SOCIOLOGY
THE ESSENTIALS

Tenth Edition

Margaret L. Andersen
University of Delaware

Howard F. Taylor
Princeton University

 CENGAGE

Australia • Brazil • Mexico • Singapore • United Kingdom • United States

Sociology: The Essentials, Tenth Edition
Margaret L. Andersen and Howard F. Taylor

Product Director: Thais Alencar

Product Manager: Ava Fruin

Content Manager: Samen Iqbal

Learning Designer: Abby Fox

Product Assistant: Megan Nauer

Marketing Manager: Tricia Salata

Digital Delivery Lead: Mike Bailey

Art Director: Marissa Falco

Manufacturing Planner: Karen Hunt

Intellectual Property Analyst: Deanna Ettinger

Cover Image Credit: iStock.com/elenabs

Text Designer & Cover Designer: Marissa Falco

Compositor: MPS Limited

For product information and technology assistance, contact us at **Cengage Customer & Sales Support, 1-800-354-9706** or **support.cengage.com.**

For permission to use material from this text or product, submit all requests online at **www.cengage.com/permissions.**

Library of Congress Control Number: 2019933416

Student Edition:
ISBN: 978-0-357-12881-7

Loose-leaf Edition:
ISBN: 978-0-357-12882-4

Cengage
20 Channel Center Street
Boston, MA 02210
USA

Cengage is a leading provider of customized learning solutions with employees residing in nearly 40 different countries and sales in more than 125 countries around the world. Find your local representative at **www .cengage.com**.

Cengage products are represented in Canada by Nelson Education, Ltd.

To learn more about Cengage platforms and services, register or access your online learning solution, or purchase materials for your course, visit **www.cengage.com**.

Printed at CLDPC, USA, 03-20

BRIEF CONTENTS

CONTENTS

9   Global Stratification   208

10   Race and Ethnicity   235

PART FOUR Social Institutions

15 Economy and Politics 391

PART FIVE Social Change

16 Environment, Population, and Social Change 425

FEATURES

maps

PREFACE

You might that think an author would get bored writing yet another edition of a book, but someone once said that if you truly understood the sociological perspective, you could never be bored.[1] For us as authors of this new edition, we are hardly bored by the tenth edition of Sociology: The Essentials. Sociology is endlessly fascinating, and we are lucky to have the opportunity to revise this book every few years so we can capture what is so compelling about the subject matter of sociology. With each new edition, we are reminded of the ever-changing nature of society, the new challenges that come from our nation's social issues, and the excitement of ongoing research on sociological subjects.

Sociology: The Essentials teaches students the basic concepts, theories, and insights of the sociological perspective. With each new edition come different challenges: new topics of study; new generations of students with different learning styles; increasing diversity among those who will read this book; and new formats for delivering course content to students. We know that how students learn and how they engage with their course material comes increasingly in the form of electronic and online learning resources.

Sociology: The Essentials, tenth edition, takes full advantage of these changes by having a fully electronic version of the book available, allowing for personalized, fully online digital learning. The platform of learning resources provided here engages students in an interactive mode, while also offering instructors the opportunity to make individualized configurations of course work. Those who want to enhance their curriculum through online resources will also be able to utilize MindTap Sociology in the way that best suits their course.

However the book is used, we have revised this edition to reflect the latest changes in society and new work in sociological scholarship. We are somewhat amazed, even as sociologists, to see how much change occurs, even in the relatively short period of time between editions. Our book adapts to these changes with each new edition. In this edition, we have maintained and strengthened the themes that have been the book's hallmark from the start: a focus on diversity in society, attention to society as both enduring and changing, the significance of social context in explaining human behavior, the increasing impact of globalization on all aspects of society, and a focus on critical thinking and an analysis of society fostered through sociological research and theory.

We know that studying sociology opens new ways of looking at the world. As we teach our students, sociology is grounded in careful observation of social facts, as well as in the analysis of how society operates. For students and faculty alike, studying sociology can be exciting, interesting, and downright fun, even though it also deals with sobering social issues, such as the growing inequality, racism, and sexism that continue to mark our time.

---

[1]Jordan, June. 1981. Civil Wars. Boston: Beacon Press, p. 100.

We try to capture the excitement of the sociological perspective, while introducing students to how sociologists do research and how they theoretically approach their subject matter. We know that most students in an introductory course will not become sociology majors, although we hope, of course, that our book and their teacher will encourage them to do so. No matter their area of study, we want to give students a way of thinking about the world that is not immediately apparent to them. We especially want students to understand how sociology differs from the individualistic thinking that tends to predominate. This is showcased in the box feature throughout the book entitled, "What Would a Sociologist Say?" Here, we take a common topic and, with informal writing, briefly discuss how a sociologist would understand this particular issue. We think this feature helps students see the unique ways that sociologists view everyday topics—things as commonplace as finding a job or analyzing sports in popular culture. We want our book to be engaging and accessible to undergraduate readers, while also preserving the integrity of sociological research and theory. Our experience in teaching introductory students shows us that students can appreciate the revelations of sociological research and theory if they are presented in an engaging way that connects to their lives. We have kept this in mind throughout this revision and have focused on material that students can understand and apply to their own social worlds.

Critical Thinking and Debunking

We use the theme of debunking in the manner first developed by Peter Berger (1963) to look behind the facades of everyday life, challenging the ready-made assumptions that permeate commonsense thinking. Debunking is a way for students to develop their critical thinking, and we use the debunking theme to help students understand how society is constructed and sustained. This theme is highlighted in the **Debunking Society's Myths** feature found throughout the book. We want students to understand the rigor that is involved with sociological research, whether quantitative research or qualitative. The box feature **Doing Sociological Research** presents a diverse array of research studies, presented to students so they can see the question being asked, the method of investigation, the research results, and the study's conclusions. This feature also includes critical thinking questions ("Questions to Consider") to help students think further about the implications of the research presented. We also include a feature to help students see the relevance of sociology in their everyday lives. The box feature **See for Yourself** allows students to apply a sociological concept to observations from their own lives, thus helping them develop their critical abilities and understand the importance of the sociological perspective.

Critical thinking is a term widely used but often vaguely defined. We use it to describe the process by which students learn to apply sociological concepts to observable events in society. Throughout the book, we ask students to use sociological concepts to analyze and interpret the world they inhabit. This is reflected in the **Thinking Sociologically** feature that is also present in every chapter.

Because contemporary students are so strongly influenced by the media, we also encourage their critical thinking through the box feature called **A Sociological Eye on the Media**. These boxes examine sociological research that challenges some of the ideas and images portrayed in the media. This not only improves students' critical thinking skills but also shows them how research can debunk these ideas and images.

A Focus on Diversity

When we first wrote this book, we did so because we wanted to integrate the then new scholarship on race, gender, and class into the core of the sociological field. We continue to see race, class, and gender—or, more broadly, the study of inequality—as one of the core insights of sociological research and theory. With that in mind, diversity, and the inequality that sometimes results, is a central theme throughout this book. A boxed theme, **Understanding Diversity**, highlights this feature, but you will find that the analysis of inequality, especially by race, gender, and class, is woven throughout the book.

Social Change

The sociological perspective helps students see society as characterized both by constant change and social stability. Throughout this book, we analyze how society changes and how events, both dramatic and subtle, influence change. We have added new material throughout the text that shows students how sociological research can help them understand that social changes are influencing their lives, even if students think of these changes as individual problems.

Global Perspective

One of the main things we hope students learn in an introductory course is how broad-scale conditions influence their everyday lives. Understanding this idea is a cornerstone of the sociological perspective. We use a global perspective to examine how global changes are affecting all parts of life within the United States, as well as other parts of the world. This means more than including cross-cultural examples. It means, for example, examining phenomena such as migration and immigration or helping students understand that their own consumption habits are profoundly shaped by global interconnections. The availability of jobs, too, is another way students can learn about the impact of an international division of labor on work within the United States. Our global perspective is found in the research and examples cited throughout the book, as well as in various chapters that directly focus on the influence of globalization on particular topics, such as work, culture, and crime.

New to the Tenth Edition

We have made various changes to the tenth edition to reflect new developments in sociological research and current social issues. These revisions should make the tenth edition easier for instructors to teach and even more accessible and interesting for students. Sociology: The Essentials is organized into five major parts: "Introducing the Sociological Imagination" (Chapter 1); "Studying Society and Social Structure" (Chapters 2 through 7); "Social Inequalities" (Chapters 8 through 12); "Social Institutions" (Chapters 13 through 15); and "Social Change" (Chapter 16).

Part I, **"Introducing the Sociological Imagination,"** introduces students to the unique perspective of sociology, differentiating it from other ways of studying society, particularly the individualistic framework students tend to assume. Within this section, **Chapter 1, "The Sociological Perspective,"** introduces students to the sociological perspective. The theme of debunking is introduced, as is the sociological imagination, as developed by C. Wright Mills. This chapter briefly reviews the development of sociology as a discipline, with a focus on the classical frameworks of sociological theory, as well as contemporary theories, including an expanded discussion of feminist theory. The tenth edition adds examples from current events to capture student interest, including new research on growing inequality, the high rate of suicide among veterans, and the influence of social media.

In **Part II, "Studying Society and Social Structure,"** students learn some of the core concepts of sociology. It begins with the study of culture in **Chapter 2, "Culture,"** that includes much discussion of social media as a force shaping contemporary culture. This includes research on social media usage both by young and older people. There is new material on the vast growth of digital viewing, but also research on body images and some of the popular titles that influence young people. **Chapter 3, "Doing Sociological Research,"** contains a discussion of the research process and the tools of sociological research—surveys, participant observation, controlled experiments, content analysis, historical research, and evaluation research. The chapter was reorganized to give better attention to the different types and tools of sociological research. As in the previous edition, we place the chapter on research methods after the chapter on culture as a way of capturing student interest early. **Chapter 4, "Socialization and the Life Course,"** contains material on socialization theory and research, including agents of socialization such as the media, family, and peers. There is more focus on the influence of social media. The chapter also emphasizes the importance of socialization in various transitions, such as the transition to adulthood.

Chapter 5, "Social Structure and Social Interaction," emphasizes how changes in the macrostructure of society influence the micro level of social interaction. We do this by focusing on technological changes that are now part of students' everyday lives and making the connection between changes at the societal level in the everyday realities of people's lives. The chapter includes material on social media, including how people create identities online and use social media websites to interact with others.

In **Chapter 6, "Groups and Organizations,"** we study social groups and bureaucratic organizations, using sociology to understand the complex processes of group influence, organizational dynamics, and the bureaucratization of society. The chapter includes a discussion of organizational culture, McDonaldization, and the significance of social networks. **Chapter 7, "Deviance and Crime,"** includes the study of sociological theories and research on deviance and crime. The core material is illustrated with contemporary events, such as police shootings of young, Black men and mass shootings. The chapter has been updated to reflect the newest research on gender-based violence and hate crime. As in previous edition, the chapter maintains a focus on race, class, and gender inequality in the criminal justice system, including mass incarceration of Black Americans and Hispanics.

In **Part III, "Social Inequalities,"** each chapter explores a particular dimension of stratification in society. Beginning with the significance of class, **Chapter 8, "Social Class and Social Stratification,"** provides an overview of basic concepts central to the study of class and social stratification. The chapter has a substantial emphasis on growing inequality with the newest research and data on this topic. The section on poverty reflects the latest research, as does material on social mobility. There is updated data throughout and new data on the likelihood of social mobility in the United States compared to other nations. **Chapter 9, "Global Stratification,"** follows with a particular emphasis on understanding the significance of global stratification, the inequality that has developed among, as well as within, various nations. Data and examples are updated throughout, and the process of globalization is a central theme in the chapter. **Chapter 10, "Race and Ethnicity,"** is a comprehensive review of the significance of race and ethnicity in society, an increasingly important topic. The chapter focuses on race as a social construction, showing the institutional basis of racial inequality. Throughout, current examples and the most recent data illustrate basic concepts about race, ethnicity, and immigration. **Chapter 11, "Gender,"** focuses on gender as a central concept in sociology closely linked to systems of stratification in society. The chapter helps students understand the distinction in sex and gender, including the social construction of gender. Research on LGBTQ people is included, as is discussion of transgender. **Chapter 12, "Sexuality,"** treats sexuality as a social construction and a dimension of social stratification and inequality. We have emphasized the influence of feminist theory on the study of sexuality. The chapter also explores research on pornography and violence against women and the sexual double standard. There is new data throughout on topics such as abortion rates, teen pregnancy, and contraception usage.

Part IV, "Social Institutions," includes three chapters, each focusing on basic institutions within society. **Chapter 13, "Families and Religion,"** maintains its inclusion of important topics in the study of families, such as interracial dating, same-sex marriage, fatherhood, gender roles within families, and family violence. We have updated material on marriage and divorce rates and also include a discussion of the impact of economic stress on families. **Chapter 14, "Education and Health Care,"** emphasizes inequality in these two important institutions. There is discussion of school segregation and its consequences and the latest material on health care, including details about the Affordable Care Act even while that is being debated in social policy. **Chapter 15, "Economy and Politics,"** analyzes the state, power, authority, and bureaucratic government. It also contains a detailed discussion of theories of power. There is current research on LGBTQ experiences in the workplace. The section on politics has been revised to show the influence of money in politics, and the chapter includes the most recent available material on the demographics of voting behavior.

Part V, "Social Change," includes **Chapter 16, "Environment, Population, and Social Change."** This chapter is framed by environmental sociology—a topic of great interest to today's students. The chapter opens with a sociological analysis of environmental change, including sustainability and climate change. There is a basic introduction to demography, as well as a discussion of the social dimensions of natural disasters. The social movements section includes an illustration from the "Black Lives Matter" movement.

Features and Pedagogical Aids

The special features of this book flow from its major themes: diversity, current theory and research, debunking and critical thinking, social change, and a global perspective. The features are also designed to help students develop critical thinking skills so that they can apply abstract concepts to observed experiences in their everyday life and learn how to interpret different theoretical paradigms and approaches to sociological research questions.

Critical Thinking Features

The feature **Thinking Sociologically** takes concepts from each chapter and asks students to think about these concepts in relationship to something they can easily observe in an exercise or class discussion. The feature **Debunking Society's Myths** takes certain common assumptions and shows students how the sociological perspective would inform such assumptions and beliefs.

An Extensive and Content-Rich Map Feature

We use the map feature that appears throughout the book to help students visualize some of the ideas presented, as well as to learn more about regional and international diversity. One map theme is **Mapping America's Diversity** and the other is **Viewing Society in Global Perspective**. These maps have multiple uses for instructional value, beyond instructing students about world and national geography. The maps have been designed primarily to show the differentiation by county, state, and/or country on key social facts.

High-Interest Theme Boxes

We use high-interest themes for the box features that embellish our focus on diversity and sociological research throughout the book. **Understanding Diversity** boxes further explore the approach to diversity taken throughout the book. In most cases, these box features provide personal narratives or other information designed to teach students about the experiences of different groups in society. Because many are written as first-person narratives, they can invoke students' empathy toward groups other than those to which they belong—something we think is critical to teaching about diversity. We hope to show students the connections between race, class, and other social groups that they otherwise find difficult to grasp. The box feature **Doing Sociological Research** is intended to show students the diversity of research questions that form the basis of sociological knowledge and, equally important, how the questions researchers ask influence the methods used to investigate the questions. We see this as an important part of sociological research—that how one investigates a question is determined as much by the nature of the question as by allegiance to a particular research method. Some questions require a more qualitative approach; others, a more quantitative approach. In developing these box features, we ask: What is the central question sociologists are asking? How did they explore this question using sociological research methods? What did they find? What are the implications of this research? We deliberately selected questions that show the full and diverse range of sociological theories and research methods, as well as the diversity of sociologists. Each box feature ends with **Questions to Consider** to encourage students to think further about the implications and applications of the research. **What Would a Sociologist Say?** boxes take a topic of interest and examine how a sociologist would likely interpret this subject. The topics are selected to capture student interest, such as a discussion of veteran suicides, hip-hop culture, and sex and popular culture. We think this box brings a sociological perspective to commonplace events.

The feature **A Sociological Eye on the Media**, found in several chapters, examines some aspect of how the media influence public understanding of some of the subjects in this book. We think this is important because sociological research often debunks taken for granted points of view presented in the media, and we want students to be able to look at the media with a more critical eye. Because of the enormous influence of the media, we think this is increasingly important in educating students about sociology.

The feature **See for Yourself** provides students with the chance to apply sociological concepts and ideas to their own observations. This feature can also be used as the basis for writing exercises, helping students improve both their analytic skills and their writing skills.

In addition to the features just described, we offer an entire set of learning aids within each chapter that promotes student mastery of the sociological concepts.

In-Text Learning Aids

Learning Objectives. We have included learning objectives to this edition, which appear near the beginning of every chapter. Matched to the major chapter headings, these objectives identify what we expect students to learn from the chapter. Faculty may choose to use these learning objectives to assess how well students comprehend the material. We tried to develop the learning objectives based on different levels of understanding and analysis, recognizing the various paths that students take in how they learn material.

Chapter Outlines. A concise chapter outline at the beginning of each chapter provides students with an overview of the major topics to be covered.

Key Terms. Key terms and major concepts appear in bold when first introduced in the chapter. A list of the key terms is found at the end of the chapter, which makes study more effective. Definitions for the key terms are found in the glossary.

Theory Tables. Each chapter includes a table that summarizes different theoretical perspectives by comparing and contrasting how these theories illuminate different aspects of different subjects.

Chapter Summary in Question-and-Answer Format. Questions and answers highlight the major points in each chapter and provide a quick review of major concepts and themes covered in the chapter. A **Glossary** and complete **References** for the whole text are found at the back of the book.

MindTap Sociology: The Personal Learning Experience

The redesigned MindTap Sociology: The Essentials, tenth edition, from Cengage represents a new approach to a highly personalized, online learning platform. A fully online learning solution, MindTap Sociology combines all of a student's learning, readings, and multimedia activities into a singular learning path that guides students through an introduction to sociology course. Three new, highly interactive activities challenge students to think critically by exploring, analyzing, and creating content, developing their sociological imagination through personal, local, and global lenses. MindTap Sociology: The Essentials, tenth edition, is easy to use and saves instructors' time by allowing them to:

- Seamlessly deliver appropriate content and technology assets from a number of providers to students, as they need them.
- Break course content down into movable objects to promote personalization, encourage interactivity, and ensure student engagement.
- Customize the course—from tools to text—and make adjustments "on the fly," making it possible to intertwine breaking news into their lessons and incorporate today's teachable moments.
- Bring interactivity into learning through the integration of multimedia assets (apps from Cengage Learning and other providers) and numerous in-context exercises and supplements; student engagement will increase, leading to better student outcomes.
- Track students' use, activities, and comprehension in real time, which provides opportunities for early intervention to influence progress and outcomes. Grades are visible and archived so students and instructors always have access to current standings in the class.
- Assess knowledge throughout each section: after readings, and in automatically graded activities and assignments.

A new digital implementation guide will help you integrate the new MindTap Learning Path into your course. Learn more at www.cengage.com/mindtap.

Instructor Resources

Sociology: The Essentials, tenth edition, is accompanied by a wide array of supplements prepared to create the best learning environment inside as well as outside the classroom for both instructors and students. All the continuing supplements for *Sociology: The Essentials*, tenth edition, have been thoroughly revised and updated. We invite you to take full advantage of the teaching and learning tools available to you.

Instructor's Resource Manual. This supplement offers instructors brief chapter outlines, student learning objectives, American Sociological Association recommendations, key terms and people, detailed chapter lecture outlines, lecture/discussion suggestions, student activities, chapter worksheets, video suggestions, video activities, and Internet exercises. The tenth edition also includes a syllabus to help instructors easily organize learning tools and create lesson plans.

Cengage Learning Testing Powered by Cognero. This flexible, online system allows teachers to author, edit, and manage test bank content from multiple Cengage Learning solutions, create multiple test versions in an instant, and deliver tests from your LMs, your classroom, or wherever you want.

PowerPoint Slides. Preassembled Microsoft® PowerPoint® lecture slides with graphics from the text make it easy for you to assemble, edit, publish, and present custom lectures for your course.

Acknowledgments

We relied on the comments of many reviewers to improve the book, and we thank them for the time they gave in developing very thoughtful commentaries on the different chapters.

We appreciate the efforts of many people who make this project possible. We are fortunate to be working with a publishing team with great enthusiasm for this project. We thank all of the people at Cengage Learning who have worked with us on this and other projects, especially Ava Fruin, who shepherded this edition through important revisions. We were also fortunate to have the guidance of Samen Iqbal, who oversaw the many aspects of production that are critical to the book's success. We especially thank Laura Sanderson for her extraordinary and careful work in the production process. Finally, our special thanks also go to our spouses Richard Morris Rosenfeld and Patricia Epps Taylor for their ongoing love and willingness to put up with us when we are frazzled by the project details!

ABOUT THE AUTHORS

Courtesy of Margaret Andersen

Margaret L. Andersen is the Edward F. and Elizabeth Goodman Rosenberg Professor Emerita of Sociology at the University of Delaware where she has held joint appointments in women's studies and Black American studies. She is the author of *Race in Society: The Enduring American Dilemma; Thinking about Women: Sociological Perspectives on Sex and Gender; Race, Class and Gender* (with Patricia Hill Collins); *Race and Ethnicity in Society: The Changing Landscape* (with Elizabeth Higginbotham); *On Land and On Sea: A Century of Women in the Rosenfeld Collection;* and *Living Art: The Life of Paul R. Jones, African American Art Collector.* She is a recipient of the American Sociological Association's Jessie Bernard Award and the Merit Award of the Eastern Sociological Society. She is the former vice president of the American Sociological Association, former president of the Eastern Sociological Society, and a recipient of the University of Delaware's Excellence in Teaching Award and the College of Arts and Sciences Award for Outstanding Teaching.

Courtesy of Howard Taylor

Howard F. Taylor was raised in Cleveland, Ohio. He graduated Phi Beta Kappa from Hiram College and has a Ph.D. in sociology from Yale University. He has taught at the Illinois Institute of Technology, Syracuse University, and Princeton University, where he is Professor of Sociology. He has published over fifty articles in sociology, education, social psychology, and race relations. His books include *The IQ Game* (Rutgers University Press), a critique of hereditarian accounts of intelligence; *Balance in Small Groups* (Van No-strand Reinhold), translated into Japanese; and the forthcoming *The SAT Triple Whammy: Race, Gender, and Social Class Bias.* He is past president of the Eastern Sociological Society, and a member of the American Sociological Association and the Sociological Research Association, an honorary society for distinguished research. He is a winner of the DuBois-Johnson-Frazier Award, given by the American Sociological Association for distinguished research in race and ethnic relations, and the President's Award for Distinguished Teaching at Princeton University.

ABOUT THE AUTHORS

THE SOCIOLOGICAL PERSPECTIVE

In this chapter, you will learn to:

Illustrate what is meant by saying that human behavior is shaped by social structure

Question individualistic explanations of human behavior

Describe the significance of studying diversity in contemporary society

Explain the origins of sociological thought

Compare and contrast the major frameworks of sociological theory

Imagine you had been switched with another infant at birth. How different would your life be? What if your accidental family was very poor . . . or very rich? How might this have affected the schools you attended, the health care you received, and the possibilities for your future career? If you had been raised in a different religion, would this have affected your beliefs, values, and attitudes? Taking a greater leap, what if you had been born another sex or a different race? What would you be like now?

We are talking about changing the basic facts of your life—your family, social class, education, religion, sex, and race. Each has major consequences for who you are and how you will fare in life. These factors play a major part in writing your life script. Your social location (meaning a person's place in society) establishes the limits and possibilities of a life.

Consider this:

- The people least likely to attend college are those most likely to benefit from it (Brand and Xie 2010).
- Neighborhoods with high concentration of African Americans and, to some extent, Latinos and Asian Americans are less likely to include health-related businesses such as fresh food markets, physical fitness facilities, and various health-providing social service organizations (Anderson 2017; Walker, Reece, and Burke 2010).
- Sons who have sisters are less likely to provide elder care to their parents than are sons without sisters. At the same time, daughters (who provide more elder care overall) provide more care to mothers than they do for fathers (Grigoryeva 2017).

These conclusions, drawn from current sociological research, describe some consequences of your particular location in society. Although we may take our place in society for granted, our social location has a profound effect on our chances in life. The power of sociology is that it teaches us to see how society influences our lives and the lives of others, and it helps us explain the consequences of different social arrangements.

Sociology also has the power to help us understand the influence of major changes on people. Currently, rapidly developing technologies, increasing globalization, a more diverse population in the United States, and growth in social inequality are affecting everyone, although in different ways. How are these changes affecting your life? Perhaps you rely on social media to keep in touch with friends. Maybe your school includes people speaking many different languages. Perhaps you see women and men in your community finding it harder to make ends meet, while people with vast sums of money shape national policy. All of these are issues that guide sociological questions. Sociology explains some of the causes and consequences of these changes.

Although society is always changing, it is also remarkably stable. People generally follow established patterns of human behavior, and you can often anticipate how people will behave in certain situations. You can even anticipate how different social conditions will affect different groups of people in society. This is what sociologists find so interesting: Society is marked by both change and stability. Societies continually evolve, creating the need for people to adapt to change while still following generally established patterns of behavior.

What Is Sociology?

Sociology is the study of human behavior in society. Sociologists are interested in the study of people and have learned a fundamental lesson: Human behavior, even when seemingly "natural" or taken for granted, is shaped by social structures—structures that have their origins beyond the immediately visible behaviors of everyday life. In other words, *all human behavior occurs in a social context*. That context—the institutions and culture that surround us—shapes what people do and think. In this book, we will examine the dimensions of society and analyze the elements of social context that influence human behavior.

Sociology is a scientific way of thinking about society and its influence on human groups. Observation, reasoning, and logical analysis are the tools of sociologists. Sociology is inspired by the fascination people have for

Sociology is the study of human behavior. What social behaviors do you see here?

David Grossman/Alamy Stock Photo

observing people, but it goes far beyond casual observations. It builds from objective analyses that others can validate as reliable.

Every day, the media in their various forms (television, film, video, digital, and print) bombard us with social commentary. Media commentators provide endless opinions about the various and sometimes bizarre forms of behavior in society. Sociology is different. Sociologists often appear in the media, and they study some of the same subjects that the media examine, such as crime, violence, or income inequality, but sociologists use specific research techniques and well-tested theories to explain social issues. Indeed, sociology can provide the tools for testing whether the things we hear about society are actually true—an increasingly important contribution in an era where many think that the media promote so-called fake news—a phenomenon that itself has social causes and consequences. Learning how to assess such claims is an important contribution of sociological research, as you will see in the feature "Debunking Society's Myths" appearing throughout this book.

The subject matter of sociology is everywhere. This is why people sometimes wrongly believe that sociology just explains the obvious. Sociologists bring a unique perspective to understanding social behavior and social change. Even though sociologists often do research on familiar topics, such as youth cultures or racial inequality, they do so using specific research tools and frames of analysis (known as sociological theory). Psychologists, anthropologists, political scientists, economists, social workers, and others also study social behavior, although each has a different perspective or "angle" on people in society.

How is sociology different from psychology? After all, both study people and both identify some of the social forces that shape our lives. There is, however, a difference. Research in psychology can inform some sociological analyses, but the focus in psychology is more on individuals—what makes individuals do what they do and how individual minds and emotions work. Increasingly, psychology is also influenced by the studies of the brain that are emerging from the techniques of neuroscience. Sociologists, on

Key Sociological Concepts

As you build your sociological perspective, you must learn certain key concepts to begin understanding how sociologists view human behavior. Social structure, social institutions, social change, and social interaction are not the only sociological concepts, but they are fundamental to grasping the sociological perspective.

- **Social Interaction.** Sociologists see **social interaction** as behavior between two or more people that is given meaning. Through social interaction, people react and change, depending on the actions and reactions of others. Because society changes as new forms of human behavior emerge, change is always in the works.
- **Social Structure.** We define **social structure** as the organized pattern of social relationships and social institutions that together constitute society. Social structure is not a "thing," but refers to the fact that social forces not always visible to the human eye guide and shape human behavior. Acknowledging that social structure exists does not mean that humans have no choice in how they behave, only that those choices are largely conditioned by one's location in society.
- **Social Institutions.** In this book, you will also learn about the significance of **social institutions**, defined as established and organized systems of social behavior with a particular and recognized purpose. Family, religion, marriage, government, and the economy are examples of major social institutions. Social institutions confront individuals at birth and transcend individual experience, but they still influence individual behavior.
- **Social Change.** As you can tell, sociologists are also interested in the process of **social change**, the alteration of society over time. As much as sociologists see society as producing certain outcomes, they do not see society as fixed, nor do they see humans as passive recipients of social expectations. Sociologists view society as stable but constantly changing.

As you read this book, you will see that these key concepts—social interaction, social structure, social institutions, and social change—are central to the sociological imagination.

the other hand, though they learn from psychological research, are more interested in the broad social forces that shape society as a whole and the people within it. (See the box "What Would a Sociologist Say?" for an example.) Together, these and other social sciences provide compelling, though different, views of human behavior.

The Sociological Perspective

Think back to the chapter opening where we asked you to imagine yourself growing up under different circumstances. Our goal in that passage was to make you feel the stirring of the *sociological perspective*—the ability to see societal patterns that influence individual and group life. The beginnings of the sociological perspective can be as simple as the pleasures of watching people or wondering how society influences people's lives. Indeed, many students begin their study of sociology because they are "interested in people." Sociologists convert this curiosity into the systematic study of how society influences different people's experiences within it.

C. Wright Mills (1916–1962) was one of the first to write about the sociological perspective in his classic book, *The Sociological Imagination* (1959). He wrote that the task of sociology was to understand the relationship between individuals and the society in which they live. He defined the **sociological imagination** as the ability to see the societal patterns that influence the individual as well as groups of individuals. Sociology should be used, Mills argued, to reveal how the context of society shapes our lives. He thought that to understand the experience of a given person or group of people, one had to have knowledge of the social and historical context in which people lived.

Think, for example, about the time and effort that many people put into their appearance. You might ordinarily think of this as merely personal grooming or an individual attempt to "look good," but this behavior has significant social origins. When you stand in front of a mirror, you are probably not thinking about how society is present in your reflection. As you look in the mirror, though, you are seeing how others see you and are very likely adjusting your appearance with that in mind, even if not consciously.

This seemingly individual behavior is actually a very social act. If you are trying to achieve a particular look, you are likely doing so because of social forces that establish particular ideals. These ideals are produced by industries that profit enormously from the products and services that people buy, even when people do so believing they are making an individual choice. Some industries suggest that you should

What Would a Sociologist Say?

Getting Pregnant: A Very Social Act

When does a woman get pregnant? Simple, you might think—it's biological. Of course, you can think of pregnancy from a biological perspective, as resulting from the process of fertilization. Or, you might think of pregnancy from a psychological perspective, analyzing the desire to have a child as rooted in an individual decision-making processes. You might even think about pregnancy from a cross-cultural or historical perspective, analyzing childbirth in different cultural contexts or analyzing historical changes in how pregnancy is managed by the medical profession. What would a sociologist say about getting pregnant?

From a sociological perspective, pregnancy is deeply social behavior. There would be many sociological angles for studying pregnancy. An example from recent research reveals the power of sociological thinking. Sociological researchers have found that the likelihood of becoming pregnant increases significantly in the two years following a friend's having had a child. As the researchers conclude, even such personal decisions as the decision to have a child result from the *web of social relationships in which people are embedded* (Balbo and Barban 2014). Pregnancy may seem like a very personal decision, but it is fertile ground for sociological study. What other social forces do you think might influence the likelihood of getting pregnant?

be thinner or curvier, your pants should be baggy or "skinny," women's breasts should be minimized or maximized—either way, you need more products. Maybe you should have a complete makeover! Many people go to great lengths to try to achieve a constantly changing beauty ideal, one that is probably not even attainable (such as flawless skin, hair always in place, with perfectly proportioned body parts). Sometimes trying to meet these ideals can even be hazardous to your physical and mental health. The ideals also vary depending on your gender.

The point is that the alleged standards of beauty are produced by social forces that extend far beyond an individual's concern with personal appearance. Appearance ideals, like other socially established beliefs and practices, are produced in particular social and historical contexts. People may come up with all kinds of personal strategies for achieving these ideals: They may buy more products, try to lose more weight or get bulked up, possibly becoming depressed and anxious if they think their efforts are failing. These personal behaviors may seem to be only individual issues, but they have basic social causes. The sociological imagination permits us to see that something as seemingly personal as how you look arises from a social context, not just individual behavior.

Sociologists are certainly concerned about individuals, but they are attuned to the social and historical context that shapes individual and group experiences. The sociological imagination distinguishes between *troubles* and *issues*. **Troubles** are privately felt problems that spring from events or feelings in a person's life. **Issues** affect large numbers of people and have their origins in the institutional arrangements and history of a society (Mills 1959). This distinction is the crux of the difference between individual experience and **social structure**, defined as the organized pattern of social relationships and social institutions that together constitute society. Issues shape the context within which troubles arise. Sociologists employ the sociological perspective to understand how issues are shaped by social structures.

Mills used the example of unemployment to explain the meaning of troubles versus issues—an example that still has resonance given people's concerns about finding work. When an individual person becomes unemployed—or cannot find work—he or she has a personal trouble, such as the worry that many college graduates have experienced in trying to find work following graduation. The personal trouble unemployment brings may include financial problems as well as the person feeling a loss of identity, becoming depressed, or having to uproot a family and move. College students may have to move back home with parents after graduation.

The problem of unemployment, however, is deeper than the experience of any one person. Unemployment is rooted in the structure of society—this is what interests sociologists. What societal forces cause unemployment? Who is most likely to become unemployed at different times? How does

Thinking Sociologically

Troubles and Issues

Personal troubles are everywhere around us: alcohol abuse or worries about money or even being upset about how you look. At an individual level, these things can be deeply troubling, and people sometimes need personal help to deal with them. But most personal troubles, as C. Wright Mills would say, also have their origins in societal arrangements. Take the example of alcohol abuse.

What are some of the things about society—not just individuals—that might influence this personal trouble? Is there a culture of drinking on your campus that generates peer pressure to drink? Do people drink more when they are unemployed? Is drinking more common among particular groups or at different times in history? Who profits from people's drinking? Thinking about these questions can help you understand the distinction that Mills makes between *personal troubles* and *social issues*.

unemployment affect an entire community (for instance, when a large plant shuts down) or an entire nation (such as when recessions hit)? Sociologists know that unemployment causes personal troubles, but understanding unemployment is more than understanding one person's experience. It requires understanding the social structural conditions that influence people's lives.

The specific task of sociology, according to Mills, is to comprehend the whole of human society—its personal and public dimensions, historical and contemporary—and its influence on the lives of human beings. Mills had an important point: People often feel that things are beyond their control, meaning that people are shaped by social forces larger than their individual lives. Social forces influence our lives in profound ways, even though we may not always know how. Consider this: Sociologists have noted a current trend, popularly labeled "the boomerang generation" or "accordion families" (Newman 2012). These terms refer to the pattern whereby many young people, after having left their family home to attend college, are returning home after graduation. Although this may seem like an individual decision to save money on housing or live "free" while paying off student loans, when a whole generation experiences this living arrangement, there are social forces at work that extend beyond individual decisions. In other words, people feel the impact of social forces in their personal lives, even though they may not always know the full dimensions of those forces. This is where sociology comes into play—revealing the *social structures* that shape the different dimensions of our day-to-day lives. Social structure is a lot like air: You cannot directly "see" it, but it is essential to living our lives.

Sociologists see social structures through careful and systematic observation. This makes sociology an **empirical** discipline. Empirical refers to careful observation, not just conjecture or opinion. In this way, sociology is very different from common sense. For empirical observations to be useful to others, they must be gathered and recorded rigorously. Sociologists are also obliged to reexamine their assumptions and conclusions constantly. Although the specific methods that sociologists use to examine different problems vary, as we will see in Chapter 3 on sociological research methods, the empirical basis of sociology is what distinguishes it from mere opinion or other forms of social commentary.

Discovering Unsettling Facts

In studying sociology, it is crucial to examine the most controversial topics and to do so with an open mind, even when you see the most disquieting facts. The facts we learn through sociological research can be "inconvenient" because the data can challenge familiar ways of thinking. Consider the following:

- Many think of the Internet as promoting more impersonal social interaction. Sociological research, however, finds that people with Internet access are actually *more likely* to have romantic partners because of the ease of meeting people online (Rosenfeld and Thomas 2012).
- Almost thirteen percent (12.7) of U.S. households are "food insecure," meaning that they do not have the money for an adequate amount of food (Oliviera 2017).

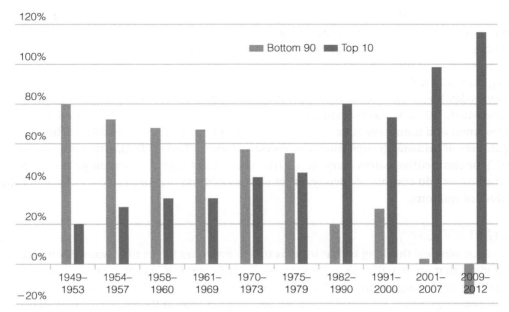

▲ **Figure 1-1   Distribution of Average Income Growth during Economic Expansions.** This figure shows how the bottom 90 percent and top 10 percent of the population experience change in their income during periods of economic expansion. What trends do you see here and how might they be affecting people's personal troubles and social issues?

Source: Tcherneva, Pavlina R. 2014. *Growth for Whom?* Levy Economics Institute of Bard College. Retrieved April 1, 2015. **www.levyinstitute.org /pubs/op_47.pdf**

- The number of women prisoners has increased at almost twice the rate of increase for men; two-thirds of women and half of men in prison are parents (Carson and Anderson 2016; Glaze and Maruschak 2008).

These facts provide unsettling evidence of persistent problems in the United States, *problems that are embedded in society, not just in individual behavior.* Sociologists try to reveal the social factors that shape society and determine the chances of success for different groups. Some never get the chance to go to college; others are unlikely to ever go to jail. These divisions persist because of people's placement within society.

▲ Figure 1-1 provides graphic evidence of how changes in society might determine the opportunities for success of different groups. This image shows what percentage of income growth went to the top 10 percent and the bottom 90 percent of the U.S. population since World War II. This was a period of great economic expansion in the United States. How was income growth distributed over this time period and who benefitted? As you can see in this image, since 2000, the bottom 90 percent of the population has experienced a dramatic decline in income growth. How does this affect opportunity for people like you?

Debunking Society's Myths

Myth: Anyone who works hard enough in the United States can get ahead.

Sociological Research: There are periods in society when some groups are able to move ahead. As examples, the Black middle class expanded following changes in civil rights laws in the 1960s; the White middle class also grew in the post-World War II period as the result of such things as GI benefits for returning vets and government support for home ownership. However, although there are exceptions, most people do not change their social class position from that in which they were born. As Figure 1.1 shows you, at times groups may even fall further behind as the result of conditions in society (Piketty 2014; Noah 2013).

How might it help explain the growing concern with class inequality? We will discuss these changes more in Chapter 8, but for now, perhaps you can begin to understand how sociologists study the broad social forces that shape people's life chances. Something as simple as being born in a particular generation can shape the course of your lifetime.

Sociologists study not just the disquieting side of society. Sociologists may also study questions that affect everyday life, such as how children's play and sports affect their developing gender identities (Messner and Musto 2016), worker–customer dynamics in nail salons (Kang 2010), or the expectations that young women and men have for combining work and family life (Gerson 2010). There are also many intriguing studies of unusual groups, such as cyberspace users (Turkel 2012), strip clubs and dancers (Barton 2017), or competitive eaters (Ferguson 2014). The subject matter of sociology is vast. Some research illuminates odd corners of society; other studies address urgent problems of society that may affect the lives of millions.

Debunking in Sociology

The power of sociological thinking is that it helps us see everyday life in new ways. Sociologists question actions and ideas that are usually taken for granted. Peter Berger (1963) calls this process "debunking." **Debunking** refers to looking behind the facades of everyday life—what Berger called the "unmasking tendency" of sociology (1963: 38). In other words, sociologists look at the behind-the-scenes patterns and processes that shape the behavior they observe in the social world.

Take schooling, for example: We can see how the sociological perspective debunks common assumptions about education. Most people think that education is primarily a way to learn and get ahead. Although this is true, a sociological perspective on education reveals something more. Sociologists have concluded that more than learning takes place in schools; other social processes are at work. Social cliques are formed where some students are "insiders" and others are excluded "outsiders." Young schoolchildren acquire not just formal knowledge but also the expectations of society and people's place within it. Race and class conflicts are often played out in schools (Lewis and Diamond 2015). Poor children seldom have the same resources in schools as middle-class or elite children, and they are often assumed to be incapable of doing schoolwork and are treated accordingly. The somber reality is that schools often stifle the opportunities of some children rather than launch all children toward success (Kozol 2012).

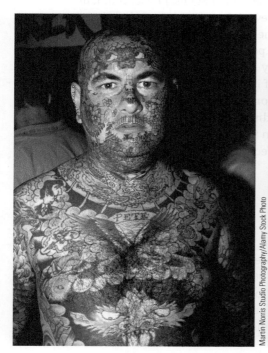

Martin Norris Studio Photography/Alamy Stock Photo

Lindsay Hebberd/Encyclopedia/Corbis

Cultural practices that seem bizarre to outsiders may be taken for granted or defined as appropriate by insiders.

Doing Sociological Research

Evicted

Research Question

Sociologist Matthew Desmond was curious about the character of poverty in the United States, but the more he thought about it, the more he realized that people's housing arrangements were a critical part of this problem. Specifically, he wanted to know more about housing eviction: How much does it happen? What are its consequences? Who gets evicted and why? He soon realized that most studies of poverty and housing focused solely on public housing, even though many of the poor are renters in the private market.

Research Method

He began his research by designing a survey of the private housing sector in Milwaukee, Wisconsin, then interviewing 1100 tenants in their homes in households across the city to be as representative as possible (see Chapter 3 for more details on research methods). At the same time, he lived in one of the city's trailer parks, taking extensive and detailed notes that allowed him to understand housing eviction through the eyes of those affected by it. His use of multiple methods, including his interviews, field notes, and formal records, provides a powerful analysis of eviction and its consequences.

Research Results

His results were published in a Pulitzer prize winning book. His results are rich and detailed, by his following the lives of some of the people he interviewed. Among other things, he found that almost half (48 percent) of forced moves were "informal evictions"—that is, not processed through the courts, but just forced by a landlord. Eviction resulted in many additional problems, including poor physical and mental health (especially for mothers), disruption in children's education, job loss, homelessness, declines in neighborhood quality, and a disruption of social bonds that stable housing provides.

Conclusions and Implications

Eviction has been an understudied problem that Desmond has exposed to the public through his research. He concludes that public initiatives that provide decent and stable, affordable housing is critical to solving this problem. Desmond also concludes that a stable home should be one of the "unalienable rights"—life, liberty and the pursuit of happiness—that are founding principles for the nation.

Questions to Consider

1. Were you to be evicted from where you live, what would be the consequences? How might they be affected by your age, your race, and your family status?
2. Are there organizations in your community that assist people who need affordable housing? If so, who do they support and how? If not, do you see a need for such assistance?

Source: Desmond, Matthew. 2016. *Evicted: Poverty and Profit in the American City.* New York: Crown Publishers.

Debunking Society's Myths

Myth: Email scams promising to deliver a large sum of cash from some African bank if you contact the email deliverer prey on people who are just stupid or old.

Sociological Research: Studies of these email scams indicate that Americans and Brits are especially susceptible to such scams because they play on widely held cultural stereotypes about Africa (that these are economically unsophisticated nations in which people are unable to manage money). These scams also exploit the American cultural belief that it is possible to "get rich quick"—reflecting a belief in individualism and the belief that anyone who tries hard enough can get ahead (Smith 2009).

Debunking is sometimes easier to do when looking at a culture or society different from one's own. Consider how behaviors that are unquestioned in one society may seem positively bizarre to an outsider. For a thousand years in China, it was usual for the elite classes to bind the feet of young girls to keep the feet from growing bigger—a practice allegedly derived from a mistress of the emperor. Bound feet were a sign of delicacy and vulnerability. A woman with large feet (defined as more than 4 inches long!) was thought to bring shame to her husband's household. The practice was supported by the belief that men were highly aroused by small feet, even though men never actually saw the naked foot. If they had, they might have been repulsed, because a woman's actual foot was U-shaped and often rotten and covered with dead skin (Blake 1994). Outside the social, cultural, and historical context in which it was practiced, footbinding seems bizarre, even dangerous. Feminists have pointed out that Chinese women were crippled by this practice, making them unable to move about freely and more dependent on men (Chang 1991).

This is an example of outsiders debunking a practice that was taken for granted by those within the culture. Debunking can also call into question practices in one's own culture that may normally go unexamined. Strange as the practice of Chinese footbinding may seem to you, how might someone from another culture view wearing shoes that make it difficult to walk? Or piercing one's tongue or eyebrow? Many take these practices of contemporary U.S. culture for granted, just as they do Chinese footbinding. Until these cultural processes are debunked, seen as if for the first time, they might seem normal.

Establishing Critical Distance

Debunking requires critical distance—that is, being able to detach from the situation at hand and view things with a critical mind. The role of critical distance in developing a sociological imagination is well explained by the early sociologist **Georg Simmel** (1858–1918). Simmel was especially interested in the role of *strangers* in social groups. Strangers have a position both inside and outside social groups. They are part of a group without necessarily sharing the group's assumptions and points of view. Because of this, the stranger can sometimes see the social structure of a group more readily than can people who are thoroughly imbued with the group's worldview. Simmel suggests that the sociological perspective requires a combination of nearness and distance. One must have enough critical distance to avoid being taken in by the group's definition of the situation, but be near enough to understand the group's experience.

Sociologists are not typically strangers to the society they study. You can acquire critical distance through a willingness to question the forces that shape social behavior. Often, sociologists become interested in things because of their own experiences. The biographies of sociologists are rich with examples of how their personal lives informed the questions they asked. Among sociologists are former ministers and nuns now studying the sociology of religion, women who have encountered sexism who now study the significance of gender in society, rock-and-roll fans studying music in popular culture, and sons and daughters of immigrants now analyzing race and ethnic relations (see the box "Understanding Diversity: Becoming a Sociologist").

The Significance of Diversity

The analysis of diversity is a central theme of sociology. Differences among groups, especially differences in the treatment of groups, are significant in any society, but they are particularly compelling in a society as diverse as that in the United States.

Defining Diversity

Today, the United States includes people from all nations and races. In 1900, one in eight Americans was not White; today, racial and ethnic minority groups (including African Americans, Hispanics, American Indians, Native Hawaiians, Asian Americans, and people of more than one race) represent 23 percent of Americans, and that proportion is growing (see ◆ Table 1-1 and ▪ map 1-1).

Table 1-1 U.S. Population Projections, 2010–2050	2010	2020	2030	2040	2050
White	79.5%	78.0%	76.6%	75.3%	74.0%
Black	12.9%	13.0%	13.1%	13.0%	13.0%
American Indian and Alaskan Native	1.0%	1.1%	1.2%	1.2%	1.2%
Asian	4.6%	5.5%	6.3%	7.1%	7.8%
Native Hawaiian and Other Pacific Islander	0.2%	0.2%	0.2%	0.3%	0.3%
Two or more races	1.8%	2.1%	2.7%	3.2%	3.7%

Note: The U.S. census counts race and Hispanic ethnicity separately. Thus, Hispanics may fall into any of the race categories. Those who identified themselves as Hispanic were 16 percent of the total U.S. population in the 2010 census.

Source: U.S. Census Bureau. 2012. National Population Projections: Summary Table. Washington, DC: U.S. Department of Commerce. **www.census.gov**

map 1-1   Mapping America's Diversity: A Changing Population

The nation is becoming increasingly diverse, but the distribution of minority groups differs in various regions of the country. Looking at this map, what factors do you think influence the distribution of the population?

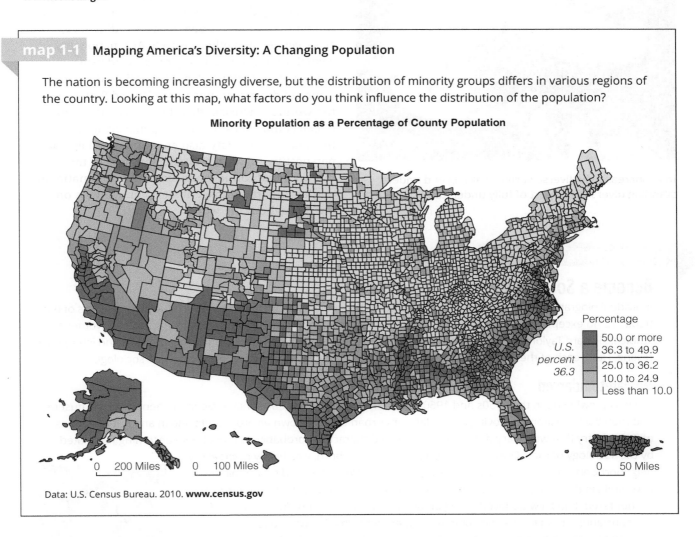

Minority Population as a Percentage of County Population

Percentage
- 50.0 or more
- 36.3 to 49.9
- 25.0 to 36.2
- 10.0 to 24.9
- Less than 10.0

U.S. percent 36.3

0 200 Miles   0 100 Miles   0 50 Miles

Data: U.S. Census Bureau. 2010. **www.census.gov**

Perhaps the most basic lesson of sociology is that people are shaped by the social context around them. In the United States, with so much cultural diversity, people will share some experiences, but not all. Experiences not held in common can include some of the most important influences on social development, such as language, religion, and the traditions of family and community. Understanding diversity means recognizing this diversity and making it central to sociological analyses.

In an increasingly diverse society, valuing and understanding diversity is a part of fully understanding society.

In this book, we use the term **diversity** to refer to the variety of group experiences that result from the social structure of society. Diversity is a broad concept that includes studying group differences in society's opportunities, the shaping of social institutions by different social factors, the formation of group and individual identity, and the process of social change. Diversity includes the study of different cultural orientations, although diversity is not exclusively about culture.

Understanding diversity is crucial to understanding society because fundamental patterns of social change and social structure are increasingly patterned by diverse group experiences. There are numerous sources of diversity, including race, class, gender, and others as well. Age, nationality, sexual orientation, and region of residence, among other factors, also differentiate the experience of diverse groups in the United States. As the world is increasingly interconnected through global communication and a global economy, the study of diversity also encompasses a global perspective—that is, an understanding of the international connections existing across national borders and the impact of such connections on life throughout the world.

Understanding Diversity

Become a Sociologist

Individual biographies often have a great influence on the subjects sociologists choose to study. The authors of this book are no exception. Margaret Andersen, a White woman, now studies the sociology of race and women's studies. Howard Taylor, an African American man, studies race, social psychology, and especially race and intelligence testing. Here, each of them writes about the influence of their early experiences on becoming a sociologist.

Margaret Andersen

As I was growing up in the 1950s and 1960s, my family moved from California to Georgia, then to Massachusetts, and then back to Georgia. Moving as we did from urban to small-town environments and in and out of regions of the country that were very different in their racial character, I probably could not help becoming fascinated by the sociology of race. Oakland, California, where I was born, was highly diverse; my neighborhood was mostly White and Asian American. When I moved to a small town in Georgia in the 1950s, I was ten years old, but I was shocked by the racial norms I encountered. I had always loved riding in the back of the bus—our major mode of transportation in Oakland—and could not understand why this was no longer allowed. Labeled by my peers as an outsider because I was not southern, I painfully learned what it meant to feel excluded just because of "where you are from."

When I moved again to suburban Boston in the 1960s, I was defined by Bostonians as a southerner and was ridiculed. Nicknamed "Dixie," I was teased for how I talked.

(*continued*)

Unlike in the South, where Black people were part of White people's daily lives despite strict racial segregation, Black people in Boston were even less visible. In my high school of 2500 or so students, Black students were rare. To me, the school seemed not much different from the strictly segregated schools I had attended in Georgia. My family soon returned to Georgia, where I was an outsider again; when I later returned to Massachusetts for graduate school in the 1970s, I worried about how a southerner would be accepted in this "Yankee" environment. Because I had acquired a southern accent, I think many of my teachers stereotyped me and thought I was not as smart as the students from other places.

These early lessons, which I may have been unaware of at the time, must have kindled my interest in the sociology of race relations. As I explored sociology, I wondered how the concepts and theories of race relations applied to women's lives. So much of what I had experienced growing up as a woman in this society was completely unexamined in what I studied in school. As the women's movement developed in the 1970s, I found sociology to be the framework that helped me understand the significance of gender and race in people's lives. To this day, I write and teach about race and gender, using sociology to help students understand their significance in society.

Howard Taylor

I grew up in Cleveland, Ohio, the son of African American professional parents. My mother, Murtis Taylor, was a social worker and the founder and then president of a social work agency called the Murtis H. Taylor Human Services Center in Cleveland, Ohio. She is well known for her contributions to the city of Cleveland and was an early "superwoman," working days and nights, cooking, caring for her two sons, and being active in many professional and civic activities. I think this gave me an early appreciation for the roles of women and the place of gender in society, although I surely would not have articulated it as such at the time.

Courtesy of Howard Taylor

My father was a businessman in a then all-Black life insurance company. He was also a "closet scientist," always doing physics experiments, talking about scientific studies, and bringing home scientific gadgets. He encouraged my brother and me to engage in science, so we were always experimenting with scientific studies in the basement of our house. In the summers, I worked for my mother in the social service agency where she worked, as a camp counselor, and in other jobs. Early on, I contemplated becoming a social worker, but I was also excited by science. As a young child, I acquired my father's love of science and my mother's interest in society. In college, the one field that would gratify both sides of me, science and social work, was sociology. I wanted to study human interaction, but I also wanted to be a scientist, so the appeal of sociology was clear.

At the same time, growing up African American meant that I faced the consequences of race every day. It was always there, and like other young African American children, I spent much of my childhood confronting racism and prejudice. When I discovered sociology, in addition to bridging the scientific and humanistic parts of my interests, I found a field that provided a framework for studying race and ethnic relations. The merging of two ways of thinking, coupled with the analysis of race that sociology has long provided, made sociology fascinating to me.

Today, my research on race, class, gender, and intelligence testing seems rooted in these early experiences. I do quantitative research in sociology and see sociology as a science that reveals the workings of race, class, and gender in society.

AP Images/Eugene Hoshiko

Globalization brings diverse cultures together, but it is also a process by which Western markets have penetrated much of the world.

Society in Global Perspective

No society can be understood apart from the global context that now influences the development of all societies. The social and economic system of any one society is increasingly intertwined with those of other nations. Coupled with the increasing ease of travel and telecommunication, a global perspective is necessary to understand change both in the United States and in other parts of the world.

To understand globalization, you must look beyond the boundaries of your own society to see how patterns in any given society are increasingly being shaped by the connections between societies. Comparing and contrasting societies across different cultures is valuable. It helps you see patterns in your own society that you might otherwise take for granted, and it enriches your appreciation of the diverse patterns of culture that mark human society and human history. A global perspective, however, goes beyond just comparing different cultures; it also helps you see how events in one society or community may be linked to events occurring on the other side of the globe.

For instance, return to the example of unemployment that C. Wright Mills used to distinguish between troubles and issues. One man may lose his job in Peoria, Illinois, and a woman in Los Angeles may employ a Latina domestic worker to take care of her child while she pursues a career. On the one hand, these are individual experiences for all three people, but they are linked in a pattern of globalization that shapes the lives of all three. The Latina domestic may have a family whom she has left in a different nation so that she can afford to support them. The corporation for which the Los Angeles woman works may have invested in a new plant overseas that employs cheap labor, resulting in the unemployment of the man in Peoria. The man in Peoria may have seen immigrant workers moving into his community. One of his children may have made a friend at school who speaks a language other than English.

Such processes are increasingly shaping many of the subjects examined in this book—work, family, education, politics, just to name a few. Without a global perspective, you would not be able to fully understand the experience of any one of the people just mentioned, much less how these processes of change and global context shape society. Throughout this book, we will use a global perspective to understand some of the developments shaping contemporary life in the United States.

The Development of Sociological Theory

Like the subjects it studies, sociology is itself a social product. Sociology first emerged in western Europe during the eighteenth and nineteenth centuries. In this period, the political and economic systems of Europe were rapidly changing. Monarchy, the rule of society by kings and queens, was disappearing, and new ways of thinking were emerging. Religion as the system of authority and law was giving way to scientific authority. At the same time, capitalism grew. Contact between different societies increased, and worldwide economic markets developed. The traditional ways of the past were giving way to a new social order. The time was ripe for a new understanding.

The Influence of the Enlightenment

The **Enlightenment** in eighteenth- and nineteenth-century Europe had an enormous influence on the development of modern sociology. Also known as the Age of Reason, the Enlightenment was characterized by faith in the ability of human reason to solve society's problems. Intellectuals believed that there were natural laws and processes in society to be discovered and used for the general good. Modern science was gradually supplanting traditional and religious explanations for natural phenomena with theories confirmed by experiments.

The earliest sociologists promoted a vision of sociology grounded in careful observation. **Auguste Comte** (1798–1857), a French philosopher who coined the term *sociology*, believed that just as science had discovered the laws of nature, sociology could discover the laws of human social behavior and thus help solve society's problems. This approach is called **positivism**, a system of thought still prominent today, in which scientific observation and description is considered the highest form of knowledge, as opposed to, say, religious dogma or poetic inspiration. The modern scientific method, which guides sociological research, grew out of positivism.

Alexis de Tocqueville (1805–1859), a French citizen, traveled to the United States as an observer beginning in 1831. Tocqueville thought that democratic values and the belief in human equality positively influenced American social institutions and transformed personal relationships. Less admiringly, he felt that in the United States the tyranny of kings had been replaced by the *tyranny of the majority*. He was referring to the ability of a majority to impose its will on everyone else in a democracy. Tocqueville also felt that, despite the emphasis on individualism in American culture, Americans had little independence of mind, making them self-centered and anxious about their social class position (Collins and Makowsky 1972).

Another early sociologist is **Harriet Martineau** (1802–1876). Like Tocqueville, Martineau, a British citizen, embarked on a long tour of the United States in 1834. She was fascinated by the newly emerging culture in the United States. Her book *Society in America* (1837) is an analysis of the social customs that she observed. This important work was overlooked for many years, probably because the author was a woman. It is now recognized as a classic. Martineau also wrote the first sociological methods book, *How to Observe Morals and Manners* (1838), in which she discussed how to observe behavior when one is a participant in the situation being studied.

As one of the earliest observers of American culture, Harriet Martineau used the powers of social observation to record and analyze the social structure of American society. Long ignored for her contributions to sociology, she is now seen as one of the founders of early sociological thought.

Classical Sociological Theory

Of all the contributors to the development of sociology, the giants of the European tradition were Emile Durkheim, Karl Marx, and Max Weber. They are classical thinkers because the ideas they offered more than 150 years ago continue to influence our understanding of society, not just in sociology but in other fields as well (such as political science and history).

Emile Durkheim

During the early academic career of the Frenchman **Emile Durkheim** (1858–1917), France was in the throes of great political and religious upheaval. Anti-Semitism (hatred of Jews) was rampant. Durkheim, himself Jewish, was fascinated by how the public degradation of Jews by non-Jews seemed to calm and unify a large segment of the divided French public. Durkheim later wrote that public rituals have a special purpose in society. Rituals create social solidarity, referring to the bonds that link the members of a group. Some of Durkheim's most significant works explore what forces hold society together and make it stable.

According to Durkheim, people in society are glued together by belief systems (Durkheim 1947/1912). The rituals of religion and other institutions symbolize and reinforce the sense of belonging. Public ceremonies create a bond between people in a social unit. Durkheim thought that such rituals as publicly punishing people sustain moral cohesion in society. Durkheim's views on this are further examined in Chapter 7, which discusses deviant behavior.

Durkheim also viewed society as an entity larger than the sum of its parts. He described this as society *sui generis* (which translates as "thing in itself"), meaning that society is a subject to be studied separately from the sum of the individuals who compose it. Society is external to individuals, yet its existence is internalized in people's minds—that is, people come to believe what society expects them to believe. Durkheim conceived of society as an integrated whole—each part contributing to the overall stability of the system. His work is the basis for *functionalism*, an important theoretical perspective that we will return to later in this chapter.

One contribution from Durkheim was his conceptualization of the *social facts*. Durkheim created the term **social facts** to indicate those social patterns that are *external* to individuals. Things such as customs and social values exist outside individuals, whereas psychological drives and motivation exist inside people. Social facts, therefore, are the proper subject of sociology; they are its reason for being.

A striking illustration of this principle was Durkheim's study of suicide (Durkheim 1951/1897). He analyzed rates of suicide in a society, as opposed to looking at individual (psychological) causes of suicide. He showed that suicide rates varied according to how clear the norms and customs of the society were, whether the norms and customs were consistent with each other and not contradictory. *Anomie* (the breakdown of social norms) exists where norms were either grossly unclear or contradictory; the suicide rates were higher in such societies or such parts of a society. It is important to note that this condition is in society—external to individuals, but felt by them (Puffer 2009). In this sense, such a condition is truly societal.

Durkheim held that social facts, though they exist outside individuals, nonetheless pose constraints on individual behavior. Durkheim's major contribution was the discovery of the social basis of human behavior. He proposed that society could be known through the discovery and analysis of social facts. This is the central task of sociology (Coser 1977; Bellah 1973; Durkheim 1950/1938).

Karl Marx

It is hard to imagine another scholar who has had as much influence on intellectual history as has **Karl Marx** (1818–1883). Along with his collaborator, Friedrich Engels, Marx not only changed intellectual history but also world history.

Marx's work was devoted to explaining how capitalism shaped society. He argued that capitalism is an economic system based on the pursuit of profit and the sanctity of private property. Marx used a class

What Would a Sociologist Say?

Suicide among Veterans

Currently, 7400 veterans commit suicide each year, accounting for 18 percent of all suicides (in 2014), even though veterans are only 8.5 percent of the population (VA Suicide Prevention Program 2016). How do sociologists explain this?

Certainly, there are psychological factors at work—post-traumatic stress, depression, and, sometimes, substance abuse—but sociological factors are at work, too. Durkheim would argue that this is a good example of suicide as a social fact. A soldier returning home is likely to encounter a far less structured environment than when in service where military life is highly structured. This can be a suicide-prone environment, especially if combined with unemployment, homelessness, or a disability. If you add to that a lack of social support services or benefits specifically to address the risk of suicide, you can have a potentially lethal social context.

Although sociologists do not ignore the psychological dimensions of behavior such as suicide, they see that there are other important social factors that produce this tragic behavior.

analysis to explain capitalism, describing capitalism as a system of relationships among different classes, including capitalists (also known as the bourgeois class), the proletariat (or working class), the petty bourgeoisie (small business owners and managers), and the *lumpenproletariat* (those "discarded" by the capitalist system, such as the homeless). In Marx's view, profit, the goal of capitalist endeavors, is produced through the exploitation of the working class. Workers sell their labor in exchange for wages, and capitalists make certain that wages are worth less than the goods the workers produce. The difference in value is the profit of the capitalist. In the Marxist view, the capitalist class system is inherently unfair because the entire system rests on workers getting less than they give.

Durkheim thought that symbols and rituals were important for producing social cohesion in society. You can witness this when shrines are spontaneously created in the aftermath of tragedies, as illustrated here after the school shooting at Sandy Hook Elementary School in Newtown, Connecticut.

Marx thought that the economic organization of society was the most important influence on what humans think and how they behave. He found that the beliefs of the common people tended to support the interests of the capitalist system, not the interests of the workers themselves. Why? The answer is that the capitalist class controls the production of goods *and* the production of ideas. It owns the publishing companies, endows the universities where knowledge is produced, and controls information industries—thus shaping what people think.

Marx considered all of society to be shaped by economic forces. Laws, family structures, schools, and other institutions all develop, according to Marx, to suit economic needs under capitalism. Like other early sociologists, Marx took social structure as his subject rather than the actions of individuals. It was the *system* of capitalism that dictated people's behavior. Marx saw social change as arising from tensions inherent in a capitalist system—the conflict between the capitalist and working classes. Marx's ideas are often misperceived by U.S. students because communist revolutionaries throughout the world have claimed Marx as their guiding spirit. It would be naive to reject his ideas solely on political grounds. Much that Marx predicted has not occurred—for instance, he claimed that the "laws" of history made a worldwide revolution of workers inevitable, and this has not happened. Still, he left us an important body of sociological thought springing from his insight that society is systematic and structural and that class is a fundamental dimension of society that shapes social behavior.

Max Weber

Max Weber (1864–1920; pronounced "vayber") was greatly influenced by and built upon Marx's work. Whereas Marx saw economics as the basic organizing element of society, Weber theorized that society had three basic dimensions: political, economic, and cultural. According to Weber, a complete sociological analysis must recognize the interplay between economic, political, and cultural institutions (Parsons 1947). Weber is credited with developing a *multidimensional* analysis of society that goes beyond Marx's more one-dimensional focus on economics.

Weber also theorized extensively about the relationship of sociology to social and political values. He did not believe there could be a value-free sociology because values would always influence what sociologists considered worthy of study. Weber thought sociologists should acknowledge the influence of values so that ingrained beliefs would not interfere with objectivity. Weber professed that the task of sociologists is to teach students the uncomfortable truth about the world. Faculty should not use their positions to promote their political opinions, he felt; rather, they have a responsibility to examine all opinions, including unpopular ones, and use the tools of rigorous sociological inquiry to understand why people believe and behave as they do.

An important concept in Weber's sociology is *verstehen* (meaning "understanding" and pronounced "vershtayen"). **Verstehen**, a German word, refers to understanding social behavior from the point of view of those engaged in it. Weber believed that to understand social behavior, one had to understand the meaning that a behavior had for people. He did not believe sociologists had to be born into a group to understand it (in other words, he didn't believe "it takes one to know one"), but he did think sociologists had to develop some subjective understanding of how other people experience their world. One major contribution from Weber was the definition of *social action* as a behavior to which people give meaning (Gerth and Mills 1946; Weber 1962/1913; Parsons 1951b), such as placing a bumper sticker on your car that states pride in U.S. military troops.

Sociology in the United States

American sociology was built on the earlier work of Europeans, but unique features of U.S. culture contribute to its distinctive flavor. In the early twentieth century, as sociology was evolving, most early sociologists in the United States took a reform-based approach, emphasizing more the importance of applying knowledge for social change. American sociologists believed that if they exposed the causes of social problems, they could alleviate human suffering. The nation in the early twentieth century was moving to a more urban society, with a new mix of immigrants and visible problems such as those we face today: urban blight, hunger, poverty, and racial segregation. Sociology, it was believed, could explain how these problems were caused and, therefore, be used to create change.

Nowhere was the emphasis on application more evident than at the University of Chicago, where a style of sociological thinking known as the Chicago School developed. The Chicago School included scholars who wanted to understand how society shapes the mind and identity of people. Sociologists such as George Herbert Mead and Charles Horton Cooley thought of society as a human laboratory where they could observe and understand human behavior to be better able to address human needs, and they used the city in which they lived as a living laboratory. You will study these thinkers more in Chapter 4.

Robert Park (1864–1944), from the University of Chicago, was a key founder of sociology. Originally a journalist who worked in several Midwestern cities, Park was interested in urban problems and how different racial groups interacted with each other. He was also fascinated by the sociological design of cities, noting that cities were typically sets of concentric circles. At the time, the very rich and the very poor lived in the middle, ringed by slums and low-income neighborhoods (Coser 1977; Collins and Makowsky 1972; Park and Burgess 1921). Today, Park would still be intrigued by how boundaries are defined and maintained in urban neighborhoods. You might notice this yourself. A single street crossing might delineate a Vietnamese neighborhood from an Italian one, an affluent White neighborhood from a barrio. The social structure of cities continues to be a subject of sociological research.

Bettmann/Corbis

Jane Addams, the only sociologist to win the Nobel Peace Prize, used her sociological skills to try to improve people's lives. The settlement house movement provided social services to groups in need, while also providing a social laboratory in which to observe the sociological dimensions of problems such as poverty.

Many early sociologists of the Chicago School were women whose work is only now being rediscovered. **Jane Addams** (1860–1935) was one of the most renowned sociologists of her day. Because she was a woman, she was never given the jobs or prestige that men in her time received. She was the only practicing sociologist ever to win a Nobel Peace Prize (in 1931), yet she never had a regular teaching job. Instead, she used her skills as a research sociologist to develop community projects that assisted people in need (Deegan 1988). She was a leader in the settlement house movement providing services and doing research to improve the lives of slum dwellers, immigrants, and other poor people.

Another early sociologist, widely noted for her work in the antilynching movement, was **Ida B. Wells-Barnett** (1862–1931). Born a slave, Ida B. Wells-Barnett learned to read and write at Rust College, a school established for freed slaves, later receiving her teaching credentials at Fisk University. She wrote numerous essays on the status of African Americans in the United States and was an active crusader against lynching and for women's rights, including the

right to vote. She was so violently attacked—in writing and in actual threats—
that she often had to write under an assumed name. Until recently, her contributions to the field of sociology have been largely unexamined. Interestingly,
her grandson, Troy Duster (b. 1936) is now a faculty member at New York
University and the University of California, Berkeley (Giddings 2008; Henry 2008;
Lengermann and Niebrugge-Brantley 1998).

W. E. B. DuBois (1868–1963; pronounced "due boys") was one of the most
important early sociological thinkers in the United States. DuBois was a prominent Black scholar, a cofounder of the NAACP (National Association for the
Advancement of Colored People) in 1909, a prolific writer, and one of the best
American minds. He received the first Ph.D. ever awarded to a Black person in
any field (from Harvard University), and he studied for a time in Germany, hearing several lectures by Max Weber (Morris 2015).

Ida B. Wells-Barnett is now
well known for her brave
campaign against the
lynching of African American people. Less known are
her early contributions to
sociological thought.

DuBois was deeply troubled by the racial divisiveness in society, writing in
a classic essay published in 1901 that "the problem of the twentieth century is
the problem of the color line" (DuBois 1901: 354). Like many of his women colleagues, he envisioned a community-based, activist profession committed to
social justice (Deegan 1988); he was a friend and collaborator with Jane Addams.
He believed in the importance of a scientific approach to sociological questions,
but he also thought that convictions always directed one's studies. Were he alive
today, he might no doubt note that the problem of the color line persists well into the twenty-first century.

Much of DuBois's work focused on the social structure of Black communities, one of his classic studies being of the city of Philadelphia. His book, *The Philadelphia Negro*, published in 1899, remains a classic
study of African American urban life and its social institutions. One of the most lasting ideas from DuBois
is his concept of "dual (or double) consciousness." DuBois saw African Americans as always having to
see themselves through the eyes of others, a response that would be typical among any group oppressed
by others. For DuBois, this dual consciousness led African Americans to always be alert to how others
see them, and at the same time, to develop a strong collective identity of themselves as "Black" or, as we
would say now, African American (DuBois 1903).

Theoretical Frameworks in Sociology

The founders of sociology have established theoretical traditions that ask basic questions about society
and inform sociological research. The idea of theory may seem dry to you because it connotes something
that is only hypothetical and divorced from "real life." Sociological theory though is one of the tools that
sociologists use to interpret real life. Sociologists use theory to organize their observations and apply them
to the broad questions sociologists ask, such as: How are individuals related to society? How is social order
maintained? Why is there inequality in society? How does social change occur? (See ◆ Table 1-2.)

Different theoretical frameworks within sociology make different assumptions and provide different
insights about the nature of society. In the realm of *macrosociology* are theories that strive to understand
society as a whole. Durkheim, Marx, and Weber were macrosociological theorists. Theoretical frameworks
that center on face-to-face social interaction are known as *microsociology*. Some of the work derived from
the Chicago School—research that studies individuals and group processes in society—is microsociological. Although sociologists draw from diverse theoretical perspectives to understand society, four theoretical traditions form the major theoretical perspectives: functionalism, conflict theory, symbolic interaction,
and, more recently, feminist theory.

Functionalism

Functionalism has its origins in the work of Durkheim, who you will recall was especially interested in
how social order is possible and how society remains relatively stable. **Functionalism** interprets each
part of society in terms of how it contributes to the stability of the whole. As Durkheim suggested,

Table 1-2   Classical Theorists Reflect on the Economic Inequality

	Major Concepts	What's the Big Idea?	An Applied Example: Economic Inequality
EMILE DURKHEIM (1858–1917)	Society sui generis Social solidarity Social facts	Social structures produce social forces that impinge on individuals even when they are not immediately visible; social solidarity is produced through identifying some as "other" or not belonging.	In times of rising economic inequality, those who are especially vulnerable tend to blame others, such as immigrants or "foreigners," for taking jobs from those perceived as more worthy. This produces solidarity among those who may even act outside of their own interests because of their perception of "others."
KARL MARX (1818–1883)	Capitalism Class conflict	Capitalism is built on the exploitation of laboring groups for the profit of others. Class conflict is embedded in the system of capitalism that then shapes other social institutions.	It is no surprise that inequality is growing; the forces of capitalism mean that the rich will amass the most resources, with everyone else becoming worse off.
MAX WEBER (1864–1920)	Multidimensional analysis Verstehen	Cultural values interact with economic and political systems to produce society; no one factor determines the character of society.	Even when the economy is stagnant, cultural beliefs in hard work and the Protestant ethic mean that people will blame individuals, not the system, for failure.
W. E. B. DUBOIS (1868–1963)	Color line Double consciousness	Racial inequality structures social institutions in the United States. Those who are oppressed by race develop a dual consciousness, ever aware of their status in the eyes of others but also having a collective identity as African American.	The "problem of the color line" extends into the twenty-first century, as African American people and other people of color are uniquely disadvantaged by economic inequality.

Bettmann/Corbis

Bettmann/Corbis

akg-images/Newscom

Historical/Corbis

Table 1-3	Manifest and Latent Functions: The Family
Manifest Functions (explicit, deliberate)	**Latent Functions (unintended, unrecognized)**
Reproduction	Sexual relations outside of the traditional family may be judged as deviant
Transmission of cultural values	Risk of intolerance of different cultures/groups
Care of the young	Neglect of public policies to support working parents
Emotional support	Silence around conflicts that occur within families
Consumption of goods	Transmission of inequality across generations as wealth and property is passed on for some and not others

functionalism conceptualizes society as more than the sum of its component parts. Each part is "functional" for society—that is, contributes to the stability of the whole. The different parts are primarily the institutions of society, each of which is organized to fill different needs and each of which has particular consequences for the form and shape of society. The parts each then depend on one another.

The family as an institution, for example, serves multiple functions. At its most basic level, the family has a reproductive role. Within the family, infants receive protection and sustenance. As they grow older, they are exposed to the patterns and expectations of their culture. Across generations, the family supplies a broad unit of support and enriches individual experience with a sense of continuity with the past and future. All these aspects of family can be assessed by how they contribute to the stability and prosperity of society. The same is true for other institutions.

The functionalist framework emphasizes the consensus and order that exist in society, focusing on social stability and shared public values. From a functionalist perspective, disorganization in the system, such as an economic collapse, leads to change because societal components must adjust to achieve stability. This is a key part of functionalist theory—that when one part of society is not working (or is *dysfunctional*, as they would say), it affects all the other parts and creates social problems. Change may be for better or worse. Changes for the worse stem from instability in the social system, such as a breakdown in shared values or a social institution no longer meeting people's needs (Eitzen and Baca Zinn 2012; Merton 1968).

Functionalism was a dominant theoretical perspective in sociology for many years, and one of its major theorists was **Talcott Parsons** (1902–1979). In Parsons's view, all parts of a social system are interrelated, with different parts of society having different basic functions. Functionalism was further developed by **Robert Merton** (1910–2003). Merton saw that social practices often have consequences for society that are not immediately apparent. He suggested that human behavior has both manifest and latent functions. *Manifest functions* are the stated and intended goals of social behavior. *Latent functions* are neither stated nor intended. The family, for example, has both manifest and latent functions, as demonstrated in ◆ Table 1-3.

Critics of functionalism argue that its emphasis on social stability understates the roles of power and conflict in society. Critics also disagree with the explanation of inequality offered by functionalism—that it persists because social inequality creates a system for the fair and equitable distribution of societal resources (discussed further in Chapter 8). Functionalists argue that it is fair and equitable that the higher social classes earn more money because they are more important (functional) to society. Critics disagree, saying that functionalism is too accepting of the status quo. From a functionalist perspective though, inequality serves a purpose in society: It provides an incentive system for people to work and promotes solidarity among groups linked by their common social standing.

Conflict Theory

Conflict theory emphasizes the role of coercion and power in society and the ability of some to influence and control others. It differs from functionalism, which emphasizes cohesion within society. Instead, conflict theory emphasizes strife and friction. Conflict theory pictures society as comprised of groups that compete for social and economic resources. Social order is maintained not by consensus but by domination, with power in the hands of those with the greatest political, economic, and social resources. When consensus exists, according to conflict theorists, it is attributable to people being united around common interests, often in opposition to other groups (Dahrendorf 1959; Mills 1956).

According to conflict theory, inequality exists because those in control of a disproportionate share of society's resources actively defend their advantages. The masses are not bound to society by their shared values but by coercion at the hands of the powerful. In conflict theory, the emphasis is on social control, not on consensus and conformity. Those with the most resources exercise power over others; inequality and power struggles are the result. Conflict theory gives great attention to class, race, gender, and sexuality in society because these are seen as the grounds of the most pertinent and enduring struggles in society.

Conflict theorists see inequality as inherently unfair, persisting only because groups who are economically advantaged use their social position to their own betterment. Their dominance even extends to the point of shaping the beliefs of other members of the society by controlling public information and holding power in institutions such as education and religion that shape what people think and know. From the conflict perspective, power struggles between conflicting groups are the source of social change. Those with the greatest power are typically able to maintain their advantage at the expense of other groups.

Conflict theory has been criticized for neglecting the importance of shared values and public consensus in society while overemphasizing inequality. Like functionalist theory, conflict theory finds the origins of social behavior in the structure of society, but it differs from functionalism in emphasizing the importance of power.

Symbolic Interaction

The third major framework of sociological theory is **symbolic interaction**. Instead of thinking of society in terms of abstract institutions, symbolic interaction emphasizes immediate social interaction as the place where "society" exists. Because of the human capacity for reflection, people give meaning to their behavior. The creation of meaning is how they interpret the different behaviors, events, or things that happen in society.

As its name implies, symbolic interaction relies extensively on the symbolic meaning that people develop and employ in the process of social interaction. Symbolic interaction theory emphasizes face-to-face interaction and thus is a form of microsociology, whereas functionalism and conflict theory are more macrosociological.

Derived from the work of the Chicago School, symbolic interaction theory analyzes society by addressing the subjective meanings that people impose on objects, events, and behaviors. Subjective

Table 1-4 Comparing Sociological Theories

Basic Questions	Functionalism	Conflict Theory	Symbolic Interaction	Feminist Theory
What is the relationship of individuals to society?	Individuals occupy fixed social roles.	Individuals are subordinated to society.	Individuals and society are interdependent.	Women and men are bound together in a system of gender relationships that shape identities and beliefs.
Why is there inequality?	Inequality is inevitable and functional for society.	Inequality results from a struggle over scarce resources.	Inequality is demonstrated through the importance of symbols.	Inequality stems from the matrix of domination that links gender, race, class, and sexuality.
How is social order possible?	Social order stems from consensus on public values.	Social order is maintained through power and coercion.	Social order is sustained through social interaction and adherence to social norms.	Patriarchal social orders are maintained by the power that men hold over women.
What is the source of social change?	Society seeks equilibrium when there is social disorganization.	Change comes through the mobilization of people struggling for resources.	Change develops from an ever-evolving set of social relationships and the creation of new meaning systems.	Social change comes from the mobilization of women and their allies on behalf of women's liberation.
Major Criticisms				
	This is a conservative view of society that underplays power differences among and between groups.	The theory understates the degree of cohesion and stability in society.	There is little analysis of inequality, and it overstates the subjective basis of society.	Feminist theory has too often been anchored in the experiences of White, middle-class women.

meanings are important because, according to symbolic interaction, people behave based on what they *believe*, not just on what is objectively true. Symbolic interaction sees society as socially constructed through human interpretation (Blumer 1969; Berger and Luckmann 1967; Shibutani 1961). Social meanings are constantly modified through social interaction.

People interpret one another's behavior; these interpretations form social bonds. These interpretations are called the "definition of the situation." For example, why would young people smoke cigarettes even though all objective medical evidence points to the danger of doing so? The answer is in the definition of the situation that people create. Studies find that teenagers are well informed about the risks of tobacco, but they also think that "smoking is cool," that they themselves will be safe from harm, and that smoking projects an image—a positive identity for boys as a "tough guy" and for girls as fun-loving, mature, and glamorous. Smoking is also defined by young women as keeping you thin—an ideal constructed through dominant images of beauty. In other words, the symbolic meaning of smoking overrides the actual facts regarding smoking and risk.

Symbolic interaction interprets social order as constantly negotiated and created through the interpretations people give to their behavior. In observing society, symbolic interactionists see not simply facts but "social constructions," the meanings attached to things, whether those are concrete symbols (like a certain way of dress or a tattoo) or nonverbal behaviors. In symbolic interaction theory, society is highly subjective—existing in the minds of people, even though its effects are very real.

Feminist Theory

Contemporary sociological theory has been greatly influenced by the development of **feminist theory**. Prior to the emergence of second-wave feminism (the feminist movement emerging in the 1960s and 1970s), women were largely absent and invisible within most sociological work—indeed, within most academic work. When seen, they were strongly stereotyped in traditional roles as wives and mothers.

Careers in Sociology

Now that you understand a bit more what sociology is about, you may ask, "What can I do with a degree in sociology?" This is a question we often hear from students. There is no single job called "sociologist" like there is "engineer" or "nurse" or "teacher," but sociology prepares you well for many kinds of jobs, whether with a bachelor's degree or a postgraduate education. The skills you acquire from your sociological education are useful for jobs in business, health care, criminal justice, government agencies, various nonprofit organizations, and other job venues.

For example, the research skills one gains through sociology can be important in analyzing business data or organizing information for a food bank or homeless shelter. Students in sociology gain experience working with and understanding those with different cultural and social backgrounds; this is an important and valued skill that employers seek. Also, the ability to dissect the different causes of a social problem can be an asset for jobs in various social service organizations.

Some sociologists have worked in their communities to deliver more effective social services. Sociologists employed in business organizations and social services use their sociological training to address issues such as poverty, crime and delinquency, population studies, substance abuse, violence against women, family social services, immigration policy, and any number of other important issues. Sociologists also work in the offices of U.S. representatives and senators, doing background research on the various issues addressed in the political process.

These are just a few examples of how sociology can prepare you for various careers. A good way to learn more about how sociology prepares you for work is to consider doing an internship while you are still in college. For more information about careers in sociology, see the booklet, "21st Century Careers with an Undergraduate Degree in Sociology," available through the American Sociological Association (www.asanet.org).

Critical Thinking Exercise

1. Read a national newspaper over a period of one week and identify any experts who use a sociological perspective in their commentary. What does this suggest to you as a possible career in sociology? What are some of the different subjects about which sociologists provide expert information?

2. Identify some of the students from your college who have finished degrees in sociology. What different ways have they used their sociological knowledge?

Flake/Alamy Stock Photo

Symbolic interaction theory can help explain why people might do things that otherwise seem contrary to what one might expect.

Feminist theory developed to understand the status of women in society and with the purpose of using that knowledge to better women's lives.

Feminist theory has created vital new knowledge about women and has also transformed what is understood about men. Feminist scholarship in sociology, by focusing on the experiences of women, provides new ways of seeing the world and contributes to a more complete view of society.

Feminist theory takes gender as a primary lens through which to view society. Beyond that, feminist theory makes the claim that without considering gender in society, one's analysis of any social behavior is incomplete and, thus, incorrect. At the same time, feminist theory purports to analyze society with an eye to improving the status of women. Men are not excluded from feminist theory. In fact, feminist theory, as we will see in various chapters that follow, also argues that men are gendered subjects too. We cannot understand society without understanding how gender is structured in society and in women's and men's lives.

Feminist theory is a now vibrant and rich perspective in sociology, and it has added much to how people understand the sociology of gender—and its connection to other social factors, such as race, sexuality, age, and class. Along with the classical traditions of sociology, feminist theory is included throughout this book.

Functionalism, conflict theory, symbolic interaction, and feminist theory are by no means the only theoretical frameworks in sociology. For some time, however, they have provided the most prominent general explanations of society. Each has a unique view of the social realm. None is a perfect explanation of society, yet each has something to contribute. Functionalism gives special weight to the order and cohesion that usually characterizes society. Conflict theory emphasizes the inequalities and power imbalances in society. Symbolic interaction emphasizes the meanings that humans give to their behavior. Feminist theory takes gender as a primary lens through which to understand society, especially in relation to other structures of inequality. Together, these frameworks provide a rich, comprehensive perspective on society, individuals within society, and social change (see ◆ Table 1-4).

Whatever the theoretical framework used, theory is evaluated in terms of its ability to explain observed social facts. The sociological imagination is not a single-minded way of looking at the world. It is the ability to observe social behavior and interpret that behavior in light of societal influences.

Chapter Summary

What is sociology?

Sociology is the study of human behavior in society. The *sociological imagination* is the ability to see societal patterns that influence individuals. Sociology is an *empirical* discipline, relying on careful observations as the basis for its knowledge.

What is debunking?

Debunking in sociology refers to the ability to look behind things taken for granted, looking instead to the origins of social behavior.

Why is diversity central to the study of sociology?

One of the central insights of sociology is its analysis of social diversity and inequality. Understanding *diversity* is critical to sociology because it is necessary to analyze *social institutions* and because diversity shapes most of our social and cultural institutions.

When and how did sociology emerge as a field of study?

Sociology emerged in western Europe during the *Enlightenment* and was influenced by the values of critical reason, humanitarianism, and positivism. *Auguste Comte*, one of the earliest sociologists,

emphasized sociology as a positivist discipline. *Alexis de Tocqueville* and *Harriet Martineau* developed early and insightful analyses of American culture.

What are some of the basic insights of classical sociological theory?

Emile Durkheim is credited with conceptualizing society as a social system and with identifying *social facts* as patterns of behavior that are external to the individual. *Karl Marx* showed how capitalism shaped the development of society. *Max Weber* sought to explain society through cultural, political, and economic factors. *W. E. B. DuBois* saw racial inequality as the greatest challenge in U.S. society.

What are the major theoretical frameworks in sociology?

Functionalism emphasizes the stability and integration in society. *Conflict theory* sees society as organized around the unequal distribution of resources and held together through power and coercion. *Symbolic interaction* emphasizes the role of individuals in giving meaning to social behavior, thereby creating society. *Feminist theory* is the analysis of women and men in society and is intended to improve women's lives.

Key Terms

conflict theory 22

debunking 8

diversity 12

empirical 6

Enlightenment 15

feminist theory 24

functionalism 19

issues 5

positivism 15

social change 3

social facts 16

social institution 3

social interaction 3

social structure 3

sociological imagination 5

sociology 2

symbolic interaction 22

troubles 5

Verstehen 18

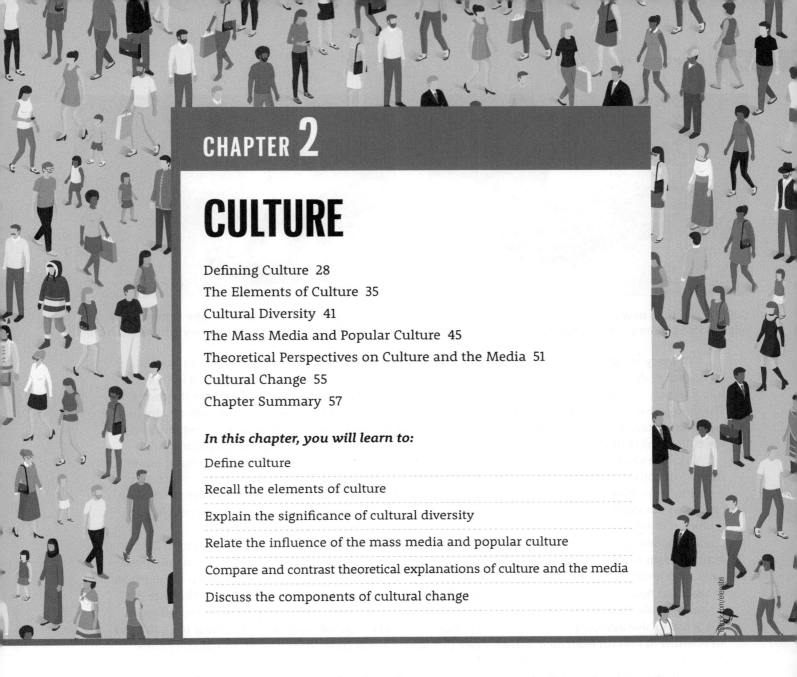

CHAPTER 2

CULTURE

In this chapter, you will learn to:

Define culture

Recall the elements of culture

Explain the significance of cultural diversity

Relate the influence of the mass media and popular culture

Compare and contrast theoretical explanations of culture and the media

Discuss the components of cultural change

iStock.com/elenabs

In one contemporary society known for its technological sophistication, people—especially the young—walk around with plugs in their ears. The plugs are connected to small wires that are themselves coated with a plastic film. These little plastic-covered wires are then connected to small devices made of metal, plastic, silicon, and other modern components, although most people who use them have no idea how they are made. When turned on, these devices put music into people's ears or, in some cases, show pictures and movies on a screen not much larger than a bar of soap. Some people who use these devices wouldn't even consider walking around without them. It is as if the devices shield them from other elements of their culture.

The same people who carry these devices around have other habits that, when seen from the perspective of someone unfamiliar with this culture, might seem peculiar and certainly highly ritualized. Apparently, when young people in this society go away to school, most take a large number of various electronic devices along with them. Many sleep with one of these devices turned on all night. It seems that everything these young people do involves looking at some kind of screen, enough so that one of the authors of this book has labeled their generation "screenagers." People in this culture now talk about "streaming" things—an innovative term that would have had no meaning not many years ago.

Cultural practices may seem strange to outsiders, but may be taken for granted by those within the culture. How might some contemporary cultural practices in the United States look strange to people from a very different culture?

Not everyone in this culture has access to all of these devices, although many want them. Indeed, having more devices seems to be a mark of one's social status, that is, how you are regarded in this culture, but very few people know where the devices are made, what they are made of, or how they work. The young also often ridicule older people for not understanding how to operate the devices or why they are so important to young people. From outside the culture, these practices seem strange, yet few within the culture think the behaviors associated with these devices are anything but perfectly ordinary.

You have surely guessed that the practices described here are taken from U.S. culture: smartphones, tablets, and other electronic devices. These devices have become so commonplace that they practically define modern American culture. As with all cultural habits, unless they are somehow interrupted, most people do not think much about their influence on society, on people's relationships, or on people's definitions of themselves.[1]

When viewed from the outside, cultural habits that seem perfectly normal often seem strange. Take an example from a different culture. The Tchikrin people—a remote culture of the central Brazilian rain forest—paint their bodies in elaborate designs. Painted bodies communicate to others the relationship of the person to his or her body, to society, and to the spiritual world. The designs and colors symbolize the balance the Tchikrin people think exists between biological powers and the integration of people into the social group. The Tchikrin also associate hair with sexual powers; lovers get a special thrill from using their teeth to pluck an eyebrow or eyelash from their partner's face (Sanders and Vail 2008; Turner 1969). To Tchikrin people, these practices are no more unusual or exotic than the daily habits we practice in the United States.

To study culture, to analyze it and measure its significance in society, we must separate ourselves from judgments such as "strange" or "normal." We must see a culture as insiders see it, but we cannot be completely taken in by that view. We should know the culture as insiders and understand it as outsiders.

Defining Culture

Culture is the complex system of meaning and behavior that defines the way of life for a given group or society. It includes beliefs, values, knowledge, art, morals, laws, customs, habits, language, and dress, among other things. Culture includes ways of thinking as well as patterns of behavior. Observing culture involves studying what people think, how they interact, and the objects they use.

In any society, culture defines what is perceived as beautiful and ugly, right and wrong, good and bad. Culture helps hold society together, giving people a sense of belonging, instructing them on how to behave, and telling them what to think in particular situations.

---

[1]This introduction is inspired by a classic article on the "Nacirema"—American, backward—by Horace Miner (1956). But it is also written based on essays students at the University of Delaware wrote regarding the media blackout exercise described later in this chapter. Students have written that, without access to their usual media devices, they "had no personality" and that the period of the blackout was the "worst forty-eight hours of my life!"

Culture is both material and nonmaterial. **Material culture** consists of the objects created in a given society—its buildings, art, tools, toys, literature, and other tangible objects, such as those discussed in the chapter opener. In the popular mind, material artifacts constitute culture because they can be collected in museums or archives and analyzed for what they represent. These objects are significant because of the meaning they are given. A temple, for example, is not merely a building, nor is it only a place of worship. Its form and presentation signify the religious meaning system of the faithful.

Nonmaterial culture includes the norms, laws, customs, ideas, and beliefs of a group of people. Nonmaterial culture is less tangible than material culture, but it has an equally strong, if not stronger, presence in social behavior. Nonmaterial culture is found in patterns of everyday life. For example, in some cultures, people eat with utensils; in others, people do not. The eating utensils are part of material culture, but the belief about whether to use them is nonmaterial culture.

Cultural patterns make humans interesting. Some animal species develop what we might call culture. Chimpanzees, for example, learn behavior through observing and imitating others, a point proved by observing different eating practices among chimpanzees in the same species but raised in different groups (Whiten et al. 1999). Elephants have been observed picking up and fondling bones of dead elephants, perhaps evidence of grieving behavior (Meredith 2003). Dolphins have a complex auditory language. Most people also think that their pets communicate with them. Apparently, humans are not unique in their ability to develop systems of communication. Are human beings different from animals? Scientists generally conclude that animals lack the elaborate symbol-based forms of knowing and communication that are common in human societies—in other words, culture.

Understanding culture is critical to knowing how human societies operate. Culture can even shape the physical and biological characteristics of human beings. Nutrition, for instance, is greatly influenced by the cultural environment. Cultural eating habits will shape the body height and weight of a given population, even though height and weight are also biological phenomena. Without understanding culture, you cannot understand such things as changes in idealized images of beauty over time, as the photos on this page show.

In the 1920s, the ideal woman was portrayed as curvaceous with an emphasis on her reproductive characteristics—wide, childbearing hips and large breasts. In more recent years, idealized images of

Thinking Sociologically

Celebrating Your Birthday!

Birthday cake, candles, family and friends singing "Happy birthday to you!" Once a year, you feel like the day is yours. Some people give you presents, send cards, and call you. If you are turning to a legal age, maybe a drinking ritual is involved. If you are older, say turning forty or fifty, perhaps people kid you about "being over the hill." Such are the cultural rituals associated with birthdays in the United States.

How would these rituals change in a different culture? Traditionally, in Vietnam everyone's birthday is celebrated on the first day of the year, and few really acknowledge the day they were born. In Russia, you might get a birthday *pie* with a birthday message carved into the crust. In Newfoundland, you might have butter rubbed on your nose for good luck—the butter is considered too greasy for bad luck to catch you. These cultural practices show how something as seemingly "normal" as celebrating your birthday has strong cultural roots.

What are the norms associated with birthday parties that you have attended? How do these reflect the values in U.S. culture?

JT Vintage/Glasshouse Images/Alamy Stock Photo

Press2000/Alamy Stock Photo

Justin de Villeneuve/Hulton Archive/Getty Images

Body size ideals, which stem from culture, have changed dramatically since the 1950s. Jayne Mansfield was a major star and sex symbol in the 1950s; she was a size 4. Marilyn Monroe was a size 8. When Twiggy became the ideal in the 1960s, she was the equivalent of a size triple zero! Kate Moss, considered now to be "average" size would wear a size 4 dress. In reality, not the ideal, the average American woman wears a size 14!

women have become increasingly thin. Body mass index (BMI) is a measure of relative size, using height and weight. In the 1950s, the body mass index of idolized women, such as Marilyn Monroe, was 20. Now models have a body mass index in the mid-teens, far below the average BMI for U.S. adult women, which is 28. The point is that the media communicate that only certain forms of beauty are culturally valued. These ideals are not "natural"; they are created within a society's culture.

The Power of Culture: Ethnocentrism, Cultural Relativism, and Culture Shock

Would you dice a jellyfish and serve it as a delicacy? Roll a cabbage through your house on New Year's Day to ensure good luck in the year ahead? Peculiar or strange as these examples may seem, from within a particular culture, each seems perfectly normal. Because culture tends to be taken for granted, it can be difficult for people within a culture to see their culture as anything but "the way things are." Seen from outside the culture, everyday habits and practices can seem bizarre, certainly unusual or quirky. Such reactions show just how deeply influential culture is.

We take our own culture for granted to such a degree that it can be difficult to view other cultures without making judgments based on one's own cultural assumptions. **Ethnocentrism** is the habit of seeing things only from the point of view of one's own group. An ethnocentric perspective prevents you from understanding the world as others experience it, and it can lead to narrow-minded conclusions about the worth of diverse cultures.

Any group can be ethnocentric. Ethnocentrism can be extreme or subtle—as in the example of social groups who think their way of life is better than that of any other group. Is there such a ranking among groups in your community? Fraternities and sororities often build group rituals around such claims; youth groups see their way of life as superior to adults; urbanites may think their cultural habits are more sophisticated than those of groups labeled "country hicks." Ethnocentrism is a powerful force because it combines a strong sense of group solidarity with the idea of group superiority.

Ethnocentrism can build group solidarity, but it can limit intergroup understanding (see, for example, ▲ Figure 2-1). Taken to extremes, ethnocentrism can lead to overt political conflict, war, terrorism, even *genocide*, the mass killing of people based on their membership in a particular group. You might wonder how people could believe so much in the righteousness of their religious faith that they would murder people. Ethnocentrism is a key part of the answer. Understanding ethnocentrism does not excuse or fully explain such behavior, but it helps you understand how such murderous behavior can occur.

Contrasting with ethnocentrism is cultural relativism. **Cultural relativism** is the idea that something can be understood and judged only in relation to the cultural context in which it appears. This does not make every cultural practice morally acceptable, but it suggests that without knowing the cultural context, it is impossible to understand why people behave as they do. For example, in the United States, burying or cremating the dead is the cultural practice. It may be difficult for someone from this culture to understand that in parts of Tibet, with a ruggedly cold climate and the inability to dig the soil, the dead

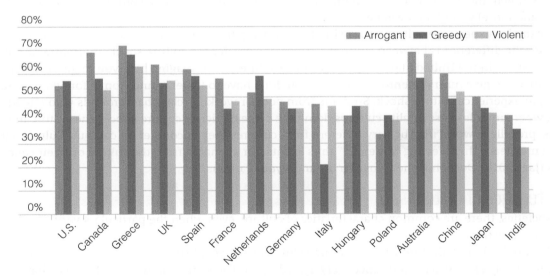

▲ **Figure 2-1   Global Views of Americans.** Most Americans see themselves as hardworking, optimistic, and tolerant of others—a view shared by many around the world. But, people from other nations also tend to see Americans as arrogant, greedy, and violent—a view that many Americans have of themselves, though not so much as in the eyes of certain nations, as this graph shows. Where do such perceptions come from? Do you think the American media has some influence on how people across the globe view American culture? If so, what does this tell you about the power of the media to shape global attitudes?

Data: Wike, Richard, Jabob Poushter, and Hani Zainulbhai. 2016 (June 28). "America's International Image." Pew Research Global Attitudes Project. Washington, DC: Pew Research Project. **www.pewglobal.org**

are cut into pieces and left for vultures to eat. Although this would be repulsive (and illegal) in the United States, this practice is understandable within parts of Tibetan culture.

Understanding cultural relativism gives insight into some controversies, such as the international debate about the practice of clitoridectomy—a form of genital mutilation. In a clitoridectomy (sometimes called female circumcision), all or part of a young woman's clitoris is removed, usually not by medical personnel, often in very unsanitary conditions, and without any painkillers. Sometimes, the lips of the vagina may be sewn together. Human rights and feminist organizations have documented this practice in some countries on the African continent, in some Middle Eastern nations, and in some parts of Southeast Asia. Around two million girls per year worldwide are at risk. This practice is most frequent in cultures where women's virginity is highly prized and where marriage dowries depend on some accepted proof of virginity.

From the point of view of Western cultures, clitoridectomy is genital mutilation—a form of violence against women. Many have called for international intervention to eliminate the practice, but there is also a debate about whether disgust at this practice should be balanced by a reluctance to impose Western cultural values on other societies. Should cultures have the right of self-determination or should cultural practices that maim people be treated as violations of human rights? This controversy is unresolved. The point is to see that understanding a cultural practice requires knowing the cultural values on which it is based. Even if you want to change a cultural practice, you will be better able to do so if you understand its origins.

The power of culture is also revealed when you are placed into a new cultural situation. The result can be **culture shock**, the feeling of disorientation when one encounters a new or rapidly changed cultural situation. The people of Puerto Rico, for example, likely experienced this in the aftermath of the devastating hurricane Maria when the entire island was without electricity for weeks and, in some parts, many months. Many also had no running water, no cell phone service, and no way to communicate with family and friends. The impact of this natural disaster on ordinary social relations—including people's ability to survive—was massive but also a stark reminder of how dependent most of us are on a functioning electrical grid. Indeed, our culture has become so dependent on electronic systems that, should the grid be disrupted, most forms of our contemporary culture (communication, commerce, transportation, banking, and more) would not even be possible.

You can also experience culture shock simply by moving from one cultural environment to another. The greater the difference is between cultural settings, the greater the culture shock. International students who study in the United States experience this routinely. Just imagine how you might feel were you to move to a foreign country to enroll in college. You don't have to travel to and from a foreign nation, however, to experience culture shock. Students from poor or working-class backgrounds who attend colleges where middle-class or elite cultures dominate can experience culture shock. The culture in such students' precollege world may be very different from cultures they encounter on campus. Culture shock for them may leave them feeling "different," thus isolated or alienated from the culture on their college campus (Jack 2014). Does this phenomenon occur on your campus?

Characteristics of Culture

Across societies, there are common characteristics of culture, even when the particulars vary. These different characteristics are:

1. **Culture is shared.** Culture would have no significance if people did not hold it in common. Culture is collectively experienced and collectively agreed upon. The shared nature of culture is what makes human society possible. The shared basis of culture may be difficult to see in complex societies where groups have diverse traditions, perspectives, and ways of thinking and behaving. In the United States, for example, different racial and ethnic groups have unique histories, languages, and beliefs—that is, different cultures. Even within these groups, there are different cultural traditions. Latinos, for example, include many groups with distinct origins and cultures. Still, there are features of Latino culture, such as values and traditions, that are shared. Latinos also share a culture that is shaped by their common experiences as minorities in the United States. Similarly, African Americans have created a rich and distinct culture that is the result of their unique experience within the United States. What identifies African American culture are the practices and traditions that have evolved from both the U.S. experience and African and

Caribbean traditions. Placed in another country, such as an African nation, African Americans would likely recognize elements of the host culture, but they would also feel culturally distinct as Americans.

Within the United States, culture varies by age, race, region, gender, ethnicity, religion, class, and other social factors. A person growing up in the South is likely to develop different tastes, modes of speech, and cultural interests than is a person raised in the West. Despite these differences, there is a common cultural basis to life in the United States. Certain symbols, language patterns, belief systems, and ways of thinking are distinctively American, forming a common culture even with great cultural diversity.

2. Culture is learned. Cultural beliefs and practices are usually so well learned that they seem perfectly natural, but they are learned nonetheless. How do people come to prefer some foods to others? How is musical taste acquired? Culture may be taught through direct instruction, such as a parent teaching a child how to use silverware or teachers instructing children in songs, myths, and other traditions in school.

Culture is also learned indirectly through observation and imitation. Think of how a person learns what it means to be a man or a woman. Although the "proper" roles for men and women may never be explicitly taught, one learns what is expected from observing others. A person becomes a member of a culture through both formal and informal transmission of culture. Until the culture is learned, the person will feel like an outsider. Sociologists refer to the process of learning culture as *socialization*, discussed in Chapter 4.

3. Culture is taken for granted. Because culture is learned, members of a given society seldom question the culture of which they are a part, unless for some reason they become outsiders or establish some critical distance from the usual cultural expectations. People engage unthinkingly in hundreds of specifically cultural practices every day. Culture makes these practices seem "normal." If you suddenly stopped participating in your culture and questioned each belief and every behavior, you would soon find yourself feeling detached and perhaps a little disoriented; you might even become ineffective within your group.

You can see this if you travel outside of your culture, such as visiting a foreign country. Even the simplest things, such as how you eat or even use the toilet, may seem strange and must be learned. As a result, tourists tend to stand out when in a foreign culture. Even when well informed, tourists typically approach the society through the vantage point of their own cultural orientation.

You do not have to leave your home country to observe this. For example, students who have been raised in a cultural group that teaches them to be quiet and not outspoken might be perceived as stupid or "slow" if they are in a classroom where they are expected to assert themselves and be aggressive in debate. Native American students, for example, may experience this. If a teacher is not aware of these cultural differences, such students may be penalized simply for observing their cultural traditions. You can probably think of many other examples in which cultural misunderstanding can lead to isolation of those perceived as different. Culture binds us together, but lack of communication across cultures can have negative consequences.

4. Culture is symbolic. The significance of culture lies in the meaning it holds for people. **Symbols** are things or behaviors to which people give meaning. The meaning in a symbol is not inherent but is bestowed by the meaning people give it. The U.S. flag, for example, is literally a decorated piece of cloth. Its cultural significance derives not from the cloth of which it is made but from its meaning as a symbol of freedom and democracy. Desecration of the flag invokes strong emotional reactions, just as flying it invokes strong feelings of patriotism and pride.

Symbols can also produce social conflict when groups define them differently. For many, especially American Indians, the Native American mascots that name and represent some sports teams is symbolic of the exploitation of Native Americans. To Native American activists and their supporters, such symbols are derogatory and extremely insulting, even when to some sports fans, these very same symbols represent tradition and team pride. (Think of the Washington Redskins, the Cleveland Indians, or the Atlanta Braves' "tomahawk chop.") The protests that have developed over controversial symbols are indicative of the enormous influence of cultural symbols.

The significance of the symbolic value of culture can hardly be overestimated. Learning a culture means not just engaging in particular behaviors but also learning the meanings that various behaviors have within a cultural context.

Tattoos: Status Risk or Status Symbol?

Research Question

Tattoos were once considered a mark of social outcasts, but now tattoos are a symbol of who is trendy and hip. How did a once stigmatized activity associated with the working class become a statement of middle-class fashion?

Research Method

Sociologist Katherine Irwin asked this when she first noticed the increase in tattooing among the middle-class. Irwin hung out in the Blue Mosque tattoo shop and began a four-year study using participant observation, also interviewing people getting their first tattoos, some of the parents of tattooees, and potential tattooees.

Research Results

Irwin found that middle-class tattoo patrons were initially fearful that their desire for a tattoo would associate them with low-status groups. They reconciled this by defining tattooing as symbolic of independence, liberation, and freedom from social constraints. Although tattoos held different cultural meanings to different groups, people getting tattooed used various techniques (what Irwin calls "legitimation techniques") to counter the negative stereotypes associated with tattooing. Women, for example, saw tattoos as symbolizing toughness and strength, thus rejecting conventional ideals of femininity.

Conclusions and Implications

Irwin concludes that people try to align their behavior with legitimate cultural values and norms even when that behavior seemingly falls outside of prevailing standards.

Questions to Consider

1. Do you think of tattoos as fashionable or deviant? What influences your judgment about this, and how might your judgment be different were you in a different culture, age group, or historical moment?
2. Are there fashion adornments that you associate with different social classes? What are they? What judgments (positive and negative) do people make about them? Where do these judgments originate? Are they associated with social class?

Source: Irwin, Katherine. 2001. "Legitimating the First Tattoo: Moral Passage through Informal Interaction." *Symbolic Interaction* 24 (March): 49–73.

5. Culture varies across time and place. Culture develops as humans adapt to the physical and social environment around them. Culture is not fixed from one place to another. Not that long ago, it would have been unimaginable to think that one could have access to one's favorite movie or television series "on demand." With the growth of technological innovation, people can now stream video and music when they wish—a cultural change that generates other adaptations, such as how furniture is arranged in people's homes. A video screen may now be the focal point for family gatherings, not the kitchen table of yesteryear. In a different cultural context, news and entertainment might be simply shared through oral traditions. Now you might get news alerts instantaneously on your smartphones or even a smart watch. These fast-paced technological changes are even penetrating remote areas of the world.

Culture also varies over time. As people encounter new situations, the culture that emerges is a mix of the past and present. Children of immigrants typically grow up with both the traditional cultural expectations of their parents' homeland and the cultural expectations of a new society. However, adapting to the new society can create conflict between generations, especially if the older generation is intent on passing along their cultural traditions. The children may be more influenced by their peers and may choose to dress, speak, and behave in ways that are characteristic of their new society but are unacceptable to their parents.

To sum up, culture is concrete because we can observe the cultural objects and practices that define human experience. Culture is abstract because it is a way of thinking, feeling, believing, and behaving. Culture links the past and the present because it is the knowledge that makes us part of human groups. Culture gives shape to human experience.

The Elements of Culture

Culture is multifaceted, consisting of material and nonmaterial things. Some parts of culture are abstract; others, more concrete. The different elements of culture include language, norms, beliefs, and values (see ◆ Table 2-1).

Language

Language is a set of symbols and rules that, combined in a meaningful way, provides a complex communication system. Human culture is made possible by language. Learning the language of a culture is essential to becoming part of society, and it is one of the first things children learn. Indeed, until children acquire at least a rudimentary command of language, it is very difficult for them to acquire other social skills. Language is so important to human interaction that it is difficult to think of life without it.

Table 2-1   **Elements of Culture**

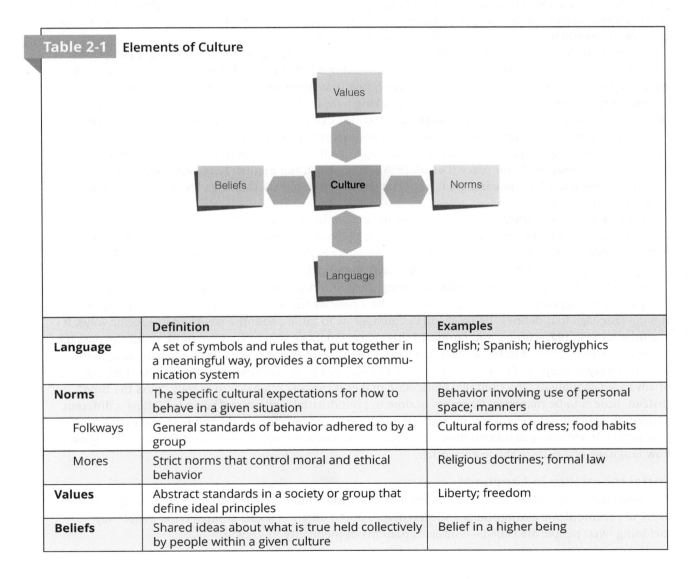

	Definition	Examples
Language	A set of symbols and rules that, put together in a meaningful way, provides a complex communication system	English; Spanish; hieroglyphics
Norms	The specific cultural expectations for how to behave in a given situation	Behavior involving use of personal space; manners
Folkways	General standards of behavior adhered to by a group	Cultural forms of dress; food habits
Mores	Strict norms that control moral and ethical behavior	Religious doctrines; formal law
Values	Abstract standards in a society or group that define ideal principles	Liberty; freedom
Beliefs	Shared ideas about what is true held collectively by people within a given culture	Belief in a higher being

Think about the experience of becoming part of a social group. When you enter a new social group (or society), you have to learn the group's language to become a member of the group. This includes any special terms of reference used by the group. Lawyers, for example, have their own vocabulary and their own way of constructing sentences called, not always kindly, "legalese." Becoming a part of any social group—a friendship circle, fraternity or sorority, or any other group—involves learning the language that group members use. Those who do not share the language of a group cannot participate fully in its culture.

Language is fluid and dynamic, evolving in response to social change. Think, for example, of how the introduction of computers has affected the English language. People now talk about "downloading apps," "hashtags," and providing "input." Only a few years ago, had you said you were going to "text" your friends, no one would have known what you were talking about. Text messaging has also introduced its own language: BFF (best friends forever), LOL (laughing out loud), and GTG (got to go)—a new language shared among those in the text-messaging culture. These expressions are now commonplace—in other words, a new form of culture. No doubt, by the time you read this, some of these examples may even feel dated, and new tech lingo will have emerged and become familiar—evidence of how culture can change, sometimes rapidly; other times, slowly.

Does Language Shape Culture?

Language is clearly a big part of culture. Edward Sapir and his student Benjamin Whorf thought that language was central in determining social thought. The **Sapir-Whorf hypothesis** asserts that language determines other aspects of culture because language provides the categories through which social reality is defined. The idea is that language determines what people think because language forces people to perceive the world in particular ways (Whorf 1956; Sapir 1921).

If Sapir and Whorf were correct, then speakers of different languages have different perceptions of reality. Whorf used the example of the social meaning of time to illustrate cultural differences in how language shapes perceptions of reality. He noted that the Hopi Indians conceptualize time as a slowly turning cylinder, whereas English-speaking people conceive of time as running forward in one direction at a uniform pace. Linguistic constructions of time shape how the two different cultures think about time and therefore how they think about reality. In Hopi culture, events are located not in specific moments of time but in "categories of being," as if everything is in a state of becoming, not fixed in a particular time and place (Carroll 1956). In contrast, the English language locates things in a definite time and place, placing great importance on verb tense, with things located precisely in the past, present, or future.

Language does not single-handedly dictate the perception of reality—but, no doubt, language has a strong influence on culture. Most scholars now see two-way causality between language and culture. Asking whether language determines culture or vice versa is like asking which came first, the chicken or the egg. Language and culture are inextricable, each shaping the other.

Consider again the example of time. Contemporary Americans think of the week as divided into two parts: *weekdays* and *weekends*, words that reflect how we think about time. When does a week end? Having language that defines the weekend encourages us to think about the weekend in specific ways. It is a time for rest, play, chores, and family. In this sense, language shapes how we think about the passage of time—we look forward to the weekend, we prepare ourselves for the work week—but the language itself (the very concept of the weekend) stems from patterns in the culture—specifically, the work patterns of advanced capitalism. The capitalist work ethic makes it morally offensive to merely "pass the time." Instead, time is to be managed. Concepts of time in preindustrial, agricultural societies follow a different pattern. In agricultural societies, time and calendars are based on agricultural and seasonal patterns; the year proceeds according to this rhythm, not the arbitrary units of time of weeks and months. This shows how language and culture shape each other.

Social Inequality in Language

The language of any culture reflects the nature of that society. In a society with inequality, language is likely to communicate assumptions and stereotypes about different social groups. What people say—including what people are called—reinforces patterns of inequality in society.

We see this in what different groups in the United States are called (see also the box later in this chapter called, "Understanding Diversity: The Social Meaning of Language"). What someone is called is significant because it imposes an identity on that person. This is why the names for various racial and ethnic groups have been so heavily debated and contested. For example, many Native Americans objected for years to being called "Indian," because White conquerors created that term about them. To emphasize their native roots in the Americas, the term *Native American* was adopted. Now, though many prefer to be called by their actual origin, *Native American* and *American Indian* are also used interchangeably. Likewise, Asian Americans are generally offended by being called "Oriental," an expression that stemmed from Western (that is, European and American) views of Asian nations.

Living in a multicultural society often juxtaposes diverse cultures, even in public places.

Language reflects the social value placed on different groups. Language also reflects power relationships, depending on who gets to name whom. Derogatory terms such as *redneck, white trash,* or *trailer park trash* stigmatize people based on regional identity and social class. This is also why it is so demeaning when derogatory terms are used to describe racial–ethnic groups. For example, throughout the period of Jim Crow segregation in the American South, Black men, regardless of their age, were routinely called "boy" by Whites. Calling a grown man a "boy" is an insult because it diminishes his status by referring to him as childlike. Calling a woman as a "girl" has the same effect. Why are young women, even well into their twenties, routinely referred to as "girls"? Just as does calling a man "boy," this diminishes women's status.

Note, however, that terms such as *girl* and *boy* are pejorative only in the context of dominant and subordinate group relationships. African American women, as an example, often refer to each other as "girl" in informal conversation. The term *girl* used between those of similar status is not perceived as derogatory, but when used by someone in a position of dominance, such as when a male boss calls his secretary a "girl," it is demeaning. Likewise, terms such as *dyke, fag,* and *queer* are terms lesbians and gay men sometimes use without offense in referring to each other, even though the same terms are offensive to lesbians and gays when others use them. By reclaiming these terms as positive within their own culture, lesbians and gays build cohesiveness and solidarity (Due 1995). These examples show that power relationships between groups supply the social context for the connotations of language.

In sum, language can reproduce the inequalities that exist in society. At the same time, changing the language that people use can, to some extent, alter social stereotypes and thereby change how people think.

Debunking Society's Myths

Myth: Bilingual education discourages immigrant children from learning English and blocks their assimilation into American culture.

Sociological Perspective: Studies of students who are fluent bilinguals show that they outperform both English-only students and students with limited bilingualism. Moreover, preserving the use of native languages can better meet the need for skilled bilingual workers in the labor market (Portes 2002).

Norms

Social norms are another component of culture. **Norms** are the specific cultural expectations for how to behave in a given situation. Society without norms would be chaos. With norms in place, people know how to act, and social interactions are consistent, predictable, and learnable. There are norms governing every situation. Sometimes norms are implicit—that is, they need not be spelled out for people to understand them. For example, when joining a line, there is an implicit norm that you should stand behind the last person, not barge in front of those ahead of you. At least this is true in the United States, not always in other cultures. Implicit norms may not be formal rules, but violation of these norms may nonetheless produce a harsh response. Implicit norms may be learned through specific instruction or by observation of the culture. Norms are part of society's (or a group's) customs. Norms are explicit when the rules governing behavior are written down or formally communicated. Typically, specific sanctions are imposed for violating explicit norms.

In the early years of sociology, William Graham Sumner (1906) identified two types of norms: folkways and mores. **Folkways** are the general standards of behavior adhered to by a group. Folkways are the ordinary customs of different group cultures. How you dress is an example of a cultural folkway. Other examples are how people greet each other, decorate their homes, and prepare their food. Folkways are loosely defined and loosely followed. Either way, they structure group customs and implicitly govern much social behavior.

Mores (pronounced "more-ays") are strict norms that control moral and ethical behavior. Mores provide strict codes of behavior, such as the injunctions, legal and religious, against killing others and committing adultery. Mores are often upheld through **laws**, the written set of guidelines that define right and wrong in society. Basically, laws are formalized mores. Violating mores can bring serious repercussions. When any social norm is violated, the violator is typically punished.

Social sanctions are mechanisms of social control that enforce folkways, norms, and mores. The seriousness of a social sanction depends on how strictly the norms or mores are held. **Taboos** are those behaviors that bring the most serious sanctions. Dressing in an unusual way that violates the folkways of dress may bring ridicule but is usually not seriously punished. In some cultures, the rules of dress are strictly interpreted, such as the requirement by Islamic fundamentalists that women who appear in public must have their bodies cloaked and faces veiled. It would be considered a taboo for women in this culture to appear in public without being veiled. The sanctions for doing so can be as severe as whipping, branding, banishment, even death.

Sanctions can be positive or negative, that is, based on rewards or punishment. When children learn social norms, for example, correct behavior may elicit positive sanctions; the behavior is reinforced through praise, approval, or an explicit reward. Early on, for example, parents might praise children for learning to put on their own clothes. Later, children might get an allowance if they keep their rooms clean. Bad behavior earns negative sanctions, such as getting spanked or grounded. In society, negative sanctions may be mild or severe, ranging from subtle mechanisms of control, such as ridicule, to overt forms of punishment, such as imprisonment, physical coercion, or death.

One way to study social norms is to observe what happens when they are violated. Once you become aware of how social situations are controlled by norms, you can see how easy it is to disrupt situations where adherence to the norms produces social order. **Ethnomethodology** is a theoretical approach in sociology based on the idea that you can discover the normal social order through disrupting it. As a technique of study, ethnomethodologists often deliberately disrupt social norms to see how people respond, thus revealing the ordinary social order (Garfinkel 1967).

See for Yourself

Identify a *norm* that you commonly observe. Construct an experiment in which you, perhaps with the assistance of others, violate the norm. Record how others react and note the sanctions engaged through this norm violation exercise. Note: Be careful not to do anything that puts you in danger or causes serious problems for others.

In a famous series of ethnomethodological experiments, college students were asked to pretend they were boarders in their own homes for a period of fifteen minutes to one hour. They did not tell their families what they were doing. The students were instructed to be polite and impersonal, to use formal terms of address, and to speak only when spoken to. After the experiment, two of the participating students reported that their families treated the experiment as a joke; another's family thought the daughter was being extra nice because she wanted something. One family believed that the student was hiding some serious problem. In all the other cases, parents reacted with shock, bewilderment, and anger. Students were accused of being mean, nasty, impolite, and inconsiderate; the parents demanded explanations for their sons' and daughters' behavior. Through this experiment, the student researchers were able to see that even the informal norms governing behavior in one's home are carefully structured. By violating the norms of the household, the norms were revealed (Garfinkel 1967).

Ethnomethodological research teaches us that society proceeds on an "as if" basis. That is, society exists because people behave as if there were no other way to do so. Usually, people go along with what is expected of them. Culture is actually "enforced" through the social sanctions applied to those who violate social norms. Usually, specific sanctions are unnecessary because people have learned the normative expectations. When the norms are violated, their existence becomes apparent (see also Chapter 5).

Beliefs

As important as social norms are, the beliefs of people in society are also important. **Beliefs** are shared ideas held collectively by people within a given culture about what is true. Shared beliefs are part of what binds people together in society. Beliefs are also the basis for many norms and values of a given culture. In the United States, belief in God or a higher power is widely shared.

Some beliefs are so strongly held that people find it difficult to cope with ideas or experiences that contradict them. Someone who devoutly believes in God may find atheism intolerable. Those who believe in magic may seem merely superstitious to those with a more scientific and rational view of the world.

Whatever beliefs people hold, they orient us to the world. They provide answers to otherwise imponderable questions about the meaning of life. Beliefs provide a meaning system around which culture is organized. Whether belief stems from religion, myth, folklore, or science, it shapes what people take to be possible and true. Although a given belief may be logically impossible, it nonetheless guides people through their lives.

Values

Deeply intertwined with beliefs are the values of a culture. **Values** are the abstract standards in a society or group that define ideal principles. Values define what is desirable and morally correct, determining what is considered right and wrong, beautiful and ugly, good and bad. Although values are abstract, they provide a general outline for behavior. Freedom, for example, is a value held to be important in U.S. culture, as is democracy. Values are ideals forming the abstract standards for group behavior, but they are also ideals that may not be realized in every situation.

The Social Meaning of Language

Language reflects the assumptions of a culture. This can be seen and exemplified in several ways:

- **Language affects people's perception of reality.**
  Example: Researchers have found that using male pronouns, even when intended to be gender neutral, produces male-centered imagery and ideas (Switzer 1990; Hamilton 1988).
- **Language reflects the social and political status of different groups in society.**
  Example: The term working man has a different connotation than the term working woman. Why?

(continued)

- **Groups may advocate changing language that refers to them as a way of asserting a positive group identity.**
  Example: Advocates for the disabled challenge the term *handicapped*, arguing that it stigmatizes people who may have many abilities, even if they are physically distinctive.
- **Language emerges in specific historical and cultural contexts.**
  Example: The naming of so-called races comes from the social and historical processes that have defined different groups as inferior or superior. The term *Caucasian*, for example, was coined in the seventeenth century by Alfred Blumenbach to refer to people from the Caucasus of Russia whom he thought were more beautiful and intelligent than any other people in the world.
- **Language can distort actual group experience.**
  Example: The terms *Hispanic* and *Latino* point to the shared experience of those from Latin cultures (Chicanos, Puerto Ricans, Hondurans, and so forth), but like the terms *Native American* and *American Indian*, the terms obscure the experiences of unique groups (such as the Lakota, Nanticoke, or Navajo).
- **Language shapes people's perceptions of groups and events in society.**
  Example: Following Hurricane Katrina in New Orleans, African American people taking food from abandoned stores were described as "looting" and White people as "finding food."
- **Terms used to define different groups change over time and can originate in movements to assert a positive identity.**
  Example: In the 1960s, Black American replaced the term Negro because the Civil Rights and Black Power movements inspired Black pride and the importance of self-naming. Earlier, Negro and colored were used to define African Americans. Now, people of color is used to refer to the solidarity of different racial-ethnic groups. Language is fraught with cultural and political assumptions.

The best way to solve this problem is for different groups to become more aware of the meaning and nuances of naming and language in order to promote better intergroup relationships.

Values can be a basis for cultural cohesion, but they can also be a source of conflict. Some of our most contested issues can often be traced to value conflicts. Should sex education be taught in schools? Should public schools allow school prayer? Should women have the right to choose to terminate a pregnancy? These and numerous other examples you can likely identify are matters of great debate—debates made more heated by the value conflicts that lie at the core of these public issues.

Values guide the behavior of people in society; they also shape social norms. An example of the impact that values have on people's behavior comes from an American Indian society known as the Kwakiutl (pronounced "kwa-kee-YOO-tal"), a group from the coastal region of southern Alaska, Washington State, and British Columbia. The Kwakiutl developed a practice known as *potlatch*, in which wealthy chiefs would periodically pile up their possessions and give them away to their followers and rivals (Wolcott 1996; Harris 1974; Benedict 1934). The object of potlatch was to give away or destroy more of one's goods than did one's rivals. The potlatch reflected Kwakiutl values of reciprocity, the full use of food and goods, and the social status of the wealthiest chiefs in Kwakiutl society. Chiefs did not lose their status by giving away their goods because the goods were eventually returned in the course of other potlatches. They would even burn large piles of goods, knowing that others would soon replace their wealth through other potlatches.

Compare this practice with patterns of consumption in the United States. Imagine the CEOs of major corporations regularly gathering up their wealth and giving it away to their workers and rival CEOs! In the United States, *conspicuous consumption* (consuming for the sake of displaying one's wealth) celebrates values similar to those of the potlatch: High-status people demonstrate their position by accumulating more material possessions than those around them (Veblen 1953/1899).

Together, norms, beliefs, and values guide the behavior of people in society. It is necessary to understand how they operate in a situation to understand why people behave as they do.

Cultural values can clash when groups have strongly held, but clashing, value systems. Values can be a source of cultural cohesion, but also of cultural conflict. What are some of the different values that are being debated in society?

Cultural Diversity

It is rare for a society to be culturally uniform. Especially as societies develop and become more complex, different cultural traditions appear. In the United States, diversity is an integral and complex part of the national history, including religious, ethnic, and racial differences, as well as regional, national origin, age, gender, and class differences. Currently, 12.9 percent of people in the United States are foreign born. Whereas earlier immigrants were predominantly from Europe, now Latin America and Asia are the greatest sources of new immigrants (U.S. Census Bureau 2017). Additionally, racial-ethnic minority groups now comprise 37 percent of the total population and that number is expected to grow to 57 percent by 2060 (U.S. Census Bureau 2012). Cultural diversity is clearly a characteristic of contemporary U.S. society. (See ■ map 2-1.)

The richness of American culture stems from the many traditions that different groups have brought to this society, as well as from the cultural forms that have emerged through their experience within the United States. Jazz, for example, is a musical form indigenous to the United States. An indigenous art form refers to something that originated in a particular region or culture. Jazz also has its roots in the musical traditions of slave communities and African cultures. Since the birth of jazz, cultural greats such as Ella Fitzgerald, Duke Ellington, Billie Holiday, and numerous others have not only enriched national music traditions but have also influenced other forms of music, including rock and roll.

Native American cultures have likewise enriched the culture of our society, as have the cultures that various immigrant groups have brought with them to the United States. With such great variety, how can the United States be called one culture? The culture of the United States, including its languages, arts, food customs, religious practices, and dress, can be seen as the sum of the diverse cultures that constitute this society.

Dominant Culture

Two concepts from sociology help us understand the complexity of culture in a given society: dominant culture and subculture. The **dominant culture** is the culture of the most powerful group in a society. Although the dominant culture is not the only culture in a society, it is commonly believed to be "the"

With increased immigration and greater diversity in the U.S. population, evidence of cultural diversity can be seen in many homes. This map shows the regional differences in the percentage of the population over age 5 who speak a language other than English at home. Twenty percent of the U.S. population do not speak English at home. Of those, 40 percent say they do not speak English very well. Contrary to what you might think, many of those who speak something other than English at home (44 percent) are not immigrants. What implications does this have for the regions most affected?

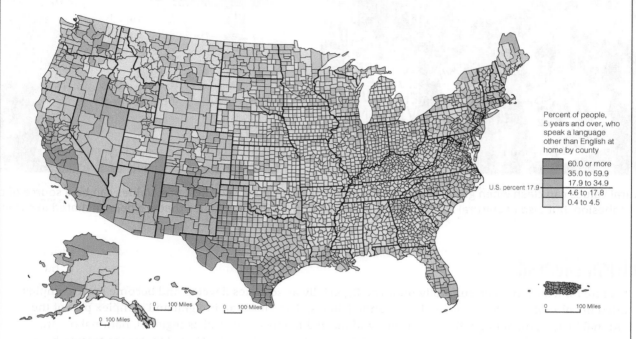

Percent of people,
5 years and over, who
speak a language
other than English at
home by county

60.0 or more
35.0 to 59.9
17.9 to 34.9
4.6 to 17.8
0.4 to 4.5

U.S. percent 17.9

Source: Zeigler, Karen, and Stephen A. Camarota. 2014 (October). "One in Five U.S. Residents Speaks Foreign Language at Home, Record 61.8 Million." Washington, DC: Center for Immigration Studies. **www.cis.org**

culture of a society, despite the other cultures present. Social institutions in the society perpetuate the dominant culture and give it a degree of legitimacy that other cultures do not share. Quite often, the dominant culture is the standard by which other cultures in the society are judged.

A dominant culture need not be the culture of the majority of people. It is simply the culture of the most powerful group in society who have the power to define the cultural framework. An example comes from the phrase Donald Trump adopted for his presidential campaign and, then, administration: "Make America Great Again." Although the reference was not specific and, in fact, has been controversial, the slogan seems to appeal to a nostalgic past where Anglo-European, Christian values dominated. Now the dominant culture is increasingly diverse and it is hard to isolate a single dominant culture, but there is a widely acknowledged "American" culture that is considered to be the dominant one. This culture is strongly influenced by the mass media, as we will see. The dominant culture includes not only various beliefs and practices, but also values that emphasize achievement and individual effort—a cultural tradition that we will later see has a tremendous impact on how many in the United States view inequality (see Chapter 8).

Sometimes, culture can be imposed on a group, as often happens in totalitarian regimes. Indigenous groups (that is, native populations) may not be allowed to speak their own language in public settings or women may not be allowed to appear in public in certain forms of dress. Even in more democratic societies, the dominant culture comes from the power of particular groups, not just the numerical proportion of a population. On a college campus, for example, even with a strong system of fraternities and sororities, the number of students belonging to the Greek system may be a numerical minority of the total student body. Still, the campus culture may be dominated by Greek life. That is, Greek life may become the dominant culture, even though only a minority of students participate in it. Indeed, one of the things that is so interesting about culture is its connection to the overall structure, including power dynamics, in a given society or group.

Subcultures

Subcultures are the cultures of groups whose values and norms of behavior differ to some degree from those of the dominant culture. Members of subcultures tend to interact frequently with one another and share a common worldview. They may be identifiable by their appearance (style of clothing or adornments) or perhaps by language, dialect, or other cultural markers. You can view subcultures along a continuum of how well they are integrated into the dominant culture. Subcultures typically share some elements of the dominant culture and coexist within it, although some subcultures may be quite separated from the dominant one. This separation occurs because they are either unwilling or unable to assimilate into the dominant culture, that is, to share its values, norms, and beliefs (Dowd and Dowd 2003).

Rap and hip-hop music first emerged as a subculture as young African Americans developed their own style of dress and music to articulate their resistance to the dominant White culture. Now, rap and hip-hop have been incorporated into mainstream youth culture. Indeed, they are now global phenomena, as cultural industries have turned hip-hop and rap into a profitable market. Even so, rap still expresses an oppositional identity for Black and White youth and other groups who feel marginalized by the dominant culture (Morgan 2010, 2009).

Some subcultures retreat from the dominant culture, such as the Amish, some religious cults, and some communal groups. In these cases, the subculture is actually a separate community that lives as independently from the dominant culture as possible. Other subcultures may coexist with the dominant society, and members of the subculture may participate in both the subculture and the dominant culture.

The Amish people form a subculture in the United States, although preserving their traditional way of life can be a challenge in the context of contemporary society.

Thinking Sociologically

Identify a group on your campus that you would call a *subculture*. What are the distinctive norms of this group? Based on your observations of this group, how would you describe its relationship to the dominant culture on campus?

Subcultures also develop when new groups enter a society. Puerto Rican immigration to the U.S. mainland, for example, has generated distinct Puerto Rican subcultures within many urban areas. Although Puerto Ricans also partake in the dominant culture, their unique heritage is part of their subcultural experience. Parts of this culture are now entering the dominant culture, such as salsa music. The themes in salsa mix the musical traditions of other Latin music, including rumba, mambo, and cha-cha. As with other subcultures, the boundaries between the dominant culture and the subculture are permeable, resulting in cultural change as new groups enter society.

Countercultures

Countercultures are subcultures created as a reaction against the values of the dominant culture. Members of the counterculture reject the dominant cultural values, often for political or moral reasons, and develop cultural practices that explicitly defy the norms and values of the dominant group. Nonconformity to the dominant culture is often the hallmark of a counterculture. Youth groups often form countercultures. Why? In part, they do so to resist the culture of older generations, thereby asserting their independence and identity. Countercultures among youth, like other countercultures, usually have a unique way of dress, their own special language, perhaps even different values and rituals.

Some countercultures directly challenge the dominant society. The white supremacist movement is an example. People affiliated with this movement have an extreme worldview, one that is anchored in racism, anti-Semitism, and overt homophobia. Within such a counterculture, members likely close themselves off from other perspectives, thus feeding their reactionary view. As we have seen in recent white supremacist actions, such as the march in Charlottesville, Virginia, in 2017, white supremacist countercultures that are so focused on hate can be very dangerous (Daniels 2016; Ferber 1998).

Countercultures may also develop in situations where there is political repression and some groups are forced "underground." Under a dictatorship, for example, some groups may be forbidden to practice their religion or speak their own language. In Spain, under the dictator Francisco Franco, people were forbidden to speak Catalan—the language of the region around Barcelona. When Franco died in 1975 and Spain became more democratic, the Catalan language flourished—both in public speaking and in the press.

The Globalization of Culture

The infusion of Western culture throughout the world seems to be accelerating as the commercialized culture of the United States is marketed worldwide. One can go to quite distant places in the world and see familiar elements of U.S. culture, whether it is McDonald's in Hong Kong, Old Navy in Japan, or Disney products in western Europe. From films to fast food, the United States dominates, largely through the influence

Cultural diffusion is occurring as U.S. culture is being exported to other nations, as well as the other way around. This photo shows the Old Navy store that opened in Tokyo, Japan.

The Asahi Shimbun/Getty Images

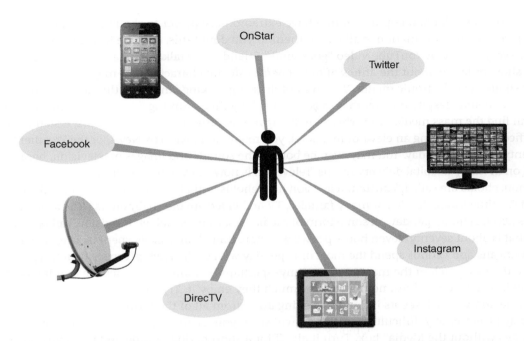

You can see how strong cultural monopolies have become if you just imagine how surrounded you are, even as an individual, by various devices (many of them owned by the same company) that deliver culture to you.

Photos: Phone, Maxx-Studio/ Shutterstock.com; Satellite Dish, Roobcio/Shutterstock.com; LCD, Oleksiy Mark/Shutterstock.com; Tablet, Telnov Oleksii/Shutterstock.com

of capitalist markets. The diffusion of a single culture throughout the world is referred to as **global culture**. Despite the enormous diversity of cultures worldwide, western and U.S. markets increasingly dominate fashion, food, entertainment, and other cultural values, thereby creating a more homogenous world culture. Global culture is increasingly marked by capitalist interests, squeezing out the more diverse folk cultures that have been common throughout the world (Steger 2009).

Does the increasing globalization of culture change traditional cultural values? Some worry that globalization imposes Western values on non-Western cultures, thus eroding long-held cultural traditions. Global economic change can also introduce more tolerant values to cultures that might have had a narrower worldview previously. As globalization occurs, *both* economic changes *and* traditional cultural values shape the emerging national culture of different societies.

The conflict between traditional and more commercial values is now being played out in world affairs. Some of the conflicts in international relations are rooted in a struggle between the values of the consumer-based, capitalist Western culture and the traditional values of local communities. As some people resist the influence of market-driven values, movements to reclaim or maintain ethnic and cultural identity can intensify, such as seen among extremist groups in the Middle East, even while pro-democratic movements also exist there.

The Mass Media and Popular Culture

Increasingly, culture in the United States and around the world is dominated and shaped by the mass media, including popular culture. Indeed, the culture of the United States is so infused by the media that, when people think of U.S. culture, they are likely thinking of something connected to the media—television, film, video, and so forth. The term **mass media** refers to the channels of communication that are available to wide segments of the population—both print and electronic. **Popular culture** refers to the beliefs, practices, and objects that are part of everyday traditions, such as music and video, mass-marketed books and magazines, newspapers, and websites.

Both the mass media and popular culture have extraordinary power to shape culture, including what people believe and the information available to them. If you doubt this, observe how much the media affect your everyday life. A YouTube video "goes viral." Friends may talk about last night's episode of a particular show or laugh about the antics of their favorite sitcom character. You may have even met your partner or spouse via electronic media. Your way of dressing, talking, and even thinking has likely been shaped by the media, despite the fact that most people deny this, claiming "they are just individuals."

You can find the mass media everywhere—in living rooms, airports, classrooms, bars, restaurants, and doctor's offices. Even entering an elevator in a hotel, you might find CNN, Fox News, or the Weather Channel on twenty-four hours a day. You may even be born to the sounds and images of television, because they are turned on in many hospital delivery rooms. Television is now so ever-present in our lives that 42 percent of all U.S. households are called "constant television households"—that is, households where television is on most of the time (Gitlin 2007). For many families, TV and video are the "babysitters." The average person spends about 11 hours per day on some form of media (smartphone, tablet, television, radio, and other devices). That is about seventy-seven hours per week—more time than is likely spent in school or at work. African Americans and Latinos spend the most time per day with such viewing, especially television.

Despite the vast reach of the mass media, many—perhaps including you—believe that it has little effect on their beliefs and values, no matter how much they enjoy it. Try to do without it for even a brief period of time and you will see its influence. Getting away from all of the forms of media that permeate daily life may be extremely difficult to do, as you will see if you try the experiment in the "See for Yourself: Two Days without the Media" box. Turn it all off for a short period of time and see if you feel suddenly left out of society. Then ask yourself how the mass media influence your life, your opinions, your values, and even how you look!

Death of a Superstar

When singer-songwriter Prince tragically died in April 2016, millions grieved his passing, and public tributes to him blanketed social media. When Aretha Franklin died in 2018, her funeral and a huge musical tribute were broadcast live on major national television networks, with millions tuning in. How can so many people be so moved by these deaths, even when they do not know the people personally?

In our celebrity culture, media coverage of the deaths of superstars tells a common tale: the tragic and premature loss of someone with enormous talent who rose from common origins to soaring heights of wealth, popularity, and power. As sociologist Karen Sternheimer writes, "Celebrity and fame are unique manifestations of our sense of American social mobility; they provide the illusion that material wealth is possible for anyone" (2011: xiii).

Emile Durkheim would say that celebrity funerals engage the *collective consciousness*, binding

Like other superstars who die while still popular, Whitney Houston's funeral in 2012 was viewed by millions on TV and brought out many memorials, including this one at the church where her singing started.

us together and reaffirming our collective beliefs and values. But you don't have to wait for a tragic death to see the cultural ideal of the American dream retold through the media. Observe celebrity culture with a sociological perspective and ask yourself where, when, and how you see the American dream replayed through various media reports.

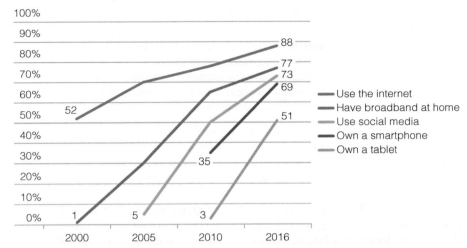

▲ **Figure 2-2   The Rapid Evolution of Electronic Media.** As you can see, the use of electronic forms of media by U.S. adults is changing rapidly. This graph shows the rapid expansion of such media in a mere sixteen years. How might the data change over the next five years? Ten?

Source: Pew Research Center. January 11, 2017. **www.research.org**

With the growth of digital viewing, the time people spend with media is increasing (see ▲ Figure 2-2). Not surprisingly, young people (those between 18 and 29) are experiencing the largest increases in time spent on the Internet. The growth of electronic media is radically changing how people communicate, including about current events. Facebook, Twitter, Instagram, Snapchat, and other electronic networks have become such a common form of interaction that they are now referred to as **social media**—the term used to refer to the vast networks of social interaction that new media have inspired. Eight in ten of Americans who have Internet access now use Facebook, an increasingly popular way that people communicate, including for political information and organization. Indeed, there are almost two billion users of Facebook worldwide—almost one-quarter of the world's population (The Statistics Portal 2017).

How does such rapid change in technology affect our social behavior? As you can see in ▲ Figures 2-3 and ▲ 2-4, social media use is influenced by such social factors as age, social class, and race/ethnicity. Altogether, two-thirds of cell phone owners say they check their phones regularly for messages, alerts, and calls—even when the phone has not rung. Ninety-three percent of young people aged 18 to 24 now use a cell phone (Pew Research Internet Project 2017; Rainie 2013). Two-thirds of all cell phone users say they check it constantly, even keeping it by their bed at night. While people say they cannot imagine living without their phone, they also think they spend too much time on their phones and the Internet. People also report that it makes it harder to separate work from personal life and to focus on a single task without being distracted (Friedman 2016; Smith 2012).

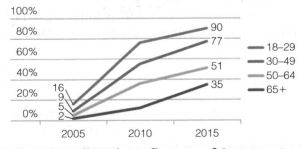

▲ **Figure 2-3   Who Uses Social Media? The Influence of Age.** Youth are often the first to pick up new social media. As you can see, usage of social media by all age groups has been increasing, but why do you think youth are the most likely users? What factors influence the change in usage over time?

Source: Pew Research Center. 2017 (January 22). *Social Media Fact Sheet*. Washington, DC: Pew Research Center. **www.pewinternet.org**

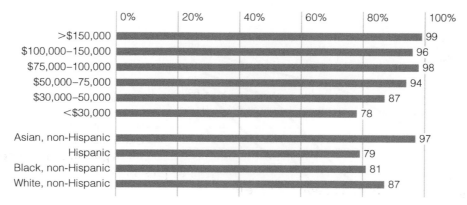

▲ **Figure 2-4   The Digital Divide.** Even with the widespread availability of the Internet, there are still significant social class and race differences in who uses the Internet. What difference do you think this makes in the daily lives of those in different income brackets and in different racial/ethnic groups? What social policies might you suggest for remedying this so that there is less of a *digital divide*?

Data: Rainie, Lee. 2016. "Digital Divides 2016." Washington, DC: Pew Research Center. **www .pewinternet.org**

The Organization of Mass Media

Mass media are not only a pervasive part of daily life; they are also a huge business. The media are owned by an increasingly small number of companies—companies that form huge media monopolies. This means that a few very powerful groups—media conglomerates—are the major producers and distributors of culture. A single corporation can control a huge share of television, radio, newspapers, music, publishing, film, and the Internet. Because the media are concentrated in the hands of a few, there may be less diversity in the content as producers try to appeal to the widest possible audience by keeping the content superficial. Moreover, advertisers are as much the customer as is the audience. Content can then be geared to those interests (Iyengar 2010).

The organization of the mass media as a system of economic interests means that there is enormous power in the hands of a few to shape the culture of the whole society. Sociologists refer to the concentration of cultural power as **cultural hegemony** (pronounced "heh-JeM-o-nee"), defined as the pervasive and excessive influence of one culture throughout society. Cultural hegemony means that people may conform to cultural patterns and interests that benefit powerful elites, even without those elites overtly forcing people into conformity. Although there seems to be enormous choice in what media forms people consume, the cultural messages are largely homogenous (meaning "the same"). Cultural monopolies are then a means by which powerful groups gain the assent of those they rule. The concept of cultural hegemony implies that culture is highly politicized, even if it does not appear to be so. Those who control cultural institutions can control people's political awareness by creating cultural beliefs that make the rule of those in power seem inevitable and right. As a result, political resistance to the dominant culture is blunted (Gramsci 1971). We explore this idea further in the discussion on sociological theories of culture.

Debunking Society's Myths

Myth: Teens are addicted to social media, isolating them from face-to-face interaction.
Sociological Perspective: People tend to misuse the term *addiction* by referring to activities that people enjoy and engage in frequently. Teens say they spend more time on social media than they would like, but policies that prevent teens from gathering in public places push them on to social media more than they actually say they would like (Boyd 2014).

Race, Gender, and Class in the Media

Many sociologists argue that the mass media can promote narrow definitions of who people are and what they can be. Even though you may think that the books you read, movies you see, and so forth are "just for fun," they can relay powerful messages about gender roles, race relations, and class ideals. Take the popular *Twilight* series of books, widely read by teen girls. Sociological analysis of the books finds that, as entertaining as they are, they reproduce stereotypes of girls as weak, passive, and needing protection. The men in the books are strong, violent, and dominating. Native Americans are portrayed as animalistic werewolves (Hayes-Smith 2011). Most likely, young girls reading these novels do not think about the messages being projected, but when popular culture is replete with such images, you cannot help but be influenced. This is why it is important for alternative images to be presented, especially to young people. The trilogy *Hunger Games* provides an example. Here the central female character is strong and self-reliant. Unlike most popular heroines, she does not wait for men to rescue her. Such images can provide young women with new models for their own leadership (McCabe et al. 2011).

Content analyses of the media (a research method discussed in the following chapter) show distinct patterns of how race, gender, class, and age are depicted in various media forms. Youth is defined as beautiful; aging, not, especially for women. Light skin is promoted as more beautiful than dark skin, although being tan is seen as more beautiful than being pale. Models in African American women's magazines are often those with Anglo features of light skin, blue eyes, and straight or wavy hair. European facial features are also pervasive in the images of Asian women and Latinas appearing in popular culture. In music videos, women wear sexy and skimpy clothing and are more often the object of another's gaze than is true for their male counterparts; music videos are especially represented in sexualized ways (Coy 2014; Collins 2004).

On prime-time television, men are still a large majority of the characters shown. Professional women in the media are generally portrayed as young, thin, and beautiful—their youth suggesting that career success comes early. Televised images of people of color are loaded with stereotypes. Latinos, though vastly underrepresented (except on Spanish Language TV) are most often portrayed as hypersexual, subservient, or stupid. South Asian Americans, if seen at all, are stereotyped as cab drivers and convenience store clerks—or somehow linked to technology (Mastro 2015; Thakore 2014). Prime-time television also vastly overrepresents White people, as if they were a larger share of the population than they actually are (see ▲ Figure 2-5).

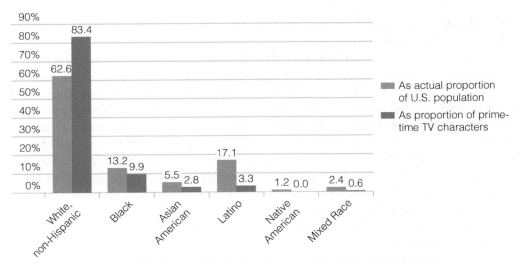

▲ **Figure 2-5   The Televised View of America: Reality versus TV**

Source: Tukachinsky, Riva, Dana Mastro, and Moran Yarchi. 2015. "Documenting Portrayals of Race/Ethnicity on Primetime Television over a 20-Year Span and Their Association with National-Level Racial/Ethnic Attitudes." *Journal of Social Issues* 71 (1): 17–38; U.S. Census Bureau. 2015. *State and County Quick Facts.* Washington, DC: U.S. Department of Commerce. **www.census.gov**

See for Yourself

Two Days without the Media

Try this experiment, created by Charles Gallagher (a sociologist at La Salle University), in which you will stage a forty-eight–hour media blackout. Begin by keeping a written log for forty-eight hours of exactly how much time you spend with some form of media. Include all time spent watching television, on the Internet, reading books and magazines, listening to music, viewing films, even using smartphones—any activity that can be construed as part of the media monopoly on people's time.

Next, eliminate all use of the media, except for that required for work and school, for a forty-eight-hour period. Keep a log as you go of what happens, what you are thinking, what others say, and how people interact with you. *Warning: If you try the media blackout, be sure to have some plan in place for having your family and/or friends contact you in case of an emergency!*

Without their connection to the media, students are likely to feel alienated, isolated, and detached. When the author of this book had her students do this experiment, many said they felt left out of conversations with friends about what they had viewed the night before on TV. Some nonetheless said they were much more reflective during this time, that they had more meaningful conversations with friends, and they had time to study more!

After trying this experiment, think about the enormous influence that the mass media have in shaping everyday life, including your self-concept and your relationship with other people. What does this exercise teach you about *cultural hegemony*? The role of the mass media in shaping society? How would each of the following theoretical frameworks explain what happened during your media blackout: functionalism, conflict theory, feminist theory, or symbolic interaction?

Source: Personal correspondence, Charles Gallagher, La Salle University.

African Americans, who watch more television than do other groups, are generally confined to a narrow variety of character types in the media. In recent years, the number of African American characters shown on television has come to match their proportion in the population, but largely because of their casting in situation comedies and in programs that are mostly minority. A recent report has found that, even though Latinos have increased as a share of the U.S. population, their presence in film and television, especially as leading actors and actresses, is shockingly low—even less than in the past. When shown, Latinos are often stereotyped as criminals, law enforcers, cheap labor, and hypersexualized or comic figures (Negrón-Muntaner 2014).

In a similar vein, African American men are most often seen as athletes and sports commentators, criminals, or entertainers. Women who work as football sports commentators are typically on the sidelines, reporting not so much on the play of the game as on human interest stories or injury reports—suggesting that women's role in sports is limited to that of nurturer. It is rare to find shows where Asians are the principal characters; with very few exceptions, they are depicted in silent roles, as sidekicks, domestic workers, or behind-the-scenes characters. Native Americans make occasional appearances, where they usually are depicted as mystics or warriors. Jewish women are generally invisible on popular TV programming, except when they are ridiculed in stereotypical roles. Arab Americans are likewise stereotyped, depicted as terrorists, rich oil magnates, or in the case of women, as perpetually veiled and secluded (Read 2003; Mandel 2001). Research documents numerous examples of stereotyped portrayals in the media—stereotypes you will see for yourself if you step outside of the taken-for-granted views with which you ordinarily observe the media.

Class stereotypes abound in the media and popular culture as well, with working-class men typically portrayed as being ineffectual, even buffoonish (Dines and Humez 2010). This has been demonstrated in research by sociologist Laura Grindstaff, who spent six months working on two popular talk shows. She did careful participant observation and interviewed the production staff and talk show guests. She found that to get airtime, guests had to enact social class stereotypes, acting vulgar and loud.

See for Yourself

Watch a particular kind of television show (situation comedy, sports broadcast, children's cartoon, or news program, for example) and make careful written notes on the depiction of different groups in this show. How often are women and men or boys and girls shown?

How are they depicted? You could also observe the portrayal of Asian Americans, Native Americans, African Americans, or Latinos. What do your observations tell you about the cultural ideals that are communicated through *popular culture*?

She concluded that, although these popular talk shows give ordinary people a place to air their problems and be heard, the shows exploit the working class, making a spectacle of their troubles (Grindstaff 2002; Press 2002).

Even a brief glance at popular television sitcoms reveals rampant homophobic joking. Recently, however, representation of gays, lesbians, and trans people has increased in the media, after years of being virtually invisible or only the subject of ridicule. As advertisers have sought to expand their commercial markets, they are showing more gay and lesbian characters on television. This makes LQBTQ more visible, although critics point out that they are still cast in narrow and stereotypical terms, or in comical roles (such as in *Modern Family*). Cultural visibility for any group is important because it validates people and can influence the public's acceptance of and generate support for equal rights protection (Gamson 1998).

Television is not the only form of popular culture that influences public consciousness, about class, gender, and race. Music, film, books, and other industries play a significant role in molding public consciousness. What images do these cultural forms produce? You can look for yourself. Try to buy a birthday card that contains neither an age nor gender stereotype. Alternatively, watch TV or a movie and see how different gender and race groups are portrayed. You will likely find that women are depicted as trying to get the attention of men; African Americans are more likely than Whites to be seen singing and dancing.

Do these images matter? Studies find that exposure to traditional sexualized imagery in music videos has a negative effect on college students' attitudes—for example, holding more adversarial attitudes about sexual relationships (Kalof 1999). Other studies find that even when viewers see media images as unrealistic, they think that others find the images important and evaluate them accordingly. Although people do not just passively internalize media images, such images form cultural ideals that have a huge impact on people's behavior, values, and self-image.

Theoretical Perspectives on Culture and the Media

Sociologists study culture and the media in a variety of ways, asking a variety of questions about the relationship of culture to other social institutions and the role of culture in modern life (see ◆ Table 2-2). Each theoretical position reveals different ways of understanding how and why the mass media influence the public.

Do the media create popular values or reflect them? The **reflection hypothesis** contends that the mass media reflect the values of the general population (Tuchman 1979). The media try to appeal to the most broad-based audience, so they aim for the middle ground in depicting images and ideas. Maximizing popular appeal is central to television program development. Media organizations spend huge amounts on market research to uncover what people think and believe and what they will like. Characters are then created with whom people will identify. Interestingly, the images in the media with which we identify are distorted versions of reality. Real people seldom live like the characters on television,

Table 2-2	Theoretical Perspectives on Culture			
According to:				
Functionalism	**Conflict Theory**	**Symbolic Interaction**	**New Cultural Studies**	**Feminist Theory**
Culture . . .				
Integrates people into groups	Serves the interests of powerful groups	Creates group identity from diverse cultural meanings	Is ephemeral, unpredictable, and constantly changing	Reflects the interests and perspectives of powerful men
Provides coherence and stability in society	Can be a source of political resistance	Changes as people produce new cultural meanings	Is a material manifestation of a consumer-oriented society	Is anchored in the inequality of women
Creates norms and values that integrate people in society	Is increasingly controlled by economic monopolies	Is socially constructed through the activities of social groups	Is best understood by analyzing its artifacts—books, films, and television images	Creates images and values that reproduce sexist and racist images

although part of the appeal of these shows is how they build upon, but then mystify, the actual experiences of people.

The reflection hypothesis assumes that images and values portrayed in the media reflect the values existing in the public, but the reverse can also be true—that is, the ideals portrayed in the media also influence the attitudes and values of those who see them. This has been illustrated in research on music videos. In a controlled experiment, the researchers exposed college men and women to hip-hop videos with high sexual content. Following their viewing, men in the sample expressed greater sexual objectification of women, more sexual permissiveness, stereotypical gender attitudes, and acceptance of rape myths; the findings did not hold for women in the sample (Kistler and Lee 2010). Although there is not a simple and direct relationship between the content of mass media images and what people think, clearly these mass-produced images can have a significant impact on who we are and what we think.

Culture and Group Solidarity

Many sociologists have studied particular forms of culture and have provided detailed analyses of the content of cultural artifacts, such as images in certain television programs or genres of popular music. Other sociologists take a broader view by analyzing the relationship of culture to other forms of social organization. Beginning with some of the classical sociological theorists (see Chapter 1), sociologists have studied the relationship of culture to other social institutions. Max Weber looked at the impact of culture on the formation of social and economic institutions. In his classic analysis of the Protestant work ethic and capitalism, Weber argued that the Protestant faith rested on cultural beliefs that were highly compatible with the development of modern capitalism. By promoting a strong work ethic and a need to display material success as a sign of religious salvation, the Protestant work ethic indirectly but effectively promoted the interests of an emerging capitalist economy. (We revisit this issue in Chapter 13.) In other words, culture influences other social institutions.

Functionalist theorists, for example, believe that norms and values create social bonds that attach people to society. Culture therefore provides coherence and stability in society. Participation in a common culture, even something as mundane as television, is an important social bond—one that unites society (Etzioni et al. 2001).

Classical theoretical analyses of culture have placed special emphasis on nonmaterial culture—the values, norms, and belief systems of society. Sociologists who use this perspective emphasize the integrative function of culture, that is, its ability to give people a sense of belonging in an otherwise complex social system (Smelser 1992). In the broadest sense, functionalism interprets culture as a major integrative force in society, providing people in society with a sense of collective identity and commonly shared worldviews.

Culture, Power, and Social Conflict

Whereas the emphasis on shared values and group solidarity drives one sociological analysis of culture, conflict and power drive another. Conflict theorists (see Chapter 1) analyze culture as a source of power in society.

Conflict theorists see contemporary culture as produced within institutions that are based on inequality and capitalist principles—that is, very powerful social institutions. The cultural values and products produced and sold within these institutions promote the economic and political interests of the few—those who own or benefit from these cultural industries. Owners of the mass media market have a vast economic stake in the distribution of images and products. The cultural products most likely to be produced are those that are consistent with the values, needs, and interests of the most powerful groups in society. The evening news, for example, is typically sponsored by major financial institutions and oil companies. Conflict theorists ask how this commercial sponsorship influences the content of the news. For example, if the news were sponsored by labor unions, would conflicts between management and workers always be defined as "labor troubles," or might newscasters refer instead to "capitalist troubles"?

Conflict theorists see culture as increasingly controlled by economic monopolies. Whether it is books, music, films, news, or other cultural forms, monopolies in the communications industry (where culture is increasingly located) have a strong interest in protecting the status quo. As media conglomerates swallow up smaller companies and drive out smaller, less-efficient competitors, the control that economic monopolies have over the production and distribution of culture becomes enormous. Megacommunications companies then influence everything—from the movies and television shows you see to the books you read in school.

Culture can also be a source of political resistance and social change. Reclaiming an indigenous culture that had been denied or repressed is one way that groups mobilize to assert their independence. An example from within the United States is the *repatriation movement* among American Indians who have argued for the return of both cultural artifacts and human remains held in museum collections. Many American Indians believe that, despite the public good that is derived from studying such remains and objects, cultural independence and spiritual respect outweigh such scientific arguments (Thornton 2001). Other social movements, such as the gay and lesbian movement, have also used cultural performance as a means of political and social protest. Cross-dressing, drag shows, and other forms of "gender play" can be seen as cultural performances that challenge homophobia and traditional sexual and gender roles (Rupp and Taylor 2003).

A final point of focus for sociologists studying culture from a conflict perspective lies in the concept of cultural capital. **Cultural capital** refers to the cultural resources that are deemed worthy (such as knowledge of elite culture) and that give advantages to groups possessing such capital. This idea has been most developed by the French sociologist Pierre Bourdieu (1984), who sees the appropriation of culture as one way that groups maintain their social power.

Bourdieu argues that members of the dominant class have distinctive lifestyles that mark their status in society. Their ability to display this cultural lifestyle signals their importance to others; that is, they possess cultural capital. From this point of view, culture has a role in reproducing inequality among dominant groups. Those with cultural capital use it to improve their social and economic position in society. Sociologists have found a significant relationship, for example, between cultural capital and grades in school. Those from the more well-to-do classes (those with more cultural capital) are able to parlay their knowledge into higher grades, thereby reproducing their social position by being more competitive in school admissions and, eventually, in the labor market (Hill 2001; Treiman 2001).

Symbolic Interaction and the Study of Culture

Symbolic interaction theory analyzes behavior in terms of the meaning people give it (see Chapter 1). The concept of culture is central to this orientation. Symbolic interaction emphasizes the interpretive basis of social behavior, and culture provides the interpretive framework through which behavior is understood.

Symbolic interaction emphasizes that culture, like all other forms of social behavior, is socially constructed. That is, culture is produced through social relationships and in social groups, such as the media organizations that produce and distribute culture. However, people do not just passively submit to cultural norms. They actively make, interpret, and respond to the culture around them. Culture is not one-dimensional; it contains diverse elements and provides people with a wide range of choices from which to select how they will behave (Swidler 1986). Culture, in fact, represents the creative dimension of human life.

In recent years, a new interdisciplinary field known as *cultural studies* has emerged that builds on the insights of the symbolic interaction perspective in sociology. Sociologists who work in cultural studies are often critical of classical sociological approaches to studying culture, arguing that the classical approach has overemphasized nonmaterial culture, that is, ideas, beliefs, values, and norms. The new scholars of cultural studies find that material culture has increasing importance in modern society (Walters 1999; Crane 1994). This includes cultural forms that are recorded through print, film, artifacts, or the electronic media. Growing from such studies is *postmodernist theory* (see Chapter 1), that is, the idea that society is not an objective thing. Rather, society is found in the words and images that people use to represent behavior and ideas. Because of this orientation, postmodernist studies have concentrated on analyzing common images and cultural products found in everyday life and, especially, through the mass media.

Classical theorists have tended to study the unifying features of culture. Cultural studies researchers tend to see culture as more fragmented and unpredictable. To them, culture is a series of images that can be interpreted in multiple ways, depending on the viewpoint of observers. From the perspective of new cultural studies theorists, the ephemeral and rapidly changing quality of contemporary cultural forms is reflective of the highly technological and consumer-based culture on which the modern economy rests. Modern culture, for example, is increasingly dominated by the ever-changing, but ever-present, images that the media bombard us with in everyday life. The fascination that cultural studies theorists have for these images is partially founded in illusions that such a dynamic and rapidly changing culture produces.

Feminist Theory and Culture

Feminist theory also adds to our understanding of culture. Feminist theory analyzes the power that men have in controlling cultural institutions. In addition, feminist theory analyses the gendered stereotypes that are culturally reproduced. It is also critical of women's exclusion from important leadership roles within cultural institutions. We explore gender and cultural stereotypes in more detail in Chapter 11, but here it is important to understand how culture reflects and reinforces gendered images that maintain gender inequality.

Feminists also note that changes are appearing in the dominant culture as women assume more significant roles in the production of culture. By and large, however, even with such changes, men still dominate both popular and elite culture. Women remain a minority on the boards of most elite cultural institutions. In film, television, and the Internet, women's bodies are routinely sexualized. When women assume positions of leadership (such as in roles as news anchors), much of the commentary about them focuses on their looks or their roles as mothers—attributes not so frequently expressed on commentaries about men.

Feminist theory analyzes cultural imagery and also criticizes the taken-for-granted nature of gender stereotyping and beliefs in the culture. Violence against women is also routinized in the media—in video games, on crime dramas, and even in comedy. On any given night, if you only listen to television in the background, you might be amazed at how frequently you will hear women screaming.

On the positive side, feminist theory also encourages the production and distribution of cultural content that challenges sexism and presents alternative images. Performance artists, for example, might transgress from dominant gendered images and roles, thereby questioning the gender binary that posits women and men as so-called opposites. *Queer theory* (examined in Chapter 13), as an example, emphasizes the social construction of gender, questioning the permanency of "man" and "woman" as social categories. As with conflict theory, feminist theory emphasizes that the transformation of culture is an important part of social movements for human liberation.

Cultural Change

In one sense, culture is a conservative force in society. Culture tends to be based on tradition and is passed on through generations, conserving and regenerating the values and beliefs of society. Culture is also increasingly based on institutions that have an economic interest in maintaining the status quo. People are also often resistant to cultural change because familiar ways and established patterns of doing things are hard to give up. But in other ways, culture is completely taken for granted, and it may be hard to imagine a society different from what is familiar.

Imagine, for example, the United States without fast food. Can you do so? Probably not. Fast food is so much a part of contemporary culture that it is hard to imagine life without it. Consider these facts about fast-food culture:

- The average person in the United States consumes three hamburgers and four orders of French fries per week.
- People in the United States spend more money on fast food than on movies, books, magazines, newspapers, videos, music, computers, and higher education combined.
- Ninety-six percent of American schoolchildren can identify Ronald McDonald—only exceeded by the number who can identify Santa Claus (Schlosser 2001).

Eric Schlosser, who has written about the permeation of society by fast-food culture, writes that "a nation's diet can be more revealing than its art or literature" (2001:3). He relates the growth of the fast-food industry to other fundamental changes in American society, including the vast numbers of women entering the paid labor market, the development of an automobile culture, the increased reliance on low-wage service jobs, the decline of family farming, and the growth of agribusiness. One result is a cultural emphasis on uniformity, not to mention increased fat and calories in people's diets.

This example shows how cultures can change over time, sometimes in ways that are hardly visible to us unless we take a longer-range view or, as sociologists would do, question that which surrounds us. Culture is a dynamic, not static, force in society, and it develops as people respond to various changes in their physical and social environments.

Culture Lag

Sometimes society adjusts slowly to changing cultural conditions, and the result can be **culture lag** (Ogburn 1922). Some parts of culture may change more rapidly than others. In other words, one aspect of culture may "lag" behind another. Rapid technological change is often attended by culture lag because some elements of the culture do not keep pace with technological innovation. In today's world, we have the technological ability to develop efficient, less-polluting rapid transit, but changing people's transportation habits is difficult. This would be an example of people's attachment to their cars (a cultural phenomenon) lagging behind the cultural need to reduce carbon emissions through the capabilities of technology.

Sources of Cultural Change

There are several causes of cultural change, including (1) a change in the societal conditions; (2) cultural diffusion; (3) innovation; and (4) the imposition of cultural change by an outside agency. Let us examine each.

1. Cultures change in response to changed conditions in the society. Economic changes, population changes, and other social transformations all influence the development of culture. A change in the makeup of a society's population may be enough by itself to cause a cultural transformation. The high rate of immigration in recent years has brought many cultural changes to the United States. Many major cities, such as Miami and Los Angeles, have a Latin feel because of the large Hispanic population. Cultural change from immigration is now apparent in locations throughout the United States. Markets selling Asian, Mexican, and Middle Eastern foods are increasingly common; school districts include students who speak a huge variety of languages; popular music bears the imprint of different world cultures. This is not the first time U.S. culture has changed because of immigration. Many national traditions stem from the patterns of immigration that marked the earlier part of the twentieth century—think of St. Patrick's Day parades, Italian markets, and Chinatowns.

2. Cultures change through cultural diffusion. Cultural diffusion is the transmission of cultural elements from one society or cultural group to another. In our world of instantaneous communication, cultural diffusion is swift and widespread. This is evident in the degree to which worldwide cultures have been Westernized. Cultural diffusion also occurs when subcultural influences enter the dominant group. Dominant cultures are regularly enriched by minority cultures. An example is the influence of Black and Latino music on other musical forms. Cultural diffusion is one thing that drives cultural evolution, especially in a society such as ours that is lush with diversity.

3. Cultures change as the result of innovation, including inventions and technological developments. Cultural innovations can create dramatic changes in society. Think, for example, of how the invention of trolleys, subways, and automobiles changed the character of cities. People no longer walked to work; instead, cities expanded outward to include suburbs. Furthermore, the invention of the elevator let cities expand not just out, but also up.

Now, the development of computer technology infiltrates every dimension of life. It is hard to overestimate the effect of innovation on contemporary cultural change. Technological innovation is so rapid and dynamic that one generation can barely maintain competence with the hardware of the next. The newest handheld computer today weighs hardly more than a few ounces, and its capabilities rival that of computers that filled entire buildings only twenty years ago. Technological innovation is now so rapid that it can leave some people in a state resembling culture shock—a phenomenon that one of the authors of this book has termed *tech-shock*, meaning that new devices and applications arrive faster than some in the public have time to learn and adjust to.

What are some of the social changes that technology change is creating? People can now work and be miles—even nations—away from their places of employment. Families can communicate from multiple sites; children can face-time; grandparents can receive live photos of a family event; criminals are tracked via cellular technology; music can be stolen without even going into a music store. Conveniences multiply with the growth of such technology, but so do the invasions of privacy and, perhaps, identity theft. In such a rapidly changing technological world, it is hard to imagine what will be common in just a few years.

New technological innovations raise interesting questions for sociological research. Studies of blogs find, for example, that women are a small proportion of bloggers—only 10 percent of the bloggers on the most widely used political sites. Some use blogs as support systems—for example, a gay person in a very traditional and isolated community may participate in a blog that provides a national community of support. One study in China found that many women are using blogs to subvert traditional concepts of womanhood (Schaffer and Xianlin 2007; Dolan 2006; Harp and Tremayne 2006).

The use of blogs is a good example of how technological innovation can create new forms of culture. Unlike traditional communities, blogging communities cross vast geographic distances, connecting people who might never meet face-to-face. Just as town meetings might have created a sense of community in the past, cyberspace communities now involve "imagined communities." Some suggest that blogs can actually create a more democratic society by directly engaging more people in political discussion and activity (Perlmutter 2008).

4. Cultural change can be imposed. Change can occur when a powerful group takes over a society and imposes a new culture. The dominating group may arise internally, as in a political revolution, or it may appear from outside, perhaps as an invasion. When an external group takes over the society of a "native," or indigenous, group—as White settlers did with Native American societies—they typically impose their own culture while prohibiting the indigenous group from expressing its original cultural ways. Manipulating the culture of a group is a way of exerting social control. Many have argued that public education in the United States, which developed during a period of mass immigration, was designed to force White, northern European, middle-class values onto a diverse immigrant population that was perceived to be potentially unruly and politically disruptive. Likewise, schools run by the Bureau of Indian Affairs have been used to impose dominant group values on Native American children (Snipp 1996).

Resistance to political oppression often takes the form of a cultural movement that asserts or revives the culture of an oppressed group. Cultural expression can be a form of political protest. Identification with a common culture can be the basis for group solidarity, as found in the example of the "Black pride" movement in the 1970s, whose influence is still felt today by having encouraged Black Americans to celebrate their African heritage with Afro hairstyles, African dress, and African awareness. Cultural solidarity has also been encouraged among Latinos through La Raza Unida (meaning "the race," or "the people, united"). Cultural change can promote social change, just as social change can transform culture.

Chapter Summary

What is culture?

Culture is the complex and elaborate system of meaning and behavior that defines the way of life for a group or society. It is shared, learned, taken for granted, symbolic, and emergent and varies from one society to another.

How do sociologists define norms, beliefs, and values?

Norms are rules of social behavior that guide every situation and may be formal or informal. When norms are violated, *social sanctions* are applied. *Beliefs* are strongly shared ideas about the nature of social reality. *Values* are the abstract concepts in a society that define the worth of different things and ideas.

What is the significance of diversity in human cultures?

As societies develop and become more complex, cultural diversity can appear. The United States is highly diverse culturally, with many of its traditions influenced by immigrant cultures and the cultures of African Americans, Latinos, and Native Americans. The *dominant culture* is the culture of the most powerful group in society. *Subcultures* are groups whose values and cultural patterns depart significantly from the dominant culture.

What is the sociological significance of the mass media and popular culture?

The *mass media* have an enormous influence on groups' beliefs and values, including images associated with racism and sexism. *Popular culture* includes the beliefs, practices, and objects of everyday traditions. Although people tend to deny its influence, the mass media and popular culture have an enormous influence on our attitudes and behaviors.

What do different sociological theories reveal about culture?

Sociological theory provides different perspectives on the significance of culture. *Functionalist theory* emphasizes the influence of values, norms, and beliefs on the whole society. *Conflict theorists* see culture as influenced by economic interests and power relations in society. *Symbolic interaction* emphasizes that culture is socially constructed. This has influenced new cultural studies, which interpret culture as a series of images that can be analyzed from the viewpoint of different observers. Feminist theory emphasizes the patriarchal control of the media and the reproduction of sexist images that pervade popular culture.

How do cultures change?

There are several sources of cultural change, including change in societal conditions, *cultural diffusion*, innovation, and the imposition of change by dominant groups. As cultures change, *culture lag* can result, meaning that sometimes cultural adjustments are out of sync with each other. People who experience new cultural situations may experience *culture shock*.

Key Terms

beliefs 39

countercultures 44

cultural capital 53

cultural diffusion 56

cultural hegemony 48

cultural relativism 31

culture 28

culture lag 55

culture shock 32

dominant culture 41

ethnocentrism 31

ethnomethodology 38

folkways 38

global culture 45

laws 38

mass media 45

material culture 29

mores 38

nonmaterial culture 29

norms 38

popular culture 45

reflection hypothesis 51

Sapir–Whorf hypothesis 36

social media 47

social sanctions 38

subculture 43

symbols 33

taboo 38

values 39

CHAPTER 3

DOING SOCIOLOGICAL RESEARCH

In this chapter, you will learn to:

Understand that sociological research is a true scientific endeavor whether it is quantitative or qualitative

Relate the steps in a research design

Identify the different research tools that sociologists use, including their relative advantages and disadvantages

Explain the role of professional ethics in the research process

iStock.com/elenabs

You have now seen some of the interesting things sociologists study through a glimpse into the sociology of culture. You also have a basic foundation in the sociological perspective and the major concepts in the field. We turn now to the tools sociologists use to study social phenomena—the methods of sociological research. These methods are varied. How you proceed in research depends on the sociological question that you are asking. Let us start with some examples.

Suppose you wanted to do some sociological research on how homeless people lived. What is life like for them? How dangerous is it? Where are the homeless to be found? Do they interact and associate with each other? Do they work at all, and if so, doing what? Do they feel rejected by society? Do they really sleep on park benches at night? Sociologist Mitch Duneier (1999) in his study entitled "Sidewalk" wanted to know all these things and more. So he decided to study a group of homeless people by living with them, and that is exactly what he did. He lived with them on their park benches and in doorways on New York City's Lower East Side. He spent four years with them. He interacted with them. He worked with them—a group consisting largely of African American men who sold books and magazines on the street. Duneier himself is White: He tells how becoming accepted into this society of African American men was itself an interesting and challenging process.

Contrary to popular belief, he discovered that these men make up a rather well-organized mini-society, with a social status structure, rules, norms, and a culture. He discovered many unknown elements of this "sidewalk society."

In this chapter we examine the participant observation method, such as Duneier used, plus other methods of sociological research. Each method is different from the others, but they all share a common goal: a deeper understanding of how society operates.

The Research Process

Sociological research is the tool sociologists use to answer questions. There are various methods that sociologists use to do research, all of which involve rigorous observation and careful analysis.

As we saw in the chapter opener, sociologist Mitch Duneier examined several questions about a group of people by living with them. He was engaged in what is called **participant observation**—a sociological research technique in which the researcher actually becomes simultaneously both participant in and observer of that which she or he studies.

In another example of participant observation, sociologist Roberto Gonzales (2016) spent twelve years conducting research with a highly vulnerable population known as "the Dreamers." Dreamers are those who came to the United States as young children with parents who are themselves undocumented. Gonzales's research was conducted with Mexican youth, though Dreamers also come from other Central American nations. As people know and as Gonzales richly details, many of the Dreamers are now young adults, but they live without U.S. citizenship even though they have lived almost their entire life knowing no other country than the United States. Right now, their future is highly uncertain, even though most have done all of the things normally expected of young people—gone to school and, in many cases, completed college. But, they are forced to live in the shadows where they cannot legally find work, vote, or otherwise enjoy either the formal or informal rights of citizenship. Without legislation to protect them, they are subject to deportation to a country that they do not know and where they might not even speak the language. Gonzales's research is further showcased in the box, "Doing Sociological Research: Lives in Limbo."

Doing Sociological Research

Lives in Limbo

Research Question
What is it like to live your life in the United States starting as a young child, only to learn once you grow a little older that you are actually not a U.S. citizen? Even though you may have come to the United States as a very young person, spoken English all your life, attended school, perhaps even college, what if you eventually learned a way that you could never get a job, apply for various benefits, vote, get a driver's license, or enjoy other rights of being a U.S. citizen? This is what sociologist Roberto Gonzales asks in his richly detailed study of undocumented young people "living in limbo" in the United States.

Research Methodology
Gonzales used multiple methods of sociological research in studying undocumented youth. He first started observing these young people when he was a youth worker in Chicago. Later, as a graduate student in Los Angeles, he continued his work with undocumented young people, interviewing them, observing them and their families, and also gathering what secondary data he could on such things as the population of undocumented workers in the Los Angeles area and the educational attainment of youth. As you might imagine, such data are not easily available because this is a population that has to live outside of the purview of the usual sources of such data. Gonzales also had to be very careful not to put his research subjects in any kind of jeopardy. His research also includes analyses of various immigration policies that have affected undocumented people.

(continued)

Research Results

The vast sweep of Gonzales research is impossible to briefly summarize, but one of his major results is the difference he finds in the experience of those youth he calls "early exiters" and "college goers." Early exiters are those who, faced with insurmountable challenges of an unequal school system, economic needs to help support their families, and the constant threat of deportation, thus leave school early, feeling disenfranchised and "outside" of the American social system. College-goers hold on to their high dreams of getting a good education and then a good career, and they are supported in these beliefs by strong social support from not only family, but also school counselors and teachers. Both groups experience huge structural disadvantages, including their undocumented status and family poverty, but the most significant differences between early exiters and college-goers is the strong support that college-goers received from strong social networks. Ironically, though, the college goers end up having the most to lose in their adult lives, as their dreams of completing their education, establishing a good career, creating a stable family are thwarted by their illegal status.

Conclusions and Implications

Although it may be easy for some to dismiss the conundrum that undocumented youth face by thinking that "illegal is illegal," remember that these are young people who have shared every dimension of the American dream. Their undocumented status severely limits their options and makes them extremely vulnerable to even minor transgressions, such as a simple traffic stop. That or a raid at a workplace can change their lives in an instant, so they must live with the constant threat of deportation to a place they do not even know. Left in complete limbo, the everyday lives of undocumented youth are constrained by the absence of immigration laws that would protect them and help them realize their dreams. Now, in a national context that is "tough on immigration" and where the stereotype of the "welfare queen" is being replaced with the caricature of the "illegal immigrant" (Gonzales 215: 219), policy changes that would help these young people seem even more unlikely than before.

Source: Gonzales, Roberto G. 2016. *Lives in Limbo: Undocumented and Coming of Age in America.* Oakland, CA: University of California Press.

You can see from the two examples above how challenging participant observation can be. It is labor intensive and can take years to complete. Although it tends to be somewhat unstructured, participant observation has led to some of the most lasting sociological insights.

There are other kinds of sociological research that sociologists do. Some approaches are more structured and focused than participant observation, such as survey research. Other methods may involve the use of official records or interviews. The different approaches used reflect the different questions asked in the first place. Some methods may require statistical analysis of a large set of quantitative information. Either way, the chosen research method must be appropriate to the sociological question being asked. (In the "Doing Sociological Research" boxes throughout this book, we explore different research projects that sociologists have done, showing what question they started with, how they did their research, and what they found.)

However it is done, research is an engaging and demanding process. It requires skill, careful observation, and the ability to think logically about the things that spark your sociological curiosity.

Sociology and the Scientific Method

Sociological research derives from what is called the *scientific method*, originally defined and elaborated by the British philosopher **Sir Francis Bacon** (1561–1626). The **scientific method** involves several steps in a research process, including observation, hypothesis testing, analysis of data, and drawing conclusions. Since its beginnings, sociology has attempted to adhere to the scientific method. To the degree that it has succeeded, sociology is a science. Yet, there is also an art to developing sociological knowledge. Sociology aspires to be both scientific and humanistic, but sociological research varies in how strictly it adheres to the scientific method. Some sociologists test hypotheses (discussed later); others use more open-ended methods, such as interviewing with open-ended questions and the examples of participant observation above.

Science is **empirical**, meaning it is based on careful and systematic observation, not just on conjecture. Although some sociological studies are highly *quantitative* and statistically sophisticated, others are *qualitatively* based, that is, based on more interpretive observations, not statistical analysis. Both quantitative and qualitative studies are empirical. Sociological studies may be based on surveys, observations, and many other forms of analysis, but they always depend on an empirical underpinning.

Sociological knowledge is not the same as philosophy or personal belief. Philosophy, theology, and personal experience can deliver insights into human behavior, but at the heart of the scientific method is the notion that a theory must be *testable*. This requirement distinguishes science from purely humanistic pursuits such as theology and literature.

Inductive and Deductive Reasoning

One wellspring of sociological insight is **deductive reasoning**. When sociologists use deductive reasoning, they create a specific research question about a focused point that is based on a more general or universal principle (see ▲ Figure 3-1). Here is an example of deductive reasoning: One might reason that because Catholic doctrine forbids abortion, Catholics would then be less likely than other religious groups to support abortion rights. This notion is "deduced" from a general principle (Catholic doctrine). You could test this notion (the research question) via a survey. As it turns out, the testing of this research question shows that it is incorrect: Surveys show that Catholics as a group are on average *more* likely to support abortion rights than are some other religious groups. That may come to you as a bit of a surprise! That is why we do research.

Inductive reasoning—another source of sociological insight—reverses this logic: That is, it arrives at general conclusions from specific observations. For example, if you observe that most of the demonstrators protesting abortion in front of a family planning clinic are evangelical Christians, you might infer that strongly held religious beliefs are important in determining human behavior. Again, referring to Figure 3-1, inductive reasoning would begin with one's observations. Either way—deductively or inductively—you are engaged in research.

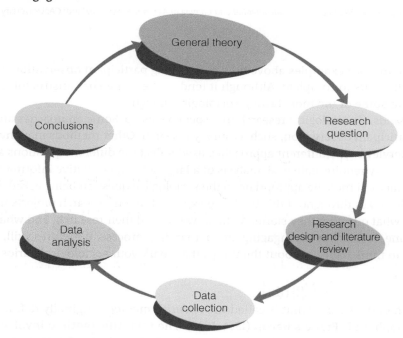

▲ **Figure 3-1   The Research Process.** Research can begin by asking a research question derived from general theory or earlier studies, but it can also begin with an observation or even from the conclusion of prior research. One's research question is the basis for a research design and the subsequent collection of data. As this figure shows, the steps in the research process flow logically from what is being asked.

Research Design

When sociologists do research, they engage in a process of discovery. They organize their research questions and procedures systematically—their research site being the social world. Through research, sociologists organize their observations and interpret them.

Developing a Research Question

Sociological research is an organized practice that can be described in a series of steps (see Figure 3-1). The first step in sociological research is to develop a research question. One source of research questions is past research. For any number of reasons, the sociologist might disagree with a research finding or wonder if it still holds and thus decide to carry out further research. A research question can also begin from an observation that you make in everyday life, such as wondering about the lives of homeless people.

Developing a sociological research question typically involves reviewing existing studies on the subject, such as past research reports or current articles on your subject matter. This process is called a *literature review*. Digital technology has vastly simplified the task of doing a literature review. Researchers who once had to burrow through paper indexes and card catalogs to find material relevant to their studies can now scan much larger swaths of material in far less time using online databases. The catalogs of most major libraries in the world are accessible on the Internet, as are specialized indexes, professional research journals, discussion groups, and other research tools developed to assist sociological researchers. Most of the many journals that report new sociological research are now available online in full-text format through university and college libraries. JSTOR (for "journal storage") and Sociological Abstracts are two of the most important for sociologists, although they are not the only sources of published research.

Of course, the vast amount of such information that is available on the Internet means it is especially important to know how to assess whether or not the information is valid. How do you know when something found on the web is valid or true? A lot of what is found on the web is of questionable accuracy, that is, unsubstantiated by accurate research or empirical study. Pay attention, for example, to what person or group has posted the website. Is it a political organization? An organization promoting a cause? A person expressing an opinion? See the box "A Sociological Eye on the Media: Fake News, Research, and the Media" on pages 63–64 for some guidelines about interpreting what you find on the web and in the media.

A Sociological Eye on the Media

Fake News, Research, and the Media

When you watch the news, read a newspaper, or search the web, you are likely to learn about various new research studies purporting some new finding. How do you know if the research results reported in the media are accurate? Just as in being careful about basing opinions on so-called fake news, you have to be learn how to interpret the legitimacy of what you hear in the media.

The following questions will help:

1. **What are the major variables in the study? Are the researchers claiming a causal connection between two or more variables?** For example, the press reported that one way parents can reduce the chances of their children becoming sexually active at an early age is for the parents to quit smoking (O'Neil 2002)—an oversimplified claim from the original research. The researcher who conducted this study actually claimed there was no direct link between parental smoking and teen sex, although she found a correlation between parents' risky behaviors—smoking, heavy drinking, and not using seat belts—and children's sexual activity. She argued that parents who engage in unsafe activities provide a model for their children's own risky behavior (Wilder and Watt 2002).

   Just because there is a link, or "correlation," between two variables does not necessarily mean one caused the other. Seeing parental behavior as a model for what children do is hardly the same thing as seeing parents' smoking as the cause of early sexual activity!

(*continued*)

2. **How have researchers defined and measured the major topics of their study?** For example, if someone claims that 10 percent of all people are gay, how is "being gay" defined? Does it mean having had only one such experience over one's entire lifetime or does it mean actually having a gay identity? Does the definition include gay, lesbian, and bisexual behavior? The difference matters because a particular definition may inflate or reduce the number reported. Different definitions and measurements can result in different conclusions.

3. **Is the research based on a representative scientific sample or is it a biased sample?** For example, headlines exclaimed "Study Links Working Mothers to Slower Learning" (Lewin 2002), but this study included only White, non-Hispanic families, thus resulting in a *biased sample* (Brooks-Gunn et al. 2002). Another study by the same research team found that there were *no significant effects* of mother's employment on children's intellectual development among African American or Hispanic children (Waldfogel et al. 2002). The point is not that the study is invalid, but that its results have more limited implications than the headlines suggest.

4. **Is there false generalization in the media report?** Often a study has more limited claims in the scientific version than what is reported in the media. Using the example just given about the connection between maternal employment and children's learning, you would make a big mistake to generalize from the study's results to all children and families. Remember that some groups were not included.

5. **Can the study be replicated?** "Replication" means "accurately repeated." Unless there is full disclosure of the research methodology (that is, how the study was conducted), this will not be possible. But you can ask yourself how the study was conducted, whether the procedures used were reasonable and logical, and whether the researchers made good decisions in constructing their research question and research design. If possible, you might be able to obtain the original study upon which the media coverage was based.

6. **Who sponsored the study and do they have a vested interest in the study's results?** For example, would you give as much validity to a study of environmental pollution that was funded and secretly conducted by a chemical company as you would to a study on the same topic conducted by independent scientists who openly report their research methods and results and who had no connection with the chemical company? Research sponsored by interested parties can raise questions about the researchers' objectivity and the standards of inquiry they used.

7. **Who benefits from the study's conclusions?** Although this question does not necessarily challenge the study's findings, it can help you think about whom the findings are likely to help.

8. **What assumptions did the researchers have to make to ask the question they did?** For example, if you started from the assumption that poverty is not the individual's fault but is the result of how society is structured, would you study the values of the poor or the values of policymakers? When research studies explore matters where social values influence people's opinions, it is especially important to identify the assumptions made by certain questions.

9. **What are the implications of the study's claims?** Thinking through the policy implications of a given result can often help you see things in a new light, particularly given how the media tend to sensationalize much of what is reported. Consider the study of maternal employment and children's intellectual development examined in question 3 above. If you take the media headlines at face value, you might leap to the conclusion that working mothers hurt their children's intellectual development, and you might then think it would be best if mothers quit their jobs and stayed at home. Is this a reasonable implication of this study? Does the study not have just as many implications for day-care policies as it does for encouraging stay-at-home mothers? Especially when reported research studies involve politically charged topics (such as issues of "family values"; or even something like "gun control"), it is important to ask questions that explore various implications of social policies.

10. **Do these questions mean you should never believe anything you hear in the media?** Of course not. Thinking critically about research does not mean being negative or cynical about everything you hear or read. The point is not to reject all claims in the media about research as out of hand. All research has limitations, but knowing how to evaluate good versus bad research can help you know when reports in the news are reliable and valid or "fake." Learning the basic tools of research can make you a better-informed citizen and prevent you from being duped by claims that are neither scientifically nor sociologically valid.

When you review prior research, you may wonder if the same results would be found if the study were repeated, perhaps examining a different group or studying the phenomenon at a different time. Research that is repeated exactly, but on a different group of people or in a different time or place, is called a **replication study**. Suppose earlier research found that women managers have fewer opportunities for promotion than do men. You might want to know if this still holds true now. You would then replicate the original study, probably using a different group of women and men managers, but asking the same questions that were asked earlier. A replication study can tell you what changes have occurred since the original study and may also refine the results of the earlier work. Research findings should be reproducible: If the research is sound, other researchers who repeat a study should get the same results, unless, of course, some identifiable change, or no identifiable change, has occurred in the interim.

Sociological research questions can also come from casual observation of human behavior. Perhaps you have observed the seating patterns in your college dining hall at lunch and wondered why people sit with the same group day after day. Does the answer point to similarity among the people on the basis of race, gender, age, or perhaps political views—or maybe any two or all of these? Answering this question would be an example of inductive reasoning—going from a specific observation (such as seating patterns at lunch) to a generalization (a theory about the effects of race and gender). Researcher Beverly Tatum (1997) found that seating patterns in a college dining room depended heavily upon race and also gender.

Creating a Research Design

A **research design** is the overall logic and strategy underlying a research project. Sociologists engaged in research may distribute questionnaires, interview people, or make direct observations in a social setting or laboratory. They might analyze cultural artifacts, such as magazines, newspapers, television shows, or Internet content. Some do research using historical records. Others base their work on the analysis of social policy. All these are forms of sociological observation. Research design consists of choosing the observational technique best suited to a particular research question.

Suppose you wanted to study the career goals of student athletes. In reviewing earlier studies, perhaps you found research discussing how athletics is related to academic achievement (Messner 2011). You might also have read an article in your student newspaper reporting that the graduation rate for women college athletes is much higher than the rate for men athletes and wondered if women athletes are better students than men athletes. In other words, are athletic participation, academic achievement, and gender interrelated, and if so, how?

Your research design would lay out a plan for investigating these questions. Which athletes would you study? How will you study them? To begin, you will need to get sound data on the graduation rates of the groups you are studying to verify that your assumption of better graduation rates among women athletes is actually true. Perhaps, you think, the differences between men and women are not so great when the men and women play the same sports. Or perhaps the differences depend on other factors, such as what kind of financial support they get or whether coaches encourage academic success. To observe the influence of coaches, you might observe interactions between coaches and student athletes, recording what coaches say about class work. As you proceed, you would probably refine your research design and even your research question. Do coaches encourage different traits in men and women athletes? To answer this question, you have to build into your research design a comparison of coaches interacting with men and with women. Perhaps you even want to compare female and male coaches and how they interact with women and men. The details of your research design flow from the specific questions you ask.

Quantitative versus Qualitative Research

A research design often involves deciding whether the research will be qualitative or quantitative or perhaps some combination of both. **Quantitative research** is that which uses numerical analysis. In essence, this approach reduces the data into numbers, for example, the percentage of teenage mothers in California. **Qualitative research** is somewhat less structured than quantitative research, yet still focuses on a central research question. Qualitative research allows for more interpretation and nuance in what people say and do and thus can provide an in-depth look at a particular social behavior. Both forms of research are useful, and both are used extensively in sociology.

Some research designs involve the testing of hypotheses. A **hypothesis** (pronounced "hy-POTH-i-sis") is a prediction or a hunch, a tentative assumption that one intends to test. If you have a research design that calls for the investigation of a very specific hunch, you might formulate a hypothesis. Hypotheses are often formulated as if–then statements. For example:

Hypothesis: If a person's parents are racially prejudiced, then that person will, on average, be more prejudiced than a person whose parents are relatively free of prejudice.

This is merely a hypothesis or expectation, not a demonstration of fact. Having phrased a hypothesis, the sociologist must then determine if it is true or false. To test the preceding example, one might take a large sample of people and determine their prejudice level by questionnaire or interviews. One would then determine the prejudice level of their parents, perhaps by interviewing their parents. (One would, of course, have to develop questions beforehand that accurately measures "prejudice.") According to the hypothesis, one would expect to find more prejudiced children among prejudiced parents and more nonprejudiced children among nonprejudiced parents. If this association is found, the hypothesis is supported. If it is not found, then the hypothesis would be rejected.

Not all sociological research follows the model of hypothesis testing, but all research does include a plan for how **data** will be gathered. (Note that *data* is the plural form; one says, "data are used . . . ," not "data is used. . . .") Data can be qualitative or quantitative; either way, they are still data. Sociologists often try to convert their observations into a quantitative form (see the "Statistics in Sociology" box later in this chapter).

Sociologists frequently design research to test the influence of one variable on another. A **variable** is a characteristic of a person or group that can have more than one value or score. The notion of "a variable" is central to sociological research. A variable can be relatively straightforward, such as age or income, or a variable may be more abstract, such as social class or degree of prejudice. In much sociological research, variables are analyzed to understand how they influence each other. With proper measurement techniques and a good research design, the relationships between different variables can be discerned. In the example of student athletes given previously, the variables you use would likely be student graduation rates, gender, and perhaps the sport played. In the hypothesis about race prejudice, parental prejudice and their child's prejudice would be the two variables you would study.

An **independent variable** is one that the researcher wants to test as the presumed cause of something else. The **dependent variable** is one on which there is a presumed effect. That is, if X is the independent variable, then X leads to Y, the dependent variable. In the previous example of the hypothesis, the amount of prejudice of the parent is the independent variable, and the amount of prejudice of the child is the dependent variable. In some sociological research, *intervening variables*—variables that fall between the independent and dependent variables (see ▲ Figure 3-2)—are also studied.

Sociological research proceeds through the study of concepts. A **concept** is any abstract characteristic or attribute that can potentially be measured. Social class and social power are concepts. These are not things that can be seen directly, although they are key concepts in the field of sociology. When sociologists want to study concepts, they must develop ways of "seeing" them.

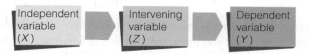

▲ **Figure 3-2   The Analysis of Variables.** Much sociological research seeks to find out whether some independent variable (X) affects an intervening variable (Z), which in turn affects a dependent variable (Y).

map 3-1   Viewing Society in Global Perspective: Human Development Index

The Human Development Index is a series of indicators developed by the United Nations and used to show the differing levels of well-being in nations around the world. The index is calculated using a number of indicators, including life expectancy, educational attainment, and standard of living. (Are these reasonable indicators of well-being? What else might you use?)

HDI: Human Development Index (HDI) Value (2014)

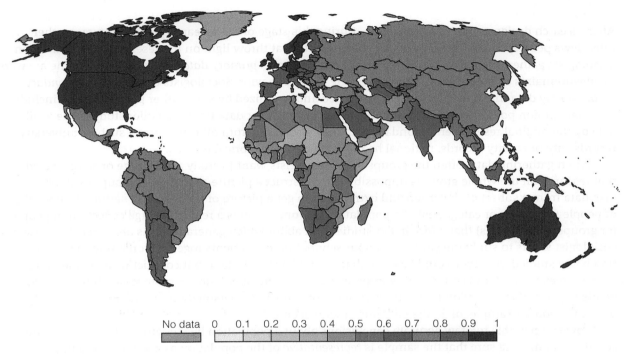

No data   0   0.1   0.2   0.3   0.4   0.5   0.6   0.7   0.8   0.9   1

Source: United Nations. 2016. Human Development Report, 2016. **hdr.undp.org/en/indicators/137506#**

Variables are sometimes used to show more abstract concepts that cannot be directly measured, such as the concept of social class. In such cases the variables studied are **indicators**—something that points to or reflects an abstract concept. An indicator is a way of "seeing" a concept. An example is shown in ▨ map 3-1 using the United Nations' Human Development Index. Here, the Human Development Index is composed of several *indicators*, including life expectancy and educational attainment, combined to show levels of well-being. "Level of well-being" is the *concept*.

The **validity** of a measurement (an indicator) is the degree to which it accurately measures or reflects a concept. To ensure the validity of their findings, researchers usually use more than one indicator for a particular concept. If two or more chosen measures of a concept give similar results, it is likely that the measurements are giving an accurate—that is, valid—depiction of the concept. For example, using a person's occupation, years of formal education, and annual earnings—namely, using three indicators of her or his social class—would likely be more valid than using only one indicator.

Sociologists also must be concerned with the **reliability** of their research results. A measurement is reliable if repeating the measurement under the same circumstances gives the same result. If a person is given a survey or a test two or three times and gets different results each time, then the reliability of the test is poor. One way to ensure that sociological measurements are reliable is to use measures that have proved sound in past studies. Another technique is to have a variety of people gather the data to make certain the results are not skewed by the tester's appearance, personality, and so forth. The researcher must be sensitive to all factors that affect the reliability of a study.

Sometimes sociologists want to gather data that would almost certainly be unreliable if the subjects (the people in the study) knew they were being studied. Knowing that they are being studied might cause people to change their behavior, a phenomenon in research known as the **Hawthorne effect**, an effect first discovered while observing work groups at a Western Electric plant in Hawthorne, Illinois. The work groups mysteriously increased their productivity (the dependent variable) right after they were observed by the researchers—an effect not noticed at first by the researchers themselves. An example of this effect would be a professor who wants to measure student attentiveness by observing how many notes are taken during class. Students who know they are being scrutinized will magically become more diligent!

Gathering Data

After research design comes data collection. During this stage of the research process, the researcher interviews people, observes behaviors, or collects facts that throw light on the research question. When sociologists gather original material, the product is known as *primary data*. Examples include the answers to questionnaires or notes made while observing group behavior. Sociologists often rely on *secondary data*, namely data that have already been gathered and organized by some other party. This can include national opinion polls, census data, national crime statistics, or data from an earlier study made available by the original researcher. Secondary data may also come from official sources, such as university records, city or county records, national health statistics, or historical records.

When gathering data, often the groups that sociologists want to study are so large or so dispersed that research on the whole group is impossible. To construct a picture of the entire group, sociologists take data from a subset of the group and extrapolate to get a picture of the whole. A **sample** is any subset of people (or groups or categories) of a population. A **population** is a relatively large collection of people (or groups or categories) that a researcher studies and about which generalizations are made. Suppose a sociologist wants to study the students at your school. All the students together constitute the population being studied. A survey could be done that reached every student, but conducting a detailed interview with every student is often highly impractical unless the population is quite small. If the sociologist wants the sort of information that can be gathered only during a personal interview, she would study only a portion, or sample, of all the students at your school.

How is it possible to draw accurate conclusions about a population by studying only part of it? The secret lies in making sure that the sample is *representative* of the population as a whole. The sample should have the same mix of people as the larger population and in the same proportions. If the sample is representative, then the researcher can *generalize* what she finds from the sample to the entire population. For example, if she interviews a sample of 100 students and finds that 80 percent of them have student debt, and if the sample is representative of the population, then she can conclude that about 80 percent of *all* the students at your school hold student debt. Note that a sample of 5 or 6 students would probably result in generalizations of poor quality, because the sample is not large enough to be representative. The size of a sample alone does not ensure its being representative. More important is trying to ensure, as best as possible, that the sample is not biased in some way. In selecting a sample, one has to be careful to avoid bias in how and who you select participants—in other words, to avoid *selection bias*. A *biased* (nonrepresentative) sample can lead to grossly inaccurate conclusions.

The best way to ensure a representative sample is to make certain that the sample population is selected randomly. A scientific **random sample** gives everyone in the population an equal chance of being selected. Quite often, striking and controversial research findings prove to be distorted by inadequate sampling. The man-on-the-street survey, much favored by TV and radio news reports, and certain other media as well, is the least scientific type of sample and the least representative. (The person-on-the-street sample includes only those who were available at that particular time and place and thus ignores those who were not there.)

Analyzing the Data

After the data have been collected, whether primary or secondary data, they must be analyzed. **Data analysis** is the process by which sociologists organize collected data to discover the patterns and uniformities that the data reveal. The analysis may be statistical or qualitative. When the data analysis is completed, conclusions and generalizations can be made.

Data analysis is labor intensive, but it is also an exciting phase of research. Here is where research discoveries are made. Sometimes while pursuing one question, a researcher will stumble across an unexpected finding, referred to by researchers as **serendipity**. A serendipitous finding is something that emerges from a study that was not anticipated, perhaps the discovery of an association between two variables that the researcher was not looking for or some pattern of behavior that was outside the scope of the research design. Such findings can be minor sidelines to the researcher's major conclusions or, in some cases, lead to major new discoveries. They are part of the excitement of doing sociological research.

One of the most significant examples of a serendipitous finding comes from science and the discovery of penicillin. Alexander Flemming, now known as the discoverer of the benefits of penicillin, discovered its

A census taker interviews a man in his home.

antiseptic influence when, while doing another study, his nasal drippings fell into a bacterial culture. When the bacteria died, Flemming realized he had discovered something that would change medical practices.

Reaching Conclusions and Reporting Results

The final stage in research is developing conclusions, relating findings to sociological theory and past research, and reporting the findings. At this stage an important questions is whether researchers' findings can be generalized. **Generalization** is the ability to draw conclusions from specific data and to apply them to a broader population. Researchers ask, do my results apply only to those people who were studied, or do they also apply to the broader population beyond? Assuming that the results have wide application, the researcher can then ask if the findings refine or refute existing theories and whether the research has direct application to practical social issues. Using the earlier example of the relationship between parent and offspring prejudice, if you found that racially prejudiced people did tend to have racially prejudiced parents (thus supporting your hypothesis), then you might report these results in a paper or research report. You might also ask, what kinds of programs for reducing prejudice do the results of your study suggest?

The Tools of Sociological Research

There are several tools or techniques sociologists use to gather data. Among the most widely used are survey research, participant observation, controlled experiments, content analysis, historical research, and evaluation research.

The Survey: Polls, Questionnaires, and Interviews

Whether in the form of a questionnaire, interview, or telephone poll, surveys are among the most commonly used tools of sociological research. Questionnaires are typically distributed to a large group of people. The *return rate* is the percentage of questionnaires returned out of all those distributed or initially requested. A low return rate introduces possible bias because the small number of responses may not be representative of the whole group.

Like questionnaires, interviews provide a structured way to ask people questions. They may be conducted face to face, by phone, by mail, or, more and more common, by using social media, such as email and Facebook. Interview questions may be open-ended or closed-ended, although the open-ended form is particularly accommodating if respondents want to elaborate.

Typically, a survey questionnaire will solicit data about the respondent (the person you are studying), such as income, occupation or employment status (employed or unemployed), years of formal education, yearly income, age, race, and gender, coupled with additional information that throws light upon a particular research question. For *closed-ended* questions, people must reply from a list of possible answers, like a multiple-choice test. For *open-ended* questions, respondents are allowed to elaborate on their answer. Closed-ended questions are generally (though not always) analyzed quantitatively, and open-ended questions are generally (though not always) analyzed qualitatively. Thus, a survey can involve both qualitative and quantitative research. Researchers may wish to analyze survey data that have already been collected by someone else. If a researcher has access to these original data, then the researcher may wish to analyze it. This is called *secondary analysis*, namely analysis of data that have already been collected. Secondary analysis has some advantages over collecting one's own data (called *primary data analysis*): It generally takes less time to do, and it can permit analysis of a very large sample of people.

As a research tool, surveys make it possible to ask specific questions about a large number of topics and then to perform sophisticated analyses to find patterns and relationships among variables. The disadvantages of surveys arise from their rigidity. Responses may not accurately capture the opinions of respondents or may fail to capture nuances in people's behavior and attitudes. Also, what people say and what they do are not always the same. Survey researchers must be persistent in order to get answers that are truthful—one reason for allowing respondents to be anonymous. Survey researchers sometimes get at this problem by asking essentially the same question in different ways. In this way, *validity* is increased.

Participant Observation

A unique and interesting way for sociologists to collect data and study society is to actually become part of the group they are studying. This is the method of **participant observation**. (*Nonparticipant observation* is also used as a technique for research. For example, one may wish to study a work group in a factory without actually participating in the group itself.) Two roles are played at the same time: subjective participant and objective observer. Usually, the group is aware that the sociologist is studying them, but not always. Participant observation is sometimes called *field research*, a term borrowed from anthropology.

Participant observation combines subjective knowledge gained through personal involvement and objective knowledge acquired by disciplined recording of what one has seen. The subjective component supplies a dimension of information that is lacking in survey data.

Street Corner Society (1943), a classic work by sociologist William Foote Whyte, documents one of the first qualitative participant observation studies ever done. Whyte studied the "Cornerville gang," a group of Italian American men whose territory was a street corner in Boston in the late 1930s and early 1940s. Although not Italian, Whyte learned to speak the language, lived with an Italian family, and then infiltrated the gang by befriending the gang's leader, whose pseudonym was "Doc." Doc was the **informant** for Whyte, a person with whom the participant observer works closely in order to learn about the group. For the duration of the study, Doc was the only gang member who knew that Whyte was doing research on his gang. This represents what is called **covert participant observation**, in which the members of the group being studied do not know that they are being researched. This is one means of trying to reduce the

Debunking Society's Myths

Myth: Public opinion polls that have sample sizes smaller than the U.S. population cannot possibly reflect people's actual opinions.

Sociological Research: A sample size smaller than the population from which it is drawn can represent the population, but only if there is no bias built into the selection of the sample. A sample should have the same mix of people as in the larger population.

Hawthorne effect. If the group is told that they are being studied and that they are the research subjects, then it is called **overt participant observation**. Sometimes the group members inadvertently find out that they are research subjects and may become angry because of the discovery. In this case, covert participant observation is by accident transformed into overt participant observation.

Most social scientists of the 1940s and 1950s thought gangs were socially disorganized, random deviant groups, but Whyte's study showed otherwise—as have participant observation studies since (Goffman 2009; Venkatesh 2008; Anderson 1976, 1990, 1999). Whyte found that the Cornerville gang, and by implication other urban street corner gangs, was a highly organized mini-society with its own social hierarchy (social stratification), morals, practices, and punishments (sanctions) for deviating from the norms of the gang.

There are a few built-in weaknesses to participant observation as a research technique. We already mentioned that it is very time-consuming. Participant observers have to cull data from vast amounts of notes. Such studies usually focus on fairly small groups, posing problems of generalization. Participant observation can also pose real physical dangers to the researcher, such as being "found out" or "outed" if one is studying a street gang using covert participant observation. Observers may also lose their objectivity by becoming too much a part of what they study. If this happens—the observer becomes so much a part of the group that she or he is no longer a scientific observer but rather a participant—it is called "going native" and is seen as one of the disadvantages of participant observation research. These limitations aside, participant observation has been the source of some of the most arresting and valuable studies in sociology.

Controlled Experiments

Controlled experiments are highly focused ways of collecting data and are especially useful for determining a pattern of cause and effect. To conduct a controlled experiment, two groups are created, an *experimental group*, which is exposed to the factor or variable one is examining, and the *control group*, which is not. In a controlled experiment, external influences are either eliminated or equalized, that is, held constant, between the experimental and the control group. This is necessary to establish cause and effect.

Suppose you wanted to study whether violent television programming causes aggressive behavior in children. You could conduct a controlled experiment to investigate this question. The behavior of children would be the dependent variable (variable Y); the independent variable (variable X) is whether or not the children are exposed to violent programming. To investigate your question, you would expose an experimental group of children (under monitored conditions) to a movie containing lots of violence (ultimate fighting, for example, or gunfighting). The control group would watch a movie that is free of violence. Beforehand, the children would be assigned randomly to the experimental group or the control group (this is called *experimental randomization*) in order to make the composition of the two groups as much alike as possible. Aggressiveness in the children (the dependent variable) would be measured twice: a *pretest* measurement made before the movies are shown and a *posttest* measurement made afterward. You would take pretest and posttest

The men in this bar, as shown by Anderson's (1976) classic participant observation study of "Jelly's Bar" in *A Place on the Corner*, have status differences among themselves that they create, such as (in descending status order) "regulars," "hoodlums," and "winos."

Mark Richards/PhotoEdit

measures on both the control and the experimental groups. Studies of this sort actually find very mixed results about whether observing violence in the media predicts youth violence.

Generally speaking, there is no consensus about whether there is a direct causal relationship between watching violence and committing it. The complexity of this connection lies in the many other factors within society that influence youth violence, including violence in the society itself, not just in the media.

Statistics in Sociology

Certain fundamental statistical concepts are basic to sociological research. Although not all sociologists do quantitative research, basic statistics are important to carrying out and interpreting sociological studies.

A **percentage** is the same as parts per hundred. To say that 22 percent of U.S. children are poor tells you that for every 100 children randomly selected from the whole population, approximately 22 will be poor. A **rate** is the same as parts per some number, such as per 10,000 or 100,000. The homicide rate in 2009 was about 7.2, meaning that for every 100,000 people in the population, approximately 7 were murdered. A rate is meaningless without knowing the numeric base on which it is founded; a rate is always the number per some other number.

A **mean** is the same as an average. Adding a list of fifteen numbers and dividing by fifteen gives the mean. The **median** is often confused with the mean but is actually quite different. The median is the midpoint in a series of values arranged in numeric order. In a list of fifteen numbers arrayed in numeric order, the eighth number (the middle number) is the median. In some cases, the median is a better measure than the mean because the mean can be skewed ("pulled" up or down) by extremes at either end. Another often-used measure is the **mode**, which is simply the value (or score) that appears most frequently in a set of data.

Let's illustrate the difference between mean and median using national income distribution as an example. Suppose that you have a group of ten people. Two make $10,000 per year, seven make $40,000 per year, and one makes $1 million per year. If you calculate the mean (the average), it comes to $130,000. The median, on the other hand, is $40,000, a figure that more accurately suggests the income profile of the group. That single million-a-year earner dramatically distorts, or skews, the picture of the group's income. If we want information about how the group in general lives, we are wiser to use the median income figure as a rough guide, not the mean. Note also that in this example the mode is the same as the median: $40,000.

Sociologists frequently examine the relationship between two variables. **Correlation** is a widely used technique for analyzing the patterns of association between pairs of variables such as income and education. We might begin with a questionnaire that asks for annual earnings (which we designate as the dependent variable, Y) and level of education (the independent variable, X). Correlation analysis delivers two types of information: It tells us the "direction" of the relationship between X and Y and also the "strength" of that relationship. The direction of a relationship is positive (that is, a positive correlation exists) if X is low when Y is low and if X is high when Y is high. But there is also a correlation if Y is low when X is high (or vice versa); this is a negative, or inverse, correlation. The strength of a correlation is simply how closely or tightly the variables are associated, regardless of the direction of correlation. With this example, you might well find a positive correlation between education (X) and annual earnings (Y), and we would also be interested in the strength of this correlation. A correlation does not necessarily imply cause and effect. A correlation is simply an association, one whose cause must be explained by means other than simple correlation analysis. A **spurious correlation** exists when there is no meaningful causal connection between apparently associated variables.

A good example of a spurious ("false") correlation is the finding that the murder rate (Y) in Denver, Colorado, dropped 50 percent as a result of the legalization of marijuana. The conclusion that the legalization of marijuana *caused* the drop in murder rate could well be—and probably is—spurious (false). Some third variable, such as a decline in unemployment or a decrease in domestic violence, could have caused a drop in the murder rate.

Another widely used method of analyzing sociological data is **cross-tabulation**, a way of seeing if two variables are related by breaking them down into categories for comparison. Take the following example. In a Pew Research poll (2016), a national sample of people was asked: "Do you favor a policy proposal to ban the sale of assault style weapons?" The following results, a cross-tabulation of answers to the question (the dependent variable) by gender (the independent variable) were obtained, with a significant gender difference appearing:

(continued)

	Favor ban	Oppose ban
Women:	60%	26%
Men:	44%	47%

Source: Pew Research Center. 2016. "Opinions on Gun Policy and the 2016 Campaign." Washington, DC: Pew Research Organization. **www.people-press.org/2016/08/26/opinions-on-gun-policy-and-the-2016-campaign**

Women are far more likely to support such policies than are men. This means that the two variables—gender and the answer to the question—are related. Gender differences on various matters about gun control have remained fairly steady over time. Public opinion has also shifted somewhat in recent years with more people favoring different forms of gun control than was true in the past, most likely the result of widely publicized mass shootings. Other factors, such as political party identification, race, education, and age are also related to support for gun control, as is whether or not someone has a gun in their household (Pew Research Organization 2017).

Statistical information is notoriously easy to misinterpret, willfully or accidentally. Examples of some statistical mistakes include the following:

- **Citing a correlation as a cause.** A correlation reveals an association between things (variables). Correlations do not necessarily indicate that one causes the other. Sociologists often say: "Correlation is not proof of causation."

- **Overgeneralizing.** Statistical findings are limited by the extent to which the sample group actually reflects or represents the population from which the sample was obtained. Generalizing beyond the population is a misuse of statistics. Studying only men and then generalizing conclusions to both men and women would be an example of overgeneralizing. This kind of mistake is fairly common in the media and also in some sociological research.

- **Interpreting probability as certainty.** Probability is a statement about chance or likelihood only. For example, in the cross-tabulation given previously, women are more likely than men to favor strict gun control. This means that women have a higher probability (a greater chance) of favoring strict gun control than men; it does not mean that all women favor strict gun control or that all men do not.

- **Building in bias.** In a famous advertising campaign, public taste tests were offered between two soft drinks. A wily journalist verified that in at least one site, the brand sold by the sponsor of the test was a few degrees colder (thus presumably better tasting) than its competitor when it was given to the people being tested, which biased the results. Bias can also be built into studies by careless wording on questionnaires.

- **Faking data.** Perhaps one of the worse misuses of statistics is actually making up, or faking, data. A famous instance of this occurred in a study of identical twins who were separated early in life and raised apart (Burt 1966). The researcher wished to show that despite their separation, the twins remained highly similar in certain traits, such as measured intelligence (IQ), thus suggesting that their (identical) genes caused their striking similarity in intelligence. It was later shown that the data were fabricated (Mackintosh 1995; Taylor 1980; Hearnshaw 1979; Kamin 1974).

- **Using data selectively.** Sometimes a survey includes many questions, but the researcher reports on only a few of the answers. Doing so makes it quite easy to misstate the findings. Researchers often do not report findings that show no association between variables, but these can be just as telling as associations that do exist. For example, researchers on gender differences typically report the differences they find between men and women, but seldom publish their findings when the results for men and women are identical. This tends to exaggerate the differences between women and men and falsely confirms certain social stereotypes about gender differences.

Among its advantages, a controlled experiment can establish causation, and it can zero in on a single independent variable. On the downside, controlled experiments can be artificial. They are for the most part performed in a contrived laboratory setting (unless it is what is called a *field experiment*), and they tend to eliminate many real-life effects. Analysis of controlled experiments includes making judgments about how much the artificial setting has affected the results (see ◆ Table 3-1).

Table 3-1   Comparison of Six Research Techniques

Technique (Tool)	Qualitative Analysis or Quantitative Analysis	Advantages	Disadvantages
The survey (polls, questionnaires, interviews)	Usually quantitative, often qualitative	Permits the study of a large number of variables; results can be generalized to a larger population if sampling is accurate	Difficult to focus in great depth on a few variables; difficult to measure subtle nuances in people's attitudes
Participant observation	Usually qualitative	Studies actual behavior in its home setting; affords great depth of inquiry	Is very time-consuming; difficult to generalize beyond the research setting
Controlled experiment	Usually quantitative	Focuses on only two or three variables; able to study cause and effect	Difficult or impossible to measure large number of variables; may have an artificial quality
Content analysis	Can be either qualitative or quantitative	A way of measuring culture	Limited by studying only cultural products or artifacts (music, TV programs, stories, other) rather than people's actual attitudes
Historical research	Usually qualitative	Saves time and expense in data collection; takes differences over time into account	Data often reflect biases of the original researcher and reflect cultural norms that were in effect when the data were collected
Secondary analysis of survey data	Can be either qualitative or quantitative	Permits the study of a large number of variables without actually collecting the primary data on these variables	Limited in the number of variables that can be measured; maintaining objectivity is problematic if research is done or commissioned by administrators of the program being evaluated
Evaluation research	Can be either qualitative or quantitative	Evaluates the actual outcomes of a program or strategy; often direct policy application	Same as a survey

Content Analysis

Researchers can learn a vast amount about a society by analyzing *cultural artifacts* such as newspapers, magazines, TV programs, Internet, or popular music. **Content analysis** is a way of measuring by examining the cultural artifacts of what people write, say, see, or hear. The researcher studies not people but the communications the people produce as a way of creating a picture of their society.

Content analysis is frequently used to measure cultural change and to study different aspects of culture (Lamont 1992). Sociologists also use content analysis as an indirect way to determine how social groups are perceived—they might examine, for example, how Asian Americans are depicted in television dramas or how women are depicted in advertisements.

Children's media have been the subject of many content analyses because of the strong impact of gender and race-based stereotypes on how children learn about their identity and those of other people. As one example, Elizabeth McDade-Montez and her colleagues studied the strong presence of sexualization in children's popular television programs (McDade-Montez, Wallander, and Cameron 2017). They conducted a systematic content analysis of episodes in the ten most popular children's television programs. They found that the sexualization of young girls was present in virtually every episode. In particular, they found extensive sexualization in how girls were dressed, their body appearance, statements the girls made, and how the young girls were objectified. The researchers also wanted to know if White characters were less sexualized than Black, Latino, Asian, and multiethnic characters. This formed one of their hypotheses. Perhaps surprisingly, they did not find significant differences in how White characters were sexualized compared to characters of color. They did however find that Latinas were the most underrepresented in children's programming relative to their numbers in the actual U.S. population.

Although some think that video games are a cause of mass shootings, research does not find a direct, causal link between viewing violent videos and the commission of violence, although it does show that media violence can desensitize children to the effects of violence.

Content analysis has the advantage of being *unobtrusive*, or "nonreactive." The research can have no effect at all on the person being studied because the cultural artifact has already been produced. Hence, content analysis will reveal very little, if any, Hawthorne effect. Content analysis is limited in what it can study, however, because it is based on mass communication—either visual, oral, or written. It cannot tell us what people think about these images or whether they affect people's behavior. Other methods of research, such as interviewing or participant observation, would be used to answer these questions. Nonetheless content analysis can be very insightful.

Historical Research

Historical research examines sociological themes over time. It is commonly done using historical archives, such as official records, church records, town archives, private diaries, or oral histories. The sources of this sort of material are critical to its quality and applicability. Oral histories have been especially illuminating, most dramatically in revealing the unknown histories of groups that have been ignored or misrepresented in other historical accounts. For example, when developing an account of the spirituality of Native Americans, one would be misguided to rely solely on the records left by Christian missionaries or U.S. Army officials. These records would give a useful picture of how Whites perceived Native American religion, but they would be a very poor source for discovering how Native Americans understood their own spirituality.

In a similar vein, the writings of a slave owner can deliver fascinating insights into slavery, but a slave owner's diary will certainly present a different picture of slavery as a social institution than will the written or oral histories of former slaves themselves.

Handled properly, comparative and historical research is rich with the ability to capture long-term social changes, and it is the perfect tool for sociologists who want to ground their studies in historical or comparative perspectives.

Evaluation Research

Evaluation research assesses the effect of policies and programs on people in society. If the research is intended to produce policy recommendations, then it is called *policy research*.

Suppose you want to know if an educational program is actually improving student performance. You could design a study that measured the academic performance of two groups of students, one that participates in the program and one that does not (a "control" group). If the academic performance of students in the program is better than that of those not in the program, and if the groups are alike in other ways (they are often matched to accomplish this), you would conclude that the program was effective. If the academic performance of the students in the program ended up being the same (or even worse) as those not in the program, then you would conclude that the program was not effective. When you do research to recommend social policy, you are doing policy research.

Research Ethics: Is Sociology Value Free?

The topics dealt with by sociology are often controversial. People have strong opinions about social questions and the settings for sociological work can be highly politicized or controversial. Imagine spending time in an urban precinct house to do research on police brutality or doing research on sex education in a conservative public school system. Under these conditions, can sociology be scientifically objective? How do researchers balance their own political and moral commitments against the need to be objective and open-minded? Sociological knowledge has an intimate connection to political values and social views. Often the very purpose of sociological research is to gather data as a step in creating social policy. Can sociology be value free? Should it be?

This is an important question without a simple answer. Most sociologists do not claim to be completely value free, but they do try as best they can to produce objective research. It must be acknowledged that researchers make choices throughout their research that can influence their results. The problems sociologists choose to study, the people they decide to observe, the research design they select, and the type of media they use to distribute their research can all be influenced by the personal values of the researcher.

Sociological research often raises ethical questions. In fact, ethical considerations of one sort or another exist with any type of research. In a survey, the person being questioned is often not told the purpose of the survey or who is funding the study. Is it ethical to conceal this type of information?

In controlled experiments, *deception* is often employed, as in the now-famous studies by Stanley Milgram, to be reviewed later in Chapter 6, where people were led to believe that they were causing harm to another, when in fact they were not. Researchers often reveal the true purpose of an experiment only after it is completed. This is called **debriefing**. The deception is therefore temporary. In the case of the Milgram experiments, some subjects felt harmed when they realized that they were easily duped into causing what they thought was serious harm to another person. In the end, researchers must consider ethical considerations, often codified in the guidelines of professional organizations, such as the American Sociological Association's Code of Ethics, which sets forth the principles and ethical standards with which sociologists pursue their work (see **www.asanet.org/membership /code-ethics**).

One of the clearest ethical violations in all of the history of science has come to be known as the *Tuskegee Syphilis Study*. The study

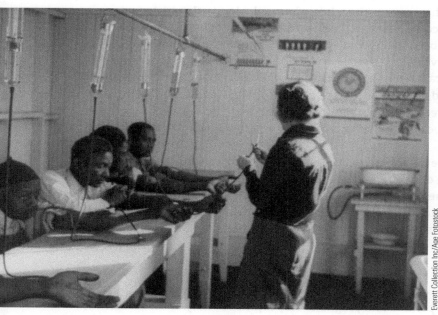

Here men with syphilis are being examined to determine the "progress" of syphilis. These unfortunate men were experimental subjects in the U.S. government's infamous Tuskegee Syphilis Study, one of the clearest ethical violations in all the history of science.

Everett Collection Inc/Age Fotostock

was conducted at the Tuskegee Institute in Macon County, Alabama, a historically Black college. For this study—begun in 1932 by the government's United States Health Service—a sample of about 400 Black males who were infected with the sexually transmitted disease syphilis (this was the "experimental" group) were allowed to go untreated medically for over forty years. Another 200 Black males who had not contracted syphilis were used as a control group. The purpose of the study was to examine the effects of "untreated syphilis in the male negro." The study was not unlike similar "studies" carried out against Jews by Hitler's Nazi regime in Germany at the same time—just before and during World War II in the 1930s and early 1940s. Jews who were injected with debilitating illnesses remained medically untreated. Untreated syphilis causes blindness, mental retardation, and death, and this is how many of the untreated Black men in the Tuskegee study fared over the period of forty-plus years.

In the 1950s, penicillin was discovered as an effective treatment for infectious diseases, including syphilis, and was widely available. Nonetheless, the scientists conducting the study decided *not* to give penicillin to the infected men in the study on the grounds that it would "interfere" with the study of the physical and mental harm caused by untreated syphilis! The U.S. government itself authorized the study to be continued until the early 1970s—that is, until quite recently. By the mid-1970s pressure from the public and the press caused the federal government to terminate the study, but by then it was too late to save approximately 100 men who had already died of the ravages of untreated syphilis, plus many others who were forced to live with major mental and physical damage.

The ethical horrors of studies such as the Tuskegee study led to the development of the American Sociological Association (ASA) professional code of ethics. The federal government also has many regulations about the protection of human subjects. Ethical researchers adhere to these guidelines and must ensure that research subjects are not subjected to physical, mental, or legal harm. Research subjects must also be informed of the rights and responsibilities of both researcher and subject. Sociologists, like other scientists, also should not involve people in research without what is called **informed consent**—that is, getting agreement to participate from the respondents or subjects after the purposes of the study are explained in detail to them. There may be exceptions to the need for informed consent, such as when observing people in public places. Sociologists also take strict measures to avoid identifying their respondents and to assure confidentiality through the use of pseudonyms or by not using names at all and by assigning random ID (code) numbers to all respondents during data analysis.

Chapter Summary

What is sociological research?

Sociological research is used by sociologists to answer questions and, in many cases, to test *hypotheses*. The research method one uses depends on the question that is asked.

Is sociological research scientific?

Sociological research is derived from the *scientific method*, meaning that it relies on empirical observation and, at times, the testing of *hypotheses*. The research process involves several steps: developing a research question, designing the research, collecting data, analyzing data, and developing conclusions. Different research designs are appropriate to different research questions, but sociologists have to be

concerned with the *validity*, the *reliability*, and the *generalization* of their results. Applying one's results obtained from a sample to a broader population is an example of generalization.

What is the difference between qualitative research and quantitative research?

Qualitative research is research that is relatively unstructured, does not rely heavily upon statistics, and is closely focused on a question being asked. *Quantitative research* is research that uses statistical methods. Both kinds of research are used in sociology.

What are some of the statistical concepts in sociology?

Through research, sociologists are able to make statements of *probability*, or likelihood. Sociologists use *percentages* and *rates*. The *mean* is the same as an average. The *median* represents the midpoint in an array of values or scores. The *mode* is the most common value or score. *Correlation* and *cross-tabulation* are statistical procedures that allow sociologists to see how two (or more) different variables are associated. There have been instances of misuse of statistics in the behavioral and social sciences, including sociology, and these have resulted in incorrect conclusions.

What different tools of research do sociologists use?

The most common tools of sociological research are surveys and interviews, participant observation, controlled experiments, content analysis, comparative and historical research, and evaluation research. Each method has its own strengths and weaknesses. You

can better generalize from surveys, for example, than participant observation, but participant observation is better for capturing subtle nuances and depth in social behavior.

Can sociology be value free?

Although no research in any field can always be value free, sociological research nonetheless strives for objectivity while recognizing that the values of the researcher may have some influence on the work. One of the worst cases of ethical violation in scientific research was the Tuskegee Syphilis Study. There are ethical dilemmas in doing sociological research, such as whether one should attempt to avoid the *Hawthorne effect* by collecting data without letting research subjects (people) know they are being observed.

Key Terms

concept 66
content analysis 74
controlled experiment 71
correlation 72
covert participant
  observation 70
cross-tabulation 72
data 66
data analysis 68
debriefing 76
deductive reasoning 62
dependent variable 66
empirical 62
evaluation research 75

generalization 69
Hawthorne effect 68
hypothesis 66
independent variable 66
indicator 66
inductive reasoning 62
informant 70
informed consent 77
mean 72
median 72
mode 72
overt participant observation 71
participant observation 70
percentage 72

population 68
qualitative research 66
quantitative research 66
random sample 68
rate 72
reliability 67
replication study 65
research design 65
sample 68
scientific method 61
serendipity 69
spurious correlation 72
validity 67
variable 66

CHAPTER 4

SOCIALIZATION AND THE LIFE COURSE

In this chapter, you will learn to:

Explain the socialization process

Identify the different agents responsible for socialization over the life course

Compare theories of socialization

Explore how socialization differs across cultures

Identify the stages of life from childhood to old age

Discuss the possibilities for resocialization throughout the life course

Do you think you could define what it means to be human? Biologists, geneticists, and many other natural scientists have attempted to identify the specific biological makeup of humans. Scientists working on the Human Genome Project have mapped the entire structure of human DNA, unlocking the genetic code of human life. The purpose of the human genome project is to see how genetics influences the development of disease, but it also raises many questions about human cloning and the possibility of creating human life in the laboratory. Is our genetic constitution what makes us human? Suppose you created a human being in the laboratory but left that creature without social contact. Would the "person" be human?

Knowing the genetic makeup of humans may mean the ability to make human beings in the laboratory, but without society, what would humans be like? What is the distinction between being "biologically" human and being "socially" human? Sociologists argue that the answer includes the influence of society. We are born with our genetic and biological makeup, but it is life experience that defines the self. As we go

Formal and informal learning, through families, schools, and other socialization agents, are important elements of the socialization process. In this photo, a mother teaches her child how to properly brush teeth.

from infancy, through the toddler years, from childhood, through adulthood and old age, our genetic DNA is unchanging. The ongoing interaction we have with other human beings over our lifetime, however, guides what it truly means to be part of human society. From the moment of birth, we are being "taught" what is expected of us to become a social being.

Consider the rare cases of *feral children*, who have been raised in the absence of human contact. Such cases, when a person has little or no social contact, provide clues about the importance of human contact to human development. One such case involved a young girl given the pseudonym of Genie. When her blind mother appeared (in 1970) in the Los Angeles County welfare office seeking assistance, caseworkers first thought the girl was six years old. In fact, she was thirteen, although she weighed only 59 pounds and was 4 feet 6 inches tall. She was small and withered, unable to stand up straight, incontinent, and severely malnourished. Her eyes did not focus, and a strange ring of calluses circled her buttocks. She could not talk. As the case unfolded, it was discovered that the girl had been kept in nearly total isolation for most of her life, never learning verbal language or any form of social interaction.

Genes may confer skin and bone and brain, but only by learning the values, norms, and roles that culture bestows on people do we become social beings—literally, human beings. Sociologists refer to this process as socialization—the subject of this chapter.

The Socialization Process

Socialization is the process through which people learn the expectations of society. To be a fully socialized member of society means to have internalized the expected *norms* of that society. Most of the time, people learn these expectations so thoroughly that they no longer question them, but simply accept them as correct and "normal." Unless something happens to disrupt what is taken for granted, people are hardly aware of how thoroughly they have been socialized.

Of course, not all human beings *act* the same way. In fact, within any one culture, not all people act the same way or believe the same things. Socialization occurs within specific contexts, and people are shaped by the many different experiences they have. Even within the same social circle, such as your college friends, there will of course be differences in how people act and what they think. Yet, socialization within such a group guides us in how to behave within our given roles.

Roles are the expected behavior associated with a given status in society. When you occupy a social role, you tend to take on the expectations of others. For example, when you make the transition from high school to college, you keenly observe the behavior, the language, the dress, perhaps even the music tastes of other college students. You likely modify your own behavior accordingly. Before you know it, you are transformed from high school student to college student. Later, you will socialize the next entering class into the same set of expectations. As you join new groups, such as a work organization, a club, or, perhaps, form a new family, your socialization continues. Roles are learned as you are socialized to a group's norms, taking on your specific role and interacting with the group in socially acceptable ways. The more important the group is to you (such as your family), the more thoroughly you are likely to internalize the expectations associated with your role in that group.

By means of socialization, people absorb their culture—customs, habits, laws, practices, and means of expression. Socialization is the basis for **identity**: how one defines oneself. Identity is both personal and social. To a great extent, our identity is bestowed by others, because we come to see ourselves as others

International Adoption

In the United States, international adoptions are not unusual, although the number of such adoptions has fallen off significantly since the mid-2000s (from 23,000 in 2014 to only about 5400 in 2016; U.S. Department of State 2017). How is socialization different in families where parents are culturally different from an adopted child?

Parents who adopt children from cultural backgrounds different from their own have to decide how to socialize their child, especially in terms of their ethnic, and possibly their racial, identity. Research studies find that the children in such families are more likely to develop a positive sense of self-esteem when parents socialize these children to have strong ethnic and racial identities. For Chinese adoptees, for example, parents might provide opportunities for the children to participate in Chinese cultural activities and perhaps learn Chinese language. For children who are racially or ethnically different from their parents, having ongoing discussions about race and discrimination is also important to the child's developing a positive identity (Mohanty 2013; Lee et al. 2006).

see us. For example, embracing the identity of a college student occurs through interaction with other college students and comes complete with new behaviors and feelings. Socialization also establishes **personality**, defined as a person's relatively consistent pattern of behavior, feelings, predispositions, and beliefs.

The socialization experience differs for individuals, depending on factors such as age, race, gender, and class, as well as more subtle aspects of personality. Women and men encounter different socialization patterns as they grow up because each gender brings with it different social expectations (see Chapter 11). Likewise, growing up Jewish, Asian, Latino, or African American involves different socialization experiences. The box "Understanding Diversity: International Adoption" also shows how socialization occurs when adopted children are from a different cultural background than their parents.

The Nature–Nurture Controversy

Examining the socialization process helps reveal the degree to which our lives are *socially constructed*, meaning that the organization of society and the life outcomes of people within it are the result of social definitions and processes. Is it "nature" (what is natural) or is it "nurture" (what is social)—or both—that makes us human? This question has been the basis for debate for many years.

From a sociological perspective, what a person becomes results more from social experiences than from *innate* (inborn or natural) traits, although innate traits do have some influence on what we become. For example, someone may have a particular gift for musical talent, but without the resources to develop that talent, may never become a good musician. Nature provides a certain stage for what is possible, but society provides the full drama of what we become. Our values and social attitudes are not inborn; they emerge through the social relations we have with others and our social position in society. Family environment, education, how people of your social group are treated, and the historic influences of the time all shape how we are nurtured by society.

Perhaps the best way to understand the nature—nurture controversy is not that one or the other fully controls who we become, but that life involves a complex interplay, or *interaction*, between "natural" (that is, biological or genetic influences) and social influences on human beings. Even neuroscientists know this as more studies of brain functioning and genetics are showing that even our biological makeup interacts with the environment (Conley and Fletcher 2017). It is the impact of the social environment that concerns sociologists.

Socialization as Social Control

Sociologist Peter Berger pointed out that not only do people live in society but society also lives in people (Berger 1963). Socialization is, therefore, a mode of social control. **Social control** is the process by which groups and individuals within those groups are brought into conformity with dominant social expectations. Sometimes a person rebels and attempts to resist this conformity, but because people generally

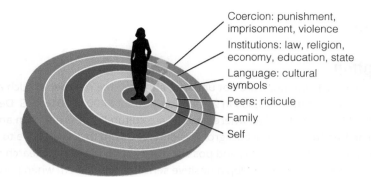

Coercion: punishment, imprisonment, violence

Institutions: law, religion, economy, education, state

Language: cultural symbols

Peers: ridicule

Family

Self

▲ **Figure 4-1   Socialization as Social Control.** Though we are all individuals, the process of socialization also keeps us in line with society's expectations. This may occur subtly through peer pressure or, in some circumstances, through coercion and/or violence.

conform to cultural expectations, socialization gives society a certain degree of predictability. Patterns are established that become the basis for social order.

To understand how socialization is a form of social control, imagine the individual in society as surrounded by a series of concentric circles (see ▲ Figure 4-1). Each circle is a layer of social controls, ranging from the most subtle, such as the expectations of others, to the most overt, such as physical coercion and violence. Coercion and violence are usually not necessary to extract conformity because learned beliefs and the expectations of others are enough to keep people in line. These socializing forces can be subtle because even when someone disagrees with others, he or she can feel pressure to conform and may experience stress and discomfort in choosing not to conform. People learn through a lifetime of experience that deviating from the expectations of others invites peer pressure, ridicule, and other social judgments that remind one of what is expected.

Conformity and Individuality

Saying that people conform to social expectations does not eliminate individuality. We are all unique to some degree. Our uniqueness arises from different experiences, different patterns of socialization, the choices we make, and the imperfect ways we learn our roles. People can resist some of society's expectations. Sociologists warn against seeing human beings as totally passive creatures because people interact with their environment in creative ways. Yet, most people conform, although to differing degrees. Socialization is profoundly significant, but this does not mean that people are robots. Instead, socialization emphasizes the adaptations people make as they learn to live in society.

Some people conform too much, for which they pay a price. Socialization into men's roles can encourage aggression and a zeal for risk-taking. Men have a lower life expectancy and higher rate of accidental death than do women, probably because of the risky behaviors associated with men's roles, that is, simply "being a man" (Kimmel and Messner 2012). Women's gender roles carry their own risks. Striving excessively to meet the beauty ideals of the dominant culture can result in feelings of low self-worth and may encourage harmful behaviors, such as smoking or severely restricting eating to keep one's weight down. Being a man or woman is not inherently bad for your health, but conforming to gender roles to an extreme can compromise your physical and mental health. Women and girls are more likely than men and boys, for example, to suffer from eating disorders or to have an unhealthy self-image (National Institute of Mental Health 2016).

The Consequences of Socialization

Socialization is a lifelong process with consequences that affect how we behave toward others and what we think of ourselves.

1. First, *socialization establishes self-concepts*. **Self-concept** is how we think of ourselves as the result of the socialization experiences we have over a lifetime. Socialization is also influenced by various social factors, such as gender. ▲ Figure 4-2 shows how students' self-concepts differ for men and women. Men rate themselves much more highly on social and intellectual self-confidence, risk-taking, and

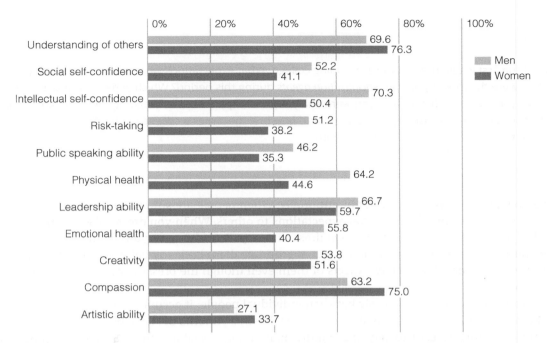

▲ **Figure 4-2   Student Self-Concepts: The Difference Gender Makes.** These data are based on a national sample of first-year college students in the fall of 2016. What patterns do you see, comparing women and men? How does socialization help you explain these data?

Source: Eagan, Kevin et al. 2016. *The American Freshman: National Norms Fall 2016.* Higher Education Research Institute. Los Angeles, CA: University of California, Los Angeles. **www.heri.ucla.edu/monographs /TheAmericanFreshman2016.pdf**

leadership, as examples. Women rate themselves higher in compassion and understanding others. These differences are largely explained by socialization.

2. Second, *socialization creates the capacity for role-taking.* As we learn societal expectations, we create the ability to see ourselves through the perspective of another. Socialization is fundamentally reflective; that is, it involves human beings seeing and reacting to the expectations of others. The capacity for reflection and the development of identity are ongoing. This is how we establish our roles in society. As we encounter new situations in life, such as going away to college or getting a new job, we are able to see what is expected. We adapt to the situation accordingly, becoming a young adult student or an employee of a company. Of course, not all people do so successfully. Unsuccessful socialization can be the basis for social deviance or other social and psychological problems.

3. Third, *socialization creates the tendency for people to act in socially acceptable ways.* Through socialization, people learn the normative expectations attached to social situations and the expectations of society in general. As a result, socialization creates some predictability in human behavior and brings some order to what might otherwise be social chaos.

4. Finally, *socialization makes people bearers of culture.* Socialization is the process by which people learn and internalize the attitudes, beliefs, and behaviors of their culture. At the same time, socialization is a two-way process—that is, a person is not only the recipient of culture but also is the creator of culture, passing cultural expectations on to others. The main product of socialization, then, is society itself.

Agents of Socialization

Socialization agents are people, or sources, or structures who pass on social expectations. Everyone is a socializing agent because social expectations are communicated in countless ways and in every

Think about the first week that you were attending college. What expectations were communicated to you and by whom? Who were the most significant *socialization agents* during this period? Which expectations were communicated formally and which informally? If you were analyzing this experience sociologically, what would be some of the most important concepts to help you understand how one "becomes a college student"? Were there expectations for dress? For transportation around campus?

interaction people have, whether or not intentionally. When people are simply doing what they consider "normal," they are communicating social expectations to others. When you dress a particular way, you may not feel you are telling others they must dress that way. Yet, when everyone in the same environment dresses similarly, some expectation about appropriate dress is clearly being conveyed. People feel pressure to become what society expects of them even though the pressure may be subtle and unrecognized.

Socialization does not occur simply between individual people; it occurs in the context of social institutions. Recall from Chapter 1 that institutions are established patterns of social behavior that persist over time. Institutions are a level of society above individuals. Many social institutions shape the process of socialization, including, as we will see, family, media, peers, religion, sports, and schools.

The Family

For most people, the family is the first source of socialization. Through families, children are introduced to the expectations of society. Children learn to see themselves through their parents' eyes. How parents define and treat a child is crucial to the development of the child's sense of self.

What children learn in families is certainly not uniform. Even though families pass on the expectations of a given culture, families within that culture may be highly diverse, as we will see in Chapter 13. Some families may emphasize educational achievement over physical activity; some may be more permissive, whereas others emphasize strict obedience and discipline. Within families, children may experience different expectations based on gender or birth order (being born first, second, or third). Living in a family experiencing the strain of social problems such as alcoholism, unemployment, domestic violence, or teen pregnancy also affects how children are socialized. The specific effects of different family structures and processes are the basis for ongoing and extensive sociological research.

Sociologists have also found that there are significant class differences in how children are socialized in families. By observing families in different social classes over an extended period of time, sociologist Annette Lareau found that children raised in upper- and middle-class families are given highly structured activities, such as music lessons, organized sports and clubs, language instruction, and other activities that both fill the family calendar and give children little free time. *Cultivated childhood* is the result as children in the upper and middle classes learn to navigate institutions and interact with authority figures in ways that will benefit them later in life. Working-class children, on the other hand, have much more unstructured childhoods, allowing them to create their own activities, but leaving them without the same skills to successfully navigate social institutions, such as school and work, later in life. In this sense, childhood socialization in families reproduces a system of inequality that advantages those who already have resources and disadvantages everyone else (Lareau 2011).

As important as the family is in socializing the young, it is not the only socialization agent. As children grow up, they encounter other socializing influences, sometimes in ways that might contradict family expectations. Even parents who want to socialize their children in less gender-stereotyped ways might be frustrated by the influence of other socializing forces, which promote highly gender-typed toys and activities to boys and girls. These multiple influences on the socialization process create what sociologist Emily Kane has called "the gender trap," meaning the expectations and structures that reproduce gender norms, even when parents try to resist them (Kane 2012).

The Media

As we saw in Chapter 2, the mass media increasingly are important agents of socialization. Television alone has a huge impact on what we are socialized to believe and become. Add to that films, music, video games, radio, Facebook, Twitter, Instagram, and other social media outlets, and you begin to see the enormous influence the media have on the values we form, our images of society, our desires for ourselves, and our relationships with others. These images are powerful throughout our lifetimes, but many worry that their effect during childhood may be particularly harmful.

There is no doubt that violence is extensive in the media. There is considerable public debate about the effect of viewing media violence on children's development. Some contend that media violence tends to desensitize children to the effects of violence, including engendering less sympathy for victims of violence. Some also draw a connection between the noted violence in video games and subsequent violent behavior. Although there may be some link between media violence and actual violent behavior, social science research finds that it is overly simplistic to see a direct causal connection between the two (Short 2015; Ferguson 2013). You cannot ignore the broader context in society where violent behavior is common and where various factors, such as family socialization, youth alienation, and a strong gun culture, influence violent behavior. Children do not watch television in a vacuum. Children live in families where they learn different values and attitudes about violent behavior. They also observe the society around them, not just the images they see in fictional representations. The sociological question is whether or not images in the media reflect societal reality or if reality is influenced by the images presented in the media.

The media expose us to numerous images that shape our definitions of ourselves and the world around us. What we think of as beautiful, sexy, politically acceptable, and materially necessary is strongly influenced by the media. Advertisers are increasingly able to target very specific images to you, perhaps based on your latest electronic search or download. If, as you look at your Facebook page or view an electronic news outlet, you might see an ad that you think will give you status and distinction. You might come to think, even if unconsciously, that your self-worth can be measured by the clothes you wear or the car you drive. If you watch sports shows, ads suggest that you will have more fun if you drink more beer. You, then, come to believe that parties are better when everyone is drinking. The values represented in the media, whether they are about violence, racist and sexist stereotypes, or any number of other social images, have a great effect on what we think and who we become.

Peers

Peers are those with whom you interact on equal terms, such as friends, fellow students, and coworkers. Among peers, there are no formally defined superior and subordinate roles, although status distinctions commonly arise in peer group interactions. Without peer approval, most people find it hard to feel socially accepted.

Peers are important agents of socialization. Young girls and boys learn society's images of what they are supposed to be through the socialization process, and peers are enormously important in that process. Peer cultures for young people often take the form of *cliques*—friendship circles where

iStock.com/PhotoEuphoria

Peers are important agents of socialization. Young girls and boys learn society's images of what they are supposed to be through the socialization process.

Children and Disasters

You might think that children who experience some horrible natural disaster—a hurricane, a flood, or a tornado—might be deeply traumatized by such an event. Psychologists might study the impact of such an event on a child's emotional state or sense of safety. Sociologists would not disagree with such an approach, but conceptualize childhood in a somewhat different manner than just focusing on emotional states or internal states of mind. Rather, sociologists see children as active agents in the construction of their own lives, even as they are influenced by the adult world around them.

This has been shown in a research study of children who experienced hurricane Katrina in 2005. Sociologists Alice Fothergill and Lori Peek spent seven years studying some of the children who witnessed Katrina first hand. They especially looked at the material artifacts (paintings, drawings, songs, games, and so forth) that children created after the disaster. They found that the things children created helped them cope with the losses they experienced and identified what had helped them cope with this natural disaster.

Fothergill and Peek's research shows that children, even when displaced, create their own cultures that help them cope with some of life's most traumatic events. You might consider how you might use children's creative works to help them adjust in the aftermath of a natural disaster.

Source: Fothergill, Alice, and Lori Peek. 2017. "Kids, Creativity, and Katrina." *Contexts* 16 (2): 65–67; Fothergill, Alice, and Lori Peek. 2015. *Children of Katrina*. Austin, TX: University of Texas Press.

members identify with each other and hold a sense of common identity. You probably had cliques in your high school and may even be able to name them. Did your school have "jocks," "goths," "tech geeks," "freaks," "stoners," and so forth? Sociologists studying cliques have found that they are formed based on a sense of exclusive membership, like in-groups and out-groups. Cliques are cohesive but also have an internal hierarchy, with certain group leaders having more power and status than other members. Interaction techniques, like inside jokes and high fives, produce group boundaries, defining who's in and who's out. The influence of peers is especially strong in adolescence, but it also persists into adulthood (Farrell et al. 2017).

As agents of socialization, peers are important sources of social approval, disapproval, and support. This is one reason groups without peers of similar status are often at a disadvantage in various settings, such as women in male-dominated professions or minority students on predominantly White campuses. Being a "token" or an "only," as it has come to be called, places unique stresses on those in settings where there are relatively few peers from whom to draw support (Thoits 2009). This is one reason those who are minorities in a dominant group context often form same-sex or same-race groups for support, social activities, and the sharing of information about how to succeed in that particular environment.

Bullying can be understood by the strong influence of peer cultures and socialization. Bullying is the verbal berating (and possible physical violence) directed at someone selected as a victim by a peer clique. Bullying can have dire consequences. A horrible example of this was the tragic death of a gay student at Rutgers University in 2010 who took his own life after his college roommate posted a video linking him romantically to another male student. The roommate who filmed the victim was later convicted on fifteen criminal counts. Even without such extremes, bullying is common and can be best understood in the context of status relationships within peer cultures. Studies find that both those who are bullied and those who do the bullying tend to be those who are disliked by peers (Sentse, Kretschmer, and Salmivalli 2015). Although this is not a direct (that is, causal) relationship, clearly the influence of peers is a strong part of adolescent socialization.

Religion

Religion is another powerful agent of socialization, and religious instruction contributes greatly to the identities children construct for themselves. Children tend to develop the same religious beliefs as their parents. Even those who renounce the religion of their youth are deeply affected by the attitudes, images, and beliefs instilled by early religious training. Very often, those who disavow religion return to their original faith at some point in their life, especially if they have strong ties to their family of origin and if they form families of their own (Wuthnow 2010).

Religious socialization influences a large number of beliefs that guide adults in how they organize their lives, including beliefs about moral development and behavior and the roles of men and women. Religious socialization also influences beliefs about sexuality, including the likelihood of tolerance for gay and lesbian sexuality (Whitehead and Baker 2012). Religion can even influence child-rearing practices, including the use of physical nurturing and strict discipline.

Sports

Most people think of sports as something that is just for fun and relaxation—or perhaps to provide opportunities for college scholarships and athletic careers—but sports are also an agent of socialization. Through sports, men and women learn concepts of self that stay with them in their later lives.

Sports are also where many ideas about gender differences are formed and reinforced (Eitzen 2016; Messner and Musto 2016). For men, success or failure as an athlete can be a major part of their identity. Even for those who have not been athletes, knowing about and participating in sports is an important source of men's gender socialization. Men learn that being competitive in sports is considered a part of manhood. Indeed, the attitude that "sports builds character" runs deep in American culture. Sports are supposed to pass on values such as competitiveness, the work ethic, fair play, and a winning attitude. Sports are considered to be where one learns to be a man.

Michael Messner's research on men and sports reveals the extent to which sports shape masculine identity. His research shows that, for most men, playing or watching sports is often the context for developing relationships with other men, especially fathers, even when the father is absent or emotionally distant in other areas of life. Through sports relationships with male peers, men shape their identities (Messner 2002).

Part of the socialization of masculine identity in sports is learning homophobic attitudes (that is, fear and hatred of homosexuals, discussed in Chapter 11). "Gay" and "athlete" are words rarely used together. The socialization of athletes includes the expectation that men are heterosexual. Jason Collins, a professional basketball player in the NBA, shocked the sports world in 2013 when he announced he was gay, the first active player to go public with his gay identity. Other professional athletes have since done the same, such as former NFL player Ray O'Callahan. The courage of these men to challenge the highly heterosexist norms of professional sports has created some dialogue within the sports media community about sexuality, gender, and sports.

Still, athletic prowess, highly esteemed in men, is more often minimized or ignored for women. Media images of women athletes tend to emphasize women athletes' family roles and personalities (Eitzen 2016). Commentary on women athletes can be downright sexist—and racist, too—such as the repeated racial slurs that have been directed against tennis superstar Serena Williams.

Research in the sociology of sports shows how activities as ordinary as shooting baskets on a city lot, playing on the soccer team for one's high school, or playing touch football on a Saturday afternoon can convey powerful cultural messages about our identity and our place in the world. Sports are a good example of the power of socialization in our everyday lives.

Debunking Society's Myths

Myth: Youth sports are simply games children play for fun.
Sociological Perspective: Although sports are a form of entertainment, playing sports, much like playing with dolls, is also a source for socialization into roles, such as gender roles (Messner and Musto 2016).

Schools

Once young people enter school, another process of socialization begins. At home, parents are the overwhelmingly dominant source of socialization cues. In school, teachers and other students are the source of expectations that encourage children to think and behave in particular ways. The expectations encountered in schools vary for different groups of students. These differences are shaped by a number of factors, including teachers' expectations for different groups and the resources that different parents can bring to bear on the educational process. The parents of children attending elite, private schools, for example, often have more influence on school policies and classroom activities than do parents in low-income communities. In any context, studying socialization in the schools is an excellent way to see the influence of gender, class, and race in shaping the socialization process.

Even as early as preschool, socialization into gender roles starts to shape different paths for boys and girls. Playing with blocks, for example, influences the development of spatial skills that later are important in mathematical reasoning. If teachers are more likely to encourage boys than girls to play with blocks, they are already laying paths that point the way to later life skills. Research finds, for example, that pre-school teachers are more likely to encourage so-called masculine activities (playing with blocks and trucks, for example) when working with all-boy groups and to encourage feminine activities (those that encourage nurturing, creativity, and domestic skills) when playing with all-girl groups (Granger et al. 2016). Social class and racial stereotypes also affect teachers' interactions with students. If teachers perceive working-class children and students of color as less bright or less motivated than middle-class White children, then these negative appraisals become *self-fulfilling prophecies. That is,* the expectations they create become the cause of actual behavior in the children, thus affecting the odds of success for these young people.

Socialization in school influences students in everything from classroom behavior to subjects they choose to study. Young girls, for example, assess their math abilities lower than do boys—even when they demonstrate similar levels of math achievement. Girls also believe they have to be exceptional to succeed in math and science careers. Such gendered socialization leaves the nation without the full range of talent needed in science, math, and engineering careers (Hill 2010).

Theories of Socialization

Knowing that people become socialized does not explain how it happens. People tend to think of socialization solely in psychological terms. The influence of **Sigmund Freud** (1856–1939), for example, permeates our culture. Perhaps Freud's greatest contribution was the idea that the unconscious mind shapes human behavior. Freud is also known for developing the technique of *psychoanalysis* to help discover the causes of psychological problems in the recesses of troubled patients' minds. Freud's approach depicts the human psyche in three parts: the id is about impulses; the superego is about the standards of society and morality; and the ego is about reason and common sense. The psychoanalytic perspective interprets human identity as relatively fixed at an early age in a process greatly influenced by one's family dynamics.

Psychological theories of socialization hold much in common with sociology but increasingly rely on studies of the brain to understand how people operate. Within sociology, socialization is explained using social learning theory, functionalism, conflict theory, and symbolic interaction theory. Each sociological perspective focuses on interactions with others and within social institutions to explain socialization and its effect on the development of the self (see ◆ Table 4-1).

See for Yourself

Visit a local day-care center, preschool, or elementary school and observe children at play. Record the activities they are involved in, and note what both girls and boys are doing. Do you observe any differences between boys' and girls' play? What do your observations tell you about *socialization* patterns for boys and girls?

Table 4-1	Theories of Socialization			
	Social Learning Theory	**Functional Theory**	**Conflict Theory**	**Symbolic Interaction Theory**
How each theory views:				
Individual learning process	People respond to social stimuli in their environment.	People internalize the role expectations that are present in society.	Individual and group aspirations are shaped by the opportunities available to different groups.	Children learn through taking the role of significant others.
Formation of self	Identity is created through the interaction of mental and social worlds.	Internalizing the values of society reinforces social consensus.	Group consciousness is formed in the context of a system of inequality.	Identity emerges as the creative self interacts with the social expectations of others.
Influence of society	Young children learn the principles that shape the external world.	Society relies upon conformity to maintain stability and social equilibrium.	Social control agents exert pressure to conform.	Expectations of others form the social context for learning social roles.

Social Learning Theory

Whereas **psychoanalytic theory** places great importance on the *internal* unconscious processes of the human mind, **social learning theory** considers the formation of identity to be a learned response to *external* social stimuli (Bandura and Walters 1963). Social learning theory emphasizes the societal context of socialization. Identity is regarded not as the product of the unconscious but as the result of modeling oneself (called role modeling) in response to the expectations of others. According to social learning theory, behaviors and attitudes develop in response to *reinforcement* and encouragement from those around us. Reinforcement comes to us as *positive reinforcement* (reward) or *negative reinforcement* (rebuke or punishment). Behavior that is positively reinforced is more likely to be repeated, whereas behavior that is negatively reinforced is not. A major tenant of social learning theory is the principle that positive reinforcement *plus* the presence of an admired role model makes the particular behavior highly likely.

Functionalism

Functionalist theory interprets socialization as integrating people into society. Seen in this way, socialization is the mechanism through which people internalize social roles and the values of society. This reinforces social consensus because it encourages at least some degree of conformity. Socialization is thus one way that society maintains its stability. Functionalism also understands that people learn identities and behaviors that are congruent with the needs of various social institutions. Schools, for example, train children, at least allegedly, to respect authority, be punctual, and follow rules—thereby preparing them for their future lives as workers in other institutions where these traits are valued.

Conflict Theory

Conflict theorists see socialization differently. Because of the emphasis in conflict theory on the role of power and coercion in society, conflict theorists thinking

Stephen Simpson/Taxi/Getty Images

Social learning theory emphasizes how people model their behaviors and attitudes on those of others.

about socialization would be most interested in how group identity is shaped by patterns of inequality in society. A person's or group's identity always emerges in a context, and if that context is one marked by different opportunities for different groups, then one's identity will be shaped by that fact. This may help you understand why, for example, women are more likely to choose college majors in areas of study that have traditionally been associated with women's traditional work opportunities.

Symbolic Interaction Theory

Recall that *symbolic interaction theory* centers on the idea that human actions are based on the meanings people attribute to behavior. These meanings emerge through social interaction (Blumer 1969). Symbolic interaction has been especially important in developing an understanding of socialization. People learn identities and values through socialization. For example, learning to become a good student means taking on the characteristics associated with that role. Because roles are socially defined, they are not real, like objects or things, but are real because of the meanings people give them.

Symbolic interaction theory sees meaning as constantly reconstructed as people act within their social environments. The **self** is what we imagine we are. It is not just an interior bundle of drives, instincts, and motives, but is imagined as people make conscious and meaningful adaptations to their social environment. From a symbolic-interactionist perspective, a person's identity is not something fixed at birth or set in stone at an early age. Because the self is socially bestowed and socially sustained, it emerges over a lifetime and, as important as one's early years are, people can adapt and change their concepts of self (Berger 1963).

Two theorists have greatly influenced the development of symbolic-interactionist theory in sociology. **Charles Horton Cooley** (1864–1929) and **George Herbert Mead** (1863–1931) were both sociologists at the University of Chicago in the early 1900s (see Chapter 1). Cooley and Mead saw the self as developing in response to the expectations and judgments of others in their social environment.

Charles Horton Cooley postulated the **looking-glass self** to explain how our conception of our self develops through considering our relationships to others (Cooley 1967/1909, 1902). The looking-glass self emerges from (1) how we think we appear to others; (2) how we think others judge us; and, (3) how the first two make us feel—proud, embarrassed, or other feelings. The looking-glass self involves perception and effect, the perception of how others see us, and the effect of others' judgment on us (see ▲ Figure 4-3).

You can see the influence of the looking-glass self in a clever study examining the time college women spend on Facebook and their body image. The researchers found that college women spend much of their time on Facebook commenting on weight, body image, exercise, and dieting and comparing it to others (Ecklet, Kalyango, and Paasch 2017). This is the looking-glass self emerging through cyberspace interaction!

Whether online or face-to-face, how others see us is fundamental to the idea of the looking-glass self. In seeing ourselves as others do, we respond to the expectations others have of us. This means that *the formation of the self is fundamentally a social process*—one based in the interaction people have with each other, as well as the human capacity for self-examination.

One unique feature of human life is the ability to see ourselves through others' eyes. People can imagine themselves in relationship to others and develop a definition of themselves accordingly. From a symbolic interactionist perspective, the reflective process is key to the development of the self. If you grow up with others who think you are smart and sharp-witted, chances are you will develop this definition of yourself. If others see you as dull-witted and withdrawn, chances are good that you will see yourself this way. George Herbert Mead agreed with Cooley that children are socialized by responding to others' attitudes toward them. According to Mead, social roles are the basis of all social interaction.

Taking the role of the other is the process of putting oneself into the point of view of another. To Mead, role-taking is a source of self-awareness. As people take on new roles, their awareness of self changes.

According to Mead, identity emerges from the roles one plays. He explained this process in detail by examining childhood socialization, which he saw as occurring in three stages: the imitation stage, the play stage, and the game stage (Mead 1934). In each phase of development, the child becomes more proficient at taking the role of the other.

▲ **Figure 4-3   The Looking-Glass Self.** The looking-glass self refers to the process by which we attempt to see ourselves as others see us. This also helps us identify what roles we play in society and in relation to others. Drawing conceptualized by Norman Andersen.

In the first stage, the **imitation stage**, children merely copy the behavior of those around them. Role-taking in this phase is nonexistent because the child simply mimics the behavior of those in the surrounding environment without much understanding of the social meaning of the behavior. Although children in the imitation stage have little understanding of the behavior being copied, they are learning to become social beings. For example, think of young children who simply mimic the behavior of people around them (such as pretending to read a book, but doing so with the book upside down).

In the second stage, the **play stage**, children begin to take on the roles of significant people in their environment, not just imitating but incorporating their relationship to the other. Especially meaningful is when children take on the role of **significant others**, those with whom they have a close affiliation. A child pretending to be his mother may talk to himself as the mother would. The child begins to develop self-awareness, seeing himself or herself as others do.

In the third stage of socialization, the **game stage**, children become capable of taking on multiple roles at the same time. These roles are organized in a complex system that gives the children a more general or comprehensive view of the self. In this stage, children begin to comprehend the system of social relationships in which they are located. The children not only see themselves from the perspective of a significant other, but also understand how people are related to each other and how others are related to them. This is the phase where children internalize (incorporate into the self) an abstract understanding of how society sees them.

Mead compared the lessons of the game stage to a baseball game. In baseball, all roles together make the game. The pitcher does not just throw the ball past the batter as if they were the only two people on the field; rather, each player has a specific role, and each role intersects with the others. The network of

Childhood Play and Socialization

The purpose of this exercise is to explain how *childhood socialization* is a mechanism for passing on social *norms* and *values*. Begin by identifying a form of play that you engaged in as a young child. What did you play? Who did you play with? Was it structured or unstructured play? What were the rules? Were they formal or informal, and who controlled whether they were observed?

1. Now think about what *norms* and *values* were being taught to you by way of this play. Do they still affect you today? If so, how?
2. How does your experience compare to those of students different from you in terms of gender, race, ethnicity, regional origin, and so forth? Are there differences in learned norms and values that can be attributed to these different social characteristics?

social roles and the division of labor in the baseball game is a social system, like the social systems children must learn as they develop a concept of themselves in society.

In the game stage, children learn more than just the roles of significant others in their environment. They also acquire a concept of the **generalized other**—the abstract composite of social roles and social expectations. In the generalized other, they have an example of community values and general social expectations that adds to their understanding of self; however, children do not all learn the same generalized other. Depending on one's social position (that is, race, class, gender, region, or religion), one learns a particular set of social and cultural expectations.

If the self is socially constructed through the expectations of others, how do people become individuals? Mead answered this by saying that the self has two dimensions: the "I" and the "me." The "I" is the unique part of individual personality, the active, creative, self-defining part. The "me" is the passive, conforming self, the part that reacts to others. In each person, there is a balance between the I and the me. To Mead, social identity is always in flux, constantly emerging (or "becoming") and dependent on social situations. Over time, identity stabilizes as one learns to respond consistently to common situations.

Social expectations associated with given roles change as people redefine situations and as social and historical conditions change; thus, the social expectations learned through the socialization process are not permanently fixed. For example, as more women enter the paid labor force and as men take on additional responsibilities in the home, the expectations associated with motherhood and fatherhood are changing. Men now experience some of the role conflicts that women have faced in balancing work and family. As the roles of mother and father are redefined, children are learning new socialization patterns; however, traditional gender expectations maintain a remarkable grip. Despite many changes in family life and organization, young girls are still socialized for motherhood and young boys are still socialized for greater independence and autonomy.

Growing Up in a Diverse Society

Understanding the institutional context of socialization is important for understanding how socialization affects different groups in society. Socialization makes us members of our society. It instills in us the values of the culture and brings society into our self-definition, our perceptions of others, and our understanding of the world around us. Socialization is not, however, a uniform process, as the different examples developed in this chapter show. In a society as complex and diverse as the United States, no two people will have exactly the same experiences. We can find similarities between us, often across vast social and cultural differences, but variation in social contexts creates vastly different socialization experiences, as shown in the box "Doing Sociological Research" on undocumented students.

Resilience among Undocumented Students

There are an estimated 11.2 million undocumented youth in the United States (Krogstad, Passel, and Cohn 2017). They face the same stresses and risk factors that all young people experience, but with the added threat of deportation, family breakup, and other forms of institutional and social exclusion.

Research Questions

How do undocumented students overcome these risks and become academically successful? What specific social and environmental factors enable their success in school, despite the risk factors they face. These are the questions examined by a team led by William Perez.

Research Method

Perez et al. sampled 110 Latino high school, community college, and college students with an equal balance of women and men. On average they had lived about thirteen years in the United States—the men having come to the United States at age seven (on average) and the women at age five. Using an online survey, the research team collected various information, including measures of educational success (academic achievement, civic leadership positions, and enrollment in advanced level academic courses), school background and demographic information, and different measures of risk and stress (including perceived societal rejection, parental and friends' valuing of school, employment while in school, family obligations, and so forth).

Research Results

The researchers found that a constellation of factors protected students from the threats posed by their undocumented status. Having positive personal and environmental supports was especially important. Living in poverty, working long hours at a job while in school, low levels of parental education, and feeling a high sense of rejection because of being undocumented put Latino students at especially high levels of risk for poor academic achievement. Resilient students, on the other hand, were more protected by a host of factors, including being identified as gifted, parental and personal valuing of school, and being engaged in extracurricular activities.

Conclusions and Implications

Higher levels of protective assets can help ensure student success, even knowing the risks for undocumented students. Understanding the impact of social and environmental factors on academic resilience can help school leaders and policy makers ensure the success of undocumented students.

Questions to Consider

1. What specific social and environmental factors put you at risk and/or what "protective resources" do you have? How might these compare to those of other students in your class?
2. What policies would you recommend specific to undocumented students that would ensure their educational success?

Source: Perez, William, Roberta Espinoza, Karina Ramos, Heidi M. Coronado, and Richard Cortes. 2009. "Academic Resilience among Undocumented Latino Students." *Hispanic Journal of Behavioral Sciences* 31 (May): 149–181.

Furthermore, current changes in the U.S population are creating new multiracial and multicultural environments in which young people grow up. Schools, as an example, are being transformed in many places by the large number of immigrant groups entering the school system. Immigrant children come into contact with native-born American children who may or may not be born to immigrant parents. This creates a new context in which children form their social values and learn their social identities.

One task of the sociological imagination is to examine the influence of different contexts on socialization. Where you grow up; how your family is structured; what resources you have at your disposal; your racial–ethnic identity, gender, and nationality—all shape the socialization experience. Socialization experiences for all groups are shaped by many factors that intermingle and intersect to form the context for socialization.

Children are socialized to behave certain ways in school. Both families and schools are agents of socialization. Social class influences parental expectations of classroom behavior and problem solving, resulting in class differences in academic performance.

African American families, as an example, have to socialize their children to be able to cope with the racism they will no doubt encounter. Asian American families, likewise, have to teach children how to cope with stereotypes of them as a "model minority"—a stereotype that can actually reduce students' academic performance. How well families are able to support their children (both financially and emotionally) as they transition into adulthood can have long-term consequences for people's well-being in adulthood. As the United States becomes more diverse in its population, differences in how well families are able to support children may shift societal norms around the transition to adulthood (Hardie and Seltzer 2016).

Aging and the Life Course

Socialization begins the moment a person is born. As soon as the sex of a child is known (which now can be even before birth), parents, grandparents, brothers, and sisters greet the infant with different expectations, depending on whether it is a boy or a girl. Socialization does not come to an end as we reach adulthood. Socialization continues through our lifetime. As we enter new situations, and even as we interact in familiar ones, we learn new roles and undergo changes in identity.

Sociologists use the term **life course** to describe and analyze the connection between people's personal attributes, the roles they occupy, the life events they experience, and the social and historical aspects of these events. The life course perspective underscores the point made by C. Wright Mills (introduced in Chapter 1) that personal biographies are linked to specific social-historical periods. Different generations are strongly influenced by large-scale events (such as war, economic depression, or events such as the 9/11 terrorist attacks, as examples).

The phases of the life course are familiar: childhood, youth and adolescence, adulthood, and old age. These phases of the life course define different generations and define some of life's most significant events, such as birth, marriage, retirement, and death. ▲ Figure 4-4 illustrates some of the ways we are socialized throughout life.

Childhood

During childhood, socialization establishes one's initial identity and values. In this period, the family is an extremely influential source of socialization. Experiences in school, peer relationships, sports, religion, and the media also have a profound effect. Children acquire knowledge of their culture through countless subtle cues that provide them with an understanding of what it means to live in society.

Socializing cues begin as early as infancy, when parents and others begin to describe their children based on their perceptions. Frequently, these perceptions are derived from the cultural expectations parents have for children. Parents of girls may describe their babies as "sweet" and "cuddly," whereas boys are described as "strong" and "alert." Even though it is difficult to physically identify baby boys and girls when they are infants, parents in this culture dress even their tiny infants in colors and styles that typically distinguish one gender from the other.

▲ **Figure 4-4   The Many Agents of Socialization.** Throughout the life course, there are many social influences on how and what we learn. We internalize the expectations of society beginning from infancy and continuing into old age.

Childhood socialization is often very subtle. Much socialization in early childhood takes place through play and games. Games that emphasize traditional gender roles will likely lead girls to grow up wanting to fulfill feminine roles and boys wanting to fulfill masculine roles. Social class, family structure, and race also affect how gender roles are taught to children. For example, research finds that children of lesbian or gay parents play games that are less gender-stereotyped and that children of heterosexual couples play more gender-specific games (Goldberg et al. 2012).

Adolescence

Only recently has adolescence been thought of as a separate phase in the life cycle. Until the early twentieth century, children moved directly from childhood roles to adult roles. It was only when formal education was extended to all classes that adolescence emerged as a particular phase in life. In adolescence, young people are regarded as no longer children, but not yet adults. There are no clear boundaries to adolescence, although it generally lasts from junior high school until the time one takes on adult roles by getting a job, forming one's own family, and financially supporting themselves.

Erik Erikson (1980), the noted psychologist, stated that the central task of adolescence is the formation of a consistent identity. Adolescents try to become independent of their families, but they have not yet moved into adult roles. Conflict and confusion can arise as adolescents swing between childhood and adult maturity. Some argue that adolescence is a period of delayed maturity. Although society expects adolescents to behave like adults, they are denied many privileges associated with adult life. Until age 18, they cannot vote, drink, or marry without permission, and they are considered too young for parenthood. They can, however, enlist in the military.

The tensions of adolescence have been blamed for numerous social problems, such as drug and alcohol abuse, youth violence, and teen pregnancy. Sociologist Michael Kimmel has examined the impact of the now-extended period of adolescence for White, educated, middle-class young men (Kimmel 2008). "Guyland" is the social space that these men inhabit—a space where young men are neither children nor quite adults. In this space, men "hang out" with other men, developing a unique culture of masculinity. By conducting over 400 interviews throughout the country, Kimmel found that women are mostly peripheral in Guyland where men seem anxious, not sure of their place in the world, especially as women have become more ascendant. In Guyland, young men drink a lot, engage in locker-room talk about women and sports, and disparage being a good student. Although not all "guys" are the same in Guyland, they play out roles of masculinity often in ill-conceived and irresponsible ways. Guyland, Kimmel argues, is a social space where men express manhood in a world when being a man is no longer a certain thing. Kimmel's research shows how different social and historical contexts shape what adolescence looks like.

Today's young people face a quite uncertain world where adult roles are less predictable than in the past and where economic inequality has created unique circumstances. Robert Putnam's research team has examined the changes for youth from different social and economic backgrounds that have come from economic transformations in the U.S. economy. The decline of a manufacturing economy and growing inequality has eroded the possibility of achieving the American Dream—that is, upward mobility, no matter one's class of origin, through work, education, and commitment. Putnam found that for working-class youth there is a decline in economic opportunity, increasing educational segregation, and a decline in social networks and a collective sense of community. The result is more difficulty in achieving the traditional markers of adulthood—steady work, stable families, and close-knit communities (Putnam 2015).

Adulthood

Work such as that of Kimmel and Putnam shows that socialization occurs in the context of social forces of the time. Moreover, socialization does not end when one becomes an adult. Building on the identity formed in childhood and adolescence, **adult socialization** is the process of learning new roles and expectations in adult life. More than at earlier stages in life, adult socialization involves learning behaviors and attitudes appropriate to specific situations and roles.

Adult life is peppered with events that may require adults to adapt to new roles. Marriage, a new career, starting a family, entering the military, getting a divorce, retirement, or dealing with a death in the family all transform an individual's previous social identity. In today's world, these transitions through the life course are not as orderly as they were in the past. Where there was once a sequential and predictable trajectory of schooling, work, and family roles through one's twenties and thirties, that is no longer the case. Younger generations now experience diverse patterns in the sequencing of work, schooling, and family formation—even returning home—than was true in the past. These changes complicate the life course, and people have to make different adaptations to these changing roles.

Becoming an adult involves many dimensions, such as financial independence, completing an education, and so forth. Conditions in society at any given time can make these aspirations difficult to achieve. Finishing an education, acquiring work, becoming independent—all of these desires can be compromised by the social conditions of the time, which is further evidence of the impact of societal forces larger than oneself on the outcome of one's life. The path to adulthood has changed significantly than in the past (see, for example, ▲ Figure 4-5). Although young men and women agree that a good job, marriage, and being able to live on their own are all desirable adult roles, achieving these milestones is not as automatic as it once was. Working-class young adults, in particular, are especially vulnerable. They find work mostly in insecure service jobs, incur large amounts of credit card debt, and have little time or other resources to build a secure life. As Jennifer Silva writes, "pervasive economic insecurity, fear of commitment, and confusion within institutions make the achievement of traditional markers of adulthood impossible and sometimes undesirable" (Silva 2014, 2013: 28).

Another part of adult socialization is **anticipatory socialization**, the learning of expectations associated with a role a person expects to enter in the future. Anticipatory socialization allows a person to foresee the expectations associated with a new role and to learn what is expected in that role in advance. Research on working-class young adults highlights that they are prepared to embrace adult roles, despite the economic obstacles they face in obtaining them.

In the transition from an old role to a new one, individuals often vacillate between their old and new identities as they adjust to fresh settings and expectations. An example is *coming out*, the process of identifying oneself as gay or lesbian. This can be either a public coming out or a private acknowledgment of sexual orientation. The process can take years and usually means coming out to a few people, perhaps

Debunking Society's Myths

Myth: Identity is more or less fixed at an early age.
Sociological Perspective: The formation of identity is a lifelong process. Though childhood and adolescence are significant times for identity formation, people learn new identities throughout the life course, including socialization into new roles as adults.

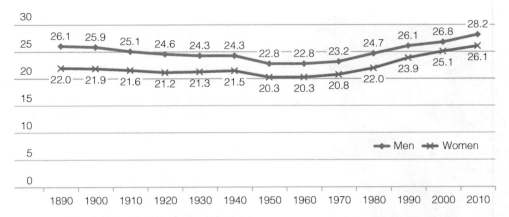

▲ **Figure 4-5   Average Age at First Marriage, 1890–2010.** You can see from this line graph that the age at first marriage for both men and women has both fallen and risen at different points in time. Women and men started marrying younger in the time period between 1940 and 1980. What social factors might explain this? And, then again, why is the age at first marriage now rising—and fairly substantially since 1980? You might also notice that the gap between men's and women's age has not changed much over time. Why?

Source: U.S. Census Bureau. 2016. Table MS-2: Estimated Median Age at First Marriage, 1890–2010. **www.census .gov/data/tables/time-series/demo/families/marital.html**

first to family members or friends who are likely to have the most positive reaction. Coming out is rarely a single event, but occurs in stages on the way to developing a new identity.

Age and Aging

Passage through adulthood involves many transitions. In our society, one of the most difficult transitions is the passage to old age. We are taught to fear aging in this society, and many people spend a lot of time and money trying to keep looking young. Unlike many other societies, ours does not revere the elderly, but instead devalues them, making the aging process even more difficult.

Despite desperate attempts to hide gray hair, eliminate wrinkles, and reduce middle-aged weight gain, aging is inevitable. The skin creases and sags, the hair thins, metabolism slows, and bones become less dense and more brittle by losing bone mass. Although aging is a physical process, the social dimensions of aging are just as important, if not more important, in determining the aging process. Just think about how some people appear to age much more rapidly than others. Some sixty-year-olds look only forty, and some forty-year-olds look sixty. These differences result from combinations of biological and social factors, such as genetics, eating, exercise, stress, smoking habits, pollution in the physical environment, and many other factors. The social dimensions of aging are what interest sociologists.

Although the physiology of aging proceeds according to biological processes, what it *means* to grow older is a social phenomenon. **Age stereotypes** are preconceived judgments about what different age groups are like. Stereotypes abound for both old and young people. Young people, especially teenagers, are stereotyped as irresponsible, addicted to loud music, lazy ("slackers"), sloppy, and so on; the elderly are stereotyped as forgetful, set in their ways, mentally dim, and unproductive. Like any stereotype, these stereotypes are largely myths, but they are widely believed.

Age stereotypes also differ for different groups. Older women are stereotyped as having lost their sexual appeal, contrary to the stereotype of older men as handsome or "distinguished" and desirable. Gender is, in fact, one of the most significant factors in age stereotypes. Age stereotypes are also reinforced through popular culture. Advertisements depict women as needing creams and lotions to hide "the telltale signs of aging." Men are admonished to cover the patches of gray hair that appear or to use other products to prevent baldness. Entire industries are constructed on the fear of aging that popular culture promotes. Facelifts, tummy tucks, and vitamin advertisements all claim to "reverse the process of aging," even though the aging process is a fact of life.

Can you tell how old this person is? Many people work hard to counter the impact of aging, even though aging is inevitable.

Age Prejudice and Discrimination

Age prejudice refers to a negative attitude about an age group that is generalized to all people in that group. Prejudice against the elderly is prominent. The elderly are often thought of as childlike and thus incapable of adult responsibility. Prejudice relegates people to a perceived lower status in society and stems from the stereotypes associated with different age groups.

Age discrimination is the different and unequal treatment of people based solely on their age. Whereas age prejudice is an attitude, age discrimination involves actual behavior. As an example, people may talk "baby talk" to the elderly. This reinforces the stereotype of the elderly as childlike and incompetent. Some forms of age discrimination are illegal. The Age Discrimination in Employment Act, first passed in 1967 but amended several times since, protects people from age discrimination in employment. It states that age discrimination is a violation of the individual's civil rights. An employer can neither hire nor fire someone based solely on age, nor segregate or classify workers based on age.

Ageism is a term sociologists use to describe age prejudice and discrimination. More than a single attitude or an explicit act of discrimination, ageism is structured into the institutional fabric of society. Like racism and sexism, ageism does not have to be intentional or overt to affect how age groups are treated. Ageism in society means that, regardless of laws that prohibit age discrimination, a person's age is a significant predictor of his or her life chances. Resources are distributed in society in ways that advantage some age groups and disadvantage others; cultural belief systems devalue the elderly; society's systems of care are often inadequate to meet people's needs as they grow old—these are the manifestations of ageism, a persistent and institutionalized feature of society.

Age Stratification

Most societies produce age hierarchies—systems in which some age groups have more power and better life chances than others. **Age stratification** refers to the hierarchical ranking of different age groups in society. Age stratification exists because processes in society ensure that people of different ages differ in their access to society's rewards, power, and privileges. In the United States and elsewhere, age is a major source of inequality. You can see in ▲ Figure 4-6 that the percentage of older Americans living

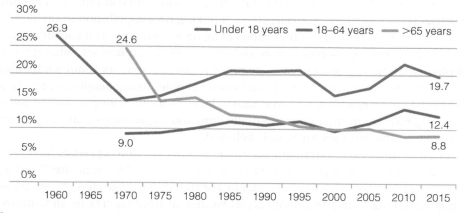

▲ **Figure 4-6   Poverty by Age, 1960–2015.** You can see from this graph that poverty among the old has declined substantially. Children (those under 18 years of age) are now the age group most likely to be poor. What implications does this have for the future?

Source: **www.census.gov**

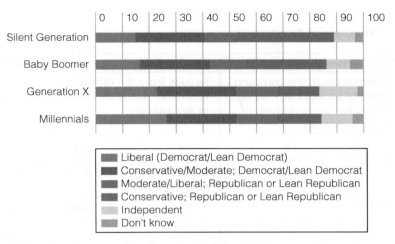

▲ **Figure 4-7   The Generation Gap in Political Views.** Millennials are those born after 1980 who in 2016, when this survey was taken, would have been between 18 and 35 years of age; Generation X, those then aged 36 to 51; baby boomers, age 52 to 70; and the Silent Generation, those 71 to 88.

Source: Maniam, Shiva, and Samanta Smith. 2016. "A Wider Partisan and Ideological Gap between Younger, Older Generations." Washington, DC: Pew Research Organization. **www.pewresearch.org**

in poverty has significantly declined since 1960, while the percentage of children living in poverty has increased since 1970.

Age is an *ascribed status*; that is, age is determined by when you were born. Different from other ascribed statuses, which remain relatively constant over the duration of a person's life, age changes steadily throughout your life. Still, you remain part of a particular generation—something sociologists call an **age cohort**—an aggregate group of people born during the same period.

People in the same age cohort share the same historical experiences—wars, technological developments, and economic fluctuations—although they might do so in different ways, depending on other life factors. Living through the Great Depression, for example, shaped an entire generation's attitudes and behaviors, as did growing up in the 1960s, as will being a member of the contemporary millennial generation. You can see in ▲ Figure 4-7 that age cohorts matter in shaping the political orientation of different generations.

Recall from Chapter 1 that C. Wright Mills saw the task of the sociological imagination as analyzing the relationship between biography and history. Understanding the experiences of different age cohorts is one way you can do this. People who live through the same historic period experience a similar impact of that period in their personal lives. The troubles and triumphs they experience and the societal issues they face are rooted in the commonality established by their age cohort. There is variation in how the old are treated. In many societies, older people are given enormous respect. There may be traditions to honor the elders, and they may be given authority over decisions in society, as they are perceived as most wise. On the other hand, among some cultures, adults who can no longer contribute to the society because of old age or illness may be perceived as extreme burdens and thus may be banished from the society altogether.

Why does society stratify people on the basis of age? Functionalist sociologists ask whether the grouping of individuals contributes in some way to the common good of society (see ◆ Table 4-2). From this perspective, adulthood is functional to society because adults are seen as the group contributing most fully to it; the elderly are not. Functionalists argue that older people are seen as less useful and are therefore granted lower status in society. According to the functionalist argument, the diminished usefulness of the elderly justifies their depressed earning power and their relative neglect in social support networks.

Conflict theory focuses on the competition over scarce resources between age groups. Among the most important scarce resources are jobs. Conflict theory interprets groups as in competition for society's

Table 4-2 Sociological Theories of Aging

	Functional Theory	Conflict Theory	Symbolic Interaction
Age differentiation	Contributes to the common good of society because each group has varying levels of utility in society	Results from the different economic status and power of age cohorts	Occurs in most societies, but the social value placed on different age groups varies across diverse cultures
Age groups	Are valued according to their usefulness in society	Compete for resources in society, resulting in generational inequities and thus potential conflict	Are stereotyped according to the perceived value of different groups
Age stratification	Results from the functional value of different age cohorts	Intertwines with inequalities of class, race, and gender	Promotes ageism, which is institutionalized prejudice and discrimination against old people

resources. Barring youth and the elderly from the labor market eliminates these groups from competition, improving the prospects for middle-aged workers. Conflict theory also helps explain the power that different age cohorts have—or do not. Older people, for example are now a political force both because of their population size and because they are less likely to be poor than in the past. The politics that emerge whenever someone threatens to reduce Social Security or Medicare shows the power that the older population has—at least for now.

Symbolic interaction theory analyzes the different meanings attributed to social entities. Symbolic interactionists ask what meanings become attached to different age groups and to what extent these meanings explain how society ranks such groups. Definitions of aging are socially constructed, as we saw in our discussion of age stereotypes. Moreover, in some societies, the elderly may be perceived as having higher status than in other societies. Symbolic interaction considers the role of social perception in understanding the sociology of age. Age clearly takes on significant social meaning—meaning that varies from society to society for a given age group and that varies within a society for different age groups.

Growing old in a society such as the United States with such a strong emphasis on youth means encountering social stereotypes about the old, adjusting to diminished social and financial resources, and sometimes living with the absence of social supports, even when facing some of life's most difficult transitions, such as declining health and the loss of loved ones. Still, many people experience old age as a time of great satisfaction and enjoy a sense of accomplishment connected to work, family, and friends. The degree of satisfaction during old age depends to a great extent on the social support networks established earlier in life—evidence of the continuing influence of socialization.

Rites of Passage

A **rite of passage** is a ceremony or ritual that marks the transition of an individual from one role to another. Rites of passage define and legitimize abrupt role changes that begin or end each stage of life. The ceremonies surrounding rites of passage are often dramatic and infused with awe and solemnity. Examples include graduation ceremonies, weddings, and religious affirmations such as the Jewish ceremony of the bar mitzvah for boys or the bat mitzvah for girls, confirmation for Catholics, and adult baptism for many Christian denominations.

Formal promotions or entry into some new careers may also include rites of passage. Completing police academy training or being handed one's diploma are examples. Such rites usually include family and friends, who watch the ceremony with pride. People frequently keep mementos of these rites as markers of the transition through life's major stages. Bridal showers and baby showers have been analyzed as rites of passage. At a shower, the person who is being honored is about to assume a new role and identity—from young woman to wife or mother. Rites of passage entail public announcement of the new status for the benefit of both the individual and those with whom the newly anointed person will interact. In the absence of such rituals, the transformation of identity would not be formally recognized, and might create uncertainty in the youngster or the community about the individual's worthiness, preparedness, or community acceptance.

Sociologists have noted that in the U.S. population as whole, there is no standard and formalized rite of passage marking the transition from childhood to adulthood. As a consequence, the period of adolescence is attended by ambivalence and uncertainty. Adolescents hover between adult and child status, so they may not have the clear sense of identity that a rite of passage can provide. Although there is no universal ceremony in our culture marking the change from childhood to adulthood, some social class and ethnic subcultures do mark the occasion. Among the wealthy, the debutante's coming-out celebration is a traditional introduction of a young woman to adult

Every culture has important rites of passage that mark the transition from one phase in the life course to another. Different cultural traditions distinguish the rites of passage associated with marriage, such as this traditional Nigerian wedding.

society. Latinos may celebrate the *quinceañera* (fifteenth birthday) of young girls. A tradition of the Catholic church, this rite recognizes the girl's coming of age, while also keeping faith with an ethnic heritage. Dressed in white, she is introduced by her parents to the larger community. Formerly associated mostly with working-class families and other Latinos, the quinceañera has also become popular among affluent Mexican Americans, who may match New York debutante society by spending as much as $50,000 to $100,000 on the event.

Resocialization

Most transitions people experience in their lifetimes involve continuity with the former self as it undergoes gradual redefinition. Sometimes, however, adults are forced to undergo a radical shift of identity. **Resocialization** is the process by which existing social roles are radically altered or replaced (Fein 1988). Resocialization is especially likely when people enter institutional settings where the institution claims enormous control over the individual. Examples include the military, prisons, monastic orders, and some cults. When military recruits enter boot camp, they are stripped of personal belongings, their heads are shaved, and they are issued identical uniforms. Although military recruits do not discard their former identities, the changes brought about by becoming a soldier can be dramatic and are meant to make the military one's *primary group*, not one's family, friends, or personal history. The military represents an extreme form of resocialization in which individuals are expected to subordinate their identity to that of the group.

Resocialization occurs again when soldiers return to civilian life. Research has found, for example, that veterans returning from the wars in Iraq and Afghanistan face challenges upon their return that make them feel alien in their own country. The contrasting identities of soldier and civilian mean they may struggle with their new autonomy, because they have become accustomed to (that is socialized into) the hierarchical authority structure of the military. Coming home means having to be resocialized into new expectations (Smith and True 2014).

Resocialization also occurs in many settings such as college sport teams, fraternities, and the military. During initiation rituals, new members may be given menial and humiliating tasks and be expected to act in a subservient manner. Such behaviors reinforce one's new identity. Although the participants may not think of this as resocialization, that is precisely what is happening.

Find an adult who has just entered a new stage of life (getting a new or first job, getting married, becoming a grandparent, or retiring). Ask them to describe any expectations that others are being communicated to them in this new role. Do they experience new expectations associated with this change? How has their own behavior changed? What do your observations tell you about *adult socialization*?

The Process of Conversion

Resocialization also occurs during what people popularly think of as conversion. A conversion is a far-reaching transformation of identity, often related to religious or political beliefs. People often think about conversion now in the context of those who become radical terrorists. What would cause someone to believe something so strongly that they would kill innocent civilians in the interest of their cause? While not excusing such behavior, sociologists would look to the social influences that would generate such resocialization.

Conversion happens in various contexts, not just among those who become radical terrorists. Conversion in cults is another example. A *cult* is a group (sometimes small) that is formed around some extremist beliefs. Those who become members of cults are typically socialized through rejection of all they believed before joining a cult (see also Chapter 13). An example is the infamous Heaven's Gate cult, in which thirty-nine members committed suicide together believing they were going to be exported by a comet to a better place (see also Chapter 6).

Conversion is not always so extreme, but might involve a major change in identity, such as a transgender person whose identity as a new gender evolves over time (Ryan 2016). Someone who converts to a new religion or disavows previously held religious beliefs also experiences resocialization as the person develops new ideas, new identities, and new relationships.

The Brainwashing Debate

Extreme examples of resocialization are seen as "brainwashing." In the popular view of brainwashing, converts have their previous identities totally stripped. The transformation is seen as so complete that only deprogramming can restore the former self. Potential candidates of brainwashing include people who enter religious cults, prisoners of war, and hostages. Sociologists have examined so-called brainwashing to illustrate the process of resocialization, but they note that even with extreme conversions, converts do not necessarily drop their former identity. Resocialization in its most extreme form can be seen in the radicalization of youth into terrorist organizations. Terrorist groups across the globe recruit children and young adults to join extremist movements, engage in violent acts, and possibly even detonate "suicide bombs" in the name of the radical group. Both political and religious doctrines can be used to resocialize people to radicalized belief systems (Nawaz 2013).

Forcible confinement and physical torture can be instruments of extreme resocialization. Under severe captivity and deprivation, a captured person may come to identify with the captor; this is known as the **Stockholm syndrome**. In such instances, the captured person has become *dependent* on the captor. On release, the captive frequently needs debriefing, or deprogramming. Prisoners of war and hostages may not lose free will altogether, but they do lose freedom of movement and association, which makes prisoners intensely dependent on their captors and therefore vulnerable to the captor's influence.

The Stockholm syndrome can help explain why some battered women do not leave their abusing spouses or boyfriends. Dependent on their abuser both financially and emotionally, battered women often develop identities that keep them attached to men who abuse them, a clear example of identification with an aggressor. In these cases, outsiders often think the women should leave instantly, whereas the women themselves may find leaving difficult, even in the most abusive situations.

The socialization process begins at birth and continues throughout life. How we are socialized as children defines much of who we are and how we interact with society. The various agents of socialization influence the roles we take on as children, young adults, and into old age.

Chapter Summary

What is socialization, and why is it significant for society?

Socialization is the process by which human beings learn the social expectations of society. Socialization creates the expectations that are the basis for people's attitudes and behaviors. Through socialization, people conform to social expectations, although people still express themselves as individuals.

What are the agents of socialization?

Socialization agents are those who pass on social expectations. They include the family, the media, peers, sports, religious institutions, and schools, among others. The family is usually the first source of socialization. The media also influence people's values and behaviors. *Peers* are an important source of individual identity; without peer approval, most people find it hard to be socially accepted. Schools also pass on expectations that are influenced by gender, race, and other social characteristics of people and groups.

What theoretical perspectives do sociologists use to explain socialization?

Psychoanalytic theory sees the self as driven by unconscious drives and forces that interact with the expectations of society. *Social learning theory* sees identity as a learned response to social stimuli such as reward–punishment and role models. *Functionalism* interprets socialization as key to social stability because socialization establishes shared roles and values. *Conflict theory* interprets socialization in the context of inequality and power relations. *Symbolic interaction theory* sees people as "constructing" the self as they interact with the environment and give meaning to their experience. Charles Horton Cooley described this process as the *looking-glass self*. Another sociologist, George Herbert Mead, described childhood socialization as occurring in three stages: imitation, play, and games.

Does socialization mean that everyone grows up the same?

Socialization is not a uniform process. Growing up in different environments and in such a diverse society means that different people and different groups are exposed to different expectations. Factors such as family structure, social class, regional differences, and many others influence how one is socialized.

Does socialization end during childhood?

Socialization continues through a lifetime, although childhood is an especially significant time for the formation of *identity*. Adolescence is also a period when peer cultures have an enormous influence on the formation of people's self-concepts. *Adult socialization* involves the learning of specific expectations associated with new roles.

What are the social dimensions of the aging process?

Although aging is a physiological process, its significance stems from social meanings attached to aging. *Age prejudice* and *age discrimination* result in the devaluation of older people. *Age stratification*—referring to the inequality that occurs among different age groups—is the result.

What does resocialization mean?

Resocialization is the process by which existing social roles are radically altered or replaced. It can take place in an organization that maintains strict social control and demands that the individual conform to the needs of the group or organization. Examples are religious conversion, excessive influence via social interaction ("brainwashing"), and the Stockholm syndrome.

Key Terms

adult socialization 96
age cohort 98
age discrimination 98
age prejudice 98
age stereotype 97
age stratification 98
ageism 98
anticipatory socialization 96
game stage 91
generalized other 92

identity 80
imitation stage 91
life course 94
looking-glass self 90
peers 85
personality 81
play stage 91
psychoanalytic theory 89
resocialization 101
rite of passage 100

roles 80
self 90
self-concept 82
significant others 91
social control 81
social learning theory 89
socialization 80
socialization agents 83
Stockholm syndrome 102
taking the role of the other 90

CHAPTER 5

SOCIAL STRUCTURE AND SOCIAL INTERACTION

In this chapter, you will learn to:

Define society and social interaction, contrasting macro- and micro-level analyses

Compare and contrast different ways society is held together

Identify the different types of society

Explain social interaction in society, including groups, status, roles, and everyday social interactions

Compare and contrast the theories used to analyze social interaction

Examine interaction in cyberspace

iStock.com/elenabs

Picture a college classroom on your campus: Students sit, and some are taking notes; most are listening; a few, perhaps, sleeping. The class period ends and students stand, then gather their books, backpacks, bags, and other gear. As they stand, many whip out their cell phones and quickly push buttons that connect them to a friend. As the students exit the room, many are engaged in social interaction—chatting with their friends: some by phone, others by texting, some by talking face-to-face. Few, if any, of them realize that their behavior is at that moment influenced by society—a society whose influence extends into their immediate social relationships, even when the contours of that society—its social structure—are probably invisible to them.

These same students might put ear buds into their ears as they move on to their next class, possibly tuning in to the latest sounds while tuning out the sounds of the environment around them. Some will return to their residences and perhaps text message friends, download some music, or send a photo via Instagram. Surrounding all of this behavior are social changes that are taking place in society, including changes in technology, in global communication, and in how people now interact with each other. How we make sense of these changes requires an understanding of the connection between society and social interaction. In this way, a sociological perspective can help you see the relationship between individuals and the larger society in which people live.

What Is Society?

In Chapter 2, we studied culture as one force that holds society together. *Culture* is the general way of life in any society, including norms, customs, beliefs, and language. Human **society** is a system of **social interaction**, typically within geographical boundaries, that includes both culture and social organization. Within a society, members have a common culture, even though there may also be great diversity within it. Members of a society think of themselves as distinct from other societies, maintain ties of social interaction, and have a high degree of interdependence. The interaction they have, whether based on harmony or conflict, is one element of society. Within society, social interaction is behavior between two or more people that is given meaning by them. Social interaction is how people relate to each other and form social bonds.

Social interaction is the foundation of society, but society is more than a collection of individual social actions. Emile Durkheim, the classical sociological theorist, described society as *sui generis*—a Latin phrase meaning "a thing in itself, of its own particular kind." To sociologists, seeing society *sui generis* means that society is more than just the sum of its parts. Durkheim saw society as an organism, something comprising different parts that work together to create a unique whole. Just as a human body is not just a collection of organs but is alive as a whole organism with relationships between its organs, society is not only a simple collection of individuals, groups, or institutions but is a whole entity that consists of all these elements and their interrelationships.

This central sociological idea, that *society is much more than the sum of individuals,* means that society takes on a life of its own. It is patterned by humans and their interactions, but it is something that endures and takes on shape and structure beyond the immediacy of any given group of people. This is a basic idea that guides sociological thinking.

You can think of it this way: Imagine how a photographer views a landscape. The landscape is not just the sum of its individual parts—mountains, pastures, trees, or clouds—although each part contributes to the whole. The power and beauty of the landscape is that all its parts *relate* to each other, some in harmony and some in contrast, to create a panoramic view. The photographer who tries to capture this landscape will likely use a wide-angle lens. This method of photography captures the breadth and comprehensive scope of what the photographer sees. Similarly, sociologists try to picture society as a whole, not only by seeing its individual parts but also by recognizing the relatedness of these parts and their vast complexity.

Macroanalysis and Microanalysis

Sociologists use different lenses to see the different parts of society. Some views are more macroscopic—that is, sociologists try to comprehend the whole of society, how it is organized, and how it changes. This is called **macroanalysis**, a sociological approach that takes the broadest view of society by studying large patterns of social interaction that are vast, complex, and highly differentiated. You might do this by looking at a whole society or comparing different total societies to each other. For example, the technology that allows you to connect to friends from long distances, through texting, photo sharing, and video calls allow for immediate social interaction with other people. This technology makes our society very different from societies before cell phones and the Internet.

Other views are more microscopic—that is, the focus is on the smallest, most immediately visible parts of social life, such as specific people interacting with each other. This is called **microanalysis**. In this approach, sociologists study patterns of social interactions that are relatively small, less complex, and less differentiated—the microlevel of society. Using the example of technology again, a sociologist might examine how people engage in social interaction through texting and social media. How are they similar or different, on the basis of age, gender, social class, or race? For example, do people text (that is, interact) with each other within racial groups more than between racial groups? Observing this would be an example of microanalysis.

A sociologist who studies social interaction via texting or on the Internet would be engaging in microanalysis but might interpret what is found in the context of macrolevel processes (such as race relations in society). Just as a photographer might use a wide-angle lens to photograph a landscape or a telephoto lens for a closer view, sociologists use both macroanalyses and microanalyses to reveal different dimensions of society.

In this chapter, we continue our study of sociology by starting with the macro level of social life (by studying total social structures), then continuing through the micro level (by studying groups and face-to-face interaction). The idea is to help you see how large-scale dimensions of society shape even the most immediate forms of social interaction.

Sociologists use the term **social organization** to describe the order established in social groups at any level. Specifically, social organization brings regularity and predictability to human behavior. Social organization is present at every level of interaction, from the whole society to the smallest groups.

Social Institutions

Societies are identified by their cultural characteristics and the social institutions that compose each society. A **social institution** (or simply an institution) is an established and organized system of social behavior with a recognized purpose. The term refers to the broad systems that organize specific activities in society. Unlike individual behavior, social institutions cannot be directly observed, but their impact and structure can still be seen. For example, the family is an institution that provides for the care of the young and the transmission of culture. Religion is an institution that organizes sacred beliefs. Education is the institution through which people learn the information and skills needed to live in the society.

The concept of the social institution is important to sociological thinking. You can think of social institutions as the enduring consequences of social behavior, but what fascinates sociologists is how social institutions take on a life of their own. For example, you were likely born in a hospital, which itself is part of the health care institution. The simple act of birth, which you might think of as an individual experience, is shaped by the structure of this social institution. You were likely delivered by a doctor, accompanied by nurses or a midwife—each of whom exists in a specific social relationship to the health care institution. Each of these people is in an *institutional* role. Moreover, this social institution also shaped the practices surrounding your birth. Thus, you might have been initially removed from your mother and examined by a doctor, which is very different from the institutional practices in other societies.

The major institutions in society include the family, education, work and the economy, the political institution (or state), religion, and health care, as well as the mass media, organized sports, and the military. These are all complex structures that exist to meet certain needs that are necessary for society to exist. *Functionalist theorists* have traditionally identified these needs (functions) as follows:

1. *The socialization of new members of the society.* This is primarily accomplished by the family, but involves other institutions as well, such as education.
2. *The production and distribution of goods and services.* The economy is generally the institution that performs this set of tasks, but this may also involve the family as an institution—especially in societies where production takes place within households.
3. *Replacement of society's members.* All societies must have a means of replacing members who die, move or migrate away, or otherwise leave the society. Families are typically organized to do this.

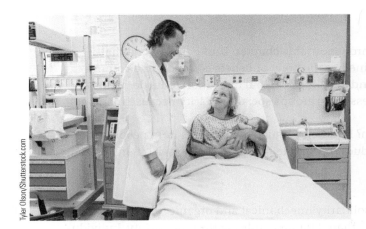
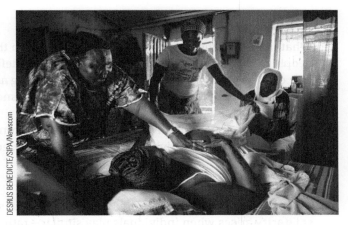

Birth, though a natural process, occurs within social institutions. These institutions vary in different societies, depending on the social organization of society. In these photos you see a new mother in the United States, holding her newborn with only the doctor nearby in a hospital room. Contrast that to a home birth in the South Sudan.

4. *The maintenance of stability and existence. Certain institutions within a society* (such as the government, the police force, and the military) contribute toward the stability and continuance of the society.

5. *Providing the members with an ultimate sense of purpose.* Societies accomplish this task by creating national anthems, for instance, and by encouraging patriotism in addition to providing basic values and moral codes through institutions such as religion, the family, and education (Parsons 1951a).

In contrast to functionalist theory, *conflict theory* further notes that because conflict is inherent in most societies, the social institutions of society do not provide for all its members equally. Some members are provided for better than others, demonstrating that institutions affect people by granting more power to some social groups than to others. The health care institution, for example, has a hierarchy of power. Nurses are generally subordinate to doctors and doctors to hospital administrators. Other hospital workers (cafeteria staff, cleaners, maintenance staff, and clerical workers) have even lower status. Beyond these specific actors within the health care institutions, different social groups in society have more or less access to health care institutions. In other words, the health care institution is patterned by elaborate and overlapping systems of inequality.

Social Structure

Sociologists use the term **social structure** to refer to the organized pattern of social relationships and social institutions that together compose society. Social structures are not immediately visible to untrained observers. They are, nevertheless, present and affect all dimensions of human experience in society. Social structural analysis is a way of looking at society whereby the sociologist analyzes the patterns in social life that produce social behavior.

Social class distinctions are an example of a social structure. Class shapes the access that different groups have to the resources of society, and it shapes many interactions people have with each other. People may form cliques with those who share similar class standing, or they may identify with certain values associated with a given class. Class then forms a social structure—one that shapes and guides human behavior at all levels, no matter how overtly visible or invisible this structure is to someone at a given time.

The philosopher Marilyn Frye aptly uses the metaphor of a birdcage to describe the concept of social structure (Frye 1983). She notes that if you look closely at only one wire in a cage, you cannot see the other wires. You might then wonder why the bird within does not fly away. Only when you step back and see the whole cage instead of a single wire do you understand why the bird does not escape. Social structure, like the birdcage, confines people. People's choices are shaped by social structures; their paths in life and their networks of social relations are shaped by social structure. Just as the birdcage is a network of wires, so is society a network of social structures, both micro and macro.

What Holds Society Together?

What holds societies together? We ask this question throughout this chapter. Durkheim argued that people in society had a **collective consciousness**, defined as the body of beliefs common to a community or society that give people a sense of belonging and a feeling of moral obligation to its demands and values. According to Durkheim, collective consciousness gives groups social solidarity because members of a group feel they are part of one society.

Where does the collective consciousness come from? Durkheim argued that it stems from people's participation in common activities, such as work, family, education, and religion—in short, society's institutions.

Mechanical and Organic Solidarity

According to Durkheim, there are two types of social solidarity: mechanical and organic. **Mechanical solidarity** arises when individuals play similar—rather than different—roles within the society. Individuals in societies marked by mechanical solidarity share the same values and hold the same things sacred. This particular kind of cohesiveness is weakened when a society becomes more complex. Contemporary examples of mechanical solidarity are rare because most societies of the world have been absorbed in the global trend for greater complexity and interrelatedness. Before European conquest, Native American nations were bound together by a kind of mechanical solidarity, as close-knit communities where values were shared by all. Indeed, many Native American nations now celebrate the mechanical solidarity on which their cultural heritage rests. The superimposition of White institutions on Native American life, however, interferes with traditional Native social structures. Individual Native groups also have their own societies with unique cultures.

In contrast, **organic** (or contractual) **solidarity** occurs when people play a great variety of roles, and unity is based on role differentiation, not similarity. Industrial societies, including the United States, are built on organic solidarity with cohesion coming from differentiation. With organic solidarity, roles are necessarily interlinked and the performance of multiple roles is necessary for the execution of society's complex and integrated functions.

Durkheim described this state as the **division of labor**, defined as the relatedness of *different* tasks that develop in complex societies. The labor force within the contemporary U.S. economy, for example, is divided according to the kinds of work people do. Within any division of labor, tasks become distinct from one another, but they are still woven into a whole.

The division of labor is a central concept in sociology because it represents how the different pieces of society fit together. The division of labor in most contemporary societies is often marked by distinctions such as age, gender, race, and social class. In other words, if you look at who does what in society, you will see that women and men tend to do different things; this is the gender division of labor. Similarly, old and young to some extent do different things; this is a division of labor by age. This is crosscut by the racial division of labor, the pattern whereby those in different racial–ethnic groups tend to do different work—or are often forced to do different work—in society. At the same time, the division of labor is also marked by class distinctions, with some groups providing work that is highly valued and rewarded and others doing work that is devalued and poorly rewarded. As you will see throughout this book, gender, race, and class intersect and overlap in the division of labor in society.

Gemeinschaft and Gesellschaft

Different societies are held together by different forms of solidarity. Some societies are characterized by what the German sociologist Ferdinand Tönnies called **gemeinschaft**, a German word that means "community"; other societies are characterized as **gesellschaft**, which literally means "society" (Tönnies 1963/1887). Each involves a type of solidarity or cohesiveness. Those societies that are *gemeinschafts* (communities) are characterized by a sense of "we" feeling, a very moderate division of labor, strong personal ties, strong family relationships, and a sense of personal loyalty. The sense of solidarity between members of the gemeinschaft society arises from personal ties; small, relatively simple social institutions; and, a collective sense of loyalty to the whole society. People tend to be well integrated into the whole, and social cohesion comes from deeply shared values and beliefs (often, sacred values). Social control need not be imposed externally because control comes from the internal sense of belonging that members share. You might think of a small community church as an example.

In contrast, in societies marked by *gesellschaft*, importance is placed on the secondary relationships people have—that is, less intimate and more instrumental relationships such as work roles instead of family or community roles. Gesellschaft is characterized by less prominence of personal ties, a somewhat diminished role of the nuclear family, and a lessened sense of personal loyalty to the total society. The solidarity and cohesion remain, and it can be very cohesive, but the cohesion comes from an elaborated *division of labor* (thus, *organic* solidarity), greater flexibility in social roles, and the instrumental ties that people have to one another.

Social solidarity under gesellschaft is weaker than in the gemeinschaft society. Although class conflict is still present in gemeinschaft, it is less prominent, making gesellschaft societies more at risk for class conflict. Racial–ethnic conflict is also more likely within gesellschaft societies because the gemeinschaft tends to be ethnically and racially very homogeneous, meaning it is often characterized by only one racial or ethnic group.

In sum, complexity and differentiation are what make the gesellschaft cohesive, whereas similarity and unity bond the gemeinschaft society. In a single society, such as the United States, you can conceptualize the whole society as gesellschaft, with some internal groups marked by gemeinschaft.

Types of Societies

In addition to comparing how different societies are bound together, sociologists are interested in how social organization evolves in different societies. Simple things such as the size of a society can also shape its social organization, as do the different roles that men and women engage in as they produce goods, care for the old and young, and pass on societal traditions. Societies also differ according to their resource base—whether they are predominantly agricultural or industrial, for example, and whether they are sparsely or densely populated.

Thousands of years ago, societies were small, sparsely populated, and technologically limited. In the competition for scarce resources, larger and more technologically advanced societies dominated smaller ones. Today, we have arrived at a global society with highly evolved degrees of social differentiation and inequality, notably along class, gender, racial, and ethnic lines.

Sociologists distinguish six types of societies based on the complexity of their social structure, the amount of overall cultural accumulation, and the level of their technology. They are *foraging, pastoral, horticultural, agricultural* (these four are called *preindustrial* societies), and then *industrial* and *postindustrial* societies (see ◆ Table 5-1). Each type of society can still be found on Earth, although all but the most isolated societies are rapidly moving toward the industrial and postindustrial stages of development.

These different societies vary in the basis for their organization and the complexity of their division of labor. Some, such as foraging societies, are subsistence economies, where men and women hunt and gather food but accumulate very little. Others, such as pastoral societies and horticultural societies, develop a more elaborate division of labor as the social roles that are needed for raising livestock and farming become more numerous. With the development of agricultural societies, production becomes more large-scale, and strong patterns of social differentiation develop, sometimes taking the form of a caste system or even slavery.

The key driving force that distinguishes these different societies from each other is the development of technology. All societies use technology to help fill human needs, and the form of technology differs for the different types of society.

Preindustrial Societies

A **preindustrial society** is one that directly uses, modifies, and/or tills the land as a major means of survival. There are four kinds of preindustrial societies, listed here by degree of technological development: foraging

Table 5-1   Types of Societies

		Economic Base	Social Organization	Examples
Preindustrial Societies	*Foraging societies*	Economic sustenance dependent on hunting and foraging	Gender is important basis for social organization, although division of labor is not rigid; little accumulation of wealth	Pygmies of central Africa
	Pastoral societies	Nomadic societies, with substantial dependence on domesticated animals for economic production	Complex social system with an elite upper class and greater gender role differentiation than in foraging societies	Bedouins of Africa and Middle East
	Horti-cultural societies	Society marked by relatively permanent settlement and production of domesticated crops	Accumulation of wealth and elaboration of the division of labor, with different occupational roles (farmers, traders, craftspeople, and so on)	Ancient Aztecs of Mexico; Inca Empire of Peru
	Agricul-tural societies	Livelihood dependent on elaborate and large-scale patterns of agriculture and increased use of technology in agricultural production	Caste system develops that differentiates the elite and agricultural laborers; may include system of slavery	American South, pre-Civil War
Industrial Societies		Economic system based on the development of elaborate machinery and a factory system; economy based on cash and wages	Highly differentiated labor force with a complex division of labor and large formal organizations	Nineteenth and most of twentieth-century United States and western Europe
Postindustrial Societies		Information-based societies in which technology plays a vital role in social organization	Education increasingly important to the division of labor	Contemporary United States, Japan, and others

(or hunting–gathering) societies, pastoral societies, horticultural societies, and agricultural societies (see Table 5-1).

In *foraging (hunting–gathering) societies*, the technology enables the hunting of animals and gathering of vegetation. The technology does not permit the refrigeration or processing of food, hence these individuals must search continuously for plants and game. Because hunting and gathering are activities that require large amounts of land, most foraging societies are nomadic, constantly traveling as they deplete the plant supply or follow the migrations of animals. The central institution is the family, which serves as the means of distributing food, training children, and protecting its members. There is usually role differentiation on the basis of gender, although the specific form of the gender division of labor varies in different societies. The pygmies of central Africa are an example of a foraging society.

In *pastoral societies*, technology is based on the domestication of animals. Such societies tend to develop in desert areas that are too arid to provide rich vegetation. The pastoral society is nomadic, necessitated by the endless search for fresh grazing grounds for the herds of their domesticated animals. The animals are used as sources of hard work that enable the creation of a material surplus. Unlike a foraging society, this surplus frees some individuals from the tasks of hunting and gathering and allows them to create crafts, make pottery, cut hair, and so forth. The surplus generates a more complex and differentiated social system with an elite class or an upper class and more role differentiation on the basis of gender. The nomadic Bedouins of Africa and the Middle East are examples of pastoral societies.

In *horticultural societies*, hand tools are used to cultivate the land, such as the hoe and the digging stick. The individuals in horticultural societies practice ancestor worship and conceive of a deity or deities (God or gods) as a creator. Horticultural societies recultivate the land each year and tend to establish relatively permanent settlements and villages. Role differentiation is extensive, resulting in different and interdependent occupational roles such as farmer, trader, and craftsperson. The ancient Aztecs of Mexico and the Incas of Peru represent examples of horticultural societies.

The *agricultural society* is exemplified by the pre-Civil War American South, a society of slavery. Such societies have a large and complex economic system that is based on large-scale farming. These societies rely on technologies such as use of the wheel and metals. Farms tend to be considerably larger than the cultivated land in horticultural societies. Large and permanent settlements characterize agricultural societies, which also create dramatic social inequalities. A rigid caste system develops, separating the peasants, or slaves, from the controlling elite caste, which is then freed from manual work, allowing time for art, literature, and philosophy, activities of which they can then claim the lower castes are incapable. The American pre-Civil War South and its system of slavery is a good example of an agricultural society.

Industrial Societies

An *industrial society* is one that uses machines and other advanced technologies to produce and distribute goods and services. The Industrial Revolution began over 250 years ago when the steam engine was invented in England, delivering previously unattainable amounts of mechanical power for the performance of work. Steam engines powered locomotives, factories, and dynamos and transformed societies as the Industrial Revolution spread. The growth of science led to advances in farming techniques such as crop rotation, harvesting, and ginning cotton, as well as industrial-scale projects such as dams for generating hydroelectric power. Joining these advances were developments in medicine, new techniques to prolong and improve life, and the emergence of birth control to limit population growth.

Unlike agricultural societies, industrial societies rely on a highly differentiated labor force and the intensive use of capital and technology. Large formal organizations are common. The task of holding society together falls more on the institutions that have a high division of labor, such as the economy and work, government, politics, and large bureaucracies.

Within industrial societies, the forms of gender inequality that we

AP Images/Sergei Grits

Andersen Ross/Stockbyte/Jupiter Images

Different types of societies produce different kinds of social relationships. Some may involve more direct and personal relationships (called gemeinschafts), whereas others produce more fragmented and impersonal relationships (called gesellschafts). Which type is depicted here?

see in contemporary U.S. society tend to develop. With the advent of industrialization, societies move to a cash-based economy, with labor performed in factories and mills paid on a wage basis and household labor remaining unpaid. This introduced what is known as the family wage economy, in which families become dependent on wages to support themselves, but work within the family (housework, child care, and other forms of household work) is unpaid and therefore increasingly devalued. In addition, even though women (and young children) worked in factories and mills from the first inception of industrialization, the family wage economy is based on the idea that men are the primary breadwinners. A system of inequality in men's and women's wages was introduced—an economic system that even today continues to produce a wage gap between men and women.

Industrial societies tend to be highly productive economically, with a large working class of industrial laborers. People become increasingly urbanized as they move from farmlands to urban centers or other areas where factories are located. Immigration is common in industrial societies, particularly because industries are forming where there is a high demand for more, cheap labor.

Industrialization has brought many benefits to U.S. society—a highly productive and efficient economic system, expansion of international markets, extraordinary availability of consumer products, and for many, a good working wage. Industrialization has, at the same time, also produced some of the most serious social problems that our nation faces: industrial pollution, an overdependence on consumer goods, wage inequality and job dislocation for millions, and problems of crime and crowding in urban areas.

Postindustrial Societies

In the contemporary era, a new type of society is emerging. Whereas most twentieth-century societies can be characterized in terms of their making of material goods, **postindustrial society** depends economically on the production and distribution of services, information, and knowledge. Postindustrial societies are information-based societies in which technology plays a vital role in the social organization. The United States has become a postindustrial society, as have Japan and other nations. In such a society most workers provide services such as administration, education, legal services, scientific research, banking, and the delivery of essential services such as food, housing, and transportation. Many also work in the development, management, and distribution of information, particularly in the areas of cybertechnology. Central to the economy of the postindustrial society are the highly advanced technologies of computers, robotics, and genetic engineering. Multinational corporations globally link the economies of postindustrial societies.

The transition to a postindustrial society has a strong influence on the character of social institutions. Educational institutions become extremely important in the postindustrial society, and science takes an especially prominent place. For some, the transition to a postindustrial society means more discretionary income for leisure activities like tourism and entertainment. Companies that specialize in relaxation and health (spas, massage centers, and exercise) become more prominent, at least for people in the upper classes. As with the United States in the last recession, the transition to postindustrialism has meant permanent joblessness for many. For others, it has meant the need to hold down more than one job simply to make ends meet.

Social Interaction and Society

You can see by now that society is an entity that exists above and beyond individuals. Also, different societies are marked by different forms of *social organization*. Although societies differ, emerge, and change, they are also highly predictable. Your society shapes virtually every aspect of your life from the structure of its social institutions to the more immediate ways that you interact with people. This is the micro level of society.

Groups

At the micro level, society is made up of many different social groups. At any given moment, each of us is a member of many groups simultaneously, and we are subject to their influence: family, friendship groups, athletic teams, work groups, racial and ethnic groups, and so on. Groups impinge on every aspect of our lives and are a major determinant of our attitudes and values regarding everything from personal issues, such as sexual attitudes and family values, to major social issues, such as the death penalty, gun control, and physician-assisted suicide.

To sociologists, a **group** is a collection of individuals who

- interact and communicate with each other;
- share goals and norms; and
- have a subjective awareness of themselves as "we," that is, as a distinct social unit.

To be a group, the social unit in question must possess all three of these characteristics. We will examine the nature and behavior of groups in greater detail in Chapter 6.

In sociological terms, not all collections of people are groups. People may be lumped together into *social categories* based on one or more shared characteristics, such as teenagers (an age category) or truck drivers (an occupational category).

Social categories can become social groups, depending on the amount of "we" feeling the group has. Only when there is this sense of common identity, as defined in the previous characteristics of groups, is a collection of people an actual group. For example, all people nationwide watching television programs at 8 o'clock Wednesday evening form a distinct social unit, an *audience*, but they are not a group because they do not interact with one another, nor do they possess an awareness of themselves as "we." However, if many viewers were to come together for a convention where they could interact and develop a "we" feeling, such as do fans of comic books who attend Comic-Con, then they would constitute a group.

We now know that people do not need to be face-to-face to constitute a group. Online communities, for example, are people who interact with each other regularly, share a common identity, and think of themselves as being a distinct social unit. On the Internet community Facebook, for example, you may have a group of "friends," some of whom you know personally and others whom you only know online. These *friends* make up a social group that might interact on a regular, indeed, daily basis—possibly even across great distances.

Groups also need not be small or "close up" and personal. *Formal organizations* are highly structured social groupings that form to pursue a set of goals. Bureaucracies such as business corporations or municipal governments or associations such as the National Rifle Association (NRA) or the American Association of Retired People (AARP) are examples of formal organizations.

Status

Within groups, people occupy different statuses. **Status** is an established position in a social structure that carries with it a degree of social rank or value. A status is a rank in society. For example, the position "vice president of the United States" is a status, one that carries relatively high prestige. "High school teacher" is another status; it carries less prestige than "vice president of the United States," but more prestige than, say, "Uber driver." Statuses occur within institutions and also within groups. "High school teacher" is a status within the education institution. Other statuses in the same institution are "student," "principal," and "school superintendent." Within a given group, people may occupy different statuses that can be dependent on a variety of factors, such as age or seniority within the group.

Typically, a person occupies many statuses simultaneously. The combination of statuses composes a **status set**, which is the complete set of statuses occupied by a person at a given time (Merton 1968). A person may occupy different statuses in different institutions. Simultaneously, a person may be an investment banker (in the economic institution), voter (in the political institution), church member (in the religious institution), and school board member (in the education institution). Each status may be associated with a different level of prestige.

Sometimes the multiple statuses of an individual conflict with one another. **Status inconsistency** exists where the statuses occupied by a person bring with them significantly different amounts of prestige and thus differing expectations. For example, someone trained as a lawyer but working as an Uber driver experiences status inconsistency. Some recent immigrants from Vietnam and Korea have experienced status inconsistency. Many refugees who had been in high status occupations in their home country, such as teachers, doctors, and lawyers, could find work in the United States only as grocers, technicians, or nail

Debunking Society's Myths

Myth: Gender is an *ascribed status* where one's gender identity is established at birth.

Sociological Perspective: Although one's biological sex identity is an ascribed status, gender is a social construct and thus is also an *achieved status*—that is, accomplished through routine, everyday behavior, including patterns of dress, speech, touch, and other social behaviors. Sex is not the same as gender (Andersen 2015).

salon operators—jobs of relatively lower status than the jobs they left behind. A relatively large body of research in sociology has demonstrated that status inconsistency—in addition to low status itself—can lead to stress and depression (Taylor et al. 2016; Thoits 2009).

Achieved statuses are those attained by virtue of individual effort. Most occupational statuses—police officer, pharmacist, or boat builder—are achieved statuses. In contrast, **ascribed statuses** are those occupied from the moment a person is born. Your biological sex is an ascribed status. Yet, even ascribed statuses are not exempt from the process of social construction. For most individuals, race is an ascribed status fixed at birth. But African American individuals with light skin may appear to be White and be treated as White people throughout their lifetime. Finally, ascribed statuses can arise long after birth, through means beyond an individual's control, such as severe disability or chronic illness.

Some ascribed statuses, such as gender, can become achieved statuses. Gender, typically thought of as fixed at birth, is a social construct. You can be born female or male (this is your sex), but becoming a woman or a man is the result of social behaviors associated with your ascribed status. In other words, gender is also achieved. Transgender people are an example of how gender is achieved, separate and apart from ascribed sex status. All people "do" gender in everyday life. They put on appearances and behaviors that are associated with their presumed gender (Andersen 2015; West and Fenstermaker 1995). If you doubt this, ask yourself what you did today to "achieve" your gender status. Did you dress a certain way? Wear "manly" cologne or deodorant? Splash on a "feminine" fragrance? These behaviors—all performed at the micro level—reflect the macro level of your gender status.

The line between achieved and ascribed status can be hard to draw. Social class, for example, is determined by occupation, education, and annual income—all of which are achieved statuses—yet one's job, education, and income are known to correlate strongly with the social class of one's parents. Hence, one's social class status is at least partly—though not perfectly—determined at birth. It is an achieved status that includes an inseparable component of ascribed status as well.

Although people occupy many statuses at one time, it is usually the case that one status is dominant, called the **master status**, overriding all other features of the person's identity. The master status may be imposed by others, or a person may define his or her own master status. A woman judge, for example, may carry the master status "woman" in the eyes of others. She is seen not just as a judge, but as a woman judge, thus making gender a master status. A master status can completely supplant all other statuses in someone's status set. Being in a wheelchair is another example of a master status. Consider, for example, the case of a person in a wheelchair who is at the same time a medical doctor, an author, and a painter. People will typically see the wheelchair, at least at first, as the most important, or salient, part of identity, ignoring other statuses that define someone as a person. The person will likely be stereotyped as the "disabled doctor" or "disabled author."

Thinking Sociologically

Make a list of terms that describe who you are. Which of these are *ascribed statuses* and which are *achieved statuses*? What do you think your *master status* is in the eyes of others? Does one's *master status* depend on who is defining you? What does this tell you about the significance of social judgments in determining who you are?

Roles

A **role** is the behavior others expect from a person associated with a particular status. Statuses are occupied; roles are acted or "played." The status of police officer carries with it many expectations; these expected behaviors comprise the role of police officer. Police officers are expected to enforce the law, pursue suspected criminals, assist victims of crime, complete forms for reports, and obey laws themselves. Usually, people behave in their roles as others expect them to, but not always. When a police officer commits a crime, such as physically brutalizing someone, he or she has violated the role expectations. Role expectations may vary according to the role of the observer—whether the person observing the police officer is a member of a minority group, for example.

People of similar status often conform to one another, especially peers. How do you see that enacted in this photograph?

As we saw in Chapter 4, social learning theory predicts that we learn attitudes and behaviors in response to the positive reinforcement and encouragement received from those around us. This is important in the formation of our own identity in society. We embrace certain statuses, and the roles associated with them, based on our interactions with others. The "Thinking Sociologically" feature suggests you consider your own status. What is your identity? Are you a college student first? Are there particular roles you feel identify you because others see you that way? These identities are often obtained through **role modeling**, a process by which we imitate the behavior of another person we admire who is in a particular role. A college freshman might admire a senior student in his dorm. The student's self-identity is influenced by his attempts to imitate the senior.

A person may occupy several statuses and roles at one time. A person's **role set** includes all the roles occupied by the person at a given time. For example, a person may be not only a student, but a store cashier, a roommate, and an admissions tour guide. Roles can also clash with each other, a situation called **role conflict**, wherein two or more roles are associated with contradictory expectations. Notice that in ▲ Figure 5-1 some of the roles diagrammed for this college student may

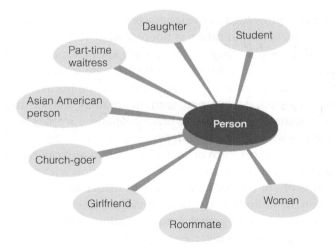

▲ **Figure 5-1   Roles in a College Student's Role Set.** Identify the different roles that you occupy and draw a similar diagram of your own role set. Then identify which roles are consistent with each other and which might produce *role conflict* and *role strain*.

conflict with others. Can you speculate about which might and which might not? Can you draw your own role set?

In U.S. society, some of the most common forms of role conflict arise from the dual responsibilities of job and family. The parental role demands extensive time and commitment, and so does the role of worker. Time given to one role is time taken away from the other. Although the norms pertaining to working women and working men have changed over time, it is still true that women are more often expected to uphold traditional role expectations associated with their gender role and are more likely responsible for tending to family issues even when job and family conflict. The sociologist Arlie Hochschild captured the predicament of today's women when she described the "second shift." An employed mother spends time and energy all day on the job, only to come home to the "second shift" of family and home responsibilities (Hochschild and Machung 1989).

Hochschild's studies point to the conflict between two social roles: family roles and work roles. This conflict also highlights the sociological concept of **role strain**, a condition wherein a single role brings conflicting expectations. Different from role conflict, which involves tensions *between* two roles, role strain involves conflicts within a single role. When considering work–family balance for women, the work role has the expectations traditionally associated with work but also the expectation that she "love" her work and be as devoted to it as to her family. The same is expected of men. The result is role strain. The role of a high school student also often involves role strain. For example, students are expected to be focused on academics and performing their best, yet students also feel pressure to be involved in sports, music, community service, or other extracurricular activities. The tension between these two competing expectations is an example of role strain.

Everyday Social Interaction

You can see the influence of society in everyday behavior, including such basics as how you talk, patterns of touch, and who you are attracted to. Although you might think these things just come "naturally," they are deeply patterned by society. The cultural context of social interaction really matters in our understanding of what given behaviors mean. An action that is positive in one culture can be negative in another. For example, shaking the right hand in greeting is a positive action in the United States, but the same action in East India or certain Arab countries might be an insult. Social and cultural contexts matter. A kiss on the lips is a positive act in most cultures, yet if a stranger kissed you on the lips, you would probably consider it an offense, perhaps even a crime.

Verbal and Nonverbal Communication

Patterns of social interaction are embedded in the language we use, and language is deeply influenced by culture and society. Furthermore, communication is not just what you say, but also how you say it and to whom. You can see the influence of society on *how* people speak, especially in different contexts. The gender of the speaker is also part of that cultural context—there are masculine and feminine styles of conversation. Japanese women, for example, are more polite and supportive when speaking to Japanese men. In conversations with English-speaking men, women are more self-assured and express their own opinions (Itakura 2014). Americans may mistakenly believe Japanese women are submissive, not realizing their conversation style changes with the context, depending on who they are talking to.

Nonverbal communication is also a form of social interaction and can be seen in various social patterns. A surprisingly large portion of our everyday communication with others is nonverbal, although we are generally not conscious of our nonverbal behavior. Consider all the nonverbal signals exchanged in a casual chat: body position, head nods, eye contact, facial expressions, touching, and so on. Studies of nonverbal communication, like those of verbal communication, show how it is influenced by social forces, including the relationships between diverse groups of people. The meanings of nonverbal communications depend heavily on race, ethnicity, social class, and gender.

For example, patterns of touch are strongly influenced by gender. Parents vary their touching behavior depending on whether the child is a boy or a girl. Boys tend to be touched more roughly; girls, more tenderly and protectively. Such patterns continue into adulthood, where women touch each other more often in everyday conversation than do men. Women are on the average more likely to touch and hug as an expression of emotional support, whereas men touch and hug more often to assert power or to express sexual interest (Baumeister and Bushman 2017). Clearly, there are also instances where women touch to express sexual interest and/or dominance, but in general, touch is a supportive activity for women and an expression of sexual interest for men. In the context of sports, however, men hug and pat other men as a show of support.

In observing patterns of touch, you can see where social status influences the meaning of nonverbal behaviors. Professors, male or female, may pat a man or woman student on the back as a gesture of approval; students will rarely do this to a professor. Male professors touch students more often than do female professors, showing the additional effect of gender. Because such patterns of touching reflect power relationships between women and men, they can also be offensive and may even involve sexual harassment (see Chapter 15).

You can also see the social meaning of interaction by observing how people use personal space. *Proxemic communication* refers to the amount of space between interacting individuals. Although people are generally unaware of how they use personal space, usually the more friendly people feel toward on another, the closer they will stand. In casual conversation, friends stand closer to each other than do strangers. People who are sexually attracted to each other stand especially close. According to anthropologist E. T. Hall (1966), we all carry around us a *proxemic bubble* that represents our personal, three-dimensional space. When people we do not know enter our proxemic bubble, we feel threatened and may take evasive action. Friends stand close; enemies tend to avoid interaction and keep far apart. According to Hall's theory, we attempt to exclude from our private space those whom we do not know or do not like, even though we may not be fully aware that we are doing so.

The proxemic bubbles of different ethnic groups on average have different sizes. Hispanic people tend to stand much closer to each other than do White, middle-class Americans; their proxemic bubble is, on average, smaller.

In a society as diverse as the United States, understanding how diversity shapes social interaction is an essential part of understanding human behavior. Ignorance of the meanings that gestures have in a society can get you in trouble. For example, some Mexicans and Mexican Americans may display the right hand held up, palm inward, all fingers extended, as an obscene gesture directed at someone in anger. This provocative gesture has no meaning at all in Anglo (White) society. Instead, extending the middle finger up, as an aggressive form of communication, is understood in many societies.

Likewise, people who grow up in urban environments learn to avoid eye contact on the streets. Staring at someone for only two or three seconds can be interpreted as a hostile act, if done man to man . If a woman maintains mutual eye contact with a male stranger for more than two or three seconds, she may be assumed by the man to be sexually interested in him. In contrast, during sustained conversation with acquaintances, women maintain mutual eye contact longer than do men .

Denise Lett/Shutterstock.com

Patterns of touch can reflect status differences. If you changed the status of one of these people, under what conditions would the touch be appropriate or not?

Interpersonal Attraction

We have already asked, "What holds society together?" This was asked at the macroanalysis level— that is, the level of society. But what holds relationships together—or, for that matter, makes them fall apart? You will not be surprised to learn that formation of relationships has a strong social structural component—that is, it is patterned by social forces.

Humans have a powerful desire to be with other human beings; in other words, they have a strong need for *affiliation*. We tend to spend about 75 percent of our time with other people when doing all sorts of activities—eating, watching television, studying, doing hobbies, working, and so on (Cassidy and Shaver 2008). People who lack all forms of human contact are very rare in the general population, and their isolation is usually rooted in psychotic or schizophrenic disorders. Extreme social isolation at an early age causes severe disruption of mental, emotional, and language development.

The affiliation tendency has been likened to *imprinting*, a phenomenon seen in newborn or newly hatched animals who attach themselves to the first living creature they encounter, even if it is of another species (Lorenz 1966). Studies of geese and squirrels show that once the young animal attaches itself to a human experimenter, the process is irreversible. The young animal prefers the company of the human to the company of its own species! A degree of imprinting may be discernible in human infant attachment, but researchers note that the process is more complex, more changeable, and more influenced by social factors in infants.

Somewhat similar to affiliation is *interpersonal attraction*, a nonspecific positive response toward another person. Attraction occurs in ordinary day-to-day interaction and varies from mild attraction (such as thinking your grocer is a "nice person") all the way to deep feelings of love. According to one view, attractions fall on a continuum ranging from hate to strong dislike to mild dislike to mild liking to strong liking to love. Another view is that attraction and love are two different feelings, able to exist separately. In this view, you can actually like someone a whole lot, but not be in love. Conversely, you can feel passionate love for someone, including strong sexual feelings and intense emotion, yet not really "like" the person.

Can attraction be scientifically predicted? Can you identify with whom you are most likely to fall in love? The surprising answer to these questions is "yes," with some qualifiers. Most of us have been raised to believe that love is impossible to measure and certainly impossible to predict scientifically. We think of love, especially romantic love, as quick and mysterious—a lightning bolt. Couples report falling in love at first sight, thinking that they were "meant for each other." Countless novels and stories support this view, but extensive research in sociology and social psychology suggests otherwise. Love can be predicted beyond the level of pure chance.

A strong determinant of your attraction to others is simply whether you live near them, work next to them, or have frequent contact with them. (This is a *proxemic* determinant.) You are more likely to form friendships with people from your own city than with people a thousand miles away. One classic study even showed that you are more likely to be attracted to someone on your floor, your residence hall, or your apartment building than to someone even two floors down or two streets over (Festinger et al. 1950). Subsequent studies continue to show this effect (Baumeister and Bushman 2017). Such is the effect of proximity in the formation of human friendships.

Now, though the general principle still holds, many people form relationships without being in close proximity, such as in online dating. In earlier societies, people would only date, fall in love, and marry people they knew from their communities. Now with social media and the ease in which we interact with one another across long distances, there is much greater likelihood to form romantic relationships with people far away. Studies of Internet dating show that people can form love relationships with people they hardly know (Rosenfeld and Thomas 2012).

We hear that "beauty is only skin deep." Apparently, that is deep enough. To a surprisingly large degree, the attractions we feel toward people of either gender are based on our perception of their physical attractiveness. Assumptions about gender differences were that men wanted beautiful women but that women cared less about attractiveness in their mate. The evidence suggests, however, that both men and women highly value attractiveness when pursuing romantic or sexual relationships (McClintock 2011). Although there are societal standards for attractiveness, there are individual preferences. "Beauty is in the eye of the beholder." Men and women see their romantic partners as attractive, even when others may not (Solomon and Vazire 2014). The point is that romantic relationships are more likely to develop between people who feel physically attracted to one another.

Of course, standards of attractiveness vary between cultures and between subcultures within the same society. What is highly attractive in one culture may be repulsive in another. In the United States, there is a maxim that you can never be too thin—a major cause of eating disorders such as *anorexia* and *bulimia*, especially among White women. The maxim is oppressive for women in U.S. society, yet it is clearly highly culturally relative, even within U.S. culture. What is considered "overweight" or "fat" is indeed a social construction. Among many African Americans, the standard of thinness is different, and larger body sizes are more ideal. Similar cultural norms often apply in certain U.S. Hispanic populations. The skinny woman is not necessarily considered attractive. Nonetheless, studies show that anorexia and bulimia are now increasing among women of color, showing how cultural norms can change (Atkins 2011; Warren et al. 2010).

Perceived physical attractiveness may predict who is attracted to whom initially, but other variables are better predictors of how long a relationship will last. So, do "opposites attract"? Not according to the research. We have all heard that people are attracted to their "opposite" in personality, social status, background, and other characteristics. Many of us grow up believing this to be true. However, if the research tells us one thing about interpersonal attraction, it is that with only a few exceptions we are attracted to people who are *similar* or *even identical* to us in socioeconomic status, race, ethnicity, religion, perceived personality traits, and general attitudes and opinions (Taylor et al. 2016). Couples tend to have similar opinions about political issues of great importance to them, such as attitudes about abortion, crime, animal rights, gun violence, and whom to vote for as president. Overall, couples tend to exhibit strong cultural or subcultural similarity, not difference.

There are exceptions, of course. We sometimes fall in love with the *exotic*—the culturally or socially different. Novels and movies return endlessly to the story of the rich young woman who falls in love with a rough and ready biker, but such a pairing is by far the exception and not the rule. That rich young woman is far more likely to fall in love with a rich young man. When it comes to long-term relationships, including both friends and lovers (whether heterosexual, lesbian, gay, or bisexual), humans vastly prefer a great degree of similarity, even though, if asked, they might deny it. In fact, the less similar a heterosexual relationship is with respect to race, social class, age, and educational aspirations (how far in school the person wants to go), then the quicker the relationship is likely to break up (Silverthorne and Quinsey 2000).

Theories about Analyzing Social Interaction

Groups, statuses, and roles form a web of social interaction. Sociologists have developed different ways of conceptualizing and understanding social interaction. Functionalist theory offers one such concept. Here we detail four others: the social construction of reality, ethnomethodology, impression management, and social exchange theory (refer to ◆ Table 5-2). The first three theories come directly from the symbolic interaction perspective.

The Social Construction of Reality

What holds society together? This is a basic question for sociologists, one that, as we have seen, has long guided sociological thinking. Sociologists note that society cannot hold together without something that is shared—a shared social reality.

Debunking Society's Myths

Myth: Love is purely an emotional experience that you cannot predict or control.
Sociological Perspective: Whom you fall in love with can be predicted beyond chance by such factors as proximity, how often you see the person (frequency, or mere exposure effect), how physically attractive you perceive the person to be, and whether you are similar (not different) to her or him in social class, race/ethnicity, religion, age, educational aspirations, and general attitudes, including political attitudes and beliefs (Taylor et al. 2016).

Table 5-2	Theories of Social Interaction			
	The Social Construction of Reality	**Ethnomethodology**	**Dramaturgy**	**Social Exchange Theory**
Interprets society as:	Organized around the subjective meaning that people give to social behavior	Held together through the consensus that people share around social norms; you can discover these norms by violating them	A stage on which actors play their social roles and give impression to those in their "audience"	A series of inter-actions that are based on esti-mates of rewards and punishments
Analyzes social inter-action as:	Based on the mean-ing people give to, or attribute to, actions in society	The encounters that hold society together based on a taken for granted sense of appropriate behaviors	Enactment of social roles played before a social audience	A rational balanc-ing act involving perceived costs and benefits of a given behavior

Some sociological theorists have argued convincingly that there is little actual reality beyond that produced by the process of social interaction itself. This is the principle of the *social construction of reality*, the idea that our perception of what is real is determined by the subjective meaning that we attribute to an experience. This is a principle central to symbolic interaction theory (Blumer 1969; Berger and Luck-mann 1967). Hence, there is no objective "reality" in itself. Things do not have their own intrinsic mean-ing. We subjectively *impose* meaning on things.

A simple example of the social construction of reality is to consider a desk and chair in a classroom. We assign meaning to these objects based on the social context within which we use them. Students sit in the chair with notebook or computer on the desk. This is a desk. Now consider these same objects, but with a tablecloth on the desk and a plate, fork, knife, and glass set up there. Now this is not a desk and chair, but a dining table and chair. The meaning assigned to these things is influenced by the interac-tion we have to them. Let's take the same desk and chair and put them upside down or balancing on the corners, bolted to a cement base, with an up-light shining on them. We can paint the surface with bright colors or add a mosaic of tiles. Now the same objects are a work of art. The social context and the social interaction people have with the object give those objects meaning.

Ethnomethodology

Our interactions are guided by rules that we follow. Sometimes these rules are nonobvious and subtle. These rules are the *norms* of social interaction. Again, what holds society together? Society cannot hold together without norms, but what rules do we follow? How do we know what these rules or norms are? An approach in sociology called *ethnomethodology* is a clever technique for finding out.

See for Yourself

Riding in Elevators

1. Try a simple experiment. Ride in an elevator and closely observe the behavior of everyone in the elevator with you. Write down in a notebook such things as how far away people stand from each other. Note the dif-ferences carefully, even in estimated inches. What do they look at? Do they tend to stand in the corners? Do they converse with strangers or the people they are with? If so, what do they talk about?

2. Now return to the same elevator and do something that breaks the usual norms of elevator behavior, such as standing too close to someone. (You will have to get up a lot of nerve to do this!) How did people react? What did they do? How did you feel? How does this experiment show how social norms are maintained through informal norms of social control?

Ethnomethodology (Garfinkel 1967), after *ethno* for "people" and *methodology* for "mode of study," is a technique for studying human interaction by deliberately disrupting social norms and observing how individuals attempt to restore normalcy. The idea is that to study such norms, one must first break them, because the subsequent behavior of the people involved will reveal just what the norms were in the first place. In the "See for Yourself" elevator example you were asked to perform previously, an application of ethnomethodology would be standing too close to someone on the elevator (this is the norm violation) and observing what that person does as a result (which would be the norm restoration behavior).

Ethnomethodology is based on the premise that human interaction takes place within a consensus, and interaction is not possible without this consensus. The consensus is part of what holds society together. According to Garfinkel, this consensus will be revealed by people's *background expectancies*, namely, the norms for behavior that they carry with them into situations of interaction. The presumption is that these expectancies are to a great degree shared, and thus studying norms by deliberately violating them will reveal the norms that most people bring with them into interaction. The ethnomethodologist argues that you cannot simply walk up to someone and ask what norms the person has and uses, because most people will not be able to articulate them. We are not wholly conscious of what norms we use even though they are shared. Ethnomethodology is designed to "uncover" those norms and typically does so through somehow violating the usual expectations. Ethnomethodology reminds us that the norms of behavior are what create the social reality. Consensus around those norms is a necessary part of this social process.

Impression Management and Dramaturgy

Another way of analyzing social interaction is to study *impression management*, a term coined by symbolic interaction theorist Erving Goffman (1959). **Impression management** is a process by which people control how others perceive them. A student handing in a term paper late may wish to give the instructor the impression that it was not the student's fault but was because of uncontrollable circumstances ("my computer crashed," "the network went down," and so on). The impression that one wishes to "give off" (to use Goffman's phrase) is that "I am usually a very diligent person, but today—just today—I have been betrayed by circumstances."

Impression management can be seen as a type of con game. We willfully attempt to manipulate others' impressions of us. Goffman regarded everyday interaction as a series of attempts to con the other. In fact, trying in various ways to con others is, according to Goffman, at the very center of much social interaction and social organization in society: Social interaction is just a big con game!

Perhaps this cynical view is not true of all social interaction, but we do present different "selves" to others in different settings. The settings are, in effect, different stages on which we act as we relate to others. For this reason, Goffman's theory is sometimes called the *dramaturgy model* of social interaction, a way of analyzing interaction that assumes the participants are actors on a stage in the drama of everyday social life. People present different faces (give off different impressions) on different stages (in different situations or different roles) with different others. To your mother, you may present yourself as the dutiful, obedient daughter, which may not be how you present yourself to a friend. Perhaps you think acting like a diligent student makes you seem like a jerk, so you hide from your friends that you are really interested in a class or enjoy your homework. Analyzing impression management reveals that we try to con others into perceiving us as we want to be perceived. The box "Doing Sociological Research: Vegetarians versus Omnivores: A Case Study of Impression Management" shows how impression management can be involved in many settings, such as conversations between vegetarians and meat eaters.

One thing that Goffman's theory makes clear is that social interaction is a very perilous undertaking. Have you ever been embarrassed? Of course you have—we all have. Think of a really big embarrassment that you experienced. Goffman defines embarrassment as a spontaneous reaction to a sudden or transitory challenge to our identity: We attempt to restore a prior perception of our "self" by others. Perhaps you were giving a talk before a class and then suddenly forgot the rest of the talk. Or perhaps you recently bent over and split your pants. Or perhaps you are a man and barged accidentally into a women's bathroom. All these actions will result in embarrassment, causing you to "lose face."

Vegetarians versus Omnivores: A Case Study of Impression Management

Research Question

Author Jessica Greenebaum is a vegan. She noticed tension between herself and her meat-eating family and friends, possibly because of stereotypes about vegetarians and vegans. She did research to ask: How do vegetarians interact with omnivores to avoid negative impressions of vegetarians? What tactics do they use in their *presentation of self*?

Research Method

Greenebaum interviewed 19 vegans and 7 vegetarians, finding her research subjects through a website for educated, upper-middle-class adults who identify as vegetarian activists. She conducted face-to-face interviews and telephone interviews, averaging about one hour per interview.

Research Results

Many of the people she interviewed spoke about avoiding confrontation. One woman explained that she used to be "in your face" with meat eaters, but changed her approach to be more gentle. A key theme from her research was how respondents timed their discussions about vegetarianism or veganism. Most respondents did not want to be the first to bring it up in conversation. Vegetarians used "face-saving" tactics to make interactions more pleasant. Respondents did not try to recruit omnivores to become vegetarian, but instead emphasized the health benefits of not eating meat. By focusing on health, they encountered fewer negative impressions of vegetarians and vegans. Face-saving also occurred by presenting a no-meat diet as easy to do and joyful. Greenebaum asserts that "If vegans are perceived as wheat grass–drinking hippies, people are less likely to keep an open mind about veganism" (Greenebaum 2012: 321).

Conclusions and Implications

Greenebaum concludes that interactions between two groups of people who have opposing views require *impression management*. Vegetarians and vegans used particular tactics to prepare themselves for conversations with omnivores, presenting themselves in a more positive way.

Questions to Consider

The next time you are talking with someone about food, diet, and overall health, observe the social interaction with particular attention to similar and differing opinions. Seek out people with different diets from your own.

1. What do you do to manage others' impressions of you and your food choices?
2. With so much media attention on the dangers of the American diet, do you worry about the impression other people get based on what you choose to eat?

Source: Greenebaum, Jessica B. 2012. "Managing Impressions: 'Face-Saving' Strategies of Vegetarians and Vegans." *Humanity & Society* 36(4): 309–325.

You will then attempt to *restore face* ("save face"), that is, eliminate the conditions causing the embarrassment. You thus will attempt to con others into perceiving you as they might have before the embarrassing incident. One way to do this is to shift blame from the self to some other. For example, you may claim that the sign saying "Women's room" was not clearly visible. This represents a deliberate manipulation (or con) to save face on your part—to restore the other's prior perception of you.

Social Exchange Theory

Another way of analyzing social interaction is through the social exchange model (see Table 5-2). The *social exchange model* of social interaction holds that our interactions are determined by the rewards or punishments that we receive from others. A fundamental principle of exchange theory is that an interaction that elicits approval from another (a type of reward) is more likely to be repeated than an interaction that incites disapproval (a type of punishment). According to the exchange principle, one can predict whether a given interaction is likely to be repeated or continued by calculating the degree of reward or punishment inspired by the interaction.

Rewards can take many forms. They can include tangible gains such as gifts, recognition, and money, or subtle everyday rewards such as smiles, nods, and pats on the back. Similarly, punishments come in many varieties, from extremes such as public humiliation, beating, banishment, or execution, to gestures as subtle as a raised eyebrow or a frown. For example, if you ask someone out for a date and the person says yes, you have gained a reward, and you are likely to repeat the interaction. You are likely to ask the person out again, or to ask someone else out. If you ask someone out, and he or she glares at you and says, "No way!," then you have elicited a punishment that will probably cause you to shy away from repeating this type of interaction with that person.

Interaction in Cyberspace

When people interact and communicate with one another by means of personal computers—through some virtual community such as email, Twitter, Facebook, or Instagram—then they are engaging in cyberspace interaction (or virtual interaction).

The character of cyberspace interaction is changing rapidly as new technologies emerge. Not long ago, nonverbal interaction was absent in cyberspace as people could not "see" what others were like. With video-based cyberspace, such as Instagram, WhatsApp, or Skype, people can display still and moving images of themselves. These images provide opportunities, as we noted previously, for what sociologists would call the presentation of self and impression management. Sometimes this comes with embarrassing consequences. A young college student who displays a seminude or nude photo of herself or himself, projecting a sexual presentation of self, may be horrified if one of the parents or a potential employer visits the Facebook site! Furthermore, the photo could be intercepted by a disgruntled boyfriend, reproduced, and made to "go viral" (seen by hundreds or thousands of people).

Cyberspace interaction is common among all age, gender, and race groups, although clear patterns are also present in who is engaged in this form of social interaction and how people use it. Two-thirds of all Americans now use a smartphone, and they rely on their phones for various services (Smith and Page 2015). Of the various social media applications, Facebook is still the most widely used. Of Americans who have online access, 79 percent use Facebook, most of them (76 percent) on a daily basis. Age and gender are good predictors of usage, with women more likely to be using Facebook than men and those under 50 using it more than older people. Younger people are more likely to use Twitter and Instagram, but older people are interacting online in ever-increasing numbers. It is unfortunate that surveys find that disabled people are much less likely to be using the Internet and electronic devices (Anderson and Perrin 2017). ◆ Table 5-3 shows that usage patterns of various social media vary by different social characteristics, such as age, income, and gender.

What Would a Sociologist Say?

Cyberbullying

With the rise of social media use, a new form of bullying has emerged—cyberbullying. Cyberbullying is unwanted aggressive behavior that is repeated through electronic communication. Although adolescents might perceive this as "just fooling around," research on cyberbullying finds that it involves power differences between the perpetrator and the victim. Particular social contexts are also significantly related to the likelihood of cyberbullying. Close parental-adolescent relationships, for example, tend to reduce young people's exposure to cyber- and other forms of bullying. Having strong friendship and peer support also makes one less likely to be bullied. Positive experiences in school also reduce the likelihood of cyberbullying. Such research results indicate the significance of social factors in explaining the likelihood of such individually-troubling behaviors as being bullied via the internet (Hong et al. 2016).

Table 5-3	Users of Social Media Sites		
Total Adults	Facebook	Twitter	Instagram
Men	75%	24%	26%
Women	83%	25%	38%
Age			
18–29	88%	36%	59%
30–49	84%	23%	33%
50–64	74%	21%	18%
65 and older	62%	10%	8%
Household Income			
Less than $30,000	84%	23%	38%
$30,000 to $49,999	80%	18%	32%
$50,000 to $74,999	75%	28%	32%
>$75,000	77%	30%	31%
Education			
High school or less	77%	26%	27%
Some college	82%	24%	37%
College degree or more	79%	24%	33%

Source: Greenwood, Shannon, Andrew Perrin, and Mauve Duggan. 2016. *Social Media Update 2016*. Washington, DC: Pew Research Organization. **www.pewresearch.org**

Perhaps even more startling is the pace of change in social media communication. As one example, in just one four-year period (from 2012 to 2016), use of Facebook by online adults (people over 18 years of age) rose from 67 percent to 79 percent—a remarkable growth in a short period of time (Greenwood, Perrin, and Duggan 2016; Duggan et al. 2015).

How people are using cyberspace interaction is also changing. People use social media for everything from keeping up with daily news to paying their bills or just avoiding boredom. Social media are, though, a powerful source of communication. Donald Trump's use of Twitter while President bypassed the traditional media but reached millions of people on a regular basis. Others have used Twitter to organize political protest. The best example comes from the Black Lives Matter movement that was originally organized through the creation of #BlackLivesMatter (Jackson 2016).

The Internet also creates more opportunity for people to misrepresent themselves or even create completely false—or even stolen—identities. Studies find that computer-mediated interactions also follow some of the same patterns that are found in face-to-face interaction. People still "manage" identities in front of a presumed audience; they project images of self to others that are consistent with the identity they have created for themselves, and they form social networks that become the source for evolving identities, just as people do in traditional forms of social interaction. The difference between LinkedIn and Facebook, for example, indicates that the professional identity is presented differently than the personal identity (van Dijck 2013).

In this respect, cyberspace interaction is an application of Goffman's principle of *impression management*. People can put forward a totally different and wholly created self, or identity. One can "give off," in Goffman's terms, any impression one wishes and, at the same time, know that one's true self is protected by anonymity. This gives the individual quite a large and free range of roles and identities from which to choose. As predicted by symbolic interaction theory, of which Goffman's is one variety, *the reality of the situation grows out of the interaction process itself*. This is a central point of symbolic interaction theory and is central to sociological analysis generally: Interaction creates reality.

Cyberspace interaction has thus resulted in new forms of social interaction in society—in fact, a new social order containing both deviants and conformists. These new forms of social interaction have their own rules and norms, their own language, their own sets of beliefs, and practices or rituals—in short, all the elements of culture, as defined in Chapter 2. For sociologists, cyberspace also provides an intriguing new venue in which to study the connection between society and social interaction.

Chapter Summary

What is society?

Society is a system of social interaction that includes both culture and *social organization*. Society includes *social institutions*, or established organized social behavior, and exists for a recognized purpose; *social structure* is the patterned relationships within a society.

What holds society together?

According to theorist Emile Durkheim, society with all its complex social organization and culture, is held together, depending on overall type, by *mechanical solidarity* (based on individual similarity) and *organic solidarity* (based on a *division of labor* among dissimilar individuals). Two other forms of social organization also contribute to the cohesion of a society: *gemeinschaft* ("community," characterized by cohesion based on friendships and loyalties) and *gesellschaft* ("society," characterized by cohesion based on complexity and differentiation).

What are the types of societies?

Societies across the globe vary in type, as determined mainly by the complexity of their social structures, their division of labor, and their technologies. From least to most complex, they are *foraging, pastoral, horticultural, agricultural* (these four constitute *preindustrial* societies), *industrial*, and *postindustrial* societies.

What are the forms of social interaction in society?

All forms of social interaction in society are shaped by the structure of its social institutions. A *group* is a collection of individuals who interact and communicate with each other, share goals and norms, and have a subjective awareness of themselves as a distinct social unit. *Status* is a hierarchical position in a structure. A *role* is the behavior others expect from a person associated with a particular status. Patterns of social interaction influence nonverbal interaction as well as patterns of attraction and affiliation.

What theories are there about social interaction?

Social interaction takes place in society within the context of social structure and social institutions. Social interaction is analyzed in several ways, including the *social construction of reality* (we impose meaning and reality on our interactions with others); *ethnomethodology* (deliberate interruption of interaction to observe how a return to "normal" interaction is accomplished); *impression management* (a person "gives off" a particular impression to "con" the other and achieve certain goals, as in *cyberspace interaction*); and *social exchange theory*.

How is technology changing social interaction?

Increasingly, people engage with each other through *cyberspace interaction*. Social norms develop in cyberspace as they do in face-to-face interaction, but people in cyberspace can also manipulate the impression that they give off, thus creating a new "virtual" self.

Key Terms

achieved status 114
ascribed status 114
collective consciousness 108
division of labor 108
ethnomethodology 121
gemeinschaft 108
gesellschaft 108
group 113
impression management 121
macroanalysis 105

master status 114
mechanical solidarity 108
microanalysis 106
nonverbal communication 116
organic solidarity 108
postindustrial society 112
preindustrial society 109
role 115
role conflict 115
role modeling 115

role set 115
role strain 116
social institution 106
social interaction 105
social organization 106
social structure 107
society 105
status 113
status inconsistency 113
status set 113

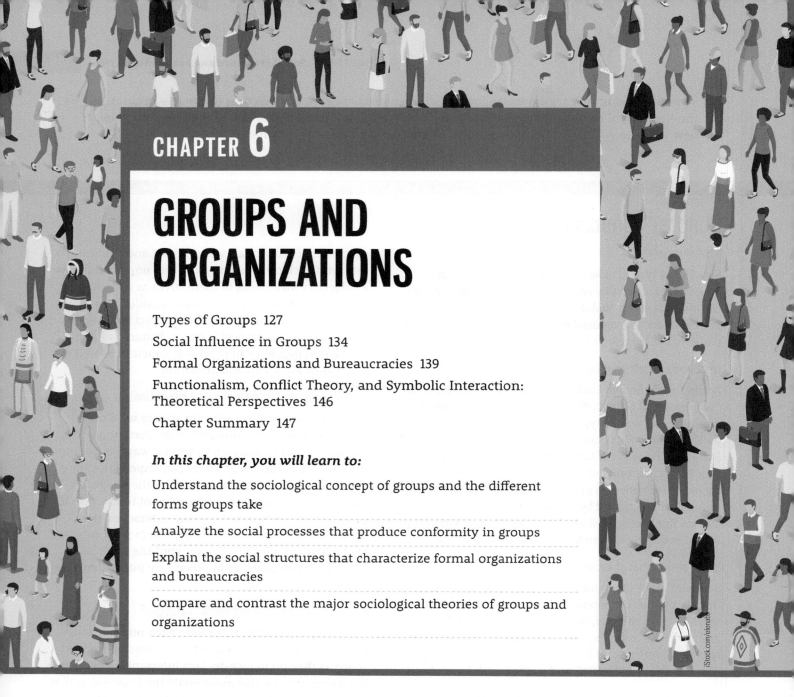

GROUPS AND ORGANIZATIONS

In this chapter, you will learn to:

Understand the sociological concept of groups and the different forms groups take

Analyze the social processes that produce conformity in groups

Explain the social structures that characterize formal organizations and bureaucracies

Compare and contrast the major sociological theories of groups and organizations

iStock.com/elenabs

It's Saturday night. You feel like staying in, perhaps to read a book, play video games, stream a movie. You're just not "up" for going out as you often do. Just as you are settling it, you get a text from a friend saying, "Hey, let's party; there's a great band at our favorite place. Let's go." Very soon, another friend texts, "I'm in! See you there." And another, "Me too." Before you know it, you are in the club, enjoying yourself but perhaps wishing you had just had a quiet night at home. The next morning, as you nurse your headache, you wonder why you went. You had really wanted a quiet night alone, but you soon found yourself surrounded by others, doing what they were doing, even though it wasn't how you had planned to spend your evening. What happened?

The answer is that you were subjected to group behavior—one of the most interesting and strongest phenomena in the social world. We like to think of ourselves as individuals and, of course, we are, but even as individuals, our behavior is strongly influenced by the groups to which we belong. At any given moment, we belong to multiple groups, some with more influence than others. Understanding group behavior is critical to understanding people's behavior.

Consider this: If someone told you that you could catch a spaceship to a next level of existence, beyond anything you had ever known on Earth, would you take a lethal combination of drugs and alcohol to get you there? Surely not, you must be thinking! But that is precisely what thirty-nine members of the Heaven's Gate cult did in 1997 in a mansion in Rancho Santa Fe, California. They were told by the group leader that a spaceship was coming, following the tail of Comet Hale-Bopp and that they would be transported to a better place. Although seemingly completely irrational, this behavior can only be understood by analyzing how these thirty-nine individuals became subject to the control of such an extremist group—in other words, succumbing to group pressure.

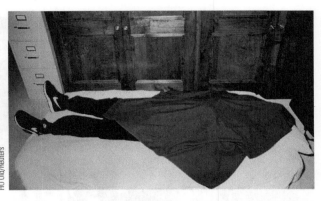

Eighteen men and twenty-one women committed mass suicide as part of the Heaven's Gate cult in 1997, all of them dressed alike in dark clothes and Nike sneakers. This is an extreme example of group conformity.

Group pressure also escalates violent behaviors. An example is a horrific rape that occurred in New Delhi, India, in 2012, when seven men gang-raped a twenty-three-year-old medical school student on a public bus. The young woman died two weeks later from the severe injuries. Rape is a violent act even when committed by one person, but research finds that rape involving more than one perpetrator—that is, group rape—is usually far more violent and involves more severe forms of violation than rape by a single perpetrator. Scholars conclude that the group behavior involved in a gang rape intensifies violence as the group members succumb to the power of a group leader and/or feel they must participate or they will be ostracized by the group (Woodhams et al. 2012).

In less dramatic and disturbing examples, group influence also shapes all kinds of ordinary behavior. Juries are groups, and jury decision making is clearly influenced by group processes. When a jury deliberates, a consensus is formed as more members of the group (that is, the jury) move to a particular verdict. Moreover, the larger the faction, the less willing an individual juror will be to defy the weight of group opinion. As we shall see, this is **group size effect**: an effect of sheer numbers in the group *independent* of the effects of individual actions and thoughts (Vidmar and Hans 2007).

You can probably think of examples in your own experience when you succumbed to group pressure, even when your individual judgment told you to do something different. Perhaps you smoke cigarettes, knowing full well that they are very harmful to your health. Maybe you have purchased something from the latest fashion trend, even though you really could not spare the money, or maybe you have gone out drinking with friends because "everybody was doing it."

People are highly subject to the social influences of groups. Whether a relatively small group—such as a jury or your friendship circle—or a large bureaucratic organization, such as the government or a work organization, people are influenced by the sociological forces of group behavior.

Types of Groups

Each of us is a member of many groups simultaneously. We have relationships in groups with family, friends, team members, and professional colleagues. Within these groups are gradations in relationships: We are generally closer to our siblings than to our cousins; we are intimate with some friends, merely sociable with others. If we count all our group associations, ranging from the powerful associations that define our daily lives to the thinnest connections with little feeling (other pet lovers, other company employees), we will uncover connections to literally hundreds of groups.

What is a group? Recall from Chapter 5, a **group** is two or more individuals who interact, share goals and norms, and have a subjective awareness as "we." To be considered a group, a social unit must have all three

José Antonio Hernaiz/Age Fotostock

The annual running of the bulls in Pamplona, Spain, is a death-defying exercise in group behavior.

characteristics, although some groups are more bound together than others. Consider two superficially similar examples: The individuals in a line waiting to board a train are unlikely to have a sense of themselves as one group. A line of prisoners chained together and waiting to board a bus to the penitentiary is more likely to have a stronger sense of common feeling.

Certain gatherings are not groups in the strict sense, but may be *social categories* (for example, teenagers, truck drivers) or *audiences* (everyone watching a movie). The importance of defining a group is not to perfectly decide if a social unit is a group—an unnecessary endeavor—but to help us understand the behavior of people in society. As we inspect groups, we can identify characteristics that reliably predict trends in the behavior of the group and even the behavior of individuals in the group.

The study of groups has application at all levels of society, from the attraction between people who fall in love to the characteristics that make some corporations drastically outperform their competitors—or that lead them into bankruptcy. The aggregation of individuals into groups has a transforming power, and sociologists understand the social forces that make these transformations possible. Within the confines of this chapter, we move from the *micro level* of analysis (the analysis of groups and face-to-face social influence) to the more *macro level* of analysis (the analysis of formal organizations and bureaucracies).

Dyads and Triads: Group Size Effects

Even the smallest groups are of acute sociological interest and can exert considerable influence upon individuals. A **dyad** is a group consisting of exactly two people. A **triad** consists of three people. This seemingly minor distinction, first scrutinized by the German sociologist **Georg Simmel** (1858–1918), can have critical consequences for group behavior (Simmel 1950/1902). Simmel was interested in discovering the effects of size on groups, and he found that the mere difference between two and three people spawned entirely different group dynamics (the behavior of a group over time).

Imagine two people standing in line for lunch. First one talks, then the other, then the first again. The interaction proceeds in this way for several minutes. Now a third person enters the interaction. The character of the interaction suddenly changes: At any given moment, two people are interacting more with each other than either is with the third. When the third person wins the attention of the other two, a new dyad is formed, supplanting the previous pairing. The group, a triad, then consists of a dyad (the pair that is interacting) plus an *isolate*.

Triadic segregation is what Simmel called the tendency for triads to segregate into a pair and an isolate (a single person). A triad tends to segregate into a **coalition** of the dyad against the isolate. The isolate then has the option of initiating a coalition with either member of the dyad. This choice is a type of social advantage, leading Simmel to coin the principle of *tertius gaudens*, a Latin term meaning "the third one gains." Simmel's reasoning has led to numerous contemporary studies of coalition formation in groups (Holyoke 2009).

For example, interactions in a triad often end up as "two against one." You may have noticed this principle of coalition formation in your own conversations. Perhaps two friends want to go to a movie you do not want to see. You appeal to one of them to go instead to a minor league baseball game. She wavers and comes over to your point of view. Now you have formed a coalition of two against one. The friend who wants to go to the movies is now the isolate. He may recover lost social ground by trying to form a new coalition by suggesting a new alternative (going bowling or to a different movie). This flip-flop interaction may continue for some time, demonstrating another observation by Simmel: A triad is a decidedly unstable social grouping, whereas dyads are relatively stable. The distinction between dyads and triads is just one person but the presence of that one person changes the character of the interaction within the group. Simmel is known as the discoverer of **group size effect**—the effects of group number on group behavior independent of the personality characteristics and opinions of the members themselves.

Primary and Secondary Groups

Charles Horton Cooley (1864–1929), a famous sociologist of the Chicago School of Sociology, introduced the concept of the **primary group**, defined as a group consisting of intimate, face-to-face interaction and relatively long-lasting relationships. Cooley had in mind the family and the early peer group. In his original formulation, "primary" was used in the sense of "first," the intimate group of the formative years (Cooley 1967/1909). The insight that there was an important distinction between intimate groups and other groups proved extremely fruitful. Cooley's somewhat narrow concept of family and childhood peers has been elaborated upon over the years to include a variety of intimate relations as examples of primary groups.

Primary groups have a powerful influence on an individual's personality or self-identity. The effect of family on an individual can hardly be overstated. The weight of peer pressure on school children is particularly notorious. Street gangs are a primary group, and their influence on individuals is significant; in fact, gang members frequently think of themselves as a family. Inmates in prison very frequently become members of a gang—primary groups perhaps based mainly upon race or ethnicity—as a matter of their own personal survival. The intense camaraderie formed among Marine Corps units in boot camp and in war is another classic example of primary group formation and the resulting intense effect on individuals and upon their survival.

In contrast to primary groups are **secondary groups**, those that are larger in membership, less intimate, and less long lasting. Secondary groups tend to be less significant in the emotional lives of people. Secondary groups include all the students at a college or university, all the people in your neighborhood, and all the people in a bureaucracy or corporation.

Primary and secondary groups serve different needs. Primary groups give people intimacy, companionship, and emotional support. These human desires are termed **expressive needs** (also called socioemotional needs). Family and friends share and amplify your good fortune, punish you when you misbehave, and cheer you up when life looks grim. Many studies have shown the overwhelming influence of family and friendship groups on religious and political affiliation, as shown in the box "Doing Sociological Research: Sharing the Journey."

Secondary groups serve **instrumental needs** (also called task-oriented needs). Athletic teams form to have fun and win games. Political groups form to raise funds and influence the government. Corporations form to make profits, and employees join corporations to earn a living. The true distinction between primary and secondary groups is in how intimate the group members feel about one another and how dependent they are on the group for sustenance and identity.

Secondary groups occasionally take on the characteristics of primary groups, even if temporarily. This is precisely what happened to a group of miners who, for nearly three months in 2011, were trapped

Thinking Sociologically

Identify a *group* of which you are a part. How does one become a member of this group? Who is included and who is excluded? Does the group share any unique language or other cultural characteristics (such as dress, jargon, or other group identifiers)? Does anyone ever leave the group, and if so, why? Would you describe this group mainly as a *primary* or a *secondary group*? Why?

Sharing the Journey

Modern society is often characterized as remote, alienating, and without much feeling of community or belonging to a group. Despite the image of society as an increasingly impersonal force, sociologist Robert Wuthnow has noticed that people in the United States are increasingly looking to small groups as places where they can find emotional and spiritual support and where they find meaning and commitment.

Research Question

Wuthnow began his research by asking: What motivates people to join support groups? How do these groups function? What do members like most and least about such groups? His broadest question, however, was to wonder how the proliferation of small support groups influences the wider society.

Research Methods

His large research team (ofifteen scholars) designed a study including both a quantitative and a qualitative dimension. They distributed a survey to a representative sample of more than 1000 people in the United States. They supplemented the survey with more than 100 interviews with support group members, group leaders, and clergy. The researchers chose twelve groups for extensive study, spending six months to three years tracing the history of the groups, meeting with members, and attending group sessions.

Research Results

Wuthnow concludes that the small group movement is fundamentally altering U.S. society. Forty percent of all Americans belong to some kind of small support group. As the result of people's participation in these groups, social values of community and spirituality are undergoing major transformation. People who join small groups are seeking community, whether the group is a recovery group, a religious group, a civic association, or some other small group.

Conclusions and Implications

Wuthnow argues that large-scale participation in small groups has arisen when the traditional support structures in U.S. society, such as the family, no longer provide the sense of belonging and social integration that they provided in the past. Geographic mobility, mass society, and the erosion of local ties all contribute to this trend. People still seek a sense of community, but they create it in groups that allow them to maintain their individuality. Wuthnow also concludes that these groups represent a quest for spirituality when, for many, traditional religious values have declined. As a consequence, support groups are redefining what is sacred. Small groups thus buffer the trend toward disintegration and isolation that people often feel in mass societies.

Questions to Consider

1. Are you a member of a voluntary small group? If so, what sense of community does the group provide for you? How do you maintain your sense of individuality at the same time?
2. What social changes do you observe in the world around you that might encourage people to join various support groups?
3. Some people join support groups in the aftermath of major life transition—a death, recovery from addiction, the desire to lose weight, and so on. What does group membership in such a situation provide for individuals?

Source: Wuthnow, Robert. 1994. *Sharing the Journey: Support Groups and America's New Quest for Community*. New York: Free Press.

a half mile below the surface in Chile's Atacama Desert. When the thirty-three miners were eventually rescued, an event that was covered live on the international news, we learned that this was a very striking example of the transition from a largely secondary group to an exceptionally close-knit primary group. A strong leader (foreman Luis Urzúa) insisted that, "It was one for all and all for one down there." As the men reported it, the experience transformed all thirty-three men into a large and very close family (primary group) even as they later coped with their newfound fame, celebrity, and requests to endorse products (Padgett et al. 2011).

Reference Groups

Primary and secondary groups are groups to which people actually belong. Both are called *membership groups*. In contrast, **reference groups** are those to which you may or may not belong but use as a standard for evaluating your values, attitudes, and behaviors (Merton and Rossi 1950). Reference groups are generalized versions of role models. They are not "groups" in the sense that the individual interacts within (or in) them. Do you pattern your behavior on that of sports stars, musicians, military officers, or business executives? If so, those are reference groups for you.

Imitation of reference groups can have both positive and negative effects. Members of a Little League baseball team may revere major league baseball players and attempt to imitate laudable behaviors such as tenacity and sportsmanship. But young baseball fans are also liable to be exposed to tantrums, fights, and tobacco chewing and spitting. This illustrates that the influence of a reference group can be both positive and negative.

Reference groups do not have to be actual people. As we have been seeing throughout this book, the media can serve as a powerful reference group, influencing how people perceive themselves. Positive representations of one's reference group can promote strong self-esteem; negative representations and negative stereotypes, such as of racial–ethnic groups, can produce diminished self-esteem. The representation of racial and ethnic groups in a society can have a striking positive effect perhaps even among children acquiring their lifetime set of group affiliations (Harris 2006; Zhou and Bankston 2000).

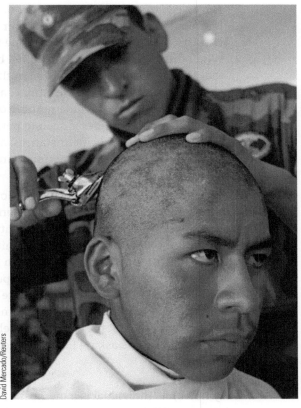

David Mercado/Reuters

Initiation into a group can mean losing one's individual identity, especially when strict conformity to the group is enforced. New initiates into military life routinely have their hair cut, symbolic of the dominance of group identity over individual identity.

In-Groups, Out-Groups, and Attribution Error

When groups have a sense of themselves as "us," they will also have a complementary sense of other groups as "them." The distinction is commonly characterized as *in-groups* versus *out-groups*. The concept was originally elaborated by the early sociological theorist **W. I. Thomas** (1863–1947) (Thomas 1931). College fraternities and sororities certainly exemplify "in" versus "out." So do families. So do gangs—especially so. The same can be true of the members of your high school class, your sports team, your racial group, your gender, and your social class.

Attribution theory is the principle that we all explain the behaviors of other people based on assumptions about their individual characteristics or their situational context. These attributions depend on whether you are in the in-group or the out-group. Thomas F. Pettigrew has summarized the research on attribution theory, showing that individuals commonly develop a distorted perception of the motives and capabilities of other people's acts based on whether that person is an in-group or out-group member (Baumeister and Bushman 2017; Gilbert and Malone 1995; Pettigrew 1992).

Pettigrew and others describe the misperception as **attribution error**, meaning errors made in crediting causes for people's behavior to their membership in a particular group, such as a racial group. Attribution error has several dimensions, all tending to favor the in-group over the out-group. All else being assumed equal, we tend to perceive people in our in-group positively and those in out-groups negatively, regardless of their actual personal characteristics:

1. When onlookers observe improper behavior by an out-group member, onlookers are likely to attribute the deviance to the disposition (the personality) of the wrongdoer. Disposition refers to the

perceived "true nature" of the person. Stereotypes work this way. A White person may observe a Black man standing on a street corner and, without further information, assume that this person is "lazy" because it fits the stereotype that Whites often hold against racial and ethnic minorities.

2. When the *same* behavior is exhibited by an in-group member, the perception is commonly held that the act is due to the *situation* of the wrongdoer, not to the in-group member's inherent disposition or personality. For example, a White person may see another White person standing on a street corner and conclude nothing other than that the person has a reason to be there.

3. If an out-group member is seen to perform in some laudable way, the behavior is often attributed to a variety of special circumstances, and the out-group member is seen as "the exception."

4. An in-group member who performs in the same laudable way is given credit for a worthy personality disposition.

Typical attribution errors include misperceptions between racial groups. In the case of race, recent events in the news tend to bear this out: When a Black person is shot and killed by a White policeman, 79 percent of Black Americans see the incident as part of a broader problem compared to 54 percent of Whites who think so. Further 4 percent of Whites see such a shooting as an isolated incident, while only 18 percent of Black Americans think so (Morin et al. 2017). Even given the numerous, well-publicized shootings of Black and Latino men by White police, surveys find a large gap in how much White and Black Americans have confidence in the police (Morin and Stepler 2016).

A related phenomenon is seen when women report that they have been sexually harassed. Many, men especially, will say that the woman somehow "asked for it." But no one asks if a wealthy man, dressed in a fine suit, was asking for it if he gets robbed.

Social Networks

As already noted, no individual is a member of only one group. Social life is far richer than that. A **social network** is a set of links between individuals, between groups, or between other social units, such as bureaucratic organizations or even entire nations. One could say that any given person belongs simultaneously to several networks. With the development of *social media*, networks that may have once been face-to-face have now developed through electronic media, such as on Facebook, Instagram, and Twitter. The development of social media brings a new dimension to the study and analysis of networks because you may be in a network with people you do not even know. Nonetheless, your group of friends, or all the people on an electronic mailing list to which you subscribe, or all of your Twitter followers, are social networks, some human, some electronic.

Let us do a bit of network analysis (including group size effects) right now. Assume first that a group consisting of only two people has by definition one two-way relationship. (Each knows the other personally). A group of three people will thus have three possible two-way relationships; and a group of four people will have six possible two-way relationships. Extending this simple counting of the number of pairs (i.e., two-way relationships) shows that a group of five people has ten possible two-way

As with the African American women's sorority, Delta Sigma Theta, groups often use clothing styles and colors to signify group belonging.

The Washington Post/Getty Images

What Would a Sociologist Say?

Finding a Job: The Invisible Hand

Is getting a job simply a matter of getting the right credentials and training? Hardly, according to sociologists. Even in a good job market, one needs the help of social networks to find a job. This has been clearly demonstrated by sociologist Deirdre Royster, who compared the experiences of two groups of men: One group was White, the other Black. Both had graduated from vocational school. The Black and White men had comparable educational credentials, the same values, and the same work ethic; yet the White men were far more likely to gain employment than were the Black men. Why? Royster's research revealed that the most significant difference between the two groups was access to job networks, just as Granovetter's work on job networks would predict (Royster 2003; Granovetter 1995).

relationships; a group of six people fifteen possible two-way relationships; and so on. With even as few as thirty "friends" on Facebook, there are actually 435 possible two-way (i.e., mutual) relationships! That is a *lot* of mutual relationships! As early as the 1950s, Robert Bales (1951) estimated that thirty is the limit for the number of people that can be in an intimate "small group" such that each person can have or recall a perception of each other member as an individual person. How "close" can your Facebook friends really be?

Networks can be critical to your success in life. Numerous research studies indicate that people get jobs via their personal networks more often than through formal job listings, want ads, or placement agencies (Granovetter 1974). Getting a job is more often a matter of whom you know than what you know. Who you know, and whom they know in turn, is a social network that may have a marked effect on your life and career.

Networks form with all the spontaneity of other forms of human interaction (Wasserman and Faust 1994). Networks evolve, such as social ties within neighborhoods, professional contacts, and associations formed in fraternal, religious, occupational, and volunteer groups. Networks to which you are only *weakly* tied (you may know only one person in your neighborhood) provide you with access to that entire network, hence the sociological paradox that there is "strength in weak ties" (Granovetter 1973).

Networks based on race, class, and gender form with particular readiness. This has been especially true of job networks. The person who leads you to a job is likely to have a similar social background. Research indicates that the "old boy network"—any network of White, male corporate executives—is less important than it used to be, although it is certainly not by any means gone as women have become increasingly prominent. Still, as we will see in various places later in this book, women and people of color remain underrepresented in corporate life, especially in high-status jobs, giving them less access to prestigious social networks (Smith 2007). Network size is a very important force in human existence. For example, network research has been used to identify populations and subpopulations that are at risk of being struck with AIDS—the human immunodeficiency virus (Gile et al. 2013).

Social Networks as "Small Worlds"

Networks can reach around the world, but how big is the world? How many of us, when we discover someone we just met is a friend of a friend, have remarked, "Wow, it's a small world, isn't it?"? Research into what has come to be known as the small world problem has shown that networks make the world a lot smaller than you might otherwise think.

Original small world researchers Travers and Milgram wanted to test whether a document could be routed via the U.S. postal system to a complete stranger more than 1000 miles away using only a chain of acquaintances (Travers and Milgram 1969). If so, how many steps would be required? The researchers organized an experiment back in 1969 in which approximately 300 senders were all charged with getting a document to one receiver, a complete stranger. (Remember that all this was well before the advent of the desktop computer in the 1980s.) The receiver was a male Boston stockbroker. The senders were one group of Nebraskans and one group of Bostonians chosen completely at random. Every sender in the study was given the receiver's name, address, occupation, alma mater, year of graduation, wife's maiden name, and hometown. They were asked to send the document directly to the stockbroker only if they knew him on a first-name basis. Otherwise, they were asked to send the folder to a friend, relative, or acquaintance known on a first-name basis who might be more likely than the sender to know the stockbroker.

How many intermediaries do you think it took, on average, for the document to get through? Most people estimate from twenty to hundreds. About one-third of the documents actually arrived at the target and, of those one-third that were completed, the average number of contacts was only 6.2! This was impressive, considering that the senders did not know the target person—hence, the current expression that any given person in the country is on average only about "six degrees of separation" from any other person. In this sense, the world is indeed "small." The sending chains tended to closely follow occupational, social class, and ethnic lines, just as general network theory would predict (Kleinfeld 1999; Wasserman and Faust 1994), suggesting that the world is especially small when within one's general social network (Ruef et al. 2003; Watts 1999).

Social networks are often organized around specific identities or group characteristics. For example, studies have found that Black leaders form a very closely knit network, one considerably more closely knit than longer-established White leadership networks (Taylor 1992). In a study of national Black leaders (including members of Congress, mayors, business executives, military generals and full colonels, religious leaders, civil rights leaders, media personalities, and entertainment and sports figures), it was found that one-fifth of the entire national Black leadership network know each other directly as a friend or close acquaintance. This is a much more closely knit social network than found among White leaders. Add only one intermediary, the friend of a friend, and the study estimated that almost *three-quarters* of the entire Black leadership network are included. Therefore, any given Black leader can generally get in touch with three-quarters of all other Black leaders in the country either by knowing them personally (a "friend") or via only one common acquaintance (a "friend of a friend").

Social networks tend to be homogenous—that is, composed of people with similar gender, ethnic, religious, and other social statuses. It has long been known that people's core networks—that is, close friends—are generally segregated by ethnicity and race. Online social networks are also highly segregated, even though they can be quite large, although not quite to the same extent as in-person networks. For example, a recent study of Facebook networks among adolescents has shown that people's online social networks are strongly segregated by ethnicity and, to some extent, by gender (Hofstra et al. 2017). All told, social networks are a significant and common component of our social lives.

Social Influence in Groups

The groups in which we participate exert tremendous influence on us. We often fail to appreciate how powerful these influences are. For example, who decides what you should wear? Do you decide for yourself each morning, or is the decision already made for you by fashion designers, role models, and your peers? Consider how closely your hair length, hair styling, and choice of jewelry have been influenced by your peers. Did you invent your skinny jeans, your dreadlocks, or your blue blazer? People who label themselves as "nonconformists" often conform rigidly to the dress code and other norms of their in-group (a type of small network).

A group such as one's family even influences your adult life long after children have grown up and formed households of their own. The choice of political party among adults (Republican, Democratic, or Independent) correlates strongly with the party of one's parents, again demonstrating the power of the primary group. Seven out of ten people vote with the political party of their parents, even though these same people insist that they think for themselves when voting (Worchel et al. 2000). Furthermore, most people share the religious affiliation of their parents, although they will insist that they chose their own religion, free of any influence by either parent.

We all like to think we stand on our own two feet, immune to a phenomenon as superficial as group pressure. The conviction that one is impervious to social influence results in what social psychologist

Philip Zimbardo calls the *not me syndrome:* When confronted with a description of group behavior that is disappointingly conforming and not individualistic, most individuals counter that some people may conform to social pressure, "but not me"; or "some people yield quickly to styles of dress, but not me"; or "some people yield to autocratic authority figures, but not me" (Zimbardo et al. 1977). Sociological experiments often reveal a dramatic gulf between what people *think* they will do and what they *actually do.* The original conformity study by Solomon Asch discussed next is a case in point.

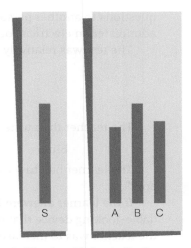

The Asch Conformity Experiment

We learned in the previous sections that social influences are quite strong. Are they strong enough to make us disbelieve our own senses? Are they strong enough to make us misperceive what is obviously objective, actual fact? In a classic piece of work known as the Asch conformity experiment, researcher Solomon Asch showed that even simple objective facts cannot withstand the distorting pressure of group influence (Asch 1955, 1951).

▲ **Figure 6-1 Lines from Asch Experiment**

Source: Asch, Solomon. 1956. "Opinion and Social Pressure." *Scientific American* 19 (July): 31–36.

Examine the two illustrations in ▲ Figure 6-1. Which line on the right is more nearly equal in length to the line on the left (Line S)? Line B, obviously. Could anyone fail to answer correctly?

In fact, Solomon Asch discovered that social pressure of a rather gentle sort was sufficient to cause an astonishing rise in the number of wrong answers. Asch lined up five students at a table and asked which line in the illustration on the right is the same length as the line on the left. Unknown to the fifth student, the first four were *confederates*—collaborators with the experimenter who only pretended to be participants. For several rounds, with similar photos, the confederates gave correct answers to Asch's tests. The fifth student also answered correctly, suspecting nothing. Then on subsequent trials the first student gave a wrong answer. The second student gave the same wrong answer. Third, wrong. Fourth, wrong. Then came the fifth student's turn. If you were the fifth student, what would you have done?

In Asch's experiment, fully *one-third* of all students in the fifth position gave the same wrong answer as the confederates at least half the time. Forty percent gave "some" wrong answers. Only one-fourth of the students consistently gave correct answers in defiance of the invisible pressure to conform.

Line length is not a vague or ambiguous stimulus. It is clear and objective, yet one-third of all subjects, a very high proportion, gave wrong answers. The subjects fidgeted and stammered while doing it, but they did it nonetheless. Those who did not yield to group pressure showed even more stress and discomfort than those who yielded to the (apparent) opinion of the group.

Would you have gone along with the group? Perhaps, perhaps not. Sociological insight grows when we acknowledge the fact that fully a third of all participants will yield to the group. The Asch experiment has been repeated many times over the years, with students and nonstudents, old and young, in groups of different sizes, and in different settings (Baumeister and Bushman 2017; Worchel et al. 2000). The results remain essentially the same! A third to a half of the participants make a judgment contrary to fact, yet in conformity with the group. Finally, the Asch findings have consistently revealed a *group size effect:* The greater the number of individuals (confederates) giving an incorrect answer (from five up to fifteen confederates), then the greater the number of subjects per group giving an incorrect answer.

The Milgram Obedience Studies

What are the limits of social group pressure? In terms of moral and psychological issues, judging the length of a line is a small matter. What happens if an authority figure demands obedience—a type of conformity—even if the task is something the test subject (the person) finds morally wrong and reprehensible? A chilling answer emerged from the now famous Milgram obedience studies done from 1960 through 1973 by Stanley Milgram (Milgram 1974).

In this study, a naive research subject entered a laboratory-like room and was told that an experiment on learning was to be conducted. The subject was to act as a "teacher," presenting a series of test questions to another person, the "learner." Whenever the learner gave a wrong answer, the teacher would administer an electric shock.

The test was relatively easy. The teacher read pairs of words to the learner, such as:

blue	box
nice	house
wild	duck

The teacher then tested the learner by reading a multiple-choice answer, such as

blue:	sky	ink	box	lamp

The learner had to recall which term completed the pair of terms given originally, in this case, "blue box."

If the learner answered incorrectly, the teacher was to press a switch on the shock machine, a formidable-looking device that emitted an ominous buzz when activated. For each successive wrong answer, the teacher was to increase the intensity of the shock by 15 volts.

The machine bore labels clearly visible to the teacher: Slight shock, moderate shock, strong shock, very strong shock, intense shock, extreme intense shock, danger: severe shock, and lastly, *XXX* at 450 volts. As the voltage rose, the learner responded with increased squirming, groans, then screams.

The experiment was rigged. The learner was a confederate. No shocks were actually delivered. The true purpose of the experiment was to see if any "teacher" would go all the way to 450 volts. If the subject (teacher) tried to quit, the experimenter responded with a sequence of prods:

"Please continue."

"The experiment requires that you continue."

"It is absolutely essential that you continue."

"You have no other choice, you must go on."

In the first experiment, fully 65 percent of the volunteer subjects ("teachers") went *all the way* to 450 volts on the shock machine!

Milgram himself was astonished. Before carrying out the experiment, he had asked a variety of psychologists, sociologists, psychiatrists, and philosophers to guess how many subjects would actually go all the way to 450 volts. The opinion of these consultants was that only one-tenth of one percent (one in one thousand) would actually do it!

What would you have done? Remember the "not me" syndrome. Think about the experimenter—a clear professorial authority figure—saying, "You have no other choice, you must go on." Most people claim they would refuse to continue as the voltage escalated. The importance of this experiment derives in part from how starkly it highlights the difference once again between what people *think* they will do and what they *actually do*.

Milgram devised a series of additional experiments in which he varied the conditions to find out what would cause subjects *not* to go all the way to 450 volts. He moved the experiment from an impressive university laboratory to a dingy basement to counteract some of the tendency for people to defer to a scientist conducting a scientific study. One learner was then instructed to complain of a heart condition with increasing trials. Still, well over half of the subjects delivered the maximum shock level! Speculating that women might be more humane than men (all prior experiments used only male subjects), Milgram did the experiment again using only women subjects (and male "learners"). The results? Exactly the same. Social class background made no difference. Racial and ethnic differences had no detectable effect on compliance rate.

At the time that the Milgram experiments were conceived, the world was watching the trial in Jerusalem of World War II Nazi Adolf Eichmann. Millions of Jews, Gypsies, gays, lesbians, and communists were murdered between 1939 and 1945 by the Nazi party, led by Adolf Hitler. As head of the Gestapo's "Jewish section," Eichmann oversaw the deportation of Jews to concentration camps and the mass executions that followed. Eichmann disappeared after the war, was abducted in Argentina by Israeli agents in 1961, and was transported to Israel, where he was tried and ultimately hanged for crimes against humanity.

The world wanted to see what sort of monster could have committed the crimes of the Holocaust, but a jarring picture of Eichmann emerged. He was slight and mild mannered, not the raging ghoul that everyone expected. He insisted that although he had indeed been a chief administrator in an organization whose product was mass murder, he was guilty only of doing what he was told to do by his superiors. He did not hate Jews, he said. In fact, he had a Jewish half-cousin whom he hid and protected. He claimed, "I was just following orders."

How different was Adolph Eichmann from the rest of us? The political theorist Hannah Arendt dared to suggest in her book *Eichmann in Jerusalem* (1963) that evil on a giant scale is banal. It is not the work of monsters, but an accident of civilization. Arendt argued that we need only look into ourselves to find the villain.

The Iraqi Prisoners at Abu Ghraib: Research Predicts Reality?

We have just learned that ordinary people will do horrible things to other humans simply because of the influence of the group, because of an authority figure, or because of a combination of both. This has been the lesson of the Asch studies and the Milgram studies. Events in the world since have once again shown vividly and clearly how accurate such sociological and psychological experiments are in the prediction of actual human behavior.

In the spring of 2004, it was revealed that American soldiers who were military police guards at a prison in Iraq (the prison was named Abu Ghraib) had engaged in severe torture of Iraqi prisoners of war. The torture included sexual abuse of the prisoners—having male prisoners simulate sex with other male prisoners, positioning their mouths next to the genitals of another male prisoner, and other such acts. Still other acts of torture involved physical abuse such as beatings, stomping on the fingers of prisoners (thus fracturing them), and a large number of other physical acts of torture, including bludgeoning, some allegedly resulting in deaths of prisoners. Such tortures are clearly outlawed by the Geneva convention and by clearly stated U.S. principles of war. Both male and female guards participated in these acts of torture. The guards later claimed that they were simply following orders, either orders directly given or indirectly assumed. Most people thought that American soldiers would never engage in such horrible acts.

Now consider what we know from research. The Milgram studies strongly suggest that many ordinary soldiers who were not at all "corrupt," at least not more than average, would indeed engage in these acts of torture, particularly if they believed that they were under orders to do so or if they believed that they would not be punished in any way if they did. Although it does not excuse such behavior, the soldiers' behavior did not stem from the personalities of a "corrupt few" (their "natures") but in the social structure and group pressures of the situation.

A now classic study of a simulated prison by Haney, Banks, and Zimbardo (1973) shows this effect quite clearly. In this study, Stanford University students were told by an experimenter to enter a dungeon-like basement. Half were told to pretend to be guards (to role-play being a guard) and half were told that they were prisoners (to role-play being a prisoner). Which students were told what was randomly determined.

After two or three days, the guards, completely on their own, began to act very sadistically and brutally toward the prisoners—having them strip naked, simulate sex, act subservient, and so on. Interestingly, the prisoners for the most part did just what the guards wanted them to do, no matter how unpleasant the requested act! The experiment was so scary that the researchers terminated the experiment after six days—more than a week early.

Remember that this study was conducted in 1973—thirty-one years *before* Abu Ghraib. Yet, this simulated prison study (as well as the Asch and Milgram studies) predicted quite precisely how both "guards" and "prisoners" would act in a real prison situation. Group influence effects uncovered by the Asch as well as the Milgram studies ruled in both the simulated prison of 1973 as well as the only too real Iraq prison of 2004.

Groupthink

Wealth, power, and experience are apparently not enough to save us from social influences. **Groupthink**, as described by I. L. Janis, is the tendency for even highly educated group members to reach a consensus opinion, even if that opinion is downright stupid (Janis 1982).

Janis reasoned that because major government policies are often the result of group decisions, it would be fruitful to analyze group dynamics that operate at the highest level of government—for instance, in the office of the president of the United States. The president makes decisions based upon group discussions with his advisers. The president is human and thus susceptible to group influence. Janis discovered a common pattern of misguided thinking in his investigations of presidential decisions. He surmised that outbreaks of this groupthink had several things in common:

1. An illusion of invulnerability;
2. A falsely negative impression of those who are antagonists to the group's plans;
3. Discouragement of dissenting opinion;
4. An illusion of unanimity.

Groupthink influences many important decisions, such as going to war, in close-knit presidential administrations. The U.S. Senate Select Committee on Intelligence fingered groupthink as leading to the intelligence failures that led government leaders to underestimate the terrorist threats to the United States prior to the terrorist attacks via airplanes on the Trade Center Towers in New York City and the Pentagon on September 11 in 2001—the infamous "9/11" attacks. Groupthink is not inevitable when a team gathers to make a decision, but it is common and appears in all sorts of groups, from student discussion groups to the highest councils of power (Tsoukalas 2007; Paulus et al. 2001). When groupthink exists, particularly in places (such as a presidential cabinet) where people make decisions that affect the lives of millions, it can be seriously harmful. Sociologists know that a diversity of expressed opinions is a far better path to decision-making, whether in presidential, corporate, or even small group settings.

Risky Shift

The term *groupthink* is commonly associated with group decision making with consequences that are not merely unexpected but disastrous. Another group phenomenon, **risky shift**, may help explain why the products of groupthink are frequently calamities. Have you ever found yourself in a group engaged in a high-risk activity that you would not do alone? When you created mischief as a child, were you not usually part of a group? If so, you might well have been in the thrall of risky shift—the general tendency for groups to be more risky than individuals taken singly.

Risky shift was first observed by MIT graduate student James Stoner (1961). Stoner gave study participants descriptions of a situation involving risk, such as one in which people seeking a job must choose between job security and a potentially lucrative but risky advancement. The participants were then asked to decide how much risk the person should take. Before performing his study, Stoner believed that individuals in a group would take less risk than individuals alone, but he found the opposite: After his groups had engaged in open discussions, they favored greater risk than they would have before discussion.

Stoner's research has stimulated literally hundreds of studies using males and females, different nationalities, different tasks, and other variables (Yardi and Boyd 2010; Worchel et al. 2000). The results are complex. Much, but not all, group discussion leads to greater risk-taking. In subcultures that value caution above daring, as in some work groups of Japanese and Chinese firms, group decisions are less risky after discussion than before. The shift can occur in either direction, driven by the influence of group discussion, but there is generally some kind of shift in one direction or the other rather than no shift at all (Kerr 1992). This is called **polarization shift**.

What causes risky shifts? The most convincing explanation is that deindividuation occurs. **Deindividuation** is the sense that one's self has merged with a group. In terms of risk-taking, one feels that responsibility

See for Yourself

Think of a time when you engaged in some risky behavior. What group were you part of, and how did the group influence your behavior? How does this illustrate the concept of *risky shift*? Is there more risky shift with more people in the group? If so, this would illustrate a group size effect.

Myth: A group of experts brought together in a small group will solve a problem according to their collective expertise.

Sociological Perspective: *Groupthink* can lead even the most qualified people to make disastrous decisions because people in groups in the United States tend to seek consensus at all costs.

(and possibly blame) is borne not only by oneself but also by the group. This seems to have happened among the American prison guards who tortured prisoners at Abu Ghraib prison: Each guard could convince himself or herself that responsibility, hence blame, was to be borne by the group as a whole. The greater the number of people in a group, the greater the tendency toward deindividuation. In other words, deindividuation is a *group size effect*. As groups get larger, trends in risk-taking are amplified.

Formal Organizations and Bureaucracies

Groups, as we have seen, are capable of greatly influencing individuals. The study of groups and their effects on the individuals represent an example of *microanalysis*, to use a concept introduced in Chapter 5. In contrast, the study of formal organizations and bureaucracies, a subject to which we now turn, represents an example of *macroanalysis*. The focus on groups drew our attention to the relatively small and less complex, whereas the focus on organizations draws our attention to the relatively large and structurally more complex.

A **formal organization** is a large secondary group, highly organized to accomplish a complex task or tasks and to achieve goals efficiently. Many of us belong to various formal organizations: work organizations, schools, and political parties, to name a few. Formal organizations are formed to accomplish particular tasks and are characterized by their relatively large size, compared with a small group such as a family or a friendship circle. Often, organizations consist of an array of other organizations. The federal government is a huge organization comprising numerous other organizations, most of which are also vast. Each organization within the federal government is also designed to accomplish specific tasks, be it collecting your taxes, educating the nation's children, or regulating the nation's transportation system and national parks.

Organizations develop routine practices that result in the production of an organizational culture. **Organizational culture** refers to the collective norms and values that shape the behavior of people within an organization; in other words, it is the environment of the organization. Organizational culture is present in any organization and involves both formal and informal norms. The culture of an organization may be reflected in certain symbols and rituals, perhaps even a certain style of dress. Organizational culture guides the behaviors of people within the organization, shaping what is perceived to be appropriate and inappropriate. Indeed, organizations appear very different depending upon their culture. Corporate organizational culture has tended to be somewhat formal and restrictive, although new and innovative companies, such as Google and Facebook, have considerably transformed these traditional organizational cultures. Even when it is informal, organizational culture guides behavior within the organization.

Organizational culture can also produce problems for organizations, as was found in the scandal at Penn State University involving the sexual abuse of young boys by assistant football coach Jerry Sandusky. The special investigative report that analyzed these incidents and made recommendations to the university specifically pointed at organizational culture as a contributing factor in the failure of Penn State University to stop Sandusky's behavior. Certainly, the individual behavior of Sandusky, now serving a prison term, is to blame, but the special counsel's report also blames the organizational culture of the university for failing to investigate when repeated reports of Sandusky's behavior came forward. The report cited the repeated failure of four university leaders (the president, the famed football coach Joe Paterno, the athletic director, and the university vice president) for participating in an organizational culture that resisted outside perspectives, had the pervasive goal of protecting the university's reputation, and had an excessive reverence for football, placing football above the protection of children and youths. An organizational culture of hierarchy and secrecy has also been identified as one of the reasons for the many coverups of childhood sexual abuse by priests in the Catholic Church.

Organizations tend to be persistent, although they are usually responsive to the broader social environment where they are located (DiMaggio and Powell 1991). Organizations are frequently under pressure to respond to changes in the society by incorporating new practices and beliefs into their structure. Business corporations, as an example, have had to respond to increasing global competition; they do so by expanding into new international markets, developing a globally focused workforce, and trimming costs by eliminating workers and various layers of management.

Organizations can be tools for innovation, depending on the organization's values and purpose. As an example, rape crisis centers originally emerged from the feminist movement at a time when there were few, if any, services for rape victims. Rape crisis centers have now changed how police departments and hospital emergency personnel respond to rape victims (Martin 2005; Schmitt and Martin 1999).

Types of Organizations

Sociologists classify formal organizations into three categories distinguished by their types of membership affiliation: normative, coercive, and utilitarian (Etzioni 1975; Blau and Scott 1974).

Normative Organizations

People join **normative organizations** to pursue goals that they consider worthwhile. They obtain personal satisfaction, but no monetary reward for membership in such an organization. In many instances, people join the normative organization for the social prestige that it offers. Many are service and charitable organizations and are often called *voluntary organizations*. They include organizations such as Kiwanis clubs, political parties, religious organizations, the National Association for the Advancement of Colored People (NAACP), B'nai B'rith, the National LGBTQ Task Force, and other similar voluntary organizations that are concerned with specific issues. Such groups have been created to meet particular needs, sometimes ones that members see as unmet by other organizations.

Gender, class, race, and ethnicity all play a role in who joins what voluntary organization. Social class is reflected in the fact that many people do not join certain organizations simply because they cannot afford to join. Membership in a professional organization, as one example, can cost hundreds of dollars each year. Those who feel disenfranchised, however, may join grassroots organizations—voluntary organizations that spring from specific local needs that people think are unmet. Tenants may form an organization to protest rent increases or lack of services, or a new political party may emerge from people's sense of alienation from existing political party organizations. African Americans, Latinos, and Native Americans have formed many of their own voluntary organizations in part because of their historical exclusion from traditional White voluntary organizations. Some of these are vibrant, ongoing organizations in their own right, such as the African American organizations Delta Sigma Theta and Alpha Kappa Alpha sororities, and the fraternities Alpha Phi Alpha, Kappa Alpha Psi, and Omega Psi Phi (Giddings 1994).

Coercive Organizations

Coercive organizations are characterized by membership that is largely involuntary. Prisons are an example of organizations that people are coerced to "join" by virtue of punishment for their crime. Similarly, mental hospitals are coercive organizations: People are placed in them, often involuntarily, for some form of psychiatric treatment. In many respects, prisons and mental hospitals are similar in their treatment of inmates or patients. They both have strong security measures such as guards, locked and barred windows, and high walls (Goffman 1961).

The sociologist Erving Goffman has described coercive organizations as total institutions. A **total institution** is an organization that is cut off from the rest of society and one in which resident individuals are subject to strict social control (Foucault 1995; Goffman 1961). Total institutions include two populations: the "inmates" and the staff. Within total institutions, the staff exercises complete power over inmates, for example, nurses over mental patients and guards over prisoners. The staff administers all the affairs of everyday life, including basic human functions such as eating and sleeping. Rigid routines are characteristic of total institutions, thus explaining the common complaint by those in hospitals that they cannot sleep because nurses repeatedly enter their rooms at night, regardless of whether the patient needs medication or treatment.

Utilitarian Organizations

The third type of organization named is **utilitarian organization**. These are large organizations, either for-profit or nonprofit, that individuals join for specific purposes, such as monetary rewards. Large business organizations that generate profits (in the case of for-profit organizations) and salaries and wages for the organization's employees (as with either for-profit or nonprofit organizations) are utilitarian organizations. Examples of large, for-profit organizations include Microsoft, Amazon, and Google. Examples of large nonprofit organizations that pay salaries to employees are colleges and universities, churches, and organizations such as the National Collegiate Athletic Association (NCAA).

Bureaucracy

As a formal organization develops, it is likely to become a **bureaucracy**, a type of formal organization characterized by an authority hierarchy, a clear division of labor, explicit rules, and impersonality. Bureaucracies are notorious for their unwieldy size and complexity as well as their reputation for being remote and cumbersome organizations that are highly impersonal and machinelike in their operation. The federal government is a good example of a cumbersome bureaucracy that many believe is ineffective because of its sheer size. Numerous other formal organizations have developed into huge bureaucracies: many universities, hospitals, and state motor vehicle registration systems, as familiar examples.

The early sociological theorist Max Weber (1947/1925) analyzed the classic characteristics of a bureaucracy. These characteristics represent what he called an **ideal type**—a model rarely seen in reality but that defines the principal characteristics of a social form. The characteristics of bureaucracies described as an ideal type are:

1. *High degree of division of labor and specialization.* The notion of the specialist embodies this criterion. Bureaucracies ideally employ specialists in the various positions and occupations, and these specialists are responsible for a specific set of duties. Sociologist Charles Perrow (2007) notes that many modern bureaucracies have hierarchical authority structures and an elaborate division of labor.
2. *Hierarchy of authority.* In bureaucracies, positions are arranged in a hierarchy so that each is under the supervision of a higher position. Such hierarchies are often represented in an *organization chart*, a diagram in the shape of a pyramid that shows the relative rank of each position plus the lines of authority between each. These lines of authority are often called the "chain of command," and they show not only who has authority, but also who is responsible to whom and how many positions are responsible to a given position.
3. *Rules and regulations.* All the activities in a bureaucracy are governed by a set of detailed rules and procedures. These rules are designed, ideally, to cover almost every possible situation and problem that might arise, including hiring, firing, salary scales, and rules for sick pay and absences.
4. *Impersonal relationships.* Social interaction in the (ideal) bureaucracy is supposed to be guided by *instrumental* criteria, such as the organization's rules, rather than by *expressive needs*, such as personal attractions or likes and dislikes. The ideal is that the objective application of rules will minimize matters such as personal favoritism—giving someone a promotion simply because you like him or her or firing someone because you do not like him or her. Of course, as we will see, sociologists have pointed out that bureaucracy has "another face"—the *informal* social interaction that actually keeps the bureaucracy working and often involves interpersonal friendships and social ties, typically among people taken for granted in these organizations, such as the support staff, traditionally consisting largely of women.

5. *Career ladders.* Candidates for the various positions in the bureaucracy are supposed to be selected on the basis of specific criteria, such as education, experience, and standardized examinations. The idea is that advancement through the organization becomes a career for the individual. Some organizations, such as some universities and some law firms, have a policy of *tenure*—a guarantee of continued employment until one's retirement from the organization, but only given after rigorous review.

6. *Efficiency.* Bureaucracies are designed to coordinate the activities of many people in pursuit of organizational goals. Ideally, all activities have been designed to maximize this efficiency. The whole system is intended to keep social-emotional relations and interactions at a minimum and instrumental interaction at a maximum.

Bureaucracy's "Other Face"

All the characteristics of Weber's "ideal type" are general defining characteristics. Rarely do actual bureaucracies meet this exact description. A bureaucracy has, in addition to the ideal characteristics of structure, an *informal structure*. This includes social interactions, even network connections, in bureaucratic settings that ignore, change, or otherwise bypass the formal structure and rules of the organization. This informal structure often develops among those who are taken for granted in organizations, such as secretaries and administrative assistants—who are most often women.

This other face is an informal culture. It has evolved over time as a reaction to the formality and impersonality of the bureaucracy. Thus, employees will sometimes "bend the rules a bit" when asked to do something more quickly than usual for someone they like and bend the rules in another direction perhaps by slowing down work or otherwise sabotaging a boss whom they dislike. As a way around the cumbersome formal communication channels within the organization, the informal network, or "grapevine" (a type of *social network*, as mentioned previously) often works better, faster, and sometimes even more accurately than the formal channels. As with any culture, the informal culture in the bureaucracy has its own norms or rules. One is not supposed to "stab friends in the back," such as by "ratting on" them to a boss or spreading a rumor about them that is intended to get them fired. Yet, just as with any norms, there is deviation from the norms, and "backstabbing" and "ratting" does happen.

Problems of Bureaucracies

In contemporary times, problems have developed that grow out of the nature of the complex bureaucracy. Two problem areas already discussed are the occurrence of risky shift in work groups and the development of groupthink. Additional problems include a tendency to *ritualism* and the potential for *alienation* on the part of those within an organization.

Ritualism

Rigid adherence to rules can produce a slavish following of them, regardless of whether it accomplishes the purpose for which the rule was originally designed. The rules become ends in themselves rather than means to an end: This is **organizational ritualism**.

A classic example of the consequences of *organizational ritualism* was the tragedy involving the space shuttle *Challenger* in 1986. Only seconds after liftoff, as hundreds watched live, the *Challenger* exploded, killing schoolteacher Christa McAuliffe and six other crewmembers. It was revealed later the failure of the essential O-ring gaskets on the solid fuel booster rockets caused the catastrophic explosion, even though the O-rings were known to become brittle at below-freezing temperatures. The managers at NASA knew about this problem with the O-rings. How could they allow the shuttle to lift off when it had been freezing at the launchpad the night before? Engineers had even warned them against the danger.

Sociologist Diane Vaughan (1996) uncovered both risky shift and organizational ritualism within the organization. The NASA insiders proceeded as if nothing was wrong when they were repeatedly faced with the evidence that something was indeed *very* wrong. They, in effect, *normalized* their own behavior so that their actions became acceptable to them, representing nothing out of the ordinary. This is an example of organizational ritualism. Unfortunately, history repeated itself on February 1, 2003, when the space shuttle *Columbia*, upon its return from space, broke up in a fiery descent into the atmosphere above Texas, killing all who were aboard. With eerie similarity to the earlier 1986 *Challenger* accident, subsequent analysis concluded that a "flawed institutional culture" and a normalization of deviance accompanying a gradual erosion of safety margins were among the causes of the *Columbia* accident (Schwartz and Wald 2003; Vaughan 1996).

You might think people would have learned from such tragedies, but tragedy struck again in 2010 when a deep-drilling oil rig in the Gulf of Mexico exploded and sank, killing eleven workers and injuring seventeen. The Deepwater Horizon rig (also known as the Macondo well) sent 210 million gallons of oil into the Gulf of Mexico, devastating Gulf shore wildlife, fisheries, and beaches. The disaster was the largest oil spill in history, at least until now.

The disaster started with two or more explosions and a subsequent fire, caused by deep drilling that brought highly charged hydrocarbons to the surface.

The 2010 explosion of the Deepwater Horizon rig in the Gulf of Mexico is an example of a disaster caused by organizational ritualism.

Frantic efforts to stop the fire and the oil spill failed. Subsequent inquiries into the cause of this horrible disaster (both in human life and environment devastation) concluded that, in addition to the scientific reasons for the explosions, human and organizational culture was to blame (Hopkins 2012). Investigators wrote that the failure to contain, control, mitigate, plan, and clean-up were "deeply rooted in a multi-decade history of organizational malfunction and shortsightedness" (Deepwater Horizon Study Group 2011: 5).

Specifically, the investigators found that people in the organization "forgot to be afraid" (Deepwater Horizon Study Group 2011: 9)—that is, *organizational ritualism* led people to underestimate the risk associated with deep water drilling. Further, management's desire to "close the competitive gap" and "improve the bottom-line" led managers away from "multiple chances to do the right things in the right ways at the right times" (Deepwater Horizon Study Group 2011: 5). As the report concluded, "The organizational causes of this disaster are deeply rooted in the histories and cultures of the offshore oil and gas industry and the governance provided by the associated public regulatory agencies" (Deepwater Horizon Study Group 2011: 9). In the end, the BP corporation pled guilty to eleven counts of manslaughter, two misdemeanors, and a felony of lying to Congress. The U.S. District Court found BP guilty *as an organization* of gross negligence and reckless conduct, and BP had to pay over $18 billion in fines, the largest corporate settlement in U.S. history to date.

In none of these accidents was a single individual at fault. The stories are not those of individual evil but rather of organizational ritualism. They are stories of rigid group conformity within an organizational setting. Organizational culture overshadows individual good judgment, creating a decrease in safety and increased risk. This is one of the hazards of organizational behavior.

Alienation

The stresses on rules and procedures within bureaucracies can result in a decrease in the overall cohesion of the organization. This often psychologically separates a person from the organization and its goals. This state of *alienation* results in increased turnover, tardiness, absenteeism, and overall dissatisfaction with an organization.

Alienation can be widespread in organizations where workers have little control over what they do, or where workers themselves are treated like machines employed on an assembly line, doing the same repetitive action for an entire work shift. Alienation is not restricted to manual labor, however. In organizations where workers are isolated from others, where they are expected only to implement rules, or where they think they have little chance of advancement, alienation can be common. As we will see, some organizations have developed new patterns of work to try to minimize worker alienation and thus enhance their productivity.

The McDonaldization of Society

Sometimes the problems and peculiarities of bureaucracy can have effects on the total society. This has been the case with what George Ritzer (2010) has called **McDonaldization**, a term coined from the well-known fast-food chain. In fact, 90 percent of U.S. children between ages 3 and 9 visit McDonald's each month! Ritzer noticed that the principles that characterize fast-food organizations are increasingly dominating more aspects of U.S. society, indeed, of societies around the world. McDonaldization refers to the increasing and ubiquitous presence of the fast-food model in most organizations that shape daily life. Work, travel, leisure, shopping, health care, politics, and even education have all become subject to McDonaldization. Each industry is based on a principle of high and efficient productivity, which translates into a highly rational social organization, with workers employed at low pay but with customers experiencing ease, convenience, and familiarity.

Ritzer argues that McDonald's has been such a successful model of business organization that other industries have adopted the same organizational characteristics, so much so that their nicknames associate them with the McDonald's chain: McPaper for *USA Today*, McChild for child-care chains like Kinder-Care, and McDoctor for the drive-in clinics that deal quickly and efficiently with minor health and dental problems.

Based in part upon Max Weber's concept of the ideal bureaucracy mentioned earlier, Ritzer identifies four dimensions of the McDonaldization process—efficiency, calculability, predictability, and control:

1. *Efficiency* means that things move from start to finish in a streamlined path. Steps in the production of a hamburger are regulated so that each hamburger is made exactly the same way—hardly characteristic of a home-cooked meal. Business can be even more efficient if the customer does the work once done by an employee. In fast-food restaurants, the claim that you can "have it your way" really means that you assemble your own sandwich or salad.

2. *Calculability* means there is an emphasis on the quantitative aspects of products sold: size, cost, and the time it takes to get the product. At McDonald's, branch managers must account for the number of cubic inches of ketchup used per day. Likewise, ice cream scoopers in chain stores measure out predetermined and exact amounts of ice cream.

3. *Predictability* is the assurance that products will be exactly the same, no matter when or where they are purchased. Eat an Egg McMuffin in New York, and it will likely taste just the same as an Egg McMuffin in Los Angeles or Paris!

4. *Control* is the primary organizational principle that lies behind McDonaldization. Behavior of the customers and workers is reduced to a series of machinelike actions. Ultimately, efficient technologies replace much of the work that humans once performed.

McDonaldization clearly brings many benefits. There is a greater availability of goods and services to a wide proportion of the population; instantaneous service and convenience to a public with less free time; predictability and familiarity in the goods bought and sold; and standardization of pricing and uniform quality of goods sold, to name a few benefits. However, this increasingly rational system of goods and services also spawns irrationalities. For example, the majority of workers at McDonald's lack full-time employment, have no worker benefits, have no control over their workplace, have no pension, and quit on average after only four or five months.

Diversity in Organizations

The hierarchical structuring of positions within organizations results in the concentration of power and influence with a few individuals at the top. Because organizations tend to reflect patterns within the broader society, this hierarchy, like that of society, is marked by inequality in race, gender, and class relations, as you can see in the box "Understanding Diversity." Although the concentration of power in organizations is incompatible with the principles of a democratic society, organizations are structured by hierarchies and discrimination is still quite pervasive, especially among the power elite.

Studies find that, even though the presence of women and people of color is growing in positions of organizational leadership, they tend to take on the same values as the dominant group, evidence once again of group conformity (Zweigenhaft and Domhoff 2006).

A classic study by Rosabeth Moss Kanter (1977) shows how the structure of organizations leads to obstacles in the advancement of groups who are underrepresented in the organization. People who are underrepresented in the organization become tokens; they feel they are under the all-too-watchful eyes

Understanding Diversity

Whitening Job Resumes

Despite the fact that racial discrimination in now against the law, it is widespread in organizations, as recently shown in a clever series of studies of job resumes. The researchers wanted to know if minority job applicants try to conceal racial clues on their resumes and whether employers discriminate based on racial clues.

The research proceeded in three stages. First, the researchers interviewed Black and Asian college students who were seeking jobs or internships. They found that 31 percent of Black students and 40 percent of Asian students "whitened" their resumes by either changing the presentation of their name to appear "white" and/or reworded and deleted experiences on their resumes that would likely identify them as Black or Asian. Obviously, these students are well aware of potential discrimination and try to position themselves more advantageously in the job market.

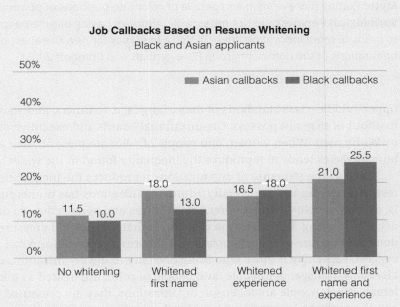

▲ **Figure 6-2 Resume Whitening.** Black and Asian applicants with "whitened resumes" (in both name and experience) were more likely to receive job callbacks than those who did not whiten their resume.

In the second phase of the research, the researchers recruited business majors to a resume writing workshop and asked them to prepare resumes for actual job postings. Researches used an experimental method to see if minority applicants were less likely to whiten their resumes if the job posting indicated that the employer was pro-diversity. Indeed, the research found that this was the case—that is, minority applicants were less likely to whiten their resumes if the employer was signaling a pro-diversity organization.

Finally, the research team used an audit study by creating fictitious Black and Asian applicants who responded to job postings. Some of the online applicants used "whitened" resumes; others did not. As you can see in ▲ Figure 6-2, Black and Asian applicants with "whitened resumes" (in both name and experience) were more likely to receive job callbacks than those who did not whiten their resume.

Further, the research design included both pro-diversity organizations and organizations that did not promote diversity. The researchers found that discrimination was just as likely in pro-diversity organizations as in others, showing how pervasive racial discrimination continues to be in work organizations.

Source: Kang, Sonia, Katherine A. DeCelles, András Tilcsik, and Sora Jun. 2016. "Whitened Resumes: Race and Self-Presentation in the Labor Market." *Administrative Science Quarterly* 61 (3):

of their superiors as well as peers. As a result—as research since Kanter's has shown—they often suffer severe stress (Smith 2007; Jackson 2000). They may be assumed to be incompetent, getting their position simply because they are women, minorities, or both—even in instances where the person had superior admissions qualifications. This is stressful for a person and shows that tokenism can have very negative consequences (Guttierez y Muhs et al. 2012).

Social class, in addition to race and gender, plays a part in determining people's place within formal organizations. Class stereotypes influence hiring practices in organizations. Personnel officers look for people with "certain demeanors," a code phrase for those who convey middle-class or upper-middle-class standards of dress, language, manners, and so on, which some people may be unable to afford or may not possess. Organizational boards and executive committees are still largely dominated by White men. When women and people of color are present, they are often tokens. Treatment in bureaucracies tends to reproduce the inequality found in the wider society.

Even as the structure of organizations reproduces the inequalities that permeate society, ample research now finds that diversity within organizations has numerous benefits. Diverse groups—that is, diverse people—bring different experiences and perspectives to organizations and to organizational decision making. Of course, the problem is that there is still pressure on such people to conform to the dominant culture of the organization. That pressure to conform can silence dissent, as groupthink would suggest, especially if those who bring diversity to the organization, such as women, gays, lesbians, bisexuals, transgender people, and people of color, are treated as tokens or silenced because they are different. When people are tokens in organizations, they are pressured not to stand out or, when they speak out, they may be ignored—or, worse, others take credit for their ideas.

Nonetheless, new research on diversity is consistently demonstrating the benefit of diversity for all kinds of organizations. In schools, all students learn more when in classrooms where there are people from different backgrounds (Gurin et al. 2002). In business organizations, racial diversity is associated with increased sales revenues, more customers, a stronger market share, and higher profits (Herring 2009). There is ample evidence that companies are now much more aware of the fact that innovation is more likely to occur in diverse work organizations (Page 2007). On college campuses, more cross-race interaction produces a more positive campus climate (Valentine et al. 2012). Studies also find that when organizations set clear expectations for how people from diverse backgrounds should interact, women and people of color are then more likely to express their ideas. Although setting such expectations within an organization leads to charges of just being "politically correct," studies document that by encouraging all people to speak up, such practices actually produce more innovation within the organization (Goncalo et al. 2015).

Functionalism, Conflict Theory, and Symbolic Interaction: Theoretical Perspectives

All three major sociological perspectives—functionalism, conflict theory, and symbolic interaction—are exhibited in the analysis of formal organizations and bureaucracies (see ◆ Table 6-1). The functional perspective, based in this case on the early writing of Max Weber, argues that certain functions characterize bureaucracies and contribute to their overall unity. The bureaucracy exists to accomplish efficiency and control, even at the cost of more personal relationships. As we have seen, however, bureaucracies develop the "other face" (informal interaction and culture, as opposed to formal or bureaucratic interaction and culture), as well as problems of ritualism and alienation of people from the organization. These latter problems are called *dysfunctions* (negative functions), which have the consequence of contributing to disunity, lack of harmony, and less efficiency in the bureaucracy. Finally, with increasing diversity in organizations, tokenism, an organizational dysfunction, may result.

The conflict perspective argues that the hierarchical or stratified nature of the bureaucracy in effect encourages rather than inhibits conflict among individuals within it. These conflicts are between superiors and subordinates, as well as between racial and ethnic groups, men and women, and people of different social class backgrounds, hampering smooth and efficient running of the bureaucracy. Furthermore,

	Functionalist Theory	Conflict Theory	Symbolic Interaction Theory
Table 6-1	**Theoretical Perspectives on Organizations**		
Central Focus	Positive functions (such as efficiency) contribute to unity and stability of the organization.	Hierarchical nature of bureaucracy encourages conflict between superiors and subordinates, men and women, and people of different racial or class backgrounds. Tokenism often results.	This theory stresses the role of self in the bureaucracy and how the self develops and changes.
Relationship of Individual to the Organization	Individuals, like parts of a machine, are only partly relevant to the operation of the organization.	Individuals are subordinated to systems of power and experience stress and alienation as a result.	Interaction between superiors and subordinates forms the structure of the organization.
Criticism	Hierarchy can result in dysfunctions such as ritualism and alienation.	This theory de-emphasizes the positive ways that organizations work.	This theory tends to downplay overall social organization.

© Cengage Learning

conflict theory helps us understand the power structures that exist in organizations—both the formal ones that come from the organizational hierarchy and the less formal ways that power is exercised between people and among groups within the organization.

Symbolic interaction theory stresses the role of the self in any group and especially how the self develops as a product of social interaction. Within organizations, people may feel that their "self" becomes subordinated to the larger structure of the organization. This is especially true in bureaucratic organizations where individuals often feel overwhelmed by the sheer complexity of working through bureaucratic structures. But symbolic interaction also emphasizes the creativity of human beings as social actors and thus would be a good perspective to use if analyzing how people change organizational structures and cultures.

Chapter Summary

What are the types of groups?

Groups are a fact of human existence and permeate virtually every facet of our lives. Group size is important, and determines quite a bit, as does the otherwise simple distinction between dyads and triads. *Primary groups* form the basic building blocks of social interaction in society. *Reference groups* play a major role in forming our attitudes and life goals, as do our relationships with in-groups and out-groups. *Social networks* partly determine things such as who we know and the kinds of jobs we get. Networks based on race or ethnicity, social class, and other social factors are extremely closely connected and are very dense. Network research has shown that network sampling (for example, sampling pairs of individuals) predicts

who is at risk for AIDS/HIV better than does traditional survey sampling.

How strong is social influence?

The social influence groups exert on us is tremendous, as seen by the Asch conformity experiments. The Milgram experiments demonstrated that the interpersonal influence of an authority figure can cause an individual to act against his or her deep convictions. The torture and abuse of Iraqi prisoners of war by American soldiers/prison guards serves as testimony to the powerful effects of both social influence and authority structures. The Iraqi tortures were in effect experimentally predicted by a simulated prison study done in the United States over thirty years earlier.

What is the importance of groupthink and risky shift?

Groupthink can be so pervasive that it adversely affects group decision making and often results in group decisions that by any measure are simply stupid. *Risky shift* (and *polarization shift*) similarly often compel individuals to reach decisions that are at odds with their better judgment.

What are the types of formal organizations and bureaucracies, and what are some of their problems?

There are several types of *formal organizations*, such as *normative*, *coercive*, or *utilitarian*. Weber typified *bureaucracies* as organizations with an efficient division of labor, an authority hierarchy, rules, impersonal relationships, and career ladders. Bureaucratic rigidities often result in organizational problems such as ritualism and resulting "normalization of deviance." The *McDonaldization* of society has resulted in greater efficiency, calculability, and control in many industries, probably at the expense of some individual creativity. Formal organizations perpetuate society's inequalities on the basis of race or ethnicity, gender, and social class. Current research finds, however, that innovation in organizations is more likely if there is greater diversity—and thus a variety of perspectives—within the organization.

What do functional, conflict, and symbolic interaction theories say about organizations?

Functional, conflict, and symbolic interaction theories highlight and clarify the analysis of organizations by specifying both organizational functions and dysfunctions (functional theory); by analyzing the consequences of hierarchical, gender, race, and social class conflict in organizations (conflict theory); and, finally, by studying the importance of social interaction and integration of the self into the organization (symbolic interaction theory).

Key Terms

attribution error 131	group 127	polarization shift 138
attribution theory 131	group size effect 127	primary group 129
bureaucracy 141	groupthink 137	reference group 131
coalition 128	ideal type 141	risky shift 138
coercive organization 140	instrumental needs 129	secondary group 129
deindividuation 138	McDonaldization 143	social network 132
dyad 128	normative organization 140	total institution 140
expressive needs 129	organizational culture 139	triad 128
formal organization 139	organizational ritualism 142	utilitarian organization 141

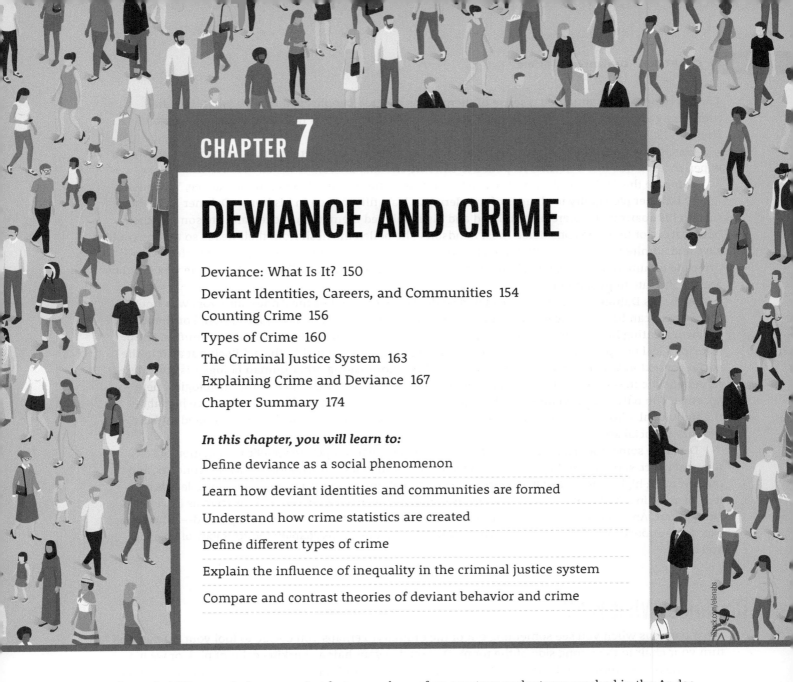

CHAPTER 7

DEVIANCE AND CRIME

In this chapter, you will learn to:

Define deviance as a social phenomenon

Learn how deviant identities and communities are formed

Understand how crime statistics are created

Define different types of crime

Explain the influence of inequality in the criminal justice system

Compare and contrast theories of deviant behavior and crime

In the early 1970s, an airplane carrying forty members of an amateur rugby team crashed in the Andes Mountains in South America. The twenty-seven survivors were stranded at 12,000 feet in freezing weather and deep snow. There was no food except for a small amount of chocolate and some wine. A few days after the crash, the group heard on a small transistor radio that the search for them had been called off.

Scattered in the snow were the frozen bodies of dead passengers. Preserved by the freezing weather, these bodies became, after a time, sources of food. At first, the survivors were repulsed by the idea of eating human flesh, but as the days wore on, they agonized over the decision about whether to eat the dead crash victims, eventually concluding that they had to eat if they were to live.

In the beginning, only a few ate the human meat, but soon the others began to eat too. The group experimented with preparations as they tried different parts of the body. They developed elaborate rules (social norms) about how, what, and whom they would eat.

After two months, the group sent out an expedition of three survivors to find help. The group was rescued, and the world learned of their ordeal. Their cannibalism (the eating of other human beings) was accepted as something they had to do to survive. Although people might have been repulsed by the story, the survivors' behavior was understood as a necessary adaptation to their life-threatening circumstances.

The survivors also maintained a sense of themselves as good people even though what they did profoundly violated ordinary standards of socially acceptable behavior in most cultures in the world (Henslin 1993; Read 1974).

Was the behavior of the Andes crash survivors socially deviant? Were the people made crazy by their experience, or was this a normal response to extreme circumstances?

Compare the Andes crash to another case of human cannibalism. In 1991, in Milwaukee, Wisconsin, Jeffrey Dahmer pled guilty to charges of murdering at least fifteen men in his home. Dahmer lured the men to his apartment, where he murdered and dismembered them, then cooked and ate some of their body parts. For those he considered most handsome, he boiled the flesh from their heads so that he could save and admire their skulls. Dahmer was seen as a total social deviant, someone who violated every principle of human decency. Even hardened criminals were disgusted by Dahmer. In fact, he was killed by another inmate in prison in 1994.

Why was Dahmer's behavior considered so deviant when that of the Andes survivors was not? The answer can be found by looking at the situation in which these behaviors occurred. For the Andes survivors, eating human flesh was essential for survival. For Dahmer, however, it was murder. From a sociological perspective, the deviance of cannibalism resides not just in the act itself but also in the social context in which it occurs. The exact same behavior—eating other human beings—is considered reprehensible in one context and acceptable in another. That is the essence of the sociological perspective. The nature of deviance is not simply about the deviant act itself, nor is deviance just about the individual who engages in the behavior. Instead, social deviance is socially constructed and is a product of social structure.

Deviance sometimes takes the form of crime—a subject that compels the public imagination. All too frequent police shootings of Black men and women; mass shootings in schools, nightclubs, concerts, and other venues; high rates of sexual violence even on our nation's college campuses; and, horrible acts of terrorism against innocent people: These events dominate the news. They can make it seem as if society is falling apart. And, yet, even the most horrid examples of crime and violence can be explained—although not excused—through sociological analysis. This is what we explore in this chapter: the sociology of deviance and crime.

Deviance: What Is It?

What happens when you see something out of the ordinary? Perhaps you decide to fool your teacher and turn your chair to the wrong side of the room. Or, you dye your hair a shocking color of pink or wear a tutu to class even though you are not a ballet dancer and it is not Halloween. These and countless other examples are acts of social deviance—that is, behaviors that violate the norms of everyday life.

See for Yourself

Perform an experiment by doing something mildly deviant for a period, such as carrying around a teddy bear doll and treating it as a live baby, or standing in the street and looking into the air, as though you are looking at something up there. Make a record of how others respond to you, and then ask yourself how labeling theory is important to the study of deviance.

How might reactions to you differ had you been of another race or gender? You might want to structure this question into your experiment by teaming up with a classmate of another race or gender. You could then compare each of your responses to the same behavior. A note of caution: Do not do anything illegal or dangerous. Even the most seemingly harmless acts of deviance can generate strong reactions, so be careful in planning your sociological exercise!

When someone sees such behavior, they likely think, "Oh, that's crazy!" Or, "That person must be sick or mentally ill." These are common responses to behavior that is bizarre or highly unusual. A sociological perspective, however, is different. Sociologists look beyond such individualistic explanations, thinking instead about the social meaning of deviant behavior, examining the context in which deviant behavior occurs, and investigating the social factors that influence such behavior.

More Than Individual Behavior

Perhaps because it violates social conventions or because it sometimes involves unusual behavior, deviance captures the public imagination. Commonly, however, the public understands deviance as the result of individualistic or personality factors. Many people see deviants as crazy, threatening, "sick," or in some other way inferior, but sociologists see deviance as influenced by society—the same social processes and institutions that shape all social behavior.

Deviance, for example, is not necessarily irrational or "sick" and may be a positive and rational adaptation to a situation. Think of the Andes survivors discussed in this chapter's opener. Was their action (eating human flesh) irrational, or was it an inventive and rational response to a dreadful situation?

To use another example, marijuana use, especially by smoking a "joint" or inhaling through a "bong," although legal in some states, is generally considered deviant behavior. Is marijuana use deviant or are some people using medical marijuana as a rational response to personal circumstances, such as illness? When would it be considered deviance and when not? Using medical marijuana to help with the pain and nausea from cancer treatments is a rational choice and blurs the lines of deviant and "normative" behavior. In fact, almost two-thirds (61 percent) of the U.S. public favor legalizing marijuana; even more approve of marijuana use if prescribed by a doctor (Geiger 2018; Pew 2010). In other words, the *use* of marijuana is only defined as deviant in the context of how it is used.

The point is that deviant behavior—or, **deviance**—is behavior that violates expected rules and norms but, more than that, deviant behavior has to be recognized as such. Thus, deviance is more than simple nonconformity. Most of us may "break the rules" now and again and, much of the time, no one notices or cares. But when people respond to unusual behavior by seeing it as deviance—either informally or formally—then we see the social reality of deviance.

Sociologists distinguish two types of deviance: formal and informal. *Formal deviance* is behavior that breaks laws or official rules. Various kinds of crime—theft, murder, hate crime, and terrorism—are examples. There are formal sanctions against formal deviance, including fines, perhaps imprisonment, or, at the extreme, the death penalty.

Informal deviance is behavior that violates customary norms, such as showing up quite drunk for a solemn ceremony or dressing too casually for a formal occasion. Although such acts of deviance are not specified by formal rules or laws, they are quite likely to be judged as deviant by others in attendance. People may try to uphold standard norms through ridicule, gossip, or even expulsion from the event. Sanctions against informal deviance may be as mild as ridicule or social isolation, but can be extreme, such as when bullying results in a criminal assault.

There are six key characteristics of deviance:

- Deviance emerges in a social context, not just in the behavior of individuals.
- The societal reaction to certain behaviors, not just the behavior itself, is what defines deviance.
- What is recognized as deviance can change over time.
- Deviance does not just violate social norms—it also reinforces them.
- The understanding of something as deviant is often the result of social movements that organize to identify behavior as deviant.
- There is a tendency to medicalize deviance—that is, define it as illness or "sick" behavior.

Let us examine each of these points in turn.

Social Context Matters

Even the most unconventional behavior can be understood if we know the context in which it occurs. Consider suicide. Are all people who commit suicide considered deviant? Might their behavior instead be explained by the social context? Think about it. There are conditions under which suicide may well be acceptable behavior—for example, when someone who voluntarily receives a lethal dose of medicine to end her life in the face of a terminal illness, compared to a despondent person who jumps from a window. Similarly, someone taking prescribed opioid drugs following a surgical procedure is considered "normal," but using opioids outside of medical supervision is defined as deviant (even if the behavior is widespread). In this context, opioid use is a form of deviance—an addictive and illegal activity.

Simply put, behavior that is deviant in one circumstance may be normal in another, or behavior may be ruled deviant only when performed by certain people. For example, whether someone who drinks is judged to be an alcoholic depends in large part on the social context in which one drinks, not solely on the amount of alcohol consumed. Drinking wine from a bottle in a brown bag on the street corner is considered highly deviant. Having martinis in a posh bar is perfectly acceptable among adults. The act of drinking alcohol is not intrinsically deviant.

In some contexts, even actions that are illegal may be commonplace and not defined as deviant. Some forms of deviance may be formally outlawed in some contexts yet perfectly ordinary in others. Gambling, for example, is illegal in most states, yet March Madness basketball pools and Sunday afternoon football bets among friends are common and are hardly considered to be a form of deviance—even though the behavior breaks the law. Among certain groups, such behavior may even be considered normative.

In some situations, deviant behavior may even be encouraged and praised. Have you ever been egged on by friends to do something that you thought was deviant, or have you done something you knew was wrong? Many argue that the reason so many college students drink excessively is that the student subculture encourages them to do so—even though students know it is harmful. Research shows that students generally perceive more drinking on campus than is actually occurring (Litt et al. 2012). The perception itself encourages people to drink, showing that perception—not just the reality—shapes deviant behavior.

The Societal Reaction to Deviance

Not only does deviance depend on context: Deviance depends on people reacting to specific behaviors (Becker 1963). The sociological study of deviance thus includes both the study of why people violate laws or norms and the study of how society reacts. Often, the most interesting part of studying deviance is not the behavior itself, but how people react to particular behaviors. Being drunk on a college campus may seem perfectly ordinary—perhaps even expected among some groups. African American students, though, are known to drink far less than White students—very likely because they perceive more risk to themselves if others see them as drunk. This is not without cause: Black students, if drunk, are far more likely to be punished than are White students. Among some groups on campus getting drunk may even be part of a ritual—such as "becoming a brother" in a fraternity. In this case, heavy drinking may even become part of an initiation ritual—a ritual that can be deadly.

Different groups might also judge the same behaviors differently. What is deviant to one group may be normative to another. People in older generations might, for example, consider tattooing and body piercing "weird" or even disgusting. Younger generations find it downright ordinary.

In other words, deviance lies not just in behavior itself but also is found in the social responses to behavior. Who is engaged in the behavior, who judges it, who punishes it and how—all of these factors are important to sociological thinking. Research has shown how even unconventional behavior is often understandable in in terms of group processes and judgments.

Debunking Society's Myths

Myth: Deviance is dysfunctional for society because it disrupts normal life.

Sociological Perspective: Deviance tends to stabilize society. By defining some forms of behavior as deviant, people are affirming the social norms of groups. In this sense, society actually *creates* deviance to some extent (Durkheim 1951/1897).

Definitions of Deviance Change Over Time

The definition of deviance can also change over time. For example, tattooing was once considered something only to be found among working-class people. Never would someone of higher status have considered covering their body with elaborate designs. Now not only is such behavior no longer deviant—it is now a mark of contemporary style and is rarely judged as somehow "lower class" (Irwin 2001).

How deviance is defined also has important social consequences. Sexual assault is a good example. Not that long ago, acquaintance rape (a form of sexual assault) was not considered social deviance. Prevailing sexual norms meant that men were expected to "seduce" women through aggressive sexual behavior, and women who said "no" were presumed to mean yes. Such beliefs continue to influence how sexual assault is perceived and prosecuted. Even now, prosecuting an offender is difficult if others do not think of a sexual assault as rape, especially under certain circumstances, such as if the woman had been drinking (Hayes, Abbott, and Cook 2016). Mobilization against sexual assault has, however, changed much about how such crimes are considered and prosecuted, including changes in how laws now define sexual assault. But perceptions persist such that somehow women who report such assaults are questioned, perhaps even vilified.

Deviance Reinstates Social Norms

No doubt, deviance violates the ordinary norms of social life, but a sociological perspective shows you that deviance also upholds social norms. How so? Deviant behavior is typically punished either through informal admonishment or, in many cases, through the administration of law. Especially when the deviance is extreme, such as a murder, punishment is typically public—either through the courts or through some other form of public display, such as a public execution.

The sociologist Emile Durkheim argued that one reason acts of deviance are publicly punished is that the social order is threatened by the deviant behavior. Judging those behaviors as deviant and punishing them confirms general social standards. Therein lies the value of widely publicized trials and public executions. Why, for example, were hangings in the Old West publicly viewed? And, why today, does administering the death penalty require public witnesses? Durkheim would say that the punishment affirms the collective beliefs of the society—that is, what he called *collective consciousness*. Public punishment of deviance, especially through public rituals, reinforces the social order, presumably to inhibit future deviant behavior.

The Influence of Social Movements

The perception of deviance is quite often influenced by *social movements*, which are networks of groups that organize to support or resist changes in society (see Chapter 16). Social movements emerge when various interest groups organize to call attention to a particular social problem. The result is often public recognition of a problem as social deviance. Even formerly acceptable behaviors may be newly defined as deviant. Cigarette smoking, for instance, was once considered glamorous, sexy, and "cool." The social climate toward smoking, however, has changed. In 1987, only 17 percent of people thought that smoking should be banned in public places. Now, over half of the public (56 percent) support a ban on smoking cigarettes in public spaces (Riffkin 2014). The increase in public disapproval of smoking results as much from social and political movements as it does from the known health risks.

Smoking for recreational purposes and using pipes or bongs is considered more deviant than using marijuana for medicinal purposes.

Thinking Sociologically

Ask people what causes domestic violence. How do they explain it? Is there evidence that the *medicalization of deviance* influences people's answers?

The antismoking movement was successful in articulating to the public that smoking is dangerous. The ability of people to mobilize against something, in this case cigarette smoking, is just as important to creating the context for deviance as is any evidence of risk. A similar movement has development regarding opioid addition. Opioids have long been known to be harmful when used without medical supervision. In fact, opioid addiction peaked in about 2010 and has been declining since, but the mobilization of groups identifying the problem of opioid addiction has been increasing. Thus, misuse of opioids is now widely known as an addiction and calls for intervention have increased (Dart et al. 2015). In other words, there has to be a social response for deviance to be defined as such; scientific evidence of harm in and of itself is not enough.

The Medicalization of Deviance

Finally, commonplace understandings of deviance often rely on what sociologists call the **medicalization of deviance** (Conrad and Schneider 1992). Commonly, people will say that someone who commits a very deviant act is "sick." Medicalizing deviance attributes deviant behavior to a "sick" state of mind, where the solution is to "cure" the deviance through therapy or other psychological treatment.

As an example, some evidence indicates that alcoholism may have a genetic basis, and certainly alcoholism must be understood at least in part in medical terms. However, understanding alcoholism *solely* from a medical perspective ignores the social causes that influence the development and persistence of this behavior. Practitioners know that medical treatment alone does not solve the problem. The social relationships, social conditions, and social habits of those with alcoholism must be altered, or the behavior is likely to recur.

Sociologists criticize the medicalization of deviance for ignoring the effects of social structures on the development of deviant behavior. Deviance, to most sociologists, is not a pathological state but an adaptation to the social structures in which people live. Factors such as family background, social class, racial inequality, and the social structure of gender relations in society produce deviance. These factors must be considered to explain it.

The point is that deviance is both created and defined within a social context. It is not just weird, pathological, or irrational behavior. Sociologists who study deviance understand it in the context of social relationships and society. They define deviance in terms of existing social norms and the social judgments people make about one another—deviance is socially constructed. Indeed, deviant behavior can sometimes be indicative of changes that are taking place in the cultural folkways, as we have noted regarding tattoos.

In sum, from a sociological perspective, deviance originates in society, not just in psychological or medical conditions. Changing the incidence of deviant behavior requires changes in society, not just changes in individuals, even though those are also important. A sociological perspective on deviance asks: Why is deviance more common in some groups than others? Why are some more likely to be labeled deviant than others, even if they engage in the exact same behavior? How is deviance related to patterns of inequality in society? Sociologists do not ignore individual psychology or medical conditions, but they focus on the social conditions surrounding the behavior, going beyond explanations of deviance that root it in the individual personality.

Deviant Identities, Careers, and Communities

Just as socialization results in the formation of individual identities, so can deviant behavior produce a definition of self as deviant—that is, a deviant identity. **Deviant identity** is the definition a person has of himself or herself as a deviant. Most often, deviant identities emerge over time (Simon 2011; Lemert 1972).

A person addicted to drugs, for example, may not think of herself as a junkie until she realizes she no longer has non-using friends. The formation of a deviant identity, like other identities, involves a process of social transformation in which a new self-image and new public definition of a person emerge. This is a process that involves how people view deviants and how deviants view themselves, as we will see in the discussion of labeling theory.

The Sociology of Stigma

A social **stigma** is an attribute that is socially devalued and discredited. Some stigmas result in people being labeled deviant—that is, perhaps carrying a deviant identity even against their will. The experiences of people who are disabled, disfigured, or in some other way stigmatized, show how such identities are formed by the social reactions of others. People with stigmas are likely to be stereotyped and defined only in terms of their presumed deviance.

Think, for example, of how someone in a wheelchair is treated in society. Their disability can become a **master status** (see Chapter 5), a characteristic of a person that overrides all other features of the person's identity (Goffman 1963). Physical disability can become a master status when other people see the disability as the defining feature of the person. People with a particular stigma are often all seen to be alike. This may explain why stigmatized individuals of high visibility are often expected to represent the whole group.

People who suddenly become disabled often have the alarming experience of their new master status rapidly erasing their former identity. People they know may treat and see them differently. A master status may also prevent people from seeing other parts of a person. A person with a disability may be assumed to have no meaningful sex life, for example, even if the disability is unrelated to sexual ability or desire. Sociologists have argued that the negative judgments about people with stigmas tend to confirm the "usualness" of others (Goffman 1963). For example, when welfare recipients are stigmatized as lazy and undeserving of social support, others are indirectly promoted as industrious and deserving. Stigmatized individuals are thus measured against a presumed norm and may be labeled, stereotyped, and discriminated against.

Sometimes, people with stigmas bond with others, maybe even strangers. This can involve an acknowledgment of kinship or affiliation that can be as subtle as an understanding look, a greeting that makes a connection between two people, or a favor extended to a stranger whom the person sees as sharing the presumed stigma. Public exchanges are common between various groups that share certain forms of disadvantage, such as people with disabilities, lesbians and gays, or members of other minority groups.

Deviant Careers

In the ordinary context of work, a career is the sequence of movements a person makes through different positions in an occupational system (Becker 1963). A **deviant career**—a direct outgrowth of the labeling process—is the sequence of movements people make through a particular subculture of deviance. Deviant careers can be studied sociologically, like any other career. Within deviant careers, people are socialized into new "occupational" roles, and are encouraged, both materially and psychologically, to engage in deviant behavior.

The concept of a deviant career emphasizes that there is a progression through deviance: Deviants are recruited, given or denied rewards, and promoted or demoted. As with legitimate careers, deviant careers involve an evolution in the person's identity, values, and commitment over time. Deviants, like other careerists, may have to demonstrate their commitment to the career to their superiors, perhaps by passing certain tests, such as when a gang expects new members to commit a crime, perhaps even to shoot someone.

Within deviant careers, rites of passage may bring increased social status among peers. Punishments administered by the authorities may even become badges of honor within a deviant community. Similarly, labeling a teenager as a "bad kid" for poor behavior in school may actually encourage the behavior to continue because the juvenile may take this as a sign of success as a deviant.

Jim West/Alamy Stock Photo

Some deviance develops within deviant communities, such as the neo-Nazis/"skinheads" shown marching here. Such right-wing extremist groups have increased significantly in recent years, as monitored by the Southern Poverty Law Center.

Deviant Communities

The preceding discussion continues to indicate an important sociological point: Deviant behavior is not just the behavior of maladjusted individuals. Deviance takes place within a group context and involves group responses. Some groups are actually organized around particular forms of social deviance; these are called **deviant communities** (Mizruchi 1983; Blumer 1969; Erikson 1966; Becker 1963).

Like subcultures and countercultures, deviant communities maintain their own values, norms, and rewards for deviant behavior. Joining a deviant community closes one off from conventional society and tends to solidify deviant careers because the deviant individual receives rewards and status from the in-group. Disapproval from the out-group may only enhance one's status within. Deviant communities also create a worldview that solidifies the deviant identity of their members. They may develop symbolic systems such as emblems, forms of dress, publications, and other symbols that promote their identity as a deviant group. Gangs wear their "colors," and skinheads have their insignia and music. Both are examples of deviant communities. Ironically, subcultural norms and values reinforce the deviant label both inside and outside the deviant group, thereby reinforcing the deviant behavior.

Some deviant communities are organized specifically to provide support to those in presumed deviant categories. Groups such as Alcoholics Anonymous, Weight Watchers, and various twelve-step programs help those identified as deviant overcome their deviant behavior. These groups, which can be quite effective, accomplish their mission by encouraging members to accept their deviant identity as the first step to recovery.

Counting Crime

Crime is one form of deviance, specifically, behavior that violates particular criminal laws. Not all deviance is crime. Deviance becomes crime when institutions of society designate it as violating a law or laws. **Criminology** is the study of crime from a social scientific perspective.

Popular accounts tend to see crime as rampant in society—and increasing at an alarming rate. One would certainly think that crime is widespread from viewing the media. Images of violent crime abound and give the impression that crime is a constant threat, can affect anyone, and is on the rise. Such sentiments, however, are rarely informed by careful knowledge about actual crime rates. Is crime increasing in the United States?

Data about crime come from two primary sources: the Federal Bureau of Investigation's (FBI) *Uniform Crime Reporting Program* and the National Victimization Surveys provided through the Department of Justice. The Uniform Crime Reporting Program publishes annual data (available online), known informally as the Uniform Crime Reports (UCR). The *Uniform Crime Reports* and are the basis for official reports about the extent of crime and its rise and fall over time. Data from the Uniform Crime Reporting Program are based on reports from police departments across the nation.

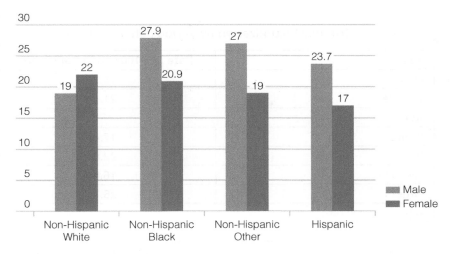

▲ **Figure 7-1   Victimization by Violent Crime by Race and Gender, 2016 (rate per 1,000 persons)**

Note: Non-Hispanic Other includes persons who report their culture or origin as something other than Hispanic without identifying race (for example, as White or Black).

Source: Bureau of Justice Statistics. "Victimization by Violent Crime, 2016." Generated using the NCVS Victimization Analysis Tool at **www.bjs.gov**

The *National Crime Victimization Surveys* are also published regularly by the Bureau of Justice Statistics in the U.S. Department of Justice. These data are based on surveys in which nationally representative samples of people are periodically asked if they have been the victims of one or more criminal acts. Different from Uniform Crime Reporting, the National Victimization Surveys are based on self-reports of crime victimization. These surveys consistently show that the likelihood of being a victim of crime is influenced by one's race, class, and gender, as you can see in ▲ Figure 7-1 and in ◆ Table 7-1. Class also predicts who most likely will be victimized by crime, with those at the highest ends of the socioeconomic scale least likely to be victims of violent crime (Barak, Leighton, and Cotton 2015). Women are somewhat less likely than men to be victimized by crime, with the exception of gender-based crimes, although this varies significantly by race and age. Black women are more likely than White women to be victims of assault; young Black women are especially vulnerable. Divorced, separated, and single women are more likely than married women to be crime victims.

The Problem with Official Statistics

Official statistics on crime are important for describing the extent of crime and various patterns in the perpetration and victimization by crime. You have to be cautious, however, in relying on these official reports because crime statistics depend on who is doing the reporting. Police reporting, for example, is highly subject to bias and even self-reports may not reveal the true extent of crime. Reported rates of crime are, like crime itself, the product of social behavior.

Official statistics are produced by people in various agencies (police, courts, and other bureaucratic organizations). These people define, classify, and record certain behaviors as falling into the category of crime—or not. Official rates of crime (and deviance) do not necessarily reflect the actual commission of crime; instead, the official rates reflect social judgments. Both of these sources of data—the *Uniform Crime Reports* and the *National Crime Victimization Surveys*—are subject to the problem of underreporting. About half to two-thirds of all crimes may not be reported to police, meaning that much crime never shows up in the official statistics. Rape is particularly known to be vastly underreported. Victims may be reluctant to report for a variety of reasons, including that the police will not take the rape seriously, especially if the assailant was known to the victim. Also, the victim may not want to undergo the continued stress of an investigation and trial.

Table 7-1	Criminal Victimization by Various Characteristics
Sex	**Rate per 100,000 in Population**
Male	15.9
Female	21.1
Race	
White	17.4
Black	22.6
Hispanic	16.8
Other	25.7
Age	
12–17	31.3
18–24	25.1
25–34	21.8
35–49	22.6
50–64	14.2
65 or older	5.2
Marital Status	
Never married	26.2
Married	9.9
Widowed	8.5
Divorced	35.3
Separated	39.5
Household Income	
$ 9,999 or less	39.2
$10,000–14,999	27.7
$15,000–24,999	25.9
$25,000–34,999	16.3
$35,000–49,999	20.5
$50,000–74,999	16.3
$75,000 or more	12.8

Source: Truman, Jennifer L., and Rachel E. Morgan. 2016. *Criminal Victimization, 2015*. Washington, DC: Bureau of Justice Statistics. **www.bjs.gov**

In an interesting example, in the aftermath of the terrorist attacks on the World Trade Center in 2001, officials debated whether to count the deaths of thousands as murder or as a separate category of terrorism. The decision would change the official murder rate in New York City that year. In the end, these deaths were not counted in the murder rate. That is a unique example, but an ongoing example is the official reporting of rape. Research finds that the police are less likely to "count" some rapes, such as those in which the victim is a prostitute, was drunk at the time of the assault, or had a previous relationship with the assailant. Rapes resulting in the victim's death are classified as homicides and thus do not appear in the official statistics on rape.

Is Crime Increasing

While headlines expose high murder rates in some of the nation's cities, crime statistics show that, despite public beliefs, crime has significantly declined since the 1990s, violent crime in particular (see ▲ Figure 7-2). For example, the murder rate in 1996 was 7.4 murders per 100,000 in the population; by

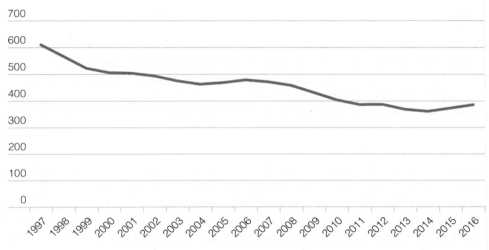

▲ **Figure 7-2   Victimization by Violent Crime, 1993–2016**

Note: Rate per 1,000 persons.

Source: Federal Bureau of Investigation. 2017. Table 1—Crime in the United States, by Volume and Rate per 100,000 Inhabitants, 1997–2016. Washington, DC: Federal Bureau of Investigation. **www.fbi.gov**

2016, the rate was 5.4. You must use caution in interpreting various crime statistics. For example, in 2015 it was widely reported that the violent crime rate (which includes murder, rape, robbery, and aggravated assault) had increased by 4 percent in one year. Although that is technically true, these data include only a one-year change. If you take a longer-range view, you will see that the violent crime rate has actually decreased, including through 2016—the most recent year that the data are available. Property crime has also decreased. Flashy media reports about murder in big cities can exaggerate the actual extent of crime. Not surprisingly, the cities with the most violent crime are also those with poverty rates that exceed 20 percent of the population (Federal Bureau of Investigation 2016).

A widespread belief about crime is that immigration is producing a rise in the crime rate. This is blatantly false. As immigration has increased in recent years, the crime rate has actually fallen. Serious studies in criminology find that recent immigrants are far less likely to commit crime than native-born people or second- and third-generation immigrants. In fact, immigrants are more likely to be victims of crime than to commit crime. Immigrant communities tend to have strong social bonds—a social structural fact that social scientists know is a strong deterrent to crime (Martinez 2014; Ousey and Kubrin 2014; Rumbaut and Ewing 2007).

Recently, women's participation in crime has been increasing, the result of several factors. Women are now more likely to be employed in jobs that present opportunities for crimes, such as property theft, embezzlement, and fraud. Violent crime by women has also increased notably since the early 1980s, possibly because the images that women have of themselves are changing, making new behaviors possible. Most significant, crime by women is related to their continuing disadvantaged status in society. Just as crime is linked to socioeconomic status for men, so it is for women (Barak, Leighton, and Cotton 2015).

Debunking Society's Myths

Myth: Immigration increases the crime rate in society.

Sociological Perspective: Recent immigrants are actually much less likely to commit crime than native-born and second and third generation immigrants (Rumbaut and Ewing 2007).

Types of Crime

When people think of crime, they may imagine a stereotypical criminal—someone who is a stranger, someone who randomly assaults you, or someone who commits a quick street crime, like a mugging. Stereotypes about crime, however, hide the many different kinds of crime committed—and the characteristics of those who commit them. The different types of crime reveal various social patterns in the commission of crime and victimization by crime, little of which is random as the stereotype suggests.

Personal and Property Crimes

The *Uniform Crime Reports* detail something called the *crime index*. The crime index includes the violent crimes of murder, manslaughter, rape, robbery, and aggravated assault, plus property crimes of burglary, larceny theft, and motor vehicle theft. The crime index includes both *personal crimes* (violent or nonviolent crimes directed against people, including murder, aggravated assault, forcible rape, and robbery) and *property crimes* (those involving theft of property without threat of bodily harm, such as burglary, larceny, auto theft, and arson). Property crimes are the most frequent criminal infractions.

Hate Crimes

Hate crime is a relatively new category of crime in the federally reported data, although hate crimes have certainly been committed throughout the nation's history. Lynching, vandalism of synagogues, and assaults on gay people are not new, but the formal reporting of hate crime did not begin until 1980. Now the U.S. Congress, via the FBI, defines **hate crime** as a criminal offense that is motivated in whole or part by bias against a "race, religion, disability, ethnic origin, or sexual orientation" (www.fbi.gov). This form of crime has been increasing in recent years, especially against gays and lesbians and against Jewish and Muslim Americans. Some of the increase is due to the ability to report and track such heinous acts, but there is also evidence that hate crimes have increased since the 2016 presidential election (Potok 2017). The vast majority of hate crimes are committed by White offenders—or, in many cases, unknown offenders. More than half of all reported hate crimes are committed against people because of race or ethnicity; 20 percent are based on sexual orientation of the victim; and another 20 percent are based on religion (Federal Bureau of Investigation 2015).

A Sociological Eye on the Media

Images of Violent Crime

The media routinely drive home two points to the consumer: Violent crime is always high and may be increasing over time, and there is much random violence constantly around us. The media bombard us with stories of "road rage," leading us to think that road rage is extensive and completely random.

No doubt there are occasions when victims are indeed picked at random. But the statistical rule of randomness cannot explain what has come to be called random violence, a vision of chaos and unpredictability that is advanced by the media. If randomness truly ruled, then each of us would have an equal chance of being a victim—and of being a criminal. This is assuredly not the case. The notion of random violence, and the notion that it is increasing, ignores virtually everything that criminologists, psychologists, sociologists, and extensive research studies know about crime: It is highly patterned and significantly predictable, beyond sheer chance, by taking into account the social structure, social class, location, race and ethnicity, gender, age, whom one's family members are, and other such variables and forces in society that affect both criminals and victims (Best 1999; Glassner 1999).

Many suggest that the portrayal of increasing crime is simply a tool to increase viewer ratings. Criminal violence is not random, but is highly patterned and predictable.

Myth: Crime is rampant in society and might victimize you at any time.
Sociological Perspective: There are clear and predictable patterns in the commission of crime. Violent crime in particular is most likely committed by someone who knows the victim, probably well.

Human Trafficking

Human trafficking has long played a role in the national and international economy. Slavery, for example, is a pernicious example of human trafficking, but this is a crime that continues in various forms. The FBI defines *human trafficking* as compelling or coercing a person to engage in some form of labor, service, or commercial sex. Sometimes the coercion is overtly physical, but it can also be psychological and subtle, such as a pimp who recruits prostitutes into sex work by initially seeming to be a boyfriend. Undocumented immigrants are particularly prone to trafficking because they are a very vulnerable population. Children are also among some of the most vulnerable, especially when coming from war-torn regions. Estimates of the extent of human trafficking are difficult to come by, because of the hidden nature of this crime. One of the problems in getting accurate data is not only the covert nature of this crime, but also lack of uniformity in how nations tabulate known cases (U.S. Department of State 2017). Human trafficking takes different forms, including sex trafficking, forced labor, and illegal recruitment of soldiers.

Gender-Based Violence

Gender-based violence is the term used to describe the various forms of violence that are associated with unequal power relationships between men and women. Gender-based violence takes many forms, including, but not limited to rape, domestic violence, sexual abuse and incest, stalking, and more. Although both men and women can be victims of gender-based violence, it far more frequently victimizes women and girls.

For all women, victimization by rape is probably the greatest fear about crime. Although rape is the most underreported crime, even with underreporting, the FBI estimates that there are about 124,000 rapes per year in the United States—a fact that the FBI itself acknowledges is very underreported (Federal Bureau of Investigation 2016).

Acquaintance rape is that committed by an acquaintance or someone the victim has just met. The extent of acquaintance rape is difficult to measure. Recently, the nation has focused its attention on the widespread phenomenon of sexual assault on college campuses. About one in five college women and one in sixteen men are sexually assaulted while in college. Researchers find that about 90 percent of sexual assaults on campus go unreported. False reporting, on the other hand, is quite rare—estimated to be between two to ten percent of all reports, even though the media focus heavily on cases where false reporting has been found (Lonsway, Archambault, and Lisak 2009; Krebs et al. 2007; Fisher, Cullen and Turner 2000).

The vast majority of sexual assaults are committed by someone who knows the victim. Numerous studies have documented the extent to which people hold various rape myths, such as believing that a woman can be considered as consenting when she is very drunk. Excessive drinking increases one's chances of being raped during campus parties. Some campus cultures and environments are especially likely to put women at risk of rape, particularly in some all-male groups and organizations, especially those organized around hierarchy, secrecy among "brothers," and loyalty, which together create an atmosphere where rape can occur. This can help you understand why different organizations, such as some fraternities, sports teams, churches, and military schools, have high rates of rape and have come to be called "rape cultures" (Langton and Sinozich 2014; Martin and Hummer 1989).

Sociologists have argued that the causes of rape lie in women's status in society—that women are treated as sexual objects for men's pleasure. The relationship between women's status and rape is also reflected in data revealing who is most likely to become a rape victim. African American women, Latinas, and poor women have the highest likelihood of being raped, as do women who are single, divorced, or separated. Younger women aged between 18 and 34 are also more likely to be rape victims than older women (U.S. Bureau of Justice Statistics 2015). Sociologists interpret these patterns to mean that the most powerless women are also most subject to this form of violence.

Cybercrime

A new type of crime has emerged in the context of the technological revolution. **Cybercrime**, earlier known as computer crime, refers to illegal activities that take place through the use of computers. Unknown not that long ago, cybercrime includes such things as security breaches of electronic information (also known as "hacking") and identity theft where someone steals personal information (such as a Social Security number) for purposes of some kind of fraud (Levi et al. 2017; Allison 2005). Not surprisingly, individuals who use the Internet for routine activities, such as banking, email, and instant messaging, are more likely to be victims of identity theft than others. Online shopping increases risk by about 30 percent. Men, older people, and those with higher incomes are those most likely to experience victimization from identity theft (Reyns 2013).

Cybercrime can be the work of common, but sophisticated criminals, but may be committed by criminal networks, organizations, or terrorists. Cyberattacks against information systems have also become the focus of national and international politics, such as in the Russian hacking of election files during the 2016 presidential election. The existence of cybercrime has led to the development of new industries designed to counter such breaches and has led to new law enforcement techniques to stop online fraud, protect computer and information security, and detect and protect national security (Payne 2016).

Victimless Crimes

Victimless crimes are those that violate laws but where there is no complainant. Victimless crimes include various illicit activities, such as gambling, illegal drug use, and prostitution. Although there is no victim per se, there is clearly some degree of victimization in such crimes: Some researchers see prostitution, in many instances, as containing at least one victim because of the consequences for one's health, safety, and well-being through participation in such activities. Enforcement of these crimes is typically not as rigorous as enforcement of crimes against people or property.

Elite and White-Collar Crime

The term *white-collar* crime refers to criminal activities by people of high social status who commit crime in the context of their occupation (Sutherland and Cressey 1978). White-collar crime includes activities such as embezzlement (stealing funds from one's employer), involvement in illegal stock manipulations (insider trading), and a variety of violations of income tax law, including tax evasion. Until very recently, white-collar crime seldom generated great public concern, far less than concerns about street crime. In terms of total dollars, however, white-collar crime is even more consequential for society. Scandals involving prominent white-collar criminals have come to the public eye more frequently. The best recent example is the recession of 2008, which many say resulted from very risky financial practices by Wall Street traders and excessive borrowing by the nation's banks. Despite its cost, white-collar crime is generally perceived as less threatening than crimes by the poor. When white-collar criminals are prosecuted and convicted, they are also likely to receive light sentences, often in minimum security prisons.

Corporate Crime

Corporate crime is wrongdoing that occurs within the context of a formal organization or bureaucracy that is actually sanctioned by the norms and operating principles of the bureaucracy (Simon 2011). This can occur within any kind of organization—corporate, educational, governmental, or religious. Sociological studies of corporate crime show that it is embedded in the ongoing and routine activities of organizations (Ermann and Lundman 2001). Individuals within the organization may participate in the behavior with little awareness that their behavior is illegitimate. In fact, their actions are likely to be defined as in the best interests of the organization—business as usual. New members who enter the organization learn to comply with the organizational expectations or leave.

Organized Crime

The structure of crime and criminal activity in the United States often takes on an organized, almost institutional character. This is crime in the form of mob activity and racketeering, known as organized crime. *Organized crime* is crime committed by structured groups typically involving the provision of illegal goods and services to others. Organized crime syndicates are typically stereotyped as the Mafia, but the term can refer to any group that exercises control over large illegal enterprises, such as the drug trade, illegal gambling, prostitution, weapons smuggling, or money laundering. These organized crime syndicates are often based on racial, ethnic, or family ties, with different groups dominating and replacing each other in different criminal "industries" at different periods in U.S. history.

A key concept in sociological studies of organized crime is that these industries are organized along the same lines as legitimate businesses; indeed, organized crime has taken on a corporate form. There are likely to be senior partners who control the profits of the business, workers who manage and provide the labor for the business, and clients who buy the services that organized crime provides. In-depth studies of the organized crime underworld are difficult to conduct, owing to its secretive nature and dangers.

Terrorism

The FBI includes *terrorism* in its definition of crime, defining it as "the unlawful use of force or violence against persons or property to intimidate or coerce a government, the civilian population, or any segment thereof, in furtherance of political or social objectives" (Federal Bureau of Investigation 2011). Terrorism crosses national borders, and to understand it requires a global perspective. Terrorism is also linked to other forms of international crime. It is suspected that profits from international drug trade fund the terrorist organization al Qaeda.

One of the most frightening things about terrorism as a crime is that its victims, unlike most other crime, may be randomly targeted. Suicide bombers or other armed attackers may select particular groups because of their identification with the West or because they are associated with Jewish people. This is what happened in Paris in 2015 when terrorists who were possibly associated with the terrorist group ISIS (Islamic State in Iraq and Syria) attacked and slaughtered at least seventeen people in a kosher market and in the offices of a satirical magazine. But terrorism also kills and maims people at random—one of the things that makes it so frightening.

The Criminal Justice System

Whether in the police station, the courts, or prison, the factors of race, class, and gender are highly influential in the administration of justice in this society. Those in the most disadvantaged groups are more likely to be defined and identified as deviant independently of their behavior and, having encountered these systems of authority, are more likely to be detained and arrested, found guilty, and punished.

The Policing of Minorities

There is little question that minority communities are policed more heavily than White communities. For middle-class Whites, the presence of the police may be reassuring, but for African Americans and Latinos, an encounter with a police officer can be terrifying. African American parents of young boys have to routinely have "the talk" to instruct young boys in protecting themselves from the dangers a police encounter can bring, even when the child is completely innocent of any wrongdoing. This has been vividly witnessed recently as the public has seen numerous incidents of police shootings of Black

Recent protests highlight the distrust between police and African American communities, especially after young Black men are killed by police officers.

men and women. In each case, serious questions about whether the shootings were justifiable have cast doubt on the fairness of the criminal justice system. The frequency and perceived injustice of police shootings have sparked the movement #BlackLivesMatter protesting this form of police brutality. Accurate data on police shootings are hard to come by, although the *Washington Post* has recently begun compiling a list of all police shootings with additional information on the victim's race, gender, and the nature of the threat. In these data, of the 962 number of police killings in 2016, over 20 percent involved Black men or women. Hispanics accounted for 17 percent (Washington Post 2017).

Police brutality, of which killing is only the most extreme form, refers to the excessive use of force by the police. Most cases of police brutality involve minority citizens, with usually no penalty for the officers involved. Sociologists have tested several hypotheses for why this occurs, reaching two conclusions: (1) The greater the proportion of minority residents in a city, the greater the use of coercive crime control, such as police force; and (2) spatially segregated minority populations are the primary targets of coercive crime control (Smith and Holmes 2014). This research also suggests that, to protect minority citizens, we should focus on reducing residential segregation, which will require better economic opportunities for Black and Latino citizens.

Racial profiling has also come to the public's attention, although it is a practice that has a long history. Often referred to half in jest by African Americans as the offense of "DWB" or "driving while Black" (also known by Latinos as "driving while 'brown'"), **racial profiling** is the use of race alone as the criterion for deciding whether to stop and detain someone on suspicion of having committed a crime. Police officers often argue that they "have no choice," claiming that racial profiling is justified because a high proportion of Blacks and Hispanics commit crimes. Although the crime rate for Blacks and Hispanics is higher than that of Whites, race is a particularly bad basis for suspicion because the vast majority of Blacks and Hispanics, like the vast majority of Whites, do not commit any crime at all. As evidence of this, studies have found that eight out of every ten automobile searches carried out by state troopers on the New Jersey Turnpike over ten years were conducted on vehicles driven by Blacks and Hispanics; the vast majority of these searches turned up no evidence of contraband or crimes of any sort (Kocieniewski and Hanley 2000; Cole 1999). ▲ Figure 7-3 highlights that minorities feel they are treated less fairly by police officers.

Arrest and Sentencing

Arrest data show a very clear pattern along lines of race, gender, and class. A key question is whether this pattern reflects actual differences in the commission of crime by different groups or whether it reflects different treatment by the criminal justice system. The answer is both. People of color are policed more heavily than other groups, so they are more likely to be arrested for crime. Poor people are also more likely than others to be arrested for crimes. Does this mean that people of color and the poor commit more crimes? To some extent, perhaps yes, as unemployment and poverty are related to crime (Reiman and Leighton 2012). Those who are economically deprived often see no alternative to crime. But prosecution by the criminal justice system is also significantly related to patterns of race, gender, and class inequality. We see this in treatment by the police in patterns of sentencing and in studies of imprisonment.

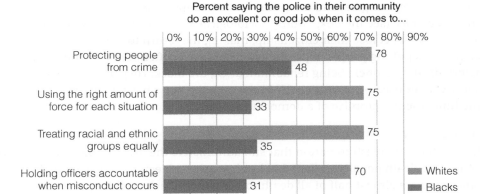

▲ Figure 7-3   A Racial Divide: Police Performance

Note: Whites and blacks include only non-Hispanics.

Source: Morin, Rich, and Renee Stepler. 2016. *The Racial Confidence Gap in Police Performance.*
Washington, DC: Pew Research Center. **www.pewresearch.org**

Racial discrimination permeates the criminal justice system. Even when convicted of the same crime as Whites, African American and Latino male defendants with the same prior arrest record as Whites are more likely to be arrested and sentenced, and they are likely to be sentenced for longer terms than White defendants (Brame et al. 2014). Bearing in mind the factors that affect the official rates of arrest and conviction—bias of official statistics, the influence of powerful individuals, discrimination in patterns of arrest, differential policing—there remains evidence that the actual commission of crime varies by race. Why? Sociologists find a compelling explanation in social structural conditions. Racial minority groups are far more likely than Whites to be poor, unemployed, and living in single-parent families. These social facts are all predictors of a higher rate of crime.

Once arrested, bail is also set higher for African Americans and Latinos than for Whites, and minorities have less success with plea bargains. And, on trial, minority defendants are more likely to be found guilty than are White defendants. At sentencing, African Americans and Hispanics get longer sentences than Whites, even when they have the same number of prior arrests and socioeconomic background as Whites. Young African American men, as well as Latinos, are sentenced more harshly than any other group, and once sentenced, they are less likely to be released on probation (Western 2014, 2007). Any number of factors influences judgments about sentencing, including race of the judge, severity of the crime, race of the victim, and the gender of the defendant, but throughout these studies, race is shown to consistently matter—and matter a lot.

Mass Incarceration

Racial minorities now account for more than half of the federal and state male prisoners in the United States. Blacks have the highest rates of imprisonment, followed by Hispanics, then Native Americans and Asians. Hispanics are the fastest-growing minority group in prison. Native Americans, though a small proportion of the prison population, are still overrepresented in prisons (Carson and Anderson 2016). In theory, the criminal justice system is supposed to be unbiased, able to objectively weigh guilt and innocence. The reality is that the criminal justice system reflects the racial and class stratification and biases in society.

The United States and Russia have the highest rate of incarceration in the world. The structure of the U.S. criminal justice system disproportionately *propels* Blacks and Hispanics into prison at a greater rate than Whites with the same criminal record. With so many people of color in prison, mass incarceration has now been called the "new Jim Crow" system (Alexander 2010).

When prisoners are released, they face new problems of re-entry in society. Many are unemployable, even when their records are clean for years after prison. Research shows that former prisoners face the stigma of imprisonment long beyond their time served. Employers are reluctant to hire them—a fact that is compounded by race (see the box, "Doing Sociological Research."). Usually denied the right to vote, former prisoners also in effect lose their citizenship, making the return to crime even more likely.

The United States is putting people in prison at a record pace, often for fairly minor drug crimes. Is crime being deterred? Are prisoners being rehabilitated? Or, are Black, Hispanic, and Native American men simply being warehoused—put on a shelf? In the end, the mass incarceration of so many citizens undermines the fundamental promise of a democratic society.

The Death Penalty and Wrongful Conviction

The United States is the only Western nation that still administers the death penalty. Race discrimination is a strong factor in the death penalty—a fact that has been the subject of intense wrangling in courts of law. African Americans are almost half of all death row inmates. Blacks and Hispanics who have already received the death penalty are even more likely to be executed, rather than being pardoned or having the execution postponed, than are Whites who have committed the same crime (Walker, Spohn, and Delone 2012; Jacobs et al. 2007). There seems to be less racial bias in death penalty decisions when the evidence is clear-cut. But, in cases that are not so clear-cut, racial disparities are higher (Baldus, cited in Barak, Leighton, and Cotton 2015: 278).

Doing Sociological Research

Race, Employment, and Prison Release

Research Question

All former inmates struggle to find work after release from prison. Sociologist Devah Pager (2007) wanted to examine whether there is additional bias against African American and Hispanic former inmates when they search for jobs.

Research Method

Pager used an *audit study* to do her research. Pre-trained role-players applied for jobs and, when interviewed, used the same preset script. The sample included Black and White role-players. Some posed as ex-cons; others posed with no criminal past on their record. The purpose of the study was to see how many of them would be invited back for another interview.

Research Results

The researchers posing as ex-cons all had trouble being invited back, but African Americans who were *not* ex-cons were *less* likely to be invited back for a job interview than were Whites who *were* ex-cons, even though White ex-cons were not invited back in large numbers. The effect of race alone actually exceeded the effect of incarceration alone. These upsetting differences could not be attributed to differences in interaction displayed during the interview, because everyone used the exact same prepared script.

Conclusions and Implications

Pager's research suggests the need for social policies such as job training programs for skilled jobs for those still in prison. Additionally, policies to combat the stigma that former prisoners, especially Black prisoners, face upon release are needed.

Questions to Consider

1. Would you support policies that would remove the requirement for job applicants to indicate if they have ever been convicted of a felony? Why or why not?
2. What policies would you suggest that would enable former prisoners of any race to find employment that would support them at a living wage?

Source: Pager, Devah. 2007. *Marked: Race, Crime, and Finding Work in an Era of Mass Incarceration*. Chicago: University of Chicago Press.

Cases of wrongful conviction also reveal deep racial inequality. In cases where subsequent DNA evidence proves the innocence of a convicted person, two-thirds are African Americans (Grimsley 2012). Eyewitness testimony is one of the main reasons for wrongful conviction. A very large portion of the cases where Black prisoners have been exonerated after serving a prison sentence are those where White eyewitnesses identified a Black person as the culprit. Death penalty cases and wrongful convictions are two more ways that racial injustice pervades the criminal justice system.

Explaining Crime and Deviance

It is easy to be fascinated by crime and deviance. Facts and data can teach you a lot, but it is just as important to be able to explain what you see. Sociologists use several theoretical frameworks to explain the many dimensions of crime and deviance, starting with understanding how, contrary to what you might think, crime and deviance actually have serve important purposes in society, as the sociological perspective of functionalism suggests.

The Functions of Crime and Deviance

Recall that functionalism is a theoretical perspective that interprets all parts of society, even those that may seem dysfunctional, as contributing to the stability of the whole. At first glance, deviance seems to be dysfunctional for society. Functionalist theorists argue otherwise (see ◆ Table 7-2). They contend that deviance is functional because it creates social cohesion. Branding certain behaviors as deviant provides a contrast with behaviors that are considered normal, giving people a heightened sense of social order. Norms are meaningless unless there is deviance from them; thus, deviance is necessary to clarify society's norms.

Group coherence then comes from sharing a common definition of legitimate, as well as deviant, behavior. The collective identity of a group is affirmed when group members ridicule or condemn others they define as deviant. Labeling someone else an "outsider" is, in other words, a way of affirming one's "insider" identity. Calling someone "gay" or using other homophobic slurs is a common practice, especially among adolescent boys. Sociological research shows that boys call others gay not because of actual sexual behavior, but because it affirms the users' sense of themselves as "masculine." In this way, homophobic slurs affirm boys' (or men's) identity as "real men" (Pascoe 2011).

Durkheim: The Study of Suicide

The functionalist perspective on deviance stems originally from the work of **Emile Durkheim** (1858–1917). One of Durkheim's central concerns was how society maintains its coherence (or social order). Durkheim saw deviance as functional for society because it produces solidarity among society's members. He developed his analysis of deviance in large part through his analysis of suicide. Through this work, he discovered a number of important sociological points. First, he criticized the usual psychological interpretations

Table 7-2	Sociological Theories of Crime and Deviance		
	Functionalist Theory	Symbolic Interaction Theory	Conflict Theory
	Deviance and crime create social cohesion.	Deviance and crime are forms of a learned behavior, reinforced through group membership.	Dominant classes control the definition of and sanctions attached to crime and deviance.
	Structural strains in society produce crime and deviance.	People can be labeled as deviant or criminal regardless of their actual behavior.	Deviance and crime result from social inequality in society.
	Crime and deviance occur when people's attachment to social bonds is diminished.	Those with the power to assign labels themselves produce deviance and identify certain groups as more likely to be criminals.	Elite and corporate crime and deviance go largely unrecognized and unpunished.

of why people commit suicide, turning instead to sociological explanations with data to back them up. Second, he emphasized the role of social structure in producing deviance. Third, he pointed to the importance of people's social attachments to society in understanding deviance. Finally, he elaborated the functionalist view that deviance provides the basis for social cohesion.

Durkheim was the first to argue that the causes of suicide were to be found in social factors, not individual personalities. Observing that the rate of suicide in a society varies with time and place, Durkheim looked for causes linked to these factors other than emotional stress. Durkheim argued that suicide rates are affected by the different social contexts in which they emerge. He looked at the degree to which people feel integrated into the structure of society and their social surroundings as social factors producing suicide.

Durkheim analyzed four types of suicide: anomic suicide, altruistic suicide, egoistic, and fatalistic suicide. **Anomie**, as defined by Durkheim, is the condition that exists when social regulations in a society break down: The controlling influences of society are no longer effective, and people exist in a state of relative normlessness. The term *anomie* refers not to an individual's state of mind, but instead to social conditions.

Anomic suicide occurs when the disintegrating forces in the society make individuals feel lost or alone. Teenage suicide is often cited as an example of anomic suicide. Feelings of depression and hopelessness can lead to suicide. The rate of suicide among returning veterans may well constitute anomic suicide, for example, if they return from war feeling as if no one understands them (Finley et al. 2015). Suicide is more likely committed by those who have been sexually abused as children and who may feel they can talk to no one (Jakubczyk et al. 2014).

Altruistic suicide occurs when there is excessive regulation of individuals by social forces. An example is someone who commits suicide for the sake of a religious or political cause. For example, after hijackers on September 11, 2001, took control of four airplanes—crashing two into the World Trade Center in New York, one into the Pentagon, and despite the intervention of passengers, one into a Pennsylvania farm field—many wondered how anyone could do such a thing, killing themselves in the process. Although sociology certainly does not excuse such behavior, it can help explain it. Terrorists and suicide bombers are so regulated by their extreme beliefs that they are willing to die and kill as many people as possible to achieve their goals. As Durkheim argued, altruistic suicide results when individuals are excessively dominated by the expectations of their social group. People who commit altruistic suicide subordinate themselves to collective expectations, even when death is the result.

Egoistic suicide occurs when people feel totally detached from society. Ordinarily, people are integrated into society by work roles, ties to family and community, and other social bonds. When these bonds are weakened through retirement, loss of family and friends, or socioeconomic hardship, the likelihood of egoistic suicide increases. Egoistic suicide is also more likely to occur among people who are not well integrated into social networks. Thus it should not be surprising that women have lower suicide rates than men (Centers for Disease Control and Prevention 2013).

Fatalistic suicide occurs when there is extreme overregulation, such as when there is total oppression in society. In such situations, people may feel that they simply have no way out of their current situation and no hope for a future. Suicide can be the result.

Durkheim's major point is that suicide is a social, not just an individual, phenomenon. Durkheim sees sociology as the discovery of the social forces that influence human behavior. As individualistic as suicide might seem, Durkheim uncovered the influence of social structure on suicide.

Applying Durkheim's Theory of Suicide: School Shootings

Durkheim's analysis of suicide can help you understand the horrific acts of mass murder rampages that have occurred in schools, movie theaters, and other places. Why would someone go into a public place, kill many people, and then shoot themselves?

Durkheim would see that there are social-structural elements that are common to school shootings and other mass murders where the perpetrator also kills himself. In the case of most school shootings, the perpetrators have been social isolates, people ridiculed by peers and without strong social networks. To Durkheim, school shootings are examples of egoistic suicide and anomic suicide, given that the shooters were socially isolated, lacked integration into society, had troubled individual histories, and wanted to "make a mark" in history by killing the largest number of individuals possible in a single attack (Newman et al. 2006).

Merton: Structural Strain Theory

The functionalist perspective on deviance has also been elaborated by the sociologist **Robert Merton** (1910–2003). Merton's **structural strain theory** traces the origins of deviance to the tensions caused by the gap between cultural goals and the means people have available to achieve those goals. Merton noted that societies are characterized by both culture and social structure. Culture establishes goals for people in society. Social structure provides, or fails to provide, the means

Durkheim would argue that public rituals, such as spontaneous memorials, help re-establish social cohesion in the aftermath of extreme deviance.

for people to achieve those goals. In a well-integrated society, according to Merton, people use accepted means to achieve the goals society establishes. In other words, the goals and means of the society are in balance. When the means are out of balance with the goals, deviance is likely to occur. According to Merton, this imbalance, or disjunction, between cultural goals and structurally available means can actually *compel* the individual into deviant behavior (see ▲ Figure 7-4; Merton 1968).

To explain further, a collective goal in U.S. society is the achievement of economic success. The legitimate means to achieve such success are education and jobs, but not all groups have equal access to those means. The result is structural strain that produces deviance. According to Merton, poor and working-class people are most likely to experience these strains because they internalize the same goals and values as the rest of society but have blocked opportunities for success. Structural strain theory therefore helps explain the high correlation that exists between unemployment and crime.

Strain between cultural goals and structurally available means can produce deviance. *Conformity* is likely to occur when the goals are accepted and the means for attaining the goals are made available to the individual by the social structure. If this does not occur, then cultural–structural strain exists, and at least one of four possible forms of deviance is likely to result: innovative deviance, ritualistic deviance, retreatism deviance, or rebellion.

Consider the case of female prostitution: The prostitute has accepted the cultural values of the dominant society—obtaining economic success. If she is poor or unable to find work, then the structural means to attain these goals are less available to her. Especially given how many women have to turn to low-wage work to support themselves and, possibly, their children, sex work may seem to be the most viable alternative.

Other forms of deviance also represent strain between goals and means. *Retreatism deviance* becomes likely when neither the goals nor the means are available. Examples of retreatism are those with severe alcoholism or people who are homeless or reclusive. *Ritualistic deviance* is illustrated in the case of college women with eating disorders, such as bulimia (purging oneself after eating). The cultural goal of extreme thinness is perceived as unattainable, even though the means for trying to attain it are plentiful, for example, good eating habits and proper diet methods. Finally, *rebellion* as a form of deviance is likely to occur when

	Cultural goals accepted?	Institutionalized means toward goal available?
Conformity	Yes	Yes
Innovative deviance	Yes	No
Ritualistic deviance	No	Yes
Retreatism deviance	No	No
Rebellion	No (old goals) Yes (new goals)	No (old means) Yes (new means)

▲ **Figure 7-4   Merton's Structural Strain Theory**

new goals are substituted for more traditional ones, and also new means are undertaken to replace older ones, as by force or armed combat. Many right-wing extremist groups, such as the American Nazi party, "skinheads," and the Ku Klux Klan (KKK), are examples of this type of deviance.

Social Control Theory

Taking functionalist theory in another direction, **social control theory** suggests that deviance occurs when a person's (or group's) attachment to social bonds is weakened (Gottfredson and Hirschi 1995, 1990; Hirschi 1969). According to this view, people internalize social norms because of their attachments to others. People care what others think of them and therefore conform to social expectations because they accept what people expect. You can see here that social control theory, like the functionalist framework from which it stems, assumes the importance of the socialization process in producing conformity to social rules. When that conformity is broken, deviance occurs.

Social control theory assumes there is a common value system within society, and breaking allegiance to that value system is the source of social deviance. This theory focuses on how deviants are (or are not) attached to common value systems and what situations break people's commitment to these values. Social control theory suggests that most people feel some impulse toward deviance at times but that the attachment to social norms prevents them from actually participating in deviant behavior. As an example, high school students who participate on an athletic team and are committed to academic success are least likely to engage in any crimes or get suspended from school (Veliz and Shakib 2012). Involvement with sports and academic success are examples of accepted social norms that help prevent deviance.

Functionalism: Strengths and Weaknesses

Functionalism emphasizes that social structure, not just individual motivation, produces deviance. Functionalists argue that social conditions exert pressure on individuals to behave in conforming or non-conforming ways. Types of deviance are linked to one's place in the social structure; thus, a poor person blocked from economic opportunities may use armed robbery to achieve economic goals, whereas a Wall Street trader may use insider trading to achieve the same. Functionalists acknowledge that people choose whether to behave in a deviant manner but believe that they make their choice from among socially prestructured options. The emphasis in functionalist theory is on social structure, not individual action. In this sense, functionalist theory is highly sociological.

Functionalists also point out that what appears to be dysfunctional behavior may actually be functional for the society. An example is the fact that most people consider prostitution to be dysfunctional behavior. From the point of view of an individual, that is true. It demeans the women who engage in it, puts them at physical risk, and subjects them to sexual exploitation. From the view of functionalist theory, however, prostitution supports and maintains a social system that links women's gender roles to sexuality, associates sex with commercial activity, and defines women as sexual objects and men as sexual aggressors. In other words, what appears to be deviant may actually serve various purposes within society.

Critics of the functionalist perspective argue that it does not explain how norms of deviance are first established. Despite its analysis of the ramifications of deviant behavior for society as a whole, functionalism does little to explain why some behaviors are defined as normative and others as illegitimate. Who determines social norms and on whom such judgments are most likely to be imposed are questions seldom asked by anyone using a functionalist perspective. Functionalists see deviance as having stabilizing consequences in society, but they tend to overlook the injustices that labeling someone deviant can produce. Others would say that the functionalist perspective too easily assumes that deviance has a positive role in society rather than studying the differential administration of justice for different groups. The tendency in functionalist theory is to assume that the system works for the good of the whole, thus too easily ignoring how inequities in society are reflected in patterns of deviance. These issues are left for sociologists who work from the perspectives of conflict theory and symbolic interaction.

Deviance, Power, and Social Inequality

Recall that conflict theory emphasizes the unequal distribution of power and resources in society. It links the study of deviance to social inequality. Based on the work of Karl Marx, conflict theory sees a dominant class as controlling the resources of society and using its power to create the institutional rules and belief systems that support its power. Like functionalist theory, conflict theory is a *macrostructural* approach; that is, both theories look at the structure of society as a whole in developing explanations of deviant behavior.

Because some groups of people have access to fewer resources in capitalist society, they are forced into crime to sustain themselves. Conflict theory posits that the economic organization of capitalist societies produces deviance and crime. The high rates of crime among the poorest groups, especially economic crimes such as theft, robbery, prostitution, and drug selling, are a result of the economic status of these groups. Rather than emphasizing values and conformity as a source of deviance as do functional analyses, conflict theorists see crime in terms of power relationships and economic inequality (Grant and Martínez 1997).

The upper classes, conflict theorists point out, can also better hide crimes they commit because affluent groups have the resources to mask their deviance and crime. As a result, a working-class man who beats his wife is more likely to be arrested and prosecuted than an upper-class man who engages in the same behavior. In addition, those with greater resources can afford to buy their way out of trouble by paying bail, hiring expensive attorneys, or even resorting to bribes.

The ruling groups in society are also in a position to develop numerous mechanisms to protect their interests. Conflict theorists argue that law, for example, is created by elites to protect the interests of the dominant class. Thus law, supposedly neutral and fair in its form and implementation, works in the interest of the most well-to-do.

Conflict theory emphasizes the significance of social control in managing deviance and crime. **Social control** is the process by which groups and individuals within those groups are brought into conformity with dominant social expectations. Social control can take place simply through socialization, but dominant groups can also control the behavior of others through marking them as deviant. An example is the historic persecution of witches during the Middle Ages in Europe and during the early colonial period in America (Ben-Yehuda 1986; Erikson 1966). Witches often were women who were healers and midwives—those whose views were at odds with the authority of the exclusively patriarchal hierarchy of the church, then the ruling institution.

One implication of conflict theory, especially when linked with labeling theory, is that the power to define deviance confers an important degree of social control. **Social control agents** are those who regulate and administer the response to deviance, such as the police and mental health workers. Members of powerless groups may be defined as deviant for even the slightest infraction against social norms, whereas others may be free to behave in deviant ways without consequence. Oppressed groups may actually engage in more deviant behavior, but it is also true that they have a greater likelihood of being labeled deviant and incarcerated or institutionalized, whether or not they have actually committed an offense. This is evidence of the power wielded by social control agents.

When powerful groups hold stereotypes about other groups, the less powerful people are frequently assigned deviant labels. As a consequence, the least powerful groups in society are subject most often to social control. You can see this in the patterns of arrest data. All else being equal, poor people are more likely to be considered criminals and therefore are more likely to be arrested, convicted, and imprisoned than middle- and upper-class people. The same is true of Latinos, Native Americans, and African Americans. Sociologists point out that this does not necessarily mean that these groups are somehow more criminally prone; rather, they take it as partial evidence of the differential treatment of these groups by the criminal justice system.

Conflict Theory: Strengths and Weaknesses

The strength of conflict theory is its insight into the significance of power relationships in the definition, identification, and handling of deviance. It links the commission, perception, and treatment of crime to inequality in society and offers a powerful analysis of how the injustices of society produce crime and result in different systems of justice for disadvantaged and privileged groups. This theory is not without its weaknesses, however, and critics point out that laws protect most people, not just the affluent, as conflict theorists argue.

In addition, although conflict theory offers a powerful analysis of the origins of crime, it is less effective in explaining other forms of deviance. For example, how would conflict theorists explain the routine deviance of middle-class adolescents? They might point out that consumer marketing drives much of middle-class deviance. Profits are made from the accoutrements of deviance—rings in pierced eyebrows, "gangsta" rap music, and so on—but economic interests alone cannot explain all the deviance observed in society. As Durkheim argued, deviance is functional for the whole of society, not just those with a major stake in the economic system.

Symbolic Interaction Theories of Deviance

Whereas functionalist and conflict theories are *macrosociological* theories, certain *microsociological* theories of deviance look directly at the interactions people have with one another as the origin of social deviance. *Symbolic interaction theory* holds that people behave as they do because of the meanings people attribute to situations. This perspective emphasizes the meanings surrounding deviance, as well as how people respond to those meanings. Symbolic interaction emphasizes that deviance originates in the interaction between different groups and is defined by society's reaction to certain behaviors.

Symbolic interactionist theories of deviance originated in the perspective of the Chicago School of Sociology. **W. I. Thomas** (1863–1947), one of the early sociologists from the University of Chicago, was among the first to develop a sociological perspective on social deviance. Thomas explained deviance as a normal response to the social conditions in which people find themselves. Thomas was one of the first to argue that delinquency was caused by the social disorganization brought on by slum life and urban industrialism. He saw deviance as a problem of social conditions, less so of individual character or individual personality.

Differential Association Theory

Thomas's work laid the foundation for a classic theory of deviance: differential association theory. **Differential association theory**, a type of symbolic interaction theory, interprets deviance, including criminal behavior, as behavior one learns through interaction with others (Sutherland and Cressey 1978; Sutherland 1940). Edwin Sutherland argued that becoming a criminal or a juvenile delinquent is a matter of learning criminal ways within the primary groups to which one belongs. To Sutherland, people become criminals when they are more strongly socialized to break the law than to obey it. Differential association theory emphasizes the interaction people have with their peers and others in their environment. Those who "differentially associate" with delinquents, deviants, or criminals learn to value deviance. The greater the frequency, duration, and intensity of their immersion in deviant environments, the more likely it is that they will become deviant.

Consider the case of cheating on college tests and assignments. Students learn from others about the culture of cheating, namely that because everyone does it, cheating is okay. Students also share the best ways to cheat without getting caught. Students who would ordinarily not engage in criminal or unethical behavior are socialized to become cheaters themselves. Sociologists found that students who were told by another student in the room how to cheat on a word memorization experiment were much more likely to do it (Paternoster et al. 2013). Differential association theory offers a compelling explanation for how deviance is culturally transmitted—that is, people pass on deviant expectations through the social groups in which they interact.

Critics of differential association theory have argued that this perspective tends to blame deviance on the values of particular groups. Differential association has been used, for instance, to explain the higher rate of crime among the poor and working class, arguing that this higher rate of crime occurs because they do not share the values of the middle class. Such an explanation, critics say, is class biased because it overlooks the deviance that occurs in the middle-class culture and among elites. Disadvantaged groups may share the values of the middle class but cannot necessarily achieve them through legitimate means.

Labeling Theory

Labeling theory is a branch of symbolic interaction theory that interprets the responses of others as the most significant factor in understanding how deviant behavior is both created and sustained (Becker 1963).

The work of labeling theorists such as Becker stems from the work of W. I. Thomas, who wrote, "If men define situations as real, they are real in their consequences" (Thomas and Thomas 1928: 572). A *label* is the assignment or attachment of a deviant identity to a person by others, including by agents of social institutions. People's reactions, not the action itself, produce deviance as a result of the labeling process.

Linked with conflict theory, labeling theory shows how those with the power to label an act or a person deviant and to impose sanctions—such as police, court officials, school authorities, experts, teachers, and official agents of social institutions—wield great power in determining societal understandings of deviance. Furthermore, because deviants are handled through bureaucratic organizations, the workers within these bureaucracies "process" people according to rules and procedures, seldom questioning the basis for those rules.

Once the label is applied, it sticks, and it is difficult for a person labeled deviant to shed the label—namely, to recover a non-deviant identity. To give an example, once a social worker or psychiatrist labels clients as mentally ill, those people will be treated as mentally ill, regardless of their actual mental state. In a kind of "catch-22," when people labeled as mentally ill plead that they are indeed mentally sound, this is taken as evidence that they are, in fact, mentally ill!

A person need not have actually engaged in deviant behavior to be labeled deviant and for that label to stick. Labeling theory helps explain why convicts released from prison have such high rates of *recidivism* (return to criminal activities). Convicted criminals are formally and publicly labeled wrongdoers. They are treated with suspicion ever afterward and have great difficulty finding legitimate employment: The label "ex-con" defines their future options.

The prison system in the United States also shows the power of labeling theory. Prisons, in effect, *train* and *socialize* prisoners into a career of secondary deviance. Reiman and Leightron (2012) argue that the goal of the prison system is not to reduce crime but to impress upon the public that crime is inevitable, originating only from the lower classes. Prisons accomplish this, even if unintentionally, by demeaning prisoners and stigmatizing them as different from "decent citizens," not training them in marketable skills. As a consequence, these people will never be able to pay their debt to society, and the prison system has created the very behavior it intended to eliminate.

The strength of labeling theory is its recognition that the judgments people make about presumably deviant behavior have powerful social effects. Labeling theory does not, however, explain why deviance occurs in the first place. It may illuminate the consequences of a young man's violent behavior, but it does not explain the actual origins of the behavior. Labeling theory helps us understand how some *are considered* deviant while others are not, but it does not explain why some people initially engage in deviant behaviors and others do not.

All told, sociological theories about crime and deviance reveal dimensions to these topics that are rarely understood in the general public. No one theory explains everything about crime and deviance, but together they provide a powerful analysis of some of the most difficult problems our society faces.

What Would a Sociologist Say?

Prison Rehabilitation

Do prisons rehabilitate people? Sociologists Jeffrey Reiman and Paul Leighton would say no. Quite the contrary, they find in their research that the prison system in the United States is in effect designed to train and socialize inmates into careers of crime. They argue that certain characteristics of prisons guarantee the further production of crime! First, label those who engage in crimes that have no unwilling victim, such as prostitution or gambling as criminals. Second, give prosecutors and judges broad discretion to arrest, convict, and sentence based on appearance, dress, race, and apparent social class. Third, treat prisoners in a painful and demeaning manner, as one might treat children. Fourth, make certain that prisoners have no training in a marketable skill that would be useful upon their release. And, finally, assure that prisoners will forever be labeled and stigmatized as different from "decent citizens," even after they have paid their debt to society. One has thus just socially constructed a U.S. prison, an institution that will continue to generate the very thing that it claims to eliminate.

Source: Reiman, Jeffrey H., and Paul Leighton. 2012. *The Rich Get Richer and the Poor Get Prison*, 10th ed. Upper Saddle River, NJ: Pearson.

Chapter Summary

How do sociologists define deviance?

Deviance is behavior that violates norms and rules of society. The definition of deviance occurs in a social context and is socially constructed, sometimes by the actions of social movements.

What are deviant identities, careers, and communities?

Deviant identities are formed as the result of labeling and occur when someone accepts such a definition of self. A *deviant career* involves a sequence of movements as people become socialized into a particular subculture of deviance. Deviance can also occur within *deviant communities*, groups that are organized around particular forms of social deviance.

How is crime measured and is it increasing?

National data on crime come from two primary sources: The *Uniform Crime Reporting* from the Federal Bureau of Investigation and the *National Victimization Surveys*. Both are subjected to some bias because of how they data are reported. Still, these are the best sources for measuring crime. Although there has been a small increase in violent crime in very recent years, overall, crime has declined significantly since the 1990s. This is contrary to widespread belief, as promulgated through the mass media.

What are the different types of crime?

Crime takes many forms, even though most people think of crime as street or interpersonal crime. Some crimes are victimless crimes. Others, such as white-collar crime are seldom punished to the same extent as crimes committed against immediate persons.

How is the criminal justice system shaped by social factors?

Class disparities exist in both arrest rates and rates of victimization. Despite public fears, middle- and upper-class Americans face lower risk of being victims of crime. Minorities and disadvantaged citizens are more likely to be both offenders and victims. Gender disparities also exist. At all stages of the criminal justice system, from racial profiling to arrest through sentencing and incarceration, Black Americans and Hispanics face a greater risk of prosecution in the criminal justice system.

What does sociological theory contribute to the study of deviance?

Functionalist theory sees both deviance and crime as functional for society because it affirms what is acceptable by defining what is not. *Structural strain theory*, a type of functionalist theory, predicts that societal inequalities actually force and compel individuals into deviant and criminal behavior. *Conflict theory* explains deviance and crime as a consequence of unequal power relationships and inequality in society. *Symbolic interaction theory* explains deviance and crime as the result of meanings people give to various behaviors. *Differential association theory*, a type of symbolic interaction theory, interprets deviance as behavior learned through social interaction with other deviants. *Labeling theory* argues that societal reactions to behavior produce deviance, with some groups having more power than others to assign deviant labels to people. Some groups suffer from *stigmas* that may define them in a *master status*.

Key Terms

altruistic suicide 168

anomic suicide 168

anomie 168

crime 156

criminology 156

cybercrime 162

deviance 151

deviant career 155

deviant community 156

deviant identity 154

differential association theory 172

egoistic suicide 168

fatalistic suicide 168

gender-based violence 161

hate crime 160

labeling theory 172

master status 155

medicalization of deviance 154

racial profiling 164

social control 171

social control agents 171

social control theory 170

stigma 155

structural strain theory 169

CHAPTER 8

SOCIAL CLASS AND SOCIAL STRATIFICATION

In this chapter, you will learn to:

Explain how class is a social structure

Describe the class structure of the United States

Identify the different components of class inequality

Analyze the extent of social mobility in the United States

Compare and contrast theoretical models of class inequality

Investigate the causes and consequences of U.S. poverty

iStock.com/filadendron

One afternoon in a major U.S. city, two women go shopping. They are friends—wealthy, suburban women who shop for leisure. They meet in a gourmet restaurant and eat imported foods while discussing their children's private schools. After lunch, they spend the afternoon in exquisite stores—some of them large, elegant department stores; others, intimate boutiques where the staff knows them by name. When one of the women stops to use the bathroom in one store, she enters a beautifully furnished room with an up-holstered chair, a marble sink with brass faucets, fresh flowers on a wooden pedestal, shining mirrors, an ample supply of hand towels, and jars of lotion and soaps. The toilet is in a private stall with solid doors. In the stall, there is soft toilet paper and another small vase of flowers.

The same day, in a different part of town, another woman goes shopping. She lives on a marginal income earned as a stitcher in a textiles factory. Her daughter badly needs a new pair of shoes because she has out-grown last year's pair. The woman goes to a nearby discount store where she hopes to find a pair of shoes for under $15, but she dreads the experience. She knows her daughter would like other new things—a bathing suit for the summer, a pair of jeans, and a blouse. But this summer, the daughter will have to wear hand-me-downs because bills over the winter have depleted the little money left after food and rent. For the mother, shopping is not recreation but a bitter chore reminding her of the things she is unable to get for her daughter.

While this woman is shopping, she, too, stops to use the bathroom. She enters a vast space with sinks and mirrors lined up on one side of the room and several stalls on the other. The tile floor is gritty and gray. The locks on the stall doors are missing or broken. Some of the overhead lights are burned out, so the room has dark shadows. In the stall, the toilet paper is coarse. When the woman washes her hands, she discovers there is no soap in the metal dispensers. The mirror before her is cracked. She exits quickly, feeling as though she is being watched.

Two scenarios, one society. The difference is the mark of a society built upon class inequality. The signs are all around you. Think about the clothing you wear. Are some labels worth more than others? Do others in your group see the same marks of distinction and status in clothing labels? Do some people you know never seem to wear the "right" labels? Whether it is clothing, bathrooms, schools, homes, or access to health care, the effect of class inequality is enormous, giving privileges and resources to some and leaving others struggling to get by.

Great inequality divides society—increasingly so, as we will see. Of late, people have also been pointing to class as a major source of people's discontent. Many working-class people feel that they have somehow been "left behind" and many people in the middle class fear that they are unprepared for emergencies or are overwhelmed by high levels of debt (including student debt). A deep tension has emerged in our society whereby some who are struggling resent others whom they perceive as somehow less deserving and yet getting more (Hochschild 2016). This dynamic has been used to explain the widespread appeal and approval of a billionaire candidate and eventual president, Donald Trump, who promised to "make America great again."

Clearly class matters, and it fundamentally shapes people's life chances even while most people still believe in the American Dream—that equal opportunity exists for all and that anyone who works hard enough can—and should—make it in the United States. This belief is so much a part of American culture that, when people fail, they tend to be blamed for their lack of success or individual achievement. Thus, many think the poor are lazy and do not value work. At the same time, the rich are admired for their supposed initiative, drive, and motivation. Neither is an accurate portrayal. There are many hardworking individuals who are poor, but they seldom get credit for their effort. At the same time, many of the richest people have inherited their wealth or have had access to resources (such as the best schools or access to elite networks) that others can barely imagine.

Observing and analyzing class inequality is fundamental to sociological study. What features of society cause different groups to have different opportunities? Why is there such an unequal allocation of society's resources? Sociologists respect individual achievements but have found that the greatest cause for disparities in material success is the organization of society. Instead of understanding inequality as the result of individual effort, sociologists study the social structural origins of inequality.

Social Differentiation and Social Stratification

All social groups and societies exhibit social differentiation. *Status*, as we have seen earlier, is a socially defined position in a group or society. Different statuses develop in any group, organization, or society. Think of a sports organization. The players, the owners, the managers, the fans, the cheerleaders, and the sponsors all have a different status within the organization. Together, they constitute a whole social system, one that is marked by social differentiation.

Status differences can become organized into a hierarchical social system. **Social stratification** is a relatively fixed, hierarchical arrangement in society by which groups have different access to resources, power, and perceived social worth. Social stratification is a system of structured social inequality. Again using sports as an example, you can see that many of the players earn extremely high salaries, although most do not. Those who do are among the elite in this system of inequality, but the owners control the resources of the teams and hold the most power in this system. Sponsors (including major corporations and media networks) are the economic engines on which this system of stratification rests. Fans are merely observers who pay to watch the teams play, but the revenue they generate is essential for keeping this system intact. Altogether, sports are systems of stratification because the

See for Yourself

Take a shopping trip to different stores and observe the appearance of stores serving different economic groups. What kinds of bathrooms are there in stores catering to middle-class clients? The rich? The working class? The poor? Which ones allow the most privacy or provide the nicest amenities? What fixtures are in the display areas? Are they simply utilitarian with minimal ornamentation, or are they opulent displays of consumption? Take detailed notes of your observations, and write an analysis of what this tells you about social class in the United States.

What Would a Sociologist Say?

Social Class and Sports

Sports are a huge part of American culture. Whether you are an athlete, a fan, or just an observer, sports are a window into how social class shapes some of our most popular activities.

Classical theorist Emile Durkheim explains how cultural symbols bind people together. Think of how many sports symbols, such as jerseys, hats, and bumper stickers, are common sights in everyday life. These symbols project an identity to others that define you as part of a collective group, They can also reflect social class. Rich people, for example, are not likely to be wearing NASCAR caps, but may well have yacht club logos on their polo shirts and ties.

As Max Weber would say, class, power, and prestige are also tangled up in sports. Some sports have more prestige than others. Sports are also intermeshed with power. During political elections, for example, you see politicians at tailgate parties and in the expensive box seats.

Sports are also big business, driven by corporate sponsors, even in college sports. Various plays in a football game might be featured as an "AT&T All-America Play of the Week!" And there is the vast amount of money spent on televised commercials during sports events.

Social class in the world of sports is everywhere, even though the workers who help put on events are often invisible. Some athletes are very highly paid, but working-class people serve the food in stadiums, clean up after the fans leave, and take out all the trash. Sports are an amazing example of a class-based social system.

groups that constitute the organization are arranged in a hierarchy where some have more resources and power than others. Some provide resources; others take them. Even within the field of sports, there are huge differences in which teams—and which sports—are among the elite.

All societies seem to have a system of social stratification, although they vary in the degree and complexity of stratification. Some societies stratify only along a single dimension, such as age, keeping the

Alex Segre/Alamy Stock Photo

AP Images/Don Ryan

Social class differences make it seem as if some people are living in two different societies.

Table 8-1	Inequality in the United States
• Among women heading their own households, 28 percent live below the poverty line (Fontenot, Semega, and Kollar 2018).
• One percent of the U.S. population controls 38 percent of the total wealth in the nation; the bottom half hold none or are in debt (Rose 2014).
• Most American families have seen their net worth decline, largely because of declines in the value of housing. Households at the bottom of the wealth distribution lost the largest share of their wealth; those at the top, the least (Pfeffer et al. 2014).
• The average CEO of a major company has total compensation of $13.1 million per year, including salary and other benefits; workers earning the minimum wage make $15,080 per year if they work 40 hours a week for 52 weeks and hold only one job (www.aflcio.org).

stratification system relatively simple. Most contemporary societies are more complex, with many factors interacting to create different social strata. In the United States, social stratification is strongly influenced by class, which is in turn influenced by matters such as one's occupation, income, and education, along with race, gender, and other influences such as age, region of residence, ethnicity, and national origin (see ◆ Table 8-1).

Estate, Caste, and Class

Stratification systems can be broadly categorized into three types: estate systems, caste systems, and class systems. In an **estate system** of stratification, the ownership of property and the exercise of power are monopolized by an elite class who have total control over societal resources. Historically, such societies were feudal systems where classes were differentiated into three basic groups—the nobles, the priesthood, and the commoners. Commoners included peasants (usually the largest class group), small merchants, artisans, domestic workers, and traders. The nobles controlled the land and the resources used to cultivate the land, as well as all the resources resulting from peasant labor.

Estate systems of stratification are most common in agricultural societies. Although such societies have been largely supplanted by industrialization, some societies still have a small but powerful land-holding class ruling over a population that works mainly in agricultural production. Unlike the feudal societies of the European Middle Ages, however, contemporary estate systems of stratification display the influence of international capitalism. The "noble class" comprises not knights who conquered lands in war, but international capitalists or local elites who control the labor of a vast and impoverished group of people, such as in some South American societies where landholding elites maintain a dictatorship over peasants who labor in agricultural fields.

In a **caste system**, one's place in the stratification system is an *ascribed status* (see Chapter 5), meaning it is a quality given to an individual by circumstances of birth. The hierarchy of classes is rigid in caste systems and is often preserved through formal law and cultural practices that prevent free association and movement between classes. The system of apartheid in South Africa was a stark example of a caste system. Under apartheid, the travel, employment, associations, and place of residence of Black South Africans were severely restricted. Segregation was enforced using a pass system in which Black South Africans could not even be present in White areas unless for purposes of employment. Those found without passes were arrested, often sent to prison without ever seeing their families again. Interracial marriage was illegal. Black South Africans were prohibited from voting; the system was one of total social control where anyone who protested was imprisoned. The apartheid system was overthrown in 1994 when Nelson Mandela, held prisoner for twenty-seven years, was elected president of the new nation of South Africa.

In **class systems**, stratification exists, but a person's placement in the class system can change according to personal achievements. That is, class depends to some degree on *achieved status*, defined as status that is earned by the acquisition of resources and power, regardless of one's origins. Class systems are more open than caste systems because position does not depend strictly on birth, although the class into which one is born can still matter. Classes are less rigidly defined than castes because class divisions are blurred when there is movement from one class to another.

Despite the potential for movement from one class to another, in the United States, class placement still depends heavily on one's social background. Although *ascription* (the designation of ascribed status according to birth) is not the basis for social stratification in the United States, the class a person is born into has major consequences for that person's life. Patterns of inheritance; access to exclusive educational resources; the financial, political, and social influence of one's family; and similar factors all shape one's likelihood of achievement. Although there are not formal obstacles to movement through the class system, one's life chances are still very much shaped by one's class of origin.

In common terms, *class* refers to style or sophistication. In sociological use, **social class** (or *class*) is the social structural position that groups hold relative to the economic, social, political, and cultural resources of society. Class determines the access different people have to these resources and puts groups in different positions of privilege and disadvantage. Each class has members with similar opportunities who tend to share a common way of life. Class also includes a cultural component in that class shapes language, dress, mannerisms, taste, and other preferences. Class is not just an attribute of individuals; it is a feature of society.

The social theorist Max Weber described the consequences of stratification in terms of **life chances**, meaning the opportunities that people have in common by virtue of belonging to a particular class. Life chances include the opportunity for possessing goods, having an income, and having access to particular jobs. Life chances are also reflected in the quality of everyday life. Whether you dress in the latest style or wear another person's discarded clothes, have a vacation in an exclusive resort, take your family to the beach for a week, or have no vacation at all, these life chances are the result of being in a particular class.

Class is a structural phenomenon; it cannot be directly observed. Nonetheless, you can "see" class through various displays that people project, often unintentionally, about their class status. Do some objects worn project higher-class status than others? What class status is displayed through the car you drive or, for that matter, whether you even have a car or use a bus to get to work? In so many ways, class is projected to others as a symbol of presumed worth in society.

Social class can be observed in the everyday habits and presentations of self that people project. Common objects, such as clothing and cars, can be ranked not only in terms of their economic value but also in terms of the status that various brands and labels carry. The interesting thing about social class is that a particular object may be quite ordinary, but with the right "label," it becomes a *status symbol* and thus becomes valuable. Take the example of Vera Bradley bags. These paisley bags are made of ordinary cotton with batting. Not long ago, such cloth was cheap and commonplace, associated with rural, working-class women. If such a bag were sewn and carried by a poor person living on a farm, the bag (and perhaps the person!) would be seen as ordinary, almost worthless. Transformed by the right label (and some good marketing), Vera Bradley bags have become status symbols, selling for a high price (often a few hundred dollars—a price one would never pay for a simple cotton purse). Presumably, having such a bag denotes the status of the person carrying it. (See also the box "See for Yourself: Status Symbols in Everyday Life.")

The early sociologist Thorstein Veblen described the class habits of Americans as **conspicuous consumption**, meaning the ostentatious display of goods to define one's social status. Writing in 1899, Veblen said, "conspicuous consumption of valuable goods is a means of respectability to the gentleman of leisure" (Veblen 1953/1899: 42). Although Veblen identified this behavior as characteristic of the well-to-do (the "leisure class," he called them), conspicuous consumption today marks the lifestyle of many. Indeed, mass consumerism is a hallmark of both the rich and the middle class, and even of many working-class people's lifestyles. What examples of this do you see among your associates?

Because sociologists cannot isolate and measure social class directly, they use other *indicators* to serve as measures of class. A prominent indicator of class is income. Other common indicators are

See for Yourself

Status Symbols in Everyday Life

You can observe the everyday reality of social class by noting the status that different ordinary objects have within the context of a class system. Make a list of every car brand you can think of—or, if you prefer, every clothing label. Then rank your list with the highest status brand (or label) at the top of the list, going down to the lowest status. Then answer the following questions:

1. Where does the presumed value of this object come from? Does the value come from the actual cost of producing the object or something more subjective?
2. Do people make judgments about people wearing or driving the different brands you have noted? What judgments do they make? Why?
3. What consequences do you see (positive and negative) of the ranking you have observed? Who benefits from the ranking and who does not?

What does this exercise reveal about the influence of status symbols in society?

education, occupation, and place of residence. These indicators alone do not define class, but they are often accurate measures of the class standing of a person or group. We will see that these indicators tend to be linked. A good income, for example, makes it possible to afford a house in a prestigious neighborhood and an exclusive education for one's children. In the sociological study of class, indicators such as income and education have had enormous value in revealing the outlines and influences of the class system.

The Class Structure of the United States: Growing Inequality

People think of the United States as a land of opportunity where one's class position matters less than individual effort. According to a recent survey, almost three-quarters of Americans think that hard work is the key to getting ahead in life. Compared to those in other countries, Americans are far more likely to believe in the importance of individual effort, a reflection of the cultural belief in individualism (Pew Research Global Attitudes Project 2014).

Despite these beliefs, class divisions in the United States are real, and inequality is growing. Perhaps this has become more apparent to people as many in the middle and working class feel that their way of life is slipping away. For the first time in our nation's history, only 37 percent of the public think that children today will be better off than their parents (Pew Research Global Attitudes Project 2017).

In the United States, the gap between the rich and the poor is greater than in other industrialized nations, and it is larger than at any time in the nation's history. Many analysts argue that this gap is the central problem of the age—contributing to crime and violence, political division, threats to democracy, and increased anxiety and frustration felt by large segments of the population (Piketty 2014; Noah 2012; Reich 2010).

Many factors have contributed to growing inequality in the United States, including the profound effects of national and global economic changes. Many think of the economic problems of the nation as stemming from individual greed on Wall Street, and this likely plays a role, but social inequality stems from systemic—that is, social structural—conditions, particularly what is called *economic restructuring*.

Economic restructuring refers to the decline of manufacturing jobs in the United States, the transformation of the economy by technological change, and the process of globalization. We examine economic restructuring more in Chapter 15 on the economy, but the point here is that these structural changes are having a profound effect on the life chances of people in different social classes. Many in the working class, for example, once largely employed in relatively stable manufacturing jobs with decent wages and good benefits, now likely work, if they work at all, in lower-wage jobs with fewer benefits, such as health care and pensions. Despite political promises to bring such jobs back, in this highly technological society where jobs once done by human hands can now be done by machines (or robots), bringing such jobs back is unlikely.

The economic problems that produce inequality are not, however, purely economic: They are social, both in their origins and their consequences. Home ownership provides an example. For most Americans, owning one's own home is the primary means of attaining economic security—a central part of the American dream. Owning a home is also the key to other resources—good schools, clean neighborhoods, and an investment in the future. Similarly, losing your home is more than just a financial crisis—it reverberates through various aspects of your life. The odds of having a home—indeed, the odds of losing your home—are profoundly connected to social factors, such as your race and your gender.

Housing foreclosure, for example, has hit some groups especially hard. The racial segregation of Hispanic and, especially, African American neighborhoods is a major contributing cause to the high rate of mortgage foreclosures (Rugh and Massey 2010). African Americans are almost twice as likely to experience foreclosure as White Americans (Bocian et al. 2010). Since the great recession of 2008, Black home-ownership, which had been rising, drastically fell and continues to do so (Potts 2012). Moreover, women are 32 percent more likely than men to have *subprime mortgages* (that is, mortgages with an interest rate *higher* than the prime lending rate). Even when they have similar middle-income earnings, Black women are nearly five times more likely to receive subprime mortgages than White men (Fishbein and Woodall 2006).

Some might argue that foreclosures occur because individual people have made bad decisions—buying homes beyond their means. But, institutional lending practices also target particular groups, making them more vulnerable to the economic forces that can shatter individual lives. Lenders may see African Americans as a greater credit risk, but they also know that the value of real estate is less in racially segregated neighborhoods. Discriminatory practices in the housing market have also been well documented (Squires 2007; Oliver and Shapiro 2006).

The sociological point is that economic problems have a sociological dimension and cannot be explained by individual decisions alone. Economic policies also have different effects for different groups—sometimes intended, sometimes not. Wealthy people, as an example, typically pay a far lower tax rate than the middle class, because much of their money comes from investments, not income, and income is taxed at a much higher rate than investment income. Various tax loopholes (such as home mortgage deductions, tax shelters on real estate investments, or even offshore banking deposits) also significantly reduce the tax burden by those with the most resources.

Corporations benefit the most from the tax structure. The corporate tax rate in the United States is 21 percent, but many corporations pay much less than that, given the various loopholes, offshore investments, and tax subsidies that lessen tax obligations. A study of the Fortune 500 companies (those companies with the highest gross revenue in a given year) has found that many of these big companies paid no tax at all in some years (McIntyre et al. 2014).

The Distribution of Income and Wealth

Understanding inequality requires knowing some basic economic and sociological terms. Inequality is often presented as a matter of differences in income, one important measure of class standing. In addition to income inequality, there are vast inequalities in who owns what—that is, the wealth of different groups.

Income is the amount of money brought into a household from various sources (wages, investment income, dividends, and so on) during a given period. In recent years, income growth has been greatest for those at the top of the population (see ▲ Figure 8-1). For everyone else, income (controlling for the value of the dollar) has either been relatively flat or grown at a far lesser rate. Indeed, for the lowest one-fifth of the U.S. population, income has actually fallen (Fontenot, Semega, and Kollar 2018).

Inequality is even more apparent when you consider both wealth and income. **Wealth** is the monetary value of everything one actually owns. Wealth is calculated by adding all financial assets (stocks,

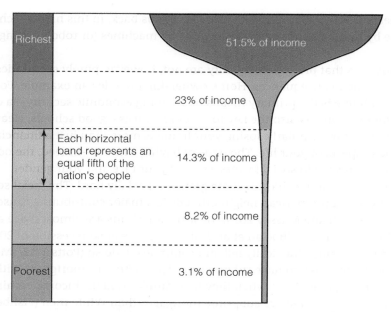

▲ **Figure 8-1   Distribution of Income in the United States.** This figure shows the percentage of all income earned by five quintiles in the U.S. population. A quintile is 20 percent of the total population and thus the figure shows the top quintile, the next to the top, the middle, second to the bottom, and the bottom quintile.

Source: Fontenot, Kayla, Jessica Semega, and Melissa Kollar. 2018. *Income and Poverty in the United States: 2017*. Washington, DC: U.S. Census Bureau. **www.census.gov**

bonds, property, insurance, savings, value of investments, and so on) and subtracting debts, resulting in one's **net worth**. Wealth allows you to accumulate assets over generations, giving advantages to subsequent generations that they might not have had on their own. Unlike income, *wealth is cumulative*—that is, its value tends to increase through investment.

To understand the significance of wealth compared to income in determining class location, imagine two college students graduating in the same year, from the same college, with the same major and same grade point average. Imagine further that both get jobs with the same salary in the same organization. In one case, parents paid all the student's college expenses and gave her a car upon graduation. The other student worked while in school and graduated with substantial debt from student loans. This student's family has no money with which to help support the new worker. Who is better off? Same salary, same credentials, but wealth (even if modest) matters, giving one person an advantage that will be played out many times over as the young worker buys a home, finances her own children's education, and possibly inherits additional assets.

Where is all the wealth? The wealthiest 1 percent own 34 percent of all net worth and have about 20 percent of all income. The bottom half hold only 1 percent of all wealth (Bricker et al. 2016; Rose 2014). Moreover, the concentration of wealth has been increasing since the 1980s, making the United States one of the most unequal nations in the world. The growth of wealth by a select few, though long a feature of the U.S. class system, has also reached historic levels. As just one example, John D. Rockefeller is typically heralded as one of the wealthiest men in U.S. history. Comparing Rockefeller with Bill Gates, controlling for the value of today's dollars, Gates has far surpassed Rockefeller's riches.

In contrast to the vast amount of wealth and income controlled by elites, a very large proportion of Americans have hardly any financial assets once debt is subtracted. ▲ Figure 8-2 shows how wealth is distributed in the population, and you can see that most of the population has very low net worth. One-fifth of the population has zero or negative net worth, usually because their debt exceeds their assets (Pew Charitable Trusts 2016). The American dream of owning a home, a new car, taking annual vacations, and sending one's children to good schools—not to mention saving for a comfortable retirement—is increasingly unattainable for many. When you see the amount of income and wealth a

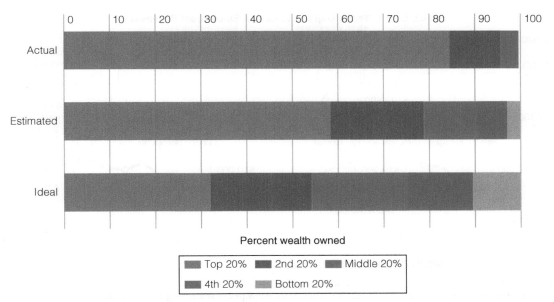

▲ **Figure 8-2   Actual, Perceived, and Ideal Distribution of Wealth.** Researchers noted the actual distribution of all wealth across different segments of the population and then asked people what they thought wealth distribution actually was and what was ideal. This figure shows you how far the actual distribution of wealth is from people's perception of reality and their perception of the ideal.

Source: Norton, Mike, and Dan Ariely. "Building a Better America—One Wealth Quintile at a Time." *Perspectives on Psychological Science* 6 (1): 9–12.

small segment of the population controls, a sobering picture of class inequality emerges, even though as Figure 8-2 shows, most people misperceive the actual distribution of wealth in America. Students themselves may be experiencing this burden, as levels of debt from student loans have escalated in recent years.

Despite the prominence of rags-to-riches stories in American legend, much of the wealth in this society is inherited. In recent years, more of those who are very rich are "self-made"—that is, starting from modest origins: Bill Gates, a Harvard dropout; Jeff Bezos, founder and CEO of Amazon; and Mark Zuckerberg, founder of Facebook come to mind. Such examples exist, although for many, if you scratch the surface of the rags-to-riches theme, you will find that they had a significant leg up. Among the now very rich, it has become more common for some to become amazingly rich, even though coming from modest origins. The technology boom has certainly helped, showing again how the historical context of one's life course can matter. Among the nation's wealthiest women, not one is self-made; all inherited their wealth. The wealthiest woman (Alice Walton) still has half the net worth of the wealthiest men—Bill Gates, Jeff Bezos, Warren Buffett, and Mark Zuckerberg (Peterson-Withorn 2016).

For most people, however, dramatically moving up in the class system remains highly unlikely. Young and minority households have been especially hard hit by these changes, in large part because of being highly "leveraged"—that is, holding too much debt on their homes (Wolff 2014; Kochhar et al. 2011). The high rate of foreclosure in Black communities during the economic recession of 2008 also had massive consequences for Black people who lost almost $200 billion in wealth, largely because of housing foreclosures. The effect, according to a national report has been "the largest loss of wealth for these communities in modern history" (Gottesdiener 2013). Given the relatively recent expansion of the Black middle class, such losses have long-term and lasting consequences.

Race strongly influences the pattern of wealth distribution in the United States. The net worth of White households is fourteen times that of Black households and ten times higher than that for Latino families—gaps that have widened since the early 1980s, as you can see in ▲ Figure 8-3 (Kochhar and Fry 2014). Further, at all levels of income, occupation, and education, Black families have lower levels of wealth than similarly situated White families.

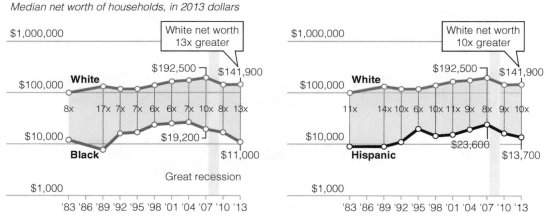

▲ FIGURE 8-3   Wealth and Race

Source: Kochhar, Rakesh, and Richard Fry. 2014. *Wealth Inequality Has Widened along Racial Ethnic Lines Since End of Great Recession*. Washington, DC: Pew Research Organization. **www.pewresearch.org/fact-tank/2014/12/12 /racial-wealth-gaps-great-recession/**

Being able to draw on assets during times of economic stress means that families with some resources can better withstand difficult times than those without assets. Even small assets, such as home ownership or a savings account, provide protection from crises such as increased rent, a health emergency, or unemployment. Because the effects of wealth are *intergenerational*—that is, they accumulate over time—providing equality of opportunity in the present does not address the differences in class status that Black and White Americans experience (Shapiro 2017; Oliver and Shapiro 2006).

What explains the disparities in wealth by race? Wealth accumulates over time. Government policies in the past have prevented Black Americans from being able to accumulate wealth. Discriminatory housing policies, bank lending policies, tax codes, and so forth have disadvantaged Black Americans, resulting in the differing assets Whites and Blacks in general hold now. Even though some of these discriminatory policies have ended, many continue. Either way, their effects persist, resulting in what sociologists Melvin Oliver and Thomas Shapiro call the *sedimentation of racial inequality*.

Understanding the significance of wealth in shaping life chances for different groups also shows how important it is to understand diversity within the different labels used to define groups. Among Hispanics, for example, Cuban Americans and Spaniards are similar to Whites in their wealth holdings, whereas Mexicans, Puerto Ricans, Dominicans, and other Hispanic groups more closely resemble African Americans in measures of wealth and class standing. Without significant wealth holdings, families of any race are less able to transmit assets from previous generations to the next generation, one main support of *social mobility* (discussed later in this chapter).

Analyzing Social Class

The class structure of the United States is elaborate, arising from the interactions of race and gender inequality with class, the presence of old mixed with new wealth, the income and wealth gap between the haves and have-nots, a culture of entrepreneurship and individualism, accelerated globalization, and immigration. Given this complexity, how do sociologists conceptualize social class?

Class as a Ladder

One way to think about the class system is as a ladder, with different class groups arrayed up and down the rungs, each rung corresponding to a different level in the class system. Conceptualized this way, social class is the common position groups hold in a status hierarchy (Lucal 1994; Wright 1979); class is indicated by factors such as levels of income, occupational standing, and educational

Understanding Diversity

The Student Debt Crisis

Numerous recent reports show that students are struggling over rising levels of debt from student loans. A record one in five households in the United States now has outstanding student debt. Not only is the number of those with student debt increasing, but so is the size of the indebtedness (see ▲ Figures 8-4 and 8-5; Lee 2013; Fry 2012). Leaving college or graduate school with large amounts of debt impedes one's ability to get financially established.

All students are at risk of accruing debt given the rising cost of education and the higher interest rates now associated with student loans. Some groups, though, are more vulnerable than others, adding to the inequalities that accrue across different groups. Among those in the bottom fifth of income earners, student debt, on average, takes up 24 percent of all income; for the top fifth of earners, only 9 percent of income. For those in the middle, student debt consumes 12 percent of income (Fry 2012).

The amount of student debt is highest among those under 35, those who are just beginning careers and, possibly, families. Race also matters. Black students are more likely to borrow money for college than other groups—and to borrow more; 80 percent of Black students have outstanding student loans, compared to 65 percent of Whites, 67 percent of Hispanics, and 54 percent of Asian students. Moreover, levels of debt are highest among Blacks—an average of $28,692, compared to $24,772 for Whites, $22,886 for Hispanics, and $21,090 for Asians (Demos 2014).

How does this reality of student debt influence the experience of those you see in your own environment? What are the sociological causes of this significant social problem?

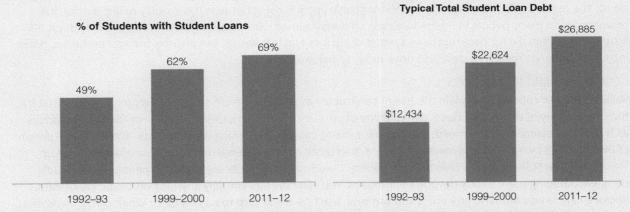

▲ FIGURE 8-4 and 8-5   **The Share of College Graduates Borrowing has Sharply Increased and the Amount a Typical Borrower Owes has more than Doubled.** Median amount borrowed by undergraduates graduating in these years who had student loan debt, in 2013 dollars.

Source: Pew Research Center. 2014. "The Changing Profile of Student Borrowers." Washington, DC: Pew Research Center. **www.pewresearch.org**

attainment. People are relatively high or low on the ladder depending on the resources they have and whether those resources are education, income, occupation, or any of the other factors known to influence people's placement (or ranking) in the stratification system. Indeed, an abundance of sociological research has stemmed from the concept of social mobility, that is, a person's or group's movement over time from one class to another. Research on social mobility, as we will see in more detail later in this chapter, describes how factors such as class origins, educational level, and occupation produce class location.

The laddered model of class suggests that stratification in the United States is hierarchical but somewhat fluid. That is, the assumption is that people can move up and down different "rungs" of the ladder—or class system. In a relatively *open class system* such as the United States, people's achievements do matter, although the extent to which people rise rapidly and dramatically through the stratification system is less than the popular imagination envisions. Some people move down in the class system, but as we will see,

The Fragile Middle Class

The hallmark of the middle class in the United States is its presumed stability. Home ownership, a college education for children, and other accoutrements of middle-class status (nice cars, annual vacations, an array of consumer goods) are the symbols of middle-class prosperity. But the American middle class is not as secure as it has been presumed to be.

Teresa Sullivan, Elizabeth Warren, and Jay Lawrence Westbrook have studied bankruptcy, and their research shows the fragility of the middle class in recent times, including an increase in personal bankruptcy.

Research Question

What is causing the rise of bankruptcy?

Research Method

Their study analyzed official records of bankruptcy in five states and included detailed questionnaires given to individuals who filed for bankruptcy.

Research Results

This team's research debunks the idea that bankruptcy is most common among poor people. Instead, bankruptcy is mostly a middle-class phenomenon representing a cross-section of those in this class (meaning that those who are bankrupt are matched on the demographic characteristics of race, age, and gender with others in the middle class). The research also debunks the notion that bankruptcy is rising because it is so easy to file. Rather, the research found many people in the middle class so overwhelmed with debt that they cannot possibly pay it off. Most people often file for bankruptcy as a result of job loss and lost wages. But divorce, medical problems, housing expenses, and credit card debt also drive many to bankruptcy court.

Conclusions and Implications

Sullivan and her colleagues explain the rise of bankruptcy as stemming from structural factors in society that fracture the stability of the middle class. The volatility of jobs under modern capitalism is one of the biggest factors, as is the "thin safety net"—no health insurance for many, coupled with rising medical costs. The American dream of owning one's own home also means many are "mortgage poor"—extended beyond their ability to keep up.

The United States is also a credit-driven society. Credit cards are routinely mailed to people in the middle class, encouraging them to buy beyond their means. You can now buy virtually anything on credit: cars, clothes, doctor's bills, entertainment, groceries. You can even use one credit card to pay off other credit cards. Increased debt is the result. Many are simply unable to keep up with compounding interest and penalty payments, and debt takes on a life of its own as consumers cannot keep up with even the interest payments on debt.

Sullivan, Warren, and Westbrook conclude that increases in debt and uncertainty of income combine to produce the fragility of the middle class. Their research shows that "even the most secure family may be only a job loss, a medical problem, or an out-of-control credit card away from financial catastrophe" (2000: 6).

Questions to Consider

1. Have you ever had a credit card? If so, how easy was it to get? Is it possible to get by without a credit card?
2. What evidence do you see in your community of the fragility or stability of different social class groups?

Source: Sullivan, T. A., E. Warren, and J. L. 2000. Westbrook, *The Fragile Middle Class: Americans in Debt*. New Haven, CT: Yale University Press.

most people remain relatively close to their class of origin. When people rise or fall in the class system, the distance they travel is usually relatively short, as we will see in a later section on social mobility.

The image of stratification as a laddered system, with different gradients of social standing, emphasizes that one's **socioeconomic status (SES)** is derived from certain factors. Income, occupational prestige, and education are the three measures of socioeconomic status that have been found to be most significant in determining people's placement in the stratification system.

The **median income** for a society is the midpoint of all household incomes. Half of all households earn more than the median income; half earn less. In 2017, median household income in the United States was $61,372 (Fontenot, Semega, and Kollar 2018). To many, this may seem like a lot of money, but consider these facts: American consumers spend about one-third of their household budgets on housing; almost another 16 percent on transportation; 13 percent on food; 24 percent on health care, and 10 percent on insurance and pensions, a total of 86 percent of income (Bureau of Labor Statistics 2017b). If you do the calculations based on the median income level, you will see there is very little left for other living expenses (clothing, education, entertainment, charities, pets, and so forth)—only $716 per month for all other expenses—hardly a lavish income, especially when you consider that half of Americans have less than this, given the definition of a median. (See also the "See for Yourself" exercise on household budgets later in this chapter.) Those bunched around the median income level are considered middle class, although sociologists debate which income brackets constitute middle-class standing because the range of what people think of as "middle class" is quite large. Nonetheless, income is a significant indicator of social class standing, although not the only one.

Occupational prestige is a second important indicator of socioeconomic status. **Prestige** is the value others assign to people and groups. **Occupational prestige** is the subjective evaluation people give to

Income Distribution: Should Grades Be the Same?

▲ Figure 8-6 shows the income distribution within the United States. Imagine that grades in your class were distributed based on the same curve. Let's suppose that after students arrived in class and sat down, different groups received their grades based on where they were sitting in the room and in the same proportion as the U.S. income distribution. Only students in the front receive A's; the back, D's and F's. The middle of the room gets the B's and C's. Write a short essay answering the following questions based on this hypothetical scenario:

1. How many students would receive A's, B's, C's, D's, and F's?

2. Would it be fair to distribute grades this way? Why or why not?

3. Which groups in the class might be more likely to support such a distribution? Who would think the system of grade distribution should be changed?

4. What might different groups do to preserve or change the system of grade distribution? What if you really needed an A, but got one of the F's? What might you do?

5. Are there circumstances in actual life that are beyond the control of people and that shape the distribution of income?

6. How is social stratification maintained by the beliefs that people have about merit and fairness?

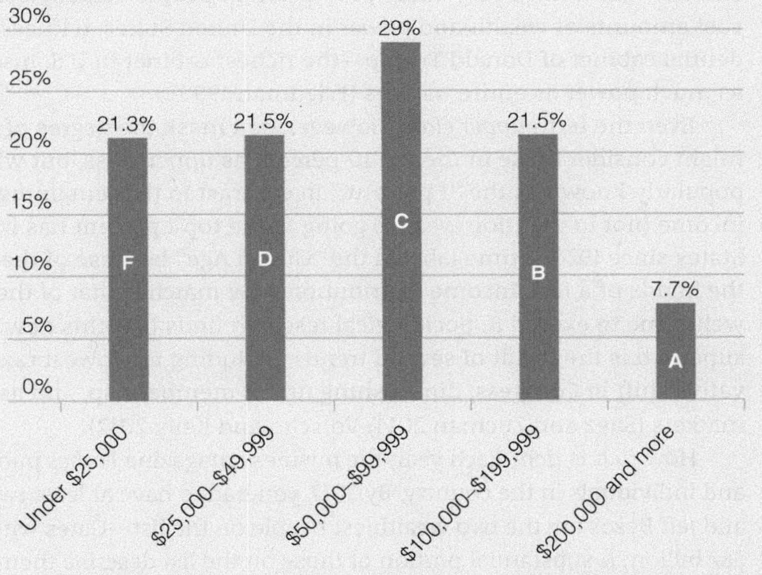

▲ **Figure 8-6   Income Distribution in the United States.** This graph shows the percentage of the total population that falls into each of five income groups. Would it be fair if course grades were distributed by the same percentages?

Source: Fontenot, Kayla, Jessica Semega, and Melissa Kollar. 2018. *Income and Poverty in the United States: 2017*. Washington, DC: U.S. Census Bureau. **www.census.gov**

Adapted from: Brislen, William, and Clayton D. Peoples. 2005. "Using a Hypothetical Distribution of Grades to Introduce Social Stratification." *Teaching Sociology* 33 (January): 74–80.

jobs. To determine occupational prestige, sociological researchers typically ask nationwide samples of adults to rank the general standing of a series of jobs. These subjective ratings provide information about how people perceive the worth of different occupations. People tend to rank professionals, such as physicians, professors, judges, and lawyers highly, with occupations such as electrician, insurance agent, and police officer falling in the middle. Occupations with low occupational prestige are maids, garbage collectors, and shoe shiners. These rankings do not reflect the worth of people within these positions but are indicative of the judgments people make about the worth of these jobs.

The final major indicator of socioeconomic status is **educational attainment**, typically measured as the total years of formal education. The more years of education attained, the higher a person's class status. As we will see, education matters a lot for people's chances for social mobility, but it comes with a price—as we have seen in the skyrocketing level of student debt. As with economic inequality, the United States is also experiencing growing inequality in the ability of children from high- and low-income families to pay for and attend college (Greenstone et al. 2013). This trend is likely to exacerbate the trend toward growing inequality overall.

Taken together, income, occupation, and education are good indicators of people's class standing. Using the laddered model of class, you can describe the class system in the United States as being divided into several classes: upper, upper middle, middle, lower middle, and lower class. The different classes are arrayed up and down, like a ladder, with those with the most money, education, and prestige on the top rungs and those with the least at the bottom.

In the United States, the *upper class* owns the major share of corporate and personal wealth. The upper class includes those who have held wealth for generations as well as those who have recently become rich. Only a very small proportion of people actually constitute the upper class, but they control vast amounts of wealth and power in the United States, witnessed of late by the billionaire-filled presidential cabinet of Donald Trump—the richest cabinet in U.S. history. Some wealthy individuals can wield as much power as entire nations (Friedman 1999).

Even the term *upper class*, however, can mask the degree of inequality in the United States. You might consider those in the top 10 percent as upper class, but within this class are the superrich, or those popularly known as the "1 percent," in contrast to the remaining 99 percent. Since about 1980, the share of income (not to mention wealth) going to the top 1 percent has increased to levels not seen in the United States since 1920, a time labeled the "Gilded Age" because of the concentration of wealth and income in the hands of a few. Income distribution now matches that of the Gilded Age and, given the trends, may well come to exceed it. Sociological research finds that this new concentration of income among the superrich is the result of several trends, including the lowest tax rates for high incomes, a more conservative shift in Congress, diminishing union membership, and asset bubbles in the stock and housing markets (Saez and Zucman 2014; Volscho and Kelly 2012).

How rich is rich? Each year, the business magazine *Forbes* publishes a list of the 400 wealthiest families and individuals in the country. By 2017, you had to have at least two billion dollars to be on the list! Bill Gates and Jeff Bezos are the two wealthiest people on the list—Gates with an estimated worth of $89 billion; Bezos, $82 billion. A substantial portion of those on the list describe themselves as "self-made," that is, living the American dream, but most of these were still able to borrow from parents, in-laws, or spouses. Although they may have built their fortunes, they did so with a head start on accumulation (Kroll and Dolan 2012). The best predictor of future wealth still remains the family into which you are born (McNamee and Miller 2009).

The upper class is overwhelmingly White, conservative, and Protestant. Members of this class exercise tremendous political power by funding lobbyists, exerting their social and personal influence on other elites, and contributing heavily to political campaigns (Domhoff 2013). They travel in exclusive social networks that tend to be open only to those in the upper class. They tend to intermarry, their children are likely to go to expensive schools, and they spend their leisure time in exclusive resorts.

Those in the upper class with newly acquired wealth are known as the *nouveau riche*. Luxury vehicles, high-priced real estate, and exclusive vacations may mark the lifestyle of the newly rich. Larry Ellison, who made his fortune as the founder of the software company Oracle, is the fifth wealthiest person in the United States. Ellis has a mega yacht that is 482 feet long, five stories high, with 82 rooms inside. The mega yacht also includes an indoor swimming pool, a cinema, a space for a private submarine, and a basketball court that doubles as a helicopter launch pad.

The *upper-middle class* includes those with high incomes and high social prestige. They tend to be well-educated professionals or business executives. Their earnings can be quite high indeed, even millions of dollars a year. It is difficult to estimate exactly how many people fall into this group because of the difficulty of drawing lines between the upper, upper-middle, and middle classes. Indeed, the upper-middle class is often thought of as "middle class" because their lifestyle sets the standard to which many aspire, but this lifestyle is actually unattainable by most. A large home full of top-quality furniture and modern appliances, two or three relatively new cars, vacations every year (perhaps a vacation home), high-quality college education for one's children, and a fashionable wardrobe are simply beyond the means of a majority of people in the United States.

The *middle class* is hard to define in part because being "middle class" is more than just economic position. Forty-four percent of Americans identify themselves as middle class (Newport 2017), even though they vary widely in lifestyle and in resources at their disposal. The idea that the United States is an open class system leads many to think that the most have a middle-class lifestyle. The "middle class" is the ubiquitous norm, even though many who consider themselves middle class have a tenuous hold on this class position. Thus, while a majority may identify as "middle class," what this means varies a lot across the population.

The *lower-middle class* includes workers in the skilled trades and low-income bureaucratic workers, some who may actually think of themselves as middle class. Also known as the *working class*, this class includes blue-collar workers (those in skilled trades who do manual labor) and many service workers, such as secretaries, hairstylists, food servers, police, and firefighters. A medium to low income, education, and occupational prestige define the lower-middle class relative to the class groups above it. The term *lower* in this class designation refers to the relative position of the group in the stratification system, but it has a pejorative sound to many people, especially to people who are members of this class, many who think of themselves as middle class.

The *lower class* is composed primarily of displaced and poor. People in this class tend to have little formal education and are often unemployed or working in minimum-wage jobs. People of color and women make up a disproportionate part of this class. The poor include the *working poor*—those who work at least twenty-seven hours a week but whose wages fall below the federal poverty level. Three percent of all people working full-time and 14 percent of those working part-time live below the poverty line, a proportion that has increased over time. Although this may seem a small number, it includes 8.6 million adults. Black and Hispanic workers are twice as likely to be among the working poor as White or Asian workers, and women are more likely than men to be so (U.S. Bureau of Labor Statistics 2017a).

The concept of the **urban underclass** has been added to the lower class (Wilson 1987). The underclass includes those who are likely to be permanently unemployed and without much means of economic support. The underclass has little or no opportunity for movement out of the worst poverty. Rejected from the economic system, those in the underclass may become dependent on public assistance or illegal activities. Structural transformations in the economy have left large groups of people, especially urban minorities, in these highly vulnerable positions. The growth of the urban underclass has exacerbated the problems of urban poverty and related social problems (Wilson 2009, 1996, 1987).

Class Conflict

A second way of conceptualizing the class system is *conflict theory*. Conflict theory defines classes in terms of their structural relationship to other classes and their relationship to the economic system. The analysis of class from this sociological perspective interprets inequality as resulting from the unequal distribution of power and resources in society (see Chapter 1). Sociologists who work from a conflict perspective see classes as facing off against each other, with elites exploiting and dominating others. Related to the work of Karl Marx (discussed in Chapter 1), the key idea in the conflict model of class is that class is not simply a matter of what individuals possess in terms of income and prestige. Class is, instead, defined by the relationship of the classes to the larger system of economic production (Vanneman and Cannon 1987; Wright 1985).

From a conflict perspective, the middle class, or the *professional–managerial class*, includes managers, supervisors, and professionals. Members of this group have substantial control over other people, primarily through their authority to direct the work of others, impose and enforce regulations in the workplace, and determine dominant social values. Marx argued that the middle class is controlled by the ruling class

Labor unions, traditionally dominated by White men in the skilled trades, are not only more diverse but also represent workers in occupations typically thought of as "white-collar" work.

but tends to identify with the interests of the elite. The professional–managerial class, however, is caught in a contradictory position between elites and the working class. Those in this class have some control over others, but like the working class, they have minimal control over the economic system (Wright 1979). Marx argued that as capitalism progresses, more and more of those in the middle class drop into the working class as they are pushed out of managerial jobs into working-class jobs or as professional jobs become organized more along the lines of traditional working-class employment.

Has this happened? Not to the extent Marx predicted, but to some extent Marx was correct. Classes have become more polarized, with the well-off accumulating even more resources and the middle and working classes seeing their income either flat or falling, measured in constant dollars. Rising levels of debt among the middle class have contributed to this growing inequality. Many now have a fragile hold on being middle class: The loss of a job, a family emergency, such as the death of a working parent, divorce, disability, or a prolonged illness can quickly leave middle- and working-class families in a precarious financial state. At the same time, high salaries for CEOs, tax loopholes that favor the rich, and sheer greed are concentrating more wealth in the hands of a few.

Members of the working class have little control over their own work lives; instead, they generally have to take orders from others. This concept of the working class departs from traditional blue-collar definitions of working-class jobs because it includes many so-called white-collar workers (secretaries, salespeople, and nurses), any group working under the rules imposed by managers. The middle class may exercise some autonomy at work, but the working class has little power to challenge decisions of those who supervise them, except insofar as they can organize collectively, as in unions, strikes, or other collective work actions.

Marx thought that, ultimately, there would only be two classes: the capitalist and the proletariat, but this has not occurred—in part because he did not see the growth of a large middle class. However, there is a great deal of class conflict, though not in the open way that Marx predicted. The class structure in the United States is quite hierarchical, giving people different access to jobs, income, education, power, and social status. And although the boundaries of class can seem somewhat blurry, you can see class conflict in the politics around issues like tax reform, health care, and various federal programs to assist those in need. The middle and working classes shoulder much of the tax burden for social programs, producing resentment by these groups toward the poor. At the same time, tax loopholes for the rich have increased, producing suspicion that the wealthy are not paying their share—witnessed particularly when Donald Trump refused to release his tax returns to the public.

Whatever perspective on the class system different sociologists employ, it is clear that class stratification is a dynamic process—one involving the interplay of access to resources, judgments about different groups, and the exercise of power by a few.

Diverse Sources of Stratification

Class is only one basis for stratification in the United States. Factors such as age, ethnicity, and national origin have a tremendous influence on stratification. Race and gender are two primary influences in the stratification system in the United States. Analyzing class without including race and gender is misleading. Race, class, and gender, as we are seeing throughout this book, are overlapping systems of stratification that people experience simultaneously. A working-class Latina, for example, does not experience herself as working class at one moment, Hispanic at another moment, and a woman the next. At any given point in time, her position in society is the result of her race, class, *and* gender status. In other words, class position is manifested differently depending on one's race and gender, just as gender is experienced differently depending on one's race and class, and race is experienced differently depending on one's gender and class. Depending on one's circumstances, race, class, or gender may seem particularly salient at a given moment in a person's life. A Black middle-class man stopped and interrogated by police when driving through a predominantly White middle-class neighborhood may at that moment feel his racial status as his single most outstanding characteristic, but at all times his race, class, and gender influence his life chances. As social categories, race, class, and gender shape all people's experience in this society, not just those who are disadvantaged (Andersen and Collins 2016).

Class also significantly differentiates group experience within given racial and gender groups. Latinos, for example, are broadly defined as those who trace their origins to regions originally colonized by Spain. The ancestors of this group include both White Spanish colonists and the natives who were enslaved on Spanish plantations. Today, some Latinos identify as White, others as Black, and others by their specific national and cultural origins. The very different histories of those categorized as Latino are matched by significant differences in class. Some may have been schooled in the most affluent settings; others may be virtually unschooled. Those of upper-class standing may have had little experience with prejudice or discrimination. Others may have been highly segregated into barrios and treated with extraordinary prejudice. Latinos who live near each other geographically in the United States and who are the same age and share similar ancestry may have substantially different experiences based on their class standing. Neither class, race, nor gender alone can be considered an adequate indicator of different group experiences. As you can see in ▲ Figure 8-7, even one's household status affects class standing.

The Race–Class Debate

The relationship between race and class is much debated among sociologists. The Black middle class goes all the way back to the small numbers of free Blacks in the eighteenth and nineteenth centuries (Frazier 1957), expanding in the twentieth century to include those who were able to obtain an education and become established in industry, business, or a profession. Likewise, Latinos include those who held land for years prior to the U.S. annexation of Mexican lands in the southwest of the country. The Latino class structure is particularly complex given the diversity in the origins of those with this designation (Vallejo 2013).

In recent years, both the African American and Latino middle classes have expanded, primarily as the result of increased access to education and middle-class occupations for people of color (Pattillo 2013; Lacy 2007). Although middle-class Blacks and Latinos may have economic privileges that others in these groups do not have, their class standing does not make them immune to the negative effects of race. Asian Americans also have a significant middle class, but they have also been stereotyped as the most successful minority group because of their presumed educational achievement, hard work, and thrift. This stereotype is

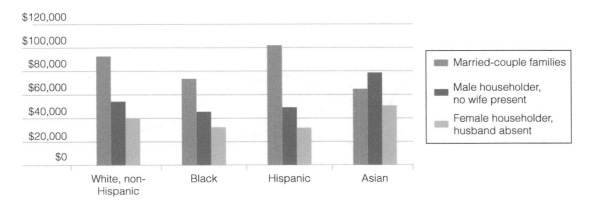

▲ Figure 8-7   **Median Income by Race and Family Structure, 2017.** As illustrated in this graph, married-couple households have the highest median income in all racial–ethnic groups; female-headed households, the least. Which groups reach median income status ($61,372 for households in 2017), and what does this tell you about the combined influence of family type, gender, and race/ethnicity?

Source: U.S. Census. 2018. *Detailed Income and Poverty Tables, FINC-02. Age of Reference Person, by Total Money Income, Type of Family, Race and Hispanic Origin of Reference Person.* Washington, DC: U.S. Census Bureau. **www.census.gov**

referred to as the *myth of the model minority* and includes the idea that a minority group must adopt alleged dominant group values to succeed. This myth about Asian Americans obscures the significant obstacles to success that Asian Americans encounter. It also ignores the hard work and educational achievements of other racial and ethnic groups. The idea that Asian Americans are the "model minority" also obscures the high rates of poverty among many Asian American groups (Lee 2009; Chou and Feagin 2008).

Despite recent successes, many in the Black middle class have a tenuous hold on this class status. The Black middle class remains as segregated from Whites as the Black poor, and continuing racial segregation in neighborhoods means that Black middle-class neighborhoods are typically closer to Black poor neighborhoods than the White middle-class neighborhoods are to White poor ones. This exposes many in the Black middle class to some of the same risks as those in poverty. This is not to say that the Black middle class has the same experience as the poor, but it challenges the view that the Black middle class "has it all" (Pattillo 2013; Lacy 2007).

The Influence of Gender and Age

Despite decades of legislation in place to protect women from discrimination and to provide equal pay for equal work, women's median income still lags behind that of men. The median income for women, even among those employed full-time, is far below the national median income level. In 2017, when median income for men working year-round and full-time was $52,146, the median income of women working year-round, full-time was $41,977—80 percent of men's income (Fontenot, Semega, and Kollar 2018). This is largely because most women work in gender-segregated jobs, a phenomenon we will explore further in Chapter 11 on gender inequality. Women have clearly improved their economic status in the United States, although a careful look at the data show that most gains have been for women at the top. As one example, between 2000 and 2016, women in the bottom 10 percent of income earners increased their wages by 2 percent; women in the top five percent, by 17 percent (Economic Policy Institute 2017).

Age, too, is a significant source of stratification. Children are the most likely age group to be poor. A whopping 17.5 percent of children live in poverty in the United States—almost thirteen million children. Although many elderly people are now poor (9.2 percent of those age 65 and over), far fewer in this age category are poor than was the case not many years ago (see ▲ Figure 8-8; Fontenot, Semega, and Kollar 2018).

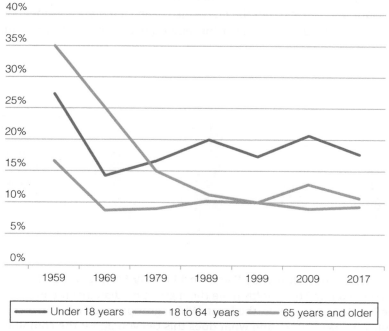

▲ **Figure 8-8   Poverty by Age 1959–2017**

Source: Fontenot, Kayla, Jessica Semega, and Melissa Kollar. 2018. *Income and Poverty in the United States: 2017*. Washington, DC: U.S. Census Bureau. **www.census.gov**

This shift reflects the greater prosperity of the older segments of the population—a trend that is likely to continue as the current large cohort of middle-class baby boomers grows older. This cohort of older Americans is also largely White—unlike the more diverse racial–ethnic composition of young people.

Social Mobility: Myths and Realities

Popular legends extol the possibility of anyone becoming rich in the United States. The rich are admired not just for their style of life but also for their supposed drive and diligence. The admiration for those who rise to the top makes it seem like anyone who is clever enough and works hard can become fabulously rich. The assumption is that the United States class system is a **meritocracy**—that is, a system in which one's status is based on merit or accomplishments, not other social characteristics. As the word suggests, people in a meritocracy move up and down through the class system based on merit, not based on other characteristics. Is this the case in the United States?

Defining Social Mobility

Social mobility is a person's movement over time from one class to another. Social mobility can be up or down, although the American dream emphasizes upward movement. Mobility can be either *inter*generational, occurring between generations, as when a daughter rises above the class of her mother or father, or *intra*generational, occurring within a generation, as when a person's class status changes as the result of business success (or disaster).

Societies differ in the extent to which social mobility is permitted. Some societies are based on *closed class systems*, in which movement from one class to another is virtually impossible. In a caste system, for

A Sociological Eye On the Media

Reproducing Class Stereotypes

The media have a substantial impact on how people view the social class system and different groups within it. Especially because people tend to live and associate with people in their own class, how they see others can be largely framed by the portrayal of different class groups in the media. Research has found this to be true and, in addition, has found that mass media have the power to shape public support for policies on public assistance.

To begin with, the media overrepresent the lifestyle of the most comfortable classes. Rare are the families that can afford the home decor and fashion depicted in soap operas, ironically most likely watched by those in the working class. Media portrayals, such as those found on television talk shows as well as sports, tend to emphasize stories of upward mobility. When the working class is depicted, it tends to be shown as deviant, reinforcing class antagonism and giving middle-class viewers a sense of moral superiority (Grindstaff 2002).

Content analyses of the media also find that the poor are largely invisible in the media (Mantsios 2010). Those poor people who are depicted in television and magazines are more often portrayed as Black, leading people to overestimate the actual number of Black poor. The elderly and working poor are rarely seen. Representations of welfare overemphasize themes of dependency, especially when portraying African Americans. Women are also more likely than men to be represented as dependent (Misra et al. 2003). How might people understand class inequality if the media routinely presented the social structural context of class differences?

Sources: Mantsios, Gregory. 2010. "Media Magic: Making Class Invisible." Pp. 386–394 in *Race, Class, and Gender: An Anthology*, edited by Margaret L. Andersen and Patricia Hill Collins. Belmont, CA: Wadsworth; Grindstaf, Laura. 2002. *The Money Shot, Trask, Class, and the Making of TV Talk Shows*. Chicago: University of Chicago Press, 269–296; Misra, Joy, Stephanie Moller, and Marina Karides. 2003. "Envisioning Dependency: Changing Media Depictions of Welfare in the 20th Century." *Social Problems* 50 (November): 482–504. See also the film, *Class Dismissed: How TV Frames the Working Class*, Media Education Foundation. **www.mediaed.org**

example, mobility is strictly limited by the circumstances of one's birth. At the other extreme are *open class systems*, in which placement in the class system is based on individual achievement, not ascription. In open class systems, there are relatively loose class boundaries, high rates of class mobility, and weak perceptions of class difference. Research on social mobility shows that, despite popular beliefs to the contrary, the mobility in the United States is actually less than that in fifteen other industrialized nations, including France, Spain, Germany, Japan, Pakistan, South Korea, Canada, and others (Corak 2016).

The Extent of Social Mobility

To what extent does social mobility occur in the United States? Social mobility is much more limited than people believe. Success stories of social mobility do occur, but research finds that experiences of mobility over great distances are rare, certainly far less than believed. Most people remain in the same class as their parents. What mobility exists is typically short in distance, and some people actually drop to a lower status, referred to as *downward social mobility*.

The fact is that those born at both the top and bottom end of the income ladder are very likely to remain there. Rates of upward social mobility are highest among White men, followed by White women, then Black men, and finally, Black women. Family social mobility has, of late, been somewhat helped by the increased participation of women in the labor force, yet research still finds that an increase in men's wages best predicts upward social mobility for families (Economic Mobility Project 2014).

Social mobility is influenced most by factors that affect the whole society, not just by individual characteristics. Just being born in a particular generation can have a significant influence on one's life chances. The fears of today's young, middle-class people that they will be unable to achieve the lifestyles of their parents show the effect that being in a particular generation can have on one's life chances. When mobility occurs, it is usually because of societal changes that create or restrict opportunities, including such changes as economic cycles, changes in the occupational structure, and demographic factors, such as the number of college graduates in the labor force (Beller and Hout 2006). Yet, mobility in the United States is not impossible. Indeed, many have immigrated to this nation with the knowledge that their life chances are better here than in their countries of origin. The social mobility that does exist is greatly influenced by education. In sum, however, social mobility is much more limited than the American dream suggests.

Class Consciousness

Because of the widespread belief that mobility is possible, people in the United States, compared to many other societies, tend not to be very conscious of class. **Class consciousness** is the perception that a class structure exists along with a feeling of shared identification with others in one's class—that is, those with whom you share life chances (Centers 1949). Notice that there are two dimensions to class consciousness: (1) the idea that a class structure exists; and (2) one's class identification.

There has been a long-standing argument that Americans are not very class conscious because of the belief that upward mobility is possible and because of the belief in individualism that is part of the culture. Images of opulence also saturate popular culture, making it seem that such material comforts are available to anyone. The faith that upward mobility is possible ironically perpetuates inequality because, if people believe that everyone has the same chances of success, they are likely to think that whatever inequality exists must be fair or the result of individual success and failure.

Class inequality in any society is usually buttressed by ideas that support (or actively promote) inequality. Beliefs that people are biologically, culturally, or socially different can be used to justify the higher position of some groups. If people believe these ideas, the ideas provide legitimacy for the system. Karl Marx used the term **false consciousness** to describe the class consciousness of subordinate classes who had internalized the view of the dominant class. Marx argued that the ruling class controls subordinate classes by infiltrating their consciousness with belief systems that are consistent with the interests of the ruling class. If people accept these ideas, which seem to justify inequality, they need not be overtly coerced into accepting the roles designated for them by the ruling class.

There have been times when class consciousness was higher, such as during the labor movement of the 1920s and 1930s. Then working-class people had a very high degree of class consciousness and mobilized on behalf of workers' rights. We see this happening again with the stagnation in income for many. Now, 28 percent of the public identify as working class (Newport 2017). The formation of a relatively large middle class and a relatively high standard of living mitigates class discontent. Racial and ethnic divisions also make strong alliances within various classes less stable. Growing inequality could result in a higher degree of class consciousness, but this has not yet developed into a significant class-based movement for change.

Why Is There Inequality?

Stratification occurs in all societies. Why? This question originates in classical sociology in the works of Karl Marx and Max Weber, theorists whose work continues to inform the analysis of class inequality today.

Karl Marx: Class and Capitalism

Karl Marx (1818–1883) provided a complex and profound analysis of the class system under capitalism—an analysis that, although more than 100 years old, continues to inform sociological analyses. Marx defined classes in relationship to the *means of production*, defined as the system by which goods are produced and distributed. In Marx's analysis, two primary classes exist under capitalism: the *capitalist class*, those who own the means of production, and the *working class* (or proletariat), those who sell their labor for wages. There are further divisions within these two classes: the *petty bourgeoisie*, small business owners and managers (those whom you might think of as middle class) who identify with the interests of the capitalist class but do not own the means of production, and the *lumpenproletariat*, those who have become unnecessary as workers and are then discarded. (Today, these would be the underclass, the homeless, and the permanently poor.)

Marx thought that with the development of capitalism, the capitalists and working class would become increasingly antagonistic (something he referred to as class struggle). As class conflicts became more intense, Marx predicted that the two classes would become more polarized, with the petty bourgeoisie becoming deprived of their property and dropping into the working class. Marx would see now that his analysis was correct as classes are becoming more polarized with an increasing gap between the very rich and everyone else.

In addition to the class struggle that Marx thought would characterize the advancement of capitalism, he also thought that capitalism was the basis for other social institutions. To Marx, capitalism is the *infrastructure* of society, with other institutions (such as law, education, the family, and so forth) reflecting capitalist interests. According to Marx, the law supports the interests of capitalists; the family promotes values that socialize people into appropriate work roles; and education reflects the interests of the capitalist class. Over time, Marx argued, capitalism would increasingly penetrate society. This can be clearly seen in the way that the profit motive permeates contemporary institutions, such as in how profits of the pharmaceutical industry permeate the delivery of health care (see also Chapter 14). You might ask yourself where you see the values of capitalism permeating other social institutions.

Why do people support such a system? Here is where ideology plays a role. **Ideology** refers to belief systems that support the status quo. According to Marx, the ruling class produces the dominant ideas of a society. Through their control of the communications industries in modern society, the ruling class is able to produce ideas that buttress their interests.

Much of Marx's analysis boils down to the consequences of a system based on the pursuit of profit. If goods were exchanged at the cost of producing them, no profit would be produced. Capitalist owners want to sell commodities for more than their actual value—more than the cost of producing them, including materials and labor. Because workers contribute value to the system and capitalists extract value, Marx saw capitalist profit as the exploitation of labor. Marx believed that as profits became increasingly concentrated in the hands of a few capitalists, the working class would become increasingly dissatisfied. The basically exploitative character of capitalism, according to Marx, would ultimately lead to its destruction as workers organized to overthrow the rule of the capitalist class. *Class conflict* between workers and capitalists, he argued, was inescapable, with revolution being the inevitable result. Perhaps the class revolution that Marx predicted has not occurred, but the dynamics of capitalism that he analyzed are unfolding before us.

At the time Marx was writing, the middle class was small and consisted mostly of small business owners and managers. Marx saw the middle class as dependent on the capitalist class, but exploited by it, because the middle class did not own the means of production. He saw middle-class people as identifying with the interests of the capitalist class because of the similarity in their economic interests and their dependence on the capitalist system. Marx believed that the middle class failed to work in its own best interests because it falsely believed that it benefited from capitalist arrangements. Marx thought that in the long run, the middle class would pay for their misplaced faith when profits became increasingly concentrated in the hands of a few and more and more of the middle class dropped into the working class. Because he did not foresee the emergence of the large and highly differentiated middle class we have today, not every part of Marx's theory has proved true. Still, his analysis provides a powerful portrayal of the forces of capitalism and the tendency for wealth to belong to a few, whereas the majority work only to make ends meet. Marx has also influenced the lives of billions of people under self-proclaimed Marxist systems that were created in an attempt, however unrealized, to overcome the pitfalls of capitalist society.

Max Weber: Class, Status, and Party

Max Weber (1864–1920) agreed with Marx that classes were formed around economic interests and that economic forces have a powerful effect on people's lives. He disagreed with Marx, however, that economic forces are the primary dimension of stratification. Weber saw three dimensions to stratification:

- *class* (the economic dimension);
- *status* (or prestige, the cultural and social dimension); and,
- *party* (or power, the political dimension).

Weber is thus responsible for a *multidimensional view* of social stratification because he analyzed the connections between economic, cultural, and political systems. Weber pointed out that, although the economic, social, and political dimensions of stratification are usually related, they are not always consistent. Someone could be high on one or two dimensions, but low on another. A major drug dealer is an example: high wealth (economic dimension) and power over others (political dimension) but low prestige (social dimension), at least in the eyes of the mainstream society.

Weber defined *class* as the economic dimension of stratification—how much access to the material goods of society a group or individual has, as measured by income, property, and other financial assets. A family with an income of $200,000 per year clearly has more access to the resources of a society than a family living on an income of $50,000 per year. Weber understood that a class has common economic interests and that economic status is the basis for one's life chances. But, in addition, he thought that people were also stratified based on their status and power differences.

Status, to Weber, is the prestige dimension of stratification—the social judgment or recognition given to a person or group. Weber understood that class distinctions are linked to status distinctions. That is, those with the most economic resources tend to have the highest status in society, but not always. In a local community, for example, those with the most status may be those who have lived there the longest, even if newcomers arrive with more money. Although having power is typically related to also having high economic standing and high social status, this is not always the case, as you saw with the example of the drug dealer.

Finally, *party* (or what we would now call power) is the political dimension of stratification. Power is the capacity to influence groups and individuals even in the face of opposition. Power is also reflected in the ability of a person or group to negotiate their way through social institutions. An unemployed Latino man wrongly accused of a crime, for instance, does not have much power to negotiate his way through the criminal justice system. By comparison, business executives accused of corporate crime can afford expensive lawyers and thus frequently go unpunished or, if they are found guilty, serve relatively light sentences in comparatively pleasant facilities. Again, Weber saw power as linked to economic standing, but he did not think that economic standing was always the determining cause of people's power.

Marx and Weber explain different features of stratification. Both understood the importance of the economic basis of stratification, and they knew the significance of class for determining the course of one's life. Marx saw people as acting primarily out of economic interests. Weber refined the sociological analyses of stratification to account for the subtleties that can be observed when you look beyond the sheer economic dimension to stratification, stratification being the result of economic, social, and political forces. Together, Marx and Weber provide compelling theoretical grounds for understanding the contemporary class structure.

Functionalism and Conflict Theory: The Continuing Debate

Marx and Weber were trying to understand why differences existed in the resources that various groups in society hold. The question persists of why there is inequality. Two major frameworks in sociological theory—functionalist and conflict theory—take quite different approaches to understanding inequality (see ◆ Table 8-2).

The Functionalist Perspective on Inequality

Functionalist theory views society as a system of institutions organized to meet society's needs (see Chapter 1). The functionalist perspective emphasizes that the parts of society are in basic harmony with each other; society is held together by cohesion, consensus, cooperation, stability, and persistence (Eitzen and Baca Zinn 2012; Merton 1957; Parsons 1951a). Different parts of the social system complement one another. To explain stratification, functionalists see the roles filled by the upper classes—such as governance, economic innovation, investment, and management—as essential for a cohesive and smoothly running society. The upper classes are then rewarded in proportion to their contribution to the social order (Davis and Moore 1945).

According to the functionalist perspective, social inequality serves an important purpose in society: It motivates people to fill the different positions in society that are needed for the survival of the whole. Functionalists think that some positions in society are more important than others and require the most

Table 8-2   Functionalist and Conflict Theories of Stratification

Interprets	Functionalism	Conflict Theory
Inequality	The purpose of inequality is to motivate people to fill needed positions in society.	Inequality results from a system where those with the most resources exploit and control others.
Reward system	Greater rewards are attached to higher positions to ensure that people will be motivated to train for functionally important roles in society.	Inequality prevents the talents of those at the bottom from being discovered and used.
Classes	Some groups are rewarded because their work requires the greatest degree of talent and training.	Classes conflict with each other as they vie for power and economic, social, and political resources.
Elites	The most talented are rewarded in proportion to their contribution to the social order.	The most powerful reproduce their advantage by distributing resources and controlling the dominant value system.
Class consciousness/ideology	Beliefs about success and failure confirm the status of those who succeed.	Elites shape societal beliefs to make their unequal privilege appear to be legitimate and fair.
Poverty	Poverty serves economic and social functions in society.	Poverty is inevitable because of the exploitation built into the system.
Social policy	Because the system is basically fair, social policies should only reward merit.	Because the system is basically unfair, social policies should support disadvantaged groups.

talent and training. The rewards attached to those positions (such as higher income and prestige) ensure that people will make the sacrifices needed to acquire the training for functionally important positions (Davis and Moore 1945). Higher class status thus comes to those who acquire what is needed for success (such as education and job training). In other words, functionalist theorists see inequality as based on a reward system that motivates people to succeed.

The Conflict Perspective on Inequality

Conflict theory also sees society as a social system, but unlike functionalism, conflict theory interprets society as being held together through conflict and coercion. From a conflict-based perspective, society comprises competing interest groups, some with more power than others. Groups struggle over societal resources and compete for social advantage. Conflict theorists argue that those who control society's resources also hold power over others. The powerful are also likely to act to reproduce their advantage and try to shape societal beliefs to make their privileges appear to be legitimate and fair. In sum, conflict theory emphasizes the friction in society rather than the coherence, and sees society as dominated by elites.

From the perspective of conflict theory, social stratification is based on class conflict and blocked opportunity. Conflict theorists see stratification as a system of domination and subordination in which those with the most resources exploit and control others. They also see the different classes as in conflict with each other, with the unequal distribution of rewards reflecting the class interests of the powerful, not the survival needs of the whole society (Eitzen, Baca Zinn and Eitzen 2016). According to the conflict perspective, inequality provides elites with the power to distribute resources, make and enforce laws, and control value systems. Elites then use these powers to reproduce their own advantage. Others in the class structure, especially the working class and the poor, experience blocked mobility.

From a conflict point of view, the more stratified a society, the less likely that society will benefit from the talents of its citizens. Inequality limits the life chances of those at the bottom, preventing their talents from being discovered and used.

The Debate between Functionalist and Conflict Theories

Implicit in the argument of each perspective is criticism of the other perspective. Functionalism assumes that the most highly rewarded jobs are the most important for society, whereas conflict theorists argue that some of the most vital jobs in society—those that sustain life and the quality of life, such as farmers, mothers, trash collectors, and a wide range of other laborers—are usually the least rewarded. Conflict theorists also criticize functionalist theory for assuming that the most talented get the greatest rewards. They point out that systems of stratification tend to devalue the contributions of those left at the bottom and to underutilize the diverse talents of all people (Tumin 1953). In contrast, functionalist theorists contend that the conflict view of how economic interests shape social organization is too simplistic. Conflict theorists respond by arguing that functionalists hold too conservative a view of society and overstate the degree of consensus and stability that exists.

The debate between functionalist and conflict theorists raises fundamental questions about how people view inequality. Is inequality inevitable? How is inequality maintained? Do people basically accept it? This debate is not just academic. The assumptions made from each perspective frame public policy debates. Whether the topic is taxation, health care, or poverty, if people believe that anyone can get ahead by ability alone, they will tend to see the system of inequality as fair and accept the idea that there should be a differential reward system. Those who tend toward the conflict view of the stratification system are more likely to advocate programs that emphasize public responsibility for the well-being of all groups and to support programs and policies that result in more of the income and wealth of society going toward the needy.

Poverty

Despite the relatively high average standard of living in the United States, poverty afflicts millions of people. There are now almost 40 million poor people in the United States—12.3 percent of the population (Proctor, Semega, and Kollar 2018). Even more startling is the large number of people living in very deep poverty, or what experts define as *extreme poverty* (the U.S. measure being living on two dollars or less per day; the world measure of extreme poverty is $1.25 per day or less; see also Chapter 9). Extreme poverty in the United States includes 3.5 million children who are living with virtually no income—a shocking fact for such a rich nation (Shaefer and Edin 2014).

Poverty deprives people of basic human needs—food, shelter, and safety from harm. It is also the basis for many of our nation's most intractable social problems. Failures in the education system; crime and violence; inadequate housing and homelessness; poor health care—all are related to poverty. Who is poor, and why is there so much poverty in an otherwise affluent society?

Defining Poverty

The federal government has established an official definition of poverty used to determine eligibility for government assistance and to measure the extent of poverty in the United States. The **poverty line** is the amount of money needed to support the basic needs of a household, as determined by government; below this line, one is considered officially poor. To determine the poverty line, the Social Security administration takes a low-cost food budget (based on dietary information provided by the U.S. Department of Agriculture) and multiplies it by a factor of three, assuming that a family spends approximately one-third of its budget on food. The resulting figure is the official poverty line, adjusted slightly each year for increases in the cost of living. In 2017, the official poverty line for a family of four (including two children) was $24,858. Although a cutoff point is necessary to administer antipoverty programs, this definition of poverty can be misleading. A person or family earning $1 above the cutoff point would not be officially categorized as poor.

There are numerous problems with the official definition of poverty. To name a few, it does not account for regional differences in the cost of living. It does not reflect changes in the cost of housing nor changes in the cost of modern standards of living that were not imagined in the 1930s, when the definition was established (Meyer and Sullivan 2012). Experts have argued that the government should develop alternative poverty measures, such as *shelter poverty*—a measure that would account for the cost of housing in different regions (Stone 1993). To date, Congress has resisted changing the official definition of poverty—a change that would likely increase the reported rate of poverty and potentially increase the cost of federal antipoverty programs.

Who Are the Poor?

Since the 1960s when poverty was more than 20 percent, poverty has declined significantly. Poverty is higher now than it was in 2000, but it has been slowly inching down since 2010. By 2017, 12.3 percent of the population was officially poor—that is, fell under the federal poverty line (Fontenot, Semega, and Kollar 2018). The majority of the poor are White, although there are disproportionately high rates of poverty among Asian Americans, Native Americans, Black Americans, and Hispanics. Twenty-one percent of African Americans are below the official poverty line; 18.3 percent of Hispanics, 9.8 percent of Asian Americans, and 8.7 percent of non-Hispanic Whites (Fontenot, Semega, and Kollar 2018). Among Hispanics, there are further differences among groups. Puerto Ricans—the Hispanic group with the lowest median income—have been most likely to suffer increased poverty, probably because of their concentration in the poorest segments of the labor market and their high unemployment rates. Asian American

See for Yourself

Using the current federal poverty line of $24,858 for a family of four, including two children), develop a monthly budget that does not exceed this income level and that accounts for all of your family's needs. Base your budget on the actual costs of such things in your locale (rent, food, transportation, utilities, clothing, and so forth). Don't forget to account for taxes (state, federal, and local), health care expenses, your children's education, and so on. What does this exercise teach you about those who live below the poverty line?

Child poverty in the United States is higher than one would expect for such an otherwise materially well-off nation.

poverty is highest among the most recent immigrant groups, including Laotians, Cambodians, Vietnamese, Chinese, and Korean immigrants; Filipino, Japanese, and Asian Indian families have lower rates of poverty (White House 2012).

The vast majority of the poor have always been women and children, but the percentage of women and children considered to be poor has increased in recent years. The term **feminization of poverty** refers to the large proportion of the poor who are women and children. This trend results from several factors, including the dramatic growth of female-headed households, a decline in the proportion of the poor who are elderly (not matched by a decline in the poverty of women and children), and continuing wage inequality between women and men. The large number of poor women is associated with a commensurate large number of poor children. By 2017, almost 18 percent of all children in the United States (those under age 18) were poor, including 9.5 percent of non-Hispanic White children, 28 percent of Black children, 25 percent of Hispanic children, and 11.3 percent of Asian American children (Fontenot, Semega, and Kollar 2018).

Families headed by women are far more likely to be poor than other families (see ▲ Figure 8-9), a fact resulting from the low wages of women, the lack of child care, and the absence of child support from men, especially when the fathers of children are young themselves. Young mothers are increasingly unlikely to have the contributing income of a spouse and are more likely than men to live with their

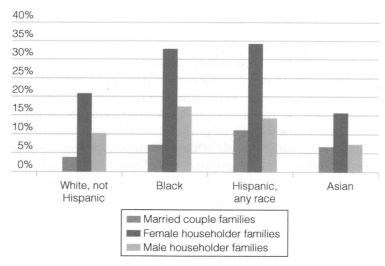

▲ **Figure 8-9   Poverty Status by Family Type and Race, 2017**

Note: Families with children under 18 present.

Source: U.S. Census Bureau. 2018. *POV-02-People in Families by Family Structure, Age and Sex, Iterated by Income-to-Poverty Ratio and Race*. Washington, DC: U.S. Census Bureau. **www.census.gov**

Debunking Society's Myths

Myth: Marriage is a good way to reduce women's dependence on welfare.
Reality: Although it is true that married-couple households are less likely to be poor than single-headed households, forcing women to marry encourages women's dependence on men and punishes women for being independent. Research indicates that poor women place a high value on marriage and want to be married, but also understand that men's unemployment and instability makes their ideal of marriage unattainable. In addition, large numbers of women receiving welfare have been victims of domestic violence (Edin and Kefalas 2005; Scott et al. 2002).

children and be financially responsible for them. Although teen birth rates have declined significantly among all races, when young women have children, the odds of their being married (and thus having two incomes available) have declined.

It is very important to remember that the poor are not a uni-dimensional group. They are racially diverse, including Whites, Blacks, Hispanics, Asian Americans, and Native Americans. They are diverse in age, including not just children and young mothers, but also men and women of all ages, and especially a substantial number of the elderly, many of whom live alone. The poor are also geographically diverse, to be found in areas east and west, south and north, urban and rural.

As ■ map 8-1 shows, poverty rates are generally higher in the South and Southwest. What the map cannot show, however, is concentrated poverty. **Concentrated poverty** occurs when 40 percent or more of those in a given census area live below the federal poverty line. Hispanic, Native American, and Black Americans are the most likely groups to live in areas with concentrated poverty, although the rate has also doubled for non-Hispanic Whites since 2000 (Jargowsky 2015; Kneebone 2014). Concentrated poverty is known to be linked to higher crime rates, more violence, poor schools, joblessness, less family stability, and other problems that plague such residential areas.

One marked change in poverty is the growth of poverty in suburban areas. One-third of the nation's poor are now found in suburbs where poverty is growing twice as fast as in center cities (Kneebone and Holmes 2014). Rural poverty also persists in the United States, even though people tend to think of poverty as an urban phenomenon (Fontenot, Semega, and Kollar 2017).

Despite the idea that the poor "milk" the system, government supports for the poor are limited. So-called welfare is now largely in the form of food stamps, not cash assistance. To be eligible for food stamps (Supplemental Nutritional Assistance Program or SNAP), a household must meet three criteria: (1) have a gross family income below 130 percent of the federal poverty line; (2) have a net income (that is, after deductions are applied) below the federal poverty line; and, (3) have total assets under $2250 ($3500 if there is an elderly or disabled person in the household); such assets only include assets that are actually accessible, such as a savings account. You can see that despite the stereotypes of people who use food stamps, the resources of such households are slim indeed. The average food stamp benefit is $125 per person per month, or $1.40 per person per meal (Center on Budget and Policy Priorities 2017). Three-quarters of households receiving food stamps include children; 12 percent have a disabled person present; and 10 percent include the elderly. Forty percent of SNAP participants are White, 26 percent are African-American, 11 percent are Hispanic, 2.4 percent are Asian, and 1 percent are Native American (U.S. Department of Agriculture 2014).

Poverty is also evident in the numbers of homeless people in the United States. Depending on how one defines and measures homelessness, estimates of the number of homeless people vary widely. If you count the number of homeless on any given night, there are estimated to be about 550,000 homeless people. The transient nature of this population makes accurate estimates of the extent of homelessness impossible, but careful monitoring has shown that the numbers of homeless people have been increasing. One-third of the homeless are young people (under age 24). Most of the homeless (65 percent) are individuals, but over a third (35 percent) are families.

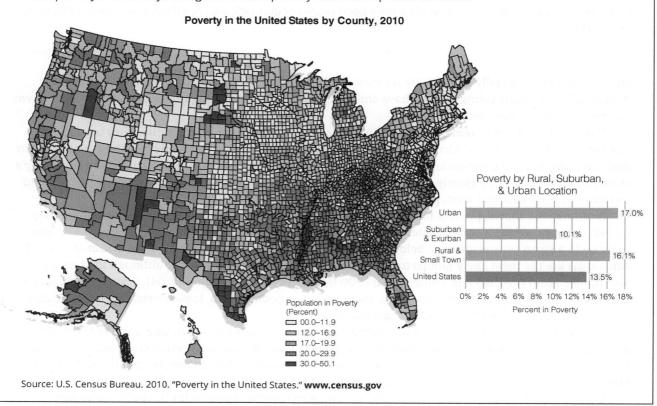

map 8-1 Mapping America's Diversity: Poverty in the United States

This map shows regional differences in poverty rates (that is, the percentage of poor in different counties). As you can see, poverty is highest in the South, Southwest, and some parts of the upper Midwest. This reflects the higher rates of poverty among Native Americans, Latinos, and African Americans, especially in rural areas. What the map does not show is the concentration of poverty in particular urban areas. According to this map, how much poverty is there in your region? Is there poverty that the map does not show?

Poverty in the United States by County, 2010

Poverty by Rural, Suburban, & Urban Location

Urban	17.0%
Suburban & Exurban	10.1%
Rural & Small Town	16.1%
United States	13.5%

0% 2% 4% 6% 8% 10% 12% 14% 16% 18%
Percent in Poverty

Population in Poverty (Percent)
- 00.0–11.9
- 12.0–16.9
- 17.0–19.9
- 20.0–29.9
- 30.0–50.1

Source: U.S. Census Bureau. 2010. "Poverty in the United States." **www.census.gov**

Family homelessness has also increased significantly in recent years, and many families have experienced eviction from their homes (National Coalition for the Homeless 2017; Desmond 2016). In fact, families are the fastest-growing segment of the homeless—40 percent—and children are also 40 percent of the homeless. Moreover, many of the women with children who are homeless have fled from domestic violence. A shocking number of the homeless are veterans (about 11 percent), including those returning from Iraq and Afghanistan (National Coalition for the Homeless 2017).

There are many reasons for homelessness. The great majority of the homeless are on the streets because of a lack of affordable housing and an increase in poverty, leaving many people with no choice but to live on the street. Add to that problems of inadequate health care, domestic violence, and addiction, and you begin to understand the factors that create homelessness. Some of the homeless have mental illness (about 16 percent of single, homeless adults); the movement to relocate patients requiring mental health care out of institutional settings has left many without mental health resources that might help them (National Coalition for the Homeless 2014).

Causes of Poverty

Most agree that poverty is a serious social problem. There is far less agreement on what to do about it. Public debate about poverty hinges on disagreements about its underlying causes. Two points of view prevail: Some blame the poor for their own condition, and some look to social structural causes to explain poverty. The first view, popular with the public and many policymakers, is that poverty is caused by the cultural habits of the poor. According to this point of view, behaviors such as crime, family breakdown, lack of ambition, and educational failure generate and sustain poverty, a syndrome to be treated by forcing the poor to fend for themselves. The second view is more sociological, one that understands poverty as rooted in the structure of society, not in the morals and behaviors of individuals.

Blaming the Victim: The Culture of Poverty

Blaming the poor for being poor stems from the myth that success requires only individual motivation and ability. Many in the United States adhere to this view and hence have a harsh opinion of the poor. This attitude is also reflected in U.S. public policy concerning poverty, which is rather ungenerous compared with other industrialized nations. Those who blame the poor for their own plight typically argue that poverty is the result of early childbearing, drug and alcohol abuse, refusal to enter the labor market, and crime. Such thinking puts the blame for poverty on individual choices, not on societal problems. In other words, it blames the victim, not the society, for social problems (Ryan 1971).

The **culture of poverty** argument attributes the major causes of poverty to the absence of work values and the irresponsibility of the poor (Lewis 1969, 1966). In this light, poverty is seen as a dependent way of life that is transferred, like other cultural values, from generation to generation. Policymakers have adapted the culture of poverty argument to argue that the actual causes of poverty are found in the breakdown of major institutions, including the family, schools, and churches.

Is the culture of poverty argument true? To answer this question, we might ask: Is poverty transmitted across generations? Researchers have found only mixed support for this assumption. Many of those who are poor remain poor for only one or two years; only a small percentage of the poor are chronically poor. More often, poverty results from a household crisis, such as divorce, illness, unemployment, or parental death. People tend to cycle in and out of poverty. The public stereotype that poverty is passed through generations is thus not well supported by the facts.

A second question is: Do the poor want to work? The persistent public stereotype that they do not is central to the culture of poverty thesis. This attitude presumes that poverty is the fault of the poor, that poverty would go away if they would only change their values and adopt the American work ethic. What is the evidence for these claims?

Detailed studies of the poor simply find no basis for the assumption that the poor hold different values about work compared to everyone else (Lakso 2013; Lee and Anat 2008). They simply find that work is difficult to find. Several other facts also refute this popular claim.

Most of the able-bodied poor *do* work, even if only part-time. As we saw previously, the number of workers who constitute the *working poor* has actually increased. You can see why this is true when you calculate the income of someone working full-time for minimum wage. Someone working forty hours per week, fifty-two weeks per year, at minimum wage will have an income far below the poverty line. This is the major reason that many have organized a living wage campaign, intended to raise the federal minimum wage to provide workers with a decent standard of living.

Debunking Society's Myths

Myth: The influx of unskilled immigrants raises the poverty level by taking jobs away from U.S. citizens who would otherwise be able to find work and lift themselves from poverty.

Sociological Perspective: A state-by-state analysis of poverty and immigration in recent years finds that immigrants actually improve local economies by increasing the supply of workers, generating labor market expansion, and promoting entrepreneurship—in other words, stimulating the economy. In some regions, immigration has actually lessened poverty (Peri 2014).

Current policies that force those on welfare to work also tend to overlook how difficult it is for poor people to retain the jobs they get. Prior to welfare reform in the mid-1990s, poor women who went off welfare to take jobs often found they soon had to return to welfare because the wages they earned were not enough to support their families. Leaving welfare often means losing health benefits, yet incurring increased living expenses. The jobs that poor people find often do not lift them out of poverty. In sum, attributing poverty to the values of the poor is both unproven and a poor basis for public policy (Greenbaum 2015).

Structural Causes of Poverty

From a sociological point of view, the underlying causes of poverty lie in the economic and social transformations taking place in the United States. Careful scholars do not attribute poverty to a single cause. There are many causes. Two of the most important are: (1) the restructuring of the economy, which has resulted in diminished earning power and increased unemployment; and, (2) the status of women in the family and the labor market, which has contributed to women being overrepresented among the poor. Add to these underlying conditions the federal policies in recent years that have diminished social support for the poor in the form of welfare, public housing, and job training. Given these reductions in federal support, it is little wonder that poverty is so widespread.

The restructuring of the economy has caused the disappearance of manufacturing jobs, traditionally an avenue of job security and social mobility for many workers, especially African American and Latino workers (Wilson 1996). The working class has been especially vulnerable to these changes. Economic decline in those sectors of the economy where men have historically received good pay and good benefits means that fewer men are the sole support for their families. Most families now need two incomes to achieve a middle-class way of life. The new jobs that are being created fall primarily in occupations that offer low wages and few benefits; they also tend to be filled by women, especially women of color, leaving women poor and men out of work. Such jobs offer little chance to get out of poverty. New jobs are also typically located in neighborhoods far away from the poor, creating a mismatch between the employment opportunities and the residential base of the poor.

Declining wage rates caused by transformations taking place within the economy fall particularly hard on young people, women, and African Americans and Latinos, the groups most likely to be among the working poor. The high rate of poverty among women is also strongly related to women's status in the family and the labor market. Divorce is one cause of poverty, because without a male wage in the household, women are more likely to be poor. Women's child-care responsibilities make working outside the home on marginal incomes difficult. Many women with children cannot manage to work outside the home, because it leaves them with no one to watch their children. More women now depend on their own earnings to support themselves, their children, and other dependents. Whereas unemployment has always been considered a major cause of poverty among men, low wages play a major role for women.

The persistence of poverty also increases tensions between different classes and racial groups. William Julius Wilson, one of the most noted analysts of poverty and racial inequality, has written, "The ultimate basis for current racial tension is the deleterious effect of basic structural changes in the modern American economy on Black and White lower-income groups, changes that include uneven economic growth, increasing technology and automation, industry relocation, and labor market segmentation" (1978: 154). Wilson's comments demonstrate the power of sociological thinking by convincingly placing the causes of both poverty and racism in their societal context, instead of the individualistic thinking that tends to blame the poor for their plight.

Welfare and Social Policy

The 1996 Personal Responsibility and Work Opportunity Reconciliation (PRWOR) Act governs current welfare policy. This federal policy eliminated the long-standing welfare program titled Aid to Families with Dependent Children (AFDC), which was created in 1935 as part of the Social Security Act. Implemented during the Great Depression, AFDC was meant to assist poor mothers and their children. This program acknowledged that some people are victimized by economic circumstances beyond their control and deserve assistance. For much of its lifetime, this law supported mostly White mothers and their children. Not until the 1960s did welfare come to be identified with Black families.

Current welfare policy gives block grants to states to administer their own welfare programs through the program called **Temporary Assistance for Needy Families (TANF)**. TANF stipulates a lifetime limit of five years for people to receive aid and requires all welfare recipients to find work within two years—a policy known as *workfare*. Those who have not found work within two years of receiving welfare can be required to perform community service jobs for free.

In addition, welfare policy denies payments to unmarried teen parents under age 18 unless they stay in school and live with an adult. It also requires unmarried mothers to identify the fathers of their children or risk losing their benefits. These broad guidelines are established at the federal level, but individual states can be more restrictive, as many have been. At the heart of public beliefs about support for the poor is the idea that public assistance creates dependence, discouraging people from seeking jobs. The very title of current welfare policy emphasizes "personal responsibility and work," suggesting that poverty is the fault of the poor. Low-income women, for example, are stereotyped as just wanting to have babies to increase the size of their welfare checks. Low-income men are also stereotyped as shiftless and irresponsible, even though research finds no support for either idea (Edin and Nelson 2013; Edin and Kefalas 2005).

Is welfare reform working? Many claim that welfare reform is working because, since passage of the new law, the welfare rolls have shrunk. Since 1996, the year that welfare reform was passed, the number receiving welfare support has declined dramatically. Having fewer people on welfare does not, however, mean that poverty is reduced. In fact, as we have seen, extreme poverty has actually increased since passage of welfare reform. Having fewer people on the rolls can simply mean that people are without a safety net. Holes in the safety net makes those already vulnerable to economic distress even more so. Single women with children and people of color have been those most negatively affected by the 1996 welfare reforms, although working families are not immune to these changes (Halpern-Meekin et al. 2015; Shaefer and Edin 2014).

Politicians brag that welfare rolls have shrunk, but reduction in the welfare rolls is a poor measure of the true impact of welfare reform because this would be true simply if people are denied benefits. Because welfare has been decentralized to the state level, studies of the impact of current law must be done on a state-by-state basis. Such studies are showing that those who have gone into workfare programs most often earn wages that keep them below the poverty line. Although some states report that family income has increased following welfare reform, the increases are slight. More people have been evicted because of falling behind on rent. Families also report an increase in other material hardships, such as phones and utilities being cut off. Marriage rates among former recipients have not changed, although more now live with nonmarital partners, most likely as a way of sharing expenses. The number of children living in families without either parent has also increased, probably because parents had to relocate to find work. Low-wage work also simply does not lift former welfare recipients out of poverty (Shaefer and Edin 2014;

Is It True?*

	True	False
1. Income growth has been greatest for those in the middle class in recent years.		
2. The average American household has most of its wealth in the stock market.		
3. Social mobility is greater in the United States than in any other Western nation.		
4. Poor teen mothers do not have the same values about marriage as middle-class people.		
5. Old people are the most likely to be poor.		
6. Poverty in U.S. suburbs is increasing.		

*The answers can be found at the end of this chapter.

Hays 2003). Forcing welfare recipients to work provides a cheap labor force for employers and potentially takes jobs from those already employed.

The public debate about welfare rages on, often in the absence of informed knowledge from socio-logical research and almost always without input from the subjects of the debate, welfare recipients themselves. Although stigmatized as lazy and not wanting to work, those who have received welfare actually believe that it has negative consequences for them, but they say they have no other viable means of support. They typically have needed welfare when they could not find work or had small children and needed child care. Most were forced to leave their last job because of layoffs or firings or because the work was only temporary. Few left their jobs voluntarily.

Welfare recipients also say that the welfare system makes it hard to become self-supporting because the wages one earns while on welfare are deducted from an already minimal subsistence. Furthermore, there is not enough affordable day care for mothers to leave home and get jobs. The biggest problem they face in their minds is lack of money. Contrary to the popular image of the conniving "welfare queen," welfare recipients want to be self-sufficient and provide for their families, but they face circumstances that make this very difficult to do. Indeed, studies of young, poor mothers find that they place a high value on marriage, but they do not think they or their boyfriends have the means to achieve the marriage ideals they cherish (Edin and Kefalas 2005; Hays 2003).

Another popular myth about welfare is that people use their welfare checks to buy things they do not need. Research finds that when former welfare recipients find work, their expenses actually go up. Although they may have increased income, their expenses (in the form of child care, clothing, transportation, lunch money, and so forth) increase, leaving them even less disposable income. Moreover, studies find that low-income mothers who buy "treats" for their children (brand-name shoes, a movie, candy, and so forth) do so because they want to be good mothers (Edin and Lein 1997).

Other beneficiaries of government programs have not experienced the same kind of stigma. Social Security supports virtually all retired people, yet they are not stereotyped as dependent on federal aid, unable to maintain stable family relationships, or insufficiently self-motivated. Spending on welfare programs is also a pittance compared with the spending on other federal programs. Sociologists conclude that the so-called welfare trap is not a matter of learned dependency, but a pattern of behavior forced on the poor by the requirements of sheer economic survival (Halpern-Meekin et al. 2015; Edin and Kefalas 2005; Hays 2003).

Chapter Summary

What different kinds of stratification systems exist?

Social stratification is a relatively fixed hierarchical ar-rangement in society by which groups have different access to resources, power, and perceived social worth. All societies have systems of stratification, although they vary in composition and complexity. *Estate sys-tems* are those in which a single elite class holds the power and property; in *caste systems*, placement in the stratification is by birth; in *class systems*, placement is determined by achievement.

How do sociologists define class?

Class is the social structural position that groups hold relative to the economic, social, political, and cultural resources of society. Class is highly significant in de-termining one's *life chances*.

How is the class system structured in the United States?

Social class can be seen as a hierarchy, like a ladder, where income, occupation, and education are indica-tors of class. *Status attainment* is the process by which people end up in a given position in this hierarchy. *Prestige* is the value others assign to people and groups within this hierarchy. Classes are also organized around common interests and exist in conflict with one another.

Is there social mobility in the United States?

Social mobility is the movement between class posi-tions. Education gives some boost to social mobil-ity, but social mobility is more limited than people believe; most people end up in a class position very close to their class of origin. *Class consciousness* is both the perception that a class structure exists and the feeling of shared identification with others in one's class. The United States has not been a particularly class-conscious society because of the belief in upward mobility.

What analyses of social stratification do sociological theorists provide?

Karl Marx saw class as primarily stemming from economic forces; Max Weber had a multidimensional view of stratification, involving economic, social, and political dimensions. Functionalists argue that social inequality motivates people to fill the different positions in society that are needed for the survival of the whole, claiming that the positions most important for society require the greatest degree of talent or training and are thus most rewarded. Conflict theorists see social stratification as based on class conflict and blocked opportunity, pointing out

that those at the bottom of the stratification system are least rewarded because they are subordinated by dominant groups.

How do sociologists explain why there is poverty in the United States?

Culture of poverty is the idea that poverty is the result of the cultural habits of the poor that are transmitted from generation to generation, but sociologists see poverty as caused by social structural conditions, including unemployment, gender inequality in the workplace, and the absence of support for child care for working parents.

What current policies address the problem of poverty?

Current welfare policy, adopted in 1996, provides support through individual states, but recipients are required to work after two years of support and have a lifetime limit of five years of support.

Is It True? (Answers)

1.	FALSE. Income growth has been highest in the top 5 percent of income groups (Fontenot, Semega, and Kollar 2018).
2.	FALSE. Eighty percent of all stock is owned by a small percentage of people. The most common financial asset is home ownership (Shapiro 2017; Oliver and Shapiro 2006).
3.	FALSE. The United States has lower rates of social mobility than Canada, Sweden, Norway, and fourteen other industrialized nations (Corak 2016).
4.	FALSE. Research finds that poor teen mothers value marriage and want to be married, but associate marriage with economic security, which they do not think they can achieve (Edin and Kefalas 2005).
5.	FALSE. Although those over age 65 used to be the most likely to be poor, poverty among the elderly has declined; the most likely to be poor are children (Fontenot, Semega, and Kollar 2018).
6.	TRUE. Poverty outside of metropolitan areas (14.8 percent) is quite close to that within principal cities (15.6 percent; Fontenot, Semega, and Kollar 2018).

Key Terms

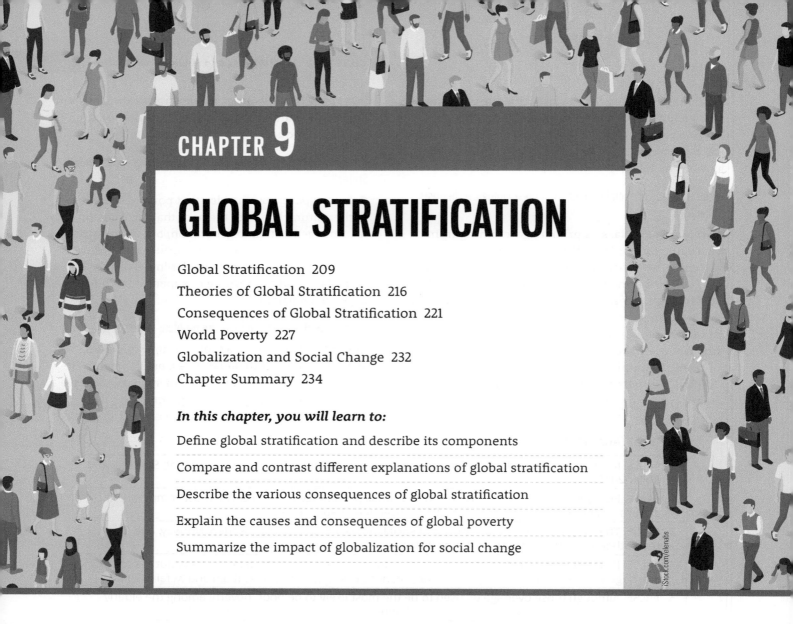

GLOBAL STRATIFICATION

In this chapter, you will learn to:

Define global stratification and describe its components

Compare and contrast different explanations of global stratification

Describe the various consequences of global stratification

Explain the causes and consequences of global poverty

Summarize the impact of globalization for social change

iStock.com/elenabs

"It takes a village to raise a child," the saying goes, but it also seems to take a world to make a shirt—or so it seems from looking at the global dimensions of the production and distribution of goods. Try this simple experiment: Look at the labels on your clothing. (If you do this in class, try to do so without embarrassing yourself and others!) What do you see? "Made in Indonesia," "Made in Vietnam," "Made in Malawi"—all indicating the linkage of the United States to clothing manufacturers around the world. The popular brand Nike, as just one example, contracts with factories all over the world. Most of Nike's products are made in hundreds of factories throughout Asia.

Taking your experiment further, ask yourself: Who made your clothing? A young person trying to lift his or her family out of poverty? Might it have been a child? The International Labour Organization (ILO) distinguishes *child labor* (those under age 17) from *employed children* (such as a teenager holding a part-time job or babysitting). *Child labor* specifically refers to "work that deprives children of their childhood, their potential, and their dignity and that is also harmful to mental and physical development" (International Labour Organization 2017). The ILO estimates that about 168 million children around the world are trapped in child labor, almost half of whom are involved in dangerous work and many of whom are separated from their families and possibly held in slavery (International Labour Organization 2017). This does not mean that a child necessarily made your clothing. In fact, most child labor occurs in agricultural work, although a significant component (25 percent) is in service and manufacturing work. The encouraging news is that the use of child labor has declined since 2000 but remains a significant part of global production.

Data on child labor indicate that our global systems of work are deeply connected to inequality. Especially in the poorest countries, trying to survive forces people into forms of work that produce some of the world's greatest injustices—both for children and for adult women and men. As we will see in this chapter, nations are interlocked in a system of global inequality, in which the status of the people in one country is intricately linked to the status of the people in others—a fact that has become apparent to workers in the United States who see their status declining as the result of jobs moving overseas. Such a recognition has even fueled anti-globalization politics in the United States, as voters have responded to political calls to bring jobs back to America.

Recall from Chapter 1 that C. Wright Mills identified the task of sociology as seeing the social forces that exist beyond individuals. This is particularly important when studying global inequality. A person in the United States (or western Europe or Japan) who thinks he or she is expressing individualism by wearing the latest style is actually part of a global system of inequality. The adornments available to that person result from a whole network of forces that produce affluence in some nations and poverty in others.

The United States and other wealthy nations are dominant in the system of global stratification. Those at the top of the global stratification system have enormous power over the fate of other nations. Although world conflict stems from many sources, including religious differences, cultural conflicts, and struggles over political philosophy, the inequality between rich and poor nations causes much hatred and resentment. One cannot help but wonder what would happen if the differences between the wealth of some nations and the poverty of others were smaller. In this chapter, we examine the dynamics and effects of global stratification.

Global Stratification

In the world today, there are not only rich and poor people but also rich and poor countries. Some countries are well off, some countries are doing so-so, and a growing number of countries are poor and getting poorer. There is, in other words, a system of **global stratification** in which the units are countries, much like a system of stratification within countries in which the units are individuals or families.

Just as we can talk about the upper-class or lower-class individuals within a country, we can also talk of the equivalent upper-class or lower-class countries in this world system. One manifestation of global stratification is the great inequality in life chances that differentiates nations around the world. Simple measures of well-being (such as life expectancy, infant mortality, access to education and health, and measures of environmental quality) reveal the consequences of global inequality. The gap between rich and poor people is also sometimes greatest in nations where poverty rates are highest. No longer can nations be understood without considering the global system of stratification of which they are a part.

The relative affluence of the United States means that U.S. consumers have access to goods produced around the world. A simple thing, such as a child's toy, can represent this global system. For many young girls in the United States, Barbie is the ideal of fashion and romance. Young girls may have not just one Barbie, but several, each with a specific role and costume.

Alice Hartley

Some nations have so much wealth that they actually serve gold as food, such as these real gold leaves on a dessert in Dubai!

Cheaply bought in the United States, but produced in China, Indonesia, or Malaysia, Barbie is made by those who are probably not much older than the young girls who play with her. Now, Barbie is played with in over 150 countries around the world. Many of the workers who make them would need a substantial part of their monthly pay to buy just one of the dolls that many U.S. girls collect by the dozens.

The manufacturing of toys and clothing is an example of the global stratification that links the United States and other parts of the world. *Global outsourcing* locates jobs overseas even while supporting U.S.-based businesses. Many of the jobs that have been outsourced in this way are semiskilled jobs, such as data entry, medical transcription, and so forth. Increasingly, outsourced jobs are also found in high-tech industries, software design, market research, and product research activities. Although it is difficult to measure the extent of global outsourcing, it has become a common phenomenon—something you experience when, for example, you engage in a telephone or Internet transaction, such as getting help for your computer or arranging a trip. India, China, and Russia have been major players in the economy of global outsourcing, but other nations, such as Ireland, South Africa, Poland, and Hungary, among others, are playing an increasingly important role. The consequences can be very positive for the economies of the host nations. The practice of outsourcing also lowers personnel costs for U.S.-based companies, given the lower wages in nations where jobs flow. Outsourcing is at the expense of jobs for workers in the United States, however. The practice of global outsourcing increasingly links the economies and social systems of nations around the world.

Rich and Poor

One dimension of stratification between countries is wealth. Enormous differences exist between the wealth of the countries at the top of the global stratification system and the wealth of the countries at the bottom. A very small proportion of the world's population receives a vast share of all income—a visual reminder of the inequality that characterizes our world. You will recall that we looked at the "champagne glass" of inequality within the United States in Chapter 8. A similar image can show you the inequality of income worldwide (see ▲ Figure 9-1). As you can see, a small percentage of the world's population has a very disproportionate share of world income. As ■ map 9-1 shows, were the world geographically arrayed by wealth, a very different looking globe would result!

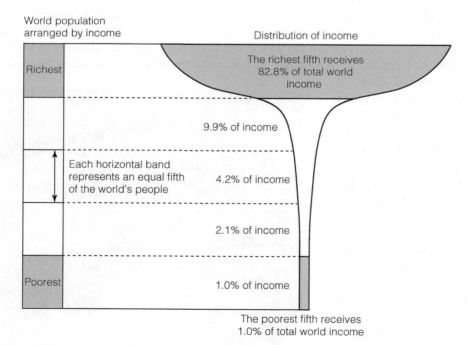

World population arranged by income

Distribution of income

Richest

The richest fifth receives 82.8% of total world income

9.9% of income

Each horizontal band represents an equal fifth of the world's people

4.2% of income

2.1% of income

Poorest

1.0% of income

The poorest fifth receives 1.0% of total world income

▲ **Figure 9-1   World Income Distribution**

Data from: Ortiz, Isabel, and Matthew Cummins. 2011. "Global Inequality: Beyond the Bottom Billion." Social and Economic Policy Working Paper. UNICEF, April. **www.worldbank.org**

map 9-1   Viewing Society in Global Perspective: The World Seen through the Distribution of Wealth

This map shows what the world would look like were geographical boundaries formed by the wealth held by different nations. What perspective does this give you on global stratification?

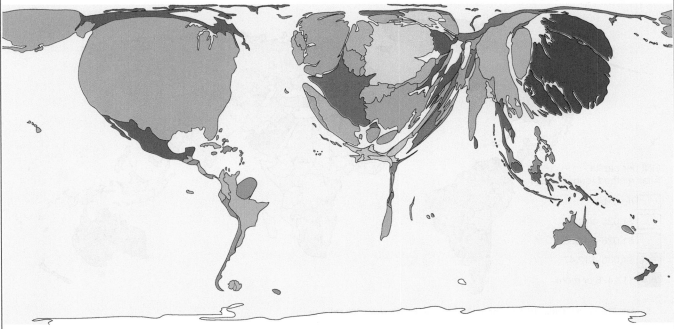

Source: © Copyright 2006 SASI Group (University of Sheffield) and Mark Newman (University of Michigan), **www.worldmapper.org/posters /worldmapper_map169_ver5.pdf**

There are different ways to measure the wealth of nations, but the most common is to use the per capita **gross national income (GNI)**. The GNI measures the total output of goods and services produced by residents of a country each year plus the income from non-resident sources, divided by the size of the population. The GNI does not truly reflect what individuals or families receive in wages or pay. It is simply each person's annual share of their country's income if income were shared equally. You can use this measure to get a picture of global stratification (see ■ map 9-2).

Per capita GNI is reliable only in countries that are based on a cash economy. It does not measure informal exchanges or bartering in which resources are exchanged without money changing hands. These noncash transactions are not included in the GNI calculation, but they are common in developing countries. As a result, measures of wealth based on the GNI, or other statistics that count cash transactions, are less reliable among the poorer countries and may underestimate the wealth of the countries at the lower end of the economic scale.

The per capita GNI of the United States, one of the wealthier nations in the world (though not the wealthiest on a per capita basis), was $56,180 in 2016. In the poorest countries of the world, the GNI per capita is quite small, as you can see in Figure 9-2. You may also notice that many of the poorest nations are in sub-Saharan Africa, the region of the world most likely to be afflicted by extreme poverty. The United States' per capita GNI shows us to be one of the most affluent nations in the world (World Bank 2017a). Which are the wealthiest nations? ▲ Figure 9-2 shows the ten richest and the ten poorest countries in the world (measured by the annual per capita GNI).

The poorest nations are largely rural, have high fertility rates and large populations, and still depend heavily on subsistence agriculture. In very poor countries, the life of an average citizen is meager. Often poor nations are rich with natural resources but are exploited for such resources by more powerful nations. Still, they rank at the bottom of the global stratification system.

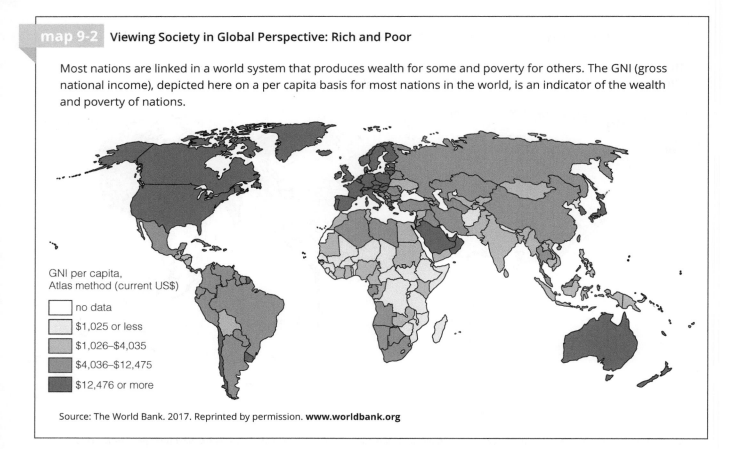

map 9-2 **Viewing Society in Global Perspective: Rich and Poor**

Most nations are linked in a world system that produces wealth for some and poverty for others. The GNI (gross national income), depicted here on a per capita basis for most nations in the world, is an indicator of the wealth and poverty of nations.

GNI per capita,
Atlas method (current US$)

- no data
- $1,025 or less
- $1,026–$4,035
- $4,036–$12,475
- $12,476 or more

Source: The World Bank. 2017. Reprinted by permission. **www.worldbank.org**

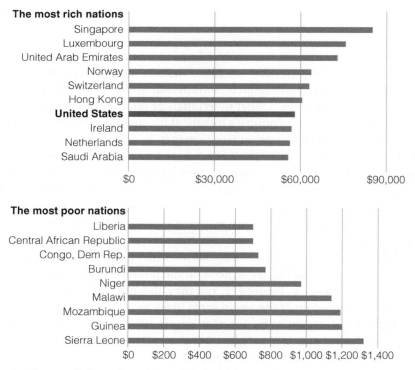

The most rich nations

Singapore
Luxembourg
United Arab Emirates
Norway
Switzerland
Hong Kong
United States
Ireland
Netherlands
Saudi Arabia

$0 $30,000 $60,000 $90,000

The most poor nations

Liberia
Central African Republic
Congo, Dem Rep.
Burundi
Niger
Malawi
Mozambique
Guinea
Sierra Leone

$0 $200 $400 $600 $800 $1,000 $1,200 $1,400

▲ Figure 9-2   **The Rich and the Poor: A World View***

*Measured by GNI per capita, in U.S. dollars, for 2011. **www.worldbank.org**

Data from: The World Bank. 2017a.

Data based on GNI per capita

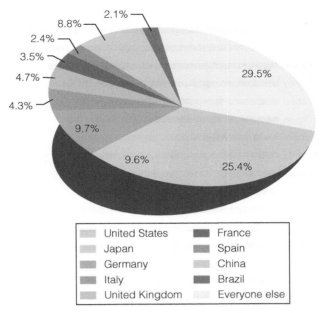

▲ Figure 9-3   Who Owns the World's Wealth?

Data from: Davies, James B., Susanna Sandstrom, Anthony Shorrocks, and Edward N. Wolff. 2008. "The World Distribution of Household Wealth." UNU-WIDER, World Institute for Development Economics Research. Helsinki, Finland.

Because the poorest nations suffer from extreme poverty, there is terrible human suffering in these places. This also produces instability and the potential for violence, as well as risks to human health. Each of these nations suffers from very high poverty rates, short life expectancy, and poor water facilities. We will look more closely at the nature and causes of such world poverty later in this chapter.

The wealthiest countries, you will see, are largely industrialized nations or those that are oil-rich (see ▲ Figure 9-3). These countries represent the equivalent of the upper class. Simply being one of the wealthiest nations in the world does not mean that all of the nation's population is well off. Some, especially the Scandinavian countries, have low degrees of inequality within. Others, including the United States, have great inequality within, as we have seen in the previous chapter on class inequality. Moreover, inequality between nations has to be seen in relative terms. For example, a very wealthy person in India may have the income of someone in the bottom 5 percent of income earners in the United States, but within India, this can afford the person an expensive mansion and a highly lavish lifestyle relative to other Indian people (Milanovic 2010).

Inequality within nations is measured by something called the Gini coefficient. The **Gini coefficient** is a measure of income distribution within a given population or nation. The figure ranges from zero to one, with zero representing a population where there is perfect equality and one indicating a population where just one person has all the money—in other words, the greatest inequality. South Africa has the highest Gini coefficient in the world; Iceland and the Scandinavian countries are among the lowest. The United States ranks very high in the degree of internal inequality among other industrialized nations, as you can see in ▲ Figure 9-4.

The International Division of Labor: A World Divided by Race and Gender

The connection between rich and poor nations is the result of an **international division of labor** that relies on cheap labor and labor. Labor in non-Western countries is generally much less expensive and, typically, the work of people of color or ethnic minority groups, sometimes children. Moreover, in this "global assembly line," it is mostly women of color who provide the most menial and under-rewarded labor.

Profits from the international division of labor accrue to wealthy owners, who are mostly White, resulting in a racially divided world. Like the division of labor within the United States (see Chapter 14), the global labor force is divided along lines of race and gender, making these social factors central to the phenomenon of global capitalism.

▲ **Figure 9-4   Gini Coefficient in OECD* Countries.** How does the United States compare to western European nations? Why do you think this has happened?

*OECD refers to the Organization for Cooperation and Economic Development, generally including the more industrialized nations.

Data from: Central Intelligence Agency. 2014. *The World Fact Book*. **www.cia.gov**

Further, in the richest nations, the dominant population is largely White. In the poorest countries of the world, mostly in Africa, the populations are people of color. The history of western capitalism has been characterized by extracting both human and natural resources from these regions. This has been the mechanism of Western imperialism and colonialism. The inequities that have resulted are enormous. And, just as "color" has tended to define race in the United States, "color" stratifies people around the world with the "global South" now a term used to describe the poor, developing continents (Africa, Asia, and Latin America). "North and South;" "rich and poor:" Just as they have described race and class stratification within the United States, they now also encompass stratification in the whole world.

The World on the Move: International Migration

The global interconnection of nations is increasingly apparent in high rates of international migration. According to the United Nations, international migration has grown rapidly since 2000, reaching 244 million migrant people in 2015 compared to 173 million in 2000. Nearly two-thirds live in Europe or Asia (counting all countries combined in both places), but the United States is, by far, the most likely destination country, followed by Germany and then the Russian Federation (United Nations 2015). Current restrictions on immigration into the United States are likely to reduce this number in years to come, unless there are significant policy changes. As you can see in ■ map 9-3, the flow of international migrants is largely out of the "global South" and into the more developed nations.

Contrary to popular belief, most migrants do not come from the poorest nations. Rather, the majority come from so-called middle-income countries—because people need some resources to leave. Most migrants are also working-age, and this provides an available and potentially less expensive labor force for the wealthier nations. Often, this means that migrants experience downward mobility in the host country. Because they are generally ethnically different from the host population, racial tensions are often the result. As we have seen in the United States, the high rate of immigration into the nation has also fueled *xenophobia* (see Chapter 2)—an ugly development that has also created ethnic tensions in western European nations.

Migration happens for a variety of reasons, including the desire for improved living conditions, fleeing from armed conflict and political turmoil, and religious and ethnic persecution. Refugees are those

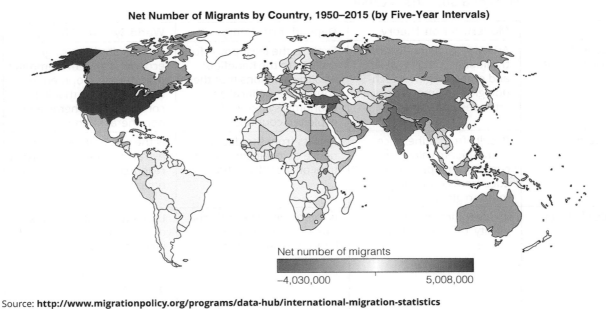

map 9-3   **Viewing Society in Global Perspective: Migration**

This map shows migration both in and out of countries around the world from 2010–2015. The nations shaded in "red" are experiencing out-migration; those in blue, in-migration. You can see that the general flow is out of parts of the global South into the nations of the global "north"—namely, the United States, Russia, and western Europe.

Net Number of Migrants by Country, 1950–2015 (by Five-Year Intervals)

Net number of migrants

−4,030,000   5,008,000

Source: **http://www.migrationpolicy.org/programs/data-hub/international-migration-statistics**

who are seeking asylum from war-torn parts of the world or where political oppression, often against ethnic minorities, is driving people away from their homeland.

International migration, sometimes legal, sometimes not, has radically changed the racial and ethnic composition of populations in the United States as well as many European nations. Some of those who have moved have done so because of war and persecution, but many move as work moves around the globe (Eitzen and Baca Zinn 2011). Most western nations that are receiving large numbers of international migrants are now more internally diverse—a development that both holds the promise of greater acceptance of difference along with the tensions that such developments also create.

One result is the development of **world cities**, that is, cities that are closely linked through the system of international commerce. Within these cities, families and their surrounding communities often form *transnational communities*, communities that may be geographically distant but socially and politically close. Linked through various communication and transportation networks, transnational communities share information, resources, and strategies for coping with the problems of international migration.

Theories of Global Stratification

How did world inequality occur? Sociological explanations of world stratification generally fall into three camps: modernization theory, dependency theory, and world systems theory (see ◆ Table 9-1).

Modernization Theory

Modernization theory views the economic development of countries as stemming from technological change. According to this theory, a country becomes more "modernized" by increased technological development.

Modernization theory sees economic development as a process by which traditional societies become more complex. For economic development to occur, modernization theory predicts, countries must change their traditional attitudes, values, and institutions. Economic achievement is thought to derive from attitudes and values that emphasize hard work, saving, efficiency, and enterprise. Modernization theory suggests that nations remain underdeveloped when traditional customs and culture discourage individual achievement and kin relations dominate.

Table 9-1 Theories of Global Stratification

	Modernization Theory	Dependency Theory	World Systems Theory
Economic Development	Arises from relinquishing traditional cultural values and embracing new technologies and market-driven attitudes and values	Exploits the least powerful nations to the benefit of wealthier nations that then control the political and economic systems of the exploited countries	Has resulted in a single economic system stemming from the development of a world market that links core, semiperipheral, and peripheral nations
Poverty	Results from adherence to traditional values and customs that prevent societies from competing in a modern global economy	Results from the dependence of low-income countries on wealthy nations	Is the result of core nations extracting labor and natural resources from peripheral nations
Social Change	Involves increasing complexity, differentiation, and efficiency	Is the result of neocolonialism and the expansion of international capitalism	Leads to an international division of labor that increasingly puts profit in the hands of a few while exploiting those in the poorest and least powerful nations

As an outgrowth of functionalist theory, modernization theory derives some of its thinking from the work of Max Weber. In *The Protestant Ethic and the Spirit of Capitalism* (1958/1904), Weber saw the economic development that occurred in Europe during the Industrial Revolution as a result of the values and attitudes of Protestantism. The Industrial Revolution took place in England and northern Europe, Weber argued, because the people of this area were hardworking Protestants who valued achievement and believed that God helped those who helped themselves.

Modernization theory can partially explain why some countries have become successful. Japan and China are examples of countries that have made huge strides in economic development, in part because of a national work ethic. Work ethic alone, however, does not explain the success of either. Modernization theory may partially explain the cultural context in which some countries become successful and others do not, but it is not a substitute for explanations that also look at the economic and political context of national development. Cultural attitudes may impede economic development in some cases, but you have to be careful not to assume that developed nations have superior values compared to others. Blaming the cultural values of a poor nation overlooks the fact that a nation's status in the world may be outside their control. Whether a country develops or remains poor is often the result of other countries exploiting the less powerful. Modernization theory does not sufficiently take into account the interplay and relationships between countries that can affect a country's economic or social condition.

Developing countries, modernization theory says, are better off if they let the natural forces of competition guide world development. Free markets, according to this perspective, will result in the best economic order. This idea still resonates in the principle of neoliberalism—an idea that guides certain politics in the United States today. **Neoliberalism** is an economic principle promoting the operation of a free market, one with limited (or no) government intervention or subsidies, deregulation, and a shift from public sector spending to the private sector. Neoliberalism is a new version of *laissez-faire* (that is, "hands off") capitalism—that is, letting the market operate on its own. Critics of neoliberalism, like the critics of modernization theory, argue that governments have responsibility to the good of the whole and that free markets primarily benefit solely wealthy nations and wealthy individuals.

Dependency Theory

Although market-oriented theories may explain why some countries are successful, they do not explain why some countries remain in poverty or why some countries have not developed. It is necessary to look at issues outside the individual countries and to examine the connections between them. Keep in mind that many of the poorest nations are former colonies of European powers. This focuses your attention on colonization and imperialism as causes of global stratification.

Dependency theory holds that the poverty of the low-income countries is a direct result of their political and economic dependence on the wealthy countries. Specifically, dependency theory argues that the poverty of many countries is a result of exploitation by powerful countries. This theory is derived from the work of Karl Marx, who foresaw that a capitalist world economy would create an exploited class of dependent countries, just as capitalism within countries had created an exploited class of workers.

Dependency theory begins with understanding the historical development of this system of inequality. As the European countries began to industrialize in the 1600s, they needed raw materials for their factories and places to sell their products. To accomplish this, the European nations colonized much of the world, including most of Africa, Asia, and the Americas. **Colonialism** is a system by which Western nations became wealthy by taking raw materials from colonized societies and reaping profits from products finished in the homeland. Colonialism worked best for the industrial countries when the colonies were kept undeveloped to avoid competition with the home country. For example, India was a British colony from 1757 to 1947. During that time, Britain bought cheap cotton from India, made it into cloth in British mills, and then sold the cloth back to India, making large profits. Although India was able to make cotton into cloth at a much cheaper cost than Britain, and very fine cloth at that, Britain nonetheless did not allow India to develop its cotton industry. As long as India was dependent on Britain, Britain became wealthy and India remained poor.

Under colonialism, dependency was created by the direct political and military control of the poor countries by powerful developed countries. Most colonial powers were European countries, but other countries, particularly Japan and China, had colonies as well. Colonization came to an end soon after the Second World War, largely because of protests by colonized people and the resulting movement for independence. As a result, according to dependency theory, the powerful countries turned to other ways to control the poor countries and keep them dependent. The powerful countries still intervene directly in the affairs of the dependent nations by sending troops or, more often, by imposing economic or political restrictions and sanctions. But other methods, largely economic, have been developed to control the dependent poor countries, such as price controls, tariffs, and, especially, the control of credit. Indeed, the level of debt that some nations accrue is a major source of global inequality.

The rich industrialized nations, according to dependency theory, are able to set prices for raw materials produced by the poor countries at very low levels so that the poor countries are unable to accumulate enough profit to industrialize. As a result, the poor, dependent countries must borrow from the rich countries. However, debt creates only more dependence. Many poor countries are so deeply indebted to the major industrial countries that they must follow the economic edicts of the rich countries that loaned them the money, thus increasing their dependency.

Multinational corporations are companies that draw a large share of their profits from overseas investments and that conduct business across national borders. They play a role in keeping the dependent nations poor, dependency theory suggests. Although their executives and stockholders are from the industrialized countries, multinational corporations recognize no national boundaries and pursue business where they can best make a profit. Multinationals buy resources where they can get them cheapest, manufacture their products where production and labor costs are lowest, and sell their products where they can make the largest profits.

Many critics fault companies for perpetuating global inequality by taking advantage of cheap overseas labor to make large profits for U.S. stockholders. Companies are, in fact, doing what they should be doing in a market system: trying to make a profit. Nonetheless, dependency theory views the practices of multinationals as responsible for maintaining poverty in the poor parts of the world.

One criticism of dependency theory is that many poor countries (for example, Ethiopia) were never colonies. Some former colonies have also done well. Two of the greatest postwar success stories of economic development are Singapore and Hong Kong. Both were British colonies—Hong Kong until 1997—and were clearly dependent on Britain, yet they have had successful economic development precisely because of their dependence on Britain. Other former colonies are also improving economically, such as India.

World Systems Theory

Modernization theory examines the factors internal to an individual country, and dependency theory looks to the relationship between countries or groups of countries. Another approach to global stratification is called **world systems theory**. Like dependency theory, this theory begins with the premise that no nation in the world can be considered in isolation. Each country, no matter how remote, is tied in many ways to the other countries in the world. However, unlike dependency theory, world systems theory argues that there is a world economic system that must be understood as a single unit, not in terms of individual countries or groups of countries. This theoretical approach derives to some degree from the work of dependency theorists and is most closely associated with the work of Immanuel Wallerstein in *The Modern World System* (1974) and *The Modern World System II* (1980). According to this theory, the level of economic development is explained by understanding each country's place and role in the world economic system.

This world system has been developing since the sixteenth century. The countries of the world are tied together in many ways, but of primary importance are the economic connections in the world markets of goods, capital, and labor. All countries sell their products and services on the world market and buy products and services from other countries. However, this is not a market of equal partners. Because of historical and strategic imbalances in this economic system, some countries are able to use their advantage to create and maintain wealth, whereas other countries that are at a disadvantage remain poor. This process has led to a global system of stratification in which the units are not people but countries.

World systems theory sees the world divided into three groups of interrelated nations: core or first-world countries, semiperipheral or second-world countries, and peripheral or third-world countries.

Servants of Globalization: Who Does the Domestic Work?

Research Question

Women are a large share of international migrants, often leaving poor nations to become domestic workers in wealthier nations. Rhacel Salazar Parreñas wanted to know about these women's experiences in the context of global stratification?

Research Method

Parreñas studied two communities of Filipina women, one in Los Angeles and one in Rome, Italy, conducting her research through extensive interviewing with Filipina domestic workers in both locations. She also used participant observation in church settings, after-work social gatherings, and in employers' homes.

Research Results

Parreñas found that Filipina domestics experienced status inconsistency. They were upwardly mobile in terms of their home country but were excluded from the middle-class Filipino communities in the host nation. Thus they experienced feelings of social exclusion in addition to being separated from their own families.

Conclusions and Implications

Women are part of a new social pattern of *transnational families*—that is, families whose members live across the borders of nations. Filipinas provide the labor for more affluent households while their own lives are disrupted by these new global forces. As global economic restructuring evolves, it may be that more families will experience this form of family living.

Questions to Consider

1. Are there domestic workers in your community who provide child care and other household work for middle- and upper-class households? What is their race, ethnicity, nationality, and gender? What does this tell you about the division of labor in domestic work and its relationship to global stratification?
2. Why do you think domestic labor is so underpaid and undervalued? Are there social changes that might result in a reevaluation of the value of this work?

Sources: Parreñas, Rhacel Salazar. 2001. *Servants of Globalization: Women, Migration and Domestic Work.* Stanford, CA: Stanford University Press.

The gap between the rich and poor worldwide can be staggering. At the same time that many struggle for mere survival, others enjoy the pleasantries of a gentrified lifestyle.

The Global Economy of Clothing

Look at the labels in your clothes. Where are they made? Who profits from the distribution of these goods? What does this tell you about the relationship of *core, semiperipheral,* and *peripheral* countries within world systems theory? What further information would reveal the connections between the country where you live and the countries where your clothing is made and distributed?

Kelsey Timmerman who wanted to know where his clothes came from. Timmerman traveled to Honduras, Bangladesh, Cambodia, and China, talking to factory workers and their families about the experiences in making the clothes that others wear. His journey teaches you a lot about global production and consumption (Timmerman 2012).

Further resources: **www.whereamiwearing.com**

The **core countries** have the most power in the world economic system. These countries control and profit the most from the world system, and thus they are the "core" of the world system. These include the powerful nations of Europe, the United States, Australia, and, increasingly, India and China.

Surrounding the core countries, both structurally and geographically, are the **semiperipheral countries** that are semi-industrialized and, to some degree, represent a kind of middle class (such as Spain, Turkey, and Mexico). They play a middleman role, and may, with further development, become core countries.

At the bottom of the world stratification system are the **peripheral countries**. These are the poor, largely agricultural countries of the world. Even though they are poor, they often have important natural resources that are exploited by the core countries. Exploitation, in turn, keeps them from developing and perpetuates their poverty. Often these nations are politically unstable. Political instability within poor nations can create a crisis for core nations that depend on their resources. Military intervention by the United States or European nations is often the result.

This world economic system has resulted in a modern world in which some countries have obtained great wealth and other countries have remained poor. The core countries control and limit the economic development in the peripheral countries so as to keep the peripheral countries from developing and competing with them on the world market; thus, the core countries can continue to purchase raw materials at a low price.

Although world systems theory was originally developed to explain the historical evolution of the world system, modern scholars now focus on the international division of labor and its consequences. This approach is an attempt to overcome some of the shortcomings in world systems theory by focusing on the specific mechanism by which differential profits are attached to the production of goods and services in the world market. A tennis shoe made by Nike is designed in the United States; uses synthetic rubber made from petroleum from Saudi Arabia; is sewn in Indonesia; is transported on a ship registered in Singapore, which is run by a Korean management firm using Filipino sailors; and is finally marketed in Japan and the United States. At each of these stages, profits are taken, but at very different rates.

World systems theorists call this global production process a **commodity chain**, the network of work processes by which a product becomes a finished commodity. By following a commodity through its production cycle and seeing where the profits go at each link of the chain, one can identify which country is getting rich and which country is being exploited. As an example, your favorite Gap hoodie could have been made from cotton grown in Uzbekistan where workers are paid about five dollars per day (half the average wage per month) to pick 120 to 154 pounds of cotton. The garment is then cut and sewn by workers perhaps in Cambodia who are paid about $153 per month. It is then distributed and sold in the United States (Khan 2016; Thui 2016; Wright 2016; Aulakh 2013). World systems theory also helps explain the growing phenomenon of international migration because the need for cheap labor in some of the industrial and developing nations draws workers from poorer parts of the globe.

It is useful to see the world as an interconnected set of economic ties between countries, and to understand that these ties often result in the exploitation of poor countries. This process of globalization means that countries that were once among the most powerful in the world system— England, for example—no longer occupy such a lofty position. Peripheral countries can improve their standard of living with investment by core countries, although the benefits do not accrue equally to groups within such nations. Investment by outsiders can also put the receiving nations in debt, thus harming them in the long run. Low-wage factories may benefit managers, but not the working class. Even core countries can be hurt by the world system, such as when jobs move overseas. Who benefits from this world system is differentiated—in all countries—by one's placement, not just in the global class system but also in the class system internal to each country within this global system. World systems theory has provided a powerful tool for understanding global inequality.

Consequences of Global Stratification

It is clear that some nations are wealthy and powerful and some are poor and powerless. What are the consequences of this world stratification system? Basic indicators of national well-being include such things as infant mortality, literacy levels, access to safe water, and the status of women. There are, as we will see, considerable differences in the quality of life based on these indicators in different places in the world.

Global stratification often means that consumption in the more affluent nations is dependent on cheap labor in other less affluent nations.

Population

One of the biggest differences in rich and poor nations is population. The poorest countries have the highest birth rates and the highest death rates. The total *fertility rate*—how many live births a woman will have over her lifetime at current fertility rates—shows that women in the poorest countries have on average almost five children. Because of this high fertility rate, the populations of poor countries are growing faster than the populations of wealthy countries. Poor countries therefore also have a high proportion of young children.

In contrast, the populations of the richest countries are not growing nearly as fast as the populations of the poorest countries. In the richest countries, women have about two children over their lifetime, and the populations of these countries are growing by only 1.2 percent. Many of the richest countries, including most of the countries of Europe, are actually experiencing population declines. With a low fertility rate, the rich countries have proportionately fewer children, but they also have proportionately more elderly, which can also be a burden on societal resources. Different from the poorest nations, the richest ones are largely urban.

Rapid population growth as a result of high fertility rates can make a large difference in the quality of life of the country. Countries with high birth rates are faced with the challenge of having too many children and not enough adults to provide for the younger generation. Public services, such as schools and hospitals, are strained in high-birth rate countries, especially because these countries are poor to begin with. Very low birth rates, as many rich countries are now experiencing, can also lead to other problems. In countries with low birth rates, there often are not enough young people to meet labor force needs, and workers must be imported from other countries.

There can be innovative solutions to reduce world poverty, such as this solar panel delivering energy in southern Mozambique.

Eric Nathan/Alamy Stock Photo

Scholars disagree about the relationship between the rate of population growth and economic development. Some theorize that rapid population growth and high birth rates lead to economic stagnation and that too many people keep a country from developing, thus miring the country in poverty (Ehrlich 1968). Yet, some countries with very large populations have become developed: China and India come to mind; both are showing significant economic development. The United States has the third largest population in the world at 318 million people, yet it is one of the richest and most developed nations in the world. Scholars now believe that although large population and high birth rates can impede economic development in some situations, fertility levels are also affected by levels of industrialization, not the other way around. That is, as countries develop, their fertility levels generally decrease and their population growth levels off (Hirschman 1994).

Health and the Environment

The basic health standards of countries also depend on where they sit in the global stratification system. The high-income countries have lower childhood death rates, higher life expectancies, and fewer children born underweight. People born today in wealthy countries can expect to live about seventy-seven years, and women outlive men by several years. Except for some isolated or poor areas of the rich countries, almost all people have access to clean water and acceptable sewer systems.

In the poorest countries, the situation is completely different. Many children die within the first five years of life, people live considerably shorter lives, and fewer people have access to clean water and adequate sanitation. In the low-income countries, the problems of sanitation, clean water, childhood death rates, and life expectancies are all closely related. In many of the poor countries, drinking water is contaminated from poor or non-existent sewage treatment. This contaminated water is then used to drink, to clean eating utensils, and to make baby formula. For adults, waterborne illnesses such as cholera and dysentery sometimes cause severe sickness but seldom result in death. Children under age 5 and, especially, those under the age of 1 are highly susceptible to the illnesses carried in contaminated water. A common cause of childhood death in countries with low incomes is dehydration brought on by the diarrhea contracted from drinking contaminated water.

Degradation of the environment is a problem that affects all nations, which are linked in one vast environmental system, but global stratification also means that some nations suffer at the hands of others. Overdevelopment is resulting in deforestation, and high population and the dependency on agriculture in the poorest nations contribute to the depletion of natural resources. In the most industrial nations where the most energy is used, the overproduction of "greenhouse gas"—emission of carbon dioxide from burning fossil fuels—is resulting in numerous threats to our environment, including climate change (see also Chapter 16).

Although high-income countries have only 15 percent of the world population, together they use more than half of the world's energy. The United States alone uses 20 percent of the world's energy, although it holds only 4 percent of the world's population (see ▲ Figure 9-5). Safe water is also crucial; 663 million people lack access to safe drinking water and the World Bank projects that as early as 2040, forty percent of the world's population will face a shortfall in the availability of water. The World Bank has, in fact, warned that we are facing a "global water crisis" (U.S. Energy Information Administration 2017; World Bank 2017b; International Energy Agency 2014). Clearly, global stratification has some irreversible environmental effects that are felt around the globe.

Education and Illiteracy

In the high-income nations of the world, education is almost universal, and the vast majority of people have attended school, at least at some level. Literacy and school enrollment are now taken for granted in the high-income nations, although people in these wealthy nations who do not have a good education stand little chance of success. In the middle- and lower-income nations, the picture is quite different. Elementary school enrollment, virtually universal in wealthy nations, is less common in the middle-income nations and even less common in the poorest nations.

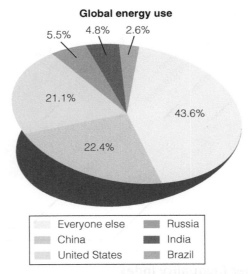

Global energy use

5.5%   4.8%   2.6%

21.1%

43.6%

22.4%

Everyone else   Russia
China   India
United States   Brazil

▲ **Figure 9-5   Who Uses the World's Energy?**

Data source: U.S. Energy Information Administration. 2017. **www.eia.gov**

How do people survive who are not literate or educated? In much of the world, education takes place outside formal schooling. Just because many people in the poorer countries never go to school does not mean that they are ignorant or that they are uneducated. Most of the education in the world takes place in family settings, in religious congregations, or in other settings where elders teach the next generation the skills and knowledge they need to survive. This type of informal education often includes basic literacy and math skills that people in these poorer countries need for their daily lives.

The disadvantage of this informal and traditional education is that, although it prepares people for their traditional lives, it often does not give them the skills and knowledge needed to operate in the modern world. In an increasingly technological world, this can perpetuate the underdeveloped status of some nations.

Gender Inequality

The position of a country in the world stratification system also affects gender relations within different countries. Poverty is usually felt more by women than by men. Although gender equality has not been achieved in the industrialized countries, compared with women in other parts of the world, women in the wealthier countries are much better off.

The United Nations (UN) is one of the organizations that carefully monitors the status of women globally. The UN has developed an index to assess the progress of women in nations around the world. Called the **gender inequality index**, the measure is a composite of three key components of women's lives: reproductive health, empowerment, and labor market status. Each of these three major components is then measured by particular facts about women's status, such as maternal mortality, educational attainment, and labor force participation (see ▲ Figure 9-6). Given how this index is computed, nations with the lowest gender inequality index have the greatest equality between women and men (see ◆ Table 9-2). Based on this index, the United Nations has concluded that, around the world, reproductive health—or lack thereof—is the greatest contributor to gender inequality (United Nations 2010).

Reports indicate mixed news with regard to women's status around the world. On the one hand, women's poverty has declined in some of the nations where it has been extreme, particularly in India, China, and some parts of Latin America. In sub-Saharan Africa, women's poverty has increased. Around the world, women have achieved near equality in levels of primary education, but large gaps remain in the secondary and higher education of women and men. This fact has huge implications for women's work, especially because the global economy increasingly demands educational skills (Inter-parliamentary Union 2012).

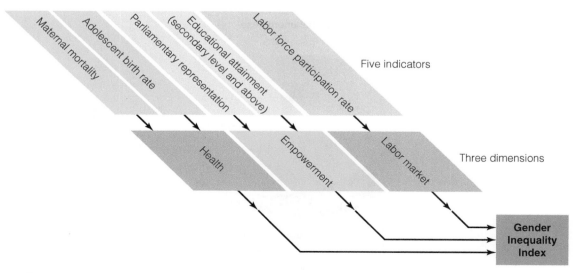

▲ **Figure 9-6** **The Gender Inequality Index**

Source: United Nations Development Programme. 2010. "Components of the Gender Inequality Index." http://hdr.undp.org. Reprinted with permission.

Table 9-2	Gender Inequality Index in Selected Countries (2015)
	Gender Inequality Index
Switzerland	0.040
Denmark	0.041
Netherlands	0.044
Sweden	0.048
Korea, Republic of	0.067
Israel	0.103
China	0.164
Libya	0.167
United States	0.203
Saudi Arabia	0.257
India	0.530
Haiti	0.593

Source: United Nations. 2017. *Human Development Data.* **http://hdr.undp.org/en/data**

Perhaps most distressing is the global extent of violence against women, as shown in the box, "Understanding Diversity." Violence takes many forms, including violence within the family, rape, sexual harassment, sex trafficking and prostitution, and state-based violence. The United Nations has concluded that "violence against women persists in every country in the world as a pervasive violation of human rights and a major impediment to achieving gender equality" (United Nations 2006: 9).

Several factors put women at risk of violence, ranging from individual-level risk factors (such as a history of abuse as a child and substance abuse) to societal-level factors, such as gender roles that entrench male dominance and societal norms that tolerate violence as a means of conflict resolution (see ◆ Table 9-3). Clearly, the inequalities that mark global stratification have particularly harmful effects for the world's women.

Understanding Diversity

Refugee Women and the Intersection of Race and Gender

About half of the world's refugees are women and children. Many, if not most, have fled from nations because of persecution based on ethnicity, nationality, religion, or race. These facts mean that gender and race are critical in developing policies to assist refugees.

Eileen Pittaway and Linda Bartolomei have studied refugees by conducting interviews in refugee camps, primarily in southeast Asia. They found that women refugees are often subjected to discrimination, partially because of their racial and ethnic identity as "different" or "other." Consequently, women refugees are highly vulnerable to rape and sexual violence. Many have already fled the threat of such violence in their nation of origin, particularly if they left because of armed conflict.

Pittaway and Bartolomei argue that policies that are intended to assist refugees are often "gender blind" and thus do not address the specific needs of women and their children. Their research shows how race and gender intersect in the social structures where refugee women and their children find themselves and they suggest social policies that are cognizant of the specific needs of these women.

Source: Pittaway, Eileen, and Linda Bartolomei, 2001. "Refugees, Race, and Gender: The Multiple Discrimination against Refugee Women." *Refuge* 19 (6): 21–32.

Table 9-3	Risk Factors for Violence against Women: A Global Analysis

The United Nations has studied the frequent use of violence against women in the world and identified the factors that put women at risk. These factors are found at various levels.

Individual Level:	Community Level:
Frequent use of alcohol and drugs	Women's isolation and lack of social support
Membership in marginalized communities	Community attitudes that tolerate and legitimate male violence
History of abuse as a child	High levels of social and economic inequality, including poverty
Witnessing marital violence in the home	
Family/Relationship Level:	**Societal Level:**
Male control of wealth	Gender roles that entrench male dominance and women's subordination
Male control of decision making	Tolerance of violence as a means of conflict resolution
History of marital violence	Inadequate laws and policies to prevent and punish violence
Significant disparities in economic, educational, or employment status	Limited awareness and sensitivity on the part of officials and social service providers

Terrorism

Terrorism can be defined as premeditated, politically motivated violence perpetrated against targets by those who try to achieve their political ends through violence (White 2013). Terrorism can be executed through violence or threats of violence and can be implemented through various means—suicide bombs, biochemical terror, cyberterrorism, or mass shootings, for example. Because terrorists operate outside the bounds of normative behavior, terrorism is very difficult to prevent. Although rigid safeguards can be put in place, such safeguards also threaten the freedoms that are characteristic of open, democratic societies. The fact that terrorism is so difficult to stop contributes to the fear that it is intended to generate.

Most people associate terrorism now with ISIS (or ISIL, the Islamic State of Iraq and the Levant)—the radical extremist group that has taken an radical view of Islam and vowed to kill so-called infidels, including citizens of the United States, who do not agree with them. All too often, we witness these violent acts both at home and abroad. The horror of beheadings, truck attacks on innocent civilians, the shooting of innocent shoppers in a Paris kosher deli, the dreadful 9/11 attacks—these and other horrific acts are all too-frequent reminders of the fragility of the world order. Terrorism, however, comes in different forms. There are terrorist acts by "lone wolfs," those who say they are inspired by ISIS (or ISIL) and terrorist acts by other radical extremist groups and/or state-sponsored violence.

There are many complex causes for terrorism. Although there is not a single explanation of terrorism, the inequities brought on by global stratification are certainly part of the problem. Global stratification can create international conflicts that bring an increased risk of terrorism by generating inequities in the distribution of power between different nations. Global stratification has also produced a world-based capitalist class with unprecedented wealth and power. This is a class that now crosses national borders—a so-called transnational capitalist class (Langman and Morris 2002). Within this class, some individuals may actually hold more wealth than entire nations. Coupled with the vast poverty in the world, the visibility of this class and its association with Western values can lead to resentment and conflict. The same power and affluence that makes the United States a leader throughout the world makes it a target by those who resent its dominance.

In the Middle East, for example, oil production has created prosperity for some and exposed people in these nations to the values of Western culture. When people from different nations, such as those in the Middle East, study at U.S. universities and travel on business or vacations, they are also exposed to Western values and patterns of consumption. The sexual liberalism of Western nations and the relative equality of women also add to the volatile mix of nations clashing (Norris and Inglehart 2002). Clashing religious values and the growth of extremist views are certainly one major cause of terrorism, but the global dominance of some nations over others is also a factor.

Inequality is also connected to the context in which terrorism emerges. For example, research on al Qaeda terrorists finds that the leaders tend to come from middle-class backgrounds, although they often recruit those who are young, poorly educated, and economically disadvantaged to carry out suicide missions. Families of suicide bombers often receive large cash payments; at the same time, they can feel they have served a sacred cause (Stern 2003). Improving the lives of those who feel collectively humiliated could provide some protection against terrorism. The complex causes of terrorism cannot, however, be explained through any singular explanation. It is important to note that an analytical view of the causes of terrorism does not, in any way, excuse the cruel and inhumane acts committed by terrorist groups or individuals.

World Poverty

One fact of global inequality is the growing presence and persistence of poverty in many parts of the world (see ■ map 9-4). There is poverty in the United States, but very few people in the United States live in the extreme levels of deprivation found in some of the poor countries of the world. We have seen in Chapter 8 how the poverty line in the United States is calculated. The definition of poverty in the United States identifies **relative poverty**, that is, a measure of poverty relative to the rest of society. Households living in poverty in the United States are poor compared with other Americans, but this would be an inaccurate measure in a worldwide context because of such huge differences in the standard of living.

The World Bank and United Nations measure world poverty in two ways: **Absolute poverty** is defined by the amount of money needed in a particular country to meet basic needs of food, shelter, and clothing. **Extreme poverty** is defined as living on less than the equivalent of $1.25 per day. Any way you measure it, it is difficult for most Americans to imagine this standard of living. Extreme poverty has been declining in the world, but 10.7 percent of the world's population—767 million people—still live below $1.90 a day (World Bank 2017c).

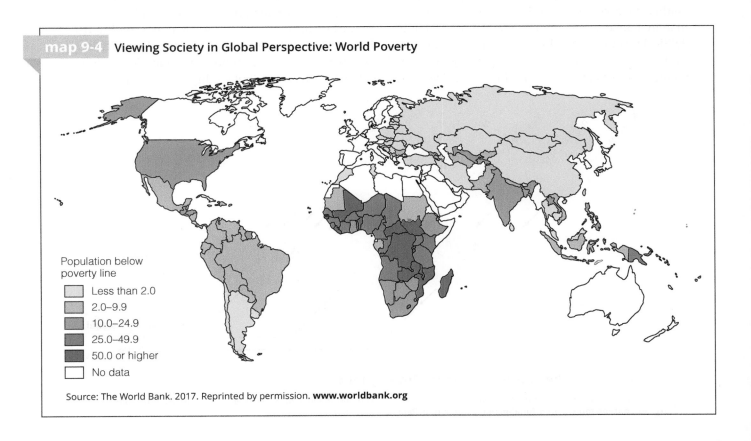

map 9-4 Viewing Society in Global Perspective: World Poverty

Population below poverty line

- Less than 2.0
- 2.0–9.9
- 10.0–24.9
- 25.0–49.9
- 50.0 or higher
- No data

Source: The World Bank. 2017. Reprinted by permission. **www.worldbank.org**

However, money does not tell the whole story because many people in poor countries do not always deal in cash. In many countries, people survive by raising crops for personal consumption and by bartering or trading services for food or shelter. These activities do not show up in calculations of poverty levels that use amounts of money as the measure. Accordingly, the United Nations also defines what it calls the multidimensional poverty index (see ▲ Figure 9-7).

The **multidimensional poverty index** measures the degree of deprivation in three basic dimensions of human life: health, education, and standard of living. These different components are then used to create a measure of poverty, including such indicators as nutrition, child mortality, educational attainment, and the availability of water, electricity, plumbing, cooking fuel—and even whether one has a floor in one's living quarters.

Measured by the multidimensional poverty index, the United Nations concludes that poverty is higher than when measured by income alone, as is the case with measuring absolute and extreme poverty. The multidimensional poverty index also points more directly to interventions to reduce poverty, such as making health clinics and running water available—projects that can significantly improve the lives of millions.

Who Are the World's Poor?

As we have seen, about 10 percent of the world's population lives in poverty, forming what the United Nations calls a *global underclass*. Although world poverty has been decreasing, it still afflicts a significant proportion of the world's people. In a world with a population over seven billion, about one billion live in extreme poverty. The decline in world poverty is largely accounted for by the economic growth in East Asia, which historically has been one of the poorest areas of the world. Now East Asia leads the world in poverty reduction. China alone has seen a decline of 400 million people moving out of poverty since the 1980s.

The causes of poverty differ around the globe. In Asia, the pressures of large population growth leave many without sustainable employment. As manufacturing has become less labor intensive with more mechanized production, the need for labor in certain industries has declined. Even though new technologies provide new job opportunities, they also create new forms of illiteracy because many people have neither the access nor the skills to use information technology. In sub-Saharan Africa, the poor live in marginal areas where poor soil, erosion, and continuous warfare have created extremely harsh conditions. Political instability and low levels of economic productivity also contribute to the high rates of poverty in some nations. Solutions to world poverty in these different regions require sustainable economic development, as well as an understanding of the diverse regional factors that contribute to high levels of poverty.

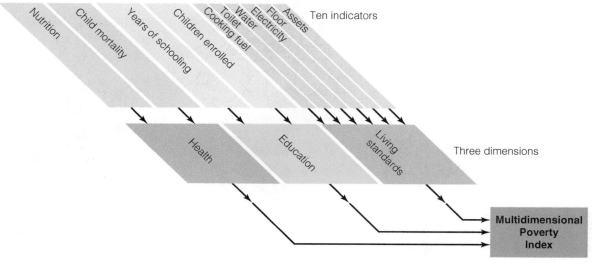

▲ **Figure 9-7** **The Multidimensional Poverty Index**

Source: United Nations Development Programme. 2016. "Components of the Multidimensional Poverty Index." **http://hdr.undp .org/en/content/multidimensional-poverty-index-mpi.** Reprinted with permission.

What Would a Sociologist Say?

Human Trafficking

The International Labour Organization estimates that there are likely as many as 21 million people worldwide enslaved in human trafficking. This includes sexual servitude, forced labor, forced child labor, and other forms of coercive treatment. Human trafficking is a modern form of slavery in which people are used for commercial gain through the use of force, coercion, or fraud (U.S. State Department 2017).

Human trafficking is a moral wrong, a criminal act by corrupt individuals, and a human rights issue. As a sociological issue, human trafficking is a complex social structure that is integrally connected to international trade, the social structure of tourism, and the racial, class, and gender inequality that crosses national borders.

Sexual trafficking is a particular form of human trafficking in which women and, often, young girls are bought and sold in an international system of prostitution. Sociologists argue that the male-dominated character of state institutions plays a part in the tolerance of sexual trafficking. Sexual trafficking and sexual tourism are part of a culture in which women's bodies are treated as a commodity. Racial and ethnic inequality also play a part as women of color are sexually exploited based on the racial/gender stereotypes that define them as exotic but also available for the pleasure of men.

An anti-trafficking movement has developed that involves a coalition of feminists, various voluntary organizations, the United Nations, some politicians, and others who have organized to stop this practice.

Women and Children in Poverty

There is no country in the world in which women are treated as well as men. As with poverty in the United States, women bear a larger share of the burden of world poverty. Some have called this *double deprivation*—in many of the poor countries, women suffer because of their gender and because they disproportionately carry the burden of poverty. For instance, in situations of extreme poverty, women have the burden of taking on much of the manual labor because the men in many cases have left to find work or food. The United Nations concludes that strengthening women's economic security through better work is essential for reducing world poverty.

Because of their poverty, women tend to suffer greater health risks than men. Although women outlive men in most countries, the life expectancy gap between women and men is *less* in the poorest countries. This is explained by several factors that put women at special risk. For one, fertility rates are higher in poor countries. Giving birth is a time of high risk for women, and women in poor countries with poor nutrition, poor maternal care, and the lack of trained birth attendants are at higher risk of dying during and after the birth.

High fertility rates are also related to the degree of women's empowerment in society—an often neglected aspect of the discussion between fertility and poverty. Societies where women's voices do not count for much tend to have high fertility rates as well as other social and economic hardships for women, including lack of education, job opportunities, and

Dan Vincent/Alamy Stock Photo

When children are poor, they may turn to child labor to help support their families. Such is the case with this child collecting trash for potential sale or use in a municipal dump in Phnom Penh.

information about birth control. Empowering women through providing them with employment, education, property, and voting rights can have a strong impact on reducing the fertility rate (Sen 2000).

Women also suffer in some poor countries because of traditions and cultural norms. Most (though not all) of the poor countries are patriarchal, meaning that men control the household. As a result, in some situations of poverty, the women eat after the men, and boys are fed before girls. In conditions of extreme poverty, baby boys may also be fed before baby girls because boys have higher status than girls. As a result, female infants have a lower rate of survival than male infants.

A distressing number of children in the world are also poor, including in the most industrialized and affluent nations. As you can see in ▲ Figure 9-8, poverty among children in the United States exceeds that of other industrialized nations and is second only to Mexico. Children in poverty seldom have the luxury of an education. Schools are usually few or nonexistent in poor areas of the world, and families are so poor that they cannot afford to send their children to school. Children from a very early age are required to help the family survive by working or performing domestic tasks such as fetching water. In extreme situations, even very young children may work as beggars (Boo 2012). Young boys and girls may end up working in sweatshops. Families may have to sell young girls into prostitution. This may seem unusually cruel and harsh by Western standards, but it is difficult to imagine the horror of starvation and the desperation that many families in the world must feel that would force them to take such measures to survive. In poor countries, families feel they must have more children for their survival, yet having more children perpetuates the poverty.

Estimates are that 156 million children under age 17 are in the labor force throughout the world. Most of the children are in Asia, though some are also in sub-Saharan Africa (Diallo, Etienne, and Mehran 2013). Many of these children work long hours in difficult conditions and enjoy few freedoms, making products (soccer balls, clothing, and toys, for example) for those who are much better off.

Another problem in the very poor areas of the world is homeless children (Mickelson 2000). In many situations, families are so poor that they can no longer care for their children, and the children must go without education or be out on their own, even at young ages. Many of these homeless children end up in the streets of the major cities of Asia and Latin America. In Latin America, it is estimated that there are thirteen million street children, some as young as six years old. Alone, they survive through a combination of begging, prostitution, drugs, and stealing. They sleep in alleys or in makeshift shelters. Their lives are harsh, brutal, and short.

Poverty and Hunger

Poverty is also directly linked to malnutrition and hunger because people in poverty cannot find or afford food. It is estimated that about 2 billion people in the world are malnourished. Experts attribute the increase to a number of factors, including poverty, lack of agricultural development, displacement and war, instability of economic markets, climate and weather, and wasting food.

Hunger stifles the mental and physical development of children and leads to disease and death. Although the food supply is plentiful in the world and is actually increasing faster than the population, the rate of malnutrition remains dangerously high.

Is there not enough food to feed all the people in the world? In fact, plenty of food is grown in the world. The world's production of wheat, rice, corn, and other grains is sufficient to adequately feed all the people in the world. Much grain grown in the United States is stored and not used. The problem is that the surplus food does not get to the truly needy. The people who are starving lack what they need for obtaining adequate food, such as arable land or a job that would pay a living wage. In the past, people in most cases grew food crops and were able to feed themselves, but much of the best land today has been taken over by

Debunking Society's Myths

Myth: There are too many people in the world, and there is simply not enough food to go around.
Sociological Perspective: Growing more food will not end hunger. If systems of distributing the world's food were more just, hunger could be reduced.

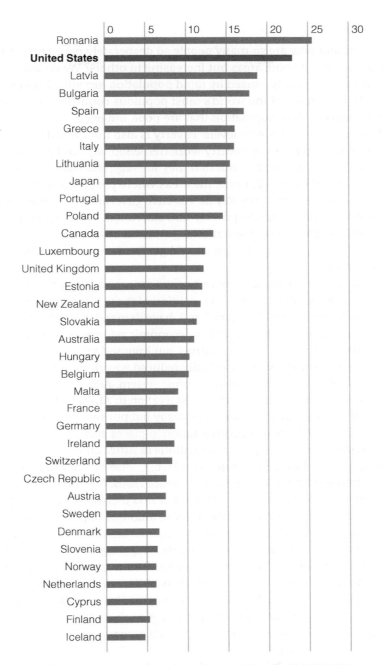

▲ Figure 9-8   Child Poverty* in the World's Richest
Nations

*Child poverty measured as living in household under 50% of national
median income.

Source: UNICEF. 2012. *Measuring Child Poverty*. Florence, Italy. **https://www
.unicef-irc.org/publications/pdf/rc10_eng.pdf**

agribusinesses that grow cash crops, such as tobacco or cotton, and subsistence farmers have been forced
onto marginal lands on the flanks of the desert where conditions are difficult and crops often do not grow.

Clearly, poverty is a cause of malnutrition, but there are other causes as well. Violence and war within
a nation can displace people, leading to large numbers of refugees crowding places where food may not
be available to all. Disasters can leave people without food and water—a situation complicated when a
nation is already poor. Even climate change can threaten to create hunger, especially if farming practices
cannot adjust to drought, floods, and extreme changes in weather patterns.

Causes of World Poverty

What causes world poverty, and why are so many people so desperately poor and starving? More to the point, why is poverty decreasing in some areas but increasing in others? We do know what does *not* cause poverty. Poverty is not necessarily caused by rapid population growth, although high fertility rates and poverty are related. In fact, many of the world's most populous countries—India and China, for instance—have large segments of their population that are poor, but even with very large populations, these countries have begun to reduce poverty levels. Poverty is also not caused by people being lazy or disinterested in working. People in extreme poverty work tremendously hard just to survive, and they would work hard at a job if they had one. It is not that they are lazy; it is that there are no jobs for them.

Poverty is a result of a mix of causes. For one, the areas where poverty is increasing have a history of unstable governments or, in some cases, virtually no effective government to coordinate national development or plans that might alleviate extreme poverty and starvation. World relief agencies are reluctant to work in or send food to countries where the national governments cannot guarantee the safety of relief workers or the delivery of food and aid to where it should go. Food convoys may be hijacked or roads blocked by bandits or warlords.

In many countries with high proportions of poverty, the economies have collapsed and the governments have borrowed heavily to remain afloat. As a condition of these international loans, lenders, including the World Bank and the International Monetary Fund, have demanded harsh economic restructuring to increase capital markets and industrial efficiency. These economic reforms may make good sense for some and can improve the standard of living in these countries, but imposed reforms have also harmed people, especially the poor, when reforms also call for drastically reduced government spending on human services.

Poverty is also caused by changes in the world economic system. Although poverty has been a long-term problem and has many causes, increases in poverty and starvation in Africa can be attributed in part to the changes in world markets that have favored Asia economically but put sub-Saharan Africa at a disadvantage. As the price of products declined with more industrialization in places such as India, China, Indonesia, South Korea, Malaysia, and Thailand, commodity-producing nations in Africa and Latin America suffered. In Latin America, the poor have flooded to the cities, hoping to find work, whereas they did the opposite in Africa, fleeing to the countryside hoping to be able to grow subsistence crops. Governments often had to borrow to provide help to their citizens. Some governments collapsed or find themselves in such great debt that they are unable to help their own people. This has created massive amounts of poverty and starvation.

Another cause of poverty is war. War disrupts the infrastructure of a society—its roads, utility systems, water, sanitation, even schools. For countries already struggling economically, this can be devastating. Food production may be disrupted and commerce can be threatened as it may be difficult, even impossible, to move goods in and out of a country. The loss of life and major injury can mean that there are fewer productive citizens who can work, thus threatening family and community well-being. Moreover, the billions of dollars spent each year on military struggle rob societies of the resources that could be used to address humanitarian needs. Add to this the fact that wars are more likely to occur in nations that are already poor, and you see the impact that war has on world poverty.

In sum, poverty has many causes. It is a major global problem that affects the billions who are living in poverty, but also affects all people in one way or another. In some areas, poverty rates are declining as some countries begin to improve their economic situation. In other areas of the world, however, poverty is increasing, and countries are sinking into financial, political, and social chaos.

Globalization and Social Change

Globalization is, in some ways, not a new thing. Nations have long been engaged through a global system of trade, travel, and tourism. What is new about globalization is the extent to which it permeates daily life for people all over the world and the pace with which globalization is developing. New technologies now allow for extraordinarily fast transactions across tremendous distances, both linking people together in new ways and transferring goods, cultural symbols, and communication systems in ways that were unimaginable not that long ago (Martell 2017; Eitzen and Baca Zinn 2011).

Globalization is thus ushering in social changes—some good, some not—that will continue to evolve in the years ahead. As we have seen, globalization has meant that many countries in the world are becoming better off, but many countries remain persistently poor, some very poor. Is the world getting better or worse? What will happen in the future? As one commentator has argued, "Perhaps the three most daunting tasks facing humanity in the 21st century are the reduction of global inequality, the preservation of our wondrous planet, and the strengthening of human security" (Steger 2017: 129).

Population overcrowding strains various natural resources.

There is some good news. In some areas of the world, particularly East Asia, but also in Latin America, many countries have shown rapid growth and are emerging as stronger nations. Examples include South Korea, Malaysia, Thailand, Taiwan, and Singapore. In these countries, the governments have invested in social and economic development, sometimes with help from other nations and corporations. Because some of these countries have large populations, their success demonstrates that economic development can occur in heavily populated countries. China, for example, has embarked on an aggressive policy of industrial growth, and India has also improved economically.

Yet, for all the success stories that globalization has generated, many nations are suffering, including nations on all continents. In many cases, governments have collapsed or are corrupt, the economy is bankrupt, the standard of living is poor, and people are starving. In many areas of the world, ethnic hatred has led to mass genocide and forced millions from their homes, creating huge numbers of refugees. Syria is a very recent example.

Globalization has also brought the expansion of the system of capitalism, including to nations once hostile to capitalist economics, such as China. This has opened new markets, increased global trade, but also expanded the reach of multinational corporations. The development of such world financial markets may bring prosperity and wealth to many nations and individuals, and it can allow some formerly poor countries to share in the world's wealth. Economic prosperity does not usually filter down to the people at the lower levels of society, and it can force nations into huge amounts of debt, thus allowing poverty and hunger to continue—or even worsen. Although market economies create opportunities for some to become wealthy, many nations and individuals do not benefit from this global transformation.

Globalization is a strong force that will continue to shape the future of most nations. Some see globalization simply as the expansion of Western markets and culture into all parts of the world. Western civilization brings positive new values (including democracy and more equality for women), but it can also bring values that may not be seen as positive changes—such as increased consumerism or a change in the nation's sexual mores. Globalization certainly brings new products to remote parts of the world (movies, clothing styles, and other commercial goods), but some see this as a form of imperialism—that is, the domination of Western nations. Resistance to Western globalization and imperialism produces some of the international problems now dominating U.S. and world history, as evidenced in the hostility felt by militant fundamentalist Islamic groups toward the United States.

Globalization has created great progress in the world, including trade, migration, the spread of diverse cultures, the dissemination and sharing of new knowledge, greater freedom for women, travel, and so forth. The technological ability for rapid and relatively easy global communication means that, for better or worse, nations are now more and more interconnected and interdependent. Globalization is a multidimensional phenomenon and will continue to influence both national and international social relations in the years to come (Martell 2017).

Chapter Summary

What is global stratification?

Global stratification is a system of unequal distribution of resources and opportunities between countries. A particular country's position is determined by its relationship to other countries in the world. The countries in the global stratification system can be categorized according to their per capita *gross national income* or wealth. The global stratification system can also be described according to the economic power countries have.

How do systems of power affect different countries in the world?

The countries of the world can be divided into three levels based on their power in the world economic system. The *core countries* are the countries that control and profit the most from the world system. *Semiperipheral countries* are semi-industrialized and play a middleman role, extracting profits from the poor countries and passing those profits on to the core countries. At the bottom of the world stratification system are the *peripheral countries*, which are poor and largely agricultural, but with important resources that the core countries exploit. Most of these nations are populated by people of color, perpetuating racism as part of the world system.

What are the theories of global stratification?

Modernization theory interprets the economic development of a country as the result of cultural attitudes and values that promote economic development. *Dependency theory* draws on the fact that many of the poorest nations are former colonies of European colonial powers that keep colonies poor and do not allow their industries to develop, thus creating dependency. *World systems theory* argues that no nation can be seen in isolation and that there is a world economic system that must be understood as a single unit.

What are some of the consequences of global stratification?

The poorest countries have more than half the world's population and have high birthrates, high mortality rates, poor health and sanitation, low rates of literacy and school attendance, and are largely rural. The richest countries have low birthrates, low mortality rates, largely urban populations, better health and sanitation, high literacy rates, and high school attendance. Although women in the wealthy countries are not completely equal to men, they suffer less inequality than do women in poor countries.

How do we measure and understand world poverty?

Relative poverty means being poor in comparison to others. *Absolute poverty* is the amount of money needed in a particular country to meet basic needs of food, shelter, and clothing. *Extreme poverty* is defined as the situation in which people live on less than $1.25 a day. The United Nations has developed a *multidimensional poverty index*—a measure that accounts for health, education, and standard of living. Poverty particularly affects women and children. Children in the very poor countries are forced to work at very early ages and do not have the opportunity for schooling. Street children are a growing problem in many cities of the world. Starvation is also a consequence of the global stratification system.

What is the future of global stratification?

The future of *global stratification* is varied and depends on the country's position within the world economic system. Some countries, particularly those in East Asia—commonly referred to as newly industrializing countries—have shown rapid growth and emerged as developed countries. Many nations, though, are not making it. Globalization is likely to continue to influence both individual nations and the world as a whole in years ahead.

Key Terms

absolute poverty 227
colonialism 217
commodity chain 220
core countries 220
dependency theory 217
extreme poverty 227
gender inequality index 224
Gini coefficient 213

global stratification 209
gross national income (GNI) 211
international division
of labor 213
modernization theory 216
multidimensional poverty
index 228
multinational corporations 218

neoliberalism 217
peripheral countries 220
relative poverty 227
semiperipheral countries 220
terrorism 226
world cities 216
world systems theory 218

CHAPTER 10

RACE AND ETHNICITY

In this chapter, you will learn to:

Define race and ethnicity as social constructions

Distinguish prejudice and discrimination

Understand the different forms of racism

Relate the brief history of diverse racial and ethnic groups

Identify current evidence of racial stratification

Compare different explanations of racial inequality

Explore and debate current strategies for achieving racial justice

Racial and ethnic inequality are deeply embedded in the structure of U.S. society. As a consequence, events such as the following are all too common:

- In 2017, a very large group of self-proclaimed white nationalists and Nazis marched in Charlottesville, Virginia. As they marched, they chanted, "Jews will not replace us" and "White lives matter," along with other Nazi slogans. During the white nationalist rally, a Nazi sympathizer White man, Christopher Cantwell, rammed his car through the crowd, killing Heather Heyer, a White woman and champion of civil rights who was there to protest the racist actions of the far-right.
- During his 2016 presidential campaign, Donald Trump riled up massive crowds of mostly White people who repeatedly chanted, "Build a wall"—a pledge Trump then signed as an Executive Order after he became the U.S. president. During the same campaign, he also called Mexican immigrants "rapists" and criminals who are "bringing drugs" and "bringing crime"

(*Washington Post*, July 8, 2015). In truth, immigrants are actually less likely to commit crime than native-born or second and third generation immigrants (Ewing, Martinez, and Rumbaut 2015).

- A sorority at a major East Coast university posted a photo of their group dressed in sombreros, ponchos, and fake mustaches, also carrying signs that said, "Will mow lawn for weed and beer." Such demeaning and offensive "racial theme parties" are common on college campuses (Cabrera 2014), including the March 2015 event in which a chapter of Sigma Alpha Epsilon at the University of Oklahoma was closed after a busload of fraternity brothers chanted a highly racist tune: "There will never be a ni**** at SAE. You can hang him from a tree, but he can never sign with me. There will never be a ni**** at SAE" (cnn.com).

- A Delta airlines employee, Rabeeya Khan—who wears a hijab—was sitting in her office in the airport lounge when Robin Rhodes, a White man, began harassing and kicking her while shouting, "Trump is here now. He will get rid of all of you." Rhodes was charged with assault, unlawful imprisonment, menacing, and harassment (*ABC7 News*, January 26, 2017). The Southern Poverty Law Center, which monitors incidents of hate, reports that hate crimes have spiked following Donald Trump's election (www.splc.org, December 16, 2016).

- On a summer day in 2015, Dylann Roof, a 21-year old self-declared white supremacist, walked into a Bible study group at the Emanuel African Methodist Episcopal Church in Charleston, South Carolina. After sitting in a Bible study group for some time, Roof pulled out a gun, then shot and killed nine African American worshippers, also wounding several others. When he confessed, he said he wanted to start a race war. Roof was convicted on 33 counts of hate crimes, including murder, attempted murder, damage to religious property, obstruction of religious belief, and various weapons charges. He was sentenced to death.

These are all ugly incidents. Of course, not all incidents of prejudice and racism are so blatant or extreme. Racial and ethnic groups do not always interact as enemies, and interracial tension is not always so obvious. It can be as subtle as a White person who simply does not initiate interaction with African Americans or Latinos. Or, it might be present in the fear a Black man feels if pulled over by a White police officer for a minor traffic violation. Interracial tension is also present in the anxiety a Black student might feel as the only person of color in classroom if she is asked to "speak for her race." The consequences of racial and ethnic inequality also show up in the persistence of racial disparities in economic status and in the high degree of segregation in our nation's schools and neighborhoods. All of this happens even while most people claim not to be racist and to hold no prejudice. How can this be?

Racial and ethnic inequality have deep roots in U.S. society. It is commonly said that this nation is strengthened by the diversity of its population and its incorporation of immigrants from so many different backgrounds and nations. Yet, immigration remains one of the most contentious issues of the time. Racial and ethnic inequality, even with the many changes that have taken place, remain hotly contested social issues (Andersen 2017).

Why does society treat racial and ethnic groups differently, and why is there inequality between the diverse groups that make up our society? As this chapter will show, race and ethnicity remain two of the most important axes of social stratification in the United States.

The Social Construction of Race and Ethnicity

It is fairly likely that when you see someone for the first time, you immediately recognize their race, probably along with their gender, age, and, perhaps, social class. The **salience principle** is the idea that we categorize people on the basis of what appears initially prominent and obvious—that is, salient—about them. Skin color is a salient characteristic and it has become how we "see" race. But what is race exactly?

Defining Race

People commonly think of race as rooted in biological differences between human groups. Consequently, noticeable physical differences between groups have come to be known as "races." But when you look deeper, you find that biological characteristics are only significant because of the social meaning they have taken on over the course of human history. Why, for example, do we differentiate people based on skin color and not some other characteristic such as height or hair color? That we use skin color and not some other observable characteristic is a manifestation of particular histories, not fixed biological differences between people.

Of course, there are differences among people, but most of the variability in almost all biological characteristics, including such basic differences as blood type, is *within* and not between racial groups. Using DNA sequencing, scientists from the human genome project have mapped the over twenty thousand human genes and convincingly shown that *there is no such thing as a race gene*. Various physical characteristics are produced through our genes, but we now know that many basic genetic features are also influenced through environmental factors (Bonham 2015).

The biological characteristics that have been used to differentiate racial groups also vary considerably both within and across groups. Many Asians, for example, are actually lighter skinned than many Europeans and White Americans but, regardless of their skin color, have been defined in racial terms as "yellow." Some light-skinned African Americans are also lighter in skin color than some White Americans. Developing racial categories overlooks the fact that human groups defined as races are, biologically speaking, much more alike than they are different (Graves 2004).

In other words, human beings do not fall into different and discretely different genetic races, as would be true of certain animal species (Graves 2004). The biological differences that are presumed to define different racial groups are somewhat arbitrary and certainly not fixed (Morning 2011, 2008; Washington 2011). Within the United States, laws that have historically defined who is Black varied from state to state. North Carolina and Tennessee law historically defined people as Black if they had even one great-grandparent who was Black (thus being one-eighth Black—called "octoroon" in the 1890 Census; see Table 10-1). In other southern states, having any Black ancestry at all defined one as a Black person—the so-called *one-drop rule*, that is, one drop of Black blood (Washington 2011; Broyard 2007; Malcomson 2000).

Society assigns people to racial categories, such as Black, White, and so on, not because of science, logic, or fact, but because of opinion and social experience. This is what it means to say that *race is a social construction*. Although the meaning of race may begin with alleged biological or genetic differences between groups (such as skin color, lip form, and hair texture), on closer examination the assumption that racial differences are biologically-based breaks down (Taylor 2012; Morning 2011; Ledger 2009; Lewontin 1996).

The social construction of race is even more apparent when we consider the meaning of race in other countries. In Brazil, a light-skinned Black person could well be considered White, especially if the person is of high socioeconomic status. One's race in Brazil is in part actually defined by one's social class. Thus, in parts of Brazil, it is often said that "money lightens" (*o dinheiro embranquence*). In fact, people in Brazil are considered Black only if they are clearly of African descent and have little or no discernible White ancestry. A large percentage of U.S. Blacks would not be considered Black in Brazil. Still, even though Brazil is often touted as being a utopia of race "mixing," light-skinned Brazilians hold a disproportionate share of the wealth and power. Brazilians of darker skin color have significantly lower earnings, occupational status, and lower access to education (Telles et al. 2011; Villareal 2010).

In the end, **race** is defined as a group treated as distinct in society based on presumed biological (and sometimes cultural) characteristics that have been assigned social importance. It is not the biological characteristics per se that define racial groups but *how groups have been treated and labeled historically and socially* (Andersen 2017; Higginbotham and Andersen 2012).

Debunking Society's Myths

Myth: Racial differences are fixed, biological categories.
Sociological Perspective: Race is a social construct, one in which certain physical or cultural characteristics take on social meanings that become the basis for racism and discrimination. The definition of race varies across cultures within a society, across different societies, and at different times in the history of a given society (Morning 2013; Graves 2004).

Racial Formation

The social construction of race has been elaborated by racial formation theory. **Racial formation** is the "process by which racial categories are created, inhabited, transformed, and/or destroyed" (Omi and Winant 1986: 64). The process of racial formation is also directly linked to the exploitation of so-called racial groups. Official social institutions, such as the law and the government, produce and maintain the meaning of race, but racial formation theory also asserts that these constructions of race can be changed—both within official institutions and because of people mobilizing to challenge racial categories.

Racial formation theory emphasizes that it matters *who* defines racial group membership. Race can be defined in a number of ways, including such as how people define themselves, how others define you, and, most importantly, how powerful interests define racial groups (Taylor 2004). In the United States, you can see the power of people to define others in how Black people were originally defined in the U.S. Constitution. With a system of slavery in place, the Constitution defined Black slaves as only three-fifths of a person. This was a compromise between northern and southern states, each vying for power in the newly forming nation. How people were counted mattered for purposes of taxation and representation in the new federal government. Such an inhumane definition directly linked concepts of race to the political and economic interests of the most powerful group in society—White people—and certain White people at that (A. L. Higginbotham 1978).

In other ways, government definitions of race continue to categorize people through the process of racial formation. American Indian people have to prove themselves as members of tribes through an elaborate set of federal regulations called the *federal acknowledgment process*. Very few Native American individuals are actually given this official status. The criteria for tribal membership for defining someone as "Indian" or "Native American" have also varied considerably throughout American history. Being defined as such matters because only those groups officially defined as Indian qualify for health, housing, and educational assistance from the Bureau of Indian Affairs (the BIA) or are able to manage the natural resources on Indian lands and maintain their own system of governance (Locklear 1999; Brown 1993; Snipp 1989).

In other words, who is defined as a race is as much a political question as a biological or cultural one. Further, it is often the state or federal government, and not the racial or ethnic group *itself*, that defines who is a member of the group and who is not.

Racial formation theory rests on understanding the process of racialization. **Racialization** is the process by which some group, perhaps an ethnic group or a religious minority, comes to be defined as a race (Omi and Winant 2014; Harrison 2000; Malcomson 2000). The experiences of Jewish people provide a good example. Jews are more accurately called an *ethnic group* because of common religious and cultural heritage (see the definition of ethnicity below), but in Nazi Germany, Hitler defined Jews as a "race." An ethnic group thus became racialized. Jews were presumed by the Nazis to be biologically inferior to the group Hitler labeled the Aryans—white-skinned, blonde, tall, blue-eyed people. On the basis of this definition—which was supported through Nazi law, taught in Nazi schools, and enforced by the Nazi military—Jewish people were segregated, brutally persecuted, and systematically murdered in the Holocaust during the Second World War.

One could say that several groups are currently undergoing the process of racialization. Despite their diversity, Hispanics are increasingly defined as "brown" people, suggesting racialization. And, although Muslims in the United States are a religious minority and, thus, an ethnic minority, increased hostility and fear of Muslims run the risk of such a group becoming racialized (Garner and Selod 2014).

Multiracial Identity

The complexity of defining race has become even more apparent with the growing number of people who identify as multiracial. Although the overall number of self-identified multiracial people is still quite small (about 4.5 percent of the U.S. population as of the 2010 U.S. census), that number has doubled since 2000 and is expected to grow (Frey 2015).

Mixed-race people defy the biological categories that are typically used to define race. Is someone who is the child of an Asian mother and an African American father Asian or Black? Reflecting this issue, the U.S. Census's current practice is for people to have the option of checking several racial categories

Table 10-1	Comparison of U.S. Census Classifications, 1890–2010				
Census Date	White	African American	Native American	Asian American	Other Categories
1890	White	Black, Mulatto, Quadroon, Octoroon	Indian	Chinese, Japanese	
1900	White	Black	Indian	Chinese, Japanese	
1910	White	Black, Mulatto	Indian	Chinese, Japanese	Other
1990	White	Black or Negro	Indian (American), Eskimo, Aleut	Chinese, Japanese, Filipino, Korean, Asian, Indian, Vietnamese	Hawaiian, Guamanian, Samoan, Asian or Pacific Islander, Other
2000 and 2010	White	Black or African American	American Indian, Alaskan Native	Chinese, Japanese, Filipino, Korean, Asian, Indian, Vietnamese	Native Hawaiian, Pacific Islander, Other

Sources: Lee, Sharon. 1993. "Racial Classification in the U.S. Census: 1890–1990." *Ethnic and Racial Studies* 16(1): 75–94; U.S. Census Bureau. 2003. "Racial and Ethnic Classification Used in Census 2000 and Beyond;" Rodriguez, Clara E. 2000. *Changing Race: Latinos, the Census, and the History of Ethnicity*. New York: New York University Press; Silver, Alexandra. 2010. "Brief History of the U.S. Census." *Time* (February 8): 16; Washington, Scott. 2011. "Who Isn't Black? The History of the One-Drop Rule." PhD dissertation, Department of Sociology, Princeton University, Princeton, NJ.

rather than just one, thus defining one's self as "biracial" or "multiracial" (Spencer 2011; Waters 1990). As ◆ Table 10-1 shows, the decennial U.S. census (taken every ten years) has dramatically changed its racial and ethnic classifications since 1890, reflecting the fact that society's thinking about racial and ethnic categorization has not remained constant through time.

The growth of people identifying as multiracial shows that the concept of race can be fluid. The black-white binary—that is, thinking of race as solely one or the other—has dissolved with much greater attention to the diversity of racial and ethnic identities. Some fear that this will dilute the civil rights agenda of Black people while others laud the development of more inclusive and complex understandings or race. Many people, however they identify, are multiracial given the mixing that has occurred.

Growing attention to multiracial identities highlights the idea that race is a socially constructed category. Racial definitions and boundaries may be created for purposes of the powerful, but as the theory of racial formation reminds us, they can also be changed through people's actions and mobilization (Rockquemore, Brunsma, and Delgado 2009).

Ethnicity

Sociologists distinguish between race and ethnicity, even though this distinction can at times be blurry. An **ethnic group** is a social category of people who share a common culture and who define themselves as having a collective identity. Italian Americans, Japanese Americans, Muslim Americans, Mexican Americans, and many other groups are examples of ethnic groups in the United States.

Of course, racial groups also share common cultures and a common identity—so much so that some have argued that we should eliminate the distinction between race and ethnicity altogether. Ethnic groups, however, are generally understood in cultural terms and only rarely are thought to be somehow biologically different from the dominant group. People can also define themselves as part of an ethnic group usually without any lasting negative consequences, although there are certainly exceptions to this fact.

Like racial groups, ethnic groups develop because of their unique historical and social experiences. These experiences become the basis for the group's *ethnic identity*, meaning the definition the group has of itself as sharing a common cultural bond. Prior to immigration to the United States, Italians, for example, did not necessarily think of themselves as a distinct group with common interests and experiences. Originating from different villages, cities, and regions of Italy, Italian immigrants identified themselves by their family background and community of origin. However, the process of immigration and the subsequent experiences of Italian Americans in the United States, including discrimination, created a new identity for the group as "Italians" (Waters and Levitt 2002; Alba 1990; Waters 1990).

Likewise, today, Latinos (also sometimes referred to as Hispanics) come from many different nations and cultures, including those who are native to what is now the American Southwest, those known as Chicano/a. Latino/a is an identity that has developed to embrace groups that are quite different in culture and history, but have some common interests. Some embrace this identity; others do not. Muslim Americans, as well, come from many different national and other ethnic backgrounds, including those who have been born and raised in the United States. Ethnic identity, as you can see, can crisscross with many other identities, just as does racial identity.

Ethnic identification tends to grow stronger when groups face prejudice or hostility from other groups. Perceived or real threats and perceived competition from other groups may unite an ethnic group around common political and economic interests. Ethnic unity can develop voluntarily, or it may be involuntarily imposed when more powerful groups exclude ethnic groups from certain residential areas, occupations, or social clubs. Such exclusionary practices strengthen ethnic identity.

Ethnic groups are also found in many other societies, such as the Shia, Sunni, Kurds, Assyrians, Turkmen, and Druze of Syria. Syria is a good example of how ethnicity can be directly connected to power and conflict in society. In Syria, the ruling autocratic government is composed of the Alawites, a fundamentalist sect that is a numerical minority and an offshoot of the Shiite. The Alawites are pitted against the nation's Sunni Muslim majority—millions of whom have fled as refugees from war and mass execution. You can see in this and many other examples that ethnic conflicts can become the source of some of the most violent and repressive acts of world history.

Panethnicity

A new term has emerged to describe how ethnicity can also evolve into a new collective identity. **Panethnicity** is what happens when multiple ethnic groups come together to forge a new collective identity for some common purpose. In doing so, they create a new name for the group, usually because of some shared political or social need (Okamoto 2014; Espiritu 1992). The label "panethnicity" brings diverse ethnic groups together under one label as a way of mobilizing their collective power.

Such is what has happened with the label Native American as well as the label Latino/a and others you can probably think of, such as "people of color." Each of these panethnic terms designates a collective group that would otherwise have a more singular identity, such as Puerto Rican, Chicana, or Cuban American. Panethnicity is often self-generated, but it can be imposed, such as when the former Soviet Union created "Russia" by incorporating people from many different nations. A panethnic identity, especially when self-generated, is strongly associated with a sense of a common history and a linked fate (Okamoto and Mora 2014: 223; Wong et al. 2011).

Panethnicity can be the result of racial and ethnic discrimination. People who migrate to the United States from Latin America, for example, are not initially likely to think of themselves as Hispanic or Latino. But, if they encounter discrimination within the United States, they are likely to develop a Hispanic/Latino identity (Golash-Boza and Darity 2008). Asian Americans, for example, emerged when groups from different Asian backgrounds forged a new common identity in the United States because of the prejudice and discrimination they experienced. The label "Asian American" then creates solidarity among the different Asian ethnic groups within the United States (Espirutu 1992). It is estimated that two-thirds of Asian Americans now use the panethnic label as "Asian American" to describe themselves (Lien, Conway, and Wong 2003). Likewise, 80 percent of Mexican Americans, Puerto Ricans, and Cuban Americans now identify with panethnic identities, whereas not that many years ago, far fewer would have done so (Fraga 2012; Jones-Correa and Leal 1996; Tienda and Ortiz 1986).

Sometimes groups adopt panethnic identities to assert what they are not. Some Arab Americans, for example, have adopted a panethnic identity because they do not want to be thought of as "white"

(Ajrouch and Jamal 2007). Likewise, some African immigrants to the United States do not define themselves as African American because of their desire to retain their specific African identity (Imoagene 2012).

Research on panethnicity, like that on multiracial identity, underscores the fluidity of racial identities. Fundamentally, panethnicity underscores the point that race is constructed via group interrelationships and boundaries (Okamoto and Mora 2014).

Minority/Majority Groups

You are likely to hear racial groups—and sometimes certain ethnic groups—referred to as "minorities." The term "minority" has a distinct meaning in a sociological usage—one that references the power of some groups over others, not the numerical proportion of a given population. Some minority groups may even constitute a large numerical majority in a given population. A classic example comes from apartheid South Africa where black South Africans were a full 90 percent of the population, but powerless to control their own destiny. Even though Blacks outnumbered Whites ten to one, Black Africans were viciously oppressed and politically excluded under the infamous apartheid (pronounced "aparthate" or "apart hite") system of government.

The specific meaning of a **minority group** is a group in society that shares common group characteristics and occupies low status in society because of prejudice and discrimination. Power, not numbers, defines whether a racial or ethnic group is a "minority group." For example, Hispanics are now nearly 40 percent of the populations of both Texas and California, but they are a sociological minority group. Minority groups are defined by their subordinate status in society because of the power that a **dominant group** holds over them.

What race is the person shown in this photo? The photo is Etta James, fabulous jazz singer who passed away in 2012. Probably best known for her rendition of "At Last," James was born to an African American mother and, people think, a White father, most likely Swiss. According to the one drop rule, she would be Black. As concepts of race change, perhaps were she still living she would be considered multiracial.

Prejudice, Discrimination, and Stereotypes

Many people use the terms *prejudice*, *discrimination*, and *racism* loosely, as if they were all the same thing. They are not. Typically, people use these terms thinking that the major problems of race results from individual people's bad will or biased ideas. Sociologists use more refined concepts to understand not just attitudes and ideas, but also the societal basis of racial and ethnic inequality. Thus, sociologists distinguish prejudice, discrimination, and racism.

Prejudice

When people think about racial inequality, they often focus on the attitudes that people hold—that is, prejudice. **Prejudice** is the evaluation of a social group and the people within it that is based on misconceptions about the group and false generalizations applied to every presumed member of the group. Prejudice involves both prejudgment and misjudgment (Jones et al. 2013; Allport 1954).[1] Prejudice is usually negative or hostile. Thinking ill of people only because they are members of a particular group is prejudice.

People are not born with stereotypes and prejudices. Research shows that prejudiced attitudes are learned and internalized through socialization, including from family, peers, teachers, and the media. Attitudes about race are formed early in childhood, (Feagin 2000; Van Ausdale and Feagin 1996; Allport 1954). If a parent complains about "immigrants taking away jobs from Americans," then the child is likely to grow up thinking negatively about immigrants. That is, the child becomes prejudiced. Research shows

---

[1] *Prejudice* is a noun; prejudiced is an adjective. Thus, the correct usage is as such: A person may have a *prejudiced* attitude or be *prejudiced*, but, if so, the person is exhibiting *prejudice*.

that the more prejudiced the parent, the more prejudiced the child will be. This is even true for individuals who insist that they can think for themselves, and who think they are not influenced by their parents' prejudice (Taylor et al. 2013).

Most people disavow racial or ethnic prejudice, yet the vast majority of us carry around some prejudices, whether about racial–ethnic groups, women and men, old and young people, the upper and working classes, or LGBTQ[2] people. Virtually no one is free of prejudice—either of harboring it and or being its recipient. Decades of research have shown that people who are more prejudiced are also more likely to stereotype and categorize others than are those who are less prejudiced (Taylor et al. 2013; 2014; Adorno et al. 1950).

Prejudice based on race or ethnicity is called *racial* or *ethnic prejudice.* If you are Latino and dislike Anglos only because they are White, this is prejudice: It is a negative judgment or prejudgment based on race and ethnicity and very little else. If a Latino individual attempts to justify these feelings by arguing that "all Whites have the same bad character," then the Latino is using a stereotype as justification for the prejudice. Note that any group can hold prejudice against another group.

Prejudice is also revealed in the phenomenon of **ethnocentrism**, the belief that one's group is superior to all other groups (see also Chapter 2). Ethnocentric people feel that their own group is moral, just, and right, and that an out-group—and thus any member of that out-group—is immoral, unjust, wrong, distrustful, or criminal. Individuals in the so-called out-group are then subject to prejudice.

Discrimination

Different from prejudice, which is an attitude, discrimination is behavior. **Discrimination** is the unequal treatment of members of some social group or stratum solely because of their membership in that group. Like prejudice, discrimination is typically negative—that is, it disadvantages one group over another. Discrimination can occur, as does prejudice, at the individual level, but it is also systemic—such as when social practices or policies exclude certain groups from the privileges and rights of others. Also like prejudice, discrimination occurs along a number of lines including race and ethnicity, but also gender, age, disability, sexual orientation, and so forth.

Discrimination is illegal under U.S. law. Nonetheless, various discriminatory processes continue and have been documented in countless research studies. One of the very effective ways of noting discrimination is the use of audit studies. **Audit studies** are research projects wherein two people identical in nearly all respects (age, education, gender, social class, dress, and other characteristics) present themselves as potential tenants or employees. If one is White and the other a person of color, audit studies routinely find that the person of color will regularly be refused housing or employment.

Sometimes audit studies are done using online research instead of people posing as live applicants or potential tenants. A recent example comes from a study where researchers submitted 9400 randomly generated resumes to online job advertisements. Each resume was identical in terms of qualifications but the researchers attached names that suggested White or Black applicants each of whom also had an identifiable female and male name. The female names were "Clare" and "May" (presumably White) and "Ebony" and "Aaliyah" (presumably Black). Likewise, the men were presented as "Cody" and "Jake" (White) and "DeShawn" and "DeAndre" (Black). In all other characteristics, the "applicants" were equal. Black applicants got 14 percent fewer callbacks but when customer interaction was involved in the job, Black applicants were 28 percent less likely to get a callback (Nunley et al. 2015). This audit study, like numerous others, shows that discrimination occurs far more frequently than most people imagine (Feagin 2007).

Stereotypes

Racial and ethnic inequality in society produce racial stereotypes, and these stereotypes become the lens through which members of different groups perceive one another. A **stereotype** is an oversimplified set of beliefs about members of a social group. Stereotypes are based on the tendency of humans to

---

[2]LGBTQ is the term now used to refer to lesbian, gay, bisexual, transgender, and queer people. Like panethnicity, it is a term forged from political interests to link the identities and experiences of those whose sexual and gender identities depart from dominant norms; see Chapters 11 and 12.

categorize a person based on a narrow range of perceived characteristics. Stereotypes are presumed to describe the "typical" member of some social group. They are usually incorrect.

In everyday social interaction, people tend to categorize other people. Fortunately or unfortunately, we all do this. The most common bases for such categorizations are race, gender, age, sexual orientation, and disability. Quick and ready categorizations help people process the huge amounts of information they receive about people with whom they come into contact. We may be taught from childhood to treat each person as a unique individual, but clearly we do not, as research shows. We process information about others quickly, assigning certain characteristics to them even with little actual knowledge of them.

Stereotypes based on race or ethnicity are called *racial–ethnic stereotypes*. Some common examples of racial–ethnic stereotypes are: Asian Americans are stereotyped as overly ambitious and academically successful; African Americans bear the stereotype of being loud but lazy; Hispanics are stereotyped as oversexed and probably illegal immigrants; Muslim Americans are stereotyped as terrorists; Jews are stereotyped as overly materialistic. Such stereotypes, presumed to describe the "typical" members of a group, are factually wrong, but wield great power in how people are perceived, understood, and treated.

No group in U.S. history has escaped the process of stereotyping. For example, Italians have been stereotyped as overly emotional and prone to crime; the Irish, as heavy drinkers. Even White people are often stereotyped as not being able to dance or, in the case of men, as unfeeling and insensitive.

Race and ethnicity are not the only basis for stereotypes. Gender stereotypes portray women as overly emotional, talkative, and probably inept at math and science. Age stereotypes depict older people as feeble-minded and asexual. People with disabilities may be stereotyped as stupid and asexual. Social class stereotypes also abound, such as in thinking that all White, working-class people are "rednecks."

Stereotypes are conveyed and supported by the cultural media—music, TV, magazines, art, and literature—and also by one's family. Even something as seemingly innocent as putting on a Halloween costume can reproduce and reinforce stereotypes within the culture (see the box, "Doing Sociological Research.").

Doing Sociological Research

Halloween Costumes: Reproducing Racial Stereotypes

Research Question

How do concepts of race develop in our minds? This question was examined by a team of sociologists who observed White students and their racial Halloween costuming.

Research Method

The faculty team analyzed 663 participant observation journals kept by undergraduate students during two fall semesters. Students were asked to write reactions and perceptions in their journals the two weeks before and after Halloween. Students recorded other student reactions to Halloween costuming, noting who, what, when, and where and also indicated the race, age, gender, and ethnicity of those they observed.

Research Results

Students reported "playing with" various cross-race identities, including not just White students costuming as Black, but also some African American and Asian students dressing as Black, often in celebrity portrayals. Many students report that Halloween freed them from fears of offending people. Yet, many (including White students) also saw this cross-race costuming as offensive to people of color.

Conclusions and Implications

The researchers conclude that many students thought that cross-race costuming at Halloween was just "for fun" despite the harm it does in reinforcing racial stereotypes. Halloween costumes trivialize race and do not subvert racism—rather they reinforce racism and harm students of color.

(*continued*)

Questions to Consider

1. Recall your last experience of Halloween. Do you see evidence of the sort that these researchers find? Why or why not?
2. Are there other sites where you might see cross-race (or cross-ethnic) costuming like that these researchers report? How is the context similar and different and what does this suggest about racial-ethnic stereotypes and their connection to power?

Source: Mueller, Jennifer C., Danielle Dirks, and Leslie H. Picca. 2007. "Unmasking Racism: Halloween Costuming and Engagement of the Racial Other." *Qualitative Sociology* 30(3): 315–335.

The Interchangeability of Stereotypes

Stereotypes, especially negative ones, are often interchangeable from one group to another—that is, from one racial or ethnic group to another, from a racial or ethnic group to a social class, or from a social class to a gender. This is the principle of **stereotype interchangeability**. Stereotype interchangeability is sometimes revealed through humor. You can think of ethnic jokes that often interchange different groups as the butt of the humor, stereotyping them as dumb and inept. Or, take the stereotype of African Americans as inherently lazy. This stereotype has also been applied in recent history to Hispanic, Polish, Irish, Italian, and other groups (illustrating interchangeability from one racial–ethnic group to another). In our society, there is a complex interplay among racial or ethnic, gender, and class stereotypes.

For example, many of the stereotypes applied to women as childlike, overly emotional, unreasonable, bad at mathematics, and so on have also been applied to African Americans, working-class people, the poor, and in the early twentieth century, even to Chinese Americans. A common theme is apparent: Through stereotype interchangeability, whatever group occupies lower social status in society at a given time (whether racial or ethnic minorities, women, or the working class), that group is negatively stereotyped. The stereotype is then used as an "explanation" for the observed behavior of a stereotyped group's members to justify their lower status in society. This in turn subjects the stereotyped group to prejudice.

Stereotype Threat

We think of stereotypes as affecting the attitudes and behaviors of those who hold them, but can stereotypes also affect the actual behavior of stereotyped individuals? An answer has been provided by social psychologist Claude Steele and his colleagues through the concept of **stereotype threat** (Steele 2010, 1997; Steele and Aaronson 1995). People know that there are stereotypes about their group and Steele has shown that being reminded of these stereotypes can actually influence people's behavior. Here is what Steele and his colleagues have shown.

When someone feels the presence of a stereotype, they will perform less well than those not subjected to stereotypes. In other words, the threat of a stereotype can produce anxiety and thus influence how well one performs. Thus, in experimental settings, when African American test takers were told in advance of the test that the test was a "true test of their ability," a stereotype of African Americans as less intelligent was invoked and, as a result, African American test takers performed less well on the exam than when, under the same conditions, the stereotype was not invoked prior to the test. To date, a large number of studies have confirmed this kind of effect.

Most research on stereotype threat has been conducted in laboratory settings, but you can imagine the impact of stereotype threat in other "real life" settings, such as how students of color might perform if a teacher thinks such students are less intelligent or less prepared for serious study. The principle of stereotype threat teaches us that even seemingly innocent comments that stereotype people can have negative effects of the stereotyped person or group. In sum, stereotypes have their consequences.

Thinking Sociologically

Identify an example of *stereotype interchangeability* that you have likely observed in your own life. What does this example reveal to you about how stereotypes reflect power dynamics between dominant and subordinated groups?

Racism: Its Many Forms

The term racism is often loosely used as if it were the same thing as prejudice. People tend to think of racism in terms of individual attitudes, like prejudice. But racism is more than prejudice. Racism is actually built into society's social institutions—at least in societies such as the United States that have been organized around racial domination and subordination. In this sense, racism is more than prejudice, although prejudice is one manifestation of societally structured racism.

We can define **racism** as the perception and treatment of a racial or ethnic group, or member of that group, as intellectually, socially, and culturally inferior to the dominant group. Racism is more than an attitude; it is institutionalized in society.

There are different forms of racism. Some is quite overt—such as building an economic system around slavery or, at the individual level, shooting someone just because they are Black or Latino. Racism can, however, also be subtle, covert, and nonobvious; this is known as **aversive racism** (Jones, Dovidio, and Vietze, 2013). A server who delays serving a Black couple is exhibiting aversive racism. This form of racism is quite common. Even when overt forms of racism dissipate, aversive racism tends to persist. Because it is less visible than overt racism, people who do not experience aversive racism can believe that racism has diminished when it has not (Gaertner and Dovidio 2005).

Another subtle form of racism, akin to aversive racism, is **implicit bias**. Implicit bias is a largely nonconscious form of racism, whereby individuals make unconscious associations, such as between race and crime. Culture forces such associations on individuals. Starting with childhood socialization, Blackness is mentally associated with criminality; Whiteness is not. Research concretely demonstrates that the association between race and crime directly impacts how individuals behave and make decisions (Eberhardt 2010).

Jennifer Eberhardt's research finds, for example, that in a court trial, holding other things constant, a defendant's skin color and hair texture correlate with the sentencing decisions of jurors: Black defendants are more likely to receive the death penalty than are otherwise similar White defendants. Eberhardt attributes such findings to implicit bias, a subtle form of racism that individuals internalize and carry with them always. This bias also carries over to police officers, who mistakenly identify Black faces as "criminal faces" relative to White faces.

Laissez-faire racism is another form of racism (Bobo 2006). Laissez-faire racism includes several elements:

1. The subtle but persistent negative stereotyping of minorities, particularly Black Americans, especially in the media;
2. A tendency to blame Blacks themselves for the gap between Blacks and Whites in socioeconomic standing, occupational achievement, and educational achievement;
3. Clear resistance to meaningful policy efforts (such as affirmative action, discussed later) designed to ameliorate racially oppressive social conditions and practices in the United States.

A close relative of laissez-faire racism is the concept of **colorblind racism**—so named because individuals affected by this type of racism prefer to ignore legitimate racial–ethnic, cultural, and other differences and insist that the race problems in the United States will go away if only race is ignored altogether. Simply refusing to perceive any differences at all between racial groups (thus being colorblind) is in itself a form of racism (Bonilla-Silva 2013; Gallagher 2013; Bonilla-Silva and Baiocchi 2001). Here is an example of colorblind racism: Have you heard someone say, as they attempt to show that they are "not racist": "I don't care if you are white, black, yellow, or green! I am not prejudiced!" Such people insist that they are only objective and fair people, people who do not notice skin color. By definition, this is colorblind racism—a form of racism, for sure. Colorblind racism ignores the reality of race and its significance in society.

Colorblindness hides White privilege. **White privilege** refers to the benefits (social, cultural, and economic) that White people receive in a society based on racial inequality, even if they are not always aware of these benefits (Andersen 2017). Even when White people define themselves as racially tolerant and believe that they do not see or judge people "by the color of their skin," they nonetheless are likely to be benefitting from White privilege. Of course, not all White people benefit equally from this system; some are likely to perceive themselves as benefitting very little and may even resent what they see as "special advantages" given to people of color. White people may think of skin color as irrelevant, but it is not. Racial

Minorities are more likely to be arrested than Whites for the same offense. This reflects institutional racism in addition to possible individual prejudice of the arresting police officer.

domination—that is, White privilege, is structured into society.

Institutional racism as a form of racism is the negative treatment and oppression of one racial or ethnic group by society's existing institutions based on the presumed inferiority of the oppressed group. Institutional racism exists at the level of social structure. In Durkheim's sense, it is external to individuals—thus institutional. It is then possible to have "racism without racists" (Bonilla-Silva 2013). Institutional racism reflects an important sociological principle: Institutional racism can exist apart from individual personalities. Key to understanding institutional racism is seeing that dominant groups have the economic and political power to subjugate minority groups, even if people in the dominant group do not have the explicit intent of being prejudiced or discriminating against others.

Consider this: Even if every White person in the country lost all of his or her personal prejudices, and even if he or she stopped engaging in individual acts of discrimination, institutional racism would still persist for some time. Over the years, institutional racism has become so much a part of U.S. institutions that discrimination can occur even when no single person is deliberately causing it. To sum up, racism is a characteristic of the institutions and not necessarily of the individuals within the institution. This is why institutional racism can exist even without prejudice being the cause of racial inequality.

Diverse Groups, Diverse Histories

The different racial and ethnic groups in the United States have arrived at their current social condition through histories that are similar in some ways, yet quite different in other respects. Their histories are related because of the common experience of White supremacy, economic exploitation, and political disenfranchisement.

Native Americans: The First of This Land

Native Americans were here tens of thousands of years before they were "discovered" by Europeans. Discovery quickly turned to conquest, and in the course of the next three centuries, the Europeans systematically drove the Native Americans from their lands, destroying their ways of life and crushing various tribal cultures. Native Americans were subjected to an onslaught of European diseases. Lacking immunity to these diseases, Native Americans suffered a population decline, considered by some to have been the steepest and most drastic of any people in the history of the world. The exact size of the indigenous population in North America at the time of the Europeans' arrival with Columbus in 1492 has been estimated at anywhere from one million

Debunking Society's Myths

Myth: The primary cause of racial inequality in the United States is the persistence of prejudice.
Sociological Perspective: Prejudice is one dimension of racial problems in the United States, but institutional racism can flourish even while prejudice is on the decline. Prejudice is an attribute of the individual, whereas institutional racism is an attribute of social structure.

to ten million people. Native American traditions have survived in many isolated places, but what is left is only an echo of the original 500 nations of North America (Snipp 2007, 1989; Nagel 1996; Thornton 1987).

At the time of European contact in the 1640s, there was great linguistic, religious, governmental, and economic heterogeneity among Native American societies. Disease and the encroachment of Europeans have ravaged these societies, leaving many American Indians among the poorest and most socially vulnerable people in the United States.

Today, the majority of American Indians and Alaska Natives (78 percent) live *outside* areas designated as American Indians or Alaskan Native lands (that is, reservations or Alaskan Native villages). Americans Indians are heavily concentrated in the West, but, as you can see in ▦ map 10-1, American Indians and Alaska Natives live in many different regions of the country, even though often overlooked or stereotyped as all living on reservations (Norris, Vines, and Hoeffel 2012).

More than one-quarter (28.5 percent) of American Indians and Alaska Natives live in poverty, far higher than the overall poverty rate for the nation. High unemployment also plagues Native Americans. As with Black Americans, Native American unemployment is twice the national average. Native Americans did not recover as well as other groups following the economic recession of 2008 (Peralta 2014). The first here in this land are now last in status, a painful irony of U.S. history.

The recent growth of casinos owned by Native Americans has improved the economic status of some groups, although it has also produced new stereotypes of Native Americans all getting rich— a stereotype obviously challenged by the data on poverty and unemployment. Research on Indian casinos finds that gaming has been a source of economic development for many tribal groups, but the improvements have not been evenly distributed. The benefit to Indian groups tends to be greatest when casinos are located in states with higher populations and a relatively high median income level (Conner and Taggart 2013).

map 10-1   Mapping America's Diversity: American Indian and Alaska Native Residence

Racial-ethnic stereotypes portray American Indians as isolated on reservations. Although many are, this map shows how widely dispersed American Indians are throughout the United States.

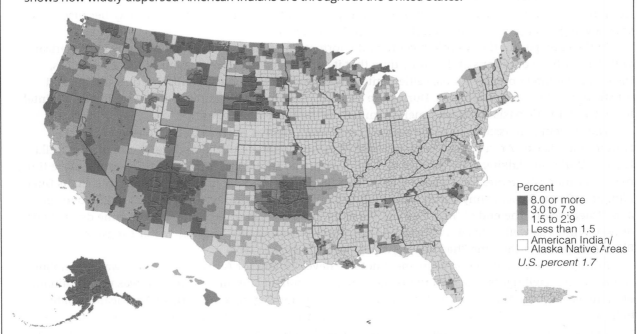

Percent
- 8.0 or more
- 3.0 to 7.9
- 1.5 to 2.9
- Less than 1.5
- American Indian/ Alaska Native Areas

U.S. percent 1.7

Source: Norris, Tina, Paula L. Vines, and Elizabeth M. Hoeffel. 2012. "The American Indian and Alaska Native Population: 2010." *U.S. Census Briefs*. Washington, DC: U.S. Census Bureau. **www.census.com**

African Americans

The development of slavery in the Americas is related to the development of world markets. Slaves were imported from Africa to provide the labor for sugar, wheat, rice, tobacco, and cotton production—a market system that enhanced the profits of slaveholders and northern traders. An estimated twelve million Africans were transported under appalling conditions to the Americas, about a quarter of whom came to the mainland United States (TransAtlantic Slave Trade Database 2014).

Slavery evolved as a form of stratification called a *caste system* (see Chapter 8). Slaveholders profited from the labor of a caste, the slaves. Central to the operation of slavery was the principle that human beings could be *chattel* (or property). Slavery was also based on the belief that Whites are superior to other races, coupled with a belief in a patriarchal social order.

The slave system also involved the domination of men over women—another aspect of the caste system. In this combination of patriarchy and White supremacy, White males presided over their "property" of White women as well as their "property" of Black men and women. This in turn led to gender stratification among the slaves themselves, reflecting the White slaveholders' assumptions about the relative roles of men and women. Black women performed domestic labor for their masters and their own families. White men further exerted their authority in demanding sexual relations with Black women (White 1999). The predominant attitude of Whites toward Blacks was paternalistic. Whites saw slaves as childlike and incapable of caring for themselves. The stereotypes of African Americans as "childlike" are directly traceable to the system of slavery.

Slaves struggled to preserve both their culture and their sense of humanity and to resist, often by open conflict, the dehumanizing effects of a system that defined human beings as mere property (Myers 1998; Blassingame 1973). Slaves revolted against the conditions of enslavement in a variety of ways, from passive means such as work slowdowns and feigned illness to more aggressive means such as destruction of property, escapes, and outright rebellion.

After slavery was ended by the Civil War (1861–1865) and the Emancipation Proclamation (1863), Black Americans continued to be exploited for their labor. In the South, the system of sharecropping emerged, an exploitative system in which Black families tilled the fields for White landowners in exchange for a share of the crop. With the onset of the First World War and the intensified industrialization of society came the Great Migration of Blacks from the South to the urban north. This massive movement, lasting from the late 1800s through the 1920s, significantly affected the status of Blacks in society because there was now a greater potential for collective action (Marks 1989).

In the early part of the twentieth century, the formation of Black ghettos victimized Black Americans with grim urban conditions at the same time that it encouraged the development of Black resources, including volunteer organizations, settlement houses, social movements, political action groups, and artistic and cultural achievements. During the 1920s, Harlem in New York City became an important intellectual and artistic oasis for Black America, known as the Harlem Renaissance.

The Harlem Renaissance gave the nation great literary figures, such as Langston Hughes, Jessie Fauset, Alain Locke, Arna Bontemps, Zora Neale Hurston, Countee Cullen, Wallace Thurman, and Nella Larsen (Marks and Edkins 1999; Gates et al. 1997; Rampersad 1988, 1986; Bontemps 1972). At the same time, many of America's greatest musicians, entertainers, and artists came to the fore, such as musicians Duke Ellington, Count Basie, Benny Carter, Billie Holiday, and Louis Armstrong, and painters Hale Woodruff and Elmer Brown. The end of the 1920s and the stock market crash of 1929 brought everyone down a peg or two, Whites as well as Blacks, although in the words of Harlem Renaissance writer Langston Hughes, Black Americans at the time "had but a few pegs to fall" (Hughes 1967).

During the twentieth century, the most notable development for African Americans was the mobilization around civil rights. The civil rights movement, examined later in this chapter, was the culmination of many years of organizing against racial discrimination and segregation. African Americans fought in both world wars, signaling their commitment as U.S. citizens but also highlighting how even with their role in these historic fights, the nation had not given Black men or women the full rights of equal belonging. That was all to change with the mass mobilization that characterized the civil rights movement of the 1950s and beyond.

Latinos/as

Latinos have recently surpassed African Americans as the largest minority population in the United States. The largest increase is that of Mexican Americans. Latinos include Chicanos/as, Mexican Americans, Puerto Ricans, Cubans, and other recent Latin and Central American immigrants to the United States. Latinos also include Latin Americans who have lived for generations in the United States. Many are not immigrants but very early settlers from Spain and Portugal in the 1400s. The terms *Hispanic* and *Latino* or *Latina* mask the great structural and cultural diversity among the various Hispanic groups.

Diverse Latino groups have been forced by institutional procedures to cause the public to perceive them "as one," for example, by the media, by political leaders, and by the U.S. Census categories (Mora 2009). The use of such inclusive terms also ignores important differences in their respective entries into U.S. society: Chicanos/as through military conquest of the Mexican–American War (1846–1848); Puerto Ricans through war with Spain in the Spanish–American War (1898); and Cubans as political refugees fleeing since 1959 from the communist dictatorship of Fidel Castro, which the U.S. government vigorously opposed (Telles et al. 2011; Glenn 2002; Bean and Tienda 1987).

Mexican Americans

Before the Anglo (White) conquest, Mexican colonists had formed settlements and missions throughout the West and Southwest. In 1834, the U.S. government ordered the dismantling of these missions, bringing them under tight governmental control and creating a period known as the Golden Age of the Ranchos. Land became concentrated into the hands of a few wealthy Mexican ranchers, who had been given large land grants by the Mexican government. This created a class system within the Chicano community, consisting of the elite ranchers, mission farmers, and government administrators at the top; *mestizos*, who were small farmers and ranchers, as the middle class; a third class of skilled workers; and a bottom class of manual laborers, who were mostly Indians (Maldonado 1997; Mirandé 1985).

With the Mexican–American War of 1846–1848, Chicanos lost claims to huge land areas that ultimately became Texas, New Mexico, and parts of Colorado, Arizona, Nevada, Utah, and California. White cattle ranchers and sheep ranchers enclosed giant tracts of land, cutting off many small ranchers, both Mexican and Anglo. Thus began a process of wholesale economic and social exclusion of Mexicans and Mexican Americans from U.S. society, much of which continues to this day (Telles and Ortiz 2008).

It was at the time of the Mexican–American war that Mexicans in the United States became defined as an inferior race that did not deserve social, educational, or political equality. This is an example of the *racial formation process* (Omi and Winant 2014). Anglos believed that Mexicans were lazy, corrupt, and cowardly, yet violent, which launched stereotypes that were used to justify the lower status of Mexicans and give Anglos control of the land that Mexicans originally occupied (Telles and Ortiz 2008; Moore 1976).

During the twentieth century, advances in agricultural technology changed the organization of labor in the Southwest and West. Irrigation allowed year-round production of crops and a new need for cheap labor to work in the fields. Migrant workers from Mexico were exploited as a cheap source of labor. Migrant work was characterized by low earnings, poor housing conditions, poor health, and extensive use of child labor. The wide use of Mexican migrant workers as field workers, domestic servants, and other kinds of poorly paid work continues even as Mexican Americans face new forms of exclusion.

Puerto Ricans

The island of Puerto Rico was ceded to the United States by Spain in 1898. In 1917, the Jones Act extended U.S. citizenship to Puerto Ricans, although it was not until 1948 that Puerto Ricans were allowed to elect their own governor. In 1952, the United States established the Commonwealth of Puerto Rico, with its own constitution. Following the Second World War, the first elected governor launched a program known as Operation Bootstrap, which was designed to attract large U.S. corporations to the island of Puerto Rico by using tax breaks and other concessions. This program contributed to rapid overall growth in the Puerto Rican economy, although unemployment remained high and wages remained low. Seeking opportunity, unemployed farmworkers began migrating to the United States. These migrants were interested in seasonal work, and thus a pattern of temporary migration characterized the Puerto Ricans' entrance into the United States (Amott and Matthaei 1996; Rodriguez 1989).

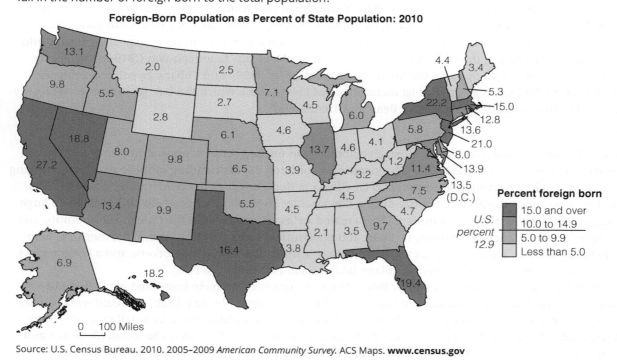

map 10-2   Mapping America's Diversity: Foreign-Born Population

This map shows the total number of foreign-born residents per state. Foreign-born includes legal residents (immigrants), temporary migrants (such as students), refugees, and illegal immigrants—to the extent they can be known. Some states have a high number of foreign-born (for example, California, Florida, and New York), and other states have fewer (for example, Wyoming, South Dakota, and Vermont). Where does your region fall in the number of foreign-born to the total population?

Foreign-Born Population as Percent of State Population: 2010

Percent foreign born
- 15.0 and over
- 10.0 to 14.9
- 5.0 to 9.9
- Less than 5.0

U.S. percent 12.9

0   100 Miles

Source: U.S. Census Bureau. 2010. 2005–2009 *American Community Survey*. ACS Maps. **www.census.gov**

Unemployment in Puerto Rico became so severe that the U.S. government even went so far as to attempt a reduction in the population by some form of population control. Pharmaceutical companies experimented with Puerto Rican women in developing contraceptive pills, and the U.S. government actually encouraged the sterilization of Puerto Rican women. More than 37 percent of the women of reproductive age in Puerto Rico had been sterilized by 1974 (Roberts 1997). More than one-third of these women have since indicated that they regret sterilization because they were not made aware at the time that the procedure was irreversible.

Cubans

Cuban migration to the United States is recent in comparison with many other Hispanic groups. The largest migration has occurred since the revolution led by Fidel Castro in 1959. Between then and 1980, more than 800,000 Cubans—one-tenth of the entire island population—migrated to the United States. The U.S. government defined this as a political exodus, facilitating the early entrance and acceptance of these migrants. Many of the first migrants had been middle- and upper-class professionals and landowners under the prior dictatorship of Fulgencio Batista, but they had lost their land during the Castro revolution. In exile in the United States, some worked to overthrow Castro, often with the support of the U.S. federal government. Many other Cuban immigrants were of more modest means who, like other immigrant groups, came seeking freedom from political and social persecution and an escape from poverty.

A second wave of Cuban immigration came in 1980, when the Cuban government, still under Castro, opened the Port of Mariel to anyone who wanted to leave Cuba. In the five months following this action, 125,000 Cubans came to the United States—more than the combined total for the preceding eight years.

The arrival of people from Mariel has produced debate and tension, particularly in Florida, a major center of Cuban migration. The Cuban government had previously labeled the people fleeing from Mariel as "undesirable"; some had been incarcerated in Cuba before leaving. They were actually not much different from previous refugees such as the "golden exiles," who were professional and high-status refugees (Portes and Rumbaut 1996). But because the refugees escaping from Mariel had been labeled (stereotyped) as undesirables, and because they were forced to live in primitive camps for long periods after their arrival, they have been unable to achieve much social and economic mobility in the United States—thus ironically reinforcing the initial perception that they were lazy and "undesirable." In contrast, the earlier Cuban migrants, who were on average more educated and much more settled, have enjoyed a fair degree of success (Portes and Rumbaut 2001, 1996; Amott and Matthei 1996; Pedraza 1996).

Asian Americans

Like Hispanic Americans, Asian Americans are from many different countries and diverse cultural backgrounds. Asian Americans include migrants from China, Japan, the Philippines, Korea, and Vietnam, as well as more recent immigrants from Southeast Asia, including India.

Chinese

Attracted by the U.S. demand for labor, Chinese Americans began migrating to the United States during the mid-nineteenth century. In the early stages of this migration, the Chinese were tolerated because they provided cheap labor. They were initially seen as good, quiet citizens, but racial stereotypes turned hostile when the Chinese came to be seen as competing with White California gold miners for jobs. Thousands of Chinese laborers worked for the Central Pacific Railroad from 1865 to 1868. They were relegated to the most difficult and dangerous work, worked longer hours than the White laborers, and were paid considerably less than the White workers.

The Chinese were virtually expelled from railroad work near the turn of the twentieth century (in 1890–1900) and settled in rural areas throughout the western states. As a consequence, anti-Chinese sentiment and prejudice ran high. This ethnic antagonism was largely the result of competition between the White and Chinese laborers for scarce jobs. In 1882, the federal government passed the Chinese Exclusion Act, which banned further immigration of unskilled Chinese laborers. Like African Americans, the Chinese and Chinese Americans were legally excluded from intermarriage with Whites (Takaki 1989). During this period, those who had been uprooted established several Chinatowns, finding strength and comfort within these enclaves of Chinese people and culture (Nee 1973).

Japanese

Japanese immigration to the United States took place mainly between 1890 and 1924, after which passage of the Japanese Immigration Act forbade further immigration. Most of these first-generation immigrants, called *issei*, were employed in agriculture or in small Japanese businesses. Many issei were from farming families and wished to acquire their own land, but in 1913, the Alien Land Law of California stipulated that Japanese aliens could lease land for only three years and that lands already owned or leased by them could not be bequeathed to heirs. The second generation of Japanese Americans, or *nisei*, were born in the United States of Japanese-born parents. They became better educated than their parents, lost their Japanese accents, and in general became more "Americanized," that is, culturally assimilated. The third generation, called *sansei*, became even better educated and assimilated, yet still met with prejudice and discrimination, particularly where Japanese Americans were present in the highest concentrations, as on the West Coast from Washington to southern California (Takaki 1989; Glenn 1986).

The Japanese suffered the complete indignity of having their loyalty questioned when the federal government, thinking they would side with Japan after the Japanese attack on Pearl Harbor in December 1941, herded them into concentration camps. By executive order of president Franklin D. Roosevelt, much of the West Coast Japanese American population (a great many—perhaps most—of them loyal second- and third-generation Americans) had their assets frozen and their real estate confiscated by the government. A media campaign immediately followed, labeling Japanese Americans "traitors" and "enemy aliens." Virtually all Japanese Americans in the United States had been removed from their homes by August 1942, and some were forced to

During World War II, Japanese Americans, who were full American citizens, were forced into concentration camps, as shown here.

stay in relocation camps until as late as 1946. Relocation destroyed numerous Japanese families and ruined them financially (Takaki 1989; Glenn 1986; Kitano 1976).

In 1986, the U.S. Supreme Court allowed Japanese Americans the right to file suit for monetary reparations. In 1987, legislation was passed, awarding $20,000 to each person who had been relocated and offering an official apology from the U.S. government. One is motivated to contemplate how far this paltry sum and late apology could go in righting what many have argued was the "greatest mistake" the United States has ever made as a government.

Filipinos

The Philippine Islands in the Pacific Ocean fell under U.S. rule in 1899 as a result of the Spanish–American War, and for a while Filipinos could enter the United States freely. By 1934, the islands became a commonwealth of the United States, and immigration quotas were imposed on Filipinos. More than 200,000 Filipinos immigrated to the United States between 1966 and 1980, settling in major urban centers on the West and East Coasts. More than two-thirds of those arriving were professional workers; their high average levels of education and skill have eased their assimilation. By 1985, more than one million Filipinos were in the United States. They are now the second largest Asian American population.

Koreans

Many Koreans entered the United States in the late 1960s, after amendments to the immigration laws in 1965 raised the limit on immigration from the Eastern Hemisphere. The largest concentration of Koreans is in Los Angeles. As much as half of the adult Korean American population is college educated, an exceptionally high proportion. Many of the immigrants were successful professionals in Korea; upon arrival in the United States, though, they have been forced to take on menial jobs, thus experiencing downward social mobility and status inconsistency. This is especially true of those Koreans who migrated to the East Coast. However, nearly one in eight Koreans in the United States today owns a business; many own small greengrocer businesses. Many of these stores are located in predominantly African American communities and have become one among several sources of ongoing conflict between some African Americans and Koreans. This has fanned negative feeling and prejudice on both sides—among Koreans against African Americans and among African Americans against Koreans (Chen 1991).

Vietnamese

Among the more recent groups of Asians to enter the United States have been the South Vietnamese, who began arriving following the fall of South Vietnam to the communist North Vietnamese at the end of the Vietnam War in 1975. These immigrants, many of them refugees who fled for their lives, numbered about 650,000 in the United States in 1975. About one-third of the refugees settled in California. Many faced prejudice and hostility, resulting in part from the same perception that has dogged many immigrant groups before them—that they were in competition for scarce jobs. A second wave of Vietnamese immigrants arrived after China attacked Vietnam in 1978. As many as 725,000 arrived in the United States, only to face discrimination throughout the country. Tensions were especially heated when the Vietnamese became a substantial competitive presence in the fishing and shrimping industries in the Gulf of Mexico on the Texas shore. Since that time, however, many communities have welcomed them, and many Vietnamese heads of households have become employed full-time (Kim 1993; Winnick 1990).

Middle Easterners

Since the mid-1970s, immigrants from the Middle East have been arriving in the United States. They have come from countries such as Syria, Lebanon, Egypt, and Iran, and more recently, especially Iraq. Contrary to popular belief, Middle Eastern immigrants follow no singular religion and thus are ethnically diverse. Some are Catholic, some are Coptic Christian, and many are Muslim. Some are from working-class backgrounds, but many were professionals—teachers, engineers, scientists, and other such positions—in their homelands. Like immigrant populations before them, Middle Easterners have formed their own ethnic enclaves in the cities and suburbs of this country as they pursue the often elusive American dream (Abrahamson 2006).

Since the terrorist attacks on the World Trade Center and the Pentagon on September 11, 2001, and terrorist attacks since, Middle Easterners of several nationalities have become unjustly suspect in this country and are subjected to severe harassment, racially motivated physical attacks, and out-and-out racial profiling. Fully 40 percent of Americans actually admit to having at least a little, if not a lot, of prejudice against Muslims (Morales 2010) even though most Muslim Americans have no connection with terrorists.

When in 2017 the Trump administration issued an Executive Order banning people from seven specific Muslim-majority countries, a full 48 percent of the public expressed support for such a ban, even though the U.S. Supreme Court later ruled the ban unconstitutional. Opponents of the can also argue that it violated core American values (Rasmussen Reports 2017).

In the heat of national fears of Islamic extremists, it is difficult for many Americans to take a calm and rational approach to understanding the life of Muslim Americans. Research finds that Muslim Americans report experiencing a lot of discrimination and Muslim Americans perceive far less support for extremist activities in the Muslim community than the general public perceives (Pew Research 2011).

White Ethnic Groups

The story of White ethnic groups in the United States begins during the colonial period. White Anglo Saxon Protestants (WASPs), who were originally immigrants from England and to some extent Scotland and Wales, settled in the New World (what is now North America). They were the first ethnic group to come into contact—most often hostile contact—on a large scale with those people already here, namely, Native American Indians. WASPs came to dominate the newly emerging society earlier than any other White ethnic group.

In the late 1700s, the WASPs regarded the later immigrants from Germany and France as "foreigners" with odd languages, accents, and customs, and applied derogatory labels ("krauts" for Germans; "frogs" for the French) to them. Again, this demonstrates that virtually every immigrant group, except WASPs themselves, was discriminated against and subject to racist name-calling.

WASPs have come to think of themselves as the "original" Americans despite the prior presence of Native American Indians, whom the WASPs in turn described and stereotyped as savages. As immigrants from northern, western, eastern, and southern Europe began to arrive, particularly during the mid- to late-nineteenth century, WASPs began to direct prejudice and discrimination against many of these newer groups.

There were two waves of migration of White ethnic groups in the mid- and late-nineteenth century. The first stretched from about 1850 through 1880, and included northern and western Europeans: English, Irish, Germans, French, and Scandinavians. The second wave of immigration occurred from 1890 to 1914, and included eastern and southern European populations: Italians, Greeks, Poles, Russians, and other eastern Europeans, in addition to more Irish. The immigration of Jews to the United States extended for well over a century, but the majority of Jewish immigrants came to the United States during the period from 1880 to 1920.

The Irish arrived in large numbers in the mid-nineteenth century and after, as a consequence of food shortages and massive starvation in Ireland. During the latter half of the nineteenth century and in the early twentieth century, the Irish in the United States were abused, attacked, and viciously stereotyped. The Irish, particularly on the East Coast and especially in Boston, underwent a period of ethnic oppression of extraordinary magnitude. A frequently seen sign posted in Boston saloons during that time proclaimed "no dogs or Irish allowed." The sign was not intended as a joke. German immigrants were similarly stereotyped, as were the French and the Scandinavians. It is easy to forget that virtually all immigrant groups have gone through times of oppression and prejudice, although these periods were considerably longer for some groups than for others. As a rule, where the population density of an ethnic group in a town, city, or region was greatest, so too was the amount of prejudice, negative stereotyping, and discrimination to which that group was subjected.

Even with a rise in prejudice against Muslim Americans, many are working to reduce such group stereotyping and hatred.

Jewish Americans

More than 40 percent of the world's Jewish population lives in the United States, making it the largest community of Jews in the world. Most of the Jews in the United States arrived between 1880 and the First World War, originating from the eastern European countries of Russia, Poland, Lithuania, Hungary, and Romania. Jews from Germany arrived in two phases; the first wave came just prior to the arrival of those from eastern Europe, and the second came as a result of Hitler's ascension to power in Germany during the late 1930s. Because many German Jews were professionals who also spoke English, they assimilated more rapidly than those from the eastern European countries. Jews from both parts of Europe underwent lengthy periods of **anti-Semitism**, defined as prejudice, hostility, and discrimination against Jewish people.

Anti-Semitism still exists in the United States—a significant proportion of American Jews say there is still a lot of discrimination against them. Perhaps surprisingly, younger Jews (those under thirty) are more likely than older Jewish people to say they have been called an offensive name, perhaps a reflection of how common such ethnic taunts are in youth cultures (Pew Research Center 2013).

Contemporary Immigration: The New Civil Rights Challenge

Other than American Indians and the indigenous Mexican people of the original American Southwest, the United States is a country of immigrants. The nation is proud of this heritage because of its incorporation of diverse people into a unified, democratic society. Immigrants bring cultural diversity, as well as innovation and diverse talents to the country. The history of immigration and current events shows, however, that immigration also reflects some of our most ugly moments. Even while immigrants have been incorporated into the nation, they have faced exclusion and discrimination. Those exclusionary actions, particularly as directed against specific groups, have shaped who enjoys the privileges and rights of citizenship.

The first law to define citizenship was the **Naturalization Act of 1790**, which established rules about who could become naturalized—that is, become a citizen. This law restricted naturalization only to "free white persons" of "good moral character." It was not repealed until 1802.

In 1924, the nation passed the **Immigration Act**, (also called the National Origins Act)—one of the most discriminatory legal actions ever taken in the United States in the area of immigration. By this act, *ethnic quotas* were established that allowed immigrants to enter the country only in proportion to their numbers already existing in 1890 as counted in the U.S. census. Thus, ethnic groups who were already here in relatively high proportions (English, Germans, French, Scandinavians, and others—that is, mostly White western and northern Europeans) could immigrate in greater numbers than those from southern and eastern Europe, such as Italians, Poles, Greeks, and other eastern Europeans. Hence, the act discriminated against southern and eastern Europeans in favor of western and northern Europeans. The European groups who were discriminated against by the National Origins Act tended to be those with darker skins on average, even though they were White and European. The law explicitly forbade immigrants from Asian countries.

The Immigration Act of 1924 also barred anyone who was classified as a convict, lunatic, "idiot," or "imbecile" from immigration. On New York City's Ellis Island, non-English–speaking immigrants, many of them Jews, were given the 1916 version of the Stanford Binet IQ test and literacy tests in English (Kamin 1974). Obviously, non-English–speaking people taking this test were unlikely to score high. On the basis of this grossly biased test, the government classified fully 83 percent of Jews, 80 percent of Hungarians, and 79 percent of Italians as "feebleminded" (Gould 1999; Taylor 1980; Kamin 1974).

The National Origins Act reduced immigration to a trickle until 1965 when the **Immigration and Nationality Act** (also known as the Hart Celler Act) eliminated the national origins quota. This law has

Debunking Society's Myths: Immigration Myths

Myth: Mexican immigrants are streaming across the border into the United States.

Fact: Immigration from Mexico has actually slowed significantly since 2000 and continues to shrink; starting around 2010, more Mexican nationals and their children started returning to Mexico in greater numbers than those who were entering the United States.

Myth: Undocumented immigrants take jobs from Americans.

Fact: States with large numbers of immigrants have low unemployment for everyone else; immigrants are also twice as likely to start businesses in the United States as are native-born citizens, thus, they create jobs for other Americans.

Myth: Immigrants come to the United States to collect welfare and other benefits.

Fact: Undocumented immigrants are not eligible for federal public benefits, nor are most legal immigrants. At the same time, about two-thirds of Mexican immigrants contribute to Social Security and pay state and federal income taxes.

Sources: Massey, Douglas S. 2005. "Five Myths about Immigration: Common Misconceptions about U.S. Border Enforcement Policy." *Immigration Policy in Focus* 4 (August): 1–11; Anti-Defamation League. 2017. "Myths and Facts about Immigrants and Immigration." **Adl.org**; Gonzalez-Barrera, Ana. 2015. "Migration Flows Between the U.S. and Mexico Have Slowed—and Turned Toward Mexico." Washington, DC: Pew Research Center. **www.pewhispanic.org**

changed the face of American society and is the law still governing the flow of immigration. In addition to eliminating the quotas of 1924, the 1965 law gave priority immigration status to those seeking family reunification, and also opened the door to highly skilled immigrants whose work has been needed in the current economy. With legal restrictions lifted, employers turned to immigrants, mostly from Mexico and Central America, to provide a supply of cheap labor. Thus, the 1965 law gave entry to less-skilled workers who fill some of the lowest wage jobs in the labor force.

As a consequence, the 1965 law has created of a bifurcated labor force: Highly skilled, well-educated immigrant workers perform much of the nation's scientific, technical, and research work. Others, including undocumented workers, perform mostly low-wage service work in a variety of industries (Portes and Rumbaut 2014).

Undocumented workers now total about 11 million in the United States, and many have to labor under clandestine arrangements. Unauthorized immigrants make up about 5 percent of the U.S. workforce (Krogstad and Passel 2015; Massey 2005). As you can see in the box "Debunking Society's Myths," the majority pay taxes and Social Security, although with only some exceptions immigrants are excluded from various federal benefits, such as food stamps, Social Security, various health care benefits, and TANF (that is, welfare; see Chapter 8).

Now immigrants to the United States come from many different parts of the world, and they are very diverse in their educational and occupational backgrounds. Many experience downward mobility once they enter the United States, perhaps coming from professional backgrounds in their country of origin but landing in various service-oriented jobs in the United States that carry lower wages and less prestige, such as by becoming nail salon workers, nannies and housekeepers, fast food workers, and other service occupations. Many immigrants come as refuges from war or other disruptions and lack of opportunity in their home nations. All come with hopes of better opportunity, but may find themselves niched in particular labor markers where upward mobility is difficult.

Unlike the European immigrants who entered around twentieth century, who no doubt experienced harsh prejudice and discrimination, new immigrants are often perceived to be unlike White Americans. As a result, the newest immigrants are quite likely to become racialized (Hirschman and Massey 2008; Alba and Nee 2003). In fact, the history of immigration in the United States has deep connections to racial inequality in that "racial divisions both shape and are shaped by immigrant arenas and experiences"—a pattern identified as the **race-immigration nexus** (Kibria, Bowman and O'Leary 2013: 5).

Most Americans welcome the cultural and social changes that immigrants bring to the county, and a large majority think that immigrants strengthen the nation through hard work and the different talents they bring. Others, however, see the new immigration as threatening the nation, either through weakening our national identity or bringing in potential terrorists (Jones 2016). The public is clearly divided on what policies should take priority in dealing with immigration, especially undocumented immigration.

This debate has culminated with strong partisanship over Donald Trump's proposed wall along the U.S.–Mexico border. Almost two-thirds of the public (59 percent) think such a wall is not important (Suls 2017). Although many think that border security is the most important policy goal (42 percent of the public), more (55 percent) think it is a bigger priority to provide a path to citizenship for those who entered illegally (Pew Research Center 2016). These contemporary conflicts beg the question of whether the United States will continue to be a melting pot or if old patterns of exclusion will prevail.

Racial Stratification: A Current Portrait

Racial stratification is a system of inequality in which race and ethnicity mark differential access to economic, social, political, and cultural resources. The important point about racial stratification is that, like other forms of social stratification, the inequality it involves is systemic. In that sense, racial stratification results from *institutional racism* (defined earlier)—that is, the sources of racial inequality are found within social institutions, not just within individual minds. Evidence of racial stratification is found within many of the chapters of this text. Here we examine two: economic inequality (also see Chapter 8) and racial segregation.

Economic Inequality

One of the starkest indications of racial stratification is the persistent income gap between White households and those of people of color. Recall from Chapter 8 that median income is the midpoint of all incomes for a given population. That means that half of the people in the population being measured earn less than the median income; half, more. In 2017, median income for White (non-Hispanic) households in the United States (that is, counting all households, regardless of family structure, size, region of residence, and so forth) was $68,145. For Black families, median income in that year was $40,258; Hispanic households, $50,496; Asian households, $81,331 (as shown in ▲ Figure 10-1; Fontenot, Semega, and Kollar 2018).

Not only is there a large income gap between Black and Hispanic households and other groups, but it is also telling that the gap between Black and White households is virtually unchanged since 1970. Black households had a median income that was 61 percent of that for White households in 1970; that gap is now *exactly the same*! There has been a substantial rise in the number of Black and Hispanic middle-class families since the 1960s and 1970s. In fact, there has always been at least a small, but significant Black middle class. But, the recent expansion of the Black and, for that matter, the Latino middle class means that middle-class people of color have a fairly tenuous hold on this economic status. Events such as the housing foreclosure crises that accompanied the economic recession of 2008 can be particularly devastating to those whose hold on middle-class status is relatively recent because they do not necessarily have the accumulated resources that could back them up during economic crises.

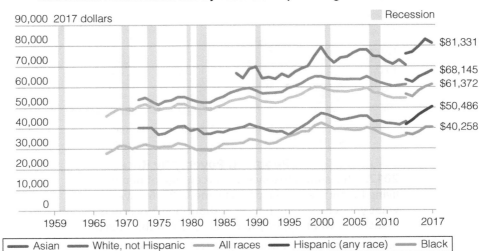

Real Median Household Income By Race And Hispanic Origin: 1967 To 2017

- Asian
- White, not Hispanic
- All races
- Hispanic (any race)
- Black

▲ **Figure 10-1   The Income Gap by Race, 1970–2017.** Be careful about interpreting this chart. The U.S. Census Bureau changed how it counted different racial–ethnic groups over this period of time. Asian Americans, for example, included Pacific Islanders prior to 2002, which inflates the overall median income. Nonetheless, you can observe persistent gaps in median income by looking at the income status of these different groups over time. Note, however, that with the changing way that the census counted race and ethnicity, data are not available for some groups at earlier points in time.

Source: Fontetnot, Kayla, Jessica Semega, Jessica, and Melissa Kollar. 2018. *Income and Poverty in the United States: 2017*. Washington, DC: U.S. Census Bureau. **www.census.gov**

As we saw earlier (see Chapter 8), even more glaring than the earnings gap is the gap in wealth between White households and those of people of color. The net worth of White families far exceeds that of Black and Latino families (see ▲ Figure 10-2). The most recent estimates are that, on average, Whites actually have *twelve times* the wealth of Black Americans and *ten times* as much as Hispanics (McKernan et al. 2015).

Poverty is also a consequence of a system of racial stratification. Poverty among Blacks has also decreased since the 1950s, but is now close to the same level as in 1970. As ▲ Figure 10-3 shows, the current

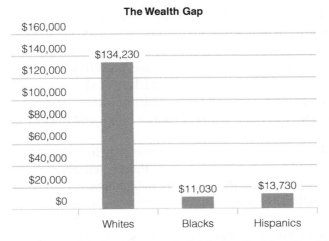

The Wealth Gap

▲ **Figure 10-2   The Racial/Ethnic Gap in Wealth**

Source: McKernan, Signe-Mary, Caroline Ratcliffe, C. Eugene Steuerle, Emma Kalish, and Caleb Quakenbush. 2015. *Nine Charts about Wealth Inequality in America*. Washington, DC: The Urban Institute. **http://datatools.urban .org/Features/wealth-inequality-charts/**

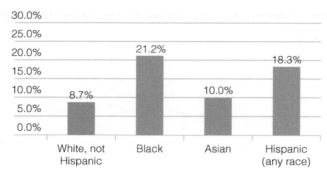

▲ **Figure 10-3   People in Poverty by Race, 2017.** This shows the overall percentage of people living below the official poverty line in 2016. Given what you learned in Chapter 8 about how poverty is measured, what might you conclude from this chart?

Source: Fontetnot, Kayla, Jessica Semega, Jessica, and Melissa Kollar. 2018. *Income and Poverty in the United States: 2017.* Washington, DC: U.S. Census Bureau. **www.census.gov**

poverty rate is highest for African Americans and Hispanics compared with Whites or Asians. Note that it is greater for Asians than for Whites. In all these racial groups, children have the highest rate of poverty.

Economic inequality based on race and ethnicity is also part of the picture. Other evidence of racial stratification is found in the nation's schools, the workplace, in health care, in people's representation in the government, even in the quality of the air someone breathes and the water they drink. But one of the starkest reminders of racial stratification is found in how segregated U.S. society is.

Racial Segregation

Segregation is the spatial and social separation of racial and ethnic groups. The nation's laws have largely outlawed racial segregation—that is, *de jure segregation* (legal segregation). Yet, *de facto segregation* (segregation in fact) still exists, most evident in housing and education. (Educational segregation is addressed in Chapter 14.) Racial segregation prevents intergroup interaction and fosters isolation form other groups. These are known factors that contribute to racial tension. Living as a person of color in a racially segregated neighborhood also lessens one's access to good jobs, ease of transportation, and the availability of healthy food options.

Levels of overall residential segregation have declined somewhat since 1990, but this general picture can be deceiving. Segregation in metropolitan areas, especially between Blacks and Whites, has actually increased. Overall, there is lower segregation between Asians and Whites, with Hispanics falling in the middle. The suburbs have become more diverse, but White flight out of cities continues even while younger White people are moving into cities (Lichter et al. 2015). The level of housing segregation is so high for some groups, especially poor African Americans and Latinos, that it has been termed **hypersegregation**—a pattern of extreme segregation (Rugh and Massey 2010; Massey 2005; Massey and Denton 1993).

Housing segregation results from many factors but is especially influenced by real estate practices. Even very recently, banks and mortgage companies have withheld mortgage loans from minorities based on "redlining," an illegal practice in which entire minority neighborhoods are designated as ineligible for loans. Predatory lending practices, such as subprime mortgages (housing loans that carry a higher interest rate than the prime lending rate), have led to higher rates of housing foreclosure for people of color, sometimes influencing entire neighborhoods and leaving people of color with higher levels of debt. *Steering* is also a practice whereby real estate agents show people of color housing properties only in neighborhoods that are already predominantly inhabited by other people of color.

Racial segregation is also highly correlated with the concentration of poverty for people of color. One-third of Black metropolitan residents now live in hypersegregated areas (Massey and Tannen 2015)—areas are usually also marked by *concentrated poverty*, that is places where 40 percent or more of the population is poor. People living in such areas are surrounded by poverty, regardless of their own social class. Concentrated poverty has more than doubled since the 1990s (Jargowsky 2015).

Residential segregation and other forms of special segregation are so pervasive that when people of color move into predominantly White areas, they have to contend with what sociologist Elijah Anderson has termed *white space*. **White space** is the perception Black people have of places where they are "typically absent, not expected, or marginalized when present" (2015:10). In white space, people of color likely feel uncomfortable, as if the place is "off limits." White people, on the other hand, rarely perceive the same thing. You can see that racial segregation shapes not just numbers, but also how people relate to and perceive one another (Andersen 2018).

Explaining Racial and Ethnic Inequality

One can fairly readily document ongoing patterns of racial and ethnic inequality. The debates come when trying to explain why this inequality occurs. Various explanations are central: (1) assimilation theory; (2) cultural versus structural explanations; (3) the class-race connection; and, (4) intersectional theory.

Assimilation Theory

Many ask why, when so many immigrant groups have prospered in this nation, others have not. "We made it; why can't they?" is a frequent refrain in conversations about racial inequality. This question assumes, of course, that different groups encounter the same context. That is actually not true. Ideas about assimilation and how much it occurs are common in the public mind, but more carefully examined through sociological theory and research.

Assimilation theory examines the process by which a minority becomes absorbed into the dominant society. Originally, this theory assumed that assimilation occurred as incoming groups adopt the values of the dominant society, especially in the second and third generation. We now know, however, that assimilation is more complex than originally assumed.

As it plays out, assimilation involves change both by the immigrant group and by the host society. Some immigrants can maintain their ethnic identity even while being incorporated into a new society. This happens to differing degrees, however, depending on the particular conditions that immigrants encounter. For example, when there is an ongoing flow of immigrants from a particular country (for example, Mexico), immigrants already present in the society may maintain, or even re-develop, a strong ethnic identity (Massey and Sanchez 2012; Jiménez 2010).

The assimilation process is also not the same for every group. Immigrants who enter with higher levels of education or particular skills, such as university professors or highly skilled technology workers, will be more easily assimilated than those who enter into the low-wage labor force (Rumbaut 1996). Immigrant groups who become racialized when they enter a new country will also find it more difficult to assimilate no matter what their values and desires for assimilation are. Especially when the host nation is in a period of intense **nativism** (that is, extreme beliefs and policies that declare the citizens of a nation superior to all others), outside groups, especially from non-Western nations, are likely to become racialized (Kibria, O'Leary and Bowman 2013).

Immigrant assimilation can also be influenced by the relationship between the host nation and the nation of origin. You can see this in the case of immigrants and refugees from predominantly Middle Eastern, predominantly Muslim countries. Fractious relationships between the United States and these nations have shaped both public policies and public attitudes toward Muslim immigrants, as witnessed by the ban proposed by the Trump administration on immigration from certain Muslim-majority countries.

To sum up, how well immigrants are assimilated into a new society depends on the structural and historical conditions that particular groups encounter. As an example, White ethnic groups entered at a time historically when the economy was growing rapidly and their labor was in high demand. Thus

Mohammed salem/Reuters

Conflict over whether to build a wall on the southern U.S. border show how deeply divided the nation has become over immigration policy.

they were able to attain education and job skills, even though they did indeed face prejudice and discrimination when they arrived in the United States. In contrast, by the time Blacks were freed from slavery and migrated in great numbers to Northern and Midwestern industrial areas during the Great Migration, Whites had already established firm control over labor and used this control to exclude Blacks from better-paying jobs and higher education.

It is also critical to point out that assimilation does not well explain the experience of African Americans in the United States. White ethnic groups came voluntarily; Blacks arrived in chains. While in many cases, Whites sought relatives in the New World; Blacks were sold and separated from close relatives. For these and other reasons, the experiences of African Americans and White ethnic groups as newcomers can hardly be compared.

The assimilation model implicitly asks whether it is possible for a society to maintain a singular national identity even while honoring **cultural pluralism**, defined as different groups in society maintaining distinctive cultures, while coexisting peacefully with the dominant group. In many ways, this is the great experiment of the United States. Current debates about immigration raise questions about the conditions under which new groups are able or not to become full-fledged citizens of a complex and democratic nation.

The Culture-Structure Debate

Debates about assimilation also reflect a debate about the role of culture in producing racial and ethnic inequality. Culture is frequently asserted in the national political discourse about causes of racial inequality and poverty as if poor people are somehow to blame for their social status. You have likely heard it said that people would not be poor if they would just adopt stronger family values or that, in the words of then Congressman Paul Ryan, "We have got this tailspin of culture, in our inner cities in particular, of men not working and just generations of men not even thinking about working or learning the value and the culture of work, and so there is a real culture problem here that has to be dealt with" (Ryan 2014).

Cultural explanations of racial inequality assume there is something wrong in the values and attitudes of those who do not succeed. This assumption has been long studied in sociology and is generally found to be completely untrue. Research finds that poor people want to work, have desires to succeed, and value marriage just as much as do others. But, as they are well aware, their ability to do so is conditioned by the many challenges they face in society (Greenbaum 2015; Edin and Kefalas 2005).

Cultural attitudes cannot, however, be completely dismissed. There can be cultural differences in how people adapt to the worlds they inhabit. That is, how people understand and interpret their situation is shaped by that very situation. This does not mean that people's cultural values *cause* inequality, but may be a result of it (Wilson 2009). All human cultures develop in the context of social structures. Cultural attitudes are not just free-floating ideas that are disconnected from the realities of people's lives. Social structures produce different outcomes for different groups, including different cultural outcomes.

Sociological research finds that there can be unique cultural beliefs and behaviors among the urban poor, even while they also share the values of the dominant group (Venkatesh 2006). For example, people in poor, racially segregated neighborhoods value "street smarts"—a unique cultural meaning system and set of behaviors, also known as "the code of the street" (Anderson 1999). The code of the street is an adaptation to the specific challenges of living in poor, urban areas where "acting cool" is about a way to stay safe and a display of masculinity otherwise denied by the dominant cultural (hooks 2003; Majors and Billson 1993). Understood in this way, culture is not a cause of poverty but is an innovative adaptation when people do not have access to the resources of the dominant culture.

In other words, culture is deeply entangled with social structure and does not exist independent of people's position in that structure. Culture is not so much a cause of poverty as an adaptation to it (see also Chapter 8). The connection between culture and social structure is thus not a question of just one or the other, but how they are intertwined.

Is It Class or Is It Race?

The debate about the relative impact of culture and structure connects to another debate—the connection between race and class. Is class more important now than race in explaining inequality? Is racial inequality simply a function of class status or does race still matter?

These questions have been most thoroughly examined by sociologist William Julius Wilson (2009, 1996, 1987, 1978). Wilson asserts that, as the economic structure in the United States has evolved, class has become increasingly important in shaping people's life chances, including those of African Americans and other people of color. Being seriously disadvantaged in the United States is now, according to Wilson, more a matter of class than of race per se, although race still matters. Others argue that race has an effect independent of class, arguing that Wilson underestimates the continuing impact of race on opportunities and people's day-to-day lives (Bonilla-Silva 2013; Feagin 2000; Willie 1979; see ◆ Table 10-2).

How does Wilson explain his argument? He shows that it is increasingly difficult for people to find the kind of work that, years ago, would have provided a solid standard of living even for those with limited education—such as factory jobs. With manufacturing jobs on the decline in the United States, people with limited education or limited technical skill have been thrown out of the labor market or shuttled into lower-age service industry positions. This transformation from a manufacturing-based economy to a service-based economy (see also Chapter 14) has harmed many workers, including White, working class workers, but for people of color, this structural change piles on top of an already existing

Table 10-2   Sociological Explanations of Racial and Ethnic Inequality

	Assimilation	Culture-Structure	Class-Race	Intersectional Theory
Social Structure:	Has social stability when diverse racial and ethnic groups are assimilated into society	Culture and structure have a dialectical relationship—that is, each shapes the other	The relationship between class and race changes with economic transformation, but both are important in shaping people's life chances	Race, class, and gender intersect in a matrix of domination in society
Minority Groups:	Are assimilated into dominant culture as they adopt cultural practices and beliefs of the dominant group	Although often blamed for holding poor cultural values, minority groups tend to share the same values for success as dominant groups	With race no longer a caste system, class can differentiate the status of people of color, but this does not make race irrelevant	Have life chances that result from the opportunities formed by the intersection of class, race, and gender
Social Change:	Is a slow and gradual process as groups adapt to the social system	Cultures, including those in poor communities, evolve as an adaptation to changed social conditions	The change to a service-based economy and away from a manufacturing-based economy makes class particularly influential in determining opportunities for various groups	Is the result of coalition as groups mobilize together to resist oppression

history of disadvantage such that the economic transformation has a particularly deleterious effect on minority communities (Wilson 1978).

Moreover, these economic transformations have created what Wilson labeled the **urban underclass**— that is, those who are stuck at the bottom of the class system, more or less permanently out of good jobs and thus subjected to some of the most disturbing social problems (including crime, joblessness, poverty, and so forth). In Wilson's argument, note that these problems arise not from people's individual bad judgments or poor values but because of structural realignments in the national economy. Thus, class and race intermingle in shaping people's life chances.

You can also see that contemporary racial tensions are then driven by these social structural changes. Racial resentment is increasingly common among many White working class people, a phenomenon you can see as having influenced the election of Donald Trump as president of the United States. Add immigration to this mix and you can understand why the nation's racial, ethnic, and class politics have become so volatile.

Wilson's argument does not mean that race is unimportant, rather that the influence of class is increasing, even though race remains extremely important. At the same time, numerous studies find that race, independent of class, still influences such things as income, wealth holdings, occupational prestige, place of residence, educational attainment, and other measures of socioeconomic well-being (Pattillo McCoy 2013; Bobo 2012; Oliver and Shapiro 2006).

The fact is that class and race combine in specific ways at different points in time. In the current period, class and race are still deeply intertwined. It would be difficult to claim that people's position in society is influenced solely by one or the other. Race and class together form a structure of opportunity that has profound effects on the life changes of various groups (Andersen 2018).

Intersectional Theory: Connecting Race, Class, and Gender

The "class versus race" controversy is connected to theories of intersectionality. **Intersectional theory** argues that class, race, and gender combine (or "intersect") to create a matrix of domination (Andersen and Collins 2016; Collins and Bilge 2016; Collins 1990). Intersectional theory points out that people are socially located in positions that simultaneously involve race, class, *and* gender. Looking at only one of them is incomplete. The effects of gender and race are intertwined, and both are intertwined with class (Crenshaw 1989).

Intersectional theory developed from the writings of women of color who, in the 1960s and 1970s, were critical of the growing feminist movement as too centered in the experiences of White women. Women of color understood their lives to be situated in a constellation of race, class, *and* gender, not just one. At any given moment, one factor might be more salient than another (such as when a White man calls a Black woman "girl" or when single Latina and African Americans mothers are called "welfare queens"), but both gender, race, and class influence these perceptions and situate women of color in a "matrix of domination" that involves all three: race, class, and gender (Collins 1990).

As intersectional theory has been elaborated, it has documented and analyzed how race, class, and gender simultaneously influence the experiences of all people—even when their influence may be invisible to more dominant groups (Andersen and Collins 2016; Baca Zinn and Dill 1996). The point is not to rank various forms of oppression, but to understand how they interlace and are together woven into the fabric of society.

Intersectional theory also explains that even as a person of color, how race is experienced also depends on one's gender and class. Similarly, how one experiences gender is influenced by one's race and so forth. Generalizations about any one of these social experiences have to be cautiously understood. Moreover, the intersection of race, class, and gender affect us all—even those in the dominant group. Masculinity, for example, may be expressed differently if one is Latino and poor compared to being Black and middle class or White and poor or White, upper class, and male. Likewise, being a woman is manifested differently depending on one's class and race position (Andersen 2018; Andersen and Collins 2016).

Intersectionality means understanding and explaining the unique and complex ways that systems of overlapping inequality shape our social and economic status, our beliefs and identities, and even how much we perceive how much race, class, and gender matter. To put it simply, race, class, and gender are not isolated one from the other. Each is manifested differently depending on its configuration with others (Andersen and Collins 2016).

Racial Justice: Changes and Challenges

Race and ethnic relations in the United States have posed a major challenge for the nation, one that is becoming even more complex as the racial–ethnic population becomes more diverse. Much progress has been made: The nation elected its first Black president, Barack Obama, in 2008; his wife, former First Lady Michelle Obama, remains one of the most popular First Ladies in recent times (Clement 2016). There is visible leadership of people of color in many places—in the halls of government, in the media, on corporate boards, and as icons in popular culture. Yet, every day we still see visible hatred, the consequence of persistent poverty, and a nation quite divided over matters of racial and ethnic justice.

How can the nation respond to its new diversity as well as to the issues faced by racial and ethnic minorities that have been present since the nation's founding? This question engages significant sociological questions and calls attention to the nation's record of social change regarding race and ethnic groups. Change comes through the mobilization of people—best evidenced by the historic civil rights movement.

The Civil Rights Movement

The civil rights movement was the single most important source of change in race relations in the twentieth century. This historic movement was initially based on the passive resistance philosophy of Martin Luther King Jr., learned from the philosophy of Satyagraha (soul "firmness and force") of the East Indian Mahatma (meaning "leader") Mohandas Gandhi. This philosophy encouraged resistance to segregation through nonviolent techniques, such as sit-ins, marches, and appealing to human conscience in the movement's calls for brotherhood, justice, and equality. Although most people date the start of the civil rights movement in the 1950s, the struggle for civil rights began many years before. The impact of the movement is unparalleled: it ended Jim Crow segregation, changed people's consciousness, and established a legal framework for equal rights. Plus, the benefits of this movement accrued to many: women, disabled people, the aged, and gays and lesbians (Andersen 2004).

The major civil rights movement in the United States intensified shortly after the 1954 *Brown v. Board of Education* decision, the famous Supreme Court case that ruled that "separate but equal" in education was unconstitutional. In 1955, African American seamstress and NAACP secretary Rosa Parks made news in Montgomery, Alabama. By prior arrangement with the NAACP, Parks bravely refused to relinquish her seat in the "White-only" section on a segregated bus when asked to do so by the White bus driver. At the time, most of Montgomery's bus riders were African American, and the action of Rosa Parks initiated the now-famous Montgomery Bus Boycott, led by the young Martin Luther King Jr. The boycott, in addition to desegregating the buses and providing more jobs for African American bus drivers hired, catapulted Martin Luther King Jr. to the forefront of the civil rights movement.

The murder of Emmett Till, a Black teenager from Chicago, who was brutally killed in Mississippi in 1954, was also a major impetus for the developing movement. After Till allegedly whistled at a White woman in a store—an accusation that his accuser only admitted in 2017 was factually untrue—a group of White men rousted him from his bed and beat him until he was dead and unrecognizable as a human being. They then tied a heavy cotton gin fan around his neck and dumped him into the nearest river. Later, his mother in Chicago allowed a picture of his horribly misshapen head and body in his casket to be published (in *Jet* magazine) so that the public could contemplate the horror vested upon her son. No one has ever been prosecuted for the murder of Emmett Till.

The civil rights movement included actions both tragic and heroic. Black and White students organized sit-ins at lunch counters in the South, insisting that Black people be served. Attempts to desegregate southern schools were often met with fierce resistance though. In some states, entire school systems were closed to avoid desegregation—in the case of Virginia, for an entire year simply to avoid desegregation.

Over the course of the movement, demonstrators, both Black and White, were beaten mercilessly—and sometimes killed. One such incident was the famous "Selma to Montgomery" march across the Edmund Pettus Bridge on March 7, 1965. Marchers were demonstrating to insist on Black voting rights. Many, including now Senator John Clark, were severely beaten, suffering bone fractures and major lacerations.

"Freedom Rides" were organized bus rides that brought Black and White northerners to the South to fight for civil rights, including voting. Numerous civil rights activists lost their lives, including among others Viola Liuzzo, a White Detroit housewife; Andrew Goodman and Michael Schwerner, two White students; and James Chaney, a Black student. The murders of civil rights workers—especially when they were White—galvanized public support for change.

Power and Militancy in the Movement for Racial Justice

While the civil rights movement developed through the 1950s and 1960s, a more radical philosophy of change also emerged, and the movement's influence spread to other racial and ethnic groups whose rights were being denied. More militant leaders grew increasingly disenchanted with the limits of the civil rights agenda, which was also perceived as moving too slowly. The militant Black Power movement had a more radical critique of race relations in the United States. The Black Power movement saw inequality as stemming not just from moral failures but also from the institutional power that Whites had over Black Americans (Carmichael and Hamilton 1967).

Before breaking with the Black Muslims (the Black Nation of Islam in the United States) and prior to his assassination in 1965, Malcolm X also advocated a form of radical pluralism whereby Black communities would take charge of their own businesses, schools, and community organization.

The Black Power movement of the late 1960s rejected assimilation, instead promoting self-determination and self-policing in Black communities. Militant groups such as the Black Panther Party also advocated fighting oppression with armed revolution, although the Panthers also launched important social service programs within Black ghettoes. Terrified by the radicalism that the Panthers espoused, the U.S. government acted vigorously and with force, imprisoning members of the Black Panther Party and other militant revolutionaries, in many cases killing them outright (Brown 1992).

Other groups were also affected by the analysis of institutional racism that the Black Power movement articulated. Protests by Mexican Americans, initially against the large number of Mexicans being killed in the Vietnam War, promoted Chicano/a solidarity through calls to "brown power" and "Chicanismo" (Escobar 1993). Likewise, the American Indian Movement (AIM) used some of the same strategies and tactics that the Black Power movement had encouraged, such as the American Indian seizure of Alcatraz Island in San Francisco Bay in 1969—an occupation that lasted for nineteen months (Smith and Warrior 1997).

Puerto Ricans, Asian Americans, and other groups were also influenced by the ideology of the Black Power and other radical movements. Racial solidarity, not just racial integration, shaped the consciousness of new generations—one of the lasting impacts of the more radical philosophy of militant movements. Overall, the more militant movements for racial justice have been suppressed by the fierce repression that they encountered. These movements, however, dramatically changed the struggle for racial justice, not only changing people's consciousness about racial identity, but forcing academic scholars to develop a deeper understanding of how fundamental racism is to U.S. social institutions (Branch 2006; Morris 1984). As you see protests in the present, such as the Standing Rock protest against the Dakota pipeline, with some historical imagination you can imagine the roots of movements in these times in earlier movements for racial justice.

#BlackLivesMatter

New conditions require new actions for change. The Black Lives Matter Movement is one of the most visible examples of a new movement in recent years. Black Lives Matter arose somewhat spontaneously out of growing concern about police shootings of Black people, particularly following the 2013 acquittal of George Zimmerman, accused of shooting and killing Trayvon Martin, a seventeen-year old Black teen.

Out of concern for providing her young son a less bleak picture of Black people than what is usually portrayed in the media, Black community activist Alicia Garza wrote what she called a "love letter to Black people" and posted it on Facebook (Cobb 2016). Joined by two of her best friends, Patrisse Cullors and Opal

Tometi, Cullors tweeted the hashtag, #BlackLivesMatter, and a new movement was born.

Black Lives Matter inspired numerous protests around the nation, especially following outrage about additional police shootings of Black people. Many of these protest actions, though not all, have been on college campuses. The movement relies on grassroots organizing and mobilization via social media. Echoing the sit-ins of the 1960s, activists staged "die-ins" as one tactic, dramatizing the deaths of so many Black men and women at the hands of the police.

Different from the earlier movements that maintained a somewhat singular focus on race, Black Lives Matter has an explicit platform of valuing *all* Black lives, specifically including love and support for queer, transgender, disabled, and undocumented people in their calls for change. You can see here the influence of intersectional theory.

Some have responded to Black Lives Matter by claiming, "All lives matter." Movement activists respond by saying that, no matter how well meaning, such claims diminish the specific value of Black lives and reinforce colorblind racism. This response to the Black Lives Matter movement reveals a fundamental tension in the push for racial change: Should change come through colorblind or color-conscious plans?

Strategies for Change: Race-Blind or Color-Conscious?

A continuing question from the dialogue between a civil rights strategy and more radical strategies for change is the debate between race-specific versus colorblind programs for change. *Race-blind* (or colorblind) *policies* are those advocating that all groups be treated alike, with no barriers to opportunity posed by race, gender, or other group differences. Equal opportunity is the key concept in colorblind policies. This is also the framework largely established through the nation's laws.

Race-specific (or color-conscious) *policies* are those that recognize the unique status of racial groups because of the long history of discrimination and the continuing influence of institutional racism. Advocates of such policies argue that color-blind strategies will not work because Whites do not start from the same position as other racial-ethnic groups.

The "Tax" on Being a Minority in America: Give Yourself a True-False Test (An Illustration of White Privilege)

On the following test, give yourself one point for each statement that is true for you personally. When you are done, total up your points. The higher your score (the more points you have), the less "minority tax" you are paying in your own life:

1. My parents and grandparents were able to purchase a house in any neighborhood they could afford.
2. I can take a job in an organization with an affirmative action policy without people thinking I got my job because of my race.
3. My parents own their own home.
4. I can look at the mainstream media and see people who look like me represented in a wide variety of roles.
5. I can choose from many different student organizations on campus that reflect my interests.
6. I can go shopping most of the time pretty well assured that I will not be followed or harassed when I am in the store.
7. If my car breaks down on a deserted stretch of road, I can trust that the law enforcement officer who shows up will be helpful.
8. I have a wide choice of grooming products that I can buy in places convenient to campus and/or near where I live as a student.
9. I never think twice about calling the police when trouble occurs.
10. The schools I have attended teach about my race and heritage and present it in positive ways.
11. I can be pretty sure that if I go into a business or other organization (such as a university or college) to speak with the "person in charge," I will be facing a person of my race.

Your total points: _____

Your racial identity: _____

Your gender: _____

How would you describe your social class? _____

Now gather results from some of your classmates and see if their total points vary according to their own race, gender, and/or social class.

Source: Adapted from the *Discussion Guide for Race: The Power of an Illusion.* **www.pbs.org**

Affirmative action is an example of a race-specific policy for reducing job and educational inequality. Affirmative action means two things: First, it means recruiting minorities from a wide base to ensure consideration of groups that have been traditionally overlooked, while not using rigid quotas based on race or ethnicity. Second, affirmative action can mean taking race into account as one factor among others that can be used in such things as hiring decisions or college admissions. Despite public misunderstandings about affirmative action, having specific quotas for minority representation is unconstitutional, but, to date, constitutional law does allow the consideration of race in such things as college admissions as long as race is not the sole factor being considered. This was most recently upheld in the Supreme Court case, *Fisher v. University of Texas* (2016).

Where Do We Go From Here?

All told, the effects of race and ethnicity are stubborn and continue to affect the distribution of social, economic, and political resources. No single strategy is likely to solve this complex social issue. How will the nation respond to an increasingly complex landscape of racial and ethnic challenges?

The United States is becoming even more diverse in its racial-ethnic composition, raising questions about the future of race and ethnicity. Latinos have now surpassed African Americans as a proportion of the U.S. population. Latinos and Asians are also expected to continue growing and, if projections are correct, will become an even larger share of the population in the not too distant future. At the same time, the White American population is expected to decline by about 6 percent by 2050 (Ortman and Guarneri 2016; Frey 2015). How will these changes affect race relations and the nation's political climate in years to come?

With more people identifying as multiracial, will the meaning of race change? Given that race is a social construction, most likely. If so, will racism diminish or simply take new forms? The traditional black-white color line that has historically defined race in the United States is no longer as sharp as it once was. Will this change our ideas about race? Will it eradicate, blur, or sharpen racial boundaries? Who will be at the top and who will be at the bottom?

Sociologist Eduardo Bonilla-Silva (2004) suggests that the United States could become a *tri-partite society*. What he means is that we could evolve into a racial system that divides people into three groups: black, white, and brown. In such a system, some people might become "honorary whites"—that is, those who are successful enough to be perceived as "white"—at least if they are light-skinned. In Bonilla-Silva's speculations, the "collective black" would include darker-skinned groups who would sit at the bottom of this racial hierarchy (Bonilla-Silva 2004; Bonilla-Silva and Glover 2004).

Only time will tell whether Bonilla-Silva's projections will happen, but there are signs of such an evolving system. Some Asian Americans have equaled or surpassed White, non-Hispanic Americans in socioeconomic status. Many middle-class and elite African Americans and Latinos/as are highly successful, in some cases even surpassing working class and poor whites who see themselves as increasingly left behind. Some Latinos, depending on their national origin and social class, already define themselves as White (Forman et al. 2004).

The current youth and millennial generations could also change the future of race—or so many believe. The millennial generation (those born after 1981) is the most diverse generation and is somewhat more racially tolerant than previous generations. Even with these vast changes in attitude, population, and people's racial identities, the truth is that race and ethnicity are hardly declining in their significance. Group boundaries and identities may no longer be so firm as the old black-white divide, but race and ethnicity still matter and they matter a lot (West 1994).

Debates about racial and ethnic inequality rage on—in dinner conversations, courtrooms, college classrooms, boardrooms, and the government. With the election of Donald Trump as president of the United States and with a mostly White male and strongly conservative administration in place, many fear that the gains of the past will be lost. The nation is already witnessing the mobilization of social movements to challenge these threats, including demonstrations to protect immigrant rights. There is the potential for a new season of social protest on behalf of race and social justice. What these movements mean for the future and where the society will go remain to be seen. What is certainly needed is continuing to sharpen our understanding of race and ethnicity and how they are produced and manifested in society.

Chapter Summary

What does it mean to say race and ethnicity are social constructions?

In virtually every walk of life, race matters. A *race* is a social construction based loosely on physical criteria, whereas an *ethnic group* is a culturally distinct group. A group is *minority* not on the basis of their numbers in a society but on the basis of which group occupies lower average social status.

What are the differences between prejudice, discrimination, and stereotypes?

Prejudice is an attitude usually involving negative prejudgment on the basis of race or ethnicity. *Discrimination* is overt, actual behavior involving unequal treatment. *Racism* involves both attitude and behavior. *Stereotypes* are oversimplified beliefs about members of a social group that have real consequences, including in such things as *stereotype threat*.

What different forms does racism take?

Racism can take on several forms, such as overt *prejudice, aversive* (subtle) *racism, implicit* (unconscious) *racism, laissez-faire racism, colorblind racism* (which masks *White privilege*), and institutional racism. *Institutional racism* is unequal treatment, carrying with it notions of cultural inferiority of a minority, which has become firmly ingrained into the economic, political, and educational institutional structure of society.

What do we learn from the histories of different racial and ethnic groups?

Historical experiences show that different groups have unique histories, although they are bound together by some similarities in the prejudice and discrimination they have experienced. These diverse histories are also linked through patterns of both exclusion and inclusion.

What is racial stratification and how is it observable?

Racial stratification is a system of inequality in which race and ethnicity mark differential access to economic, social, political, and cultural resources. It is especially apparent in economic inequality and racial segregation but is manifested in many ways within all social institutions.

What different explanations are there for racial and ethnic inequality?

Assimilation theory examines the process by which new groups become incorporated into a host society. Debates about the relative influence of *culture and social structure* show that culture, in and of itself, does not determine group success. Social structures both facilitate and block group life chances. *Class and race* are very much interconnected though the relative influence of one or the other varies depending on the social structural conditions at given points in time. *Intersectional theory* explores the simultaneous influence of race, class, and gender.

What are the current challenges in attaining racial and ethnic equality?

The civil rights movement expanded opportunities for African Americans, as well as other groups who benefitted from this historic movement. The movement put a legal framework of equal rights into the law. Subsequent movements took a more militant approach based on the argument that racism was embedded in dominant institutions. Contemporary challenges rest on whether *race-blind* or *color-conscious strategies* for change are the most effective.

Key Terms

affirmative action 266
anti-Semitism 254
assimilation theory 259
audit study 242
aversive racism 245
colorblind racism 245
cultural pluralism 245
discrimination 242
dominant group 241
ethnic group 239
ethnocentrism 242
hypersegregation 258
Immigration Act 254

Immigration and Nationality
Act 254
implicit bias 245
institutional racism 245
intersectional theory 262
laissez-faire racism 245
minority group 241
nativism 259
Naturalization Act of 1790 254
panethnicity 240
prejudice 241
race 237
race-immigration 255

racial formation 238
racial stratification 256
racialization 239
racism 245
residential segregation 259
salience principle 236
segregation 258
stereotype 242
stereotype interchangeability 244
stereotype threat 244
urban underclass
White privilege 245
white space 259

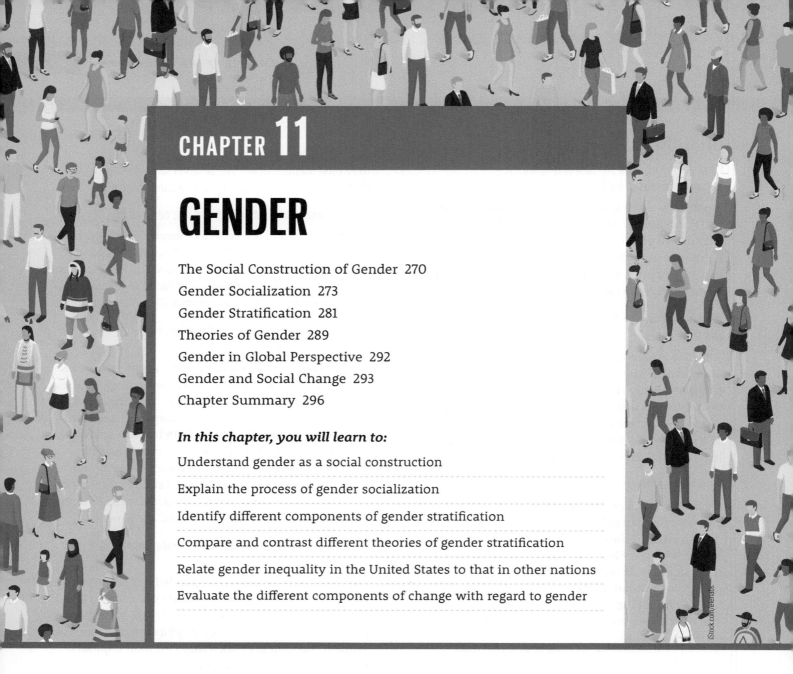

CHAPTER 11

GENDER

In this chapter, you will learn to:

Understand gender as a social construction

Explain the process of gender socialization

Identify different components of gender stratification

Compare and contrast different theories of gender stratification

Relate gender inequality in the United States to that in other nations

Evaluate the different components of change with regard to gender

Imagine suddenly becoming a member of the other sex. What would you have to change? First, you would probably change your appearance—clothing, hairstyle, and any adornments you wear. You would also have to change some of your interpersonal behavior. Contrary to popular belief, men talk more than women, are louder, are more likely to interrupt, and are less likely to recognize others in conversation. Women are more likely to laugh, express hesitance, and be polite. Gender differences also appear in nonverbal communication. Women use less personal space, touch less in impersonal settings (but are touched more), and smile more, even when they are not necessarily happy (Wood 2013). Researchers even find that men and women write email in a different style, women writing less opinionated email than men and using it to maintain rapport and intimacy (Colley and Todd 2002; Sussman and Tyson 2000). Finally, you might have to change many of your attitudes because men and women differ significantly on many, if not most, social and political issues (see ▲ Figure 11-1 and ◆ Table 11-1 where you can quiz yourself on some common assumptions about gender).

If you are a woman and became a man, perhaps the change would be worth it. You would probably see your income increase (especially if you became a White man). You would have more power in virtually every social setting. You would be far more likely to head a major corporation, run your own business, or

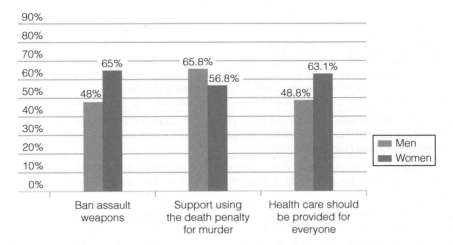

▲ **Figure 11-1   The Gender Gap in Attitudes.** When people use the term *gender gap*, they are often referring to pay differences between women and men. As the data here show, there is also a significant gender gap on some of the important issues of the day.

Data Source: Pew National Opinion Research Center, General Society Survey. 2016. Chicago, IL: University of Chicago. **www.norc.org**; Pew Research Center. 2015. "Continued Bipartisan Support for Expanded Background Checks on Gun ales." Washington; DC: Pew Research Organization. **www.people-press.org/2015/08/13/continued-bipartisan-support-for-expanded-background-checks-on-gun-sales/#total**

be elected to a political office—again, assuming that you are White. Would it be worth it? As a man, you would be far more likely to die a violent death and would probably not live as long as a woman (National Center for Health Statistics 2016).

If you are a man who became a woman, your income would most likely drop significantly. More than fifty years after passage of the Equal Pay Act in 1963, men still earn 20 percent more than women, even comparing those working year-round and full-time (Fontenot, Semega, and Kollar 2018). You would probably become resentful of a number of things because poll data indicate that women are more resentful than men about things such as the amount of money available for them to live on, the amount of help they get from their mates around the house, how much men share child care, and how they look. Women also report being more fearful on the streets than men. Women are, however, more satisfied than men with their role as parents and with their friendships outside of marriage.

For both women and men, there are benefits, costs, and consequences stemming from the social definitions associated with gender. As you imagined this experiment, you may have had difficulty trying

Table 11-1	Is It True?*	True	False
1. Men are more aggressive than women.			
2. Parents have the most influence on children's gender identities.			
3. Most women hold feminist values.			
4. Being a "stay-at-home" mom is the most satisfying lifestyle for women.			
5. In all racial–ethnic groups, women earn less on average than men.			
6. The wage gap between women and men has closed since the 1970s, largely as the result of women being more likely to enter the labor force.			
7. In terms of wages, middle-class women have most benefited from antidiscrimination policies.			

*The answers can be found on the following page.

Is It True ? (Answers)

1. FALSE. Generalizations such as this ignore variation occurring within gender categories. Aggression is also a broad term that can have multiple meanings.
2. FALSE. There are numerous sources of gender socialization; even parents who try to raise their children not to conform too strictly to gender norms will find that peers, the media, schools, and other socialization agents all push people into the expected behaviors associated with gender (Kane 2012).
3. TRUE. Although many women do not use the label *feminist* to define themselves, surveys show that the majority of women agree with basic feminist principles. Self-identification as a feminist is most likely among well-educated, urban women (McCabe 2005).
4. FALSE. Surveys show that "stay-at-home" moms experience far more depression, worry, anxiety, and anger than employed women (Mendes et al. 2012); half of women say they would prefer to have a job outside the home (Saad 2012a).
5. TRUE. But the earnings gap differs depending a lot on who you compare to whom. Considering only year-round, full-time workers, White women earn 80 percent of what White men earn. But, comparing racial–ethnic women to White men, the earnings gap is much wider. Black women earn 61 percent of White men's earnings; Hispanic women, 52 percent of White men's earnings; and Asian women, 84 percent of White men's earnings. White and Asian American women, on average, earn more than Black and Hispanic men (U.S. Census Bureau 2017).
6. FALSE. The most significant reason for the decline in the wage gap between women and men is the decline in men's wages; a smaller portion of this closing gap is attributed to changes in women's wages (Economic Policy Institute 2017).
7. FALSE. Although all women do benefit from equal employment legislation, wage data indicate that the group whose wages have increased the most since the 1970s are women in the top 20 percent of earners. Middle- and working-class women have seen far lower gains, and poor women's wages have been relatively flat over this period of time (Economic Policy Institute 2017).

to picture the essential change in your biological identity: Is this the most significant part of being a man or woman? Nature determines whether you are male or female, but society gives significance to this distinction. Sociologists see gender as a social fact, because who we become as men and women is largely shaped by cultural and social expectations.

The Social Construction of Gender

From the moment of birth, gender expectations influence how boys and girls are treated. Now that it is possible to identify the sex of a child in the womb, gender expectations may begin even before birth. Parents and grandparents might select pink clothes and dolls for baby girls, sports clothing and brighter colors for boys. Even if they try to do otherwise, it will be difficult because baby products are so typed by gender. Much research shows how parents and others continue to treat children in stereotypical ways. Girls may be expected to cuddle and be sweet, whereas boys are handled more roughly and given greater independence.

See for Yourself

Changing Your Gender

Try an experiment based on the example of changing gender that opens this chapter.

1. First, list everything you think you would have to do to change your behavior if you were the opposite gender. Separate your list items according to whether they are related to such factors as appearance, attitude, or behavior.
2. Second, for twenty-four hours, try your best to change any of these things that you are willing to do. Record how others react to you during this period and how the change makes you feel.
3. When your experiment is over, write a report on what your brief experiment tells you about how gender identities are (or are not) supported through social interaction.

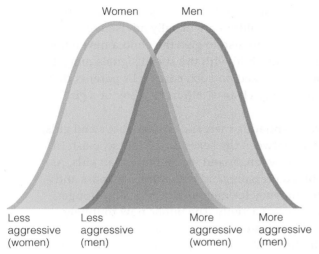

▲ **Figure 11-2   Gender Differences: Aggression.** Even when men and women as a whole tend to differ on a given trait, within-gender differences can be just as great as across-gender differences. Some men, for example, are less aggressive than some women.

Defining Sex and Gender

Sociologists use the terms *sex* and *gender* to distinguish biological sex identity from learned gender roles. **Sex** refers to biological identity, being male or female. For sociologists, the more significant concept is **gender**—the socially learned expectations, identities, and behaviors associated with members of each sex. This distinction emphasizes that behavior associated with gender is culturally learned. Gender is a "system of social practices" (Ridgeway 2011: 9) that creates categories of people—men and women—who are defined in relationship to each other on unequal terms.

The definitions that surround these categories stem from culture—made apparent especially by looking at other cultures. Across different cultures, gender expectations associated with men and women vary considerably. In Western industrialized societies, people tend to think of men and women (and masculinity and femininity) in dichotomous terms, even defined as "opposite sexes." The views from other cultures challenge this assumption. Historically, the *berdaches* (pronounced berdash) in Navajo society were anatomically normal men defined as a third gender between male and female. Berdaches, considered ordinary men, married other men who were not berdaches. Neither the berdaches nor the men they married were considered homosexuals, as they would be considered in other places (Nanda 1998; Lorber 1994).

There are also substantial differences in the construction of gender across social classes and within subcultures in a given culture. Within the United States, there is considerable variation in the experiences of gender among different racial and ethnic groups (Andersen and Collins 2016; Baca Zinn et al. 2015). Differences within a given gender can be greater than differences between men and women. That is, the variation on a given trait, such as aggression or competitiveness, can be as great within a given gender group as the difference across genders. Thus, some women are more aggressive than some men, and some men are less competitive than some women (see ▲ Figure 11-2).

Sex Differences: Nature or Nurture?

Is gender a matter of nature or nurture? Obviously, there are some biological differences between women and men, but looking at gender sociologically quickly reveals the extraordinary power of social and cultural influences on things often popularly seen as biologically fixed. Let's first examine some basic biological facts about sex and gender.

A person's sex identity is established at the moment of conception. The mother contributes an X chromosome to the embryo; the father, an X or Y. The combination of two X chromosomes makes a female; the combination of an X and a Y makes a male. Usually, chemical events directed by genes on the sex-linked chromosomes lead to the formation of male or female genitalia.

Intersexed people are those born with a combination of characteristics that are usually assumed to mean either male or female (Davis 2015). For example, an intersexed infant may be born with ovaries or testes but with ambiguous or mixed genitals, or may be born a chromosomal male but have an incomplete penis and no urinary canal. In the past, this condition was defined as hermaphroditism, but intersexed people have argued that such a term is too highly medicalized and tends to define people as somehow deficient, thus most now use the term intersex.

Case studies of intersexed people reveal the extraordinary influence of social factors in shaping a person's identity, including the language we use to describe people. Intersexed people have frequently been described through medical language or terms that describe them as somehow deficient or abnormal, an indication of how we so rigidly and, too often, judgmentally, think of gender solely in "either/or" terms (Davis 2015; Preves 2003). Parents of intersexed children are usually advised to have their child's genitals surgically assigned to either male or female and also to give the child a new name, a different hairstyle, and new clothes—all intended to provide the child with the social signals judged appropriate to a single gender identity. One physician who has worked on such cases tells parents that they "need to go home and do their job as child rearers with it very clear whether it's a boy or a girl" (Kessler 1990: 9).

Why, you might ask, are we as a culture so fixated on differentiating men and women, boys and girls, gender "insiders" and gender "outsiders?" Physical differences between the sexes do, of course, exist. In addition to differences in anatomy, boys at birth tend to be slightly longer and weigh more than girls. As adults, men tend to have a lower resting heart rate, higher blood pressure, and higher muscle mass and muscle density. These physical differences contribute to the tendency for men to be physically stronger than women, but this can be altered, depending on level of physical activity. The public now routinely sees displays of women's athleticism. Women can achieve a high degree of muscle mass and muscle density through bodybuilding and can win over men in activities that require high levels of endurance, such as the four women who have won the Iditarod—the Alaskan dog sled race considered one of the most grueling competitions in the world. In other words, until men and women compete equally in activities from which women have historically been excluded, we may not know the true extent of physical differences between women and men. The point here is that, despite the cultural focus on physical sex traits, as we will see, *gender is a social construction.*

Biological determinism refers to explanations that attribute complex social phenomena to physical characteristics. The argument that men are more aggressive because of hormonal differences (in particular,

---

Women in the Media: Where Are Women's Voices?

Who reports the news? Even with the increased presence of women as news reporters, anchors, editors, and writers, the news media continue to be dominated by men—White men at that. Women are the vast majority (74 percent) of those graduating from college with majors in journalism and mass communications (Becker et al. 2010), but their voices and words do not narrate the news of the nation. Some facts from recent research include:

- Women are underrepresented as reporters, writers, and anchors in every form of media (see the accompanying figure);
- Men are quoted three and a half times as often as women, a finding based on a detailed three-month analysis of the front page of *The New York Times*;
- Men write 82 percent of film reviews;
- Men vastly outnumber women as experts and guests on TV talk shows.

Gender Gap in the News Media

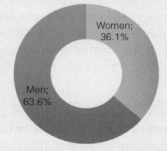

Women; 36.1%

Men; 63.6%

Note: Percentage of women and men with bylines and on-camera appearances in major TV networks, newspapers, news wires, and Internet news services, 2013

Women are most likely to report on lifestyle, education, culture, and health, not politics, technology, or criminal justice. Women are most underrepresented as reporters of world affairs. Even when the subject matter is of deep concern to women—abortion and birth control—80 percent of those reporting on these issues are men (4th Estate 2014).

Try observing these patterns yourself: Who writes the bylines in Internet news you read? Who is cited as an expert? Who covers what issues? Although sociologists would argue that the mere presence of women does not necessarily change the presentation of ideas, do you think that it matters that men are the ones most likely to narrate our nation's news?

Source: *The Status of Women in the U.S. Media 2017*. 2017. New York: Women's Media Center. **www.womensmediacenter.com**

Pete Saloutos/Getty Images

Title IX, passed in 1972, has opened up athletic opportunities to women that were not available to earlier generations.

the presence of testosterone) is a biologically determinist argument. Despite popular belief, studies find only a modest correlation between aggressive behavior and testosterone levels. Furthermore, changes in testosterone levels do not predict changes in men's aggression (such as by "chemical castration," the administration of drugs that eliminate the production or circulation of testosterone). What's more, there are minimal differences in the levels of sex hormones between girls and boys during early childhood, yet researchers find considerable differences in the aggression exhibited by boys and girls as children (Fausto-Sterling 2000, 1992).

Arguments based on biological determinism assume that differences between women and men are "natural" and, presumably, resistant to change. Like biological explanations of race differences, biological explanations of inequality between women and men tend to flourish during periods of rapid social change. They protect the status quo (existing social arrangements) by making it appear that the status of women is "natural" and therefore should remain as it is. If social differences between women and men were biologically determined, there would also be no variation in gender relations across cultures, but extensive differences are well documented.

In sum, we would not exist without our biological makeup, but we would not be who we are without society and culture. As sociologist Cecilia Ridgeway puts it, gender is a "substantial, socially elaborated edifice constructed on a modest biological foundation" (2011: 9). Culture defines certain behaviors as appropriate (or not) for women and for men; these are learned throughout life, beginning with the earliest practices by which people are raised. Understanding the process of *gender socialization* is then a key part of understanding the formation of gender as a social and cultural phenomenon.

Gender Socialization

As we saw in Chapter 4, socialization is the process by which social expectations are taught and learned. Through **gender socialization**, men and women learn the expectations and identities associated with gender in society. The rules of gender extend to all aspects of society and daily life. Gender socialization affects the self-concepts of women and men, their social and political attitudes, their perceptions about other people, and their feelings about relationships with others. Although not everyone is perfectly socialized to conform to gender expectations, socialization is a powerful force directing the behavior of men and women in gender-typical ways.

Even people who set out to challenge traditional expectations often find themselves yielding to the powerful influence of socialization. Women who consciously reject traditional women's roles may still find themselves inclined to act as hostess or secretary in group settings. Similarly, men may accept equal responsibility for housework, yet they fail to notice when the refrigerator is empty or a child needs a bath—household needs they have been trained to let someone else notice (DeVault 1991). Gender expectations are so pervasive that it is also difficult to change them on an individual basis. If you doubt this, try buying clothing or toys for a young child without purchasing something that is gender-typed, or talk to parents who have tried to raise their children without conforming to gender stereotypes and see what they report about the influence of such things as children's peers and the media.

The Formation of Gender Identity

One result of gender socialization is the formation of **gender identity**, which is one's definition of oneself as a woman or man, or perhaps as transgender. Gender identity is basic to our self-concept, shaping our expectations for ourselves, our abilities and interests, and how we interact with others. Gender identity shapes not only how we think about ourselves and others but also influences numerous behaviors, including the likelihood of drug and alcohol abuse, violent behavior, depression, or even how aggressive you are in driving (Andersen 2015).

One area in which gender identity has an especially strong effect is in how people feel about their appearance. Studies find strong effects of gender identity on body image. Concern with body image begins mostly during adolescence. Studies of young children (that is, preschool age) find no gender differences in how boys and girls feel about their bodies (Hendy et al. 2001), but clear differences emerge by early adolescence (Saunders and Frazier 2017). Adolescent girls report comparing their bodies to others of their sex more often than boys do, and girls report more negativity about their body image than do boys, thus reducing their self-esteem (Jones 2001; Polce-Lynch et al. 2001). Idealized images of women's bodies in the media, as well as peer pressures, have a huge impact on young girls' and women's gender identity and feelings about their appearance.

Sociologist Debra Gimlin argues that bodies are "the surface on which prevailing rules of a culture are written" (Gimlin 2002: 6). You see this especially with regard to gender. Men and women alike practice elaborate rituals to achieve particular gender ideals, ideals that are established by the dominant culture.

Understanding gender socialization helps you see that gender, like race, is a social construction. We "do" gender—as you will see in the section near the end of this chapter on theories of gender. Understanding gender as a social construction means thinking about the many ways that gender is produced through social interaction instead of seeing it as a fixed attribute of individuals (Connell 2009; West and Zimmerman 1987).

The experiences of transgender people make this point even more obvious. **Transgender** people are those who live as a gender different from that to which they were assigned at birth (Schilt 2011). Transgender individuals experience great pressure to fit within the usual expectations. Because they do not necessarily display the expected routines of gender, they may be ridiculed, shunned, discriminated against, or even violently assaulted. Their experiences show how powerful the social norms are to conform to gender expectations (Westbrook and Schilt 2014; Connell 2010, 2012).

Similarly, those who undergo sex changes as adults report enormous pressure from others to be one sex or the other. Managing such identities can be stressful, largely because of expectations that others have about what are appropriate categories of gender identity. Anyone who crosses these or in any way appears to be different from dominant expectations is frequently subject to exclusion and ridicule, showing again just how strong gender expectations are (Stryker and Whittle 2006).

Sources of Gender Socialization

As with other forms of socialization, there are different agents of gender socialization: family, peers, children's play, schooling, religious training, mass media, and popular culture, to name a few. Gender socialization is reinforced whenever gender-linked behaviors receive approval or disapproval from these multiple influences.

Parents are one of the most important sources of gender socialization. Parents may discourage children from playing with toys that are identified with the other sex, especially when boys play with toys meant for girls. Research finds that parents are more tolerant of girls not conforming to gender roles than they are for boys (Kane 2012). Fathers, especially, discourage sons from violating gender norms

Thinking Sociologically

Look at the products sold to men and women used as part of their daily grooming. How are they packaged? What color are they? What are their names? How do these artifacts of everyday life reflect *norms* about gender?

(Martin 2005). Although fathers are now more involved than in the past in children's care, they are less likely to provide basic care and more likely to be involved with discipline (LaFlamme et al. 2002).

Expectations about gender are changing, although researchers suggest that the cultural expectations about gender may have changed more than people's behavior. Mothers and fathers now report that fathers should be equally involved in child rearing, but the reality is different. Mothers still spend more time in child-related activities and have more responsibility for children. Furthermore, the gap that mothers perceive between fathers' ideal and actual involvement in child rearing is a significant source of mothers' stress (Milkie et al. 2002).

Gender socialization patterns also vary within different racial–ethnic families. Latinas, as an example, have generally been thought to be more traditional in their gender roles, although this varies by generation and by the experiences of family members in the labor force. Within families, young women and men learn to formulate identities that stem from their gender, racial, and ethnic expectations.

Peers strongly influence gender socialization—sometimes more so than one's immediate family. Peer relationships shape children's patterns of social interaction. Young people's play also shapes analytical skills, values, and attitudes. Studies find that boys and girls often organize their play in ways that reinforce not only gender but also race and age norms (Moore 2001). Peer relationships often reinforce the gender norms of the culture—norms that are typically even more strictly applied to boys than to girls. Thus, boys who engage in behavior that is associated with girls are likely to be ridiculed by friends—more so than are girls who play or act like boys. Homophobic attitudes, routinely expressed among peers, also reinforce dominant attitudes about what it means to "be a man" (Pascoe 2011). Although girls may be called "tomboys," boys who are called "sissies" are more harshly judged. Note, though, that tomboy behavior among girls beyond a certain age may result in the girl being labeled a "dyke."

Children's play is another source of gender socialization. You might think about what you played as a child and how this influenced your gender roles. Typically, though not in every case, boys are encouraged to play outside more; girls, inside. Boys' toys are more machinelike and frequently promote the development of militaristic values; they tend to encourage aggression, violence, and the stereotyping of enemies—values rarely associated with girls' toys. Children's books in schools also communicate gender expectations. Even with publishers' guidelines that discourage stereotyping, textbooks still depict men as aggressive, argumentative, and competitive. Men and boys are also more likely to be featured in children's books, although systematic analyses of children's books show that, even now, fathers are not very present and, when they do appear, they are most often shown as ineffectual (Anderson and Hamilton 2005).

Schools are particularly strong influences on gender socialization because of the amount of time children spend in them. Teachers often have different expectations for boys and girls. Studies find, for example, that teachers hold gender stereotypes that women are not as capable in math as men

Even with social changes in gender roles, boys and girls tend to engage in play activities deemed appropriate for their gender.

(Riegle-Crumb and Humphries 2012). We know from other research that when such stereotypes are present, student performance, such as on standardized tests, is negatively affected by the presence of what is called *stereotype threat* (Steele 2010). Earlier studies have also shown that when teachers respond more to boys in school, even if negatively, they heighten boys' sense of importance (American Association of University Women 1998; Sadker and Sadker 1994). Gender inequality is pervasive in schools and at all levels. Even in college, course-taking patterns, selection of majors, and teachers' interaction with students are shaped by gender (Mullen 2010).

Religion is an often overlooked but significant source of gender socialization. The major Judeo-Christian religions in the United States place strong emphasis on gender differences, with explicit affirmation of the authority of men over women. In Orthodox Judaism, men offer a prayer to thank God for not having created them as a woman or a slave. The patriarchal language of most Western religions as well as the exclusion of women from positions of religious leadership in some faiths signify the lesser status of women in religious institutions. Any religion interpreted in a fundamentalist way can be oppressive to women. Indeed, the most devout believers of any faith tend to hold the most traditional views of women's and men's roles. The influence of religion on gender attitudes cannot be considered separately from other factors, however. For many, religious faith inspires a belief in egalitarian roles for women and men. Christian, Islamic, and Jewish women, along with other religious women, have often organized to resist fundamentalist and sexist practices (Messina-Dysert and Ruether 2014).

The *media* in their various forms (television, film, magazines, music, and so on) communicate strong—some would even say cartoonish—gender stereotypes. Despite some changes in recent years, television and films continue to depict highly stereotyped roles for women and men. In family films, prime-time television, and even children's TV shows, women and girls are shown more so than men and boys wearing sexy attire, with exposed skin, and with thin bodies, reinforcing gendered images of both women and men, even at an early age (Smith et al. 2013). Men are also seen as more formidable, stereotyped in strong, independent roles. Women are more likely now to be portrayed as employed outside of the home and in professional jobs, but it is still more usual to see women depicted as sex objects. In fact, the sexualization of women is so extensive in the media that the American Psychological Association has concluded that there is "massive exposure to portrayals that sexualize women and girls and teach girls that women are sexual objects" (American Psychological Association 2007: 5).

Do people believe what they see on television? Research with children reveals that they identify with the television characters they see. Both boys and girls rate the aggressive toys that they see on television commercials as highly desirable. They also judge them as more appropriate for boys' play, suggesting that something as seemingly innocent as a toy commercial reinforces attitudes about gender and violence (Klinger et al. 2001). Even with adults, researchers find that there is a link between viewing sexist images and having attitudes that support sexual aggression, antifeminism, and more traditional views of women (American Psychological Association 2007).

It is easy to observe the pervasiveness of gendered images by just watching the media with a critical eye. Women in advertisements are routinely shown in poses that would shock people if the characters were men. Consider how often women are displayed in ads dropping their pants, skirts, or bathrobe, or are shown squirming on beds. How often are men shown in such poses? Men are sometimes displayed as sex objects in advertising, but not nearly as often as women. The demeanor of women in advertising—on the ground, in the background, or looking dreamily into space—makes them appear subordinate and available to men.

The Price of Conformity

A high degree of conformity to stereotypical gender expectations takes its toll on both men and women. One of the major ways to see this is in the very high rate of violence against women—both in the United States and worldwide. Too frequently, men's power in society is manifested in physical and emotional violence. Violence takes many forms, including rape, sexual abuse, intimate partner violence, stalking, genital mutilation, and honor killings. Around the world, the United Nations is working to reduce violence against women, including some initiatives to help men examine cultural assumptions about masculinity that promote violence. Violence against women stems from the attitudes of power and control that gender expectations produce and that can lead *some* men to engage in violent behavior—toward women, as well as toward other men.

Violence by men is only one of the harms to women stemming from dominant gender norms. Adhering to gender expectations of thinness for women and strength for men is related to a host of negative health behaviors, including eating disorders, smoking, and for men, steroid abuse. The dominant culture promotes a narrow image of beauty for women—one that leads many women, especially young women, to be disturbed about their body image. Striving to be thin, millions of women engage in constant dieting, fearing they are fat even when they are well within or below healthy weight standards. Many develop eating disorders by purging themselves of food or cycling through various fad diets—behaviors that can have serious health consequences (see Table 11-1). Many young women develop a distorted image of themselves, thinking they are overweight when they may actually be dangerously thin.

Additionally, despite the known risks of smoking, young women who smoke think that it will keep them thin. Eating disorders can be related to a woman having a history of sexual abuse, but they also come from the promotion of thinness as an ideal beauty standard for women—a standard that can put girls' and women's health in jeopardy (Alexander et al. 2010).

Men also pay the price of conformity if they too thoroughly internalize gender expectations that say they must be independent, self-reliant, and unemotional. Although many men are more likely now than in the past to express intimate feelings, gender socialization discourages intimacy among them, affecting the quality of men's friendships. Conformity to traditional gender roles denies women access to power, influence, and achievement in the public world, but it also robs men of the more nurturing and other-oriented relationships that women have customarily had. Men's physical daring

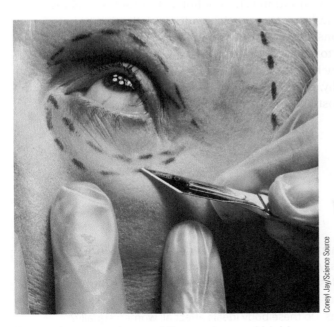

Coneyl Jay/Science Source

Cosmetic surgery is a rapidly growing, and highly profitable, industry. Some do it to try to eliminate signs of aging, others for aesthetic purposes. The pervasive influence of Western images means that some women in other nations pay for procedures to make them look more "Western."

Table 11-2	Facts about Eating Disorders
Men and women do not differ in their degree of dissatisfaction with their bodies, but women are more preoccupied with their weight than are men. White Americans are more preoccupied with weight, dieting, and body appearance than are African Americans (Dye 2016).	
Over half (58 percent) of women on college campuses report feeling pressure to lose weight; nearly half (44 percent) of them were of normal weight (Malinauskis 2006).	
The average woman fashion model weighs 23 percent less than the average woman; most runway models meet the body mass index criteria for anorexia (Jones 2012).	
The weight loss industry is worth an estimated $60 billion per year (LeTourneau 2016).	

Sources: Malinauskas, Brenda, et al. 2006. "Dieting Practices, Weight Perceptions, and Body Composition: A Comparison of Normal Weight, Overweight, and Obese College Females." *Nutrition Journal* 5 (March 31): 5–11; Spitzer, Brenda L., Katherine A. Henderson, and Marilyn T. Zivian. 1999. "A Comparison of Population and Media Body Sizes for American and Canadian Women." *Sex Roles* 700 (7/8): 545–565. Jones, Madeline. 2012. "Plus Size Bodies, What is Wrong with Them Anyway?" *Plus Model.* **www.plus -model-mag.com/2012/01/plus-size-bodies-what-is-wrong-with-them-anyway/**; LeTourneau, Nancy. 2016 (May 2). "What the $60 Billion Weight Loss Industry Doesn't Want you to Know." *Washington Monthly.* **http://washingtonmonthly.com/2016/05/02 /what-the-60-billion-weight-loss-industry-doesnt-want-you-to-know/**; Dye, Heather. 2016. "Are There Differences in Gender, Race, and Age Regarding Body Dissatisfaction?" *Journal of Human Behavior in the Social Environment* 26 (6): 499–508.

and risk-taking leaves them at greater risk of early death or injury from accidents. On many college campuses, for example, men showing bravado through heavy drinking can result in death or injury—and, as campus leaders know, heavy drinking is also strongly correlated with sexual assault against women on campus (Insight 2013). The strong undercurrent of violence in today's culture of masculinity can in many ways be attributed to the learned gender roles that put men and women at risk.

Gender Socialization and Homophobia

One of the primary ways that gender is produced during socialization is homophobia. **Homophobia** is the fear and hatred of gays and lesbians. Homophobia plays an important role in gender socialization because it encourages stricter conformity to traditional expectations, especially for men and young boys. Slurs directed against individuals who are gay encourage boys to act more masculine as a way of affirming for their peers that they are not gay (Pascoe 2011). Through such ridicule, homophobia discourages so-called feminine traits in men, such as caring, nurturing, empathy, emotion, and gentleness. The consequence is not only conformity to gender roles, but also a learned hostility toward gays and lesbians.

Doing Sociological Research

Eating Disorders: Gender, Race, and the Body

Research Question

An obsession with weight and body image permeates the dominant culture, especially for girls and women. If you glance at the covers of popular magazines for women and girls, you will very likely find article after article promoting new diet gimmicks, each bundled with a promise that you will lose pounds in a few days if you only have the proper discipline or use the right products. The models on the covers of such magazines are likely to be thin, often danger-ously so because being too thin causes serious health problems. Do these body ideals affect all women equally?

Research Method

Meg Lovejoy wanted to know if the drive for thinness is unique to White women and how gendered images of the body might differ for African American and White women in the United States. Her research is based on review-ing the existing research literature on eating disorders, which has generally concluded that, compared with White women, Black women are less likely to develop eating disorders.

(continued)

Research Results

Black women are less likely than White women to engage in excessive dieting and are less fearful of fat, although they are more likely to be obese and experience compulsive overeating. White women, on the other hand, tend to be very dissatisfied with their body size and overall appearance, with an increasing number engaging in obsessive dieting. Black and White women also tend to distort their own weight in opposite directions: White women are more likely to overestimate their own weight (that is, saying they are fat when they are not); Black women are more likely to underestimate their weight (saying they are average when they are overweight by medical standards). Why?

Conclusions and Implications

Lovejoy concludes that you cannot understand eating disorders without knowing the different stigmas attached to Black and White women in society. She suggests that Black women develop alternative standards for valuing their appearance as a way of resisting mainstream, Eurocentric standards. Black women who do so are then less susceptible to the controlling and damaging influence of the institutions that promote the ideal of thinness as feminine beauty. On the other hand, the vulnerability that Black women experience in society can foster mental health problems that are manifested in overeating. Eating disorders for Black women can also stem from the traumas that result from racism.

Lovejoy concludes that eating disorders must be understood in the context of social structures—gender, race, class, homophobia, and ethnicity—that affect all women, although in different ways. The cultural meanings associated with bodies differ for different groups in society but are deeply linked to our concepts of ourselves and the basic behaviors—like eating—that we otherwise think of as "natural."

Questions to Consider

1. What cultural messages are being sent to different race and gender groups about appropriate appearance in music and visual images in popular culture? How do these messages affect people's body image and self-esteem?

2. Lovejoy examines eating disorders in the context of gender, race, class, and ethnicity. What cultural meanings are broadcast with regard to age?

3. Is there a "culture of thinness" among your peers? If so, what impact do you think it has on people's self-concept? If not, are there other cultural meanings associated with weight among people in your social groups?

Source: Lovejoy, Meg. 2001. "Disturbances in the Social Body: Differences in Body Image and Eating Problems among African American and White Women." *Gender & Society* 15 (April): 239–261.

Homophobia is a learned attitude, as are other forms of negative social judgments about particular groups. Homophobia is also deeply embedded in people's definitions of themselves as men and women. Boys are often raised to be manly by repressing so-called feminine characteristics in themselves. Being called a "fag" or a "sissy" is one of the peer sanctions that forces conformity to gender roles. Similarly, pressures on adolescent girls to abandon tomboy behavior teach girls to adopt behaviors and characteristics associated with womanhood. Being labeled a lesbian may cause those with a strong attraction to women to repress this emotion and direct love only toward men. Although homophobic ridicule may seem like play and joking, it has serious consequences for both heterosexual and homosexual men and women. Homophobia not only socializes people into expected gender roles, it also produces numerous myths about gays and lesbians—examined in more detail in the following chapter.

Race, Gender, and Identity

Attitudes about gender roles are influenced by a large number of social factors. Studies have found some differences in attitudes comparing racial–ethnic groups. African Americans, for example, have more egalitarian gender role beliefs than do Whites, although women in both groups are more egalitarian in their beliefs than men (Vespa 2009). Traditionally, African American women have had more egalitarian beliefs about gender roles, which most interpret as the result of Black women's having been more commonly in the labor force than White women. The difference in Black and White attitudes toward gender roles is now diminishing as White women's labor force participation has increased (Carter et al. 2009).

Generally, Latinos and Latinas are more conservative in their gender role beliefs than are White men and women. Studies of Mexican American attitudes, however, find that, among those who have immigrated, these differences in gender role attitudes diminish in the second and third generations (Su et al. 2010). Native Americans also tend not to differ in their gender beliefs from other groups, despite stereotypes that they are more egalitarian in their outlooks (Harris et al. 2000). Factors such as the work experience of family members, religion, and other social experiences have more influence on gender attitudes than race per se. In addition to attitudes, race shapes people's identities. African American women, for example, are socialized to become self-sufficient, aspire to an education, desire an occupation, and regard work as an expected part of a woman's role (Thomas, Hacker, and Hoxha 2011; Collins 1990).

Men's gender identity is also affected by race. Latino men, for example, bear the stereotype of *machismo* and Black men, "the player"—both stereotypes about exaggerated masculinity and sexuality. Latinos and African American men adopt different strategies for defining their identities in the context of such stereotypes. Latinos associate machismo not with sexism but with honor, dignity, and respect (Hurtado and Sinha 2008; Baca Zinn 1995; Mirandé 1979).

African American men may adopt what has come to be called a "cool pose" to assert a positive presentation of self where they, not others, control their identity. Other Black men may assert an identity of "respectability"—that is, distancing themselves from negative stereotypes of Black men (Wilkins 2012; Young 2003). Identities are negotiated in a context where gender, race, and class frame how both women and men construct their identities.

The Institutional Basis of Gender

Gender is not just a matter of identity: *Gender is embedded in social institutions*. This means that institutions are patterned by gender, resulting in different experiences and opportunities for men and women. Sociologists analyze gender not just as interpersonal expectations but also as characteristic of institutions. This is what is meant by the term *gendered institution*.

Gendered institutions are the total pattern of gender relations that structure social institutions, including the stereotypical expectations, interpersonal relationships, and the different placement of men and women that are found in institutions. Schools, for example, are not just places where children learn gender roles but are gendered institutions because they are structured on specific gender patterns. Seeing institutions as gendered reveals that gender is not just an attribute of individuals but is also "present in the processes, practices, images and ideologies, and distributions of power in the various sectors of social life" (Acker 1992: 567).

As an example of the concept of gendered institution, think of what it is like to work as a woman in a work organization dominated by men. Women in this situation report that men's importance in the organization is communicated in subtle ways, whereas women are made to feel like outsiders. Important career connections may be made in the context of men's informal interactions with each other—both

Myth: Women of color are taking a lot of jobs away from White men.

Sociological Perspective: Sociological research finds no evidence of this claim. Quite the contrary, women of color work in gender- and race-segregated jobs and only rarely in occupations where they compete with White men in the labor market (Branch 2011).

inside and outside the workplace. Women may be treated as tokens or may think that company policies are ineffective in helping them cope with the particular demands in their lives.

Thinking of institutions as gendered is to think about the gendered characteristics of the institution itself. Work institutions have been structured on the old assumption that men work and women do not. Such institutions demand loyalty to work, not family. Even when men and women try to integrate work and family in their lives, gendered institutional practices make this difficult. Gendered institutions thus affect men as well as women, especially if they try to establish more balance between their work and personal lives. To say that institutions are gendered means that, taken together, there is a cumulative and systematic effect of gender throughout the institution.

Gender is not just a learned role: Gender is a social structure, just as class and race are structural dimensions of society. Notice that people do not think about the class system or racial inequality in terms of "class roles" or "race roles." It is obvious that race relations and class relations are far more than matters of interpersonal interaction. Race, class, *and* gender inequalities are experienced within interpersonal relationships, but they extend beyond relationships. Just as it would seem strange to think that race relations in the United States are controlled by race–role socialization, it is also wrong to think that gender relations are the result of gender socialization alone. Like race and class, gender is a system of privilege and inequality in which women are systematically disadvantaged relative to men. There are institutionalized power relations between women and men, and men and women have unequal access to social and economic resources.

Gender Stratification

Gender stratification refers to the hierarchical distribution of social and economic resources according to gender. Most societies have some form of gender stratification, although the specific form varies from country to country. Comparative research finds that women are more nearly equal in societies characterized by the following traits:

- Women's work is central to the economy.
- Women have access to education.
- Ideological or religious support for gender inequality is not strong.
- Men make direct contributions to household responsibilities, such as housework and child care.
- Work is not highly segregated by sex.
- Women have access to formal power and authority in public decision making (Chafetz 1984).

In Sweden, where there is a relatively high degree of gender equality, the participation of both men and women in the labor force and the household (including child care and housework) is promoted by government policies. Women also have a strong role in the political system, although women still earn less than men in Sweden. In many countries, women and girls have less access to education than men and boys, but that gap is closing. Still, in most countries, the illiteracy rate among women is much higher than among men (United Nations 2014).

Gender stratification is multidimensional. In some societies, women may be free in some areas of life but not in others. In Japan, for example, women tend to be well educated and they participate in the labor force in large numbers. Within the family, however, Japanese women have fairly rigid gender roles. Yet the

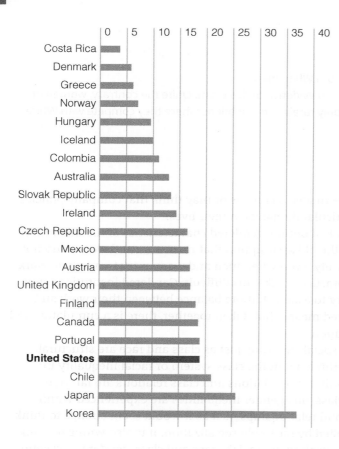

▲ **Figure 11-3** **The Gender Wage Gap in OECD Countries, 2015.** This chart shows the percentage gap in men's and women's wages in the most economically developed countries. Those with the smallest gap in women's and men's wages are at the top of the graph. Note the OECD average for all countries and where the United States falls relative to the rest of the world. How do you explain this?

Note: OECD is the Organisation for Economic Co-operation and Development.

Data: From OECD Employment Outlook. **www.oecd.org**

rate of violence against women in Japan is quite low in relation to other nations, even though women are widely employed as "sex workers" in hostess clubs, bars, and sex joints (Allison 1994). Patterns of gender inequality are most reflected in the wage differentials between women and men around the world, as ▲ Figure 11-3 shows.

Gender stratification can be extreme. The public witnessed this in 2012, when Malala Yousafzai, a fourteen-year-old Pakistani girl, was shot in the face by Taliban extremists simply because she advocated girls' rights to an education. She has since won the Nobel Peace Prize for her work promoting girls' education, the youngest person ever to receive this coveted prize. The Taliban, an extremist militia group, strips women and girls of basic human rights. When the Taliban took control of Afghanistan in 1996, women were banished from the labor force, expelled from schools and universities, and prohibited from leaving their homes unless accompanied by a close male relative. The windows of houses where women lived were painted black to keep women completely invisible to the public. This extreme segregation and exclusion of women from public life has been labeled **gender apartheid**. Gender apartheid is also evident in other nations, even if not as extreme as it was under Taliban rule: In Saudi Arabia, women were not even allowed to drive until 2017.

Sexism: The Biased Consequences of Beliefs

Gender stratification is supported by beliefs that treat gender inequality as "natural." Sexism defines women as different from and inferior to men. Sexism can be overt, but can also be subtle. Like racism, sexism makes gender roles seem natural when they are actually rooted in entrenched systems of power and privilege. In this sense, sexism is a belief that is anchored in social institutions. For example, the idea that men should be paid more than women because they are the primary breadwinners is a sexist idea, but that idea is also embedded in the wage structure. Because of this institutional dimension to sexism, people no longer have to be individually sexist for sexism to have consequences. Increasing amounts of research are, in fact, showing that *unintentional bias*—both based on gender and race—has enormous negative consequences for women and people of color (Eberhardt et al. 2004; Valian 1999).

Like racism, sexism generates social myths that have no basis in fact but support the continuing advantage of dominant groups over subordinates. Take the belief that women of color are being hired more often and promoted more rapidly than others. This belief misrepresents the facts. Women rarely take jobs away from men because most women of color work in gender- and race-segregated jobs. Women of color are especially burdened by obstacles to job mobility that are not present for men, particularly White men (Andersen 2015; Padavic and Reskin 2002). The myth that women of color get all the jobs makes White men seem to be the victims of race and gender privilege. Although there may be occasional cases where a woman of color (or a man of color, for that matter) gets a job that a White man also applied for, gender and race privilege usually favor White men.

Sexist beliefs also devalue the work that women do—both in dollar terms and in more subjective perceptions. To give a historical example, in jobs that were once held by men but became dominated by women, wages declined as women became more numerous. When this happens, the prestige of the occupation also tends to fall (Andersen 2015). You can also see this tendency in something labeled the *mommy tax*, referring to the loss of income women experience if they reenter the labor market after staying home to raise children. According to the author who coined this term, a college-educated woman with one child will lose about a million dollars in lifetime earnings as the result of the mommy tax (Crittenden 2002). Although the mommy tax afflicts all employed mothers, it has been shown to be especially severe among those who can afford it the least—low-income women (Budig and Hodges 2010).

Sexism emerges in societies structured by **patriarchy**, referring to a society or group in which men have power over women. It can be present in the private sector, such as in families in which husbands have authority over their wives, but patriarchy also marks public institutions when men hold all or most of the powerful positions. Forms of patriarchy vary from society to society, but it is common throughout the world. In some societies, it is rigidly upheld in both the public and private spheres, and women may be formally excluded from voting, holding public office, or working outside the home. In societies like the contemporary United States, patriarchy may be somewhat diminished in the private sphere (at least in some households), but the public sphere continues to be based on patriarchal relations.

Matriarchy has traditionally been defined as a society or group in which women have power over men. Anthropologists have debated the extent to which such societies exist. Matriarchies do exist, though not in the form the customary definition implies. Based on her study of the Minangkabau matriarchal society in West Sumatra (in Indonesia), anthropologist Peggy Sanday argues that scholars have used a Western definition of power to define matriarchy that does not apply in non-Western societies. The Minangkabau define themselves as a matriarchal society, meaning that women hold economic and social power. The Minangkabau are not, however, ruled by women. The people believe that rule should be by consensus, including that of men and women. Matriarchy exists but not as a mirror image of patriarchy (Sanday 2002).

In sum, gender stratification is an institutionalized system that rests on specific beliefs and attitudes that support, even inadvertently, the inequality of men and women. Although one could theoretically have a society stratified by gender where women hold power over men, gender stratification has not evolved in this way, as we will see in the next section.

Women's Worth: Still Unequal

Gender stratification is especially obvious in the persistent earnings gap between women and men (see ▲ Figure 11-4). Although the gap has closed somewhat since the 1960s, when women earned 59 percent of what men earned, women who work year-round and full-time still earn, on average, only 80 percent of what men earn. Women with bachelor degrees earn the equivalent of men with associate's degrees (see ▲ Figure 11-5). In 2017, the median income for women working full-time and year-round was $41,977; for men, it was $52,146 (Fontenot, Semega, and Kollar 2018). Perhaps more telling is that in the period of time since 1980, women in the top 10 percent of women earners have increased their earning power by 60 percent; *women in the bottom 10 percent have actually seen their income remain flat, as* measured in constant dollars (Economic Policy Institute 2017).

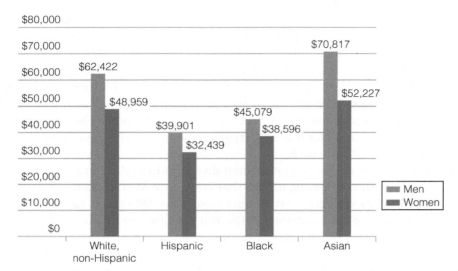

▲ **Figure 11-4   Median Income by Race and Gender, Full-Time Workers.** You will often hear that there is a 20 percent gap between women and men wage earners, but the wage gap varies depending on who you compare. For example, the gap between Hispanic women and White men is 52 percent and the gap between Hispanic women and Hispanic men is quite small. Given the data depicted here, what gaps do you see?

Note: These data are for year-round, full-time workers.

Data: U.S. Census Bureau. 2018. *Detailed Income Tables (2017): Selected Characteristics of People 15 Years and Over, by Total Money Income, Work Experience, Race, Hispanic Origin, and Sex, PINC-01,.* Washington, DC: U.S. Census Bureau. **www.census.gov**

The income gap between women and men persists despite the increased participation of women in the labor force. The **labor force participation rate** is the percentage of those in a given category who are employed either part-time or full-time. Fifty-four percent of all women are in the paid labor force compared with 66 percent of men. As recently as 1973, only 44 percent of women were in the labor force, so you can see the rather dramatic change that has occurred (Bureau of Labor Statistics 2017a). Since 1960, married women with children have also significantly increased their likelihood of employment. Seventy-one percent of women with children are in the labor force; 59 percent of women with children under one year of age are employed. Current projections indicate that women's labor force participation will continue to rise, and men's will decline slightly (Bureau of Labor Statistics 2017b; Toossi 2015).

Women of color have long been in the labor market, but this now also characterizes the experience of White women. The labor force participation rates of White women and women of color have, in fact, converged. More women in all racial groups are also now the sole supporters of their families.

Why do women continue to earn less than men, even when laws prohibiting gender discrimination have been in place for more than fifty years? The **Equal Pay Act of 1963** was the first federal law to require that men and women receive equal pay for equal work, an idea that is supported by the majority of Americans. The **Lily Ledbetter Fair Pay Act of 2009** extended the protections of the Civil Rights Bill of 1964. This act states that discrimination claims on the basis of sex, race, national origin, age, religion, and disability accrue with every

Debunking Society's Myths

Myth: Latinos are the least likely to support feminist ideas.
Sociological Perspective: Research on educated Latino men finds more identification with feminism that the myth would suggest. Latinos reject rigid definitions of masculinity and construct a definition of manhood that stems from their particular race, class, and sexual status in society. (Hurtado and Sinha 2008).

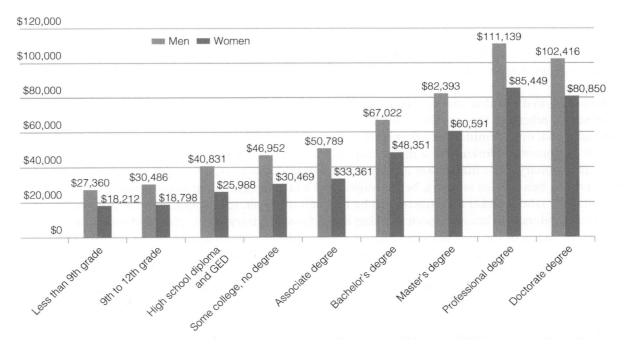

▲ Figure 11-5   Education, Gender, and Income, 2017. Although the "economic return" on education is high for both men and women at every level of educational attainment, men's income, on average, exceeds that of women. According to this national data, how much education would the average woman have to get to exceed the income of men with a high school diploma? To exceed men with a college degree?

Note: These data include all people 25 years and order, including those who may not be working year–round and full–time.

Data: U.S. Census Bureau. 2018. *Detailed Income Tables: PINC-03. Educational Attainment—People 25 Years Old and Over, by Total Money Earnings in 2016, Work Experience in 2017, Age, Race, Hispanic Origin, and Sex.* Washington, DC: U.S. Census Bureau. **www.census.gov**

paycheck, giving workers time to file claims of discrimination and eliminating earlier incentives for employers to cloak discrimination and then claim the employee filed their claims after the fact.

Wage discrimination, however, is rarely overt. Most employers do not even explicitly set out to pay women less than men. Despite good intentions and legislation, however, differences in men's and women's earnings persist. Research reveals four strong explanations for this: human capital theory, dual labor market theory, gender segregation, and overt discrimination, examined as follows.

Human Capital Theory

Human capital theory explains gender differences in wages as resulting from the individual characteristics that workers bring to jobs. Human capital theory assumes that the economic system is fair and competitive and that wage discrepancies reflect differences in the resources (or *human capital*) that individuals bring to their jobs. Factors such as age, prior experience, number of hours worked, marital status, and education are human capital variables. Human capital theory asserts that these characteristics will influence people's worth in the labor market. For example, higher job turnover rates or work records interrupted by child rearing and family responsibilities could negatively influence the earning power of women. There is a significant earnings penalty for women because of motherhood (Abendroth et al. 2014).

Education, age, and experience do influence earnings, but when you compare men and women who have the same level of education, previous experience, and number of hours worked per week, women still earn less than men (see Figure 11-5). Although human capital theory explains some of the difference between men's and women's earnings, it does not explain it all. Sociologists look to other factors to complete the explanation of wage inequality.

The Dual Labor Market

Dual labor market theory contends that women and men earn different amounts because they tend to work in different segments of the labor market. The dual labor market reflects the devaluation of women's work because women are most concentrated in low-wage jobs. Although it is hard to untangle cause and effect in the relationship between the devaluation of women's work and low wages in certain jobs, once such an earnings structure is established, it is difficult to change. Although equal pay for equal work may hold in principle, it applies to relatively few people because most men and women are not engaged in equal work.

According to dual labor market theory, the labor market is organized in two different sectors: the *primary market* and the *secondary market*. In the primary labor market, jobs are relatively stable, wages are good, opportunities for advancement exist, fringe benefits are likely, and workers are afforded due process. Working for a major corporation in a management job is an example of this. Jobs in the primary labor market are usually in large organizations where there is greater stability, steady profits, benefits for workers, better wages, and a rational system of management. In contrast, the secondary labor market is characterized by high job turnover, low wages, short or nonexistent promotion ladders, few benefits, poor working conditions, arbitrary work rules, and capricious supervision. Many of the jobs students take—such as waiting tables, bartending, or cooking and serving fast food—fall into the secondary labor market. For students, however, these jobs are usually short term, not lifelong.

Within the primary labor market, there are two tiers. The first consists of high-status professional and managerial jobs with potential for upward mobility, room for creativity and initiative, and more autonomy. The second tier comprises working-class jobs, including clerical work, skilled, and semi-skilled blue-collar work. Women and minorities in the primary labor market tend to be in the second tier. Although these jobs may be more secure than jobs in the secondary labor market, they are more vulnerable and do not have as much mobility, pay, prestige, or autonomy as jobs in the first tier of the primary labor market.

There is, in addition, an informal sector of the market where there is even greater wage inequality, no benefits, and little, if any, oversight of employment practices. Individuals may hire such workers as private service workers or under-the-table workers who perform a service for a fee (painting, babysitting, car repairs, and any number of services). Although there are no formal data on the informal sector because much of it tends to be in an underground economy, women and minorities form a large segment of this market activity, particularly given their role in care work (Duffy 2011).

Gender Segregation

Dual labor market theory explains wage inequality as a function of the structure of the labor market, not the individual characteristics of workers as suggested by human capital theory. Because of the dual labor market, men and women tend to work in different occupations and, when working in the same occupation, in different jobs. This is referred to as **gender segregation**, a pattern in which different groups of workers are separated into occupational categories based on gender. There is a direct association between the number of women in given occupational categories and the wages paid in those jobs. In other words, the greater the proportion of women in a given occupation, the lower the pay (U.S. Bureau of Labor Statistics 2017a). Segregation in the labor market can also be based on factors such as race, class, age, or any combination thereof.

Despite several decades of legislation prohibiting discrimination against women in the workplace, most women and men still work in gender-segregated occupations. That is, the majority of women work in occupations where most of the other workers are women, and the majority of men work mostly with men. Women also tend to be concentrated in a smaller range of occupations than men. To this day, more than half of all employed women work as clerical workers and sales clerks or in occupations such as food service workers, maids, health service workers, hairdressers, and child-care workers. Men are dispersed over a much broader array of occupations. Women make up 79 percent of elementary and middle school teachers, 89 percent of secretaries, 89 percent of bookkeepers, 90 percent of registered nurses, and 94 percent of child-care workers—stark evidence of the persistence of gender segregation in the labor force (Bureau of Labor Statistics 2017a).

Gender segregation also occurs within occupations. Women usually work in different jobs from men, but when they work within the same occupation, they are segregated into particular fields or job types.

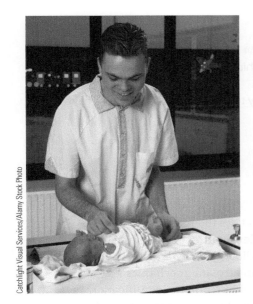

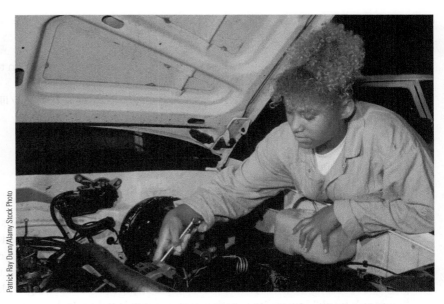

Because gender segregation is so pervasive in the workplace, people may still be surprised when they see women and men in nontraditional occupations.

For example, in sales work, women tend to do noncommissioned sales or to sell products that are of less value than those men sell. Even in fast-food restaurants, where you would expect wages to be routinized, women tend to work in those establishments with lower wages, fewer benefits, and less scheduling control (Haley-Lock and Ewert 2011).

Immigrant women are especially prone to being niched into highly segregated areas of the labor market. Many constraints shape the experience of immigrant women, including their legal status. Immigrant Latinas, for example, routinely find themselves in low-wage jobs regardless of their prior level of education, skill, or experience (Flippen 2014).

Overt Discrimination

A fourth explanation of the gender wage gap is discrimination. **Discrimination** refers to practices that single out some groups for different and unequal treatment. Despite the progress of recent years, overt discrimination continues to afflict women in the workplace. Some argue that men (White men in particular) have an incentive to preserve their advantages in the labor market. They do so by establishing rules that distribute rewards unequally. Women pose a threat to traditional White male privileges, and men may organize to preserve their own power and advantage (Reskin 1988). Historically, White men used labor unions to exclude women and racial minorities from well-paying, unionized jobs, usually in the blue-collar trades. A more contemporary example is seen in the efforts of some groups to dilute legislation that has been developed to assist women and racial–ethnic minorities. These efforts can be seen as an attempt to preserve group power.

Another example of overt discrimination is the harassment that women experience at work, including *sexual harassment* and other means of intimidation. Sociologists see such behaviors as ways for men to protect their advantages in the labor force. No wonder that women who enter traditionally male-dominated professions suffer the most sexual harassment. The reverse seldom occurs for men employed in jobs historically filled by women. Although men can be victims of sexual harassment, this is rare. Sexual harassment is a mechanism for preserving men's advantage in the labor force—a device that also buttresses the belief that women are sexual objects for the pleasure of men.

Each of these explanations—human capital theory, dual labor market theory, gender segregation, and overt discrimination—contributes to an understanding of the continuing differences in pay between women and men. Wage inequality between men and women is clearly the result of multiple factors that together operate to place women at a systematic disadvantage in the workplace.

Debunking Society's Myths

Myth: African American men are less likely than African American women to think that sexism *and* racism are equally important.

Sociological Perspective: African American men and women are equally likely to see racism and sexism as linked together and to see the need to address both (Harnois 2010).

The Devaluation of Women's Work

Across the labor market, women tend to be concentrated in those jobs that are the most devalued, causing some to wonder if the fact that women hold these jobs leads to devaluation of the jobs. Why, for example, is pediatrics considered a less prestigious specialty than cardiology? Why are preschool and kindergarten teachers (98 percent of whom are women) paid less than airplane mechanics (98 percent of whom are men)? The association of preschool teaching with children and its identification as "women's work" lowers its prestige and economic value. Indeed, if measured by the wages attached to an occupation, child care is one of the least prestigious jobs in the nation—paying on average only $452 per week in 2016, which would come out to an income below the federal poverty line if you worked every week of the year. Male highway maintenance workers, roofers, and construction workers all make more (Bureau of Labor Statistics 2017a).

The representation of women in skilled blue-collar jobs has increased, but it is still a very small fraction (typically less than 2 percent) of those in skilled trades such as plumbers, electricians, and carpenters (U.S. Bureau of Labor Statistics 2017a). Likewise, very few men work in occupations historically considered to be women's work, such as nursing, elementary school teaching, and clerical work. Interestingly, men who work in occupations customarily thought of as women's work tend to be more upwardly mobile within these jobs than are women who enter fields traditionally reserved for men (Budig 2002; Williams 1992).

Gender segregation in the labor market is so prevalent that most jobs can easily be categorized as men's work or women's work. Occupational segregation reinforces the belief that there are significant differences between the sexes. Think of the characteristics of a soldier. Do you imagine someone who is compassionate, gentle, and demure? Similarly, imagine a secretary. Is this someone who is aggressive, independent, and stalwart? The association of each characteristic with a particular gender makes the occupation itself a gendered occupation.

For all women, perceptions of gender-appropriate behavior influence the likelihood of success within institutions. Even something as simple as wearing makeup is linked to women's success in professional jobs (Dellinger and Williams 1997). When men or women cross the boundaries established by occupational segregation, they are often considered to be gender deviants. Women who work in the skilled trades, for example, are routinely assumed by others to be lesbians, whether or not they are (Denissen and Saguy 2014). Men who are nurses may be stereotyped as effeminate or gay; women marines may be stereotyped as "butch." Social practices like these serve to reassert traditional gender identities, perhaps softening the challenge to traditionally male-dominated institutions that women's entry challenges (Williams 1995).

As a result, many men and women in nontraditional occupations feel pressured to assert gender-appropriate behavior. Men in jobs historically defined as women's work may feel compelled to emphasize their masculinity, or if they are gay, they may feel even more pressure to keep their sexual orientation secret. Such social disguises can make them seem unfriendly and distant, characteristics that can have a negative effect on performance evaluations. Lesbian, gay, and bisexual workers face the added stress of having to "manage" their identities at work. Whether or not they are comfortable disclosing their identities is related to a number of factors, including the organizational climate and policies related to LGBT employees (King et al. 2014).

Balancing Work and Family

As the participation of women in the labor force has increased, so have the demands of keeping up with work and home life. Research finds that young women and men now want a good balance between work and family life, but they also find that institutions are resistant to accommodating these ideals (Gerson 2010). Men are also more involved in housework and child care than has been true in the past, although the bulk of this work still falls to women—a phenomenon that has been labeled "the second shift" (Hochschild and Machung 1989).

The social speedup that comes from increased hours of employment for both men and women (but especially women), coupled with the demands of maintaining a household, are a source of considerable stress (Jacobs and Gerson 2004). Women continue to provide most of the labor that keeps households running—cleaning, cooking, running errands, driving children around, and managing household affairs. Although more men are engaged in housework and child care, a huge gender gap remains in the amount of such work women and men do. Women are also much more likely to be providing care, not just for children, but also for their older parents. The strains these demands produce have made the home seem more and more like work for many. A large number of women and men report that their days at both work and home are harried and that they find work to be the place where they find emotional gratification and social support. In this contest between home and work, simply finding time can be an enormous challenge (Hochschild 1997). It is not surprising then that women report stress about household finances and family responsibilities more than men (American Psychological Association 2008).

Theories of Gender

Why is there gender inequality? The answer to this question is important, not only because it makes us think about the experiences of women and men, but also because it guides attempts to address the persistence of gender injustice. The major theoretical frameworks in sociology provide some answers, but feminist scholars have developed new theories to address women's lives more directly (◆ Table 11-3).

Functionalist theory traditionally purported that men fill instrumental roles in society whereas women fill expressive roles. Feminist scholars criticize functionalism for interpreting gender as a fixed role and one that is functional for society. Although few contemporary functionalist theorists would argue that now, functionalism does emphasize gender socialization as the major impetus behind gender inequality. Choices women make because of their gender socialization can make them shy away from leadership roles in public institutions, thus also reproducing traditional gender arrangements (Sandberg 2013).

Conflict theorists, in contrast, see women as disadvantaged by power inequities between women and men that are built into the social structure. This includes economic inequality and women's disadvantages in political and social systems. Wage inequality, for example, is produced by men who hold power in social institutions. Men also benefit as a group from the services that women's labor provides. Conflict theorists have been much more attuned to interactions of race, class, and gender inequality because they see all forms of inequality as stemming from power that dominant groups in society have.

Conflict theory also interprets women's inequality as stemming from capitalism. Influenced by the work of Karl Marx, some feminist scholars argue that women are oppressed because they have historically constituted a cheap supply of labor. Women provide a reserve supply of labor and are pulled into the labor market when underpaid workers are needed. Women also do much of the work that is essential to life, but not paid—that is, housework and child care. Conflict theorists understand women's inequality as stemming from economic exploitation and the power that men hold in virtually all social institutions.

Feminist Theory and the Women's Movement

As you will recall, *symbolic interaction theory* focuses on the immediate realm of social interaction. Feminist sociologists have developed an adaptation of symbolic interaction to explain the social

Table 11-3 Theorizing Gender: Sociological and Feminist Perspectives

	Functionalism	Conflict Theory	Symbolic Interaction Theory	Doing Gender/ Queer Theory	Liberal Feminism	Radical Feminism	Multiracial Feminism
Gender Identity	Gender roles are learned through socialization.	Conflict theory focuses on social structures, not individual identities.	Identity is constructed through ongoing social interaction and "doing gender."	Identity is constructed as people "perform" expected gender displays.	Except for the differences created by gender socialization, women and men are really no different.	Gender identity is produced by the power of social institutions that are controlled by men.	Gender identity is manifested differently depending on other social factors, such as race, class, and age, among others.
Status of Women	Women's status stems from the social roles of women and men in the family.	Women's status stems from their position as a cheap (or free) source of labor and their relative lack of power compared to men.	Women's status stems from the enactment of gender in social interaction.	Women's status is produced through the constant reenactment of gendered expectations.	Liberal feminism advocates equal rights for women.	Men's power is the basis for women's subordination.	Women of color are caught in a matrix of domination based on the intersection of race, gender, and other social factors.
Status of Men	Men hold instrumental roles; women, expressive roles.	Men hold economic advantages in the labor market and hold power in social institutions.	Masculinity is a learned identity that is created and sustained through social interaction.	Men's status is a performance.	Men's status derives from traditions that are learned and can be changed.	Men use their power to maintain their advantage.	Men's status is not monolithic as there are significant differences in men's power depending on their location in the matrix of domination and the intersection of race and gender.
Social Change	Social change emerges when social institutions become dysfunctional.	Social change comes from transformation of economic institutions and change in power structures that advantage men.	Social change comes when men and women disrupt existing gender displays.	Performances can disrupt existing expectations that redefine gender.	Equal rights is the framework for change most likely to generate equality.	Only the elimination of patriarchy will produce women's liberation.	Social change must come from challenging race, gender, and other forms of oppression simultaneously and as intersecting social systems.

construction of gender. An approach known as **doing gender** interprets gender as something accomplished through the ongoing social interactions people have with one another (West and Fenstermaker 1995; West and Zimmerman 1987). Seen from this framework, people produce gender through the interaction they have with one another and through the interpretations they have of certain actions and appearances.

Doing gender sees gender not as a fixed attribute of people, but as constantly made up and reproduced through social interaction. When you act like a man or a woman, you are confirming gender and reproducing existing social order. People act in gender-appropriate ways, displaying gender to others. Others then interpret the gender display that they see and assign people to gender categories, thereby reinforcing the gender order. People can disrupt this taken-for-granted behavior though. From the view of doing gender, gender structures would change if large numbers of people behaved differently. This is one reason the theory has been criticized by those with a more macrosociological point of view. Critics argue that the doing gender perspective ignores the power differences and economic differences that exist based on gender, race, and class. In other words, it does not explain the structural basis of women's oppression (Collins et al. 1995).

Related to doing gender is queer theory, a theoretical perspective that has emerged from gay and lesbian studies. **Queer theory** challenges the idea that sex and gender are binary opposites—that is, either/or categories. Instead, queer theory sees that dichotomous sex and gender categories are enforced by the power of social institutions and those who control them. People who disrupt these categories, such as transgender people, show us how gender is determined through such enforcement practices (Jenness and Fenstermaker 2014; Westbrook and Schilt 2014). Briefly put, queer theory interprets society as forcing people into presumed gender and sexual identities and behaviors. We examine queer theory further in Chapter 12 on sexuality.

Doing gender and queer theory are forms of feminist theory. **Feminist theory** has emerged from the women's movement and refers to analyses that seek to understand the position of women in society for the explicit purpose of improving their position in it.

The feminist movement has fostered widespread changes in and has transformed how people understand women's and men's lives. Simply put, **feminism** refers to beliefs and action that seek a more just society for women. Feminism is not a single way of thinking, as feminists understand the position of women in society in different ways. Different forms of feminism that emerged in the women's movement also provide different theoretical ways of viewing women's status in society.

Liberal feminism emerged from a long tradition that began among British liberals in the nineteenth century. Liberal feminism emphasizes individual rights and equal opportunity as the basis for social justice and reform. From this perspective, inequality for women originates in past and present practices that pose barriers to women's advancement, such as laws that historically excluded women from certain areas of work. From a liberal feminist framework, *discrimination* (that is, the unequal treatment of women) is the major source of women's inequality. Removing discriminatory laws and outlawing discriminatory practices are primary ways to improve the status of women. Calls for equal rights for women are the hallmark of a liberal feminist perspective.

Liberal feminism has broad appeal because it is consistent with American values of equality before the law. More **radical feminists** think that liberal feminism is limited by assuming that social institutions are basically fair were women and men treated the same within them. Radical feminists argue that there cannot be justice for women as long as men hold power in social institutions. As one example, men control the laws that govern women's reproductive lives. In this sense, *patriarchy* (that is, the power of men) is the major source of women's oppression. Patriarchy is found at the institutional level where men have control and at the individual level where men's power is manifested in high rates of violence against women. Indeed, many radical feminists see violence against women as mechanisms that men use to assert their power in society. Radical feminists think that change cannot come about through the existing system because men control and dominate that system.

Multiracial feminism has also opened new avenues of theory for guiding the study of gender and its relationship to race and class (Andersen and Collins 2016; Baca Zinn and Dill 1996; Collins 1990). Multiracial feminism evolves from studies pointing out that earlier forms of feminist thinking excluded women of color from analysis, which made it impossible for feminists to deliver theories that informed people about the experiences of all women. Multiracial feminism examines the interactive influence of gender, race, and class, showing how they together shape the experiences of all women and men.

From this perspective, gender is not a singular or uniform experience, but rather intersects with race and class in shaping the experience of women and men. Gender is thus manifested differently, depending on the particular location of a given person or group in a system shaped by gender, race, and class, along with other social identities, such as sexual orientation, ability/disability status, age, nationality, and so forth. Also known as *intersectional theory*, analyses that are situated in multiracial feminist thinking have opened up sociological theory to new ways of thinking that include the multiplicity of experiences that people have in a society as diverse as the United States.

Gender in Global Perspective

Increasingly, the economic condition of women and men in the United States is linked to the fortunes of people in other parts of the world. The growth of a global economy and the availability of a cheaper industrial labor force outside the United States mean that U.S. workers have become part of an international division of labor. U.S.-based multinational corporations looking around the world for less expensive labor often turn to the developing nations and find that the cheapest laborers are women or children. The global division of labor is thus acquiring a gendered component, with women workers, usually from the poorest countries, providing a cheap supply of labor for manufacturing products that are distributed in the richer industrial nations.

Worldwide, women work as much as or more than men. It is difficult to find a single place in the world where the workplace is not segregated by gender. On a global scale, women also do most of the work associated with home, children, and the elderly. Although women's paid labor has been increasing, their unpaid labor in virtually every part of the world exceeds that of men. The United Nations estimates that women's unpaid work (both in the home and in the community) is valued as at least $11 trillion (**www.un.org**). Generally speaking, women's status in the world is improving, but slowly and unevenly in different nations (see ■ map 11-1).

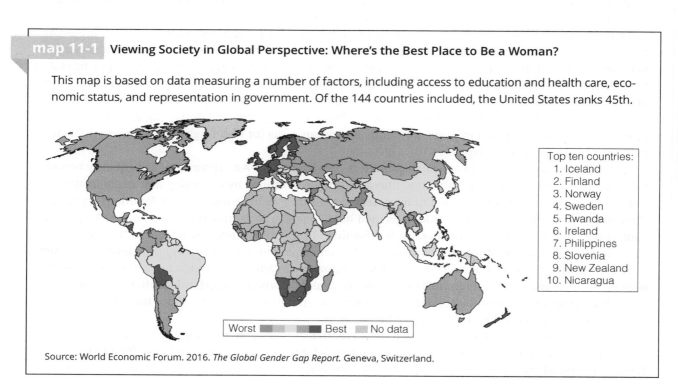

map 11-1   **Viewing Society in Global Perspective: Where's the Best Place to Be a Woman?**

This map is based on data measuring a number of factors, including access to education and health care, economic status, and representation in government. Of the 144 countries included, the United States ranks 45th.

Top ten countries:
1. Iceland
2. Finland
3. Norway
4. Sweden
5. Rwanda
6. Ireland
7. Philippines
8. Slovenia
9. New Zealand
10. Nicaragua

Worst ▬▬ Best ▬ No data

Source: World Economic Forum. 2016. *The Global Gender Gap Report.* Geneva, Switzerland.

Work is not the only measure by which the status of women throughout the world is inferior to that of men. Women are vastly underrepresented in national parliaments (or other forms of government) everywhere. As late as 2014, only 24 nations had women heads of state or government. Worldwide, women hold only 22 percent of all parliamentary seats. The United States, presumed to be the most democratic nation in the world, ranks 99th of 193 nations in the representation of women in national parliaments (Inter-parliamentary Union 2017).

The United Nations has also concluded that violence against women and girls is a global epidemic and one of the most pervasive violations of human rights (United Nations 2012a). Violence against women takes many forms, including rape, domestic violence, infanticide, incest, genital mutilation, and murder (including so-called honor killings, where a woman may be killed to uphold family honor if she has been raped or accused of adultery). Although violence is pervasive, some specific groups of women are more vulnerable than others—namely, minority groups, refugees, women with disabilities, elderly women, poor and migrant women, and women living in countries with armed conflict. Statistics on the extent of violence against women are hard to report with accuracy, both because of the secrecy that surrounds many forms of violence and because of differences in how different nations might report violence. Nonetheless, the United Nations estimates that between 20 and 50 percent of women worldwide have experienced violence from an intimate partner or family member.

As we saw in Chapter 9, many factors put women at risk of violence, including cultural norms, women's economic and social dependence on men, and political practices that either provide inadequate legal protection or provide explicit support for women's subordination.

Gender and Social Change

Few lives have not been touched by the transformations that have occurred in the wake of the feminist movement. The women's movement has opened work opportunities, generated laws that protect women's rights, spawned organizations that lobby for public policies on behalf of women, and changed public attitudes. Many young women and men now take for granted freedoms struggled for by earlier generations. These include access to birth control, equal opportunity legislation, laws protecting against sexual harassment, increased athletic opportunities for women, more presence of women in political life, and greater access to child care, to name a few changes. These impressive changes occurred in a relatively short period of time.

Indeed, many believe that the gender revolution is over and that there is no further need for feminist change. Some say that the nation is *postfeminist*, a term that means different things to different people. For some, it simply means that the women's movement is over because feminism has outlived the need. For others, it means that second-wave feminism (that of the 1970s and 1980s) does not meet the needs of new generations of women.

The growth of the contemporary #MeToo movement also shows that there remains a serious need to address the high rates of violence against women that continue to threaten women's safety and well-being. Sexual assault and sexual harassment are all too common (see Chapters 9 and 14) and reflect the ongoing connection between gender and power in that gender-based violence is explained as a reflection of men's power over women. As long as sexual assaults against women are dismissed as either women's fault or the expected behaviors of men, little will be done to reduce these high rates of violence.

Even with persistent reminders of the unequal status of women in society, it is also true that women have reached pinnacles of power and influence unprecedented in U.S. history. Women are highly visible as CEOs of major corporations, as Supreme Court justices, as presidential cabinet members, and as extraordinary and highly paid athletes—all of which would have been highly unusual not that many years ago. Women have also risen to positions of political influence, perhaps signaling a new era for women in the political realm, although many see the defeat of Hillary Clinton for president of the United States in 2016 as a major setback. Even with this rise, women have also been a part in the new conservative movement, and they have long played significant roles in neo-Nazi and white nationalist movements (Blee 2008).

No doubt there has been substantial progress for women, but the tensions between progressive and conservative politics on women's issues indicate that women's rights are hardly a settled issue. Despite the greater visibility of women, most women still struggle with low wages, managing both work and family, or perhaps struggling alone to support a family. On the conservative side, people feel that the value of women as traditional homemakers is being eroded, perhaps even threatened by women's independence because women in that position need men's economic support. On the progressive side, feminists perceive constant threats to women's reproductive rights and think that the nation has not come far enough in protecting and supporting women's rights. There is currently a complex mix of progressive and conservative politics surrounding gender in the contemporary world. Yet, the changes the feminist movement has inspired have completely transformed many dimensions of women's and men's lives, especially apparent in changed public attitudes.

Contemporary Attitudes

Surveys of public opinion about women's and men's lives are good indicators of the changes we have witnessed as the result of the feminist movement. Now, only a small minority of people disapproves of women being employed while they have young children, and both women and men say it is not fair for men to be the sole decision maker in the household.

Young women and men have different ideals for their future lives than was true for earlier generations. Kathleen Gerson's research finds that most young adults (those she calls "children of the gender revolution") want a lifelong partner and shared responsibilities for work and family. Yet, Gerson also finds a strong gender divide in how men and women imagine what they would do as a backup plan if their ideals were not realized. Men, much more than women, think that if balancing work and family does not work out in their future, they would fall back on traditional arrangements with wives staying home and husbands working. Women disagree, understanding that they have to be self-reliant in the event that their ideals are not met (Gerson 2010).

Gerson also found that, although both men and women want to share work and family life, institutions have not adjusted to this reality. In other words, resistant institutions have not adjusted to the attitudinal changes and desires for new lifestyles that most women and men embrace.

Attitudinal changes are not, however, complete. Many people want more flexible gender arrangements, but traditional gender norms also remain. Beliefs that there are basic differences between women and men and support for traditional gender norms in many aspects of personal life still prevail. In short, change in gender has been uneven—revolutionary in some regards and stagnant in others (Prokos 2011; England 2010).

Legislative Change

Attitudes are only one dimension of social change. Some of the most important changes have come from laws that now protect against discrimination—laws that have opened new doors, especially for

Thinking Sociologically

Is your pet gendered? People typically think of the *social construction of gender* in the context of human relations. What about pets? Almost two-thirds of all U.S. households have at least one pet. A recent study of dog owners found that people also gender their dogs, selecting dogs that reflect the owner's gender identity and describing them in gendered terms. Female pets have names with more syllables and that are more likely to have a diminutive ending. You can find lots of advice how to appropriately gender your pet on the Internet. Try to observe this yourself. Apparently, gender norms pervade human relationships with animals as well as with each other!

Sources: Ramirez, Michael. 2006. "'My Dog's Just Like Me': Dog Ownership as a Gender Display." *Symbolic Interaction* 29 (3): 373–391; Abel, Ernest L., and Michael L. Kruger. 2007. "Gender Related Naming Practices: Similarities and Differences between People and Their Dogs." *Sex Roles: A Journal of Research* 57 (1–2): 15–19.

professional, well-educated women. The *Equal Pay Act of 1963* was one of the first pieces of legislation requiring equal pay for equal work. Following this was the **Civil Rights Act of 1964**, adopted as the result of political pressure from the civil rights movement. You will be interested to learn that adding "sex" to this law actually protects women by accident. A southern congressman added the term, thinking that the idea of giving women these rights was such a joke that the addition of "sex" would prevent passage of the civil rights law.

The Civil Rights Act bans discrimination in hiring, promoting, and firing. This law also created the Equal Employment Opportunity Commission, an arm of the federal government that enforces laws prohibiting discrimination on the basis of race, color, national origin, religion, or sex.

The passage of the Civil Rights Act opened new opportunities to women in employment and education. This was further supported by **Title IX**, adopted as part of the educational amendments of 1972. Title IX forbids gender discrimination in any educational institution receiving federal funds.

Most people think of Title IX as only protecting women from discrimination in athletics, but Title IX also defines various forms of sexual violence (sexual harassment, sexual assault, and other forms of gender-based violence) as a violation of law. Title IX requires schools and colleges to be free of gender discrimination. The bill also requires that schools and colleges have established procedures for handling complaints of sex discrimination, sexual harassment, and sexual violence. With one in four college women reporting sexual assault while in college, this is clearly a huge problem. Title IX, however, also protects men and gender-nonconforming students, as well as faculty and staff on campus.

Adoption of Title IX, although not perfect in its implementation, has radically altered the opportunities available to female students and has laid the foundation for many coeducational programs that are now an ordinary part of college life. This law has been particularly effective in opening up athletics to women.

Has equality been achieved as the result of Title IX and other laws? In college sports, men still outnumber women athletes by more than two to one, and there is still more scholarship support for male athletes than for women. Title IX allows institutions to spend more money on male athletes if they outnumber women athletes, but the law stipulates that the number of male and female athletes should be closely proportional to their student body representation. Studies of student athletes show that although support for women's athletics has improved since implementation of Title IX, there is still a long way to go toward equality in women's sports. Some argue that Title IX has reduced opportunities for men in sports. Proponents of maintaining strong enforcement of Title IX counter this by noting that budget reductions in higher education, not Title IX per se, are responsible for any reduction in athletic opportunities for men. Furthermore, they point out that men still greatly predominate in school sports.

In the workplace, a strong legal framework for gender equity is in place, yet equity has not been achieved. The United States has never, as an example, approved the **Equal Rights Amendment**, which would provide a constitutional principle that "equality of rights under the law shall not be denied or abridged by the United States or by any state on the basis of sex" **(www.equalrightsamendment.org)**.

It is clear that the passage of antidiscrimination policies is essential, but not a guarantee that discrimination will end. One solution to the problem of gender inequality is to have more women in positions of public power. Is increasing the representation of women in existing situations enough? Without reforming sexism in institutions, change will be limited and may generate benefits only for groups who are already privileged. Feminists advocate restructuring social institutions to meet the needs of all groups, not just those who already have enough power and privilege to make social institutions work for them. The successes of the women's movement demonstrate that change is possible, but change comes only when people are vigilant about their needs.

Feminism has become a global movement as women around the world have organized for their rights.

Chapter Summary

How do sociologists distinguish sex and gender?

Sociologists use the term *sex* to refer to biological identity and *gender* to refer to the socially learned expectations associated with members of each sex. *Biological determinism* refers to explanations that attribute complex social phenomena entirely to physical or natural characteristics.

How is gender identity learned?

Gender socialization is the process by which gender expectations are learned. One result of socialization is the formation of *gender identity*. Overly conforming to gender roles has a number of negative consequences for both women and men, including eating disorders, violence, and poor self-concepts. *Homophobia* plays a role in gender socialization because it encourages strict conformity to gender expectations.

What is a gendered institution?

Gendered institutions are those where the entire institution is patterned by gender. Sociologists analyze gender both as a learned attribute and as an institutional structure.

What is gender stratification?

Gender stratification refers to the hierarchical distribution of social and economic resources according to gender. Most societies have some form of gender stratification, although they differ in the degree and kind. Gender stratification in the United States is obvious in the differences between men's and women's wages.

How do sociologists explain the continuing earnings gap between men and women?

There are multiple ways to explain the pay gap. *Human capital theory* explains wage differences as the result of individual differences between workers. *Dual labor market theory* refers to the tendency for the labor market to be organized in two sectors: the primary and secondary markets. *Gender segregation* persists and results in differential pay and value attached to men's and women's work. *Overt discrimination* against women is another way that men protect their privilege in the labor market.

Are men increasing their efforts in housework and child care?

Many men are now more engaged in housework and child care than was true in the past, although women still provide the vast majority of this labor. Balancing work and family has resulted in social speedup, making time a scarce resource for many women and men.

What is feminist theory?

Different theoretical perspectives help explain the status of women in society. *Functionalist theory* emphasizes how gender roles that differentiate women and men work to the benefit of society. *Conflict theory* interprets gender inequality as stemming from women's status as a supply of cheap labor and men's greater power in social institutions. *Feminist theory*, originating in the women's movement, refers to analyses that seek to understand the position of women in society for the explicit purpose of improving their position in it. *Doing gender* and *queer theory* interpret gender as accomplished in social interaction and enforced through powerful institutions. *Liberal feminism* is anchored in an equal rights framework. *Radical feminism* sees men's power as the primary force that locates women in disadvantaged positions in society. *Multiracial feminism*, or intersectional theory, emphasizes the linkage between gender, race, and class inequality.

When seen in global perspective, what can be observed about gender?

The economic condition of women and men in the United States is increasingly linked to the status of people in other parts of the world. Women provide much of the cheap labor for products made around the world. Worldwide, women work as much or more than men, though they own little of the world's property and are underrepresented in positions of world leadership.

What are the major social changes that have affected women and men in recent years?

Public attitudes about gender relations have changed dramatically in recent years. Women and men are now more egalitarian in their attitudes, although women still perceive high degrees of discrimination in the labor force. A legal framework is in place to protect against discrimination, but legal reform is not enough to create gender equality.

Key Terms

biological determinism 272
Civil Rights Act of 1964 295
discrimination 287
doing gender 291
dual labor market theory 285
Equal Pay Act of 1963 284
Equal Rights Amendment 295
feminism 291
feminist theory 291
gender 271

gender apartheid 282
gender identity 274
gender segregation 286
gender socialization 273
gender stratification 281
gendered institution 280
homophobia 278
human capital theory 285
intersexed 271
labor force participation rate 284

liberal feminism 291
Lily Ledbetter Fair Pay Act 284
matriarchy 283
multiracial feminism 291
patriarchy 283
queer theory 291
radical feminism 291
sex 271
Title IX 295
transgender 274

SEXUALITY

In this chapter, you will learn to:

Understand the social basis of human sexuality

Identify current attitudes and behaviors involving sexuality

Comprehend how sexuality is linked to other forms of inequality

Compare and contrast theoretical perspectives on sexuality

Be able to define homophobia and heterosexism and their influence on lesbian and gay experience

Comprehend sociological underpinnings of current issues regarding sexuality

Assess how sexuality is being affected by social change

A visitor from another planet might conclude that people in the United States are obsessed with sex. Young people watch videos where women gyrate in sexual movements. A stroll through a shopping mall reveals expensive shops selling delicate, skimpy women's lingerie. Popular magazines are filled with images of women in seductive poses trying to sell every product imaginable. Even bumper stickers brag about sexual accomplishments. People dream about sex, form relationships based on sex, fight about sex, and spend money to have sex. On the one hand, the United States appears to be a very sexually open society; however, sexual oppression still exists. Gay men, lesbians, and bisexuals are viewed with prejudice and are discriminated against—that is, treated like minority groups (see Chapter 10).

Sexuality, usually thought to be a most private matter, has taken on a public life by being at the center of some of our most heated public controversies. Should young people be educated about birth control or only encouraged to abstain from sex? Should government require employers to provide insurance coverage for birth control? Who decides whether a woman can choose to have an abortion and under what circumstances? Should someone because of their religious beliefs be allowed to discriminate against someone based on their sexual orientation?

Sexuality and the concerns it spins off clearly polarize the public on a range of social issues. Sexuality is seen as a private matter, but it is also very much on the public agenda. Studying sexuality reveals how deeply it is entrenched in social norms, values, and social inequalities. Human sexuality, like other forms of social behavior, is shaped by society and culture.

Sex and Culture

Sexual behavior would seem to be utterly natural. Pleasure and sometimes the desire to reproduce are reasons people have sex, but sexual relationships and identities develop within a social context. Social context establishes what sexual relationships mean, how we define our sexual identities, and what social supports are given (or denied) to people based on their sexual identity. Sexuality is socially defined and patterned.

Sex: Is It Natural?

From a sociological point of view, little in human behavior is purely natural, as we have learned in previous chapters. Behavior that appears to be natural is the behavior accepted by cultural customs and sanctioned by social institutions. People engage in sex not just because it feels good, but also because it is an important part of our social identity. Sexuality creates intimacy between people. Void of a cultural context and the social meanings given to sexual behavior, people might not attribute the emotional commitments, spiritual meanings, and social significance to sexuality that it has in different human cultures.

Sexual orientation refers to the attraction that people feel for people of the same or different sex. The term *sexual orientation* implies something deeply rooted in a person. **Sexual identity** is the definition of oneself that is formed around one's sexual relationships. Sexual identity is learned in the context of our social relationships and the social structures in which we live. Because sexual identity emerges in a social context, *sexual identity, like other social identities, is a social construction.* How sexual identities are understood and defined thus can change over time—witnessed by the fact that the label LGBTQ is a relatively new concept referring to lesbian, gay, bisexual, transgender, queer (or, sometimes, questioning) people. The term did not emerge until the late 1980s as a way to recognize diversity among LGBTQ people.

Debates about what to call people of different sexual orientations show how fundamentally social sexual identity is. The language used to describe different sexual identities has changed a lot over time—and continues today. Not that many years ago, the term "homosexual" was the most widely used nomenclature, still used by many. Advocates for LGBTQ rights argued that this term had a clinical sound, thus implying that being gay or lesbian was a medical condition, an abnormality, or something that needed to be "cured." (Note the discussion of the medicalization of deviance in Chapter 7.) Only with political organizing was the term dropped from the diagnostic manual of the American Psychological Association (in 1973) and, today, many find the term offensive when used to describe LGBTQ people. You might say this is just "political," and it is, but it shows how the social context of language—indeed, the social context of sexual identity—emerges within a social and political environment.

Although the term sexual orientation and sexual identity are sometimes used interchangeably, they are not the same thing, nor is one's sexual identity simply based on one's sexual practices. For example, a man may have sex with other men—perhaps even on a regular basis, but not have a sexual identity as being gay. A person may have a sexual identity as heterosexual even in the absence of actual sexual relationships.

Sexual identity as gay, lesbian, heterosexual, or bisexual emerges in a social context, as we will see in the following discussion on the social construction of sexual identity. As an example, recall from the previous chapter that *transgender* people are those who construct a gender identity different from their biological identity. Being transgender does not, however, predict a particular sexual orientation (Schilt 2011). A transgender person may be straight, lesbian, gay, or bisexual. The point here is that sexual identity is not necessarily simply based on biological or "natural" states, even though there is debate about whether there is a biological basis to sexual orientation.

Is there a biological basis to sexual identity? This question is debated in both popular and scientific literature. Gay, lesbian, transgender, and bisexual people often say that they do not choose their sexual orientation and that it is something they just "are," as if it were a biological imperative. The debate about a biological basis to sexual orientation—heated at times—comes from rejecting the idea that being gay is a choice, as if people could change their sexual orientation at will. There are political reasons for rejecting the idea of sexual orientation as a choice because, if it is something inherent in people, then perhaps others might be more accepting of gay, lesbian, and bisexual people.

Perhaps there is some biological basis to sexual orientation, but the evidence is not yet there. Further, even if a biological influence exists, social experiences are far more significant in shaping sexual identity. The social dimensions of sexual identity are, however, rarely reported in the media and never with as much acclaim as claims about a biological bases to human sexuality. Whatever the origins of sexual orientation, social influences are no doubt a very significant part of all people's sexual identity.

Periodically, there is a public claim that scientists have discovered a so-called gay gene. Interestingly, there are never claims about a so-called heterosexual gene because the implicit assumption seems to be that heterosexuality is the natural state, and gay or lesbian behavior is somewhat a mutant form. Despite the frequency of such claims, scientific experts are skeptical, usually pointing to very small samples, lack of control groups, or other weak standards of evidence (Tanner 2014).

Even if there is some yet undiscovered basis for sexual orientation, there is extensive evidence that social and cultural environments play a huge part in creating sexual identities. What interests sociologists is how sexual identity is constructed through social relationships and in the context of social institutions.

The Social Basis of Sexuality

We can see the social and cultural basis of sexuality in numerous ways:

1. *Human sexual attitudes and behavior vary in different cultural contexts.* If sex were purely natural behavior, sexual behavior would also be uniform among all societies, but it is not. Sexual behaviors considered normal in one society might be seen as peculiar in another. Think about this: In some cultures, women do not believe that orgasm exists, even though it does biologically. In the eighteenth century, European and American writers advised men that masturbation robbed them of their physical powers and that instead they should apply their minds to the study of business. These cultural dictates encouraged men to conserve semen on the presumption that its release would lessen men's intelligence or cause insanity (Rutter and Schwartz 2011; Freedman and D'Emilio 1988).

2. *Sexual attitudes and behavior change over time.* Fluctuations in sexual attitudes are easy to document. For example, in 1968, 68 percent of the American public thought premarital sex was morally wrong, compared to 20 percent who think so now (Riffkin 2014; Smith and Son 2013). Teens, as well, have changed their attitudes about sex. In 1977, one-third of teens thought it was morally wrong to have sex outside of marriage; more recently, 60 percent of young women and 71 percent of young men (age 15 to 24) think it is okay for 18-year-olds to have sex if they have strong affection for each other. Only a small percentage think you should abstain from sex until marriage, suggesting that teen tolerance of casual sex has increased (Daugherty and Copen 2016).

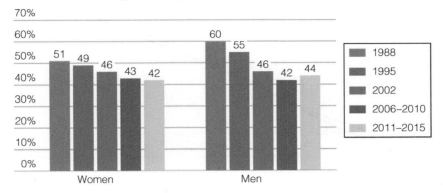

Sexual Activity Among Never Married Women and Men, Aged 15–19 (1988–2015)

▲ Figure 12-1   **Sex among Teenagers: A Change over Time.** This chart shows the change in the percentage of unmarried teen men and women (aged 15 to 19) who have ever had sex. What do you observe in these data? What sociological factors explain what you observe? Were you to develop a research project to explore this subject, how might you design it?

Source: National Vital and Health Statistics. 2017. *Sexual Activity and Contraceptive Use among Teenagers in the United States, 2011–2015*. National Survey of Family Growth. Washington, DC: National Center for Health Statistics, U.S. Department of Health and Human Services. **www.cdc.gov/nchs/data/nhsr/nhsr104.pdf**

Of course, attitudes do not necessarily predict behavior. Despite what teens say, by age 18, 58 percent of men and 56 percent of women have actually had sex; the percentage of teens having sex has also declined since the late 1980s (see ▲ Figures 12-1 and 12-2; Martinez and Amba 2015). Also, although a vast majority of Americans (84 percent) think extramarital affairs are immoral, (Poushter 2014; Wolfinger 2017), the public also seems endlessly fascinated by the affairs and sexual escapades of celebrities and high-profile people. At the same time, attitudes toward being unfaithful to one's spouse have become less tolerant. People's reports of actually being unfaithful to their spouse have wavered over time—with about 25 percent of men and 15 percent of women now admitting they have had marital affairs (Carr 2010).

3. *Sexual identity is learned.* Like other forms of social identity, sexual identity is acquired through socialization and ongoing relationships. Information about sexuality is transmitted culturally and becomes the basis for what we know about ourselves and others. Where did you first learn about sex? What did you learn? For some, parents are the source of information about sex and sexual behavior. For many, peers have the strongest influence on sexual attitudes. Long before young people become sexually active, they learn **sexual scripts** that teach us what is appropriate sexual behavior for each gender (Rutter and Schwartz 2011).

   Children learn sexual scripts by playing roles—playing doctor as a way of exploring their bodies or hugging and kissing in a way that can mimic heterosexual relationships. The roles learned in youth profoundly influence our sexual attitudes and behavior throughout life.

4. *Social institutions channel and direct human sexuality.* Social institutions, such as religion, education, or the family, define some forms of sexual expression as more legitimate than others. Debates about same-sex marriage illustrate this concept. Without being able to marry, gay and lesbian couples lose some of the privileges that married couples receive, such as employee benefits and the option to file joint tax returns. Social institutions, such as the media, also influence sexuality through the production and distribution of images that define cultural meanings attributed to sexuality. Even children's G-rated films have been shown to contain strong and explicit messages that enforce heterosexual romance (Martin and Kazyak 2009).

5. *Sex is influenced by economic forces in society.* Sex sells. In the U.S. capitalist economy, sex appeal is used to hawk everything from cars and personal care products to stocks and bonds. In this sense,

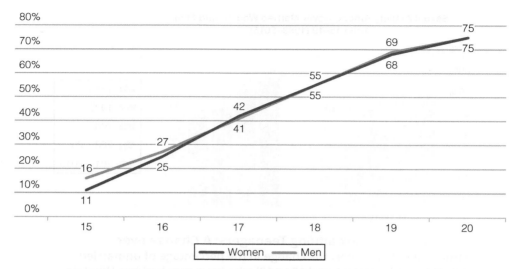

▲ Figure 12-2   **Teen Sex: When Does It Start?** This chart shows the probability of teen men and women having had sex by a certain age (the figures are percentages of those who have had sex). What facts stand out to you, and how would you explain what you see in sociological terms?

Source: National Vital and Health Statistics. 2017. *Sexual Activity and Contraceptive Use among Teenagers in the United States, 2011–2015.* National Survey of Family Growth. Washington, DC: National Center for Health Statistics, U.S. Department of Health and Human Services. **www.cdc.gov/nchs/data/nhsr/nhsr104.pdf**

sex has become a commodity—something bought and sold in the marketplace of society. Sex also can be big business. By one estimate, Americans spend over $10 billion per year in the sex indus-try, including strip bars, peep shows, phone sex, sex acts, sex magazines, and pornography rentals—and this is likely to be an underestimate (Barton 2006). The business of sex means some people may actually be "bought and sold." Think of women who may feel that selling sexual services is their best option for earning a living wage. Sex workers are among some of the most exploited and misunderstood workers. Global sex trafficking is also more extensive than commonly thought.

6. *Public policies regulate sexual and reproductive behaviors.* In many ways, the government and other social policies intervene in people's sexual and reproductive decision making. The existing prohibition on using federal funds to pay for abortion eliminates reproductive choices for women who are depen-dent on state or federal aid. Government decisions about which reproductive technologies to endorse influence what birth control is available to men and to women. Government funding, or lack thereof, for sex education can influence how people understand sexual behavior. These facts challenge the idea that sexuality is a private matter, showing how social institutions can direct sexual behavior.

To summarize, human sexual behavior occurs within a cultural and social context. That context defines certain sexual behaviors as appropriate or inappropriate.

See for Yourself

Children's Media and Sexuality

Carefully watch two movies that are marketed to a specific population of young people (children, teens, or young adults), and list any comments or behaviors you see that suggest *sexual scripts* for men and women. Note the assumptions in this script about heterosexual and homosexual behavior. What scripts did you see, and who is the intended audience? What do your observations suggest about how sexual scripts are promoted, overtly or not, through popular culture?

Contemporary Sexual Attitudes and Behavior

Sexual attitudes and behaviors in the United States are a mix of ideas and practices, both of which vary depending on the social factors that shape people's experiences. The growth of a conservative movement has, for example, shaped the sexual values of American society. Differences around sexual values have also been at the heart of highly contested issues shaping U.S. politics in recent years, such as debates about same-sex marriage, sex education in the schools, and policies about reproductive and contraceptive health. This makes sociological research on sexuality all the more fascinating as public attitudes and behaviors shift with changes in society.

Changing Sexual Values

Public opinion is now a mix of both liberal and conservative values about sexuality. Only 20 percent of Americans now think that premarital sex is always wrong, compared to 34 percent in 1972 (Smith and Son 2013). Americans are also more conservative on this matter than people in western Europe. In Germany and France, as examples, a large majority of people (in Germany, 86 percent; France, 88 percent) say that they find premarital sexual morally acceptable (Rheault and Mogahed 2008).

On gay rights, more Americans than ever before think that "gay/lesbian relations are morally acceptable" (63 percent in 2017, compared to 40 percent in 2001). Eighty-nine percent think that "homosexuals should have equal rights in terms of job opportunities." Change in the public's opinion about same-sex marriage has been remarkable and has happened over a quite short period of time. As recently as 1996, 68 percent of Americans opposed legal recognition of same-sex marriage; now 34 percent oppose such rights, a change in attitude that began even before the historic Supreme Court decision that now requires states to license marriages between same-sex couples. By the time of the Court's decision (in 2015), 60 percent of the public supported same-sex marriage; 37 percent were opposed (Gallup 2017). The historic Court decision to recognize same-sex marriage (*Obergefell v. Hodges*) also means that same-sex marriages made in one state must be recognized in any other state. This gives same-sex couples the same benefits of marriage enjoyed by heterosexual couples. Even with the majority of the public supporting this decision, resistance to same-sex marriage remains fierce among some. Only time will tell how attitudes will evolve as same-sex marriage becomes more familiar.

Attitudes about sex vary significantly depending on various social characteristics. For example, men are more likely than women to think that gay/lesbian relations are morally wrong. Sexual attitudes are also shaped by age. Younger people are more likely than older people to think that gay/lesbian relations are morally acceptable. These differences likely reflect not only the influence of age, but also historical influences on different generations. Religion also matters. Those who attend church weekly are far less likely to support gay rights compared to those who worship less often (Pelham and Crabtree 2009). Public opinion on matters about sexuality taps underlying value systems, thus generating public conflict. In general, sexual liberalism is associated with greater education, youth, urban lifestyle, and political liberalism on other social issues.

Lazyllama/Alamy Stock Photo

Gays, lesbians, and their allies have mobilized for social change, fostering pride and celebration as well as a reduction over time in homophobic attitudes.

Sexual Practices of the U.S. Public

Sexual practices are difficult to document. What we know about sexual behavior is typically drawn from surveys. Most of these surveys ask about sexual attitudes, not actual behavior. What people say they do may differ significantly from what they actually do.

As much as sex is in the news, national surveys of sexual practices are rare. Those that have been conducted tell us the following:

- Contrary to public opinion, teens are waiting longer to have sex than was true in the past, though seven in ten teens have had sex by age 19 (Guttmacher Institute 2017).
- Having only one sex partner in one's lifetime is rare (Laumann et al. 1994).
- A significant number of people have extramarital sex; estimates are that 20 to 40 percent of all marriages experience at least one instance of infidelity (Marín, Christensen, and Atkins 2014).
- A small percentage of the public say they personally identify as gay, lesbian, or bisexual (4 percent), but among millennials (those born between 1980 and the beginning of the new century; see chapter 16), about 20 percent now say they are LGBTQ—a fact that is interpreted as this generation rejecting binary categories of sexual identity (Gonella 2017). Over three-quarters of the public also now say they have a family member, coworker, or personal acquaintance who they know is gay (Gallup 2017).

Sex and Inequality: Gender, Race, and Class

When you take a sociological view of sexuality, you see how sexuality is linked to other social identities and social systems. Start with gender. Our gender identities link with sexuality in many ways. Indeed, men learn to be men by dissociating themselves from anything that seems "gay" or "sissy" (Pascoe 2011). Being called a "fag" or a "sissy" is one peer sanction that socializes children to conform to particular gender roles. Boys are raised to be manly by repressing any so-called feminine characteristics. Similarly, verbal attacks on lesbians by using the term *butch* are a mechanism of social control because ridicule encourages social conformity to the presumed "normal" gender roles (Pascoe 2011). Homophobia produces many misunderstandings about LGBTQ people, such as that gays have a desire to seduce straight people. There is little evidence that this is true.

Gender also shapes the so-called *double standard* for men and women. The double standard is the idea that different standards for sexual behavior apply to men and women. The double standard has weakened somewhat over time, particularly in the context of "hooking up" among young people. *Hooking up* is the term used to describe casual sexual relations, ranging from kissing to sexual intercourse, that occur without any particular commitment. The double standard, though, is still very much present. Men are more often encouraged to have casual sex than women, who pay a higher price for engaging in casual sex than do men. A national survey of American youth finds that men who report higher numbers of sexual partners are more popular than men with fewer partners. The opposite is true for women—that is, women with high numbers of sexual partners are less popular (Rudman et al. 2017; Kreager and Staff 2009).

Although the popular image of hooking up is that it is totally free and without constraint, this behavior in fact has very gendered norms. Women who hook up too frequently or with too many different partners are likely to be judged as "slutty." Men who do the same things are not judged in the same way. They may be seen as "players," but they are not subjected to the same shame or attribution of guilt that is targeted at women (England and Bearak 2014).

The sexual double standard for men and women has many consequences, both for people's identities and reputation, and in people's reactions to sexual violence. An extensive body of research shows that the stereotype of women as sexual temptresses is highly correlated with the acceptance of rape myths—such as that women's provocative dress encourages rape. The sexual script that men cannot stop once they become sexually aroused is also related to whether people believe rape myths. This has serious consequences because

Sex and Popular Culture

Imagine that some of the classical sociological theorists were reincarnated and observed sexuality and references to it in everyday current life. What observations and comments might they make?

Emile Durkheim would remind us that marking some behaviors as deviant is how people in society also define what is considered "normal." He might observe young boys calling each other "faggots." Interestingly, were he watching carefully, he would also see that the person being targeted by such comments, including by adults, is probably *not* gay. But the homophobic banter among boys and men is a way of asserting the dominant (and somewhat narrow) norms of what masculinity is presumed to be.

Max Weber would see something else. He would notice that sexuality, particularly heterosexuality, is rampant in popular culture. Clothing styles, popular lingo, and styles of dance are all marked by overt displays of sexuality. He would argue that such cultural displays go hand in hand with an economy that treats sexuality as part of the marketplace. Moreover, the interplay between culture and the economic marketplace leads to social judgments about some sexual styles and identities being more highly valued than others. In other words, Weber would emphasize the multidimensional connection between the economy, culture, and social judgments.

Karl Marx, on the other hand, would be intrigued by the commercial exploitation of sex, or "sexploitation." Sexuality has become a commodity in modern society. It is used to sell things for the benefit of those who own the various industries where fashion, personal care products, and even sex itself profit some while exploiting others. Marx might even note, were he paying attention to the treatment of women, that sex workers are among the most exploited workers, actually selling their bodies for the benefit of others.

W. E. B. DuBois would add that sexual exploitation is particularly harsh for Black people. The popular practice of college students partying as "pimps and hos" rests on a stereotype of Black men as sexual predators and Black women as promiscuous. These stereotypes have a long history stemming from racism. Black women are among the most sexually exploited. Their bodies are abused, they are portrayed in sexualized stereotypes in popular culture, and they are most victimized by sexual violence.

Together, these classical theorists provide sociological perspectives on images and practices common in everyday life, but rarely challenged or questioned, unless someone is on the receiving end of narrow, sexist, and racist understandings of sexuality.

women who are raped and believe such myths themselves are far less likely to report rape and may believe they somehow encouraged the rape (Deming et al. 2013; Davies et al. 2012; Ryan 2011).

Sexuality is also integrally tied to race and class inequality in society—a fact that is apparent in sexual stereotypes associated with race and class. Latinas are stereotyped as either "hot" or "virgins"; Latino men are stereotyped as "hot lovers." African American men are stereotyped as overly virile; Asian American women, as compliant and submissive, but passionate. Class relations also produce sexual stereotypes of women and men. Working-class and poor men may be stereotyped as dangerous, whereas working-class women may be disproportionately labeled "sluts."

Class, race, and gender hierarchies historically depict people of color and certain women as sexually promiscuous and uncontrollable (Nagel 2003). During slavery, for example, the sexual abuse of African American women was one way that slave owners expressed their ownership of African American people. Slaveowners thought they had rights to women slaves' sexuality. Under slavery, racist and sexist images of Black men and women were developed to justify the system of slavery. Black men were stereotyped as lustful beasts whose sexuality had to be controlled by the "superior" Whites. Black women were also depicted as sexual animals who were openly available to White men.

A Black man falsely accused of having had sex with a White woman could be murdered (that is, lynched) without penalty to his killers (Genovese 1972; Jordan 1968). Sexual abuse was also part of the

White conquest of American Indians. Historical accounts show that the rape of Indian women by White conquerors was common (Freedman and D'Emilio 1988; Tuan 1984). These patterns also can be seen in the extensive rape of women that often accompanies war and military conquest. This has been witnessed in various places—Darfur, the Democratic Republic of the Congo, Sudan, and other places where war rages. In the aftermath of war, victors may see raping women as their claim to power, making it clear that horrific acts of sexual violence against women—and also children—are acts of domination.

Poor women and women of color are the groups most vulnerable to sexual violence and exploitation. Becoming a prostitute or otherwise working in the sex industry (as a topless dancer, striptease artist, pornographic actress, or other sex-based occupation) are often last resorts for women with limited options to support themselves. Women who sell sex also are condemned for their behavior more so than their male clients—further illustration of how gender stereotypes mix with race and class exploitation. Why, for example, are women, and not their male clients, arrested for prostitution? Although data from the Uniform Crime Reports do not report the arrest rates of customers, prostitutes claim that only about 10 percent of those arrested for prostitution are customers. They also say that women of color are more likely to be arrested for prostitution than are White women, even though they are a smaller percentage of all prostitutes (Prostitutes Education Network 2009). Although these are not scientific data, they suggest the role that gender and race play in how laws against prostitution are enforced.

How do sociologists frame these complex connections between sexual identities, sexual stereotypes, and the various forms of social inequality that permeate society? For answers, we turn to sociological theory.

Sexuality: Sociological and Feminist Theory

How are sexual identities formed? Does the government have a role in regulating sexual relationships? What role does power have in sexual relationships? Where does the right to privacy in sexual relations begin and end? Although these and other questions are debated as moral issues in society, sociologists see them as subjects for sociological study. Sociological theory puts an analytical framework around the study of sexuality, examining its connection to social institutions and current social issues. How do the major sociological theories frame an understanding of sexuality?

Sex: Functional or Conflict-Based?

The three major sociological frameworks—functionalist theory, conflict theory, and symbolic interaction—take divergent paths in interpreting the social basis of human sexuality (See ◆ Table 12-1). Feminist theory and gay/lesbian/transgender studies have also very much transformed sociological understanding of sexuality and its connection to society.

Functionalist theory, with its emphasis on the interrelatedness of different parts of society, tends to depict sexuality in terms of its contribution to the stability of social institutions. Norms that restrict sex to marriage encourage the formation of families. Similarly, beliefs that give legitimacy to heterosexual behavior but not homosexual behavior maintain a particular form of social organization in which gender roles are easily differentiated and the nuclear family is defined as normative. From this point of view, regulating sexual behavior is functional for society because it prevents the instability and conflict that more liberal sexual attitudes supposedly generate. You can see that functionalist theory relates to conservative views about sexuality, but remember that theory is an analytic, not a moral, approach to understanding social issues.

Conflict theorists see sexuality as part of the power relations and economic inequality in society. *Power* is the ability of one person or group to influence the behavior of another, including the power that some sexual groups have over others and power within sexual relationships. Conflict theorists argue that sexual relations are linked to other forms of stratification, namely, race, class, and gender inequality. According to this perspective, sexual violence (such as rape or sexual harassment) is the result of power imbalances, specifically between women and men.

Table 12-1	Theoretical Perspectives on Sexuality			
Interprets	Functionalism	Conflict Theory	Symbolic Interaction	Feminist Theory
Sexual norms	As functional for society because they produce stability in institutions such as the family	As often contested by those who are subordinated by dominant and powerful sexual groups	As emerging and reinforced through social interaction	As established in a system of male domination, producing narrow definitions of women's and men's sexuality
Sexual identity	As learned in the family and other social institutions, with deviant sexual identities contributing to social disorder	As regulated by individuals and institutions that enforce only some forms of sexual behavior as desirable, thus enforcing heterosexism	As socially constructed when people learn the sexual scripts produced in society	As acknowledging that multiple forms of sexual identity are possible with some people crossing the ordinarily assumed boundaries
Sex and social change	As regulating sexual values and with norms being important for maintaining traditional and social stability, and too much change resulting in social disorganization	As coming through the activism of people who challenge dominant belief systems and practices	As evolving as people construct new beliefs and practices over time	As sexual values being changed through disrupting taken-for-granted categories of the dominant culture

At the same time, conflict theorists see economic inequality as a major basis for social conflict. Conflict theory, for example, examines how the global sex trade is linked to poverty, the status of women in society, and the economics of international development and tourism (Altman 2001; Enloe 2001). In connecting sexuality and inequality, conflict theorists use a structural analysis of sexuality.

Because both functionalism and conflict theory are macrosociological theories (that is, they take a broad view of society, seeing sexuality in terms of the overall social organization of society), they do not tell us much about the social construction of sexual identities. This is where the sociological framework of symbolic interaction is valuable.

Doing Sociological Research

Is Hooking Up Bad for Women?

Research Question

The presence of a hookup culture on college campuses is a relatively recent phenomenon and reflects changes in sexual attitudes and behaviors among, especially, young people. Some argue that hooking up liberates women from traditional sexual values that have constrained women's sexuality. Others argue that this culture is harmful to women, making them sexual objects for men's pleasure (for example, the practice of women making out with other women in public settings, such as bars and campus parties). Is hooking up harmful to women or is it a sign of their sexual liberation?

Research Method

Several sociologists have examined this question, some using national surveys, others using a more qualitative approach. Paula England and her colleagues, for example, studied sexual activity in a survey of over 14,000 students at 18 different campuses in the United States, exploring students' experiences with hooking up, dating, and relationships. Laura Hamilton and Elizabeth Armstrong used a more qualitative approach, actually residing among students in a so-called party dorm, observing as well as interviewing students for a full year. Leila Rupp and Verta Taylor had their undergraduate students interview other students about the party scene on their campus.

(continued)

Research Findings

Research finds that both arguments about the effect of hookup cultures on women are true: Some parts of this culture are harmful to women, but women's experiences within the hookup culture also vary and are not uniformly negative.

England, for example, found that hooking up is not as wildly rampant as assumed. She found that 72 percent of both men and women participated in at least one hookup, but 40 percent had engaged in three or fewer hookups and only 20 percent of students had engaged in ten or more. England concludes that the image of "girls gone wild" in popular culture is simply not true. She also found that hooking up has not replaced committed relationships. As Hamilton and Armstrong found, the hookup culture allows women (and men) a chance for sexual exploration. At the same time, the hookup culture does present risks to women—risks of being pushed to drink too much, risks of sexual violence, and loss of self-esteem. Some women say that the hookup culture frees them to pursue education and careers without the emotionally consuming pressures of committed relationships (Hamilton and Armstrong 2009).

Rupp and Taylor have similarly found that the college party scene commonly includes women making out with other women. Unlike those who argue that this practice sexually objectifies women for the pleasure of men, Rupp and Taylor argue that women who do so are exploring sexuality. But they do so within social boundaries and heterosexual norms. Even though women may engage in same-sex sexual practices, they do not necessarily develop a lesbian identity, although the lines of sexual identity are expanding for women.

Conclusions and Implications

There is not a simple or single answer to the question of whether the hookup culture harms or liberates women. Taken together, these studies reveal a complex portrait of young women's sexuality today. Sexual boundaries are perhaps more fluid than they once were, although they are still marked by sexual double standards and norms of heterosexuality.

Questions to Consider

1. Is there a hookup culture on your campus? If so, are there both positive and negative consequences of the hookup culture for women on your campus? What are they? If there is not such a culture on your campus, why not?

2. How is the hookup culture shaped by such social factors as age, social class, race, or gender? How might it change over time—both as history evolves and as the current generation ages?

Sources: Armstrong, Elizabeth A., Laura Hamilton, and Paula England. 2010. "Is Hooking Up Bad for Young Women?" *Contexts* 9 (Summer): 23–27; Hamilton, Laura, and Elizabeth A. Armstrong. 2009. "Gendered Sexuality in Young Adulthood: Double Binds and Flawed Options." *Gender & Society* 23 (October): 589–616; Rupp, Leila J., and Verta Taylor. 2010. "Straight Girls Kissing." *Contexts* 9 (Summer): 28–33.

Symbolic Interaction and the Social Construction of Sexual Identity

Symbolic interaction theory uses a **social construction perspective** to interpret sexual identity as learned, not inborn. Symbolic interaction interprets culture and society as shaping sexual experiences. Social approval and social taboos make some forms of sexuality permissible and others not (Seidman 2014; Lorber 1994).

The social construction of sexual identity is revealed by **coming out**—the process of defining oneself as gay or lesbian. The process is a series of events and redefinitions in which a person comes to see herself or himself as having a gay identity. While coming out, a person consciously adopts a gay identity either to himself or herself or to others (or both). This is usually not the result of a single experience. If it were, there would be far more self-identified gays and lesbians, because researchers find that a substantial portion of both men and women have some form of homosexual experience at some time in their lives. Typically, a person who is coming out will do so gradually, disclosing their identity in the context of specific social relationships.

Developing and disclosing a sexual identity is not necessarily a linear or unidirectional process, with people moving predictably through a defined sequence of steps or phases. Although there may be milestones in a person's identity development, people also experience periods of ambivalence about their identity and may switch back and forth between lesbian, heterosexual, and bisexual identity over time (Rust 1995, 1993). Some people may engage in lesbian or gay behavior but not adopt an identity as lesbian or gay. Certainly, many gays and lesbians never adopt a public definition of themselves as gay or lesbian, instead remaining "closeted" for long periods, if not for their entire lifetime.

One's sexual identity may also change. For example, a person who has always thought of himself or herself as heterosexual may conclude at a later time that he or she is gay, lesbian, or possibly bisexual. In more unusual cases, people may undergo a sex change operation, perhaps changing their sexual identity in the process.

Although most people learn stable sexual identities, sexual identity evolves over the course of one's life. Change is, in fact, a normal outcome of the process of identity formation. Changing social contexts (including dominant group attitudes, laws, and systems of social control), relationships with others, political movements, and even changes in the language used to describe different sexual identities all affect people's self-definition.

Feminist Theory: Sex, Power, and Inequality

Feminist theory has had a tremendous impact on the study of sexuality, as has the theoretical work that has emerged from gay/lesbian/bisexual/transgender studies. Feminist theory emphasizes that sexuality, like other forms of social identity, exists within a sex/gender system of structured inequality. Dominant institutions define heterosexuality as the only legitimate form of sexual identity, and enforce heterosexuality through social norms and sanctions, including peer pressure, socialization, law and other social policies, and, at the extreme, violence (Rich 1980). In other words, sexuality exists within a context of power relationships within society.

Sexual politics refers to the link between sexuality and power, not just within individual relationships. The feminist movement has linked sexuality to the status of women in society, including how women are treated within sexual relationships. High rates of violence against women as well as hate crimes against sexual minorities are evidence of the degree some go to in exercising power over others.

The feminist and gay and lesbian liberation movements have put sexual politics on the public agenda by challenging gender *and* sexual oppression (D'Emilio 1998). Such thinking has profoundly changed public knowledge of gay and lesbian sexuality. Gay, lesbian, and feminist scholars have convincingly argued that being lesbian or gay is *not* the result of psychological deviance or personal maladjustment. The political mobilization of many lesbian, gay, bisexual, and transgender people and the willingness of many to make their sexual identity public have also raised public awareness of the civil and personal rights of gays and lesbians. Indeed, one of the most striking and seemingly rapid changes in public opinion has been in the number of those who now support the freedom of gays and lesbians to marry (see ▲ Figure 12-3). It is hard to think of another social issue where attitudes have changed so quickly in such a short period of time.

One of the new perspectives that sexual liberation movements have generated is *queer theory* (see also Chapter 11). **Queer theory** has evolved from recognizing the socially constructed nature of sexual identity and the role of power in defining only some forms of sexuality as "normal"—that is, socially legitimate. Instead of seeing heterosexual or homosexual attraction as fixed, queer theory interprets society as forcing sexual boundaries, or dichotomies, on people. By challenging the "either/or" thinking that one is either gay or straight, queer theory disputes the idea that only one form of sexuality is normal and all other forms are deviant or wrong. Queer theory has opened up fascinating new studies of gay, straight, bisexual, and transsexual identities and introduced the idea that sexual identity is a continuum of different possibilities for sexual expression and personal identity (Wilchins 2014; Sullivan 2003).

Queer theory has also linked the study of sexuality to the study of gender, showing how transgressing (or violating) fixed gender categories can reconstruct the possibilities of how all people—men and women, gay, bisexual, transgender, or straight—construct their gender and sexual identity. Transgressing gender categories shows how sex and gender categories are usually constructed in dichotomous ways (that is, opposite or binary types). By violating these constructions, people are liberated from the social constraints that so-called fixed categories of identity create. Queer theory also encourages playing with

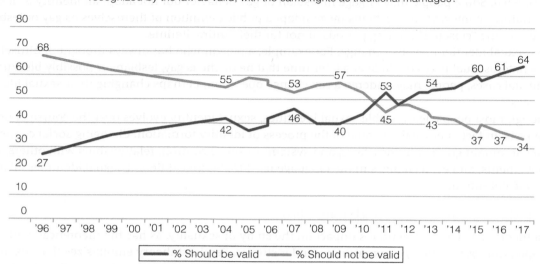

▲ **Figure 12-3** **Support for Same-Sex Marriage.** Change in public attitudes toward same-sex marriage has been unusually rapid. What do you think explains this fast rate of change?

Note: Trend shown for polls in which same-sex marriage question followed questions on gay/lesbian rights and relations 1996–2005 wording: "Do you think marriages between homosexuals..."

Data: McCarthy, Justin. 2017 (May 15). "US Support for Gay Marriages Edges to New High." *Gallup News.* Princeton, NJ: Gallup Organization. **http://news.gallup.com/poll/210566/support-gay-marriage-edges-new-high.aspx**

gender categories through various forms of performance as a political tool for deconstructing fixed sex and gender identities (Rupp and Taylor 2003).

A Global Perspective on Sexuality

Cross-cultural studies of sexuality show that sexual norms, like other social norms, develop differently across cultures. Sexual norms have deep roots in the socio-cultural contexts of different nations. In the most male-centered cultures, there are strong differences in the sexual well-being of men and women, with men reporting far more satisfactory sexual relationships than women. And, yet, cross-national research also finds that even in societies where there is more gender equality, men still report greater satisfaction in their sexual well-being than do women, although not to the same degree as in the more male-centered cultures (Laumann et al. 2006).

Cultures also vary considerably on how they view teen sexuality. Parents in the United States mostly discourage sex between teenagers, but in some other nations parents see sexuality for teens as part of the normal course of adult development. A study in the Netherlands, for example, found that two-thirds of Dutch teens (age 15 to 17) are allowed to sleep over with steady girlfriends or boyfriends—with parental approval. Interestingly, the Netherlands also has the lowest rate of teen pregnancy in the world, even with the age at first intercourse there having dropped over the years.

Compare this to the United States where the age at first intercourse has actually been increasing, but where the teen pregnancy rate is one of the highest among industrial nations. What makes the difference? The answer is cultural values around sexuality. In the United States, teen sexuality is discouraged and seen as risky. In the Netherlands, to the contrary, cultural morals see sexuality as part of developing self-determination, and it is treated with frank discussion, a strong place in public policy for sex education, and an idea that mutual respect is part of healthy sexual relationships (Schalet 2010).

Likewise, tolerance for gay and lesbian relationships varies significantly in different societies around the world. Germany has legalized gay and lesbian relationships, allowing them to register same-sex partnerships and have the same inheritance rights as heterosexual couples. The new law does not, however, give them the same tax advantages, nor can same-sex couples adopt children. Cross-cultural studies can make someone more sensitive to the varying cultural norms and expectations that apply to sexuality in different contexts. Different cultures simply view sexuality differently. In Islamic culture, for example, women and men are viewed as equally sexual, although women's sexuality is seen as potentially disruptive and needing regulation (Mernissi 2011). Not surprisingly, countries with the greatest gender equality also have the most positive public attitudes toward lesbians and gay men and the strongest legislative protections (Henry and Wetherell 2017).

Sex is also big business, and it is deeply tied to the world economic order. As the world has become more globally connected, an international sex trade has flourished—one that is linked to economic development, world poverty, tourism, and the subordinate status of women in many nations.

Sex trafficking refers to the use of women and girls worldwide as sex workers in an institutional context in which sex itself is a commodity. Sex is marketed in an international marketplace. As sex workers, women are used to promote tourism, cater to business and military men, and support a huge industry of nightclubs, massage parlors, and teahouses (Bales 2010; Sara 2010; Shelley 2010). Through sex trafficking, women—usually very young women—are forced by fraud or coercion into commercial sex acts. Sometimes identified as a form of slavery, sex trafficking can involve a system of debt and bondage, where young women (typically under age 18) are obligated to provide sexual services in exchange for alleged debt for the price of their housing, food, or other living expenses. Sometimes these young women are actually kidnapped; other times, they may simply be duped, initially lured into sex work by promises of marriage, large incomes, or the glamour of travel, but are soon trapped in a cycle of debt and/or actual captivity.

Related to sex trafficking is **sex tourism**, referring to the practice whereby people travel to particular parts of the world specifically to engage in commercial sexual activity. "Sex capitals" are places where prostitution openly flourishes, such as in Thailand and Amsterdam. Sex is an integral part of the world tourism industry. In Thailand, for example, men as tourists outnumber women by a ratio of three to one. Although certainly not all men—or even a majority—go to Thailand solely to explore sex tourism, some men do go explicitly to Thailand and other destinations to buy sexual companionship. For example, hostess clubs in Tokyo cater to corporate men. Although these clubs are not houses of prostitution, scholars who have studied them say that they are based on an environment in which women's sexuality is used as the basis for camaraderie among men (Allison 1994).

Sex tourism is such a profitable enterprise that the International Labour Organization estimates that somewhere between 2 and 14 percent of the gross domestic product in Thailand, Indonesia, Malaysia, and the Philippines is derived from sex tourism. Much of this lucrative business involves the exploitation of children (U.S. Department of State 2017).

Sex tourism and sexual trafficking are now part of the global economy, contributing to the economic development of many nations and supported by the economic dominance of certain other nations. As with other businesses, sex industry products may be produced in one region and distributed in others. Think, for example, of the pornographic film industry centered in southern California, but distributed globally. The sex trade is also associated with world poverty. Sociologists have found that the weaker the local economy, the more important the sex trade. The international sex trade is also implicated in problems such as the spread of AIDS (Kloer 2010; Altman 2001).

The international trafficking of women for sex exploits women—and often children—and puts them at risk for disease and violence.

Understanding Gay and Lesbian Experience

Sociological understanding of sexual identity has developed largely through new studies of lesbian, gay, bisexual, and transgender experiences. Long thought of only in terms of social deviance (see Chapter 7), LGBTQ people have long been negatively ga stereotyped in traditional social science. The feminist and gay liberation movements have discouraged this approach, arguing that diverse sexual identities are part of the broad spectrum of human sexuality. While there is resistance in the mainstream culture to this idea, it is now supported by strong social science research.

The institutional context for sexuality within the United States, as well as other societies, is one in which homophobia permeates the culture. **Homophobia** is the fear and/or hatred of lesbians and gays. Homophobia is manifested in prejudiced attitudes toward gays and lesbians, as well as in overt hostility and violence against people suspected of being gay. Like other forms of negative social judgments about particular groups, homophobia is a learned attitude.

Homophobia produces numerous fears and misunderstandings about LGBTQ people, including fears about anyone who lives outside of dominant, heterosexual norms. For example, there is the widespread, though incorrect, belief that children raised in gay and lesbian households will be negatively affected. This is not true. Sociological research has shown that the ability of parents to form good relationships with their children is far more significant in children's social development than is their parents' sexual orientation (Stacey and Biblarz 2001).

Other myths include the false idea that gay men are mostly White, well-off financially, and work mostly in artistic areas and personal service jobs (such as hairdressing). This stereotype prevents people from recognizing that gays and lesbians come from all racial–ethnic groups, may be working class or poor, and are employed in a wide range of occupations (Connell 2014). Stereotypes also define LGBTQ people as mostly young or middle-aged when, in fact, many are elderly (Cronin and King 2014). Homophobia, not truth, drives these attitudes.

Related to homophobia is heterosexism. **Heterosexism** refers to the institutionalization of hetero-sexuality as the only socially legitimate sexual orientation. Heterosexism is rooted in the belief that heterosexual behavior is the only natural form of sexual expression and that anything else is a perversion of "normal" sexual identity. Heterosexism is reinforced through institutional mechanisms that project the idea that only heterosexuality is normal. Some businesses, for example, may actively discriminate against people presumed to be gay.

Heterosexism is an institutional structure (just as racism is an institutional structure and sexism is an institutional structure). Because institutions distribute privileges differently based on heterosexist structures, it is possible for an individual to be accepting of gay and lesbian people (that is, not be homo-phobic) while still benefitting from being heterosexual. At the behavioral level, heterosexist practices can exclude lesbians and gays, such as when coworkers talk about dating, assuming that everyone is inter-ested in a heterosexual partner.

In the absence of institutional supports from the dominant culture, lesbians and gays have had to invent their own institutional support systems. Gay communities and gay rituals, such as gay pride marches, affirm LGBTQ identities and provide a support system that is counter to the dominant hetero-sexual culture. Those who remain "in the closet" deny themselves this support system.

The absence of institutionalized roles for lesbians and gays affects the roles they adopt within relationships. Despite popular stereotypes, gay partners typically do not assume roles as a dominant or submissive sexual partner. They are more likely to adopt roles as equals. Gay couples and lesbian couples are also more likely than heterosexual couples to both be employed, another source of greater equality

Thinking Sociologically

Keep a diary for one week and write down as many examples of *homophobia* and *heterosexism* as you observe in routine social behavior. What do your observations tell you about how heterosexuality is enforced?

Understanding Diversity

Sexuality and Disability: Understanding "Marginalized" Masculinity

What does it mean to be a "real man"? In a gender-stratified world, masculinity is typically perceived as something one either has or does not. Beliefs about masculinity define "real men" as those who are strong, straight, and sexually powerful. Thus, socially constructed beliefs about manhood are deeply tied to assumptions about male sexuality. What happens to men who do not fit this narrowly constructed definition of manhood—men who are marginalized in a social system that only privileges those with particular social characteristics?

Idealized definitions of masculinity shape experiences of disabled men who may be negatively stereotyped by others as "asexual," "impotent," or "not real men." Discrimination and prejudice against disabled people is pervasive in society. As a result, men with disabilities have to resist the perception that they somehow do not meet social standards of masculinity. Some may respond by acting "hypermasculine"—that is, working to exaggerate their strength and endurance. Others create different sets of standards for themselves, thus reformulating ideas about masculinity.

Examining experiences of disabled men with regard to sexuality and gender shows how male privilege is not a universal experience—that is, men's gender privilege is conditioned upon other forms of privilege or lack thereof, such as one's status as abled or disabled. A sociological perspective that recognizes the diversity of social experiences provides a multidimensional lens through which men's sexuality can be understood.

Sources: Coston, Beverly M., and Michael Kimmel. 2012. "Seeing Privilege Where It Isn't: Marginalized Masculinities and the Intersectionality of Privilege." *Journal of Social Issues* 68: 97; Gerschick, T. J., and A. S. Miller. 1995. "Coming to Terms: Masculinity and Physical Disability." Pp. 183–204 in *Men's Health and Illness: Gender, Power, and the Body, Research on Men and Masculinities Series*, vol. 8, edited by D. Sabo and D. F. Gordon. Thousand Oaks, CA: Sage Publications.

within the relationship. Researchers have also found that the quality of relationships among gay men is positively correlated with social support the couple receives from others (Lyons et al. 2013).

Lesbians and gays are a minority group in our society, denied equal rights and singled out for negative treatment in society. *Minority groups* are not necessarily numerical minorities; they are groups with similar characteristics (or at least perceived similar characteristics) who are treated with prejudice and discrimination (see Chapter 10). As a minority group, gays and lesbians have organized to advocate for their civil rights and to be recognized as socially legitimate citizens. Some organizations and municipalities have enacted civil rights protections on behalf of gays and lesbians, typically prohibiting discrimination in hiring. The historic Supreme Court ruling in 2015 that supported same-sex marriage was based on the constitutional principle of equal protection under the law, as mandated by the Fourteenth Amendment to the U.S. Constitution. How this will play out in protecting LGBTQ people from job or other forms of discrimination is yet to be seen.

Sex and Social Issues

In studying sexuality, sociologists tap into some highly contested social issues of the time. Birth control, reproductive technology, abortion, teen pregnancy, pornography, and sexual violence are all subjects of public concern and are important in the formation of social policy. Debates about these issues hinge in part on attitudes about sexuality and are shaped by race, class, and gender relations. These social issues can generate personal troubles that have their origins in the structure of society—recall the distinction C. Wright Mills made between personal troubles and social issues (see Chapter 1).

Birth Control

The availability of birth control is now less debated than it was in the not-too-distant past. Still, specific forms of birth control, such as the so-called morning-after pill, are the subjects of intense public debate, especially regarding young and poor women. To this day, mostly men define laws and make scientific decisions about what types of birth control will be available. Women, though, are seen as being responsible for reproduction because it is a presumed part of their traditional role. Changes in birth control technology have broken the link between sex and reproduction, freeing women from some traditional constraints.

The right to birth control is a fairly recently won freedom. In 1965, the U.S. Supreme Court, in *Griswold v. Connecticut,* defined the use of birth control as a right, not a crime. This ruling originally applied only to married people. Unmarried people were not extended the same right until the 1972 Supreme Court decision *Eisenstadt v. Baird.* Today, birth control is routinely available by prescription, but there is heated debate about whether access to birth control should be curtailed for the young—at a time when youths are experimenting with sex at younger ages and risking teenage pregnancy, sexually trans-mitted diseases, and AIDS. Some argue that increasing access to birth control will only encourage more sexual activity among the young, though there is little evidence that this is true.

Class and race relations also have had a role in shaping birth control policy. In the mid-nineteenth century, increased urbanization and industrialization ended the necessity for large families, especially in the middle class, because fewer laborers were needed to support the family. Early feminist activists such as Emma Goldman and Margaret Sanger also saw birth control as a way of freeing women from unwanted pregnancies and allowing them to work outside the home if they chose. As the birthrate fell among White upper- and middle-class families during this period, these classes feared that immigrants, the poor, and racial minorities would soon outnumber them (Gordon 1977).

The *eugenics* movement of the early twentieth century grew from fear of domination by immigrant groups. **Eugenics** sought to apply scientific principles of genetic selection to "improve" offspring of the human race. It was explicitly racist and class-based, calling for, among other things, the compulsory sterilization of those who eugenicists thought were unfit. Eugenicist arguments appeal to a public that fears social problems that emerge from race and class inequality. Instead of attributing these problems (such as crime) to the structure of society, eugenicists blame genetic composition of the least powerful groups in society.

In contemporary society, attitudes about the use of birth control have evolved and have also been shaped by greater awareness about risks of HIV/AIDS infection. Now two-thirds of women who are sexually active and aged 15 to 44 use contraceptives, a large increase since 1995 (Guttmacher Institute 2017). Among women users, the pill is the most frequent type of birth control used, followed by tubal ligation (sterilization), and then condoms. Similar data are not reported for men—itself a reflection of the belief that contraception is primarily a woman's responsibility.

Among teen women (who are most at risk of unintended pregnancy), 82 percent now use contraceptives. Still, one in five teen women do not use contraception the first time they had sex, leaving them at significant risk of an unplanned pregnancy (Guttmacher Institute 2017; see also ▲ Figure 12-4).

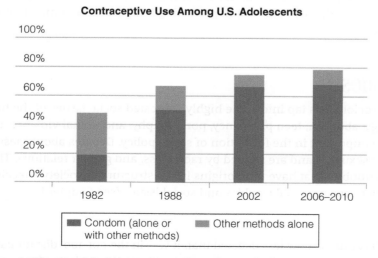

▲ **Figure 12-4 Contraceptive Use at First Sexual Intercourse, 15- to 19-year-olds**

Source: **www.guttmacher.org**

New Reproductive Technologies

Practices such as surrogate mothering, in vitro fertilization, and new biotechnologies of gene splicing, cloning, and genetic engineering mean that reproduction is no longer inextricably linked to biological parents. A child may be conceived through means other than sexual relations between one man and one woman. One woman may carry the child of another. Offspring may be planned through genetic engineering. A sheep can be cloned (that is, genetically duplicated). So can monkeys. Are humans next? With such developments, those who could not otherwise conceive children (infertile couples, single women, or lesbian couples) are now able to do so, raising new questions: To whom are such new technologies available? Which groups are most likely to sell reproductive services? Which groups are most likely to buy? What are the social implications of such changes?

These questions have no simple answers, but sociologists would point first to the class, race, and gender dimensions of these issues (Roberts 2012, 1997). Poor women, for example, are far more likely than middle-class or elite women to sell their eggs or offer their bodies as biological incubators. Groups that can afford new, costly methods of reproduction may do so at the expense of women whose economic need places them in the position of selling themselves for financial necessity.

Breakthroughs in reproductive technology raise especially difficult questions for makers of social policy. Developments in the technology of reproduction have ushered in new possibilities and freedoms but also raise questions for social policy. With new reproductive technologies, there is potential for a new eugenics movement. Sophisticated prenatal screenings make it possible to identify fetuses with pre-sumed defects. Might society then try to weed out those perceived as undesirables—the disabled, certain racial groups, certain sexes? Will parents try to produce "designer children"? If boys and girls are differently valued, one sex may be more often aborted, a frequent practice in India and China—two of the most populous nations on earth. Because of population pressures, state policy in China, for example, encourages families to have only one child. Because girls are less valued than boys, aborting and selling girls is common and has created a U.S. market for adoption of Chinese baby girls.

There are no traditions to guide us on such questions. Although the concept of reproductive choice is important to most people, choice is conditioned by constraints of race, class, and gender inequalities in society. Like other social phenomena, sexuality and reproduction are shaped by their social context.

Abortion

Abortion is one of the most seriously contested political issues. The public is sharply divided on whether they identify as "pro-choice" (49 percent) or "pro-life" (46 percent), but this varies according to various social factors. College graduates and those with postgraduate degrees are more likely to iden-tify as pro-choice. Women, young people, and those identifying as Democrats are also more likely to identify as pro-choice (see ▲ Figure 12-5). Furthermore, even among those self-identified as "pro-life,"

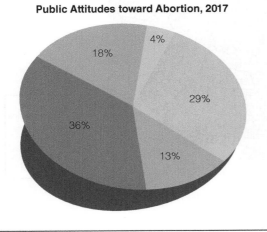

Public Attitudes toward Abortion, 2017

4%
18%
29%
36%
13%

Legal in all circumstances
Legal in most circumstances
Legal in some circumstances
Illegal in all circumstances
No opinion

▲ **Figure 12-5   Attitudes toward Abortion.**
Attitudes toward the legality of abortion have been fairly steady over time. As you can see, the majority of Americans support abortion rights, at least under some circumstances.

Data: Saad, Lydia. 2017 (June 9). "US Abortion Attitudes Stable; No Consensus on Legality." *Gallup News.* Princeton, NJ: Gallup Organization. **http://news.gallup.com/poll/211901/abortion-attitudes-stable-no -consensus-legality.aspx?g_source=ABORTION&g_medium=topic&g _campaign=tiles**

69 percent support abortion when the woman's life or physical health is endangered. Half of those identifying as pro-choice think abortion should be illegal in the second trimester (Gallup 2017; Saad 2014, 2011). Clearly, more complex views are involved in thinking about abortion than simple labels can suggest.

The right to abortion was first established in constitutional law by the *Roe v. Wade* decision in 1973. In *Roe v. Wade,* the Supreme Court ruled that at different points during a pregnancy separate but legitimate rights collide—the right to privacy, the right of the state to protect maternal health, and the right of the state to protect developing life. To resolve this conflict of rights, the Supreme Court ruled that pregnancy occurred in trimesters: In the first, women's right to privacy without interference from the state prevails; in the second, the state's right to protect maternal health takes precedence; in the third, the state's right to protect developing life prevails. In the second trimester, the government cannot deny the right to abortion, but it can insist on reasonable standards of medical procedure. In the third, abortion may be performed only to save the life or health of the mother. More recently, the Supreme Court has allowed states to impose restrictions on abortion, but it has not, to date, overturned the legal framework of *Roe v. Wade.*

Data on abortion show that it occurs across social groups, although certain patterns emerge. The abortion rate has declined since 1980, from a rate of 25.1 per 1000 women in 1990 to 14.6 by 2014 (among women aged 15 to 44). Increased contraceptive use (especially by younger women) and its availability are a major factor, although, no doubt, the increased restrictions placed on abortion services explain some of this decline. As you can see in ■ map 12-1, the distance that a significant number of women have to travel to reach an abortion clinic is quite significant. Policies, such as twenty-four hour waiting periods, can mean that women's options are quite restricted, especially for low-income women (Bearak, Burke, and Jones. 2017).

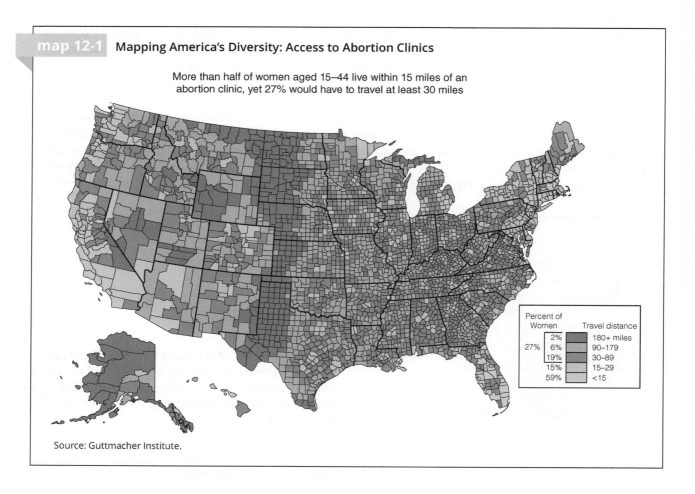

map 12-1 **Mapping America's Diversity: Access to Abortion Clinics**

More than half of women aged 15–44 live within 15 miles of an abortion clinic, yet 27% would have to travel at least 30 miles

Percent of Women		Travel distance
	2%	180+ miles
27%	6%	90–179
	19%	30–89
	15%	15–29
	59%	<15

Source: Guttmacher Institute.

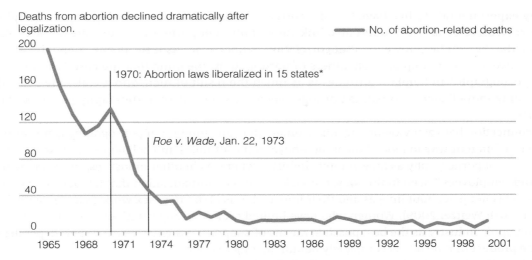

Deaths from abortion declined dramatically after legalization.

━━━ No. of abortion-related deaths

1970: Abortion laws liberalized in 15 states*

Roe v. Wade, Jan. 22, 1973

▲ **Figure 12-6 Deaths from Abortion: Before and After *Roe v. Wade***

*By the end of 1970, four states had repealed their antiabortion laws, and eleven states had reformed them.

Source: Boonstra, Heather, Rachel Benson Gold, Cory L. Richards, and Lawrence B. Finer. 2006. *Abortion in Women's Lives*. New York: Guttmacher Institute, p. 13.

Public controversy over the right to choose an abortion continues, even though the number of people identifying as pro-choice has inched up in recent years (Gallup 2017). Pro-choice advocates note that serious health risks to women when abortion is illegal (see ▲ Figure 12-6), as the number of deaths from illegal abortions plummeted in the years following the *Roe v. Wade* decision.

The abortion issue provides a good illustration of how sexuality has entered the political realm. Abortion rights activists and antiabortion activists hold very different views about sexuality and the roles of women. Antiabortion activists tend to believe that giving women control over their fertility breaks up the stable relationships in traditional families. They tend to view sex as something that is sacred, and they are disturbed by changes that loosen sexual norms. This belief has been fueled by the activism of the religious right, where strong passions against abortion have driven issues about sexual behavior directly into the political realm. Abortion rights activists, on the other hand, see women's control over reproduction as essential for women's independence. They also tend to see sex as an experience that develops intimacy and communication between people who love each other. The abortion debate can be interpreted as a struggle over the right to terminate a pregnancy as well as a battle over differing sexual values and a referendum on the nature of men's and women's relationships (Luker 1984).

Pornography and the Sexualization of Culture

Little social consensus has emerged about the acceptability and effects of pornography. Part of this debate is about defining what is obscene. The legal definition of obscenity is one that changes over time and in different political contexts. Public agitation over pornography has divided people into those who think it is solidly protected by the First Amendment, those who want it strictly controlled, those who think it should be banned for moral reasons, and those who think it must be banned because it harms women.

An ongoing question in research is whether viewing pornography promotes violence against women. This is a complex question with mixed answers in research because the connection is difficult to observe directly. Verbal and physical abuse of women in pornography is shockingly common. One analysis of popular pornographic videos found that 88 percent of scenes in such videos include physical aggression against women—hitting, gagging, and slapping, for example. Moreover, these aggressive scenes depict women as either enjoying it or responding neutrally (Bridges et al. 2010).

Many experimental studies have found a correlation between men's viewing of pornography and sexually aggressive behavior. Studies also link consumption of pornography to a greater acceptance of rape myths. Remember, however, that correlation is not cause. As it turns out, men who are sexually aggressive are more frequent consumers of pornography, meaning that the correlation between viewing pornography and violent attitudes is as much the result of men who are already sexually aggressive being more likely to consume pornography as it is the use of pornography per se (Malamuth et al. 2011).

The connection between violence against women and consumption of pornography is not settled, but there is a different way to look at the issue—beyond the attitudes and behaviors of those who view it. Instead, think of pornography as an economic industry where, as in other industries, there are owners, bosses, and "employees." Seen in this way, the violence and dehumanization that happens to women in pornography is not just about images and their impact, but is about how sex workers are exploited (Voss 2012; Boyle 2011; Weitzer 2009). Women who have limited economic opportunities may turn to work in the pornographic industry as the best possible means of supporting themselves and their dependents, even if the product that results is completely dehumanizing to them.

One thing that is certain about pornography is that it permeates contemporary culture. Once available only in more "underground" places like X-rated movie houses, pornography is now far more public than in the past. Hotel rooms have a huge array of pornographic films available on television; pornographic spam appears regularly in people's email inboxes; casual references to pornography are made in popular shows on prime-time TV; and images that once would have seemed highly pornographic are now commonly found on widely distributed magazines such as *Maxim, Cosmopolitan,* and others.

Pornography has also infiltrated the web. There are now over 4.2 million pornographic websites, a 2.8-billion-dollar industry. Researchers estimate that worldwide, men spend an average of $3000 per second purchasing pornographic material (Johnson 2011; Ropelato 2007). These figures do not account for women's use of pornography, nor do they capture the extent to which pornography is viewed on the web for free.

The United States is both the largest producer and exporter of "hard-core" porn on the web. Studies of Internet porn also find that the most popular website searches are for Black and Asian porn sites; Blacks and Asians have the most porn sites devoted to them, followed by Latinos—a reflection of the sexualized imagery of racism. Sites devoted to "teen porn" also tripled between 2005 and 2013 (Ruvolo 2011).

Pornography clearly reflects the racism and sexism that are endemic in society. Highly sexualized expressions and images are so widely seen throughout society that one commentator has said we are experiencing the *pornification of culture* (Levy 2005). Does this indicate that society has become more sexualized?

According to a major report from the American Psychological Association (APA), the answer is yes. The APA defines *sexualization* as including any one of the following conditions:

- People are judged based only on sexual appeal or behavior, to the exclusion of other characteristics.
- People are held to standards that equate physical attractiveness with being sexy.
- People are sexually objectified—meaning made into a "thing" for others' use. Sexuality is inappropriately imposed on a person (American Psychological Association 2007).

The APA report then details the specific consequences, especially for young girls, of a culture marked by sexualization. This report shows that overly sexualizing young girls harms them in psychological, physical, social, and academic ways. Young girls may spend more time tending to their appearance than to their academic studies; they may engage in eating disorders to achieve an idealized, but unattainable image of beauty; or, they may develop attitudes that put them at risk of sexual exploitation. Although the focus of this report is on young girls, one cannot help but wonder what effects the "pornification of culture" also has on young boys, as well as adult women and men.

Despite public concerns about pornography, most people believe that pornography should be protected by the constitutional guarantees of free speech and free press. Yet people also believe that pornography dehumanizes women; women especially think so. Public controversy about pornography is not likely to go away because it taps so many different sexual values among the public.

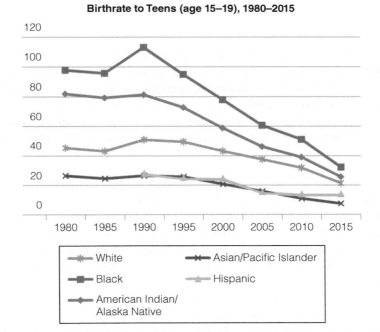

Birthrate to Teens (age 15–19), 1980–2015

Legend:
- ——✳—— White
- ——✕—— Asian/Pacific Islander
- ——■—— Black
- ——▲—— Hispanic
- ——◆—— American Indian/ Alaska Native

▲ **Figure 12-7   Teen Birthrate (per 1000 Women) by Race/Ethnicity, 1980–2015.** Despite public beliefs to the contrary, teen pregnancy among all racial/ethnic groups has declined substantially. What factors explain this decline?

Source: Martin, Joyce A., Brady E. Hamilton, Michelle J. K. Osterman, Anne K. Driscoll, and T. J. Mathews. 2017. "Births: Final Data for 2015." *National Vital Statistics Reports* 66 (1). **www.cdc.gov/nchs/data/nvsr /nvsr66/nvsr66_01.pdf**

Teen Pregnancy

Each year about 230,000 teenage girls (under age 19) have babies in the United States. The United States has the highest rate of teen pregnancy among developed nations, even though levels of teen sexual activity around the world are roughly comparable. Teen pregnancy has declined since 1990, a decline caused almost entirely from the increased use of birth control. Contrary to popular stereotypes, the teen birthrate among African American women has declined more than for White women (see also ▲ Figure 12-7). Most teen pregnancies are unplanned, due largely to inconsistent use of birth control (Guttmacher Institute 2017; Martin et al. 2017).

Beginning in the early 1980s, the federal government encouraged abstinence policies, putting money behind the belief that encouraging chastity was the best way to reduce teen pregnancy. Under programs that encourage abstinence, young people are encouraged to take "virginity pledges," promising not to have sexual intercourse before marriage. Do such pledges work?

In one of the most comprehensive and carefully controlled studies of abstinence, researchers compared a large sample of teen "virginity pledgers" and "nonpledgers" who were matched on social attitudes such as religiosity and attitudes toward sex and birth control. The study compared the two groups over a five-year period. Results showed that over time, there were *no differences* in the number of times those in each group had sex, the age of first sex, or the practice of oral or anal sex. The main difference between the two groups was that pledgers were less likely to use birth control when they had sex. Also, five years after having taken an abstinence pledge, pledgers denied having done so. The researchers concluded that not only are abstinence pledges ineffective (supporting other research findings) but that taking the pledge makes pledgers less likely to protect themselves from pregnancy and disease when having sex (Rosenbaum 2009). Consistent with these findings are other studies that find that abstinence policies account for a very small portion of the decline in teen pregnancy—probably only about 10 percent of the difference (Santelli et al. 2007; Boonstra et al. 2006).

Myth: Providing sex education to teens only encourages them to become sexually active.

Sociological Perspective: Comprehensive sex education actually delays the age of first intercourse; abstinence-only education has not been shown to be effective in delaying intercourse. Furthermore, of the 12 percent of young women who pledge abstinence, most break their pledge. Because they are less likely to be using birth control than nonpledgers, pledgers who beak their pledge are at high risk for unwanted pregnancy and contraction of the human papillomavirus (HPV) (Palk, Sanchagrin, and Heimer 2016; Risman and Schwartz 2002).

Although the rate of teen pregnancy has declined, so has the marriage rate for teens who become pregnant. Now, most babies born to teens will be raised by single mothers—a departure from the past when teen mothers often got married. What concerns people about teen parents is that teens are more likely to be poor than other mothers, although sociologists have cautioned that this is because teen mothers are more likely poor *before* getting pregnant (Luker 1996). Teen parents are among the most vulnerable of all social groups.

Teenage pregnancy correlates strongly with poverty, lower educational attainment, joblessness, and health problems. Teen mothers have a greater incidence of problem pregnancies and are most likely to deliver low-birth-weight babies, a condition associated with myriad other health problems. Teen parents face chronic unemployment and are less likely to complete high school than those who delay childbearing. Many continue to live with their parents, although this is more likely among Black teens than among Whites.

Although teen mothers feel less pressure to marry now than in the past, if they raise their children alone, they suffer the economic consequences of raising children in female-headed households—the poorest of all income groups. Teen mothers report that they do not marry because they do not think the fathers are ready for marriage. Sometimes their families also counsel them against marrying precipitously. These young women are often doubtful about men's ability to support them. They want men to be committed to them and their child, but they do not expect their hopes to be fulfilled (Edin and Kefalas 2005). Research shows that low-income single mothers are distrustful of men, especially after an unplanned pregnancy. They think they will have greater control of their household if they remain unmarried. Many teen mothers also express fear of domestic violence as a reason for not marrying (Edin 2000).

Why do so many teens become pregnant given the widespread availability of birth control? Teens often delay the use of contraceptives until several months after they become sexually active. Teens who do not use contraceptives when they first have sex are twice as likely to get pregnant as those who use contraception. In recent years, the percentage of teens using birth control has increased, although patterns of contraceptive use change over time (Guttmacher Institute 2017).

Sociologists have argued that the effective use of birth control requires a person to identify himself or herself as sexually active (Luker 1975). Teen sex, however, tends to be episodic. Teens who have sex on a couple of special occasions may not identify themselves as sexually active and may not feel obliged to take responsibility for birth control. Despite many teens initiating sex at an earlier age, social pressure continues to discourage them from defining themselves openly, or even privately, as sexually active.

Teen pregnancy is integrally linked to gender expectations of men and women in society. Some teen men consciously avoid birth control, thinking it takes away from their manhood. Teen women often romanticize motherhood, thinking that becoming a mother will give them social value they do not otherwise have. For teens in disadvantaged groups, motherhood confers a legitimate social identity on those otherwise devalued by society. Although their hopes about motherhood are not realistic, they indicate how pessimistic the teenagers feel about their lives that are often marked by poverty, a lack of education, and few good job possibilities. For young women to romanticize motherhood is not surprising in a culture where motherhood is defined as a cultural ideal for women, but the ideal can seldom be realized when society gives mothers little institutional or economic support.

Sexual Violence

Before the development of the feminist movement, sexual violence was largely hidden from public view. One great success of the women's movement has been to identify, study, and advocate better social policies to address the problems of rape, sexual harassment, domestic violence, incest, and other forms of sexual coercion. Sexual coercion is not just a matter of sexuality; it is also a form of power relations shaped by social inequality between women and men. In the forty years or so that these issues have been identified as serious social problems, volumes of research have been published on these different subjects, and numerous organizations and agencies have been established to serve victims of sexual abuse and to advocate reforms in social policy.

Rape and sexual violence were covered in Chapter 7 on deviance and crime, in keeping with the argument that these are forms of deviant and criminal behavior, not expressions of human sexuality. Here we point out that various forms of sexual coercion (rape, domestic violence, and sexual harassment) can best be understood (and therefore changed) by understanding how social institutions shape human behavior and how social interactions are influenced by social factors such as gender, race, age, and class.

Take, for example, the phenomenon now known as *acquaintance rape* (sometimes also called *date rape*). Acquaintance rape is forced and unwanted sexual relations by someone who knows the victim (even if only a brief acquaintance). This kind of rape is common on college campuses, although it is also the most underreported form of rape. Exact measurement of the extent of sexual assault on campus is notoriously difficult to determine given the reluctance of victims to report. Some surveys have estimated that between 15 to 25 percent of college women experience some form of acquaintance rape (Fisher et al. 2000), but how one defines sexual assault, how a researcher asks the question, and the context in which studies of sexual assault are done can all affect how sexual assault is "counted." What is clear is that women aged 18 to 24 and women enrolled in college are more likely to experience sexual assault than are other age groups. Students are more likely victims of sexual assaults than non-students, at least as measured by national victimization surveys (Sinozich and Langton 2017; see also Chapter 7).

Studies show that although rape is an abuse of power, it is related to people's gender attitudes. Holding stereotypical attitudes about women is strongly related to adversarial sexual beliefs, accepting rape myths, and tolerating violence against women (Deming et al. 2013).

Violence against women is more likely to occur in some contexts than others, especially in organizations that are set up around a definition of masculinity as competitive, where alcohol abuse occurs, and where women are defined as sexual prey. This is one explanation given for the high incidence of rape in some college fraternities (Martin 2015; Armstrong et al. 2006; Stombler and Padavic 1997; Martin and Hummer 1989).

Research on violence against women also finds that Black, Hispanic, and poor White women are more likely to be victimized by various forms of violence, including rape. African American and White women report the highest incidence of intimate partner violence; Latinas, Asian Americans, and Pacific Islanders have the lowest incidence (Truman and Morgan 2016; Catalano 2012).

Warren Goldswain/Shutterstock.com

The feminist movement has called attention to the once-private phenomenon of domestic violence.

Myth: Campus parties are inherently dangerous because they encourage violence against women.
Sociological Perspective: Not all parties are dangerous for women, but certain individual and organizational characteristics make some parties risky places for sexual violence. At the individual level, concerns about one's social status and holding traditional sexual values put one at risk in the party scene. At the organizational level, parties where men have a "home turf" advantage and women do not, where excessive drinking is encouraged, where men know each other more than the women know each other, and where traditional sexual/gender scripts are played out make women especially prone to sexual violence (Martin 2015; Armstrong et al. 2006).

In sum, sociological research on sexual violence shows how strongly sexual coercion is tied to the status of diverse groups of women in society. Rather than explaining sexual coercion as the result of maladjusted men or the behavior of victims, feminists have encouraged a view of sexual coercion that links it to an understanding of dominant beliefs about the sexual dominance of men and sexual passivity of women. Researchers have shown that those holding the most traditional gender role stereotypes are most tolerant of rapists and least likely to give credibility to victims of rape (Deming et al. 2013). Understanding sexual violence requires an understanding of the sociology of sexuality, gender, race, and class in society.

Sex and Social Change

As with other forms of social behavior, sexual behavior is not static. Sexual norms, beliefs, and practices emerge as society changes. Some major changes affecting sexual relations come from changes in gender roles. Technological change and emphasis on consumerism in the United States also affect sexuality. As you think about sex and social change, you might try to imagine what other social factors influence human sexual behavior.

The Sexual Revolution: Is It Over?

The **sexual revolution** refers to the widespread changes in men's and women's roles and the greater public acceptance of sexuality as a normal part of social development. Many changes associated with the sexual revolution have been changes in women's behaviors. The sexual revolution has narrowed the differences in sexual experiences of men and women. The feminist and gay and lesbian movements have put the sexual revolution at the center of public attention by challenging gender role stereotyping and sexual oppression, profoundly changing our understanding of gay and lesbian sexuality. The sexual revolution has meant greater sexual freedom, especially for women, but it has not eliminated the influence of gender in sexual relationships.

Some argue that the sexual revolution has created some new challenges, especially for women, whose gender and sexual roles have changed so much. Sociologist Leslie Bell calls this the "paradox of sexual freedom"—meaning that women, especially younger women, are now expected to build successful careers while at the same time meeting the cultural ideals that are laid out for them as sexually fulfilled women (Bell 2013). Women are now told they should be successful and independent but also have fulfilling, intimate relationships. The messages that permeate popular culture, as we have seen throughout this book, make it all seem easy if you are just sexy enough, thin enough, attractive to men, but not too pushy, sexually active, or demanding. This sexual revolution may have changed people's attitudes and expectations, but it has also produced new dilemmas as people navigate this changed social environment.

Technology, Sex, and Cybersex

Technological change has also brought new possibilities for sexual freedom. One significant change is the widespread availability of the birth control pill. Sex is no longer necessarily linked with reproduction; new sexual norms associate sex with intimacy, emotional ties, and physical pleasure (Freedman and

D'Emilio 1988). These sexual freedoms are not equally distributed among all groups, however. For women, sex is still more closely tied to reproduction than it is for men because women are still more likely to take the responsibility for birth control. Contraceptives are not the only technology influencing sexual values and practices. Now the Internet has introduced new forms of sexual relations as many people seek sexual stimulation from pornographic websites or online sexual chat rooms. *Cybersex,* as sex via the Internet has come to be known, can transform sex from a personal, face-to-face encounter to a seemingly anonymous relationship with mutual online sex. This introduces new risks, such as by sexual predators who use the Internet. The Internet has introduced new forms of deviance that are very difficult to control.

Commercializing Sex

At the same time that there are new sexual freedoms for women, many worry that this will only increase their sexual objectification. Furthermore, sexuality is becoming more and more of a commodity in our highly consumer-based society. Girls are being sexualized at younger ages, evidenced by the marketing of thongs to very young girls, the promotion of "sexy" dolls sold to young girls, and the highly sexualized content of media images that young boys and girls consume (Levy 2005). Often, these images are extremely violent and depict women as hypersexual victims of men's aggression, such as the popular video game, *Grand Theft Auto.*

Definitions of sexuality in the culture are heavily influenced by the advertising industry, which narrowly defines what is considered "sexy." Thin women, White women, and rich women are all depicted as more sexually appealing in mainstream media. Images defining "sexy" also are explicitly heterosexual. The commercialization of sex uses women and, increasingly, men in demeaning ways. Although the sexual revolution has removed sexuality from many of its traditional constraints, inequalities of race, class, and gender still shape sexual relationships and values.

The combination of sexualization and commoditization means that people become "made" into things for others' use. When people are held to narrow definitions of sexual attractiveness or are seen as valuable solely for their sexual appeal, you have social conditions that are ripe for exploitation and that damage people's sense of self-worth and value (American Psychological Association 2007). Even in what seems to be an increasingly "free" sexual society, sexuality is still nested in American culture within a system of power relations—power relations that, despite the sexual revolution, continue to influence how different groups are valued and defined.

Chapter Summary

In what sense is sexuality, seemingly so personal an experience, a part of social structure?

Sexual relationships develop within a social and cultural context. Sexuality is learned through socialization, is channeled and directed by social institutions, and reflects the race, class, and gender relations in society.

What evidence is there of contemporary sexual attitudes and behavior?

Contemporary sexual attitudes vary considerably by social factors such as age, gender, race, and religion. Sexual behavior has also changed in recent years, with mixed trends in both liberal and conservative sexual values. In general, attitudes on issues of premarital sex and gay and lesbian rights have become more liberal,

though this depends on social characteristics such as age, gender, and degree of religiosity, among others.

How is sexuality related to other social inequalities?

Sexuality intertwines with gender, race, and class inequality. This is especially revealed in the sexual stereotypes of different groups, as well as in the double standard applied to men's and women's sexual behaviors, such as in the *hooking up* culture.

What does sociological theory have to say about sexual behavior?

Functionalist theory depicts sexuality in terms of its contribution to the stability of social institutions. *Conflict theorists* see sexuality as part of the power relations and economic inequality in society. *Symbolic interaction* focuses on the social construction of sexual identity.

Feminist theory uncovers the power relationships that frame different sexual identities and behaviors, as well as linking sexuality to other forms of inequality.

How do homophobia and heterosexism influence lesbian and gay experience?

Homophobia is the fear and hatred of gays and lesbians. *Heterosexism* refers to institutional structures that define heterosexuality as the only socially legitimate sexual orientation. Both produce relationships of power that define gays and lesbians as a social minority group.

How is sexuality related to contemporary social issues?

Sexuality is related to some of the most difficult social problems, including birth control, abortion, reproductive technologies, teen pregnancy, pornography, and sexual violence. Such social problems can be understood by analyzing the sexual, gender, class, and racial politics of society.

How is sex related to social change?

The *sexual revolution* refers to widespread changes in the roles of men and women and a greater acceptance of sexuality as a normal part of social development. The sexual revolution has been fueled by social movements, such as the feminist movement and the gay and lesbian rights movement. Technological changes, such as development of the pill, have also created new sexual freedoms. Now, sexuality is influenced by the growth of cyberspace and its impact on personal and sexual interactions. At the same time, sex is treated as a commodity in this society, bought and sold and used to sell various products.

Key Terms

coming out 308
eugenics 314
heterosexism 312
homophobia 312
queer theory 309

sex tourism 311
sex trafficking 311
sexual identity 299
sexual orientation 299
sexual politics 309

sexual revolution 322
sexual scripts 301
social construction
　perspective 308

FAMILIES AND RELIGION

In this chapter, you will learn to:

Understand the features that define different kinship systems

Compare and contrast sociological theories of families

Relate the characteristics that define different types of contemporary family experiences

Refute some of the myths about contemporary marriage and divorce

Be able to identify social structures that affect family violence

Analyze current social policies that affect family experience

Define how religion is a social institution

Relate the different measures of religiosity

Compare and contrast sociological theories of religion

Identify the sources of religious diversity

Define the different organizational structures of religion

Explain how religion is related to social change and vice versa

iStock.com/elenabs

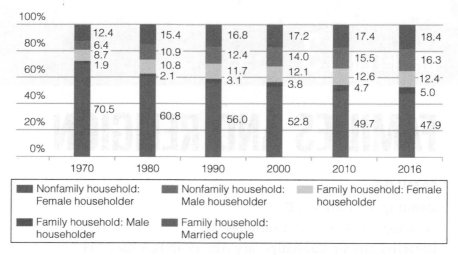

Data: U.S. Census Bureau. 2017. *Households by Type, 1940–Present, Table HH-1*. Washington, DC: U.S. Department of Commerce. **www.census.gov**

▲ **Figure 13-1   The Changing Character of U.S. Families, 1970–2016.**
Over the family household period from 1970 to 2016, the composition of U.S. families has changed considerably. By the U.S. Census Bureau definitions, a *family household* consists of a householder and others related to the householder by birth, marriage, or adoption; family household may also include unrelated people. *Nonfamily households* consist of a person living alone or with nonrelatives only. What are the major changes that you can identify from this figure? What sociological factors do you think help explain this change and, similarly, what other changes in society do such changes in the composition of families then create?

Suppose you were to ask a large group of people in the United States to describe their families. Many would describe divorced families. Some would describe single-parent families. Some would describe stepfamilies with new siblings and a new parent stemming from remarriage. Others would describe gay or lesbian households, perhaps with children present. Also included would be adoptive families and families with foster children. Others would describe the so-called traditional family with two parents living as husband and wife in the same residence as their biological children. Families have become so diverse that it is no longer possible to speak of "the family" as if it were a single thing.

The traditional *family ideal*—a father employed as the breadwinner and a mother at home raising children—has long been the dominant cultural norm, communicated through a variety of sources, including the media, religion, and the law. Few families now conform to this ideal (see ▲ Figure 13-1), and the number of families that ever did is probably fewer than generally imagined (Coontz 1992). Regardless of their form, families now face new challenges, such as managing the demands of family plus work or struggling to meet family needs when work disappears. Many families feel that they are under siege by changes in society that are dramatically altering all family experiences. Some of these changes are immediate, such as the loss of work in economically hard times or the strain on families from a family member's illness. Other changes are long-term changes in the social structure of society, such as women's increased work roles and accompanying changes in men's

Family diversity is the norm in American society, with no one type of family shaping people's experience.

What are some of the popular television shows about family life? Spend some time systematically observing one or more of these shows. What do they communicate about the *family ideal*? Are the images consistent with the actual data on families shown in this chapter?

roles; population changes (such as aging and immigration); and even things like the increased reliance on technology, which can alter communication patterns in families.

Many view the changes taking place in families as positive. Women have new options and greater independence. Fathers are discovering that there can be great pleasure in domestic and child-care responsibilities. Change, however, also brings difficulties: balancing the demands of family and employment, coping with the interpersonal conflicts caused by changing expectations, and striving to make ends meet in families without sufficient financial resources. These changes bring important questions to the sociological study of families.

Family affairs are believed to be private, but as an institution, the family is very much part of the public agenda. Many people believe that "family breakdown" causes society's greatest problems—thus the intense national discussion around so-called family values. Public policies shape family life directly and indirectly, and family life is now being openly negotiated in political arenas, corporate boardrooms, and courtrooms, as well as in the bedrooms, kitchens, and "family" rooms of individual households.

Defining the Family

The family is a *social institution*, that is, an established social system that emerges, changes, and persists over time. Institutions are "there;" we do not reinvent them every day, although people adapt in ways that make institutions constantly evolve, such as is the case with how families have changed over time. The **family** refers to a primary group of people—usually related by ancestry, marriage, or adoption—who form a cooperative economic unit to care for offspring and each other and who are committed to maintaining the group over time (adapted from Lamanna, Riedmann, and Stewart 2017: 10).

Families are part of what are more broadly considered to be kinship systems. A **kinship system** is the pattern of relationships that define people's relationships to one another within a family. Kinship systems vary enormously across cultures and over time. In some societies, marriage is seen as a union of individuals. In others, marriage is seen as creating alliances between groups. In some kinship systems, marriages are arranged, possibly even involving a broker whose job is to conduct the financial transactions and arrange marriage ceremonies (Lu 2005). In still other kinship systems, maintaining multiple marriage partners may be the norm. Kinship systems can generally be categorized by:

- how many marriage partners are permitted at one time;
- who is permitted to marry whom;
- how descent is determined;
- how property is passed on;
- where the family resides; and
- how power is distributed.

Polygamy is the practice of men or women having multiple marriage partners. Polygamy usually involves one man having more than one wife, technically referred to as *polygyny*. *Polyandry* is the practice of a woman having more than one husband, an extremely rare custom. Within the United States, polygamy is commonly associated with Mormons, even though only a few Mormon fundamentalists (estimated to be only about 2 percent of the state population of Utah) now practice polygamy; those who practice polygamy do so without official church sanction (Brooke 1998).

Monogamy is the practice of a sexually exclusive marriage with one spouse at a time. Monogamy is the most common form of marriage in the United States and other Western industrialized nations. In the United States, monogamy is a cultural ideal that is prescribed through law and promoted through religious teachings. Lifelong monogamy is not always realized, however, as evidenced by the high rate of divorce and extramarital affairs. Many sociologists characterize modern marriage as *serial monogamy* in which individuals may, over a lifetime, have more than one marriage, but only one spouse at a time (Lamanna, Riedmann, and Stewart 2018).

In addition to defining appropriate marriage partners, kinship systems shape the distribution of property in society by determining descent. **Patrilineal kinship** systems trace descent through the father; **matrilineal kinship** systems, through the mother. **Bilateral kinship** (or bilineal kinship) traces descent through both. You can see the continuing influence of patrilineal kinship in contemporary society by noting how children are typically given the name of the father, not the mother. Even though practices are evolving, it is still quite common for women to take men's names when they marry. **Matrilocal** and **patrilocal** are terms used to describe where a married couple resides. In some societies, married couples are expected to move into the husband's residence—or even the husband's family residence.

Kinship systems also determine who one can marry. Even without specific laws, society establishes normative expectations about appropriate marriage partners. In general, people in the United States marry people with very similar social characteristics, such as class, race, religion, and educational backgrounds. Interracial marriage is increasing, though, as the public has become more accepting. Although, overall, interracial marriages are 6 percent of all married couples, by 2013 a record high of 12 percent of new marriages were between people of a different race. Interracial marriage is most common among American Indians, half of whom married someone not Indian (in 2013). Asian Americans are the next most likely to marry interracially, followed by Black Americans at 17 percent. Whites are the least likely to intermarry (Wang 2015; Note: These data do not include marriages between Hispanics and non-Hispanics).

Although interracial marriages have not historically been common, a tremendous amount of energy has been put into preventing them. Laws have prohibited marriage between various groups, including between Whites and African Americans and between Whites and Chinese, Japanese, Filipinos, Hawaiians, Hindus, and Native Americans. Not until 1967 were laws prohibiting interracial marriage declared unconstitutional by the U.S. Supreme Court in the significant case, *Loving v. Virginia* (Kennedy 2003; Takaki 1989).

Extended Families

Extended families are the whole network of parents, children, and other relatives who form a family unit. Sometimes extended families, or parts thereof, live together, sharing their labor and economic resources. In some contexts, "kin" may refer to those who are not related by blood or marriage but who are intimately involved in the family support system and are considered part of the family (Stack 1974). *Othermothers* may be a grandmother, sister, aunt, cousin, or a member of the local community, but she

See for Yourself

Analyzing Family Structures

What is the kinship structure of your family? Make a list of all the people you consider to be part of your family. List them by name and then by your relationship (that is, father, sister, partner, othermother, and so forth). Then, using the concepts that describe kinship systems, identify the specific form of kinship that describes your family (nuclear, extended, matrilineal, patriarchal, and so forth). Are there characteristics of your family that suggest any need to revise these concepts? How does your family structure compare to that of other students in your class?

is someone who provides extensive child care and receives recognition and support from the community around her (Collins 1990: 119). This term emerged from the experience of African American women, whose historically dual responsibilities in the family and work have meant that they have a history of creating alternative means of providing family care for children. Now this is a common practice for many families, including the presidential "first family" during former President Obama's two terms: Upon moving to the White House, the Obamas brought Michelle Obama's mother, Marian Robinson, with them to assist with raising their then young daughters, Malia and Sasha.

Diversity can occur within families, as with this mixed-race family.

The system of *compadrazgo* among Chicanos is another example of an extended kinship system. In this system, the family is enlarged by the inclusion of godparents, to whom the family feels a connection that is the equivalent of kinship. The result is an extended system of connections between "fictive kin" (those who are not related by birth but are considered part of the family) and actual kin that deeply affects family relationships among Chicanos (Baca Zinn, Eitzen, and Wells 2015).

Nuclear Families

In the **nuclear family**, a married couple resides together with their children. Like extended families, nuclear families develop in response to economic and social conditions. The origin of the nuclear family in Western society is tied to industrialization. Before industrialization, families were the basic economic unit of society. Large household units produced and distributed goods, whether in small communities or large plantation or feudal systems where slaves and peasants provided most of the labor. Production took place primarily in households, and all family members were seen as economically vital. Household and production were united, with no sharp distinction between economic and domestic life. Women performed and supervised much of the household work, engaged in agricultural labor, and produced clothes and food. The work of women, men, and children was also highly interdependent. Although the tasks each performed might differ, together they were a unit of economic production.

With industrialization, paid labor was performed mostly away from the home in factories and public marketplaces. The transition to wages for labor created an economy based on cash rather than domestic production. Families became dependent on the wages that workers brought home. The shift to wage labor was accompanied by an assumption that men should earn the "family wage" (that is, be the breadwinner). The family wage system paid men as laborers more than women, leaving women more economically dependent on men. A man's status was also enhanced by having a wife who could afford to stay at home—a privilege seldom accorded to working-class or poor families. The *family wage system* is still reflected in the unequal wages of men and women and in the belief of some that men should be the primary breadwinner.

The unique social conditions that racial–ethnic families have faced also affect the development of family systems. Disruptions posed by the experiences of slavery, migration, and urban poverty affect how families are formed, their ability to stay together, the resources they have, and the problems they face. For example, historically, Chinese American laborers were explicitly forbidden to form families by state laws designed to regulate the flow of labor. Only a small number of merchant families were exempt from the law.

During the westward expansion of the United States, many Mexicans who had settled in the southwest were displaced. The loss of their land disrupted their families and kinship systems. Some then found work in the mines that needed labor as the nation was industrializing. Employers thought they had better control over workers if they did not have their families with them, so families were typically prohibited from living with the working member (usually a man). One result was the development of prostitution camps that followed workers from place to place (Dill 1988).

Families continue to be influenced by social structural forces. Some families, particularly those with marginal incomes, find it necessary for the entire family to work to meet the economic needs of a household. Migration to a new land and exposure to new customs also disrupt traditional family values. Now, the nation is witnessing families being broken up at the U.S.-Mexico border as families are seeking asylum in the United States as they flee from violence in their home nations. The ability to form and sustain nuclear families is directly linked to the economic, political, and racial organization of society.

Understanding Diversity

Interracial Dating and Marriage

Picture this: A young couple, stars in their eyes, holding hands, intimacy in their demeanor. Newly in love, the couple imagines a long and happy life together. When you visualize this couple, who do you see? If your imagination reflects the sociological facts, odds are that you did not imagine this to be an interracial couple. Although interracial couples are increasingly common (and have long existed), people are more likely to form relationships with those of their same race—as well as social class, for that matter. What do sociologists know about interracial dating and marriage?

First, patterns of interracial dating are influenced by race, gender, and ethnicity. Among college students, for example, Black men and women are least likely to date (or even hook up) with a person of a different race, although Black men are more likely to do so than Black women. Hispanic and Asian students are more likely to date outside their group than Black students, and White students are least likely to do so of all (McClintock 2010).

These patterns continue in marriage. Interracial marriages are on the rise, although they are still a small percentage of marriages formed. Hispanics and Whites are the most likely interracial couples; Native Americans, Asians and Hispanics are those most likely to marry someone outside of their own group. Whites are the least likely to do so (Pew Research Center 2010; see also ▲ Figure 13-2).

People in interracial relationships report negative reactions from their families; most reactions are not extremely hostile but are strong enough to put pressure on the interracial couple (Childs 2005; Dalmage 2000). Although most Blacks and Whites profess to have a color-blind stance toward interracial marriages, when pressed, they raise numerous qualifications and concerns about such pairings (Bonilla-Silva 2013).

Regardless of these attitudes, interracial marriage is on the rise, and attitudes are changing. But the facts about interracial dating and marriage show how something seemingly "uncontrollable," such as love, is indeed shaped by many sociological factors.

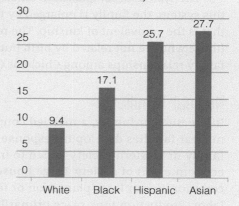

Percent of newlyweds married to someone of different race or ethnicity

White 9.4
Black 17.1
Hispanic 25.7
Asian 27.7

▲ **Figure 13-2 Intermarriage Rates by Race and Ethnicity.** A small percentage of all marriages are interracial/interethnic (6 percent in 2013), but the percentage of newlyweds who married someone of another race or ethnicity is higher, especially among some groups. How would you explain the sociological factors that might affect these group differences in rates of intermarriage?

Source: Pew Research Center. 2010. *Marrying Out: One-in-Seven New U.S. Marriages is Interracial or Interethnic*. Washington, DC: Pew Research Center. **www.pewsocialtrends.org**

Sources: Childs, Erica Chito. 2005. *Navigating Interracial Borders: Black-White Couples and Their Social World*. New Brunswick, NJ: Rutgers University Press; Dalmage, Heather M. 2000. *Tripping on the Color Line: Black-White Multiracial Families in a Racially Divided World*. New Brunswick, NJ: Rutgers University Press.

Sociological Theory and Families

Is the family a source of stability or change in society? Are families organized around harmonious interests, or are they sources of conflict and differential power? How do new family forms emerge, and how do people negotiate the changes that affect families? These questions and others guide sociological theories of the family (see ◆ Table 13-1).

Functionalist Theory and Families

Functionalist theorists interpret the family as filling particular societal needs, including socializing the young, regulating sexual activity and procreation, providing physical care for family members, and giving psychological support and emotional security to individuals. According to functionalism, families exist to meet these needs and to ensure a consensus of values in society. In the functionalist framework, the family was traditionally conceptualized as a mutually beneficial exchange, wherein women receive protection, economic support, and status in return for emotional and sexual support, household maintenance, and the production of offspring (Glenn 1986). At the same time, men in such marriages get the services that women provide—housework, nurturing, food service, and sexual partnership. Functionalists also see families as providing care for children, who are taught the values that society and the family support. In addition, functionalists see the family as regulating reproductive activity, including cultural sanctions about sexuality.

According to functionalist theory, when societies experience disruption and change, institutions such as the family become disorganized, weakening social cohesion. Currently, some analysts interpret the family as "breaking down" under societal strains. Functionalist theory suggests that this breakdown is the result of the disorganizing forces that rapid social change has fostered.

Functionalists also note that, over time, other institutions have begun to take on some functions originally performed solely by the family. For example, as children now attend school earlier in life and stay in school for longer periods of time, schools (and other caregivers) have taken on some of the functions of physical care and socialization originally reserved for the family. Functionalists would say that the diminishment of the family's functions produces further social disorganization because the family no longer carefully integrates its members into society. To functionalists, the family is shaped by the template of society. Such things as the high rate of divorce and the rising numbers of female-headed and single-parent households are the result of social disorganization.

Table 13-1 | **Theoretical Perspectives on Families**

	Functionalism	Conflict Theory	Symbolic Interaction	Feminist Theory
Families	Meet the needs of society to socialize children and reproduce new members	Reinforce and support power relations in society	Emerge as people interact to meet basic needs and develop meaningful relationships	Are gendered institutions that reflect the gender hierarchies in society
	Teach people the norms and values of society	Inculcate values consistent with the needs of dominant institutions	Are where people learn social identities through their interactions with others	Are a primary agent of gender socialization
	Are organized around a harmony of interests	Are sites for conflict and diverse interests of different family members	Are places where people negotiate their roles and relationships with each other	Involve a power imbalance between men and women
	Experience social disorganization ("breakdown") when society undergoes rapid social changes	Change as the economic organization of society changes	Change as people develop new understandings of family life	Evolve in new forms as the society becomes more or less egalitarian

Conflict Theory and Families

Conflict theory interprets the family as a system of power relations that reinforces and reflects the inequalities in society. Conflict theorists also are interested in how families are affected by class, race, and gender inequality. This perspective sees families as the units through which the advantages, as well as the disadvantages, of race, class, and gender are acquired. Conflict theorists view families as essential to maintaining inequality in society because they are the vehicles through which property and social status are acquired (Baca Zinn, Eitzen, and Wells 2015).

The conflict perspective also emphasizes that families in the United States are shaped by capitalism. The family is vital to capitalism because it produces the workers that capitalism requires. Accordingly, personalities within families are shaped to the needs of a capitalist system; thus, families socialize children to become obedient, subordinate to authority, and good consumers. Those who learn these traits become the kinds of workers and consumers that capitalism needs. Families also serve capitalism in other ways. As an example, giving a child an allowance teaches a child capitalist values about money.

Whereas functionalist theory conceptualizes the family as an integrative institution (meaning it has the function of maintaining social stability), conflict theorists depict the family as an institution subject to the same conflicts and tensions that characterize the rest of society. Families are not isolated from the problems facing society as a whole. The struggles brought on by racism, class inequality, sexism, homophobia, and other social conflicts are played out within family life.

Symbolic Interaction Theory and Families

Symbolic interaction emphasizes that meanings people give to their behavior and that of others is the basis of social interaction. Those who study families from this perspective tend to take a more microscopic view of families. Symbolic interaction theorists might ask how different people define and understand their family experience. Symbolic interaction studies how people negotiate family relationships, such as deciding who does what housework, how they will arrange child care, and how they will balance the demands of work and family life.

To illustrate, when people get married, they form a new relationship and new identities with specific meanings within society. Some changes may seem very abrupt—a change of name certainly requires adjustment, as does being called a husband or wife. Other changes are more subtle—how one is treated by others and the privileges couples enjoy (such as being a recognized legal unit). Symbolic interaction theory interprets the marriage relationship as socially constructed; that is, it evolves through the definitions that others in society give it as well as through the evolving definition of *self* that married partners make for themselves.

The symbolic interaction perspective understands that roles within families are not fixed, but rather evolve as participants define and redefine their behavior toward each other. Symbolic interaction is especially helpful in understanding changes in the family because it supplies a basis for analyzing new meaning systems and the evolution of new family forms over time.

Feminist Theory and Families

Feminist theory has contributed new ways of conceptualizing the family by focusing sociological analyses on women's experiences in the family and by making gender a central concept in analyzing the family as a social institution. Feminist theories of the family emerged initially as a criticism of functionalist theory. Feminist scholars argued that functionalist theory assumed that the gender division of labor in the household is functional for society. Feminists have also been critical of functional theory for assuming an inevitable gender division of labor within the family. Feminist critics argue that, although functionalists may see the gender division of labor as functional, it is based on stereotypes about men's and women's roles.

Influenced by the assumptions of conflict theory, feminist scholars do not see the family as serving the needs of all members equally. Quite the contrary, feminists have noted that the family is one of the primary institutions producing the gender relations found in society. Feminist theory conceptualizes the family as a system of power relations and social conflict. In this sense, it emerges from conflict theory, but adds that the family is a *gendered institution* (see Chapter 11). Each of the different theoretical perspectives illuminates different features of family experiences.

Thinking Sociologically

Apply sociological theory to thinking about your own family. What do each of the theories identified here (*functionalism, conflict theory, symbolic interaction, feminist theory*) reveal about your family?

Diversity among Contemporary American Families

Today, the family is one of the most rapidly changing of all of society's institutions. Families are systems of social relationships that emerge in response to social conditions that, in turn, shape the future direction of society. There is no static or natural form for the family. Change and variation in families are social facts.

Among other changes, families today are smaller than in the past. There are fewer births, and they are more closely spaced, although these characteristics of families vary by social class, region of residence, race, and other factors. With longer life expectancy, childbearing and child rearing now occupy a smaller fraction of parents' adult life. During earlier periods, death (often from childbirth) was more likely to claim the mother than the father of small children. Men in the past would have been more likely than now to raise children on their own after the death of a spouse. That trend is now reversed: Women are now more likely to be widowed with children, and death, once the major cause of early family disruption, has been replaced by divorce (Cherlin 2010; Rossi and Rossi 1990).

Demographic and structural changes have resulted in great diversity in family forms. Married couples now make up a smaller proportion of households than was true even thirty years ago. Single-parent households have increased dramatically. Divorced and never-married people make up a larger proportion of the population. Overall, married-couple families make up about half of all households. Single-parent households (typically headed by women), post-childbearing couples, gay and lesbian couples, childless households, and single people are increasingly common (U.S. Census Bureau 2016a). Now people may also spend more years caring for elderly parents than they did raising their children.

Female-Headed Households

One of the greatest changes in family life is the increase in the number of families headed by women. One-quarter of all children live with one parent, the vast majority of whom (85 percent) live with their mother. Although a large number of those living with one parent do so because of divorce, the largest number live with one parent who has never been married. One-quarter of all households are headed by women, although the number of households headed by single fathers has also increased. The odds of living in a single-parent household are even greater for African American and Latino children (U.S. Census Bureau 2016b).

The two primary causes for the growing number of women heading their own households are the high rate of pregnancy among unmarried teens and the high divorce rate, with death of a spouse also contributing. As discussed in Chapter 12, even though the rate of pregnancy among teenagers has declined, the proportion of teen births occurring outside marriage has increased.

Many people see the increase of female-headed households as representing a breakdown of the family and a weakening of social values. An alternative view, however, is that the rise of female-headed households reflects the growing independence of women, some of whom are making decisions to raise children on their own. Not all female-headed households are women who have never married; many are divorced and widowed women whose circumstances may be quite different from those of a younger, never-married woman.

Some claim that female-headed households are linked to problems such as delinquency, the school dropout rate, children's poor self-image, and other social problems. Sometimes the cause of these troubles is attributed explicitly to the absence of men in the family. Sociologists, however, have not found the absence of men as the basis for such problems. Rather, the presence of economic pressure faced by female-headed households, compared with that of male-headed households, puts female-headed households under great strain of the threat of poverty. Add to this fact that of the mothers who are supposed to receive child support, fewer than half actually get the full amount and one-quarter never receive any at all (Grall 2016). Among female-headed households with children, one-third live below the poverty line, with the rates of poverty highest among Black and Hispanic female-headed households (Semega, Fontenot, and Kollar 2017). It is not the makeup of households headed by women that is a problem but the fact that they are most likely to be poor. This phenomenon was discussed in Chapter 8 as the *feminization of poverty*.

Although women head the majority of single-parent families, families headed by a single father are increasing. Male-headed households are, however, less likely than female-headed households to experience severe economic problems. Unlike female-headed households where a man is not present to help with housework and children, single fathers commonly get domestic help from women—either girlfriends, daughters, or mothers (Popenoe 2001).

Married-Couple Families

Married-couple families, long the social norm in U.S. society, are rapidly changing. No longer does marriage refer only to heterosexual couples, as same-sex marriage is increasingly common. One of the biggest changes affecting married couples, especially heterosexual couples, is the increased participation of women in the paid labor force. This change has added new challenges to family life. Families now are able to sustain a median income level only by having both partners in the paid labor force. As a result, families are experiencing substantial *social speedup*, a term reflecting the common feeling among working parents that there is too much to do and too little time to do it.

Women's labor force participation has created other changes in family life. One is the number of married couples who have *commuter marriages*, when work requires one partner in a dual-career couple to reside in a different city, separated by jobs too distant for a daily commute. The common image of a commuter marriage is one consisting of a prosperous professional couple, each holding important jobs, flashing credit cards, and using airplanes like taxis. However, working-class and poor couples also do their share of long-distance commuting: Agricultural workers follow seasonal work; skilled laborers sometimes have to leave their families to find jobs; and many families cross national borders in search of work, often separating one or both parents and children. Although their commute may be less glamorous than that of professional spouses, they are commuting nonetheless. When all types of commuter marriages are included, this form of marriage is more prevalent than is typically imagined.

Stepfamilies

Because of the rise in divorce and remarriage, stepfamilies are now fairly common in the United States. They take numerous forms, including married adults with stepchildren, cohabiting stepparents, and stepparents who do not reside together (Stewart 2001). About 40 percent of marriages involve stepchildren. Stepfamilies may face a difficult period in the transition when two families blend, introducing people to a "new" family. Parents and children may have to learn new roles when they become part of a stepfamily. Children accustomed to being the oldest child in the family, or the youngest, may find that their status in the family group is suddenly transformed. New living arrangements may require children to share rooms, toys, and time with people they perceive as strangers.

In stepfamilies, the parenting roles of mothers and fathers suddenly may expand to include more children, each with his or her needs. Jealousy, competition, and demands for time and attention can make the relationships within stepfamilies tense. The problems are compounded by the absence of norms and institutional support systems for stepfamilies. Without norms to follow, people have to adapt by creating new language to refer to family members and new relationships. Many develop strong relationships within this new kinship system; others find the adjustment extremely difficult, resulting in a high probability of divorce among remarried couples with children (Baca Zinn, Eitzen, and Wells 2015).

Same-Sex Families

The increased visibility of gay and lesbian families has challenged the traditional understanding of families as only heterosexual. The landmark Supreme Court case *Obergefell v. Hodges* (2015) declaring marriage a constitutional right legalized gay marriage in every state. Like other families, gay and lesbian couples share living arrangements, make decisions as partners, and often raise children. They can now do so legally—that is, as long as this case continues to be constitutional law (Mezey 2008).

As we saw in Chapter 12, the number of Americans who support same-sex marriage has increased quite dramatically—even in a short period of time (from 37 percent in 2007 to 62 percent in 2017; Masci, Brown, and Kiley 2017). Same-sex marriage tends to be more acceptable in the eyes of younger people than older groups, raising the question of whether social support for gay marriages will increase over time or whether young people will shift their values as they age.

Public opinion polls do not tell the whole story about acceptance of same-sex couples. Although more people now express support for the formal rights of same-sex couples, fewer people accept the informal rights that heterosexual couples enjoy—such as showing affection in public (Doan et al. 2014). Gains in the approval of formal rights for same-sex couples are important, but on a day-to-day basis, the less formal ways that couples experience acceptance are just as significant. Gays and lesbians who form strong and lasting relationships sometimes do so without support from families and other institutions. Without the full benefits of citizenship—both formal and informal—same-sex couples have had to be innovative in producing new support systems.

Gay and lesbian couples also tend to be more flexible and less gender-stereotyped in their household roles than heterosexual couples. Lesbian households, in particular, are more egalitarian than are either heterosexual or gay male couples. Money also has less effect on the balance of power in lesbian relationships than is true for heterosexual couples. However, where one partner is the primary breadwinner and the other the primary caregiver for children, the partner staying at home becomes economically vulnerable and less able to negotiate her needs, just as in heterosexual relationships (Sullivan 1996).

The new family forms that lesbian and gay couples are creating mean that they have to actively construct new meanings of such things as housework, parenting, and family relationships. Research finds that they do so in collaborative ways, including elaborate networks of family and friends (Dalton and Bielby 2000; Dunne 2000). To date, most gay fathers are those who have children from a previous heterosexual marriage, although the number of gay men adopting children is increasing.

Public debate about gay marriage often centers on the implications for children raised in gay and lesbian families. Research on children in gay and lesbian households finds that, for the most part, there is little difference in outcomes for children raised in gay and lesbian households compared to those raised in heterosexual households. What differences are found result from other factors—not just the parents' sexual orientation. The greatest differences are the result of the homophobia that is directed against children in lesbian and gay families, who are very likely to be stigmatized by others. Such children are also less likely to develop stereotypical gender roles and are more open-minded about sexual matters, although they are no more likely to become gay themselves (Stacey and Biblarz 2001). If we lived in a society more tolerant of diversity, the differences that emerge might be viewed as strengths, not deficits.

Debunking Society's Myths

Myth: The extent to which the public accepts same-sex relationships is best indicated by the increase in public support for the right of same-sex couples to marry.

Sociological Perspective: Sociologists make a distinction in *formal rights* and *informal privileges*. Formal rights would include the legal right to marry; informal privileges mean such things as being able to demonstrate affection in public without ridicule or consequence. Sociological research finds that even though more people are accepting of formal rights for same-sex couples, they are less accepting of the informal privileges that heterosexual couples enjoy (Doan et al. 2014).

Single People

Single people, including those never married, widowed, divorced, and separated, today constitute half of the population (of those over age 15; U.S. Census Bureau 2016). Some of the increase is a result of the rising number of divorced people, but there has also been an increase in the number of those never married (32 percent of the population over age 15 and one-third of those aged 30 to 34). Men and women are also marrying at a later age—at age 27.4 on average for women, age 29.5 for men, compared with age 21 for women and age 24 for men in 1980 (U.S. Census Bureau 2016).

Among singles, patterns of establishing intimate relationships have changed significantly. "Hooking up"—a phrase referring to a casual sexual alliance between two people—has supplanted dating as the pattern by which many young people get to know each other. Courtship no longer follows pre-established norms. Hooking up is widespread on college campuses and influences the campus culture, although only a minority of students engages in it (about 40 percent of college women). Hooking up carries multiple meanings. For some, it means kissing; for others, it means sexual-genital play, but not intercourse; for some, it means sexual intercourse. The vagueness of the term contributes to its becoming a shared cultural phenomenon. A majority engage of college women say that hooking up makes them feel desirable, but also awkward, and they are wary of getting a bad reputation from hooking up too often. The majority of college women still want to meet a spouse while at college (Hamilton and Armstrong 2009; Bogle 2008).

The path to a committed relationship, possibly marriage, involves increasing phases of commitment and sexual exclusivity. Many people find the same sexual and emotional gratification in single life as they would in marriage. Being single is also no longer the stigma it once was, especially for women. Increasing numbers of single people are also forming new types of families. As one example, sociologist Rosanna Hertz has studied women who become mothers by choice outside of marriage. In some cases, these are women who have not found marriage partners or do not want one. Some are lesbian; many are not. Hertz has studied how these women form new family structures, drawing, for example, on kin and friendship networks for familial support (R. Hertz 2006).

Such changes in family patterns reflect an important sociological conclusion: The particular form of the family per se does not predict happiness so much as other social factors, including financial resources, the presence of conflict and violence, and the presence of stressful life problems.

Cohabitation (living together) has become increasingly common. Some of the increase is the result of better census taking, but the increase is also real. Almost two-thirds of married couples now live together before marriage, compared to only 10 percent in 1970. Although some cohabit because they are critical of the existing norms surrounding marriage (Elizabeth 2000), living together has become a common pattern prior to marriage, leading many to ask if living together before marriage stabilizes or destabilizes marriage.

Comparing those who lived together before marriage and those who did not, current research finds that co-habiting couples report little difference in the quality of marriage compared to couples who did not live together before marriage. The greatest predictor of marital quality among those who lived together seems to be the presence of children. That is, those couples who had children prior to marriage report lower-equality marriages than those who did not have children prior to marriage. Regardless of cohabitation prior to marriage, however, researchers also find that all couples report lower quality in their marriages over time (Tach and Halpern-Meekin 2009).

In addition to couples living together, a growing number of single people are remaining in their parents' homes for longer periods of time. Known as the *boomerang generation*, or "accordion families," young people in their twenties are returning home when they would normally be expected to live independently. A much higher proportion of young adults now live with their parents than at any time since the 1940s. Nearly one-third (three in ten) young adults between ages 25 and 34 now live with their parents, compared to only 11 percent in 1980 (Parker 2012). For most, this is for financial reasons, although the particulars differ across social class. Working-class young adults tend to have never left while young people in middle- and upper-middle class families are more likely to "boomerang"—that is, return home to save rent money while perhaps establishing a career or attending school (Klinenberg 2012; Newman 2012).

Marriage and Divorce

Even with the extraordinary diversity of family forms in the United States, the majority of people will still marry at some point in their lives (see ▲ Figure 13-3). Indeed, the United States has the highest rate of marriage of any Western industrialized nation, as well as a high divorce rate.

Marriage

The picture of marriage as a consensual unit based on intimacy, economic cooperation, and mutual goals is widely shared, although marital relationships also involve a complex set of social dynamics, including cooperation and conflict, different patterns of resource allocation, and a division of labor. Sociologists must be careful not to romanticize marriage to the point that they miss other significant social patterns within marriage.

Gender roles are a significant reality of family life, shaping power dynamics within marriage, as well as the allocation of work, the degree of marital happiness, the likelihood of marital violence, and even the leisure time that each partner has. Although people do not like to think of marriage as a power relationship, gender shapes the power that men and women have within marriage, as it does in other relationships. For one thing, sociologists have long found that the amount of money a person earns establishes that person's relative power within the marriage, including the ability to influence decisions, the degree of autonomy and independence held by each partner, and the control of expectations about family life. Although husbands remain the sole or major earner in the majority of heterosexual marriages, there has been a substantial increase in the number of couples in which wives are the sole or major earner (Raley, Mattingly, and Bianchi 2006). This matters because numerous researchers find that women are more able to negotiate marital decisions when they are significant contributors to the household earnings (Tichenor 2005).

Within marriage, gender also shapes the division of household labor. Women do far more work in the home and have less leisure time (Sarkasian and Gerstel 2012). Most employed mothers do two jobs—the so-called *second shift* of housework after working all day in a paid job (Hochschild and Machung 1989)—a pattern that exists not just in the United States, but in other nations as well (see ▲ Figure 13-4). Indeed, sociologists have now even identified the *third shift* of women's work—that is, the greater amount of help that women compared to men give to family and friends, such as assisting those who are sick, preparing holiday celebrations, planning family visits, and so forth (Gerstel 2000). It is little wonder that people are feeling that they have less and less time.

Are men more involved in housework than in the past? Yes and no. Men report that they do more housework, but they devote only slightly more of their time to housework than in the past. Estimates vary regarding the amount of housework that men do, but studies generally find a large gap between the number of hours women give to housework and child care and the hours men give (Offer 2014; Sayer and Fine 2011). Even among couples where both partners are employed, women still do more of the housework in most cases. Fathers do more when there is a child under two in the house, but the increase is mostly accounted for by the amount of child care men provide, not the housework they do. The end result is that men have more time for leisure than do women (Craig and Mullan 2013). Sociologists have found that the allocation of housework is greatly affected by men's and women's experience in their own families of origin; those from households with a more egalitarian division of labor are likely to carry this into their own relationships (Cunningham 2001).

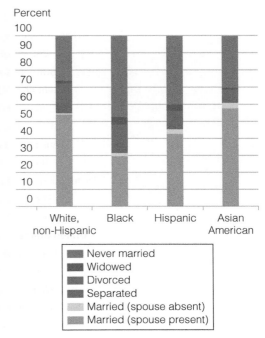

▲ **Figure 13-3 Marital Status of the U.S. Population by Race.** What differences in marital status do you observe here, comparing different racial–ethnic groups? What might explain some of the differences you see?

Data: U.S. Census Bureau. 2013. *America's Families and Living Arrangements.* Washington, DC: U.S. Department of Commerce. **www.census.gov**

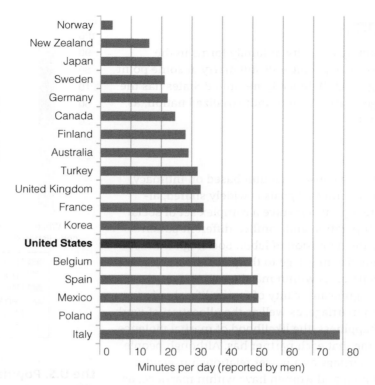

▲ **Figure 13-4** **Men's and Women's Leisure Worldwide.** This graph shows the greater number of leisure minutes per day that men report relative to women. What social factors do you think affect men's greater leisure in these different countries?

Source: OECD. 2009. *Society at a Glance—OECD Social Indicators.* **www .oecd.org/els/social/indicators/SAG**

Despite a widespread belief that young professional couples are the most egalitarian, studies find that there is little difference across social class in the amount of housework that men do. Even women who are earning more than their husbands do more of the housework (Tichenor 2005). Hispanic and Asian women tend to devote more time to housework than do White or Black women, though there are smaller gender differences in the housework men and women do among African American couples, possibly because of the long pattern of Black women's likelihood of employment (Craig and Mullan 2013). Within all families, housework and other forms of family care are negotiated, but typically within the confines of existing gender ideologies (Legerski and Cornwall 2010).

Although marriage can be seen as a romantic and intimate relationship between two people, it can also be seen within a sociological context. Marriage relationships are shaped by a vast array of social factors, not just the commitment of two people to each other. You see this especially when examining marital conflicts. Life events, such as the birth of a child, job loss, retirement, and other family commitments, such as elder care or caring for a child with special needs, all influence the degree of marital conflict and stability (Moen et al. 2001). As conditions in society change, people make adjustments within their relationships, but how well they can cope within a marriage depends on a large array of sociological, not just individual, factors.

Divorce

The United States leads the world not only in the number of people who marry, but also in the number of people who divorce. More than sixteen million people have divorced but not remarried in the population today; more women are in this group than men because women are less likely to remarry following a divorce. Since 1960, the rate of divorce has more than doubled, although it has declined recently since its all-time high in 1980.

Number per 1,000 people

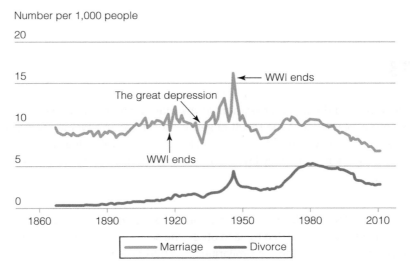

▲ **Figure 13-5   Marriage and Divorce in the United States: Historical Trends.** As you can see, rates of marriage and divorce respond to major events in society. How do you explain the peaks and valleys that you see in this graph?

You will often hear that one in every two marriages ends in divorce, but this is a misleading statistic. The marriage rate is 6.9 marriages per 1000 people and the divorce rate, 3.2 per 1000 people (National Center for Health Statistics 2014). At first glance, it appears that there are half as many divorces as marriages. The marriage rate, however, measures only the number of marriages formed in a year and does not include the number of continuing marriages; thus, divorce is not as widespread as one in every two of all marriages.

As you can see in ▲ Figure 13-5, rates of marriage and divorce fluctuate over time and with major changes in society. Events such as major wars, economic recessions, and other less dramatic changes influence even these most intimate life events.

As you might expect using a sociological lens, the likelihood of divorce is also not equally distributed across all social groups. Divorce is more likely for couples who marry young, while in their teens or early twenties. Second marriages are more likely than first marriages to end in divorce. Divorce is somewhat higher among low-income couples, a fact reflecting the strains that financial problems create. Divorce is also somewhat higher among African Americans than among Whites, partially because African Americans make up a disproportionate part of lower-income groups. Hispanics have a lower rate of divorce than either Whites or Blacks, probably the result of religious influence. It has long been the case that marriages in which the wife has more education than the husband were more likely to end in divorce, but that tendency has disappeared among more recently married couples. Sociologists interpret this as indicative of a shift away from rigid gender expectations in marriage (Schwartz and Han 2014).

A number of factors contribute to the rate of divorce in the United States. Demographic changes (shifts in the composition of the population) are part of the explanation. The rise in life expectancy, for example, has an effect on the length of marriages. In earlier eras, people died younger, and thus the average length of marriages was shorter. Some marriages that earlier would have ended with the death of a spouse may now be dissolved by divorce. Still, cultural factors also contribute to divorce.

In the United States, individualism is a cultural norm, placing a high value on a person's satisfaction within marriage. The cultural orientation toward individualism may predispose people to terminate a marriage in which they are personally unhappy. In other cultural contexts (including this society years ago), marriage, no matter how difficult, may be seen as an unbreakable bond, regardless of whether one is unhappy. Even with the American belief in individualism, people still value the ideal of lifelong romantic love and the security of a long-term partner, whether or not they are able to actually achieve this (Hull et al. 2010).

Men's Caregiving

Research Question

Much research has documented the fact that women do the majority of the housework and child care within families. Why? Many have explained it as the result of gender socialization—women learn early on to be nurturing and responsible for others, whereas men are less likely to do so. Yet, things are changing, and some men are more involved in the "care work" of family life. What explains whether men will be more engaged in family care work?

Research Method

Sociologists Naomi Gerstel and Sally Gallagher studied a sample of 188 married people. They interviewed ninety-four husbands and ninety-four wives, married to each other; the sample was 86 percent White and 14 percent African American but was too small to examine similarities or differences by race.

Research Results

You might expect that men who had attitudes expressing support for men's family responsibilities would be more involved in family care (defined by Gerstel and Gallagher to include elder care, child care, and various household tasks). This is not what Gerstel and Gallagher found. Gender attitudes did not influence men's involvement in caregiving.

Rather, the characteristics of the men's families were the most influential determinant of their engagement in housework and child care. Men whose wives spent the most time helping kin and men who had daughters were more likely to help kin. Having sons had no influence. In addition, men with more sisters tended to spend less time helping with elder parents than men with fewer sisters. Furthermore, men's employment (measured as hours employed, job flexibility, and job stability) did not affect their involvement in care work.

Conclusions and Implications

It is the social structure of the family, not gender beliefs, that shapes men's involvement in family work. As the researchers put it, "It is primarily the women in men's lives who shape the amount and types of care men provide" (Gerstel and Gallagher 2001: 211). This study shows a most important sociological point: Social structure, not just individual attitudes, is the most significant determinant of social behavior.

Questions to Consider

1. Who does the work in your family? Is it related to the social organization of your family, as Gerstel and Gallagher find in other families?
2. Do you think that men's gender identity changes when they become more involved in care work? What hinders and/or facilitates men's engagement in this kind of work?

Source: Gerstel, Naomi, and Sally Gallagher. 2001. "Men's Caregiving: Gender and the Contingent Character of Care." *Gender & Society* 15 (April): 197–217.

Is It True?*

	True	False
1. Half of all marriages end in divorce.		
2. Children who grow up in gay or lesbian families are likely to become gay.		
3. Single people are a larger proportion of the population than was true in the past.		
4. Children are better off growing up in a home where mothers are not employed.		
5. Putting in long hours at work is bad for the health of families.		

*The answers can be found on the next page.

Is It True? (Answers)

1. FALSE. The divorce rate is based on the number of divorces in one year per 1000 people in the population; the marriage rate, the number of marriages in one year per 1000 people. This does NOT mean that half of all marriages end in divorce because marriages made in one year can last many years beyond, and thus not all marriages are counted.
2. FALSE. There is little difference in outcomes for children growing up in gay or lesbian households relative to those living in heterosexual households, including their later sexual orientation (Stacey and Biblarz 2001).
3. TRUE. The number of never-married people has increased substantially since the 1970s, as have the number of divorced people in the population (U.S. Census Bureau 2013).
4. FALSE. Researchers find no negative influence of parental employment per se on children; far more significant are other conditions, especially the economic stability of the family and other stresses for children that emerge from conflict and violence (Perry-Jenkins et al. 2000).
5. FALSE. Although balancing the demands of work and family is stressful for women and men, various factors shape whether family members' health is negatively affected. Men's longer working hours actually reduce stress for the wife, mainly because of the increased resources the increased work brings to the family. Wives' longer working hours, on the other hand, predict worse health for husbands, explained by their lessening of time for exercise as family demands increase (Kleiner and Pavalko 2014).

Changes in women's roles also are related to the rate of divorce. Women today are now less financially dependent on husbands than in the past, even though they still earn less. As a result, the economic interdependence that once bound women and men as a marital unit is no longer as strong. Although most married women would be less well off without access to their husband's income, they could probably still support themselves. This can make it possible for people to end marriages that they find unsatisfactory.

For people in unhappy marriages, divorce, though painful and financially risky, can be a positive option (Kurz 1995). The belief that couples should stay together for the sake of the children is now giving way to a belief, supported by research, that a marriage with protracted conflict is more detrimental to children than divorce. Although there are periodic public outcries about the negative effect of divorce on children, many other factors influence their long-term psychological and social adjustment. Few children feel relieved or pleased by divorce; feelings of sadness, fear, loss, and anger are common, along with desires for reconciliation and feelings of conflicting loyalties. Most children adjust to divorce reasonably well after a year or so. Moreover, children's adjustment is influenced most by factors that precede the divorce.

The single most important factor influencing children's poor adjustment to divorce is marital violence and prolonged discord (Amato et al. 2007; Amato 2010). The emotional strain on children is significantly reduced if the couple remains amicable. If both parents remain active in the upbringing of the children, the evidence shows that children do not suffer from divorce. Especially important is the ability of the mother to be an effective parent after a divorce. Her ability to be effective can be influenced by the resources she has and her ongoing relationship with the father (Baxter et al. 2011).

In the aftermath of divorce, many fathers become distant from their children. Sociologists have argued that the tradition of defining men in terms of their role as breadwinners minimizes the attachment they feel for their children. If the family is then disrupted, they may feel that their primary responsibility, as financial provider, is lessened, leaving them with a diminished sense of obligation to their children.

Family Violence

Generally speaking, the family is depicted as a private sphere where members are nurtured and protected, existing away from the influences of the outside world.

Although this is the experience of many, families also can be locales for violence, disruption, and conflict. Family violence, hidden for many years, is a phenomenon that has recently been the subject of much sociological research.

Domestic Violence and Abuse

Estimates of the extent of domestic violence are hard to come by and notoriously unreliable because the majority of cases of domestic violence go unreported. The National Centers for Disease Control estimates that 33 percent of women will be raped, physically assaulted, or stalked by an intimate partner in their lifetime (Black et al. 2011). Men also experience partner violence, although far less frequently. Women who experience violence are also twice as likely as men to be injured. Violence also occurs in gay and lesbian relationships, although silence around the issue may be even more pervasive given the marginalized status of gays and lesbians. Men living with male partners are just as likely to be raped, assaulted, or stalked as are women living with men, but the incidence of violence against women by women partners is about half as likely as heterosexual violence. Researchers conclude that this is because most domestic violence is committed by men. Violence is usually accompanied by emotionally abusive and controlling behavior (Renzetti et al. 2010; Tjaden and Thoennes 2000).

One of the most common questions asked about domestic violence is why victims stay with their abuser. First, despite the belief that battered women do not leave their abusers, the majority do leave—at least for a period of time—and they seek ways to prevent further victimization. Some do not leave, and others leave and then return. Why? The answers are complex and stem from sociological, psychological, and economic problems. Victims tend to believe that the batterer will change, but they also find they have few options. Victims may also perceive that leaving will be more dangerous, because violence can escalate when an abuser thinks he (or she) has lost control. Many women are unable to support their children and meet their living expenses without a husband's income. Mandatory arrest laws in cases of domestic violence can exacerbate this problem because they may, despite their intentions, discourage a woman from reporting violence for fear her batterer will lose his job (Miller 1997). Sociological analyses of violence in the family have led to the conclusion that women's relative powerlessness in the family is at the root of high rates of violence against women. Because most violence in the family is directed against women, the imbalance of power between men and women in the family is the source of most domestic violence. Because women are relatively powerless within the society, they may not have the resources to leave their marriage.

Child Abuse

Violence within families also victimizes many children who experience child abuse. Not all forms of child abuse are alike. Some people consider repeated spanking to be abusive; others think of this as legitimate behavior. Child abuse, however, is behavior that puts children at risk and may include physical violence and neglect. As with battering, the exact incidence of child abuse is difficult to know. States vary in how they define and count child abuse and neglect. Most consider physical abuse, neglect, sexual abuse, and psychological maltreatment as constituting abuse. At the national level, the best estimates are that about three and a half million children are reviewed by child protection services (cite). Many cases, however, are likely hidden from public scrutiny.

The most frequent reports involve children under one year of age. Neglect is the most common form of abuse. Parents and relatives are the most common perpetrators. Girls are 63 percent of reports. White children are 54 percent of reports; African American children, 20 percent; and Hispanic children, 14 percent. The remaining cases are of other racial–ethnic groups and those not identified by race. These data must be interpreted with caution, however, because higher rates of reported victimization among African Americans or other groups may simply reflect the greater likelihood of some communities and families being scrutinized by social service agencies and the police, thus leading to higher levels of reporting (U.S. Administration for Children and Families 2015).

Whatever the actual incidence of child abuse, research finds a number of factors associated with abuse, including chronic alcohol use by a parent, unemployment, and isolation of the family. The absence of social supports—in the form of social services, community assistance, and cultural norms about the primacy of motherhood—is related to child abuse, because most abusers are those with weak community ties and little contact with friends and relatives (Baca Zinn, Eitzen, and Wells 2015).

Incest

Incest is a particular form of child abuse involving sexual relations between people who are closely related. A history of incest has been related to a variety of other problems, such as drug and alcohol abuse, runaways, delinquency, and various psychological problems, including the potential for violent partnerships in adult life. Studies find that fathers and uncles are the most frequent incestuous abusers. Incest is most likely in families where mothers are debilitated (such as by mental illness or alcoholism). In such families, daughters often take on the mothering role, being taught to comply with men's demands to hold the family together. Scholars have linked women's powerlessness within families to the dynamics surrounding incest (Fischer 2003; Herman 1981).

Elder Abuse

The National Center on Elder Abuse estimates that about 10 percent of elders in the United States are abused, but it is difficult to gauge the true extent of the problem. Elder abuse tends to be hidden in the privacy of families, and victims are reluctant to talk about their situations, so estimates are only approximations. What is known is that reports of elder abuse have increased. Whether this reflects an actual increase or more reporting is open to speculation (Lachs and Pillemer 2015; National Center on Elder Abuse 2015).

Why are the elderly abused? One explanation is that caring for the elderly is very stressful for a caregiver—usually a daughter who may be employed in addition to caring for the elderly person. Abusers are most likely to be middle-aged women and (sadly) the daughter of the victim—the person most likely to be caring for the older person. Sons, however, are most likely to be engaged in direct physical abuse, accounting for almost half of the known physical abuses. Sometimes the physical abuser is a husband, where the abuse is a continuation of abusive behavior in the marriage. The same factors that affect family life in any generation contribute to the problem of elder abuse (Teaster 2000).

Changing Families in a Changing Society

Like other social institutions, the family is in a constant state of change, particularly as new social conditions arise and as people in families adapt to the changed conditions of their lives. Some changes affect only a given family—the individual changes that come from the birth of a new child, the loss of a partner, divorce, migration, and other life events. These changes are what C. Wright Mills referred to as "troubles" (see Chapter 1). Some may even be happy events; the point is that they are changes that happen at an individual level, as people adjust to the presence of a new child, adjust to a breakup with a long-term partner, or grieve the loss of a spouse.

As Mills would have noted, many microsociological events that people experience in families have their origins in the broader macrosociological changes affecting society as a whole (see ▲ Figure 13-6 as

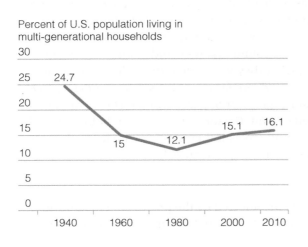

Percent of U.S. population living in multi-generational households

▲ **Figure 13-6   Multigenerational Households over Time.** Living in an extended family household was common before World War II, but, as you can see, the trend was a decline in this form of living arrangement until fairly recently. Sociologists explain the decline as the result of several changes in society, including the growth of nuclear family-centered suburbs, a decline in immigration in the post–World War II period, and a general rise in the affluence of the population. This trend reversed around 1980, as you can see in this line graph. What changes in society do you think have been influencing the rise in multigenerational households? Is there evidence of these trends in your own family? Why?

Source: Pew Research Center. 2010. "The Return of the Multi-Generational Family Household. Data from U.S. Decennial Census Data, 1940–2000." Washington, DC: Pew Research Center. **www.pewsocialtrends.org**

an example). These may be long-term changes (such as changes in women's roles we have been noting), or they may be particular to a given time in history. For example, what impact do difficult economic times have on families?

This is a question that has been examined through sociological research. Economic problems can lead to a broken family, but research has also found patterns by which families adapt to stress. One example comes from sociological work by Jennifer Sherman. Sherman studied a rural community where there was massive job loss when the local plant shut down. She found that, even under financial strain, families were more likely to stay together when men in the family were flexible in their gender roles, such as men taking on additional household work and child care while mothers take on more employment (Sherman 2009). Those who are rigid in their outlook and roles may have more trouble adjusting to social change.

A project by sociologist Marianne Cooper also looked at how families adapted to the stress generated from the economic recession of 2008. She found that how families adapt to economic insecurity varies across the three social class groups she studied. Upper-income families actually dealt with economic insecurity by what she called "upscaling"—that is, yearning for more than what they already had. Their worry that they did not have enough paradoxically increased their stress, even though they were the more economically privileged group in her research. Middle-income families, on the other hand, downscaled, resigning themselves to living with less and trying to convince themselves that they could do just fine with fewer material resources. Low-income families were the most likely group to turn to religion as a way of easing their insecurity, although, like middle-class families, they too downscaled their material expectations. Cooper's work focuses on the emotion work that families do when faced with insecure times. She also found that women were more likely than men to do the worrying (Cooper 2014).

Immigration is another source of change for families. Immigration can unite families, but it can also break them up. The societal conditions that generate immigration are macrosociological—that is, a family's or person's decision to migrate may be the result of blocked opportunities in one's homeland and perceived opportunities—or even safety—in another. But macrosociological forces are experienced by families and individuals at the microsociological level—that is in our personal relationships, emotional state, and other individual experiences.

The linkage between macro- and microsociological conditions and their impact on families is especially apparent in the debate about "dreamers"—very young children who came to the United States with someone in their family and remained for years, even though under illegal entry. Now grown, unless the nation adopts policies to protect them legally, dreamers are threatened with deportation to a country they have never known, as their entire experience has been within the United States. Dreamers then "live in limbo" as sociologist Roberto Gonzales writes after having studied these young people (Gonzales 2015). Their lack of legal identity leaves them without an anchor in either country and can override all of their other achievements, identities, and dreams.

Immigration and economic conditions are two sources of family change. Other long-term structural changes are also changing people's expectations about families. Young people, in particular, have new expectations, particularly about men's and women's roles in the family. The influence of the feminist movement, along with the greater employment of women, has particularly led younger people to have more egalitarian expectations about family life. Sociologist Kathleen Gerson has extensively studied young adults and their expectations for family and work life. She has found that young adults want relationships where both partners combine work and family, but also found that men and women differ significantly in what she calls their "fallback position"—that is, what they would do if their ideal egalitarian arrangements either are not realized or fail. Men, Gerson found, are far more likely than women to say they would "fall back" on a traditional arrangement, that is, men working and women staying home. On the other hand, women say that their fallback is to be self-reliant. Gerson concludes that both men and women will need to be more flexible in their gender and family roles if they are to weather the changes that will likely impact families in the future (Jacobs and Gerson 2015; Gerson 2010).

Global Changes in Family Life

Changes in the institutional structure of families are also being affected by the process of globalization. The increasing global basis of the economy means that people often work long distances from other family members—a phenomenon that occurs at all points on the social class spectrum, although the experience of such global mobility varies significantly by social class. A corporate executive may accumulate thousands—even millions—of first-class flight miles, crossing the globe to conduct business. A regional sales manager may spend most nights away from a family, likely staying in modestly priced motels and eating in fast-food franchises along the way. Truckers may sleep in the cabs of their tractor trailers after logging extraordinary numbers of hours of driving in a given week. Laborers may move from one state or country to the next, following the pattern of the harvest, living in camps away from families, and being paid by the amount they pick.

Global patterns of work and migration have created a new family form, the **transnational family**, defined as families where one parent (or both) lives and works in one country while his or her children remain in the country of origin. A good example is found in Hong Kong, where most domestic labor is performed by Filipina women who work on multiple-year contracts managed by the government, typically on a live-in basis. They leave their children in the Philippines, usually cared for by a relative, and send money home; the meager wages they earn in Hong Kong far exceed the average income of workers in the Philippines. This pattern is so common that the average Filipino migrant worker supports five people at home; one in five Filipinos directly depends on migrant workers' earnings (Parreñas 2001).

One need not go to other nations to see such transnational patterns in family life. In the United States, Caribbean women and African American women have had a long history of having to leave their children with others while they sought employment in different regions of the country. Central American and Mexican women may come to work in the United States while their children stay behind. Mothers may return to see their children whenever they can, or alternatively, children may spend part of the year with their mothers, part with other relatives.

Mothers in transnational families have to develop new concepts of their maternal role, because their situation means giving up the idea that biological mothers should raise their own children. Many have expanded their definition of motherhood to include breadwinning, traditionally defined as the role of fathers. Transnational women also create a new sense of home, one not limited to the traditional understanding of "home" as a single place where mothers, fathers, and their children reside (Parreñas 2005; Hondagneu-Sotelo 2001).

Families and Social Policy

Family social policies are the subject of intense national debate. How people think about families—and whether they consider certain forms of relationships as family—very much shapes public debates about family policies (Powell et al. 2010). What responsibility does society have to help parents balance the demands of work and family? Is educational failure a result of social problems in the family? Many issues on the front lines of national social policy engage intense discussions of families. Some claim the family is breaking down. Others celebrate the increased diversity among families. Many blame the family for the social problems our society faces. Drugs, low educational achievement, crime, and violence are often attributed to a crisis in "family values," as if rectifying these attitudes is all it will take to solve our nation's difficulties.

The family is the only social institution that typically takes the blame for all of society's problems. Is it reasonable to expect families to solve social problems? Families are afflicted by most of the structural problems that are generated by racism, poverty, gender inequality, and class inequality. Expecting families to solve the problems that are the basis for their own difficulties is like asking a poor person to save us from the national debt.

Balancing Work and Family

Balancing the multiple demands of work and family is one of the biggest challenges for most families. With more parents employed, it is difficult to take time from one's paid job to care for newborn or newly adopted children, tend to sick children, or care for elderly parents or other family members. As more families include two earners, more people feel pulled in multiple directions, always strategizing to find the time to get everything done. Work institutions are structured on a gendered model of the male bread-winner, where family and work are assumed to be separate, nonintersecting spheres. Now though there is significant "spillover" between family and work—work seeping into the home and home also affecting people's work (Moen 2003).

The **Family and Medical Leave Act (FMLA)**, adopted by Congress in 1993, is meant to provide help for these conflicts. The FMLA requires employers to grant employees a total of twelve weeks in unpaid leave to care for newborns, adopted children, or family members with a serious health condition. The FMLA is the first law to recognize the need of families to care for children and other dependents. A number of conditions, however, limit the effectiveness of the FMLA, not the least of which is that the leave is unpaid, making it impossible for many employed parents. Many workers in firms where there are family-friendly policies worry that taking advantage of these policies will harm their prospects for career advancement (Blair-Loy and Wharton 2002). Currently, only 11 percent of workers have child-care benefits available to them from employers (Bureau of Labor Statistics 2016). Among industrial-ized nations, the United States provides the least in support for maternity and child-care policies (see ◆ Table 13-2).

Child Care

Family leave policies, much as they are needed, also do not address the ongoing needs for child care. Half of all working families have child-care expenses; the other half either have unpaid relatives or friends providing care or they arrange their work schedules to coincide with school hours. The average family spends about $18,000 per year on child care. Because the U.S. Department of Health and Human Services recommends that a family should spend about 10 percent of their household budget on child care, this means that a family would need to earn $180,000 to meet this guideline—a figure far above the median family income (Backman 2016).

Many parents struggle to find good and affordable child care for their children, some relying on rela-tives for care; others, on paid providers; and some, a combination of both. In the United States, one-half

Table 13-2	Maternity Leave Benefits: A Comparative Perspective		
Country	Length of Maternity Leave	Percentage of Wages Paid in Covered Period	Provider of Coverage
Zimbabwe	90 days	100%	Employer
Cuba	18 weeks	100%	Social Security
Iran	90 days	67% for 16 weeks	Social Security
China	90 days	100%	Employer
Saudi Arabia	10 weeks	50 or 100%	Employer
Canada	17–18 weeks	55% for 15 weeks	Employment insurance
Germany	14 weeks	100%	Social Security to a ceiling; employer pays difference
France	16–26 weeks	100%	Social Security
Italy	5 months	80%	Social Security
Japan	14 weeks	60%	Social Security or health insurance
Russian Federation	140 days	100%	Social Security
Sweden	14 weeks	450 days, 100% paid	Social Security
United Kingdom	26 weeks	90% for 6 weeks; flat rate thereafter	92% public funds; remainder from employer
United States	12 weeks	0%	Solely at discretion of employer

Source: United Nations. 2015. *The World's Women 2015: Trends and Statistics.* New York: United Nations.

of three-year-olds and two-thirds of four-year-olds now spend much of their time in child-care centers. But the national approach is one of patching together different programs and primarily relying on private initiatives for care. Compare this with France. Although participation is voluntary, almost all parents in France enroll young children in the *école maternelle* system, where a place is guaranteed to every child aged 3 to 6. These child-care centers are integrated with the school system and are seen as a form of early education. Moreover, in the United States, child-care costs match tuition costs at public universities, but child care in France is seen as a social responsibility and is paid by the government. National norms about whether families are a private or public responsibility clearly shape social policy (Clawson and Gerstel 2002; Folbre 2001).

Care work—work that sustains life, including child care, elder care, housework, and other forms of household labor—is increasingly provided to middle- and upper-class families by women of color and immigrant women (Duffy 2011). Nannies, cleaners, and personal attendants now do much of the domestic work that wives and mothers once provided. These trends raise many new questions for sociologists, such as how work is negotiated outside of the public labor market, how "mothering" is defined when it is provided by multiple people, how domestic workers care for their own families, and what work conditions exist in the lives of domestic laborers (Hondagneu-Sotelo 2001). Clearly, new social policies are needed to address the needs of diverse families.

Elder Care

The shrinking size of families means that the proportion of elderly people is growing faster than the number of younger potential caretakers. As life expectancy has increased and people live longer, elder care becomes a greater and greater need. Family members provide almost all long-term care for the elderly—work that is often taken for granted.

Women, who shoulder much of the work of elder care, can now expect to spend more years as the child of an elderly parent than as the mother of children under eighteen. The effects of the burden of care are apparent in the stress that women report from this role. Women also believe they are better at elder care than their husbands and brothers, but with the rapid increase in the older population that lies ahead, these social norms may have to change. As the U.S. population ages, social policies will likely need to respond to this growing need.

Because families are so diverse, different families need different social supports. Family leave policies that give parents time off to care for their children or sick relatives are helpful but of little use to people who cannot afford to take time off work without pay. Greater employer support for child care can help men and women meet family needs. Some policies will benefit some groups more than others—one reason why policymakers need to be sensitive to the diversity of family experiences. Social policies cannot solve all the problems that families face, but they can go a long way toward creating the conditions under which diverse family units can thrive.

Defining Religion

Religion has a profound effect on society and human behavior. This is easily observed in daily life outside the family. Church steeples dot the landscape. Invocations to a religious deity occur at the beginning of many public gatherings. The news frequently reports on events generated by religious conflict.

Religious beliefs have led to conflict, but religious beliefs have also been the soul of some of the most liberating social movements, including the civil rights movement and other human rights movements around the world. Some of life's sweetest moments are marked by religious celebration, and some of its most bitter conflicts persist because of unshakable religious conviction. Religion is both an integrative force in society and the basis for many of our most deeply rooted social conflicts.

Sociologists study religion as both a belief system and a social institution. The belief systems of religion have a powerful hold on what people think and how they see the world. The patterns and practices of religious institutions are among the most important influences on people's lives. Sociologists

Religious socialization is a powerful source of people's values; most of the time, children take on the religious orientation of their family of origin.

Sally and Richard Greenhill/Alamy Stock Photo

are interested in several questions about religion: How are religious belief and practice related to other social factors, such as social class, race, age, gender, and level of education? How are religious institutions organized? How does religion influence social change? In using sociology to understand religion, what is important is not what one believes about religion, but one's ability to examine religion objectively in its social and cultural context.

What is religion? Most people think of it as a category of experience separate from the mundane acts of everyday life, perhaps involving communication with a deity or communion with the supernatural (Johnstone 1992). Sociologists define **religion** as an institutionalized system of symbols, beliefs, values, and practices by which a group of people interprets and responds to what they feel is sacred and that provides answers to questions of ultimate meaning (Johnstone 1992; Glock and Stark 1965). The elements of this definition bear closer examination:

1. **Religion is institutionalized.** Religion is more than just beliefs: It is a pattern of social action organized around the beliefs, practices, and symbols that people develop to answer questions about the meaning of existence. As an institution, religion presents itself as larger than any single individual; it persists over time and has an organizational structure into which members are socialized.

2. **Religion is a feature of groups.** Religion is built around a community of people with similar beliefs. Religion is a cohesive force among believers because it is a basis for group identity and gives people a sense of belonging to a community or organization. Religious groups can be formally organized, as in the case of bureaucratic churches, or they may be more informally organized, ranging from prayer groups to cults. Some religious communities are extremely close-knit, as in convents. Other religious communities are more diffuse, such as people who identify themselves as Protestant but attend church only on Easter.

3. **Religions are based on beliefs that are considered sacred.** The **sacred** is that which is set apart from ordinary activity for worship, seen as holy, and protected by special rites and rituals. The sacred is distinguished from the **profane**, which is of the everyday world and specifically not religious (Chalfant et al. 1987; Durkheim 1947/1912). Each religion defines what is to be considered sacred; most religions have sacred objects and sacred symbols. The holy symbols are infused with special religious meaning and inspire awe.

   A **totem** is an object or living thing that a religious group regards with special reverence. A statue of Buddha is a totem and so is a crucifix hanging on a wall. Among the Zuni (a native American group), fetishes are totems; these are small, intricately carved animal objects representing different dimensions of Zuni spirituality. A totem is important not for what it is, but for what it represents. To a Christian taking communion, a piece of bread is defined as the flesh of Jesus; eating the bread unites the communicant mystically with Christ. To a nonbeliever, the bread is simply that—a piece of bread (McGuire 2008). Likewise, Native Americans hold certain ground to be sacred and are deeply offended when the holy ground is disturbed by industrial or commercial developers who see only potential profit.

4. **Religion establishes values and moral proscriptions for behavior.** A proscription is a constraint imposed by external forces. Religion typically establishes proscriptions for the behavior of believers, some of them quite strict. For example, the Catholic Church defines living together as sexual partners outside marriage as a sin. Often religious believers come to see such moral proscriptions as simply "right" and behave accordingly. At other times, individuals may consciously reject moral proscriptions, although they may still feel guilty when they engage in a forbidden practice.

In an ancient Indian ritual, devotees of Jainism soak in orange vermillion. Durkheim would have seen this sacred ritual as promoting cohesion and identity.

Of course, what people believe and what they do can be very contradictory, perhaps best exemplified by the various scandals recently reported involving sexual abuse of young boys by Catholic priests.

5. **Religion establishes norms for behavior.** Religious belief systems establish social norms about how the faithful should behave in certain situations. Worshippers may be expected to cover their heads in a temple, mosque, or cathedral, or to wear certain clothes. Such behavioral expectations may be quite strong. The next time you are at a gathering where a prayer is said before a meal, note how many people bow their heads, even though some of those present may not believe in the deity being invoked.

6. **Religion provides answers to questions of ultimate meaning.** The ordinary beliefs of daily life are secular beliefs and may be institutionalized, but they are specifically not religious. Science, for example, generates secular beliefs based on particular ways of thinking—logic and empirical observations are at the root of scientific beliefs. Religious beliefs, in contrast, often have a supernatural element. They emerge from spiritual needs and may provide answers to questions that cannot be probed with the profane tools of science and reason. Think of the difference in how religion and science explain the origins of life. Whereas science explains this as the result of biochemical and physical processes, different religions have other accounts of the origin of life.

See for Yourself

For one week, keep a daily log, noting every time you see an explicit or implicit reference to religion. At the end of the week, review your notes and ask yourself how religion is connected to other social institutions. Based on your observations, how do you interpret the relationship between the sacred and the secular in this society?

Mario Tama/Getty Images News/Getty Images

The Significance of Religion in the United States

The United States is one of the most religious societies in the world. Religion has a firm grip on the nation's culture, even while the number of people who identify themselves as religious is on the decline, at least as measured in public opinion surveys. About half (53 percent in 2016) of the public say that religion is very important in their own life; another 22 percent say it is fairly important. Fifty-five percent of Americans think religion can solve all or most of society's problems, a decline from 82 percent who said so in the late 1950s (Swift 2017; Gallup 2016). Religion is, for millions of people, the strongest component of their individual and group identity. Much of the world's most celebrated art, architecture, and music has its origins in religion, whether in the classical art of western Europe, the Buddhist temples of the east, or the gospel rhythms of contemporary rock.

Religion is also strongly related to a number of social and political attitudes. Religious identification is a good predictor of how traditional a person's beliefs will be. People who belong to religious organizations that encourage intolerance of any form are most likely to be racially prejudiced. However, there is not a simple relationship between religious belief and prejudice, because religious principles are also often the basis for lessening racial prejudice. Those with deeper religious involvement tend to have more traditional gender attitudes. And, certainly not all religious people are homophobic, but religion does play a significant role in shaping attitudes about such things as same-sex marriage and other gay rights (Olson, Cadge, and Harrison 2006).

The Dominance of Christianity

Despite the U.S. Constitution's principle of the separation of church and state, Christian religious beliefs and practices dominate U.S. culture. Indeed, Christianity is often treated as if it were the national religion. It is commonly said that the United States is based on a Judeo-Christian heritage, meaning that our basic cultural beliefs stem from the traditions of the pre-Christian Old Testament of the Bible (the Judaic tradition) and the gospels of the New Testament. The dominance of Christianity is visible everywhere. State-sponsored colleges and universities typically close for Christmas break, not Yom Kippur. Christmas is a national holiday, but not Ramadan, the most sacred holiday among Muslims. Despite the dominance of Christianity, however, the pattern of religion in the United States is a mosaic one.

Measuring Religious Faith

Religiosity is the intensity and consistency of practice of a person's (or group's) faith. Sociologists measure religiosity both by asking people about their religious beliefs and by measuring membership in religious organizations and attendance at religious services (see ▲ Figure 13-7).

The majority of people in the United States identify themselves as Protestant (49 percent) or Catholic (23 percent), though there is great religious diversity within the nation. Moreover, the number of Americans who say they are religious but do not have a particular denomination has grown. This change reflects the growth of nondenominational congregations that are only loosely affiliated with traditional denominations; most of these are evangelical (Newport 2012b).

The other change you can see is the number of Americans who describe themselves as evangelical Christians—those who take their religions (almost always Protestants) very seriously and believe that they have a religious calling. Forty percent of Americans describe themselves as evangelical; 70 percent of Black Americans say they are evangelical. The significance of evangelical religion extends beyond religious faith, as being evangelical is significantly related as well to a host of conservative political beliefs (Newport 2016; Newport and Carroll 2005).

Forms of Religion

Religions can be categorized in different ways according to the specific characteristics of faiths and how religious groups are organized. In different societies and among different religious groups, the

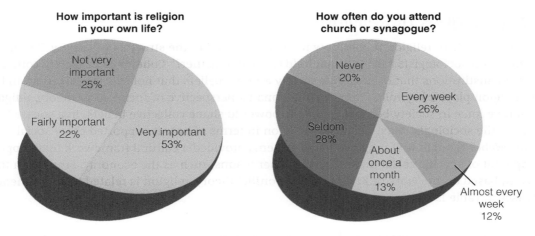

How important is religion
in your own life?

- Not very important 25%
- Fairly important 22%
- Very important 53%

How often do you attend
church or synagogue?

- Never 20%
- Every week 26%
- Seldom 28%
- About once a month 13%
- Almost every week 12%

▲ **Figure 13-7   Measuring Religiosity.** There are different ways of measuring religiosity. Public opinion polls provide one way to do so, as noted in the following two questions.

Source: Gallup Poll. 2016. "Religion." Princeton, NJ: *The Gallup Poll.* **http://news.gallup.com/poll/1690/religion.aspx**

form religion takes reflects differing belief systems and reflects and supports other features of the society. Believing in one god or many, worshipping in small or large groups, and associating religious faith with gender roles all contribute to the social organization of religion and its relationship to the rest of society.

Religiosity varies significantly among different groups in society. Church membership and attendance is higher among women than men and more prevalent among older than younger people. African Americans are more likely than Whites to belong to and attend church. On the whole, church membership and attendance fluctuate over time; membership has decreased slightly since 1940, but attendance has remained largely the same since. Large, national religious organizations, such as the mainline Protestant denominations, have lost many members, whereas smaller, local congregations have increased membership.

In recent years, there has been a decrease in the number of people who think that religion can answer all or almost all of today's problems. Changes in immigration patterns have also affected religious patterns in the United States, with Muslims, Buddhists, and Hindus now accounting for several million believers. One of the greatest changes has been a tremendous increase in the number identifying as evangelical Protestants, but Islam is now one of the fastest-growing religions in the United States (Lipka 2017; Pew Research Center 2015).

A basic way to categorize religions is by the number of gods or goddesses adherents worship. **Monotheism** is the worship of a single god. Christianity and Judaism are monotheistic in that both Christians and Jews believe in a single god who created the universe. Monotheistic religions typically define god as omnipotent (all-powerful) and omniscient (all-knowing). **Polytheism** is the worship of more than one deity. Hinduism, for example, is extraordinarily complex, with millions of gods, demons, sages, and heroes—all overlapping and entangled in religious mythology. Within Hinduism, the universe is seen as so vast that it is believed to be beyond the grasp of a single individual, even a powerful god (Grimal 1963).

Religions may also be patriarchal or matriarchal. *Patriarchal religions* are those in which the beliefs and practices of the religion are based on male power and authority. Christianity is a patriarchal religion; the ascendancy of men is emphasized by the role of women in the church, the instruction given on relations between the sexes, and even the language of worship itself. *Matriarchal religions* are based on the centrality of female goddesses, who may be seen as the source of food, nurturance, and love, or who may serve as emblems of the power of women. In societies based on matriarchal religions, women are more likely to share power with men in the society at large. Likewise, in highly sexist, patriarchal societies, religious beliefs are also likely to be patriarchal.

Sociological Theories of Religion

The sociological study of religion probes how religion is related to the structure of society. Recall that one basic question sociologists ask is, "What holds society together?" Coherence in society comes from both the social institutions that characterize society and the beliefs that hold society together. In both instances, religion plays a key role. From the functionalist perspective of sociological theory, religion is an integrative force in society because it has the power to shape collective beliefs. In a somewhat different vein, the sociologist Max Weber saw religion in terms of how it supported other social institutions. Weber thought that religious belief systems provided a cultural framework that supported the development of specific social institutions in other realms, such as the economy. From yet a third point of view, based on the work of Karl Marx and conflict theory, religion is related to social inequality in society (see ◆ Table 13-3).

Emile Durkheim: The Functions of Religion

Emile Durkheim argued that religion is functional for society because it reaffirms the social bonds that people have with each other, creating social cohesion and integration. Durkheim believed that the cohesiveness of society depends on the organization of its belief system. Societies with a unified belief system are highly cohesive; those with a more diffuse or competing belief system are less cohesive.

Religious **rituals** are symbolic activities that express a group's spiritual convictions. Making a pilgrimage to Mecca, for example, is an expression of religious faith and a reminder of religious belonging. In Durkheim's view, religious rituals are vehicles for the creation, expression, and reinforcement of social cohesion. Groups performing a ritual are expressing their identity as a group. Whether the rituals of a group are highly elaborated or casually informal, they are symbolic behaviors that sustain group awareness of unifying beliefs. Lighting candles, chanting, or receiving a sacrament are behaviors that reunite the faithful and help them identify with the religious group, its goals, and its beliefs. Durkheim believed that religion binds individuals to the society in which they live by establishing what he called a **collective consciousness**, the body of beliefs common to a community or society that gives people a sense of belonging. In many societies, religion establishes the collective consciousness and creates in people the feeling that they are part of a common whole.

Durkheim's analysis of religion suggests some of the key ideas in symbolic interaction theory, particularly in the significance he gave to symbols in religious behavior. Symbolic interaction theory sees

Table 13-3	Theoretical Perspectives on Religion		
	Functionalism	**Conflict Theory**	**Symbolic Interaction**
Religion and the social order	Is an integrative force in society	Is the basis for intergroup conflict; inequality in society is reflected in religious organizations, which are stratified by factors such as race, class, or gender	Is socially constructed and emerges with social and historical change
Religious beliefs	Provide cohesion in the social order by promoting a sense of collective consciousness	Can provide legitimation for oppressive social conditions	Are socially constructed and subject to interpretations; can also be learned through religious conversion
Religious practices and rituals	Reinforce a sense of social belonging	Define in-groups and out-groups, thereby defining group boundaries	Are symbolic activities that provide definitions of group and individual identity

religion as a socially constructed belief system, one that emerges in different social conditions. From the perspective of symbolic interaction, religion is a meaning system that gives people a sense of identity, defines one's network of social belonging, and confers one's attachment to particular social groups and ways of thinking.

Emile Durkheim theorized that public rituals provide cohesion in society.

Max Weber: The Protestant Ethic and the Spirit of Capitalism

Theorist Max Weber also saw a fit between the religious principles of society and other institutional needs. In his classic work, *The Protestant Ethic and the Spirit of Capitalism*, Weber argued that the Protestant faith supported the development of capitalism in the Western world. He began by noting a seeming contradiction: How could a religion that supposedly condemns extensive material consumption coexist in a society (such as the United States) with an economic system based on the pursuit of profit and material success?

Weber argued that these ideals were not as contradictory as they seemed. As the Protestant faith developed, it included a belief in predestination—one's salvation is predetermined and a gift from god, not something earned. This state of affairs created doubt and anxiety among believers, who searched for clues in the here and now about whether they were among the chosen—called the "elect." According to Weber, material success was taken to be one clue that a person was among the elect and thus favored by god, which drove early Protestants to relentless work as a means of confirming (and demonstrating) their salvation. As it happens, hard work and self-denial—the key features of the **Protestant ethic**—lead not only to salvation but also to the accumulation of capital. The religious ideas supported by the Protestant ethic therefore fit nicely with the needs of capitalism. According to Weber, these austere religionists stockpiled wealth, had an irresistible motive to earn more (that is, eternal salvation), and were inclined to spend little on themselves, leaving a larger share for investment and driving the growth of capitalism (Weber 1958/1904).

Karl Marx: Religion, Social Conflict, and Oppression

Durkheim and Weber concentrated on how religion contributes to the cohesion of society. Religion can also be the basis for conflict, as we see in the daily headlines of newspapers. In the Middle East, differences between Muslims and Jews have caused decades of political instability. These conflicts are not solely religious, but religion plays an inextricable part. Certainly religious wars and other religious conflicts have contributed to some of the most violent and tragic episodes of world history. The image of religion in history has two incompatible sides: piety and contemplation on the one hand, battle flags on the other. In the United States, domestic conflicts over ethical issues such as abortion, assisted suicide, and school prayer evolve from religious values even though they are played out in the secular world of politics and public opinion. Conflict theory illuminates many of the social and political conflicts that engage religious values.

The link between religion and social inequality is also key to the theories of Karl Marx. Marx saw religion as a tool for class oppression. According to Marx, oppressed people develop religion, with the urging of the upper classes to soothe their distress (Marx 1972/1843). The promise of a better life hereafter makes the present life more bearable, and the belief that "god's will" steers the present life makes it easier for people to accept their lot. To Marx, religion is a form of *false consciousness* (see Chapter 8) because it prevents people from rising up against oppression. He called religion the "opiate of the people" because he thought it encouraged passivity and acceptance.

Marx saw religion as supporting the status quo and being inherently conservative (that is, resisting change and preserving the existing social order). To Marx, religion promotes stratification because it supports a hierarchy of people on earth and the subordination of humankind to divine authority. Christianity, for example, supported the system of slavery. When European explorers first encountered African people, they regarded them as godless savages, and they justified the slave trade by arguing that slaves were being converted to the Christian way of life. Principles of Christianity thus legitimated the system of slavery in the eyes of the slave owners and allowed them to see themselves as good people, despite their enslavement of other human beings.

At the same time, religion can be the basis for liberating social change. In the civil rights movement in the United States and in Latin American liberation movements, the words and actions of religious organizations have been central in mobilizing people for change. This does not undermine Marx's main point, however, because there remains ample evidence of the role of religion in generating social conflict and resisting social change.

Symbolic Interaction: Becoming Religious

Recall that symbolic interaction theory states that people act toward things on the basis of the meaning things have for them and that those meanings emerge through social interaction. This can explain much about human behavior that is based in religious ideas. Seen from outside the faith, religious practices (kneeling in church, wearing a yarmulke, making a pilgrimage to Mecca, or chanting) may seem peculiar or different, but within the faith, these and other religious practices carry meaning— meaning that is deeply important to religious believers. Not only does symbolic interaction explain particular religious behaviors, it also explains how other behaviors may be based in the meanings that religion holds for people. For example, we have earlier said that even something as reprehensible as suicide bombings can be understood if you understand how religious zealots interpret the meaning of religious texts.

Symbolic interaction theory can also help you understand how people become religious, a process sociologists call *religious socialization*. Religious socialization may be a slow and gradual process, such as in how children learn religious values over time. Religious socialization can also be more dramatic, as when a person joins a cult or some other extreme religious group. People may also reinterpret religious beliefs when they question their religious faith or even switch to a new religion, such as when a Christian converts to Judaism. The emphasis on meaning that is typical of symbolic interaction helps explain how the same religion can be interpreted differently by different groups or in different times. For example, how could Christian beliefs be used by some to make slavery seem legitimate, while at the same time they provided the belief system that helped others survive slavery and to fight against it?

Symbolic interaction thus sees religious belief—and its meaning to different people—as essential for understanding many forms of social behavior. Symbolic interaction also helps explain how different religious beliefs and practices emerge in social and historical contexts—contexts that shape what religion means to people.

Diversity and Religious Belief

The world is marked by diverse religious beliefs (see ■ map 13-1). Christianity has the largest membership, followed by Islam. But Hindus, Jews, Confucianists, Buddhists, and observers of folk religions also comprise the world's religions. In the United States, religious identification varies with a number of social factors, including age, income level, education, and political affiliation. Younger people are more likely than older people to express no religious preference. Those in higher income brackets are more likely to identify as Catholic or Jewish than those in lower income brackets; Fundamentalist Protestants are most likely to come from lower-income groups.

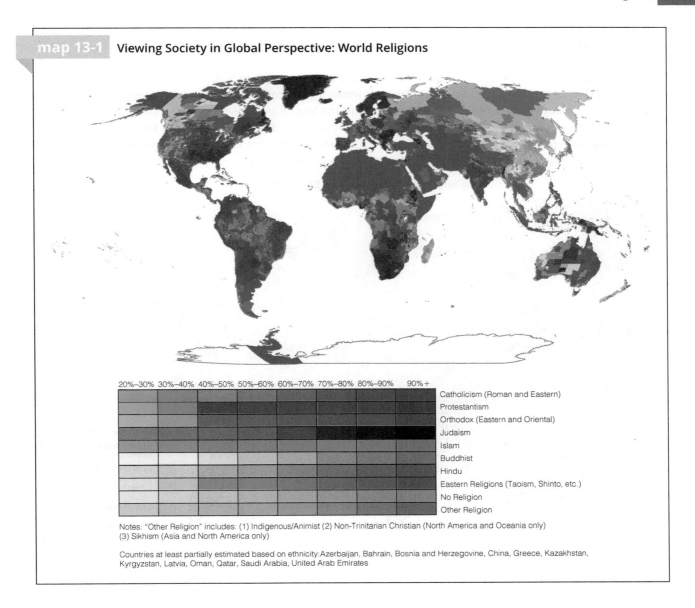

map 13-1 Viewing Society in Global Perspective: World Religions

	20%–30%	30%–40%	40%–50%	50%–60%	60%–70%	70%–80%	80%–90%	90%+	
									Catholicism (Roman and Eastern)
									Protestantism
									Orthodox (Eastern and Oriental)
									Judaism
									Islam
									Buddhist
									Hindu
									Eastern Religions (Taoism, Shinto, etc.)
									No Religion
									Other Religion

Notes: "Other Religion" includes: (1) Indigenous/Animist (2) Non-Trinitarian Christian (North America and Oceania only) (3) Sikhism (Asia and North America only)

Countries at least partially estimated based on ethnicity:Azerbaijan, Bahrain, Bosnia and Herzegovine, China, Greece, Kazakhstan, Kyrgyzstan, Latvia, Oman, Qatar, Saudi Arabia, United Arab Emirates

The Influence of Race and Ethnicity

Religious diversity also marks the United States (see ■ map 13-2). Race is one of the most significant indicators of religious orientation. African Americans are much more likely than Whites, Hispanics, or Asian Americans to say that religion is very important in their lives. Although most African Americans identify as Protestant (75 percent), a small, but growing number are Catholic (5 percent; Pew Research Center 2009).

Many urban African Americans have also become committed Black Muslims, which involves strict regulation of dietary habits and prohibition of many activities, such as alcohol use, drug use, gambling, use of cosmetics, and hair straightening. The emphasis among Black Muslims on self-reliance and traditional African identity has earned it a fervent following, although the actual number of Black Muslims in the United States is relatively small. For many African Americans, religion has been a defense against the damage caused by racism. Churches have served as communal centers, political units, and sources of social and community support, making churches among the most important institutions within the African American community (Gilkes 2000). Religion also has been a strong force in Latino communities, with map 13-2 **Mapping America's Diversity: Religious Diversity in the United States**

The Simpson's Diversity Index is a measure developed by the Association of Statisticians of American Religious Bodies. It calculates the likelihood of two individuals within a given county belonging to different religious groups. The lower the index, noted in the darkest shade, the less religious diversity in that county; conversely, the higher the index, noted in the lightest shade, the more religious diversity exists in that location. According to this map, where is there the most and least religious diversity? Do you think that affects such things as political values and other social issues in these different regions of the country?

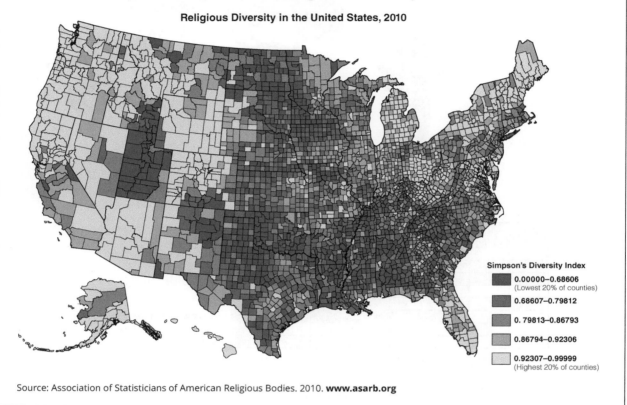

Religious Diversity in the United States, 2010

Simpson's Diversity Index

- 0.00000–0.68606 (Lowest 20% of counties)
- 0.68607–0.79812
- 0.79813–0.86793
- 0.86794–0.92306
- 0.92307–0.99999 (Highest 20% of counties)

Source: Association of Statisticians of American Religious Bodies. 2010. **www.asarb.org**

the largest number identifying as Catholic. There are now, however, a growing number of Latino Protestants, both in mainstream Protestant denominations and in fundamentalist groups (Mulder, Ramos, and Martí 2017).

Asian Americans have a great variety of religious orientations, in part because the category "Asian American" is constructed from so many different Asian cultures. Hinduism and Buddhism are common among Asians, but so is Christianity. As with all groups whose family histories include immigration, religious belief and practice among Asian Americans frequently changes between generations. The youngest generation may not worship as their parents and grandparents did, although some aspects of the inherited faith may be retained. Within families, the discontinuity with a religious past brought on by cultural assimilation can be a source of tension between grandparents, parents, and children. Within the United States, Asian Americans often mix Christian and traditionally Buddhist, Confucian, or Hindu beliefs, resulting in new religious practices (Pew Forum on Religion & Public Life 2012).

Muslims are a growing segment of U.S. society. Of the estimated 2 million Muslim adults in the United States, 58 percent are immigrants. The remainder are native born, either African American or others who have converted to this faith or were raised in a Muslim household. Because of their higher birth rate and the relative youth of the population, Muslims are predicted by 2050 to surpass the number of people in the U.S. population who identify as Jewish. Despite stereotypes about Muslim conservatism, Muslim Americans are actually more liberal than the general public on many issues—for example, they are more likely to vote Democratic. Muslim Americans are, though, generally less tolerant of gays and lesbians than the public at large. Studies find that younger Muslims (those under thirty) tend to be more observant than older Muslims—perhaps explained by the heightened identity that has emerged since 9/11 and in the aftermath of the election of Donald Trump as president (Lipka 2017; Pew Research Center 2007).

Among Latinos, religious identity tends to be strong. Slightly more than half of Latinos (55 percent) are evangelical. Among Latinos of all faiths, people tend to worship in places that are primarily comprised of other Hispanics. This makes religion for Latinos a strong ethnic, as well as religious, identity (Pew Research Religion & Public Life Project 2014).

Religious Organizations

Sociologists have organized their understanding of the various religious organizations into three types: churches, sects, and cults. These are *ideal types* in the sense that Max Weber used the term. That is, the ideal types convey the essential characteristics of some social entity or phenomenon, even though they do not explain every feature of each entity included in the generic category.

Churches are formal organizations that tend to see themselves, and are seen by society, as the primary and legitimate religious institutions. The term can be broadly applied to formal religious organizations, including temples and mosques. Such religious organizations tend to be integrated into the secular world to a degree that sects and cults are not. They are sometimes closely tied to the state. Many churches are organized as complex bureaucracies with a division of labor and different roles for groups within, including a formally trained clergy and professional staff. Some churches may be smaller, less formal, with devoted, but less formally trained, clergy. A new phenomenon for churches is the development of *megachurches*—those with memberships numbering into the thousands. These are increasingly common. Not only do megachurches have huge attendance but they also may broadcast on huge screens, possibly even televising church services.

Sects are groups that have broken off from an established church. They emerge when a faction within an established religion questions the legitimacy or purity of the group from which they are separating. Many sects form as offshoots of existing religious organizations. Sects tend to place less emphasis on organization (as in churches) and more emphasis on the purity of members' faith. The Shakers, for example, were formed by departing from the Society of Friends (the Quakers). They retain some Quaker practices, such as simplicity of dress and a belief in pacifism, but have departed from Quaker religious philosophy. The Shakers believe that the second coming of Christ is imminent, but that Christ will appear in the form of a woman (Kephart 1993).

Sects tend to admit only truly committed members, refusing to compromise their beliefs. Some sects hold emotionally charged worship services, although others, like the Amish, are more stoic. The Shakers, for example, have such emotional services that they shake, shout, and quiver while "talking with the lord," earning them their name. The only bodily contact permitted among the Shakers is during the unrestrained religious rituals; they are celibate (do not have sexual relations) and gain new members only through adoption of children or recruitment of newcomers (Kephart 1993).

Cults, which are like sects in their intensity, are religious groups devoted to a specific cause or charismatic leader. Many cults arise within established religions and sometimes continue to peaceably reside within the parent religion simply as a fellowship of people with a particular, often mystical, dogma. As they are developing, it is common for tension to exist between cults and the society around them. Cults tend to exist outside the mainstream of society, arising when believers think that society is not satisfying their spiritual needs and attracting those who feel a longing for meaningful attachments. Internally, cults seldom develop an elaborate organizational structure but are instead close-knit communities held together by personal attachment and loyalty to the cult leader.

The Rise of Religious Fundamentalism

Religious fundamentalism is an increasing force in today's world. Why is this happening now? Max Weber thought that the process of modernization would lead to a more secular society but, in fact, religious fundamentalism has been on the rise in recent years—both in the United States and abroad. How do you explain this? Is it contrary to sociological logic?

Religious fundamentalists are those who are "true believers." They have very literal interpretations of religious texts and tend to be highly certain of their religious worldview. Religious fundamentalists tend to see the world in simplistic either/or terms—dividing people into either good or evil, godly or demonic. Such divisive imagery reduces the complexity of human life into simplistic categories—categories that can fuel hate and conflict (Anthony et al. 2002). When such religious fanaticism is intertwined with the power of a state government or an armed militia, religiously inspired leaders can become extremely dangerous.

Religious fundamentalists tend to hold their beliefs as absolute—that is, having little doubt about the truth of their beliefs. Often, religious fundamentalists have a messianic and millennial view, believing that a moment will come when all truth is revealed—or a messiah appears. Religious fundamentalism is intolerant of other beliefs and can provoke extreme violence. Religious extremism has fueled horrendous acts in various parts of the world, including mass executions and genocide, enslavement, and other heinous crimes against humanity. Any religion, taken to an extreme, can be dangerous because extreme adherents think they are doing sacred work even when they are engaging in violent, murderous behavior.

It is easy to see the acts of religious extremists as the work of misled individuals, but those who study religious extremism know that it has social origins. Religious extremism is learned, usually within a narrowly circumscribed social world, such as the madrassas—religious camps in Pakistan (and other areas) where young boys are taught a strict interpretation of Islam. For young boys uprooted from families by war, detached from other social contacts, and with no other education, it is easy to be socialized into a narrow worldview that gives them a cause to fight for (Rashid 2000).

Fundamentalist religious organizations tend to have charismatic leadership and to make sharp boundaries between believers and nonbelievers. Fundamentalists typically enforce strict behavioral rules, such as dress, comportment, perhaps hairstyle, or eating and drinking habits. Religious fundamentalists are morally fervent in trying to achieve their particular worldview. In most cases, religious fundamentalists are also very conservative on matters involving the family and appropriate gender roles—stemming largely from their literal interpretation of religious texts (Emerson and Hartman 2006).

Religious fundamentalism in the United States surged starting in the 1970s, very likely as a counterreaction to the marginalization of religion that had emerged during the social movements of the 1960s. It has been fueled as well by a counterreaction to increasing secularization and the increasing trend for commercialism to shape social relations. Religious fundamentalism can reflect a search for a firmly anchored identity and sense of community. In other parts of the world, where people perceive that their traditional way of life is being overtaken by Western influences, religious extremism can come from trying to defend a traditional way of life (Stern 2003; Pain 2002; Khashan and Kreidie 2001).

Seen in this light, perhaps Weber was partially right—that the tendency of modernization is toward secularization, but secularization then can produce a counterreaction in the form of a rise of religious fundamentalism.

Cults form around leaders with great **charisma**, a quality attributed to individuals believed by their followers to have special powers (Johnstone 1992). Typically, followers are convinced that the charismatic leader has received a unique revelation or possesses supernatural gifts. Although there are exceptions, cult leaders are usually men, probably because men are more likely to be seen as having the characteristics associated with charismatic leadership.

Myth: People who join extreme religious cults are maladjusted and have typically been quickly brainwashed by cult leaders.

Sociological Perspective: Conversion to a religious cult usually involves a gradual process of resocialization wherein the convert voluntarily develops new associations with others and develops a new worldview based on these new relationships.

Religion and Social Change

What is the role of religion in social change? Durkheim saw religion as promoting social cohesion; Weber saw it as culturally linked to other social institutions; Marx assessed religion in terms of its contribution to social oppression. Is religion a source of oppression, or is it a source of personal and collective liberation from worldly problems? There is no simple answer to this question. Religion has had a persistent conservative influence on society, but it has also been an important part of movements for social justice and human emancipation.

One of the major changes in society is the increase in both membership and influence of evangelical groups. Such groups are affiliated with conservative political causes, dramatically increasing the influence of religion on politics. At the same time, there has been a decrease in the importance of religion to many people. As a social institution, religion is in transition. Religion, like other aspects of society, is also becoming more commercialized. A large self-help industry has developed in religious publication, and religious music is increasingly successful as a form of enterprise. All sorts of religious products are bought and sold in what sociologists now call a "spiritual marketplace" (Wuthnow 1998). Clearly, religion influences social change, but it is also influenced by the same changes that affect other social institutions.

At the same time, religion continues to have an important role in liberation movements around the world. Throughout the world, liberation theologians have used the prestige and organizational resources of the Catholic Church to develop a consciousness of oppression among poor peasants and working-class people. Likewise, in the United States, churches have had a prominent role in the civil rights movement (Morris 1984; Marx 1967/1867). Churches supplied the infrastructure of the developing Black protest movements of the 1950s and 1960s, and the moral authority of the church was used to reinforce the appeal to Christian values as the basis for racial justice. Now they continue to be important places for the mobilization of Black politics and provide an important source of community support—often when other institutions have abandoned the Black community (Zuckerman 2002).

The role of women is also changing in most religious organizations. Women have long been denied the right to full participation in many faiths. Some religions still refuse to ordain women as clergy, but the public generally supports the ordination of women. Women now make up a large portion of divinity students. Whereas traditional religious images of women have provided the basis for the subordination of women, those stereotypes are eroding. In sum, religion is a force of both social change and social stability.

Chapter Summary

How are different kinship systems defined?

All societies are organized around a *kinship system*, varying in how many marriage partners are allowed, who can marry whom, how descent is determined, family residence, and power relations within the family. *Extended family* systems develop when there is a need for extensive economic and social cooperation. The *nuclear family* is the result of the rise of Western industrialization that separated production from the home.

What does sociological theory contribute to our understanding of families?

Functionalism emphasizes that families have the function of integrating members to support society's needs. *Conflict theorists* see the family as a power relationship, related to other systems of inequality. *Symbolic interaction* takes a more microscopic look at families, emphasizing how different family members experience and define their family experience. *Feminist theory* emphasizes the family as a gendered institution and is critical of perspectives that take women's place in the family for granted.

What changes characterize the diversity in contemporary families?

One of the greatest changes in families has been the increase in female-headed households, which are most likely to live in poverty. The increase in women's labor force participation has also affected families, resulting in dual roles for women. Stepfamilies face unique problems stemming from the blending of two households. Gay and lesbian households are also more common and challenge traditional heterosexual definitions of the family. Single people make up an increasing portion of the population, due in part to the later age when people marry.

Is marriage declining?

The United States has both the highest marriage rate and the highest divorce rate of any industrialized nation. The high divorce rate is explained as the result of a cultural orientation toward individualism and personal gratification, as well as structural changes that make women less dependent on men within the family.

Why is family violence such a problem?

Family violence takes several forms, including partner violence, child abuse, incest, and elder abuse. Power relationships within families, as well as gender differences in the division of labor, help explain domestic violence.

What major changes are affecting contemporary families?

Changes at the global level are producing new forms of families—*transnational families*—where at least one parent lives and works in a nation different from the children. Social policies designed to assist families should recognize the diversity of family forms and the interdependence of the family with other social conditions and social institutions.

What are the elements of a religion?

Sociologists are interested in religion because of the strong influence it has in society. *Religion* is an institutionalized system of symbols, beliefs, values, and practices by which a group of people interprets and responds to what they feel is sacred and that provides answers to questions of ultimate meaning.

How do sociologists measure the significance of religion for people, and what forms does religion take?

The United States is a deeply religious society. Christianity dominates the national culture, even though the U.S. Constitution specifies a separation between church and state. *Religiosity* is the measure of the intensity and practice of religious commitment.

How do the different sociological theories analyze religion?

Durkheim understood religions and religious rituals as creating social cohesion. Weber saw a fit between the ideology of the *Protestant ethic* and the needs of a capitalistic economy. Religion is also related to social conflict. Marx saw religion as supporting societal oppression and encouraging people to accept their lot in life. *Symbolic interaction theory* focuses on the process by which people become religious. Religious conversion involves a dramatic transformation of religious identity and involves several phases through which individuals learn to identify with a new group and lose other existing social ties.

What diversity exists in religious faith and practice?

The United States is a diverse religious society. Protestants, Catholics, Jews, and, increasingly, Muslims make up the major religious faiths in the United States. Religious extremism can emerge in any religion and is generated by certain societal characteristics.

How is a religion organized?

Churches are formal religious organizations. They are distinct from *sects*, which are religious groups that have withdrawn from an established religion. *Cults* are groups that have also rejected a dominant religious faith, but they tend to exist outside the mainstream of society.

How has religion been affected by social change?

In recent years, there has been an enormous growth in conservative religious groups. Religion is a conservative influence in society, but religion also has an important part in movements for human liberation, including the civil rights movement and the move to ordain women in the church.

Key Terms

bilateral kinship 328

charisma 358

churches 357

collective consciousness 352

cults 357

extended families 328

family 327

Family and Medical Leave Act
    (FMLA) 346

kinship system 327

matrilineal kinship 328

matrilocal 328

monogamy 328

monotheism 351

nuclear family 329

patrilineal kinship 328

patrilocal 328

polygamy 327

polytheism 351

profane 348

Protestant ethic 353

religion 348

religiosity 350

rituals 352

sacred 348

sects 357

secular 349

totem 348

transnational family 345

CHAPTER **14**

EDUCATION AND HEALTH CARE

In this chapter, you will learn to:

Understand the role of schools in society

Explain the connection between education and social mobility

Outline the race and class inequalities within education

Compare and contrast theoretical perspectives on education

Summarize the principal ideas guiding debates about educational reform

Explain what it means to say that health and illness are conditioned by social factors

Report the factors that influence health disparities in the United States

Describe the social organization of health care in the United States

Compare and contrast theoretical perspectives on health care

Summarize the principal ideas regarding contemporary health care reform

The United States is thought to have the best education system in the world. Increasingly, though, people wonder if schools are living up to their promise. Do all Americans get a high quality public education? Many think schools are failing—failing their students, not supporting their teachers, and deteriorating because of their crumbling infrastructure. For some, schools have become more like prisons—the place where dreams are shattered and students experience schools as just another place that batters their self-confidence and limits their life opportunities. Some may fear that schools are not even safe, as people have witnessed horrific mass shootings in schools. School lockdowns because of the threat of violence have become more common.

Beyond public schooling, higher education has become so expensive that many now question whether a college degree is worth the price of tuition. Student debt has reached record levels, leaving even those who complete higher education with loans that can exceed the price of a home. For-profit education is more common and in too many cases people with the most resources are abandoning public education—thus further reproducing social inequality.

These challenges ask you to examine the character of education as a social institution—one that, like other social institutions, has a social structure organized around inequality. The social structure of educational institutions means that, for some, schooling provides a path to a good job and a good future. For others, schools only reproduce the inequalities that are pervasive throughout society, leaving many, especially poor and minority children, with less opportunity for success.

Public debate around education rages on. Teachers' unions are under fire. Urban schools in many American cities are facing serious budgetary and performance problems. Higher education is changing in the character of the student body and how educational content is delivered. Sociologists ask: How has the role of formal education in society changed over time? Are we providing a good education for all Americans? If not, who is being left behind? And, is education meeting the needs of a highly technological and global economy?

Schooling and Society

Education in any society is about the transmission of knowledge. In some societies, such as the United States, education is highly formalized—indeed, even regulated by government (at least for public institutions). In other societies, education may be less formal, perhaps provided solely through the transmission of knowledge by elders or family members (for example, home schooling is the norm in some protected religious communities). In the United States, education teaches formal knowledge, such as reading, writing, and arithmetic, as well as cultural knowledge, such as morals, values, and ethics. Education prepares the young for entry into society and is thus a form of socialization. Sociologists refer to the more formal, institutionalized aspects of education as **schooling**.

Early on, in the nation's history, education was considered a luxury, available only to White, male children of the upper classes. But by the nineteenth century, industrialization and urbanization were changing American society. With those changes, education changed. Public education expanded as schools were formed to inculcate middle-class values into an increasingly diverse society. Education was increasingly seen as a necessary skill in the national work force. Teacher training programs were introduced and public high schools were created to try to standardize what students learned, making them compliant with the newly emerging social order (Rury 2015).

By the beginning of the twentieth century, federal guidelines made education compulsory. Yet, state laws requiring attendance were generally enforced only for White Americans and then only through eighth grade. As you can see in ▲ Figure 14-1, high school graduation rates have increased steadily over the years. Most recently, the greatest increase in high school completion has been among Latino students. Although educational attainment has expanded dramatically over the years, as we will see great inequalities in education still remain, especially for racial–ethnic minorities.

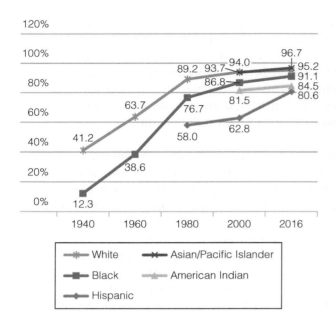

▲ **Figure 14-1 Percentage of People (age 25–29) Completing High School or More.** As you can see from this graph, the percentage of people completing high school has increased significantly, especially in the middle of the twentieth century. Although there has been much improvement for Black, Hispanic, Asian, and American Indian students, the current gap tells us there is still a need for improving the education of all citizens.

Note: Data for American Indians is for 1995, the first year tabulated for this group in these data. You can also see that data for Hispanics and Asian/Pacific Islanders did not routinely begin until the early 1990s.

Source: National Center for Education Statistics. 2017. *Digest of Education Statistics*. Washington, DC: U.S. Department of Education. **www.nces.ed.gov**

Traditionally, there have been three kinds of education in the United States: public education, private education, and homeschooling. In the mid-nineteenth century, slightly under two-thirds of the population aged 5 to 17 was enrolled in school—a figure that has grown to a full 93 percent now. Not only has the number of students enrolled increased, but so has the average length of the school year (National Center for Educational Statistics 2016). Still, public schools struggle for finances and, in many districts, teachers have to buy their own supplies.

New to public schools are *charter schools*—those that receive public funds but are not subject to the same rules and regulations as other public schools. Since 2000, the number of charter schools in the United States has tripled as people select them as an alternative to what they perceive as a weak public education. Likewise, *magnet schools*—those that typically focus on a themed curriculum (such as the arts or a strong science program)—have more than doubled since 2000. In some school districts, charter and magnet schools are attended by choice. Sometimes, enrollment is through a lottery system. Although charter and magnet schools are still accountable to local and state school authorities, many think they offer an alternative to failing public schools. Although charter and magnet schools appeal to "school choice," they add to the divide between "good" and "bad" public schools, often giving White families more privileged access to a better public education and leaving poor students and many students of color behind in increasingly class- and race-segregated schools (Parcel, Hendrix, and Taylor 2016; Pattillo 2015; Quiroz and Lindsay 2015; Roda and Wells 2013).

Approximately 5.7 million students are currently enrolled in private schools (National Center for Education Statistics 2016), although that number has declined somewhat in recent years (since the mid-1990s), likely the result of two factors: (1) the growth of charter and magnet schools, but also (2) a large decline in the number of students attending parochial (that is, Catholic) schools. Over the same time period, the percentage of students attending conservative Christian schools or other schools with a specific religious orientation (but not Catholic) has inched up (National Center for Educational Statistics 2016).

When looking at the enrollment of students in private schools, you cannot help but see the influence of the 1954 Supreme Court decision, ***Brown v. Board of Education***. This landmark decision ruled separate schools for Black and White children unconstitutional. As a result, at the time many school districts were under court order to desegregate public education. Many White parents took children out of public schools and more private schools developed. Since the *Brown* decision,

You can see the inequality in education by seeing the contrast in schools that look like prisons versus those with excellent facilities.

the number of private schools has actually doubled (National Center for Education Statistics 2016). This leaves many public school systems poorly funded with large minority student populations—creating a cycle of school failure for some and success for others. As we will see, public schools are now also re-segregating along racial and class lines.

Another form of private education is *homeschooling*, where education is overseen through online or occasional in-school programs, usually by a mother. Homeschooling now educates about 3 percent of school-aged children. It is more common for girls than boys and when there are three or more children in the household. Homeschooling is far more common among White and Asian students; it is rare for those who are poor or close to being poor (National Center for Education Statistics 2016). The most common reasons given for homeschooling are criticisms of the environment and academic instruction in the public schools and the desire for more parental involvement in education. Research studies vary in what they find regarding the academic achievement of homeschooled children compared to other students, although some say there is little difference in college preparedness. (Murphy 2014).

Higher education is also facing many challenges. Among other issues, the cost of attending a four-year college or university is higher now than in any time in history. Although federal loans can help students attend college, outstanding debt on these loans now tops $1.45 trillion, more than the total amount people hold on car loans (Federal Reserve 2017). Students are graduating with more debt so that even if they find a high-paying job, they are unable to achieve financial independence. Higher education in the United States is facing a crisis as more parents are asking if attending college is worth it. The answer is still yes because research still finds a high "return" on education—in terms of material wellbeing over time; this is especially true for women (DiPrete and Buchmann 2014).

As college has become important for preparing for and finding a good career, many students have turned to community colleges to pursue their education. Community college enrollment soared between 2000 and 2010, but has, of late, declined somewhat. Still, these institutions serve very large numbers of students and are more diverse than most other higher education institutions. Now, 42 percent of all undergraduate students and 25 percent of all full-time undergraduates are enrolled in community colleges. Community colleges serve a very large proportion of minority, first-generation, low-income, and adult students. Data show that Latino students, relative to their proportion in the total U.S. population, are disproportionately enrolled in community colleges. Black students, on the other hand, are disproportionately enrolled in private, for-profit institutions. Asian and White students are much more likely to be enrolled in four-year institutions (Ma and Baum 2016).

Almost two million students are now enrolled in for-profit institutions of higher education, a six-fold increase since the 1980s. Unlike other higher education institutions, for-profit institutions must be self-sustaining. They typically do so through high tuition and a preponderance of online instruction. While this provides convenience and flexibility for students, especially if they have work and family commitments, critics ague that such institutions are "diploma mills," pushing students through by providing programs of dubious quality. At the same time, however, such institutions provide opportunities to non-traditional students that they would not otherwise get. Studies have found that the rate of student loan default is much higher in private for-profit institutions, both because students in such institutions tend to come from relatively poorer households and because of the high cost of the college itself. Students in for-profit institutions also have a greater likelihood of dropping out (Cellini and Darolla 2017; Bennett, Lucchesi, and Vedder 2010).

Does Schooling Matter?

In a highly technological society such as the United States, education is increasingly necessary for future opportunities. Why then do some of our schools resemble prisons where entering students are searched and the physical environment is dilapidated and bleak? Other schools look like beautiful campuses, places with modern facilities and sophisticated scientific equipment.

How much does schooling really matter? Does more schooling actually lead to a better job, more annual income, and enhanced opportunities? Parents and students who invest money and time in higher education want to see the payoff at the end, but how much schooling matters, and for whom, is not just about individual success or failure. How education is organized as a social institution and how education is related to systems of inequality in society are larger sociological questions.

One way that sociologists measure a person's social class or *socioeconomic status (SES)* is to determine the person's amount of schooling, income, and type of occupation. Sociologists call these the *indicators* of SES. In the general population, there is a strong relationship between formal education and occupation. Although the relationship is not perfect, it is true that the higher a person's occupational status, the more formal education he or she is likely to have received. Thus, on average, doctors, lawyers, professors, and nuclear physicists spend many more years in school than garbage collectors and shoe shiners. This relationship is strong enough that you can often, although not always, guess a person's level of educational attainment just by knowing his or her occupation. There are indeed instances of laborers, such as Uber drivers, who have PhDs, but they are relatively rare. Also rare is the reverse: the self-educated, self-made individual who completed only high school and is now the CEO of a major corporation.

Schools are stratifying institutions; that is, they sort people into categories based on social factors such as one's social class, race, gender, and even perceived social worth. Schools themselves also build on the stratification that exists in the society at large. This is the case not just in the United States, but in other societies as well.

In England, for example, the education curriculum is divided into four key stages until age 16 when students pursue General Certificate of Secondary Education. In China, the educational system is about subject-specific knowledge. The "gaokao" system is the college entrance exam program for all Chinese students. How students perform on the exam is nearly the only determinant for how and where students are placed. And in Japan, education is only compulsory through middle school although almost all students continue through high school. High school entry, however, requires an entrance examination, which determines a child's subsequent educational opportunities.

In the United States, educational stratification is not so explicit and, until recently, was not based solely on test scores—an issue discussed later in this chapter. But education is a stratified system, and it produces social mobility at the same time that it reproduces inequalities otherwise found in society.

Look, for example, at the connection between income, education, and social factors such as gender. Although, the higher one's education, the higher one's income on the whole, it is nonetheless true that the median income for women is less than the median income for men at *every* education level. Some of this difference is because women tend to get degrees in different fields than men. Still, the connection between education, income, and gender holds across all levels of education.

Education and Social Mobility

Education has traditionally been viewed in the United States as a way out of poverty and low social standing—that is, as the main route to upward *social mobility*. The assumption has been that a person can overcome modest beginnings by staying in school.

There is some truth to this. Those with more education do have higher earnings, better jobs, and more perceived social worth. Much sociological research, however, has demonstrated that the effect of education on a person's eventual job and income greatly depends on the *social class* that the person was born into (Bowles et al. 2005).

Class and race also work together to "protect" the upper classes from downward social mobility. Education is used by the upper classes to avoid downward mobility by such means as sending their children to elite private secondary schools. Among middle-class Whites, education considerably improves the chances of getting middle-class jobs, yet access to upper-class positions is limited. Among those of the working class, getting a good education is not impossible, but it involves a lot of social support, financial aid, and, sometimes, just plain luck. For the chronically unemployed—the underclass—chances of getting a good education are minimal. In sum, education is strongly affected by social class origins. Occupation and income are heavily influenced by social class and by education. These interrelationships are summarized in ▲ Figure 14-2, which shows that social class origin affects occupation and income both directly and indirectly by way of education.

Understanding Diversity

Social Class and the College Party Scene

Would it have ever occurred to you that the party scene on many college campuses contributes to class inequality? How so? Sociologists Elizabeth A. Armstrong and Laura T. Hamilton studied the college party scene and its ramifications by actually living in a college residence hall for an extended period of time and observing and interviewing the women who lived there. They lived with the women for a full year but continued interviewing them through their graduation when they then followed up with them.

Their book, *Paying for the Party*, highlights how different the experience of college is for middle-class and working-class students. Even seemingly small differences between women from privileged families and women from solidly middle-class families led to different outcomes for students as they progressed through college and into the job market.

Armstrong and Hamilton found that women with more financial resources are able to graduate without debt, go on spring break trips, join sororities, buy alcohol, and fully participate in the party culture on campus. Women with fewer financial resources are less likely to experience the college culture this way. These differences extend to their postgraduate lives, altering opportunities for employment and graduate school. The researchers identified three "pathways" that students experience in college. One focuses on partying, another on professional development, and a third focuses on improving class status with a college education. College, particularly college parties, provides opportunities to meet peers for career advancement, for marriage, and for friendship networks. The students with fewer financial resources do not have the same opportunities to participate in the college culture that will create networks and open doors for them. Working-class students are actually disadvantaged by a party culture that, to many, just seems like a lot of fun. In the end, as Armstrong and Hamilton conclude, the party culture on college campuses is one way that the educational system reproduces class inequality.

Is there a party culture on your campus? If so, are there differences between students who frequently go out for parties, often order pizza, or can participate in other college activities that require money? Armstrong and Hamilton make the argument that more privileged students have a different orientation or attitude toward college than other students. Do you agree?

Source: Armstrong, Elizabeth A., and Laura T. Hamilton. 2013. *Paying for the Party: How College Maintains Inequality.* Cambridge, MA: Harvard University Press.

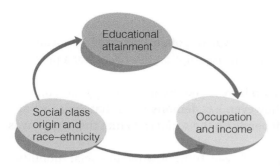

▲ **Figure 14-2 Relationship of Social Class, Race–Ethnicity, Education, Occupation, and Income.** This figure shows how social class origin affects occupation and income both directly and indirectly by way of education.

Education and Inequality

Education has reduced many inequalities in society. More high school diplomas are awarded to all race and class groups, and more minorities and women attend and graduate from two- and four-year colleges. Nonetheless, many inequalities still exist in U.S. education. These can be shown in numerous ways. Studies of inequality and poverty, specifically in U.S. urban centers, highlight how schools are organized, often along race and class lines, leaving many of our most marginalized students without quality education (Kozol 2006).

Segregation and Resegregation

As noted before, *Brown v. Board of Education* ruled that "separate but equal" in all public facilities, including schools, was unconstitutional. Although it took years before school districts actually began implementing this decision, and only then with substantial pressure from the federal government, some measure of school desegregation followed the *Brown* decision. Although school desegregation was not as great as was hoped, progress was made. By the 1980s, many school districts, particularly in the American South, had made significant gains in integrating schools by race.

Now, however, that historic change is reversing, and the nation is retreating to highly segregated schools. Researchers have found that American schools are now more racially segregated by race and by class than was the case in the 1980s, in every part of the country for both African Americans and Latinos. Communities have reversed the desegregation orders of the 1970s to create neighborhood schools. Residential segregation by race means that schools are then also segregated. More than half of African American and Latino students now attend schools that are "majority minority"—that is, more than half of the students in the schools are minority students. This would not be problematic in and of itself were it not for the fact that "majority minority" schools tend to have none of the resources of predominantly white schools (Orfield et al. 2014).

School segregation is problematic on many counts, one of which is the isolation of groups from one another and the resulting loss of friendship, interracial understanding, and co–mingling. Segregated schools that are heavily minority or poor are also generally of very poor quality—as the *Brown* decision noted. In other words, segregation breeds inequality, and even a cursory look at segregated schools that are predominantly minority and/or poor will reveal this. Unqualified teachers, ill-equipped science labs, a weak curriculum, and a prison-like atmosphere prevail in such schools, thus denying students, who likely are perfectly smart and capable, from achieving the kind of education that will lead to a good job (Kozol 2006). This unequal, subpar education creates a situation in which young African American men are more likely to end up in prison than to graduate from college (National Center for Education Statistics 2013; Carson and Sabol 2012).

The current criticism of the American educational system centers on these disparate conditions between wealthy, predominantly White, suburban schools and poor, predominantly minority, urban schools. Communities with a greater percentage of high-income families simply have better schools. Students raised in low-income communities are denied access to these schools and the strong education provided by them. Despite the perception that the United States is a fair and equitable nation, millions of U.S. children are lacking opportunity to live up to their potential. The educational structure of U.S. schools is not fair and equitable.

Testing and Accountability

The education system in the United States has relied heavily upon the idea that intelligence, or ability, or potential is a single trait—one that can be gauged according to the numerical results of standardized tests. Whether standardized tests are a strong measure of ability—and, in some forms, achievement—has become increasingly important as testing has become the major measure of school success under various educational reforms.

There are three major criticisms regarding the use of standardized tests. First, the tests tend to measure only limited ranges of abilities (such as quantitative aptitude or verbal aptitude) while ignoring other cognitive endowments such as creativity, musical ability, spatial perception, or even political skill and athletic ability (Taylor 2017; Zwick 2004; Freedle 2003).

Second, the tests possess at least some degree of cultural and gender bias—and also a strong social class bias (Taylor 2017). As a result, they may perpetuate rather than reduce inequality between different cultural, racial, gender, and class groups. Many studies show that although standardized ability tests are somewhat capable of predicting future school performance for White men, most studies show less accurate forecasts for the success of minorities, especially Latinos, African Americans, and American Indians; they also predict school performance less accurately for women than for men. In other words, the extent to which the tests accurately predict later college grades is compromised for minorities, women, and people of working-class origins.

Third, SATs actually do not predict school performance very well for all groups. For example, SAT scores are only modestly accurate predictors of college grades even for White students (Zwick 2004). This fact is not well known. Grade point average in high school (and school class rank as well) is also only a modestly accurate predictor of success in college. High school grades are about as accurate as the SATs in predicting college grades—maybe even a little better (Alon and Tienda 2007).

In general, average scores for tests such as the SAT differ across different groups. Whites score higher on average than minorities, and the higher a person's social class, the higher his or her test score is likely to be. This is where the intelligence debate begins. The segregation of schools discussed previously indicates clear reasons for poor test scores among some students. Lack of resources in schools, inadequate teacher training, and unsafe conditions are environmental factors that likely contribute to below-average academic performance.

Still, occasional claims are made that differences in test scores are somehow genetically inherited. A notorious example was the publication of a book, *The Bell Curve*, in 1994. The book caused a major stir, one that is still ongoing among educators, lawmakers, teachers, public officials, policymakers, and the general public. Authors of *The Bell Curve*, Herrnstein and Murray, argued that the distribution of intelligence in the general population closely approximates a bell-shaped curve (called the *normal distribution*). They also asserted that there is one basic, fundamental kind of intelligence and that it is genetically inherited.

Herrnstein and Murray's research was widely criticized for arguing that intelligence is determined primarily by one's genes rather than by one's social and educational environment. Although there is some small genetic basis to intelligence, as long as society is marked by the inequalities that we can sociologically observe, then group differences in ability must be seen within that context (Taylor 2017). Understanding the drastic race and class differences in schools across this country highlights access to educational opportunity as opposed to intellectual ability (Fischer et al. 1996).

Consider the standardized tests commonly used to apply for college admission. On average, students from lower-income families have lower scores on exams such as the Scholastic Assessment Test (SAT)

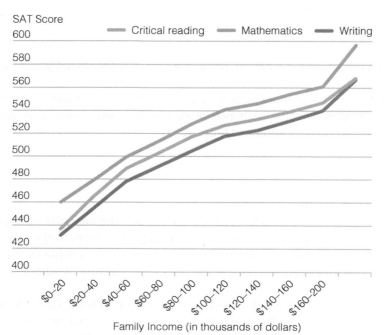

▲ **Figure 14-3** **SAT Scores and Family Income**

Source: The College Board. Copyright © 2010. National Report on College-Bound Seniors. Reproduced with permission. **www.collegeboard.com**

and the American College Testing (ACT) program. As shown in ▲ Figure 14-3, there is a smooth and dramatic increase in average (mean) SAT score as family income increases, for both SAT verbal and math scores. In this sense, a student's SAT score is a proxy, or substitute, measure of that student's social class: Within a certain range, you can guess someone's likely SAT score from knowing only the income and social class of his or her parents! As you can see from Figure 14-3, each additional $10,000 in family income is worth about 10 to 15 points on either the SAT verbal or the SAT math tests.

One possible reason for this is access to test preparation courses. These courses typically cost money and may be inaccessible to some students. Devine-Eller (2012) finds that as household income goes up, the likelihood of participating in test preparation courses also increases for ninth- through eleventh-graders. This research also shows that students are much more likely to participate in a test preparation course if they are active in school activities, have parents who are highly educated themselves, and have parents who are actively involved in school (Devine-Eller 2012). The idea of **cultural capital** in this con-text suggests that certain types of parents will have access to knowledge and information about preparing their student for college entrance exams. Beyond simply the ability to pay for test preparation courses, parents with knowledge and experience regarding college admissions are able to provide better opportu-nity for their children.

Less help preparing for standardized tests may diminish a student's chance of getting into the best colleges or universities. The intersection of race and class also contributes to the inequality of educational attainment. Statistics about SAT scores indicate that White students typically score higher in critical reading, mathematics, and writing than Blacks and Latinos (see ◆ Table 14-1). These patterns indicate that college entrance exams continue to stratify student access to educational success.

The educational system in the United States appears to allow for some social mobility as the result of education, but clearly not as much as people believe. A good education is essential for a good job, but the odds of getting such an education are considerably shaped by one's social class of origin, one's race, and, to a lesser extent, one's gender. Too many of the poorest children in the United States lack the very basic necessities for a good education (Kozol 2006). Support and scholarship programs intended to aid those with greater disadvantages in education help, but without such intervention, the forces of social stratification are reproduced in the educational system.

Table 14-1	Average SAT Scores by Ethnicity and Gender					
	Critical Reading		Mathematics		Writing	
	Men	Women	Men	Women	Men	Women
American Indian or Alaska Native	482	480	502	472	456	466
Asian, Asian American, or Pacific Islander	522	520	611	584	522	532
Black or African American	427	433	436	423	408	426
Mexican or Mexican American	453	446	481	450	438	445
Puerto Rican	458	454	468	441	440	449
Other Hispanic, Latino, or Latin American	456	446	480	446	440	445
White	530	525	552	519	508	521
Other	491	493	539	498	483	496

Source: The College Board. Copyright © 2013. Total Group Profile Report. Reproduced with permission. **www.collegeboard.com**

School Tracking

Tracking (also called *ability grouping*) is the separating of students within schools according to some measure of ability (Oakes 2005). Tracking has taken place for more than seventy years. As early as first grade, children are likely divided into high-track, middle-track, and lower-track groups, or some variation thereof. Perhaps you were assigned to one of these tracks in elementary, junior high, or high school. In high school, the high-track students take college preparatory courses in math and science and read Shakespeare. Middle-track students take courses in business administration and typing. Lower-track students take vocational courses in auto mechanics, masonry, or dental hygiene. Although this kind of tracking is now on the decline in the United States, versions of it still exist.

The original idea behind tracking is that students would get a better education and be better prepared for life after high school if they are grouped early according to ability. Theoretically, students in all tracks learn faster because the curriculum is tailored to their ability level, and the teacher can concentrate on smaller, more homogenous groups.

Advocates of *detracking* give the opposite argument. Detracking is based on the belief that combining students of varying cognitive abilities benefits the students more than tracking, especially by the time students get to junior high and high school. Students of high and low ability can thus learn from each other; the high-ability students are not seen to be "held back" by students with less ability, but are enriched by their presence.

Tracking also affects students with learning disabilities or special educational requirements. American students with disabilities have often been isolated from other students and given an entirely separate curriculum. The Individuals with Disabilities Education Act, however, gives federal guidelines for providing quality education for students with disabilities. Since the act was amended in 1997, new trends focus on the need for **individualized education programs** (IEPs), which outline specific types of learning that target specific needs.

Most researchers and educators who have studied tracking agree that not all students should be mixed together in the same classes. The differences between students can be too great and their needs too dissimilar. Some degree of tracking has always had advocates based on its presumed benefits for all students. This presumption is under attack. One of the most consistent research findings on tracking is that students in the higher tracks receive positive effects, but that the lower-track students suffer negative effects. To begin with, students in the lower tracks learn less because they are, quite simply, taught less. They are asked to read less and do less homework. High-track students are taught more and are consistently rewarded by teachers and administrators for their academic

Aveiage SA Scoies by Eihiicity and Genuei

Thinking Sociologically

Were you in a *tracked* elementary school? What were the tracks? Did you get the impression that teachers devoted different amounts of actual time to students in different tracks? Did teachers "look down" on those in the lower tracks? What about the students—did they treat some tracks as "better" or "worse" than others (were they perceived as differing in prestige)? Based on your recollections, what does this tell you about tracking and social class?

abilities (Oakes 2005). At the elementary level, mixed-ability classrooms are more common. As students progress through middle and high schools, IEPs are developed to address needs for each student and for each subject in school. Structurally, this can be very challenging for teachers and school administrators. Individualized lesson plans could benefit all students, not just those with learning disabilities. Teachers, however, must find ways to teach the same material to a classroom full of students, all of whom learn differently.

Who gets assigned to which tracks? Research shows that track assignment is not solely based on the performance in cognitive ability tests. Social class and race are involved. Students with the same test scores often get assigned to different tracks because of differences in their social class and race. Few administrators or teachers consciously and deliberately assign students to tracks based on these criteria, but it occurs nevertheless. Researchers have consistently found that when following two students with identical scores on cognitive ability tests, the student of higher social class is more likely than the student of lower social class status to get assigned to the higher track.

This inequality is at the root of the American education debate. A core American value states that all people are created equally. Through a fair and equitable educational system, all students would have equal access to opportunity and success. Sociological research examines the social institution of education to better understand the consequences for students and how to improve those consequences.

Education in Global Perspective

For years, the United States has been heralded as having the best school system in the world. Once at the top of the list, the United States now ranks thirty-eighth out of seventy-one nations in student math and twenty-fourth in science, as measured by international achievement tests (DeSilver 2017).

What explains this decline in the nation's standing? Most of it has to do with the enormous inequality that characterizes U.S. schools and the failure of American education to address the needs of poor, African American, and Latino students. Unlike other nations, the United States spends more on students from higher socioeconomic backgrounds, leaving those in poor and racially segregated schools disadvantaged on educational achievement (Darling-Hammond 2010).

These facts mean that inequality in education is strongly linked to the nation's standing in the global community. In other nations, students spend more time in school, raising questions about whether the school year should be extended for U.S. students. An extended school year is one possible solution. Other recommendations have been made, including funding for early education, increased teacher training, and better resources for all students across race and class lines. One thing is clear: The under-education of minority students explains a large portion of the academic achievement gap (Darling-Hammond 2010).

Sociological Theories of Education

As with other social institutions, sociological theory provides perspectives that illuminate public concerns about education. Questions about the purposes of education, how education is organized, and who education serves are addressed in different ways by the major theoretical perspectives in sociology.

Myth: Intelligence is mostly determined by genetic inheritance.
Sociological Perspective: Intelligence is a complex concept not easily measured by one thing and is likely shaped as much by environmental factors as by genetic endowment (Zwick 2004).

Functionalist Theory

Functionalist theory in sociology argues that education accomplishes the following consequences, or "functions," for a society. First is *socialization*. As we have already seen, socialization takes place in the family, but the family is not the sole location of socialization. Schools also have a socializing influence through passing on "book knowledge" in the form of information and skills. Schools pass on cultural heritage and history, including values, beliefs, habits, and norms—in short, culture. Some of this is explicit in the schools, such as learning a language or the music and art of one's culture. Learning culture can also be implicit, such as guiding students through norms around punctuality, discipline, and manners.

Occupational training is another function of education, especially in an industrialized society such as the United States. In less technologically based societies, jobs and training may be passed from parent to child. A significant number of occupations and professions today are still passed on from parents to offspring, particularly among the upper classes (such as a father passing on a law practice to his son) or among certain highly skilled occupations (plumbers, ironworkers, and electricians). In today's highly technological society, though, higher levels of education are increasingly necessary to secure a job with a livable wage.

Social control is also a function of education, although a less obvious one. Such indirect, subtle consequences emerging from the activities of institutions are called latent functions of the institution. Increased urbanization and immigration beginning in the late nineteenth century were accompanied by rises in crime, overcrowding, homelessness, and other urban ills. One perceived benefit of compulsory education (that is, one latent function) was that it kept young people off the streets and out of trouble. There is a *hidden curriculum* in schools—a latent function of education; that is, schools not only "function" to give skills and training, but they also teach students norms, identities, and other forms of social learning that are not part of the formal curriculum.

Conflict Theory

In contrast to functionalist theory, which emphasizes how education unifies and stabilizes society, conflict theory emphasizes the power and inequality that are part of education as a social institution. Inequality in education occurs along numerous lines, with class, race, and gender among the most significant. The higher one's social class, the more likely one will have higher educational attainment. Racial differences in education have also produced what is called the *achievement gap*—with volumes of research and heated public debate about the causes and consequences of this gap. Women now also outpace men in completing

PhotoAlto/Frederic Cirou/Getty Images

Contrary to the impression given in this photo, girls are frequently underrepresented in scientific and technical classes in school, particularly in upper grades and in college.

college (DiPrete and Buchmann 2014), although a gender gap persists is in science, technology, engineering, and mathematic majors, or STEM disciplines. Men still outnumber women in these fields, which are widely believed to be important for global competition and employment.

These facts support the argument of conflict theorists that educational institutions are a site for producing and reproducing inequality in our society. Educational and behavioral expectations are reinforced in schools, from preschool through college and beyond. Education is designed to produce workers for the continued growth of a capitalist economy. Those people in society, given the opportunity for educational advancement, are the same ones awarded opportunities within the economic structure of the United States.

Schools are also systems of power—not just on the level of teacher–student relationships, but also as a social system. School boards, principals, parents, teachers, and unions all vie for power and control in a system that ties them together—not just in stability and cooperation as functionalist theory presumes, but also in conflict and through power dynamics, the insight of conflict theory.

Symbolic Interaction Theory

Symbolic interaction theory focuses on how people interpret social interaction—in other words, some of the subjective dimensions of education and schooling. This is well illustrated in what has come to be known as the **teacher expectancy effect**—the effect of teacher expectations on a student's actual performance. When students and teachers interact, certain expectations arise on the part of both. The teacher may expect or anticipate certain behaviors, good or bad, from students. Through the operation of the teacher expectancy effect, these expectations can actually create the very behavior in question. Thus fulfilled, the behavior is actually caused by the expectation rather than the other way around. For example, if a White teacher expects Latino boys to perform below average on a math test relative to White students, over time the teacher may act in ways that encourage the Latino boys to get below-average math test scores. The point is that what the teacher expects students to do affects what they will do. Teachers' expectations can dramatically influence how much students learn independent of their actual ability.

Insights into the teacher expectancy effect come from symbolic interaction theory. In a classic study, Rosenthal and Jacobson (1968) told teachers of several grades in an elementary school that certain children in their class were academic "spurters," who would increase their performance that year. The rest of the students were called "nonspurters." The researchers selected the "spurters" list completely at random, unbeknownst to the teachers. The distinction had no relation to an ability test the children took early in the school year, although the teachers were told (falsely) that it did. At the end of the school year, although all students improved somewhat on the **achievement test**, those labeled "spurters" made greater gains than those designated "nonspurters." Although more recent sociological research has attempted to replicate this study, true replication is nearly impossible. Studies continue to show evidence that teacher expectations influence outcomes, but this is mediated by many factors beyond labeling of students (Jussim and Harber 2005).

How are expectations converted into performance? The powerful mechanism of the **self-fulfilling prophecy**, in which merely applying a label has the effect of justifying it, affects performance (Ballantine and Spade 2015). In other words, if a student is defined (labeled) as a certain type, the student often becomes that type. You can see how such a process might also be deeply affected by race, class, and gender stereotypes.

A very good example of this is the concept of *stereotype threat* (see also Chapter 10). This refers to the fact that perceived negative stereotypes about one's group can actually affect one's academic performance. A series of experiments have repeatedly shown that when students perceive that they are being judged by a negative stereotype, their performance, such as on tests, actually declines. This holds true for both students of color and women (Steele 2010; Steele and Aronson 1995). Stereotypes about women and minorities regarding educational abilities in math and science lead to anxiety for

Homeroom Security

Research Question

Random searches, zero-tolerance policies, metal detectors, surveillance cameras, security guards—the presence of these things in our public schools makes it seem as though the school system is a police state. What effect does such a strong security system have on today's students? Is this excessive discipline necessary and effective? These are the questions that drove sociologist Aaron Kupchik to study the climate of punishment in today's schools.

Research Method

Kupchik and his research assistants spent two years observing classrooms, hallways, and disciplinary meetings in four high schools located in two different states. In each state, two high schools were largely White and middle class; two, mostly poor with students predominantly from racial–ethnic minority backgrounds. In addition to participant observation, Kupchik interviewed 100 people, including students, parents, teachers, security staff, police, and administrators, and analyzed data from questionnaires that were given to all juniors in each of the four schools. This was a very comprehensive research design.

Research Results

Kupchik found that an atmosphere of punishment, not of learning, predominates in each of these schools and guides the interaction between students, faculty, and staff. Although poor, minority students are most likely subjected to such punishment, this climate also characterizes the treatment of White, middle-class schools. Students in each school perceived that the rules were unfair. The primary finding coming from this research is that the practices of punishment and discipline far outweigh the threat of actual wrongdoing. No doubt, Kupchik argues, high schools are sites of bullying, crime, and victimization, but the level of security in schools now is disproportionate to the actual risk of crime and wrongdoing.

Conclusion and Implications

Ironically, Kupchik concludes that increased security and surveillance in schools actually increases student wrongdoing. Why? Because students will follow rules if they think they are fair, but will thwart them if they perceive the rules as unfair. Moreover, the fixation on rules and punishment overlooks the real problems students face—the context in which student wrongdoing actually emerges. With the priority given to punishment in schools, students' actual needs are not being met, and taxpayer dollars may not be used to greatest effect.

Questions to Consider

1. What kind of security was in place in the high school you attended? Was it effective? Did it decrease or increase misbehavior?

2. Why has there been such an emphasis on what Kupchik calls "homeroom security" in recent years? Do enhanced security practices in the schools address student needs?

Source: Kupchik, Aaron. 2010. *Homeroom Security: School Discipline in an Age of Fear.* New York: New York University Press.

the students and can also affect their choice of college majors, leaving science, engineering, and math courses, for example, if stereotypes about them are invoked. One of the brilliant insights of symbolic interaction theory is that the meaning attributed to a behavior can be a powerful predictor of what a person becomes. Each of the three core sociological theories offers an important perspective on education (see ◆ Table 14-2).

Table 14-2	Sociological Theories of Education		
	Functionalism	**Conflict Theory**	**Symbolic Interaction**
Education in society	Fulfills certain societal needs for socialization and training; "sorts" people in society according to their abilities	Reflects other inequities in society, including race, class, and gender inequality, and perpetuates such inequalities by tracking practices, for example	Emerges depending on the character of social interaction between groups in schools
Schools	Inculcate values needed by the society	Are hierarchical institutions reflecting conflict and power relations in society	Are sites where social interaction between groups (such as teachers and students) influences chances for individual and group success
Social change	Means that schools take on functions that other institutions, such as the family, originally fulfilled	Threatens to put some groups at continuing disadvantage in the quality of education	Can be positive as people develop new perceptions of formerly stereotyped groups

Educational Reform

There are clearly major challenges facing the educational system in the United States. Calls for reform are many and are coming from parents, communities, teachers, and administrators, as well as presidents and politicians.

Most programs for reform now focus on "accountability"—that is, measured assessments of where and how schools are succeeding or failing. This has resulted in what is referred to as *high-stakes testing*. Under such a "testocracy" (Guinier 2016), students cannot graduate or move on to the next grade without reaching certain levels of proficiency, as measured on standardized exams. In some places, teachers are held accountable for their students' test scores and can be replaced if the scores are low. Schools that are underachieving, as measured by test scores, are also often threatened with budget reductions or outright closure. This is in stark contrast to other nations where it is more common for additional resources to be provided to schools that are struggling (Darling-Hammond 2010).

Intense political controversy centers on how to fund education, where best to spend tax dollars, and what policies would be most successful. Key issues in the education debate are:

1. What standards and assessments will prepare students to succeed and to compete in a global economy?
2. What are the best measures of student success and how should they be used to improve instruction and student learning?

Schools in the United States are rapidly resegregating by race.

3. How should we recruit, reward, and retain the best teachers and principals?
4. How can low-achieving schools be improved?
5. How can we ensure the most success for all students, regardless of their socio-economic standing?

Educational reform is difficult to implement. It must begin with a clear understanding of education as an institution, including how schools create and reinforce inequality. Continued research and governmental commitment will help create a more balanced, fair, and successful model for educating Americans.

Health Care in the United States

Like education, health care in the United States is a social institution. Although you might think of health and illness as purely physiological phenomena, sociologists know that there are deep and complex social dimensions to each. Put simply, explaining health and illness, treating people who need care, and understanding how health care is socially organized require not just an understanding of the physiology of health, but also of how social factors influence the health of any population.

From a sociological perspective, several themes guide the sociology of health care. They include:

- Health and illness, although physiologically rooted, are strongly influenced by social factors.
- Systems of inequality in society produce various health disparities, including both the likelihood of illness and the treatment people receive.
- How the system of health care is organized stems from its social and historical context.
- Health care policies and reforms reflect complex social attitudes and behaviors.

We examine each of these points in the remainder of this chapter, but let's start with a brief explanation of each.

(1) Health and illness are strongly influenced by social factors. How likely are you to have a nutritious diet? Scientific knowledge tells us what foods produce good health and which ones do not. Focusing on food value per se, however, does not solve the problem of poor nutrition or problematic eating. There are strong social and cultural dimensions to eating, as well as to physical activity. Recently, for example, American culture has emphasized fitness and healthy eating. Whole industries have developed that provide various apps and gadgets that are designed to help people monitor their physical activity and food intake. Why then is there a national problem of obesity?

We know that obesity is a major contributing factor for heart disease, stroke, diabetes, and some cancers. Recently, the Centers for Disease Control and Prevention even classified obesity as an epidemic, with about one-third of adults and 17 percent of children classified as obese, costing the United States nearly $150 billion annually in additional health care expenditures (Ogden et al. 2015). Many would say that obesity is an individual problem stemming from someone lacking self-control. No doubt, individual decisions guide what and how much we eat, but when an entire culture markets fast food, makes the least healthy food the most convenient, and systematically markets over-indulging in food and alcohol, then

See for Yourself

Mapping Food

Identify two neighborhoods in your community that differ by their social class and/or racial composition. Draw a map of each neighborhood and then take a drive through each with your map in hand. Mark every place where you see some kind of food outlet, and mark whether it is a major grocery chain, a convenience store, fast-food outlet, or other provider of meals. You might also note what kind of transportation is needed to get to each location. When you have finished, what patterns do you see about the availability of healthy food in each neighborhood? If you lived in either, how far would you have to go to purchase fresh, good-quality food? Can you get there without a car? What does your experiment suggest about class and race disparities in health outcomes?

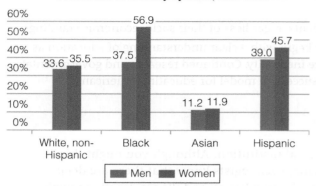

Prevalence of Obesity by Race, 2011–2014

▲ **Figure 14-4   Prevalence of Obesity, 2011–2014 (adults aged 20 and over).** In each racial–ethnic group, women tend to be more obese than men, although the difference is very slight among Asian Americans. You can also see that there are large differences by race and ethnic status. How would sociologists explain these facts?

Source: Ogden, Cynthia L., Margaret D. Carroll, Cheryl D. Fryar, and Katherine M. Flegal. 2015. *Prevalence of Obesity among Adults and Youth: United States, 2011–2014.* National Center for Health Statistics Brief, No. 219, November. Atlanta, GA: Centers for Disease Control. **www.cdc.gov/nchs/data/databriefs/db219.pdf**

clearly social and cultural factors are a significant part of the obesity problem. Further, when you can document that obesity is more frequent among Black Americans and Latinos, then you must ask about the social dimensions of this unhealthy condition.

Sociologists look at the particular environments in which people live as influencing their health. For example, many people in low-income inner-city neighborhoods have to go miles before reaching a grocery store with fresh, affordable, and healthy food—so-called *food deserts*. Convenience stores, vending machines, and fast-food restaurants dominate the city landscape, especially certainly in poorer areas. High-priced, up-scale markets are mostly found in wealthier, predominantly White neighborhoods (Cannuscio et al. 2014; Block, Scriber, and DeSalvo 2004). It is little surprise then that some racial and ethnic groups are more likely than others to be obese (see ▲ Figure 14-4; National Center for Health Statistics 2017).

(2) Systems of inequality result in significant health disparities. Generally speaking, the citizens of the United States are quite healthy in relation to the rest of the world, but there are very great discrepancies in the United States in terms of how healthy certain groups are compared to others and which groups have the best health care.

Race, gender, class, age, disability, and even immigrant status are well known to influence the likelihood of good or poor health. Even how long you live and how you will die are highly correlated with race, gender, and ethnicity. White women, for example, live ten years longer on average than Black men. Women are more likely to die from some form of cancer; men from heart disease. Death rates by opioids or heroin are much higher among White men and women than any other groups. And, although disability can affect people at any income level, people below or near the poverty level are more likely to have at least one disabling condition. Even something as basic as vision limitations is highly correlated with social class (National Center for Health Statistics 2017). Facts such as these point to the very strong connection between health, illness, and social systems of inequality.

(3) Social and historical contexts shape the social organization of health care. Health care in the United States is now a vast institution, including not only hospitals and doctors but also many auxiliary sectors, such as nursing homes, rehabilitation centers, drop-in clinics, and various "alternative" health care services, such as homeopathy, wellness centers, and even exercise and nutrition centers. Health is big business. How health care is delivered is thus massively influenced by a for-profit system, including hospitals and other treatment centers as well as the colossal pharmaceutical industry. The for-profit sector in health care also has a massive influence on social policies about health care as lobbyist groups spend enormous amounts of money to influence health care policy and even national elections. In 2016, for example, the pharmaceutical industry spent over $250 million in just this one year to influence national policies (Center for Responsive Politics 2017).

Critics of the health care system contend that the for-profit character of health care produces too strong a focus on treatment, not prevention. Medicine in this country follows mostly a *disease model* in which patients are first diagnosed and then treated for the illness. Despite evidence that prevention of many illnesses is possible, most health insurance does not usually, for example, reimburse a health club or gym membership even though doing so would likely reduce the cost of medical care.

(4) *Understanding health care reform requires also analyzing complex social attitudes and behaviors.* The vast majority of Americans have great confidence in health care providers. A significant majority (60 percent) also think that it is the government's responsibility to provide health care coverage to all Americans (Kiley 2017). Yet, there is growing concern about the high cost of health care and intense debate about what to do about it.

The United States still has some of the most sophisticated health care treatment in the world, but is it affordable? Who has access? Why are costs for medical insurance so high? What role should the government play in providing health care to its citizens? These questions are at the core of current political struggles about health care, and they point to the social structures and tensions in health care as a social institution.

The Sociology of Health and Illness

When you get sick, what do you do? Call a doctor? Visit a clinic? Go to the emergency room? Use over-the-counter drugs or some herbal or holistic treatment? Years ago, you might have taken some cod liver oil or a tonic provided by someone we would now consider a "quack." In the nineteenth century, for example, middle- and upper-class women were frequently treated for "hysteria," which was believed to be a "disease of the uterus." Most likely the so-called symptoms were simply women rebelling from the confinement in roles for bourgeois women (Ehrenreich and 2011). During the same period, men who masturbated were considered mentally ill, sometimes even confined to mental institutions (Stengers and Van Neck 2001).

Your health or illness—what you do about it, how it is defined, who treats it and how, indeed, whether it gets treated at all—are social phenomena. That is, you might make an individual decision about your health, but the choices you have and how you think about your condition are all shaped by society. This includes the basic fact of whether, when sick, you can stay home or whether you have to go to work no matter what for fear of losing your job.

Sociologists focus on the social contexts that influence health and illness. Obviously, physiological factors are not to be ignored in studying and treating health conditions, but a focus on the social and cultural environment in which disease is produced, identified, and treated lies at the core of the sociological perspective. Social factors are also increasingly being recognized by medical professionals as an important part of responding to patients' needs.

Consider how any particular disease is ultimately defined and understood. Is a disease identified simply because of the numbers of people afflicted or the deadliness of the disease? Not necessarily. How illness is understood does not stem from medical knowledge alone.

The history of HIV/AIDS provides a good example of how social attitudes and behaviors influence our understanding of health and illness. When AIDS first appeared in the early 1980s, it was defined as solely a disease of gay men. Not only were those who were infected heavily stigmatized, but little was done to study and try to stop the disease. For years, while thousands died, federal policy downplayed the use of condoms, instead moralizing about gay sex. Hardly any money was provided for research. Had there not been such "fatal inaction" (Lawson 2015), many of these deaths could have been avoided. Only with the mobilization of the gay community did the public finally demand more research and better treatment such that today HIV/AIDS is no longer the death sentence that it was when homophobia framed the public response (Shilts 2007). Of course, advances in medical knowledge about HIV/AIDS mattered, but social action, not just medical science, was critical to our current understanding of HIV/AIDS.

The current heroin and "opioid epidemic" provides another example of how social contexts define health. Until recently, heroin use was largely associated with poor, inner city, largely Black populations. Throughout the 1960s, 1970s, and 1980s, heroin and crack/cocaine were also associated with the rise in violent street crime. Policies to address this problem were thus focused on punishing and criminalizing users. Now, however, heroin and opioid abuse (and death) has become largely a White problem (see ▲ Figure 14-5). What happened?

The pharmaceutical company Purdue Pharma introduced oxycontin to the market in 1996. When they did so, they began an aggressive marketing campaign even declaring that this painkilling drug was non-addictive. Prescriptions for the drug soared, as did those for other opioids. Soon the drug became

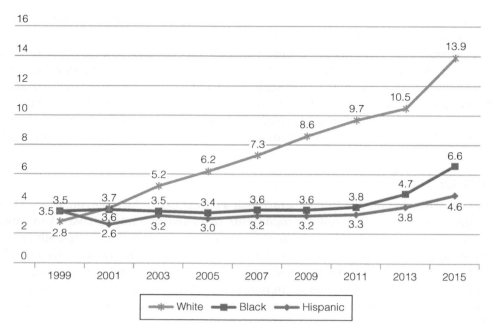

▲ **Figure 14-5** **Opioid Deaths by Race and Ethnicity.** As you can see, opioid deaths became defined as an epidemic when the rate of death among White Americans soared. How do you explain the trends and facts that you see in this line graph?

Note: Rate is per 100,000 people

Source: Centers for Disease Control and Prevention. 2017. *Multiple Causes of Death, 1999–2015*. Atlanta, GA: National Center for Health Statistics. **http://wonder.cdc.gov/mcd-icd10.html**

commonly used outside of prescribed usage and the abuse of prescription drugs became defined as a social problem. As medical professionals started to limit opioid prescriptions, addicted people then looked for substitutes, which they could find in heroin. As a result, not only did the nation declare an "opioid epidemic," but heroin use among White people also surged (Cohen 2015). But, unlike the earlier concerns about heroin in mostly poor, Black communities, the nationally declared opioid epidemic has been defined as a problem of public health—referred as a *manufactured illness* (Weiss and Lonnquist 2017).

In other words, when heroin use was perceived as mostly a problem for poor, inner city Black people, it was never defined as a health problem. Instead heroin was defined as a criminal problem and social policies followed suit. This illustrates how social attitudes and behaviors define "illness" and how social conditions shape how we think about disease.

Health Disparities and Social Inequality

Scientific breakthroughs in the natural sciences have brought us to a remarkable time in Western medicine when Americans have access to diagnosis, treatment, and cures for so many diseases once believed to be fatal. In many ways, the modern American system of health care and medicine is a model of great success. Yet, much like education as a social institution, health care involves social structures that create different experiences for different groups of people. The social reality of health and illness is especially apparent when examining known health disparities—disparities that can only be understood in the context of understanding the various forms of social inequality. Something as basic as how long one waits for treatment in an emergency room has been shown to be related to the social status of the patient, not just the seriousness of their condition (Lara-Millan 2014).

Health care is more readily available and more readily delivered to White people than to others. Yet, as late as 2014, almost 8 percent of children in the United States had no usual source of health care.

Myth: Once people know that there is a health risk associated with a particular activity, they will stop doing it.

Sociological Perspective: People can easily dismiss or ignore evidence of health risks, especially when there are institutions with an interest in maintaining the status quo by not addressing the problem. The known risk of brain injury from repeated hits to the head is a case in point. Although the National Football League has developed new protocols for possible concussions, what institutional changes would be needed to make football a non-violent sport? How likely is such change? Whose interests would be at stake and what does this tell you about the social organization of health—and football, for that matter (Coakley 2014)?

Both social class and race influence the likelihood of access to care. Slightly more than 10 percent of Latino and Black children had no usual source of care, compared to 6 percent of White children. Children in low-income families were much less likely to have a usual source of care than children in middle- and high-income families (Zewde and Berdahl 2017).

Even the region where you live can have a huge influence on your health. Each year, many in the United States die because they live too far away from good health care. Doctors and hospitals are concentrated in urban areas; they are less likely to be situated in rural areas, where health is actually worse (Sun 2017; Hartley 2014). In the remainder of this section we briefly review some of the major sources of health disparities.

Race, Ethnicity, and Health Care

On virtually every measure, racial–ethnic minorities in the United States have poorer health than Whites. There are racial disparities in rates of mortality, the early onset of disease, and in the severity and consequences of disease. Furthermore, health disparities for people of color have persisted over time and at all levels of income and education (Williams and Mohammed 2013). African Americans are more likely than Whites to fall victim to various diseases, including cancer, heart disease, stroke, and diabetes. Although the occurrence of breast cancer is lower among African American women than White women, the *mortality rate* (death rate) for breast cancer in African American women is considerably higher than it is for White women (National Cancer Institute 2017). African American men and Mexican-origin men are more likely than others to live with hypertension, a fact interpreted as reflecting the stress that living with racism produces (Sternthal, Slopen, and Williams 2011).

Surprisingly Hispanics actually have a longer life expectancy than do other groups and on many measures their health is better than that of White, non-Hispanics. Researchers refer to this as the *Hispanic paradox* because, given the similarities in Hispanic socioeconomic status to that of Black Americans, you would expect Hispanics to be less healthy. It seems that one explanation of the Hispanic paradox is the lower rate of smoking among Hispanics, but, in general, scholars have been unable to fully explain this surprising finding (Lariscy et al. 2016; Markides and Eschback 2011).

In addition to differences in patterns of health, it is also well established that African Americans, Latinos, and Native Americans simply do not receive medical attention as early as Whites. When they do get treatment, the stage of their illness is often more advanced and the treatment they receive is not of the same quality. African Americans and Hispanics, especially when they are poor, are also less likely than Whites to have a regular source of medical care (see ▲ Figure 14-6). Native Americans are the group least likely to use health services, such as doctors' offices and clinics (National Center for Health Statistics 2017).

Patterns of health among immigrants also tell you a lot about the impact of socioeconomic conditions on health and illness. Generally speaking, foreign-born people in the United States have better health than their native-born counterparts. But, the longer an immigrant remains in the United States, the worse their health becomes. Sociologists explain this as the result of residential segregation and the decline in socioeconomic status that tends to occur following immigration. The better health of first-generation immigrants is also partially explained by the fact that only those in relatively good health are likely to make the journey, but the overall pattern is clear: One's health status is quite dependent on the context in which one lives (Centers for Disease Control and Prevention 2015).

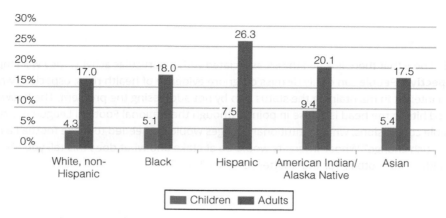

▲ **Figure 14-6** **Children and Adults without a Usual Source of Health Care, 2014–2015**

Source: National Center for Health Statistics. 2017. *Health United States 2016*. Atlanta, GA: Centers for Disease Control and Prevention. **www.cdc.gov/nchs/data/hus/hus16.pdf**

Social Class and Health Care

Social class has a pronounced effect on health and the availability of health services. Simply put, the lower one's social class, the less long one will live. The effects of social class are nowhere more evident than in the distribution of health and disease, showing up dramatically in the rates of infant mortality, tuberculosis, heart disease, cancer, stroke, diabetes, and a variety of other illnesses. The reasons lie partly in personal habits that are themselves somewhat dependent on one's social class. For example, those with lower socioeconomic status smoke more often, and smoking is the major cause of lung cancer and a significant contributor to cardio-vascular disease (National Center for Health Statistics 2017), but social circumstances also have a direct effect on health. Poor living conditions, elevated levels of pollution in low-income neighborhoods, and lack of access to health care facilities all contribute to the high rate of disease among low-income people.

Research has consistently shown correlations between stress and physical illness (Sternthal, Slopen, and Williams 2011). Most people experience stress on a regular basis, but the poor and people of color are more subject to chronic stress than are other groups. Stress shows up in their comparatively higher levels of illness. Sociologists and health care researchers also find a correlation between financial problems and the likelihood of illness. People who experience job loss typically see their health—both physical and mental—decline. When people are out of work, such as during recessions, without good health insurance, they may simply forego care.

For those who work in low-income jobs, however, the simple fact of working affects one's health, especially if the working conditions are poor. Research has also found that during times of economic recession, people's health is affected, particularly if, as low-income people do, they rely on public services for health care. During economic recessions, budgets reductions are likely to reduce or eliminate such services (Burgard and Kalouska 2015; Towne et al. 2017).

Medicaid is the government program that provides health care for poor and low-income people. Medicaid is funded through tax revenues. The costs covered per individual vary from state to state because the state must provide funds to the individual in addition to the funds that are provided by the federal government. Now, with political unrest over so-called Obamacare, the future of Medicaid is quite uncertain, as we will see in the section on health care reform.

Gender and Health Care

Race and social class are not alone in influencing people's health. Gender matters too, although gender, race, and class intersect in predicting and shaping people's health (Mullings and Schulz 2005). Gender, though, also has an influence all its own, indicated by the different health outcomes for women and men.

Ironically, women in general live longer than men, but they are sick more often and are more likely than men to suffer from chronic conditions. Common mental health problems also differ based on gender. Women, for example, are more likely to experience depression and anxiety. Men, on the other hand, engage in more risky behaviors that threaten their health. Men are more likely to smoke, use controlled substances, and engage in other risky behaviors, such as carrying a weapon and fighting. Women report more serious psychological distress and are more likely to seriously consider suicide (National Center for Health Statistics 2017).

What explains the gender disparities in health? Sociologists look to the social structures of gender that organize women's and men's lives to explain gender disparities in health—and health care. Social roles have profound social and physiological effects on women's and men's health. Women experience what some have called "constrained choices" (Bird and Rieker 2008: 5). This means that the individual choices women and men make occur in the context of social forces that shape our lives in so many ways. As just one example, media images of the ideal woman encourage excessive dieting and other eating behaviors that are known to harm women's health. Women's health is also affected by the treatment they receive in society. Sexual harassment, as just one example, has profound effects on women's physical and mental health. Finally, women's individual choices are constrained by the politics of reproduction choices and how women's and men's bodies are defined by society (Lorber and Moore 2010).

LGBTQ Health

Long ignored, but also important as influencing people's health, is a person's LGBTQ status—that is, one's identity as lesbian, gay, bisexual, and/or transgender. The silence around the experiences of LGBTQ people has meant that, until now, there has been hardly any data collected on LGBTQ health. National surveys of people's health rarely, if ever, ask about people's sexual orientation or gender identity other than the binary categories "male" or "female." Some new projects have been designed to address this issue (see, for example, pridestudy.org) so we are beginning to understand the social conditions that influence LGBTQ health and wellbeing (National Academies of Sciences, Engineering and Medicine 2011).

We know, for example, that LGBTQ youth are at an increased risk for suicide, and it seems they may be more likely to smoke and drink as responses to stress. LQBTQ people, especially trans youth, are also subject to elevated levels of violence and victimization of various forms. Sociologists also know the influence of stigmas—defined as when someone is socially devalued by others because of an identifiable characteristic. Because of the stigma LGBTQ people face, they may utilize health services less frequently for fear of being misunderstood, ridiculed, or badly treated. Indeed, sociologists know that being a target for stigma is risky for one's health.

As in other social groups, LGBTQ people differ considerably in their age, race, and educational and income levels. In order to fully understand LGBT health, scholars need to understand these overlapping social factors (Institute of Medicine 2011).

Health and Disability

The *disability rights movement* has transformed how people think about disability, challenging many preconceived ideas. For example, people tend to see someone with a disability solely in terms of that social status—a stigma. As with LGBTQ people, when someone is stigmatized, that identity tends to override all other identities, and the person is treated accordingly.

The disability rights movement has fundamentally changed how disability is now understood and treated. The movement has called attention to the social realities of disabilities, even questioning the very language used to identify people with disabilities—for example, using the term *physically challenged* rather than the more negative connotation of *disabled*.

One of the most significant achievements of the disability rights movement is the Americans with Disabilities (ADA) Act, passed by Congress in 1990. This law prohibits discrimination against people with disabilities. The ADA legislates that people with disabilities may not be denied access to public facilities—thus the presence of such things as ramps, wheelchair access on buses and stairways, handicapped parking spaces, and chirping sounds in crosswalk lights for blind pedestrians, all social changes that are now so prevalent that you might even take them for granted. They have resulted, however, from the mobilization of those who saw a need for social change.

The Americans with Disabilities Act also requires employers and schools to provide "reasonable accommodations" such that those with disabilities are not denied access to employment and education.

Peter Hvizdak/The Image Works

The disability rights movement has opened up new opportunities for those who face the challenge of disability.

For many students with various learning disabilities, this has meant making accommodations for taking tests with extended time or in settings where the test taker is not subject to as much distraction as in a crowded classroom. The increased awareness of disability rights has transformed society in ways that have opened up new opportunities for those who, years ago, would have found themselves with less access to education and jobs and, therefore, more isolated in society.

Age and Health Care

As people age, their health care needs are no doubt likely to increase. Not surprisingly, age is related to most indicators of health, although in general people are living longer. How well one ages, however, is clearly linked to other social factors including—as you would predict from above—one's race, ethnicity, social class, gender, LGBTQ identity, and ability status. *Gerontologists*, those who study aging, know that aging cannot be understood simply as a biological process. Rather, it is a "significantly social phenomenon influenced by wider social, political, and economic factors" (Matthews 2015).

The national **Medicare** system is the major source of health care for older people. Medicare was begun in 1965, under the administration of President Lyndon Johnson. It provides medical insurance, including hospital care, prescription drug plans, and other forms of medical care for all individuals age 65 or older. Medicare is partially funded through payroll taxes whereby both employees and employers pay a small percentage of employee wages to cover some of the cost of this large (and costly) federal program. With so many people in the population now living longer and the now aging baby boomer population being such a large share of the total population, many wonder if Medicare can be sustained in the near future. The number of workers paying payroll taxes is shrinking and the cost of health care is rising, twp factors that, in addition to the large elder population, mean that Medicare mau not be sustainable. Although not the sole basis for the nation's challenges in health care, the health needs of the older population are clearly a major problem.

The Social Organization of Health Care

The sociology of health and illness also calls attention to how health care systems are organized. Like other social institutions, health care is stratified in a way that reflects the social inequalities in society at large. There is a hierarchy that puts medical doctors at the top and medical assistants, nursing staff, orderlies, food service workers, and others at the bottom. Patients have the lowest status in this system, sometimes being treated as children with little agency in determining how treatment in administered. In other words, health care institutions recreate the structural inequality of society.

Unlike in the past when a doctor might have made a home visit or midwives delivered babies in the home, health care today is big business and tends to be organized in a highly bureaucratized system. There are many components of this huge system, including hospitals, and also powerful pharmaceutical companies, nursing homes and rehabilitation centers, addiction treatment facilities, insurance companies, clinics, and more. Most of this system in the United States is structured as for-profit businesses with enormous ramifications for how people are treated. Insurance companies try to minimize people's hospital stays as a way of cutting costs. Pharmaceutical companies are occasionally charged with gouging the public with the cost of drugs. Physicians may have to raise their rates to cover the sky-rocketing cost of insurance policies. At the core of this system is the need to make money, and some argue that this can at times impact the quality of care that patients receive. Within such a system, pharmaceutical companies, hospital conglomerates, and vast insurance companies can actually have more power to determine patient care than an individual physician.

The pharmaceutical industry is one of the largest components of this elaborate system. Prescription drug prices in the United States are among the highest in the world (Olson and Sheiner 2017)!. Spending for prescription drugs in the United States has increased from $40 billion in 1990 to a whopping $325 billion in 2015! There is little sign that this spending will do anything but go further up. In fact, prescription drugs are one of the fastest-growing components of health care costs. The rise in the cost of drugs is partially attributed to the high cost of research and development, but is mostly because of the money for-profit drug companies spend on aggressive marketing (Kaiser Family Foundation 2010). Marketing drugs directly to consumers has become increasingly common, something you can see this yourself as hardly an hour goes by on television without an advertisement for some kind of prescription drug. Spending on drug advertisements has in fact grown by 62 percent just since 2012, now exceeding $6 billion per year (Horovitz and Appleby 2017).

Bureaucratization is also part of the social organization of health care, and it adds to the alienation that patients can experience when confronting this complex system. There are seemingly endless forms and reports both for physicians and patients, including paperwork to enter people's records into the system, authorize procedures, dispense medicines, monitor progress, and process payments. Physicians increasingly complain about the large percentage of their time spent with insurance paperwork. Some-times it even seems that bureaucracy stands in the way of people receiving good care.

Technological advances in medical science have made astounding progress in treating many illnesses. Cancers that years ago would have inevitably led to death are now treated with sophisticated surgical procedures and/or drug "cocktails" that are specifically tailored to given forms of the disease. Even with such advanced care, however, many feel that medical model of treating disease is insufficient and have thus turned to alternative (and, often, longstanding traditional) practices of medicine. Acupuncturists are successfully treating people for everything from chronic back pain to breach pregnancies. Holistic health care providers are proliferating, sometimes used complementary to medical treatment. The most recent survey finds that Americans paid at least $30 billion a year on alternative medicines some of which, such as yoga and meditation, can have positive health effects, but others of which have found to be completely ineffective. Some physicians now incorporate some alternative practices into their patient care, though they are rarely covered by insurance.

Theoretical Perspectives on Health Care

The sociology of health is anchored in the same major theoretical perspectives that we have studied through-out this book: functionalist theory, conflict theory, and symbolic interaction theory (see ◆ Table 14-3).

Functionalist Theory

Functionalism argues that any institution, group, or organization can be interpreted by looking at its positive and negative functions in society. Positive functions contribute to the harmony and stability of society. The positive functions of the health care system are the prevention and treatment of disease. Ide-ally, this would mean the delivery of health care to the entire population without regard to race, ethnicity, social class, gender, age, or any other characteristic. At the same time, the health care system is notable for a number of negative functions, those that contribute to disharmony and instability of society.

Functionalism also emphasizes the systematic way that various social institutions are related to each other, together forming the relatively stable character of society. You can see this with regard to how the health care system is entangled with government through such things as federal regulation of new drugs and procedures. The government is also deeply involved in health care through scientific institutions such as the National Institutes of Health, a huge government agency that funds new research on various

Table 14-3	Theoretical Perspectives on the Sociology of Health		
	Functionalism	**Conflict Theory**	**Symbolic Interaction**
Central point	The health care system has certain functions, both positive and negative.	Health care reflects the inequalities in society.	Illness is partly socially constructed.
Fundamental problem uncovered	The health care system produces some negative functions.	Excessive bureaucratization of the health care system and privatization lead to excess cost.	Patients and health professionals serve specific roles. What is determined as illness is specific to cultural context.
Policy implications	Policy should decrease negative functions of health care system for minority groups, the poor, and women.	Policy should improve access to health care for minority racial–ethnic groups, the poor, and women.	Determining something as disease will make insurance reimbursement more likely.

matters of health and health care policy. As a social institution, health care is also one of the nation's largest employers and thus is integrally tied to systems of work and the economy.

Conflict Theory

Conflict theory stresses the importance of social structural inequality in society. From the conflict perspective, the inequality inherent in our society is responsible for the unequal access to medical care. We have seen that people of color and the poor usually get less quality care than the middle and upper classes, and the middle-aged. Restricted access is exacerbated by the high costs of medical care.

Also important in conflict theory is the study of power. In health care, the large for-profit industries that comprise this institution have enormous power over the kind of care that Americans receive. Power operates at the institutional level, but it also operates at the group level. That is to say that certain groups exercise more power over others within a system that is hierarchically structured. Doctors hold power over nurses; nurses have the power to shape some parts of patient care; hospital administrators have power over lower-level workers; insurance companies have the power to determine what treatment someone will get; and policy makers have the power to shape national health care policy. From a conflict perspective, you cannot understand the nation's health care system without an analysis of these power relationships.

Symbolic Interaction Theory

Symbolic interaction theory holds that illness is partly socially constructed. The definitions of illness and wellness are culturally relative—the social context of a condition partly determines whether or not it is sickness. Consider the example of alcoholism and other addictions. During the era of prohibition, people who drank were considered deviants and lacking moral fortitude. Now, however, alcoholism is a diagnosable disease, listed in the *Diagnostic Statistical Manual* as an illness. The medicalization of alcoholism refers to how Americans culturally and socially label abuse of alcohol as a disease that requires treatment. This has profound consequences for how people with alcoholism are treated. People who are ill receive more sympathy and more care than those who are labeled *deviant*.

The symbolic interaction approach to studying health care institutions also focuses on the roles of the patient and medical professionals and on the cultural context within which disease is labeled and treated. Furthermore, symbolic interaction emphasizes that illness can become an identity—one a person adopts for himself or herself, such as in playing a "sick role" when one wants to be excused from school or work. Others may also impose an illness identity on someone else, such as when being gay or lesbian was socially defined as pathological and people were treated according. Symbolic interaction theory reminds us of the statement that "situations defined as real are real in their consequences" (Thomas 1928). There are countless examples throughout history of how defining a presumed health condition had dire consequences for many people.

Health Care Reform

Currently, the cost of health care in the United States is approximately 18 percent of our gross domestic product, the highest in the world. Health care is also the nation's leading industry, at least as measured by their share of gross domestic product, total revenues, and the highest profits. The United States also tops the list of all countries in per person expenditures for health care (Schriever 2017; The World Bank 2017). Other countries spend considerably less money and deliver a level of health care at least as good. For example, Sweden and the United Kingdom spend roughly half as much per capita as the United States, and Turkey spends a bit more than one-third as much.

The health care crisis in the United States is largely a question of cost, but it also entails a debate over the nation's responsibility for the health of its citizens. Who should pay for the soaring costs of health care? Who receives the benefits of such sophisticated medicine? Should there be universal health care for all, like we are seeing through the Affordable Care Act? These questions are at the heart of the current national debate about health care reform.

The debate over affordable health care and equality of care has dominated the recent political land-scape. Medicaid, Medicare, and now the **Affordable Care Act** (also referred to as Obamacare) are as close as the United States has come to the ideal of universal health insurance.

The United States is one of the few industrialized nations that does not provide universal health care to its citizens. The passage of the Affordable Care Act in 2012 addressed the problem of there being too many uninsured Americans. Now, though, it is threatened with being repealed, although to date there has been no political agreement on the future of this law. Health care in the United States is a labyrinth of health care deliverers, for-profit insurance companies, and government programs that provide health care for the aged and for the poor. Entitlement programs like Medicare and Medicaid also are the subject of political debate.

Several key provisions of the Affordable Care Act addressed some of the earlier problems of a lack of health care insurance, including:

1. Expansion of the availability of health care insurance;
2. Insurance companies cannot deny coverage to people because of preexisting conditions;
3. Lifetime coverage limits on insurance coverage were eliminated, although there could be annual limits;
4. Insurance plans must cover preventative care, such as mammograms and colonoscopies, without charging deductibles and co-pays;
5. Young adults are allowed to stay on parents' plan until age 26;
6. Early retirees keep their employer-sponsored benefits until they are eligible for Medicare.

The Affordable Care Act clearly helped reduce the number of uninsured people. Despite threats to repeal it, as of 2018 nearly 12 million Americans have selected Obamacare policies through the system

See for Yourself

Youth and Health Insurance

Identify a group of young people you know and ask them if they are covered by health insurance. If they are insured, where does their insurance come from? Who pays? Did they use the Affordable Care Act marketplace to find insurance? If they are not insured, ask them why not and whether they think this is important. Do they support a national health insurance program?

Having conducted your interviews, ask yourself how social factors such as the age, race, ethnicity, gender, and educational/occupational status of those you interviewed might have affected what people say about their insurance. Do you think any or all of these social characteristics are related to the likelihood that people are covered by health insurance and whether these characteristics are related to their attitudes about coverage? What are the implications of your results for public support for new health care policies?

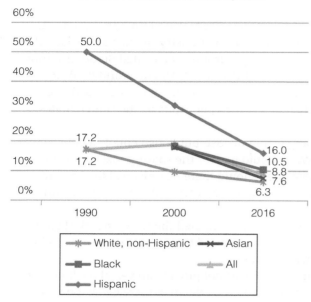

Uninsured Americans, 1990–2016 by Race

Legend:
—※— White, non-Hispanic   —✖— Asian
—■— Black   —▲— All
—◆— Hispanic

▲ Figure 14-7   People without Health Insurance, 1990–2016. As you can see, the passage of the Affordable Care Act in 2012 had a huge impact on reducing the number of uninsured Americans.

Source: Barnett, Jessica C., and Edward R. Berchick. 2017. *Health Insurance Coverage in the United States: 2016.* Washington, DC: U.S. Census Bureau. **www.census.gov/content/dam/Census /library/publications/2017/demo/p60-260.pdf**; National Center for Health Statistics. 2017. *Health United States 2016.* Atlanta, GA: Centers for Disease Control and Prevention; U.S. Census Bureau. 2000. Health Insurance Detailed Tables. Washington, DC: U.S. Census Bureau. **www2.census.gov /programs-surveys/demo/tables/p60/215/hi00ta.txt**

of federal exchanges that the law provides. This has been especially important as the number of people covered by employer-based health insurance has been falling steadily since the late 1970s. With the Affordable Care Act in place, the number of uninsured Americans fell from 16.3 percent in 2010 to only 10 percent in 2016 (DeNavas-Walt, Proctor, and Smith 2011; Collins, Gunja, and Bhupal 2017; see also ▲ Figure 14-7).

Despite the success of the Affordable Care Act in getting more Americans insured, in 2017, the government voted to repeal the law. The majority of Americans did not want the law repealed without a replacement, and their primary concern was the high cost of insurance (Kaiser Family Foundation 2017).

Despite the political division over so-called Obamacare, public opinion is strongly in favor of universal health care coverage.

Why is the United States so out of line with other Western nations in providing health care to its citizens? Sociologists offer several explanations. First, the antigovernment attitude among many in the United States fuels resistance to a national health care system. Many also think that the government should not *force* people to buy health insurance. Second, unlike in other Western nations, there is a relatively weak labor movement in the United States, and thus more limited state-based benefits for workers. Third, racial politics have shaped the nation's health care system; many people associate federal programs with racial groups, and this, too, fuels the politics of health care reform. Finally, the health care system in this country is fundamentally structured based on private, for-profit interests (Quadagno 2005).

Without universal health care coverage, millions of Americans will be uninsured. This creates vulnerability for people, especially people in poor communities, when they get sick. For Americans without insurance, the main source for medical care is a hospital emergency room, often called the "doctor's office of the poor." This is a very expensive way to deliver routine health care—and there is rarely any follow-up care or comprehensive and preventative treatment. The confusion and frustration in managing care is challenging even in the best of cases. For fully insured Americans with high education and good incomes, navigating through the health care system is often complicated and difficult. For the millions of Americans who are not insured, have less education, and are financially vulnerable, an illness can be devastating in more ways than one.

Chapter Summary

What is the importance of the education institution?

Education is the social institution that is concerned with the formal transmission of society's knowledge. It is therefore part of the socialization process. Although the U.S. education system has long produced students at the top of the world's educational achievements, the United States is falling behind other nations on standardized test scores.

How does education link to social mobility?

The number of years of formal education for individuals has important effects on their ultimate occupation and income. Social class origin affects the extent of educational attainment (the higher the social class origins, the more education is ultimately attained), as well as occupation and income (higher social class origin likely means a more prestigious occupation and more income).

Does the educational system perpetuate or reduce inequality?

Although the education system in the United States has traditionally been a major means for reducing racial, gender, and class inequalities among people, the education institution has perpetuated these inequalities. Segregation of schools and communities keep minority and poor children in schools that lack resources for success.

How does sociological theory inform our understanding of education?

Functionalism interprets education as having various purposes for society, such as socialization, occupational training, and social control. *Conflict theory* emphasizes the power relationships within educational institutions, as well as how education serves the powerful interests in society. *Symbolic interaction theory* focuses on the subjective meanings that people hold. These meanings influence educational outcomes.

What current reforms are guiding education?

The No Child Left Behind Act program emphasized accountability in the schools, largely through testing. Current educational reforms focus on achieving educational standards, assessing school progress, and developing strong measures of student and teacher success. Free community college is also an educational reform idea.

How do social factors influence health and illness?

There are strong social and cultural dimensions to how health and illness are defined. These can show up in cultural differences, but also in how illness is defined over time.

How is social inequality manifested in health care?

Health disparities occur along the same lines as other forms of inequality. Some of the most significant are based on race, social class, gender, age, LGBTQ status, and disability. The disparities show up in patterns of disease but also in patterns of care.

How is health care organized as a social institution?

In the United States, health care is a social institution mostly organized through a complex system of for-profit businesses. At the same time, the institution is heavily bureaucratized, which can lead to patient alienation.

How does sociological theory inform our understanding of health and health care?

Functionalism interprets the health care system in terms of the systematic way that health care institutions are related to each other. Conflict theory addresses the inequalities and power relations that occur within the health care system. Symbolic interaction analyzes the interpretations that can affect people's health care, such as the tendency to place patients in a sick role and in how different ailments are defined and understood.

What is driving current health care policies and reform?

The United States is only recently providing nearly universal health care for its citizens through the Affordable Care Act. Nonetheless, political debates about this law make its future uncertain even though more Americans have health insurance with the law in place.

Key Terms

achievement test 374
Affordable Care Act 387
Brown v. Board of Education 364
cultural capital 370

individualized education
  programs 371
Medicaid 382
Medicare 384

schooling 363
self-fulfilling prophecy 374
teacher expectancy effect 374
tracking 371

CHAPTER 15

ECONOMY AND POLITICS

In this chapter, you will learn to:

Compare different types of economic systems

Identify the components of change in the contemporary global economy

Identify the different organizational components of the workplace

Explain the conditions affecting diverse groups in the workplace

Compare and contrast sociological theories of work

Compare different types of political systems

Define different forms of power and authority and the organization of bureaucracies

Analyze patterns of political participation

Comprehend the organization of the military as a social institution

iStock.com/elenabs

Because you are reading this book, chances are that you are in school, probably seeking a college degree that will help you find a good job. Your hopes and dreams likely rest on getting a decent education, discovering what you plan to do in a job, and gaining the skills you need to succeed. You might be wondering how your education will prepare you for work and how you will match your interests to the current job market.

On one level, these are individual issues that involve your interests and talents, your motivation, and how well prepared you are to study and work. But behind these individual matters lie other social structures—structures that shape the options you face, the resources you have to pursue your dreams, and whether good jobs are available to you. The background structures that frame your individual life are the result of social institutions and how those institutions are organized. Finding work situates you within economic institutions—that is, the economy, which, like other social institutions, has a particular social structure that includes an organized system of social roles, norms, and values. Social institutions extend beyond us, but they shape day-to-day life.

Typically, people do not think about such structures and may not even be aware of the institution that shape their daily lives. But in today's world, individual experiences about work are being radically transformed by changes in the nation's—indeed, the world's—economy. Every day, news reports note the impact of globalization on work and workers; new technologies are transforming work for everyone—those at the top and those at the bottom. Some even fear that robots and other technologies will completely displace whole classes of workers—drivers, as an example, should self-driving cars become the norm.

In other words, the work you have—or do not—is shaped not just by individual decisions, but also by large-scale sociological and economic forces that may feel far beyond your own control. Think about finding a job. The likelihood of doing so depends not only on your individual attributes (level of education, skills, region where you live, and so forth) but also on the economic conditions of the time. For some, these conditions may mean long-term unemployment. Young people entering the labor market for the first time, even as college graduates, may find themselves facing possible joblessness, through no fault of their own but merely because they exited school at a time when broader social forces were shaping their life opportunities. Although understanding the institutional forces at work at any given time does not reduce the suffering that people may experience, such an understanding can help you think about social changes that shape people's experience in society. In this chapter, we look at the sociological analysis of the economy and work in society.

Economy and Society

All societies are organized around an economic base. The **economy** of a society is the system by which goods and services are produced, distributed, and consumed. The historic transformation from an agriculturally based society to an industrial society and now a post-industrial society is the foundation that will guide an understanding of the economy and work.

The Industrial Revolution

In Chapter 5, we discussed the evolution of different types of societies. Recall that one of the most significant of these changes was, first, the development of agricultural societies and, later, the far-ranging impact of the Industrial Revolution. Now, the Industrial Revolution is giving way to a postindustrial revolution—a development with far-reaching consequences for how work in society is organized.

The Industrial Revolution is usually pinpointed as beginning in mid-eighteenth-century Europe, soon thereafter spreading throughout other parts of the world. The Industrial Revolution led to numerous social changes because Western economies became organized around the mass production of goods. The Industrial Revolution created factories, separating work and family by relocating the place where most people were employed.

We still live in a society that is largely industrial, but that is quickly giving way to a new kind of social organization: the postindustrial society. Whereas industrial societies are primarily organized around the production of goods, **postindustrial societies** are organized around the provision of information and services. The United States has thus moved from being a manufacturing-based economy to an economy centered on the provision of services. Service is a broad term meant to encompass a wide range of economic activities now common in the labor market. It includes banking and finance, retail sales, hotel and restaurant work, and health care. It also includes parts of the information technology industry—not electronics assembly, but areas such as software design and the exchange of information (through the Internet, publishing, video production, and the like). Some service jobs, such as software engineers, are at the high end of the labor market; others, such as fast food cooks, are among the lowest paid and least well-regarded.

Comparing Economic Systems

Economic systems are characterized by different forms: The three major forms are *capitalism*, *socialism*, and *communism*. These are not totally distinct, that is, many societies have a mix of these economic systems. **Capitalism** is an economic system based on the principles of market competition, private property, and the pursuit of profit. Within capitalist societies, stockholders own corporations—or a share of the corporations' wealth. Under capitalism, owners keep a surplus of what is generated by the economy; this is their *profit*, which may be in the form of money, financial assets, or other commodities.

Socialism is an economic institution characterized by state (government) ownership and management of the basic industries; that is, the means of production are the property of the state, not of individuals. Modern socialism emerged from the writings of Karl Marx, who predicted that capitalism would give way to egalitarian, state-dominated socialism, followed by a transition to stateless, classless communism. Many European nations, for example, have strong elements of socialism that mix with the global forces of capitalism. Sweden supports an extensive array of state-run social services, such as health care, education, and social welfare programs, but Swedish industry is capitalist. Other world nations are more strongly socialist, although they are not immune from the penetrating influence of capitalism. China was formerly a strongly socialist society that is currently undergoing great transformation to capitalist principles, including state encouragement of a market-based economy, the introduction of privately owned industries, and increased engagement in the international capitalist economy.

Communism is sometimes described as socialism in its purest form. In pure communism, industry is not the private property of owners. Instead, the state is the sole owner of the systems of production. Communist philosophy argues that capitalism is fundamentally unjust because powerful owners take more from laborers (and society) than they give and use their power to maintain the inequalities between the worker and owner classes. Communist theorists in the nineteenth century declared that capitalism would inevitably be overthrown as workers worldwide united against owners and the system that exploited them. Class divisions were supposed to be erased at that time, along with private property and all forms of inequality. History has not borne out these predictions.

The Changing Global Economy

One of the most significant developments of modern times is the creation of a global economy, affecting work in the United States and worldwide. The concept of the **global economy** acknowledges that all dimensions of the economy now cross national borders, including investment, production, management, markets, labor, information, and technology. Economic events in one nation now can have major reverberations throughout the world. When the economies of any major nation are unstable, the effects are felt worldwide.

Multinational corporations—those that draw a large share of their revenues from foreign investments and conduct business across national borders—have become increasingly powerful, spreading their influence around the globe. The global economy links the lives of millions of Americans to the experiences of other people throughout the world. You can see the globalization of the economy in everyday life: Status symbols such as high-priced sneakers are manufactured for just a few cents in China. The Barbie dolls that young girls accumulate are inexpensive by U.S. standards, yet it would require one month's wages for an Indonesian or Chinese worker who makes the doll to buy it for her child.

In the global economy, the most developed countries control research and management, and assembly-line work is performed in nations with less privileged positions in the global economy. A single product, such as an automobile, may be assembled from parts made all over the world—the engine assembled in Mexico, tires manufactured in Malaysia, and electronic parts constructed in China. The relocation of manufacturing to wherever labor is cheap has led to the emergence of the **global assembly line**, an international division of labor in which research and development is conducted in the United States, Japan, Germany, and other major world powers, and the assembly of goods is done primarily in underdeveloped and poor nations—mostly by women and, all too often, children.

Related to the global assembly line is the phenomenon known as outsourcing. **Outsourcing** is the transfer of a specialized task from one organization to another that occurs for cost saving; often, the work is transferred to a different nation, as you have likely witnessed when calling someone for help with your computer. The person who answers may well be working in India or somewhere else and is part of an economy that is deeply entangled with that in the United States.

Within the United States, the development of a global economy has created anxieties about the movement of jobs overseas. This was vividly seen during Donald Trump's ascendancy to the presidency where he wooed working-class voters by promising to "bring their jobs back," even though his own branded products were typically made abroad. Anxieties about losing jobs to foreign workers, including immigrants, produce **xenophobia**, the fear and hatred of foreigners. Campaigns to "buy American" reflect this trend, although the concept of buying American is becoming increasingly antiquated in a global economy.

When buying a product from a U.S. company, it is likely that the parts, if not the product itself, were built overseas. In a global economy, distinctions between U.S. and foreign businesses blur. Moreover, the label "made in U.S.A." does not necessarily mean that well-paid workers in the United States made the product. In the garment industry, sweatshop workers—many of whom are recent immigrants and primarily women—are likely to have stitched the clothing that bears such a label. Moreover, these workers are likely to be working under exploitative conditions.

The development of a global economy is part of the broad process of **economic restructuring**, which refers to the contemporary transformations in the basic structure of work that are permanently altering the workplace. This process includes the changing composition of the workplace, deindustrialization, and use of enhanced technology. Some changes are *demographic*—that is, resulting from changes in the population. The labor force is becoming more diverse, with women and people of color becoming the majority of those employed. Other changes are driven by *technological developments*. For example, the economy is based less on its earlier manufacturing base and more on service industries.

A More Diverse Workplace

A more diverse workplace is becoming a common result of economic restructuring. Today's workforce is older, includes more women, and is more racially and ethnically diverse than ever before. People in the older age group (age 55 and older) have lower labor force participation rates than middle-aged people, yet they comprise, at least for the time being, a somewhat larger percentage of workers than was true in the past—due in large part to the size of the baby boomer population.

Thinking Sociologically

Identify a job you once held (or currently hold) and make a list of all the ways that workers in this segment of the labor market are being affected by the various dimensions of *economic restructuring*: demographic changes, globalization of the economy, and technological change. What does your list tell you about how social structure shapes people's individual work experiences?

Think about the labor market in the region where you live. What racial and ethnic groups have historically worked in various segments of this labor market, and how would you now describe the racial and ethnic *division of labor*?

As the U.S. population becomes more diverse, diversity in the labor market will continue. In fact, the White, non-Hispanic share of the labor market has fallen to 60 percent of the labor force, compared to 76 percent as recently as 1994. White, non-Hispanics in the labor market are expected to fall below 60 percent by the year 2030. Meanwhile, Asian and Hispanic workers—and all women—are expected to become an increased proportion of the labor market. The growth for young people (aged 16 to 24) in the labor market is also predicted to decline. The workforce is also growing more slowly than in the past, even though the U.S. population is growing (Toossi 2015). These basic facts about population change in the workforce will shape the experience of generations to come.

These changes in the social organization of work and the economy are creating a more diverse labor force, but much of the growth in the economy is projected to be in service industries, where education and training are required for the best jobs. People without these skills will not be well positioned for success. Manufacturing industries, where racial minorities and less-educated young people have previously gained a foothold on employment, are now in decline. New technologies and corporate layoffs have reduced the number of entry-level corporate jobs that recent college graduates have always used as a starting point for career mobility. Many college graduates are employed in jobs that do not require a college degree. College graduates, however, do still have higher earnings than those with less education.

Deindustrialization

Deindustrialization refers to the transition from a predominantly goods-producing economy to one based on the provision of services. This does not mean that goods are no longer produced, but that fewer workers in the United States are needed to produce them. Machines can do the work people once did, and many goods-producing jobs have moved overseas.

Deindustrialization is most easily observed by looking at the decline in the number of jobs in the manufacturing sector of the U.S. economy since the World War II. The manufacturing sector includes workers who actually produce goods. At the end of the war in 1945, the majority of workers (51 percent) in the United States were employed in manufacturing-based jobs. Now, manufacturing accounts for only about 7 percent of the total labor force (Henderson 2015).

The *service sector* employs the other 93 percent, including two segments: the actual delivery of services (such as food preparation, cleaning, or child care) and the transmission and processing of information (such as banking and finance, computer operation, clerical work, and even education workers, such as teachers). Parts of the service sector are higher-wage and prestigious jobs, such as physicians, lawyers, financial professionals, and so forth, but huge parts of the service sector—and where there is the largest occupational growth—are low-wage, semiskilled, and unskilled jobs. This lower end of the service sector employs many women, people of color, and immigrants.

The human cost of deindustrialization can be severe. Deindustrialization has led to *job displacement*, the permanent loss of certain job types that occurs when employment patterns shift. When a manufacturing plant shuts down, many people may lose their jobs at the same time, and whole communities can be affected. Communities that were heavily dependent on a single industry, such as steel towns or automobile-manufacturing cities such as Detroit, are among the areas hardest hit by deindustrialization. Rural communities, too, can be harmed by deindustrialization because they often have only one major employer, such as a coal mine or textile plant.

Job displacement hits people in both rural areas and inner cities hard because emerging new industries tend to be located in suburban, not urban or rural, areas—a phenomenon called *spatial mismatch*. It is then no surprise that poverty is highest in central cities and rural America. Unless young people have the educational and technical skills for employment in a new economy, or if they live in areas hard hit by job displacement, they have little opportunity for getting a good start in the now global economy. You see this in the extremely high unemployment rates for Black and Hispanic teens (see Figure 15-3 later in this chapter; Wilson 2009, 1996).

Technological Change

Coupled with deindustrialization, rapidly changing and developing technologies are bringing major changes in work, including how it is organized, who does it, and how much it pays. One of the most influential technological developments of the twentieth century has been the invention of

Doing Sociological Research

Precarious Work: The Shifting Conditions of Work in Society

Historically, once in the labor market, an American could count on a relatively stable job over the course of a lifetime, often in the same company. Now this is rare. Rather, as sociologist Arne Kalleberg labels it, precarious work is defined as work that is uncertain, unpredictable, and risky from the point of view of a worker (Kalleberg 2013).

Research Question

What are the social conditions that have made work more precarious and made workers less secure in their employment?

Research Methods

Kalleberg has studied these questions using a macro-level approach, drawing from secondary data sources, such as information from the U.S. Department of Labor, U.S. Department of Education, the General Social Survey, the Economic Policy Institute, and his own analyses of other published research studies.

Research Results

Kalleberg finds that the growth of precarious work produces increased stress for workers, thus also affecting families and personal relationships. He documents several reasons for this particular transformation in employment:

1. The expansion and institutionalization of nonstandard employment relations, such as temporary work and contract labor;
2. A general decline in job stability, meaning that people do not remain with the same employer over time;
3. An increasing tendency for employers to hire workers from outside of the work organization, rather than developing skills and talents from within;
4. Growth in involuntary job loss, especially among prime-age White men in white-collar occupations;
5. Growth in long-term unemployment;
6. A shift of risk from employers to employees, particularly in the decline of employee benefits;
7. Decline of unions and worker protections as employers have sought greater flexibility in management practices and the protection of profit margins.

Conclusions and Implications

Work is central to people's identity. As these various and multiple social, economic, and political forces have aligned, work has become more precarious and thus eroded people's sense of security. Kalleberg argues that new forms of work arrangements will be needed to ensure that employees, not just employers, have a commitment to the economy and society.

Questions to Consider

1. Is the work you plan to do precarious? Why or why not?
2. How do you see the social factors that Kalleberg identifies as influencing the work of people in your social networks?

Sources: Kalleberg, Arne. 2013. *Good Jobs, Bad Jobs: The Rise of Polarized and Precarious Employment Systems in the United States, 1970s to 2000s.* New York: Russell Sage Foundation.

the semiconductor. Computer technology has made possible workplace transactions that would have seemed like science fiction not that many years ago. Electronic information can be transferred around the world in less than a second. Employees can provide work for corporations located on another continent. A woman in Southeast Asia or the Caribbean can download and produce a book manuscript for a publishing house in New York. Some argue that the computer chip has as much significance for social change as the earlier inventions of the wheel and the steam engine.

Increasing reliance on the rapid transmission of electronic data has produced *electronic sweatshops*, a term referring to the back offices found in many industries, such as airlines, insurance firms, and mail-order houses, where workers at computer terminals process thousands of transactions in a day. Computers may monitor workers, conjuring up images of "Big Brother" invisibly watching. Computers can measure how fast cashiers ring up groceries and how fast ticket agents book reservations. Records derived from computer monitoring then become the basis for job performance evaluation.

Technological innovation in the workplace is a mixed blessing. **Automation**—the process by which human labor is replaced by machines—eliminates many repetitive and tiresome tasks, and it makes rapid communication and access to information possible. Our increasing dependence on technology may make workers subservient to machines, though. Robots can do the spot welding on automobiles; robots are even used for human surgery. Sophisticated robots are capable of highly complex tasks, enabling them to assemble finished products or flip burgers in fast-food restaurants. Will robots replace human workers? Robots are expensive to buy, but employers can see other savings because "the robot hamburger-flipper would need no lunch or bathroom breaks, would not take sick days, and most certainly would neither strike nor quit" (Rosengarten 2000: 4).

Deskilling is a process whereby the level of skill required for performing certain jobs declines over time. Deskilling may result when a job is automated or when a more complex job is divided into a sequence of easily performed units. With deskilling, workers are paid less and have less control over their tasks. Jobs may become routine and boring. Deskilling contributes to polarization of the labor force. The best jobs require increasing levels of skill and technological knowledge, whereas people at the bottom of the occupational hierarchy may become stuck in dead-end positions and become alienated from their work (Apple 1991).

Along with deskilling has come an increasing reliance on temporary or contingent workers. **Contingent workers** are those who do not hold regular jobs, but whose employment is dependent on demand. This includes contract workers, temporary workers, on-call workers (those called only when needed), the self-employed, and day laborers. Women are more likely than men to be employed in these jobs, and women are concentrated in the least desirable jobs—those with the lowest pay and least likelihood of providing benefits. Considering race, Whites are more likely to be independent contractors or self-employed, whereas Blacks and Hispanics are more likely to be found in temporary and part-time contingent work (Kalleberg 2013).

Immigration

One of the most significant changes in the U.S. labor force has been the increased presence of immigrant labor. There are approximately 40 million foreign-born people in the United States—13 percent of the U.S. population; almost half (44 percent) of the foreign born have become U.S. citizens. There are also about 11 million undocumented immigrants, so-called illegal immigrants. Immigrants constitute more than one-third of the labor force in fields such as building cleaning and maintenance; agriculture; meat, poultry, and fish production; and construction. Eighty-two percent of immigrant households have at least one worker present, more than is true for native-born households (where the figure is 73 percent; U.S. Census Bureau 2014).

Immigrants have far lower wages than native-born citizens, and they also experience higher rates of poverty. Although immigrants have been concentrated in particular states (California, Florida, Arizona, and Texas, among others), the recent trend has been that immigrants have moved into other geographic regions and smaller towns, thus transforming the character of hundreds of American communities (Jimenez 2009).

Myth: Immigrants take jobs away from American citizens.
Sociological Perspective: Immigrants and native workers do not tend to compete for the same jobs; studies also find that the average U.S. workers' wages actually increase because of immigration (Peri and Sparber 2009; Cortes 2008).

Immigration has not only changed the composition of the workforce, but it has also stimulated intense political debate. Should the nation restrict immigration? Should a guest worker program be created? What rights do immigrants have? These issues engage different interest groups—including immigrants themselves and the organizations that support them, business leaders, nonimmigrant workers, and local communities where immigrants are employed. These interests often diverge, creating conflict and certainly raising questions for policymakers about the best way to handle immigration.

Popular wisdom holds that the bulk of new immigrants, particularly Latino immigrants, are illegal, poor, and desperate, but the facts show otherwise. In fact, the proportion of professionals and technicians among legal immigrants exceeds the proportion of professionals in the labor force as a whole. This conclusion, though, is based on formal immigration data that exclude undocumented immigrants, most of whom are from the working class. Still, those who migrate are usually not the most downtrodden in their home country. The poor are seldom those able to migrate. Even undocumented immigrants tend to have higher levels of education and occupational skills than the typical workers in their homeland. Immigrants include both the most educated and the least educated segments of the population.

Some European nations have created *guest worker programs* to support the labor that immigrants provide. Guest worker programs allow immigrants to work in a nation for a limited period of time without fear of deportation, but they must return to their nation of origin at the end of their time limit. Critics argue that such programs threaten jobs for U.S. workers and allow employers to exploit this unequal class of workers. Supporters say guest worker programs allow for better regulation of immigration and avoid some of the problems of illegal immigration. It is unclear what direction the United States will take in immigration policy, but what is clear is that the U.S. economy is highly dependent on immigrant labor.

Social Organization of the Workplace

Most people think of work as an activity for which a person is paid, but work also includes the labor that people do without pay. Unpaid jobs such as housework, child care, and volunteer activities make up much of the work done in the world. Sociologists define **work** as productive human activity that creates something of value, either goods or services. Given this definition of work, housework, though unpaid, is defined as work, even though it is not included in the official measures of productivity used to indicate national work output.

Arlie Hochschild (1983) has introduced the concept of emotional labor to address some forms of work that are common in a service-based economy. **Emotional labor** is work specifically intended to produce a desired state of mind in a client and often involves putting on a false front before clients. Many jobs require some handling of other people's feelings, and emotional labor is performed where inducing or suppressing a feeling in the client is one of the primary work tasks. Airline flight attendants perform emotional labor—their job is to please the passenger and, as Hochschild suggests, to make passengers feel as though they are guests in someone's living room.

The Division of Labor

The **division of labor** is the systematic *interrelatedness* of different tasks that develop in complex societies. When different groups engage in different economic activities, a division of labor is said to exist. In a relatively simple division of labor, one group may be responsible for planting and harvesting crops, whereas another group is responsible for hunting game. As the economic system becomes more complex, the division of labor becomes more elaborate.

In the United States, gender, race, class, and age—the major axes of stratification—shape the division of labor. The *class division of labor* can be observed by looking at the work done by people with different educational backgrounds, because education is a fairly reliable indicator of class. People with

Fritz Hoffmann/The Image Works

Race and gender segregation in the labor market mean that women of color are concentrated in occupations where most other workers are also women of color.

more education tend to work in higher-paid, higher-prestige occupations. Class also leads to perceived distinctions in the value of manual labor versus mental labor. Those presumed to be doing mental labor (management and professional positions) tend to be paid more and have more job prestige than those presumed to be doing manual labor. Class thus produces stereotypes about the working class; manual labor is presumed to be the inverse of mental labor, meaning it is presumed to require no thinking. By extension, workers who do manual labor may be incorrectly assumed not to be very smart, regardless of their intelligence.

The *gender division of labor* refers to the different work that women and men do in society. In societies with a strong gender division of labor, the belief that some activities are women's work (for example, secretarial work) and other activities men's work (for example, construction) contributes greatly to the propagation of inequality between women and men, especially because cultural expectations usually place more value (both social and economic) on men's work. This helps explain why librarians and social workers, most of whom are women, are typically paid less than electricians despite the likelihood that both professions require higher education than that for electricians. Clearly all of these jobs require skill and training, but what determines the actual value of the work done? Sociologists argue that the structure of stratification, not the actual value of the job, is the answer.

See for Yourself

Anatomy of Working-Class Jobs

Identify an occupation that you think of as a working-class job. Then, find someone who does this kind of work and who would be willing to talk with you. Ask this person to tell you exactly what he or she does and what is most difficult and most rewarding about the work. Where does this job fit in the *division of labor*? Who does the job, and how are they rewarded (or not)? How would each of the three major theoretical perspectives in sociology (functionalism, conflict theory, and symbolic interaction) interpret this work in the context of the broader society?

Also determining the division of labor is race and ethnicity—the *racial–ethnic division of labor*. Even with civil rights and equal employment opportunity laws in place, race and ethnicity structure how people are distributed in jobs. Although racial–ethnic groups are distributed over a wide array of jobs, including as professional workers, there remains a concentration of people of color and immigrant labor in the lower-wage, low-status positions in the division of labor. This work is as necessary for society's survival as high-end jobs, but it is devalued and, often, underappreciated labor.

The **glass ceiling** is the term used to describe the limits to advancement that women, as well as racial–ethnic people and minorities, experience at work. Many barriers to the advancement of women and minorities have been removed, yet invisible barriers still persist. The existence of the glass ceiling is well documented. Although there has been an increase in the number of managers who are women and minorities, most top management jobs are still held by White men, who are far more likely than other groups to control a budget, participate in hiring and promotion, and have subordinates who report to them. Women and racial minorities remain clustered at the bottom of managerial hierarchies (see ▲ Figure 15-1).

The Occupational System and the Labor Market

Jobs are organized into an *occupational system*—the array of jobs that together constitute the labor market. Within the occupational system, people are distributed in patterns that reflect the race, class, and gender organization of society. Jobs vary in their economic rewards, their perceived value and prestige, and the opportunities they hold for advancement.

The Dual Labor Market

The labor market can be seen as comprising two major segments, the *primary labor market* and the *secondary labor market*, a phenomenon known as the **dual labor market** (see also Chapter 11).

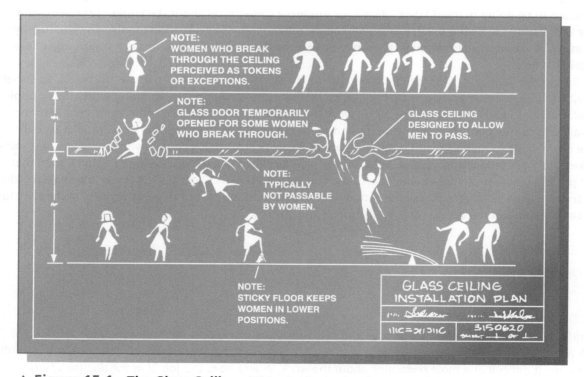

▲ **Figure 15-1   The Glass Ceiling.** Whimsically depicted as an architectural drawing by Norman Andersen, the *glass ceiling* refers to the structural obstacles still inhibiting upward mobility for women workers. Most employed women remain clustered in low-status, low-wage jobs that hold little chance for mobility. Those who make it to the top often report being blocked and frustrated by patterns of exclusion and gender stereotyping.

The primary labor market offers jobs with relatively high wages, benefits, stability, good working conditions, opportunities for promotion, job protection, and due process for workers (meaning workers are treated according to established rules and procedures that are allegedly fairly administered). Blue-collar and service workers in the primary labor market are often unionized, which leads to better wages and job benefits. High-level corporate jobs and unionized occupations fall into this segment of the labor market.

The secondary labor market is characterized by low wages, few benefits, high turnover, poor working conditions, little opportunity for advancement, no job protection, no retirement plan, and perhaps arbitrary treatment of workers. Many service jobs, such as waiting tables, nonunionized assembly work, and domestic work, are in the secondary labor market. Women and minority workers are the most likely groups to be employed in the secondary labor market. This particular structure in the labor market can explain much about race, gender, and class inequalities in work.

Occupational Distribution

Occupational distribution describes the pattern by which workers are located in the labor force. Workers are dispersed throughout the occupational system in patterns that vary greatly by race, class, and gender, revealing a certain **occupational segregation** on the basis of such characteristics. Women are most likely to work in technical, sales, and administrative support, primarily because of their heavy concentration in clerical work. This is now true for both White women and women of color. White men are most likely found in managerial and professional jobs, whereas African American and Hispanic men are most likely employed in some of the least well paid and least prestigious jobs in the occupational system (U.S. Bureau of Labor Statistics 2017).

Changes in occupational segregation are noticeable over time. For example, in 1960, 38 percent of all Black women were employed as private domestic workers; by the 1990s, this had declined to less than 2 percent. Over time, there has also been some increase in the number of women employed in working-class jobs traditionally held only by men. Today, women comprise 3 percent of construction workers, a small increase from the past, but a very small proportion of such workers (U.S. Bureau of Labor Statistics 2017).

Earnings

Sociologists have extensively documented that earnings from work are highly dependent on race, gender, and class, as shown in ▲ Figure 15-2. White men earn the most, with a gap between men's and women's earnings among all groups. African American women and Hispanic men and women earn the least. Occupations in which White men are the numerical majority tend to pay more than occupations in which women and minorities are a majority of the workers (U.S. Bureau of Labor Statistics 2018).

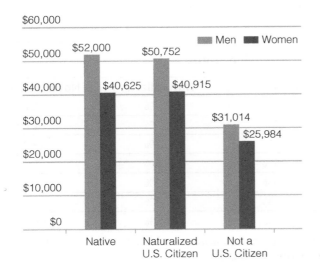

▲ **Figure 15-2 The Income Gap: U.S.-Born, Naturalized, and Non–U.S. Citizen Workers, 2012.** These median income figures are for full-time workers only. Does anything surprise you about what you see in the graph? What social factors do you think influence these earning differences?

Data: U.S. Census Bureau. 2013. *Total Money Income of Full-Time, Year-Round Workers 15 Years and Over with Earnings by Sex, Nativity, and U.S. Citizenship Status: 2012.* Washington, DC: U.S. Department of Commerce. **www.census.gov**

Diverse Groups/Diverse Work Experiences

Data on characteristics of the U.S. labor force typically are drawn from official statistics reported by the U.S. Department of Labor. The labor force now includes approximately 153 million people (U.S. Bureau of Labor Statistics 2018). Who works, where, and how varies considerably for different groups in the population, however.

One of the most dramatic changes in the labor force since the Second World War has been the increase in the number of women employed. Since 1976, the employment of women has increased from 43 to 55 percent of all women by 2017. Women now constitute almost half (47 percent) of all workers. Other changes in the labor force include that racial minorities (Hispanics, Asians, and African Americans) are the fastest-growing segment of the labor force, although White, non-Hispanic workers still make up 70 percent of the workforce. Hispanics and Asians have the fastest growth in the labor force, largely because of immigration (U.S. Bureau of Labor Statistics 2018).

These trends, however, do not mean—as popularly believed—that minorities and women are routinely taking jobs from White men. Jobs where White men have predominated are precisely those that have been declining because of economic restructuring, such as in the manufacturing sector. Jobs in areas of the labor market that are race- and gender-segregated are increasing, such as fast-food work and other low-wage service jobs. The conflicts that exist about work—such as the belief that immigrant and foreign workers are taking U.S. jobs or that women and minorities are taking away White men's jobs—stem from social structural transformations in the economy, not the individual behaviors of people affected by these changes (see ◆ Table 15-1).

Unemployment and Joblessness

The U.S. Department of Labor regularly reports the **unemployment rate**, defined as the percentage of those not working but officially defined as looking for work. Full employment is considered to be around 4 percent. During the recession that hit the nation in 2010, unemployment was as high as 9.6; it has been dropping below 4 percent through 2018 (U.S. Bureau of Labor Statistics 2018).

Even with this low rate of unemployment, however, many are without work. How can there be a decline in the unemployment rate while so many people are still out of work? The official unemployment rate does not include all people who are jobless. It includes only those who meet the official definition of unemployment—those who do not have a job and who have looked for work in the period being reported. Thus the official measure excludes people who earned money at any job during the time prior to the data being collected; it excludes people who have given up looking for full-time work (so-called *discouraged workers*); those who have settled for part-time work; people who are ill or disabled; those who cannot afford the child care, transportation, or other necessities for getting to work; and those who work only a few hours a week even though their economic position may differ little from that of someone with no work at all. Joblessness, very difficult to measure, actually exceeds the formal unemployment rate.

Table 15-1	Occupational Status by Nativity and Citizenship Status		
	Native	**Naturalized**	**Not a U.S. Citizen**
Management, business, financial	16.7%	16.4%	8.1%
Professional	23%	23.4%	14.7%
Service	16.5%	20.7%	39.2%
Sales	11%	9.6%	7.6%
Farming, fishing, forestry	0.4%	0.5%	2.5%
Construction, extraction, maintenance	7.6%	6.9%	14.4%
Production, transportation, material moving	11%	12.5%	17.1%

Source: U.S. Census Bureau. 2013. "Occupation of Employed Civilian Workers 16 Years and Over by Sex, Nativity, and U.S. Citizenship Status: 2013." Washington, DC: U.S. Department of Commerce. **www.census.gov**

Debunking Society's Myths

Myth: The unemployment rate is a good measure of the nation's joblessness.
Sociological Perspective: Unemployment is not the same as joblessness. The unemployment rate only counts those who are actively seeking work and excludes those who are working part-time, those who are so discouraged they have quit looking, and those who are underemployed, among others. The unemployment rate is one indicator of the state of the economy, but joblessness is typically higher.

Migrant workers and other transient populations are also undercounted in the official statistics. Workers on strike are also counted as employed, even if they receive no income while they are on strike. Because so many people are excluded from the official definition of unemployment, the official unemployment rate seriously underestimates actual joblessness. The people most likely to be left out of the unemployment rate are those for whom unemployment runs the highest—the youngest and oldest workers, women, and racial minority groups. These groups are also those most likely to have left jobs that do not qualify them for unemployment insurance, because to be eligible for unemployment you have to have worked a certain period of time and earned enough wages to qualify for a claim.

Official unemployment rates also ignore **underemployment**—the condition of being employed at a skill level below what would be expected given a person's training, experience, or education. This condition can also include working fewer hours than desired. A laid-off autoworker flipping hamburgers at a fast-food restaurant is underemployed as is a person with a college degree cleaning houses for a living.

Even given the problems in measuring unemployment, the highest rates of unemployment are among Black men, followed by Hispanic men (see ▲ Figure 15-3). When the government reports the national unemployment rate, it is a safe bet that unemployment among African Americans will be at least twice the national rate—a pattern that has persisted over time. Although not regularly reported by the Department of Labor, unemployment among Native Americans is also staggeringly high—around 11 percent. The unemployment rate that is found among Native Americans, African Americans, and Hispanics (with the exception of Cuban Americans) exceeds the unemployment rate that was considered a national crisis during the Great Depression of the 1930s.

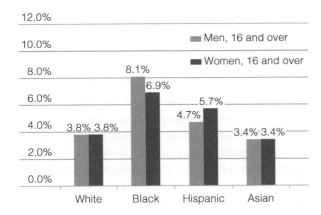
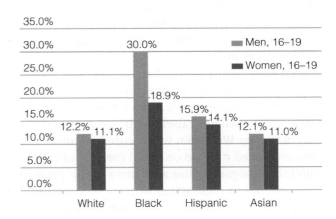

▲ **Figure 15-3   Unemployment by Race, Age, and Gender, 2016.** Unemployment varies significantly among different groups. The graph on the left shows unemployment for all people over age 16; on the right, for teens between age 16 and 19. What patterns do you see depicted in this graph? What does this suggest about the need for new social policies?

Data: Bureau of Labor Statistics. 2017. *Employment and Earnings Online.* Washington, DC: U.S. Department of Labor. **www.bls.gov**

People often attribute unemployment to the individual failings of workers, claiming that unemployed people do not try hard enough to find jobs or prefer a welfare check to hard work and a paycheck. This leads some to attribute unemployment to the "laziness" of unemployed individuals rather than to actual, factual structural conditions in society. This viewpoint reflects the common myth that *anyone* who works hard enough and puts forward sufficient effort can succeed.

Economic and structural conditions beyond an individual's control usually result in unemployment and joblessness. During the recession, hundreds of thousands of workers either lost jobs or found themselves in highly precarious working situations. In times like a major recession, it is easier to see how structural changes in the economy affect people at an individual level. (Recall the distinction between *troubles* and *issues* that C. Wright Mills made and that was examined in Chapter 1.)

Sociologists examine how changes in the social organization of the economy trickle into the day-to-day reality of people's lives. Although people do not ordinarily interpret their situations as shaped by processes such as deindustrialization, corporate downsizing, or spatial mismatch, the fact is that these structural conditions have enormous influence on how we live. Hence, unemployment is rarely caused by personal "laziness."

Sexual Harassment

Workplaces, like other social institutions, are shaped by power relationships among workers. One consequence for women workers is the possibility of sexual harassment. **Sexual harassment** is legally defined as unwanted physical or verbal sexual behavior that occurs in the context of a relationship of unequal power and that is experienced as a threat to the victim's job or educational activities (Zippel 2006; Saguy 2003).

Sexual harassment is of two forms, according to the law. *Quid pro quo sexual harassment* forces sexual compliance in exchange for an employment or educational benefit. A professor who suggests to a student that going out on a date or having sex would improve the student's grade is engaging in quid pro quo sexual harassment.

The other form of sexual harassment recognized by law is the creation of a *hostile working environment*, in which unwanted sexual behaviors are a continuing condition of work. This kind of sexual harassment may not involve outright sexual demands, but includes unwanted behaviors such as touching, teasing, sexual joking, and other kinds of sexual behavior and comments.

Sexual harassment was first made illegal by Title VII of the 1964 Civil Rights Act, which identifies sexual harassment as a form of sex discrimination. Title IX of the Educational Amendments of 1972 further defined sexual harassment as a form of discrimination in education. Regarding employment, in 1986, the U.S. Supreme Court upheld the principle that sexual harassment violates federal laws against sex discrimination (*Meritor Savings Bank v. Vinson*). Same-sex harassment also falls under the law, as does harassment directed by women against men. The law also makes employers liable for financial damages if they do not have policies appropriate for handling complaints or have not educated employees about their paths of redress.

Fundamentally, sexual harassment is an abuse of power when perpetrators use their position to exploit subordinates. The phenomenon of sexual harassment is not new and has been in the experience of countless numbers of women for years. But recent scandals and publicity about sexual harassment by high-profile men has raised public awareness of the extent of this problem. As an example, when movie mogul Harvey Weinstein was accused of numerous cases of sexual harassment and assault, millions of women posted #metoo on social media, indicating that they had experienced sexual harassment. Such public campaigns to raise awareness of this problem are important, but they can also obscure sexual harassment of less high-status women who are just as likely to experience this problem and are less likely to have the resources or institutional supports that allow them to pursue complaints.

The true extent of sexual harassment is difficult to estimate because it tends to be underreported. Surveys indicate though that as many as half of all employed women experience some form of sexual harassment at some time in their working lives. Men are sometimes the victims of sexual harassment, although far less frequently than women—about 3 percent of all cases. There is some evidence that women of color are more likely to be harassed than White women (Rospenda et al. 2009). Women are more likely to be harassed when they work in male-dominated occupations. The likelihood of harassment is also higher among single women (McDonald 2012; McLaughlin, Uggen, and Blackstone 2012).

Research finds that the negative effects of sexual harassment are numerous, including psychological distress, but also impediments to women's advancement in workplaces. There is even a financial cost to women from sexual harassment because it can precipitate a job change or be deleterious to one's career if a victim starts skipping work, withdrawing from colleagues, or otherwise losing commitment to her job (McLaughlin, Uggen, and Blackstone 2017).

Most studies find that neither women nor men are typically aware of the proper channels for reporting sexual harassment. Victims of sexual harassment often do not report it (except perhaps to friends), primarily because they believe that nothing will be done to stop the behavior. When employers have strong policies in place that are widely communicated, victims are more likely to report (Vijayasiri 2008; Pershing 2003).

Gays and Lesbians in the Workplace

The increased willingness of lesbians and gay men to be open about their sexual identity has resulted in more attention paid to their experience in the workplace. Surveys find that a large majority of the U.S. population (89 percent) endorse the general concept that lesbians and gay men should have equal rights in job opportunities, but when asked about specific occupations, a significant proportion hold prejudices that gays should not be elementary school teachers (43 percent), high school teachers (36 percent), or clergy (47 percent) (Gallup Poll 2008).

Some employers have been developing more LGBTQ-friendly policies that are shown to have a positive effect on the workplace experiences of LGBTQ workers (Lloren and Porini 2017; Williams and Giuffre 2011). Even with these advances, however, LGBTQ employees are all too frequently subjected to stereotyping and discrimination and thus must develop coping strategies for managing stress (Buddel 2011). LGBTQ employees may feel compelled to downplay their identity at work and to feel constrained by stereotypes about how gay people are supposed to act, play, and work. This can force people to remain closeted at work (Williams et al. 2009). When people feel they must conceal a particular identity at work, they experience a reduced sense of belonging and, thus, experience negative work outcomes (Newheiser, Barreto, and Tiemersma 2017).

Disability and Work

Not too many years ago, people with disabilities were not thought of as a social group; rather, disability was thought of as an individual frailty or perhaps a stigma. Sociologist **Irving Zola** (1935–1994) was one of the first to suggest that people with disabilities face issues similar to those of minority groups. Instead of using a medical model that treats disability like a disease and sees individuals as impaired, conceptualizing people with disabilities as a minority group enabled people to think about the social, economic, and political environment that this population faces. Instead of seeing people with disabilities as pitiful victims, this approach emphasizes the group rights of the disabled, illuminating things such as access to employment and education (Zola 1993, 1989).

Now those with disabilities have the same legal protections afforded to other minority groups. Key to these rights is the **Americans with Disabilities Act (ADA)**, adopted by Congress in 1990. Building on the Civil Rights Act of 1964 and earlier rehabilitation law, the Americans with Disabilities Act protects people with disabilities from discrimination in employment. The law stipulates that employers and other public

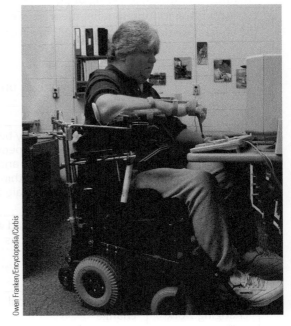

Owen Franken/Encyclopedia/Corbis

The Americans with Disabilities Act provides legal protection for workers with disabilities, giving them rights to reasonable accommodations by employers and access to education and jobs.

entities (such as schools and public transportation systems) must provide "reasonable accommodation" to people with disabilities when they are otherwise qualified for the jobs or activities for which they seek access. This means that the person so accommodated must be able to perform the essential requirements of the job or program. For students, reasonable accommodation includes provision of adaptive technology, exam assistants, and accessible buildings.

The law prohibits employers with fifteen or more employees from discriminating against job applicants who are disabled or current employees who become disabled, including in their job application, wages and benefits, advancement, and employer-sponsored social activities. The law applies to state and local governments, as well as employers. The ADA also legislates that public buses, trains, and light rail systems must be accessible to riders with disabilities; airlines are excluded from this requirement. The law also requires businesses and public accommodations to be accessible and requires telephone companies to provide services that allow people with speech or hearing impairments to communicate by telephone.

Sociological Theories of Economy and Work

Sociological research continues to find large differences in the experiences that different people have at work. Why? We have to turn to theory to answer such a question. The major theoretical perspectives identified in this book also provide the frameworks for the social structure of work. Each viewpoint—functionalism theory, conflict theory, and symbolic interaction—offers a unique analysis of work and the economic institution of which it is a part (see ◆ Table 15-2).

Functionalism

Functionalism interprets work and the economy as a functional necessity for society. Functionalists argue that society "sorts" people into occupations, with the more able sorted into prestigious occupations that pay more because they are more valuable—more "functional"—for society. Functionalist theory also explains that when society changes too rapidly, work institutions generate social disorganization—perhaps creating **alienation**, a feeling of powerlessness and separation from society (Chapter 6).

According to functionalist theories, workers are paid according to their value, which is derived from the characteristics they bring to a job: education, experience, training, and motivation to work. As we saw

Table 15-2	Theoretical Perspectives on Work		
	Functionalism	**Conflict Theory**	**Symbolic Interaction**
Defines work as:	Functional for society because work teaches people the values of society and integrates people within the social order; more "talented" people rank higher	Generating class conflict because of the unequal rewards associated with different jobs	Organizing social bonds between people who interact within work settings
Views work organizations as:	Functionally integrated with other social institutions	Producing alienation, especially among those who perform repetitive tasks	Interactive systems within which people form relationships and create beliefs that define their relationships to others
Interprets changing work systems as:	An adaptation to social change	Based in tensions arising from power differences between different class, race, and gender groups	The result of the changing meanings of work resulting from changed social conditions
Explains wage inequality as:	Motivating people to work harder	Reflecting the devaluation of different classes of workers	Producing different perceptions of the value of different occupations

in Chapter 8, functionalist theorists see inequality as what motivates people to work. From this point of view, the high wages and other rewards associated with some jobs are the incentives for people to spend long years in training and garnering experience; otherwise, the jobs would go unfilled. To functionalists, then, differential wages are a source of motivation and a means to ensure that the most talented workers fill jobs essential to society and that different wages reflect the differently valued characteristics (education, years of experience, training, and so forth) that workers bring to a job.

Conflict Theory

Conflict theorists strongly disagree with the functionalist point of view, arguing that many talented people are thwarted by the systems of inequality they encounter in society. Far from ensuring that the most talented will fill the most important jobs, conflict theorists see that some of the most essential jobs are, in fact, the most devalued and under-rewarded. From a conflict perspective, wage inequality is one way that systems of race, class, and gender inequality are maintained. Factors such as the dual labor market, overt race and gender discrimination, and persistent and unequal judgments about what work is appropriate for what groups shape the inequalities that are found in work.

Conflict theorists view the transformations taking place in the workplace as the result of inherent tensions in the social systems, tensions that arise from the power differences between groups vying for social and economic resources. Class conflict is then a major element of the social structure of work. Conflict theorists see the class division of labor as the source of unequal rewards for workers. Conflict theorists analyze the fact that some forms of work are more highly valued than others, both in how the work is perceived by society and how it is rewarded. As noted long ago by social-economic theorist Thorstein Veblen (1994/1899), mental labor has always been more highly valued than manual labor.

Symbolic Interaction Theory

Symbolic interaction theory brings a different perspective to the sociology of work. *Symbolic interaction theorists* would be interested in what work means to people and how social interactions in the workplace form social bonds. Some symbolic interaction studies examine how new workers learn their roles and how workers' identities are shaped by the social interactions in the workplace. Symbolic interaction theorists also note the creative ways that people deal with routinized jobs, perhaps performing exaggerated displays of routine tasks to humanize otherwise boring work (Leidner 1993).

Power, Politics, and Government

Throughout this book we have emphasized that social structures extend beyond our immediate day-to-day life, influencing the things we do and how we behave every day. Systems of power and authority are no exception. The government, for example, as an official system of power and authority, reaches far into people's daily lives. You cannot drive your car without a license and registration provided by a government agency; you pay taxes according to government rules; you need a government-issue license to marry; even going to the beach may depend on the government having set aside certain land for public use. The range of things regulated by systems of power and authority—both formal and informal—is enormous, even though many may be barely noticed.

The role of these systems of power and authority is also hotly debated in national politics. How should we manage the national debt? What is the proper balance between protecting civil liberties and protecting the nation from harm? Is government too big or should it do more to support those in need? Where should we draw the line on the authority of the police? These and other questions frame current political debates, but they point to the importance of a sociological analysis of the state.

State and Society

The term **state** refers to the organized system of power and authority in society. The state is an abstract concept that includes the institutions that represent official power in society, including the government, the legal system (law, courts, and the prison system), the police, and the military. Theoretically, the state exists to regulate social order, ranging from individual behavior and interpersonal conflicts to international affairs. Less powerful groups in the society may see the state more as an oppressive force than as a protector of individual rights. However, they may still turn to the state to rectify injustice, for example, by advocating for civil rights or seeking state-based rights and protections for people with disabilities.

The state has a central role in determining the rights and privileges of various groups. The state determines who is a citizen and who is not. A case in point is the ongoing battle over whether United States-born children of foreign-born and undocumented immigrant parents are full-fledged citizens. Furthermore, the state may be called upon to resolve conflicts between management and labor (such as in airline strikes), and the state may pass legislation determining the benefits of different groups or make decisions that extend rights to various groups, such as the right to same-sex marriage.

Numerous institutions make up the state, including the government, the legal system, the police, and the military. The *government* creates laws and procedures that regulate and guide a society. The *military* is the branch of government responsible for defending the nation against domestic and foreign conflicts. The *court system* is designed to punish wrongdoers and adjudicate disputes. Court decisions also determine the guiding principles or laws of human interaction. *Law* is a fundamental type of formal social control that outlines what is permissible and what is forbidden. The *police* are responsible for enforcing law in the community and for maintaining public order. The *prison system* is the institution responsible for punishing those who have broken the law. Under the U.S. Constitution, these state institutions treat people equally, although sociologists have documented how often this is not the case (see Chapter 7).

Sociological analyses of the state focus on several different questions. An important issue is the relationship between the state and inequality in society. State policies can have very different impacts on different groups, as we will see later in this chapter. Another issue explored by sociological theory is the connection between the state and other social institutions—the state and religion or the state and the family. States can, for example, regulate the inclusion or exclusion of religious practices in the schools or can regulate the family forms allowed under law. All in all, the state has enormous power in shaping the freedoms—or lack thereof—that people and groups in society hold.

The State and Social Order

Throughout this book, we have seen that a variety of social processes contribute to order in society, but none so explicitly and unambiguously as the official system of power and authority in society. In making laws, the state decrees which actions are or are not legitimate. Punishments for illegitimate actions are enforced, and systems for administering punishment are maintained. The state also influences public opinion through its power to regulate the media.

Some states, such as the United States, are **democracies**—that is, there is a representative government with elections by the population and, typically, a multiparty political system. Democracy can be compared to states that are **authoritarian**—that is, where power is concentrated in the hands of a very few individuals who rule through centralized power and control. An authoritarian state can become **totalitarian**, an extreme form of authoritarianism where the state has total control over all aspects of public and, to the extent possible, private life. Such was the case under Nazi Germany where the Nazi Party and Hitler, in particular, had absolute power. Under such repressive states as authoritarian and totalitarian states, if there are elections, they are often corrupt. Such states are also likely to circulate **propaganda**, disseminated with the intention to justify the state's power. *Censorship* is another means by which the state can direct public opinion and try to enforce a singular way of thinking—that of the dominant group.

The state's role in maintaining public order is also apparent in how the state manages dissent. Having a free press, for example, is critical for democratic states. The government's shutting down the press is one of the first clues of state repression. Protest movements that challenge state authority or disrupt society may be repressed through state action, such as through police or military force. Even in democratic societies, social control can be exercised in multiple ways, including electronic surveillance. Should the state be allowed to intercept email, track phone calls, or shut down social media during times of political unrest? Signs of such increased surveillance by the state are everywhere, such as the increasingly common cameras at traffic intersections. New forms of technological surveillance have, in fact, raised important questions about rights to public access to information, the protection of civil rights, and the need to protect the citizenry from acts of terrorism. Fundamentally, these questions revolve around the power of the state to intervene in the affairs of its citizens.

Privacy concerns about the state have become especially heightened as electronic technologies make surveillance of the public so easy. Businesses now use sophisticated software called "beacons" to track what people are viewing on the Internet. Businesses can then profile a person's income, interests, and location—even their medical condition. Such information is being traded on Wall Street to advertisers who can then target ads specifically to just one person, all the while with this tracking being invisible to the person browsing the web (Angwin 2010).

Such information can also be used by the state to track people's movements, listen to their conversations, and analyze their social and political networks. The reach of the state into people's lives is vast, but would there be social order without this power? There is not a single or definitive answer to such a question, but it depends on the balance between civil liberties and the need for states to regulate society. On a global scale, such questions become increasingly complex.

Global Interdependence and the State

The global character of modern society means that political systems, like economic systems, are now elaborately entangled. You can observe this in the daily news. News of massive U.S. protests over police shootings of young African American men has quickly spread around the world via social media. Pro-democracy protestors in Hong Kong adopted the "hands up, don't shoot" tactic used by U.S. protestors; Palestinians tweeted U.S. demonstrators advising them about how to deal with tear gas (McGrath 2014). Increasingly, the use of social media such as twitter, Facebook, and other social media platforms has enabled fast, global communication that facilitates the development of protest movements (Lee 2017; Carney 2016).

World interdependence is occurring at a time when there is increasing nationalism in some nations. **Nationalism** is the strong identity associated with an extreme sense of allegiance to one's culture or nation, often to the exclusion of interdependent relations with others. Nationalism can become a political movement, such as when groups subordinated by external nations use their original national culture as the basis for resisting oppression. It can also be a political movement when a group identifying itself as "a nation" (regardless of its official status as such) tries to become a dominant force in the world, such as the militant terrorist group known as ISIS (called the Islamic State in Iraq and Syria). Although nationalism may seem to be the same thing as patriotism, when taken to extremes, nationalism has historically been the basis for some of the world's worst violence.

Power, Authority, and Bureaucracy

The concepts of power and authority are central to sociological analyses of the state. **Power** is the ability of one person or group to exercise influence and control over others. The exercise of power can be seen in relationships ranging from the interaction of two people (husband and wife, police officer and suspect) to a nation threatening or dominating other nations. Sociologists are most interested in how power is structured in society: who has it, how it is used, and how it is built into institutions such as the state. In the United States, a society that is heavily stratified by race, class, and gender, power is structured into basic social institutions in ways that reflect these inequalities. Moreover, institutionalized power in society influences the social dynamics within individual and group relationships.

The exercise of power may be persuasive or coercive. For example, a strong political leader may persuade the nation to support a military invasion or a social policy through popular appeal. Power can also be exerted by sheer force. Generally speaking, groups with the greatest material resources have the greatest power, but not always. A group may by sheer size be able to exercise power, through, for example, organized social protests. Smaller groups may also be able to exercise power, such as in armed uprisings.

Power can be legitimate—accepted by the members of society as right and just—or it can be illegitimate. **Authority** is power perceived by others as legitimate and formal. Authority emerges not from the exercise of power, but from the belief of constituents that the power is legitimate. In the United States, the source of the president's domestic power is not just because of his status as commander of the armed forces but also because under normal circumstances most people believe that his power is legitimate. The law is also perceived by most as a legitimate system of authority.

In contrast, *coercive power* is achieved through force, often against the will of the people being so forced. A dictatorship typically relies on its ability to exercise coercive power through its control of the military or state police, at least until the military (or the police), sometimes in conjunction with a popular social movement, overthrows the dictator.

Types of Authority

Max Weber (1864–1920), the German classical sociologist, postulated that three types of authority exist in society: traditional, charismatic, and rational–legal (Weber 1978/1921). **Traditional authority** stems from long-established patterns that give certain people or groups legitimate power in society. A monarchy is an example of a traditional system of authority. Within a monarchy, kings and queens rule, not necessarily because they have won elections, but because of long-standing tradition.

Charismatic authority is derived from the personal appeal of a leader. Charismatic leaders are often believed to have special gifts, even magical powers, and their presumed personal attributes inspire devotion and obedience. Charismatic leaders often emerge from religious movements, but they come from other realms also. Donald Trump is a charismatic leader to some. His personal attributes, even though they offend many, had strong appeal to those in his so-called "base." President Obama was also a charismatic leader, admired not only for being the first African American president of the United States but also for his ability to inspire so many people.

Rational–legal authority stems from rules and regulations, typically written down as laws, procedures, or codes of conduct. This is the most common form of authority in the contemporary United States. People obey not because national leaders are charismatic or because of social traditions, but because there is a legal system of authority established by formalized rules and regulations.

The Growth of Bureaucracies

According to Weber, rational–legal authority leads inevitably to the formation of bureaucracies. As we noted in Chapter 6, a **bureaucracy** is a formal organization characterized by an authority hierarchy, a clear division of labor, explicit rules, and impersonality. Bureaucratic power comes from the accepted legitimacy of the rules, not personal ties to individuals. The rules may change, but they do so through formal, bureaucratic procedures. People who work within bureaucracies are selected, trained, and promoted based on how well they apply the rules. Those who establish the rules are unlikely to be the same people who administer them. Bureaucracies are hierarchical, and the bureaucratic leadership may be quite remote. Power in bureaucracies is dispersed downward through the system to those who actually carry out the bureaucratic functions. It is an odd feature of bureaucracy that those with the least power to influence how the rules are formulated—those at the bottom of the hierarchy—are often the most adamant about strict adherence to the rules; their job evaluation may rest on their enforcement.

Thinking Sociologically

Observe the national news for one week, noting the people featured who have some kind of authority. List each of them and note their area of influence. What form of authority would you say each represents: traditional, charismatic, or rational–legal? How is the kind of authority that a person has reflected in his or her position in society (that is, race, class, gender, occupation, education, and so on)?

Within bureaucracies, personal temperament and individual discretion are not supposed to influence the application of rules. Bureaucracy has another face though, as we saw in Chapter 6. Rank-and-file bureaucratic workers frequently exercise discretion in applying rules and procedures, "working the system," perhaps by personalizing the interaction or dodging bureaucratic stipulations. Most of the time, though, dealing with an elaborate bureaucracy—even an electronic one like voice mail—can be very frustrating.

Theories of Power

Does the state act in the interests of its different constituencies or does it merely reflect the needs of the most powerful? In other words, how is power exercised in society? This question has spawned much sociological study and debate and has resulted in several theoretical models of state power. Sociologists have developed four theoretical models to answer this question: the *pluralist model*, the *power elite model*, the *autonomous state model*, and *feminist theories of the state*. Each begins with a different set of assumptions and arrives at different conclusions (see ◆ Table 15-3).

The Pluralist Model

The **pluralist model** interprets power in society as derived from the representation of diverse interests of different groups in society. This model assumes that in democratic societies, the system of government works to balance the different interests of groups in society. An **interest group** can be any constituency in society organized to promote its own agenda, including large, nationally based groups such as the American Association of Retired Persons (AARP) and the National Rifle Association (NRA). Some interest groups are organized around professional and business interests, such as the American Medical Association (AMA), PhRMA (representing the interest of the pharmaceutical industry), and the Tobacco Institute. Others concentrate on one political or social goal, such as the National Organization for the Reform of

Table 15-3   **Theories of Power in Society**

	Pluralism	Power Elite	Autonomous State	Feminist Theory
Interprets the state as:	Representing diverse and multiple groups in society	Representing the interests of a small but economically dominant class	Taking on a life of its own, perpetuating its own form and interests	Masculine in its organization and values (that is, based on rational principles and a patriarchal structure)
Interprets political power as:	Derived from the activities of interest groups and as broadly diffused throughout the public	Held by the ruling class	Residing in the organizational structure of state institutions	Emerging from the dominance of men over women
Interprets social conflict as:	The competition between diverse groups that mobilize to promote their interests	Stemming from the domination of elites over less powerful groups	Developing between states, as each vies to uphold its own interests	Resulting from the power men have over women
Interprets social order as:	The result of the equilibrium created by multiple groups balancing their interests	Coming from the interlocking directorates created by the linkages among those few people who control institutions	The result of administrative systems that work to maintain the status quo	Resulting from the patriarchal control that men have over social institutions

The pluralist model of the state sees diverse interest groups as mobilizing to influence policy and gain political power.

Marijuana Laws (NORML). According to the pluralist model, interest groups achieve power and influence through their organized mobilization of concerned people and groups.

The pluralist model has its origins in functionalist theory. The pluralist model sees power as broadly diffused across the public with people who want to effect a change or express their points of view needing only to mobilize to do so. The pluralist model suggests that members of diverse groups can participate equally in a representative and democratic government. This model underscores how various special interest groups compete for government attention and action. The pluralist model sees special interest groups as an integral part of the political system, even though they are not an official part of government. In the pluralist view, special interest groups make government more responsive to the needs and interests of different people, an especially important function in a highly diverse society.

The pluralist model helps explain the importance of **political action committees (PACs)**, groups of people who organize to support candidates they feel will represent their views. In 1974, Congress passed legislation enabling employees of companies, members of unions, professional groups, and trade associations to support political candidates with money they raise collectively. There are now thousands of PACs, most of which spend millions of dollars, giving them enormous power to influence the political process (Center for Responsive Politics 2017).

Super PACs are a new kind of political action committee. These enormously powerful groups are allowed to spend unlimited money on individual political candidates. Their influence was strengthened by a U.S. Supreme Court decision in 2010, *Citizens United v. Federal Election Commission*. The Court's decision was that the First Amendment prohibits the government from restricting spending by corporations. Widely interpreted as defining corporations to be "people," this decision has given increased power to wealthy people and corporations to influence national elections.

PACs and super PACs are now so powerful that they have a huge impact on elections. In the 2016 election, super PACs spent over $1 billion to influence voters. Voters may not even know who is sponsoring the political ads these outside groups disseminate, a fact revealed when it was learned that a foreign nation, namely Russia, purchased many Facebook ads designed to influence the election of Donald Trump. The influence of Russian meddling in the 2016 election is yet another example of how easily, even in a democratic government, a particular group, individual, or even nation can influence election outcomes.

The Power Elite Model

The **power elite model** is linked to the framework of conflict theory. Karl Marx early argued that the dominant or ruling class controls all the major institutions in society. The state in this framework is the instrument by which the ruling class exercises its power. Conflict theory emphasizes the power of the upper class over the lower classes, the small group of elites over the rest of the population. From this perspective, the state is not a representative, rational institution, but an expression of the will of the ruling class.

C. Wright Mills (1956) popularized the term *power elite*. Mills attacked the pluralist model, arguing that the true power structure consists of people well positioned in three areas: the economy, the government, and the military. These three institutions are considered the bastions of the power elite, wielding extraordinary power (Domhoff 2013). Although sharing common beliefs and goals, the power elite shape political agendas and outcomes in the society along the narrow lines of their particular collective interests.

The political influence of wealthy individuals, such as Charles and David Koch—popularly known as the "Koch brothers"—is hard to overestimate. The Koch brothers are worth at least $40 billion, and they use their money to fund conservative causes and candidates. Their fortune is from oil and natural gas and thus is often used to fight regulation of such industries. They, along with other extremely wealthy people, have long used their power to influence national politics (Dickinson 2014; Mayer 2016).

The influence of money in politics raises serious concerns about how representative the U.S. political system really is.

The power elite model posits a strong link between government and business, a view supported by the strong hand government takes in directing the economy and by the role of military spending as a principal component of U.S. economic affairs. The power elite model also emphasizes how power overlaps between influential groups.

Interlocking directorates are organizational linkages created when the same people sit on the board of directors for numerous corporations. People in elite circles may serve on the boards of several major companies, universities, and foundations at the same time. People drawn from the same elite group receive most of the major government appointments; thus, the same relatively small group of people tends to the interests of all these organizations and the interests of the government. These interests naturally overlap and reinforce each other.

Members of the upper class do not need to occupy high office themselves to exert their will, as long as they are in a position to influence people who are in power (Domhoff 2013). White men are the majority of the power elite, which means that their interests and outlooks dominate the national agenda.

The Autonomous State Model

A third view of power developed by sociologists, the **autonomous state model**, interprets the state as its own major constituent. From this perspective, the state develops interests of its own, which it seeks to promote independently of other interests and the public that it allegedly serves. The state does not reflect the needs of the dominant groups, as Marx and power elite theorists would contend. It is an administrative organization with its own needs, such as maintenance of its complex bureaucracies and protection of its special privileges (Rueschmeyer and Skocpol 1996; Skocpol 1992).

The huge government apparatus now in place in the United States is a good illustration of autonomous state theory. The government provides a huge array of social support programs, including Social Security, unemployment benefits, agricultural subsidies, public assistance, and other economic interventions intended to protect citizens from the vagaries of a capitalist market system. The purpose of these programs is to serve people in need. Autonomous state theory argues that the government has grown into a massive, elaborate bureaucracy, run by bureaucrats more absorbed in their own interests than in meeting the needs of the people. As a consequence, government can become paralyzed in conflicts between revenue-seeking state bureaucrats and those who must fund them. You can imagine autonomous state theory as well explaining what many now perceive as a completely stalled federal government.

Feminist Theories of the State

Feminist theorists diverge from the preceding theoretical models by seeing men as having the most power in society. The pluralists see power as widely dispersed through the class system, power elite theorists see political power directly linked to upper-class interests, and autonomous state theorists see the state as relatively independent of class interests.

Some feminist theorists argue that all state institutions reflect men's interests; they see the state as fundamentally patriarchal, its organization embodying the fixed principle that men are more powerful than women. Feminist theories of the state conclude that despite the presence of a few powerful women, the state is devoted primarily to men's interests. Moreover, the actions of the state will tend to support gender inequality (Haney 1996; Blankenship 1993). One historical example would be laws denying women the right to own property once they married. Such laws protected men's interests at the expense of women.

Evidence that "the state is male" (MacKinnon 2006, 1983) is easy to observe by looking at powerful political circles. Despite the inclusion of more women in powerful circles and the presence of some notable women as major national figures, most of the powerful are men. Both the U.S. Senate and House of Representatives, despite recent gains for women, are still 80 percent men. Groups that exercise state power, such as the police and military, are predominantly men. Moreover, these institutions are structured by values and systems that can be described as culturally masculine—that is, based on hierarchical relationships, aggression, and force. Feminist theory begins with the premise that an understanding of power cannot be sound without a strong analysis of gender (Haney 1996).

Government: Power and Politics in a Diverse Society

The terms *government* and *state* are often used interchangeably. More precisely, the government is one of several institutions that make up the state. The **government** includes those institutions that represent the population, making rules that govern the society. The government of the United States is a *democracy*; therefore, it is based on the principle of representing all people through the right to vote.

The actual makeup of the government, however, is far from representative of society. Not all people participate equally in the workings of government, neither as elected officials nor as voters. Women, the poor and working class, and racial–ethnic minorities are less likely to be represented by government than are White middle- and upper-class men. Sociological research on political power has concentrated on inequality in government affairs and has demonstrated large, persistent differences in the political participation and representation of various groups in society.

Diverse Patterns of Political Participation

One would hope that all people in a democratic society would be equally eager to exercise their right to vote and be heard. That is far from the case. Among democratic nations, the United States has one of the *lowest* voter turnouts (see ▲ Figure 15-4). In the 2016 presidential election, the percentage of eligible voters who went to the polls was only 58 percent of the population, less than the all-time high of 62 percent in the 2008 presidential election. A turnout of 50 percent or less is more typical of U.S. national elections. Voter turnout in congressional and local elections is even lower.

Generally, older, better-educated, and financially better-off people are the most likely to vote. One of the biggest changes in voting is the change stemming from diversity within the U.S. population. Historically, racial–ethnic minority groups have had lower voter turnout than White Americans. In the 2012 presidential election, more African Americans, Hispanics, and Asian Americans voted than ever

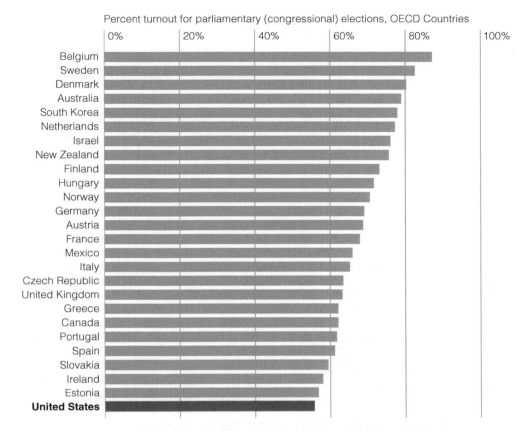

▲ Figure 15-4   International Voter Turnout, 2013–2017. As you can see, the United States has lower voter turnout in national elections than other industrialized nations. How would you explain this in sociological terms?

Source: DeSilver, Drew. 2017 (May 25). "U.S. Trails Most Developed Countries in Voter Turnout." Washington, DC: Pew Research Center. **www.pewresearch.org/fact-tank/2017/05/15/u-s-voter-turnout-trails-most -developed-countries/**

before, but in the 2016 election, voter turnout among African American and young people was markedly low. Voter suppression has also become an increasing concern as organized groups have tried to influence election outcomes by preventing the votes especially of African Americans and Latinos. The practice of *gerrymandering*, referring to manipulating voting districts to favor one party, also means that there is basically no contest between party candidates in some areas. Gerrymandering has particularly suppressed minority voting by concentrating minority votes in certain regions where they are unlikely to influence the broader election.

Some claim that there is also significant voter fraud, although rigorous research has found that the extent of voter fraud is quite minimal—less than 0.0025 percent of all votes (Levitt 2017). Voter suppression tactics are far more prevalent and can influence the outcome of elections. Voter apathy, especially when people become cynical about politics, also significantly depresses voting turnout.

Not only do social factors influence the likelihood of voting, but they also influence how people vote (see ▲ Figure 15-5). African Americans, Asian Americans, and Latinos, with the exception of Cuban Americans, tend to be markedly Democratic. Women tend to vote Democratic; men, Republican—at least among White men. In the 2016 presidential election, the majority of women voted for Hillary Clinton (54 percent), whereas men were more likely to vote for Trump (53 percent). Although women tend to be more liberal than men, women are also politically active in conservative movements (Tyson and Maniam 2016). African Americans, Latinos, and Asians overwhelmingly supported Hillary Clinton in the 2016 election while Whites overwhelmingly favored Trump (Roper Center 2017).

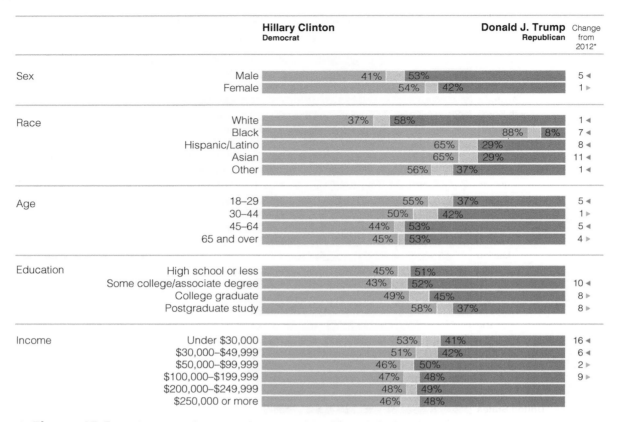

		Hillary Clinton Democrat	Donald J. Trump Republican	Change from 2012*
Sex	Male	41% / 53%		5 ◄
	Female	54% / 42%		1 ►
Race	White	37% / 58%		1 ◄
	Black	88% / 8%		7 ◄
	Hispanic/Latino	65% / 29%		8 ◄
	Asian	65% / 29%		11 ◄
	Other	56% / 37%		1 ◄
Age	18–29	55% / 37%		5 ◄
	30–44	50% / 42%		1 ►
	45–64	44% / 53%		5 ◄
	65 and over	45% / 53%		4 ►
Education	High school or less	45% / 51%		
	Some college/associate degree	43% / 52%		10 ◄
	College graduate	49% / 45%		8 ►
	Postgraduate study	58% / 37%		8 ►
Income	Under $30,000	53% / 41%		16 ◄
	$30,000–$49,999	51% / 42%		6 ◄
	$50,000–$99,999	46% / 50%		2 ►
	$100,000–$199,999	47% / 48%		9 ►
	$200,000–$249,999	48% / 49%		
	$250,000 or more	46% / 48%		

▲ **Figure 15-5   Who Voted How? The 2016 Presidential Election**

Source: Huang, Jon, Samuel Jacoby, Michael Strickland, and K.K. Rebecca Lai. 2016. "Election 2016: Exit Polls." *The New York Times*, November 6. **www.nytimes.com**

Income and education also affect voter behavior, as does region (see ■ maps 15-1 and 15-2). In the 2016 presidential election, people making under $50,000 per year were more likely to vote for Clinton than Trump. Despite the tendency to explain Trump's election as primarily the result of disaffected working-class voters, income was actually a rather poor predictor of the 2016 outcome and those with higher incomes were evenly split between Trump and Clinton. Education and race were far greater influences on voting behavior in the 2016 election (see Figure 15-5).

Political Power: Who's in Charge?

A democratic government is supposed to be representative of the people in the nation. Is this the case in the United States? Hardly. Women and racial–ethnic minorities are vastly underrepresented in our government. Most members of the U.S. Senate and House of Representatives are from upper-middle-class or upper-class backgrounds. Donald Trump's presidential cabinet is the wealthiest in our nation's history. Very few were blue-collar workers before coming to Congress.

Simply getting into politics requires a substantial investment of money. The total cost of the 2016 elections was a record $6.8 billion! The two presidential candidates (Donald Trump and Hillary Clinton) together spent $2.6 billion. Over five billion dollars was spent in the 2018 midterm elections, when only 36 percent of eligible voters turned out—the lowest voter turnout in years (Center for Responsive Politics 2018).

Candidates depend on contributions from individuals and groups to finance their election campaigns, with wealthy individuals among the largest campaign contributors, especially to presidential elections (Center for Responsive Politics 2012). The largest contributors to political campaigns are typically PACs. Much of the money given by individuals and PACs goes to incumbents, who have an overwhelming edge

maps 15-1 and 15-2   Mapping America's Diversity: Electoral Vote by State and County

The electoral vote is usually only reported state by state (blue for Democrats, red for Republicans), as you can see in the top map. But, as you see in the bottom map, if you shade the outcome by proportion of the vote at a county-by-county level, you get a somewhat different picture of the U.S. electorate. What do these maps teach you about the social-demographic basis of people's voting decisions?

2016 Electoral Vote by State

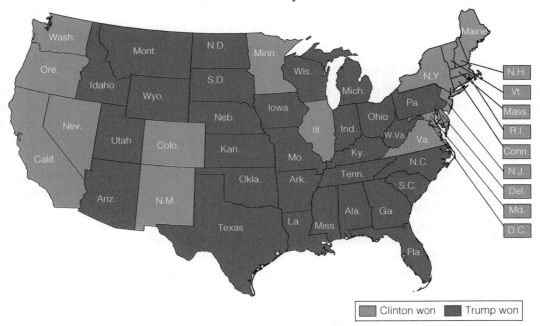

2016 Electoral Vote by County

Source: Top map © Cengage Learning®; lower map from Professor Mark Newman, University of Michigan.

in elections and already sit on the committees where public policy is hammered out. This picture of elites and business interests funneling money to candidates, who return to the same donors for more money when the next campaign rolls around, has shaken the faith of many Americans in the political system.

The amount of money flowing into political races is one reason that people have become so cynical about politics. Few think that the political process is a democratic and populist mechanism by which the "little people" can select political leaders to represent them. National surveys show that only 11 percent of U.S. citizens have a great deal or quote a lot of confidence in Congress, compared to 47 percent with a great deal of confidence in the Supreme Court and 37 percent in the presidency. By way of comparison, 74 percent have confidence in the military and 54 percent, the police (Gallup Poll 2018).

Women and Minorities in Government

There have been some gains in the number of women and minorities in government, but they are still underrepresented—both at the federal and state levels, even internationally (see ■ map 15-3). A record one hundred women now serve in the House of Representatives (out of 435 in the 116th Congress of 2019). There are 25 women in the U.S. Senate (out of 100). In the 116th Congress, women and people of color made huge gains in their representation.

There is very little religious diversity. In the 115th Congress only thirty identified as something other than Christian (Jewish, Buddhist, Hindu, and Muslim), and only nine did not identify as Christian in the Senate (one Buddhist and eight Jewish members; Marcos 2016). There is a long way to go before Congress truly represents the diversity in the population. Without greater inclusion of diverse groups, does democracy represent the diversity of the citizenry? The absence of greater representation may also explain some of the voter apathy that depresses voter turnout.

Researchers offer several explanations for why women and racial–ethnic minorities continue to be underrepresented in government. Certainly, prejudice plays a role. It was not long ago, in the 1960

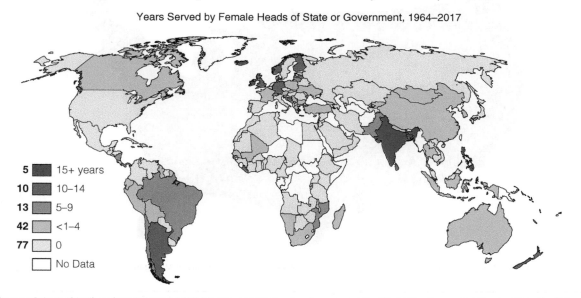

map 15-3 **Viewing Society in Global Perspective: Women Heads of State**

Does it surprise you that the United States fares poorly, relative to much of the rest of the world, when it comes to women's political leadership at the national level? What do you think explains this?

Years Served by Female Heads of State or Government, 1964–2017

5	15+ years
10	10–14
13	5–9
42	<1–4
77	0
	No Data

Source: Geiger, Abigail, and Lauren Kent. 2017. "Number of Women Leaders around the World Has Grown, But They're Still a Small Group." Washington, DC: Pew Research Organization. **www.pewresearch.org/fact-tank/2017/03/08/women-leaders-around-the-world/**

Diversity in the Power Elite

Various groups—women, racial–ethnic groups, lesbians, and gays—have vied for more representation in the halls of power, but have their efforts succeeded? If they make it to power, does this change the corporations, military, or government—the major institutions composing the power elite?

Sociologists Richard L. Zweigenhaft and G. William Domhoff examined these questions by analyzing the composition of boards of directors and chief executive officers (CEOs) of the largest banks and corporations in the United States, as well as analyzing Congress, presidential cabinets, and the generals and admirals who form the military elite. In addition, they examined the political party preferences and the political positions of people found among the power elite. Do women and minorities bring new values into power, thereby changing society as they move into powerful positions, or do their values match those of the traditional power elite or become absorbed by a system more powerful than they are?

They find that women, Jews, gays, lesbians, Black Americans, and Hispanics have become more numerous within the power elite, but only to a small degree. The power elite is still overwhelmingly White, wealthy, Christian, and male. Women and other minorities who make it into the power elite tend to come from already privileged backgrounds, as measured by their social class and education. Furthermore, Zweigenhaft and Domhoff find that the perspectives and values of women and minorities who rise to the top do not differ substantially from their White male counterparts. Some of this is explained by the common class origins of those in the power elite.

Zweigenhaft and Domhoff conclude that "the irony of diversity" is that greater diversity may have strengthened the position of the power elite because its members appear to be more legitimate through their inclusion of those previously left out. By including only those who share the perspectives and values of those already in power, little is actually changed.

Sources: Zweigenhaft, Richard L, and G. William Domhoff. 2011. *The New CEOs: Women, African Americans, Latino, and Asian American Leaders of Fortune 500 Companies*. Lanham, MD: Rowman and Littlefield; Zweigenhaft, Richard L., and G. William Domhoff. 2006. *Diversity in the Power Elite: Have Women and Minorities Reached the Top?* New Haven: Yale University Press; Domhoff, G. William. 2002. *Who Rules America?* New York: McGraw-Hill.

Kennedy–Nixon election, that Kennedy became the first Catholic president elected. Joseph Lieberman was the first Jewish candidate to appear on a major national ticket. Mitt Romney in 2012 was the first Mormon candidate to appear on a national ballot. Even then, 18 percent of the public said they would not vote for a Mormon as president, even if qualified (Newport 2012a). We also witnessed much prejudice directed toward former President Obama when he was accused by Donald Trump of not being a U.S. citizen, even though he was born in Hawaii.

Gender and racial prejudice run deep in the public mind. Although almost all Americans say they would vote for a woman president if their party nominated a woman, far fewer actually do and the implicit bias of sexism is known to be a factor in explaining this (Case-Levine 2016; Fiske and Durante 2016). Prejudice does not, however, fully account for the lack of a more representative government. Social structural causes are a major factor in the successful elections of women and people of color. Women and minority candidates receive a great deal of political support from local groups, but at the national level, they do not fare as well. The power of incumbents, most of whom are White men, is also a disadvantage to any new office seeker.

The Military as a Social Institution

Social institutions are stable systems of norms and values that fulfill certain functions in society. The military is a social institution whose function is to defend the nation against external (and sometimes internal) threats. A strong military is often considered an essential tool for maintaining peace. The military arm of the state is among the most powerful and influential social institutions in almost all societies. In the United States, the military is the largest single employer. Approximately 2.4 million men and women serve in the U.S. military, 1.3 million on active duty, and the rest in the reserves. This does not include the many hundreds of thousands who are employed in industries that support the military, nor does it include civilians who work for the Department of Defense and other military-affiliated agencies (Department of Defense 2017).

The military is one of the most hierarchical social institutions, and its hierarchy is extremely formalized. People who join the military are explicitly labeled with rank, and if promoted, they pass through a series of additional well-defined levels (ranks), each with clearly demarcated sets of rights and responsibilities. An explicit line exists between officers (lieutenants and higher ranks) and enlisted personnel, and officers have many privileges that others do not. Higher ranks are also entitled to absolute obedience from the ranks below them, with elaborate rituals created to remind both dominants and subordinates of their status.

As in other social institutions, military enlistees are carefully socialized to learn the norms of the culture they have joined. Military socialization places a high premium on conformity and eliminates individuality. All new recruits are issued identical uniforms and are allowed to retain very few of their personal possessions. They are endlessly harangued by the infamous "D.I." (drill instructor). They must quickly learn new, strictly enforced codes of behavior.

Most of the military is a part of the institution of government, but there has also been *privatization of the military*—meaning that an increasing number of military functions have been paid on a contract basis to private, for-profit employers. Under this development, the military becomes like a business, with people and corporations reaping profits on activities that once would have been not for profit. The privatization of the military can include companies that provide specific services (such as security), as well as engineering and building contracts. Critics of this trend warn that it will sacrifice safety and national security for the sake of corporate and individual profit and could lure the brightest people away from traditional military service if they see economic gains from private military service (Singer 2007).

Race and the Military

The greatest change in the military as a social institution is the representation of racial minority groups and women within the armed forces. Picture a U.S. soldier. Whom do you see? At one time, you would have almost certainly pictured a young White male, possibly wearing army green camouflage and carrying a weapon. Today, the image of the military is much more diverse. Drawing on the cultural images you have stored in your mind, you are just as likely to picture a young African American man in a military dress uniform with a stiffly starched shirt and a neat and trim appearance or perhaps a woman wearing a flight helmet in the cockpit of a fighter plane.

African Americans have served in the military for almost as long as the U.S. armed forces have been in existence. Except for the Marines, which desegregated in 1942, the armed forces were officially segregated until 1948, when President Harry Truman signed an executive order banning discrimination in the armed services. Although much segregation continued after this order, the desegregation of the armed forces is often credited with promoting more positive interracial relationships and increased awareness among Black Americans of their right to equal opportunities than has been the case in society at large. Until that time, the widespread opinion among Whites was that to allow Black and White soldiers to serve side by side would destroy soldiers' morale.

Currently, 18 percent of active military personnel are African American and 11 percent are Hispanic (who fall into various other racial–ethnic categories). Almost 3.6 percent are Asian Americans, 1.7 percent American Indian or Alaskan Native, 3.2 percent multiracial, and 1.1 percent native Hawaiian or Pacific Islander (U.S. Department of Defense 2017). Enlistees have many reasons to join the military, but the desire for education and job training is certainly among the strongest motivators, along with wanting to serve one's country.

Within the military today, there is a policy of equal pay for equal rank. African Americans and Latinos, however, are overrepresented in the lowest ranking support positions. Often, they are excluded from the higher-status, technologically based positions—those most likely to bring advancement and higher earnings both in the military and beyond. Most minorities remain in positions with little supervisory responsibility, such as service and supply jobs. Although the number of racial minorities in officers' positions has been increasing, they are still underrepresented and are less likely to get there via the route of military academies, as is the case for White officers (Segal and Segal 2004). Still, for both Whites and racial minorities, serving in the military leads to higher earnings relative to one's nonmilitary peers.

Women in the Military

The military academies did not open their enrollment to women until 1976. Since then, the armed services have profoundly changed their admission policies, and in 1996, the Supreme Court ruled (in *United States v. Virginia*) that women cannot be excluded from state-supported military academies such as the Citadel and the Virginia Military Institute (VMI). This was a landmark decision that opened new opportunities for women who want the rigorous physical and academic training that military academies provide (Kimmel 2000).

The involvement of women in the military has reached an all-time high in recent years. Women are 16 percent of enlisted military personnel. Now, almost 204,000 women are on active duty in the United States. The Army has the highest proportion of women, followed by the Air Force, the Navy, and the Marines. Women now comprise 16 percent of military officers and 27 percent of military academy cadets and midshipmen (U.S. Department of Defense 2017).

The former exclusion of women from military service was rationalized by the popular conviction that women should not serve in combat. Despite this attitude, women have been fighting in active combat to defend the nation and were officially made eligible for combat roles in 2013.

The presence of women in the military has transformed the armed forces, but it also has raised new issues for military personnel. Slightly more than half (56 percent) of military personnel are married, and 6 percent of active-duty members are in dual-military marriages. Family separations, frequent moves (on average every three years), risk of injury or death, and living in a foreign country are only some of the challenges that military personnel face in trying to manage their lives (Segal and Segal 2004).

For women in the military, the highly gendered organization of which they are a part is also a challenge. Indeed, recent reports have documented an alarmingly high rate of sexual assault and sexual harassment against women in the military—by other military personnel. Reports from the Pentagon have found that one-third of women in the military (and 6 percent of men) experienced sexual harassment, including unwanted crude and offensive behavior, unwanted sexual attention, and sexual coercion; the same report found that 5 percent of women in the military experienced some form of unwanted sexual contact, such as rape, unwanted sodomy, or indecent assault (U.S. Department of Defense 2012). Periodic scandals involving rape, sexual harassment, and other forms of intimidation against women in the military (including the military academies) reveal that, although certainly not all military men engage in these behaviors, institutions organized around such masculine characteristics as aggression, domination, and hierarchy put women at risk.

Women are an increasing presence in the U.S. military.

Gays and Lesbians in the Military

Gays and lesbians have long served in military duty, despite the policies that have attempted to exclude them. The military has admitted that there always have been gays and lesbians in all branches of the U.S. armed forces, but homophobia is a pervasive part of military culture (Becker 2000; Myers 2000).

The Obama administration ended the "don't ask, don't tell" policy by which recruiting officers could not ask about sexual preference. It remains unclear whether gays and lesbians will be permitted to live openly as gay while also pursuing careers in the armed services. The Trump administration has tried to ban transgender people from military service, although it is unclear how the Department of Defense will implement this discriminatory proposal.

Supporters of the ban on LGBTQ people in the military often use arguments similar to the arguments used before 1948 to defend the racial segregation of fighting units. As in 1948, detractors claim that the morale of soldiers will drop if forced to serve alongside gay men and women, national security will be threatened, and known homosexuals serving in the military will upset the status quo and destroy the fighting spirit of military units. Seeing these arguments in historical perspective helps you see through some of the myths perpetuated by such attitudes.

Military Veterans

Now, almost two million veterans have returned from the wars in Iraq, Afghanistan, and other current world conflicts. Add to that the veterans of the Gulf War, Vietnam War, Korean War, and the living veterans of World War II, and it totals 22 million veterans living in the United States (U.S. Department of Veteran's Affairs 2016). For all veterans, the return home—though joyful—also has risks, risks that result from social, as well as physical, needs for recovery and adaptation.

The changed nature of combat in the two Iraq and Afghanistan wars has meant that returning veterans have more complex forms of physical and emotional injury. Exposure to repeated blasts of IEDs (improvised explosive devices) has resulted in more traumatic brain injuries. Having an all-volunteer army has also produced more frequent redeployment, resulting in longer-term exposure to war trauma, as well as greater exposure to blasts and other forms of violence. Changes in military and medical technology have also increased survival rates from injuries that would have killed military personnel in the past.

All of these factors have meant increased risks for returning veterans, including not only difficult recoveries from physical injuries, but also high rates of mental health disorders, a high risk of suicide, depression, and/or drug and alcohol addiction. In addition to these social problems, veterans face various adaptation challenges as they transition back into the civilian workforce—that is, if they find work. Veterans and their families and partners also have to adapt to new family roles, perhaps even including a readjustment as parents because their children will have matured in their absence. Adding to this complexity is the fact that there may be a significant readjustment to a new physical or mental disability. Managing the health problems that may have developed during deployment produces new forms of stress on preexisting relationships (Institute of Medicine of the National Academies 2010).

When veterans return home, often the social supports they need are not strong. One consequence has been an increase in the number of homeless veterans—a figure that has doubled since 2010 (Zoroya 2012). African American veterans, for whom the military has been a path for social mobility, may face the additional fact that social institutions fail them again in the form of unemployment and persistent racism (Finkel 2014; Fleury-Steiner 2012).

The situation for U.S. veterans shows how critical social institutions are in the lives of these men and women. For all members of society, the support—or lack thereof—provided by social institutions is a critical backdrop to the character of everyday life.

Chapter Summary

How are societies economically organized?

Societies are organized around an economic base. The *economy* is the system on which the production, distribution, and consumption of goods and services are based. *Capitalism* is an economic system based on the pursuit of profit, market competition, and private property. *Socialism* is characterized by state ownership of industry; *communism* is the purest form of socialism.

How has the global economy changed?

As capitalism has spread throughout the world, *multinational corporations* conduct business across national borders. A number of countries have undergone *deindustrialization*, or changeover from a goods-producing economy to a services-producing one. This has caused many heavy-industry jobs in U.S. cities to vanish, thus increasing the unemployment rate in those cities. Changes in information technology, plus increased *automation*, have resulted in the further elimination of jobs in both the United States and abroad.

What is the social organization of work?

Sociologists define *work* as human activity that produces something of value. Some work is judged to be more valuable than other work. *Emotional labor* is work that is intended to produce a desired state of mind in a client. The *division of labor* is the differentiation of work roles in a social system. In the United States, there is a class, gender, and racial division of labor. The labor market in the United States is described as a *dual labor market*. Jobs in the primary sector of the labor market carry better wages and working conditions, whereas those in the secondary labor market pay less and have fewer job benefits. Women and minorities are disproportionately employed in the secondary labor market. Patterns of occupational distribution also show tremendous segregation by race and gender in the labor market. Race and gender also affect the occupational prestige, as well as the earnings, of given jobs.

How is diversity reflected in the workplace?

The workplace is becoming more diverse with greater numbers of racial–ethnic groups, women, and an older workforce. Official *unemployment rates* underestimate the actual extent of joblessness. Women and minorities often encounter the *glass ceiling*—a term used to describe the limited mobility of women and minority workers in male-dominated organizations. In addition, women more often than men face *sexual harassment* at work—defined as the unequal imposition of sexual requirements in the context of a power relationship. Homophobia in the workplace also negatively affects the working experience of gays and lesbians. New protections are in place for disabled workers through the *Americans with Disabilities Act (ADA)*.

What is the state?

The *state* is the organized system of power and authority in society. It comprises different institutions, including the government, the military, the police, the law and the courts, and the prison system. The state is supposed to protect its citizens and preserve society, but it often protects the status quo, sometimes to the disadvantage of less powerful groups in the society. States can also be organized as *democracies*, as *authoritarian*, or as *totalitarian*.

How do sociologists define power and authority?

Power is the ability of a person or group to influence another. *Authority* is power perceived to be legitimate and formal. There are three kinds of authority: *traditional authority*, based on long-established patterns; *charismatic authority*, based on an individual's personal appeal or charm; and *rational–legal authority*, based on the authority of rules and regulations (such as law).

What theories explain how power operates in the state?

Sociologists have developed four theories of power. The *pluralist model* sees power as operating through the influence of diverse interest groups in society. The *power elite model* sees power as based on the interconnections between the state, industry, and the military. *Autonomous state theory* sees the state as an entity in itself that operates to protect its own interests. *Feminist theorists* argue that the state is patriarchal, representing primarily men's interests.

How well does the government represent the diversity of the U.S. population?

An ideal democratic government would reflect and equally represent all members of society. The makeup of the U.S. government does not reflect the diversity of the general population. African Americans, Latinos, Native Americans, Asians, and women are underrepresented within the government. Political participation also varies by a number of social factors, including income, education, race, gender, and age. African Americans and Latinos, however, are overrepresented in the military, in part because of the opportunity the military purports to offer groups otherwise disadvantaged in education and the labor market; however, both are underrepresented at the levels of high-level commissioned officers. There is an increased presence of women in the military; however, prejudice and discrimination continue against lesbians and gays in the military.

Key Terms

alienation 406

Americans with Disabilities Act (ADA) 405

authoritarian 408

authority 410

automation 397

autonomous state model 413

bureaucracy 410

capitalism 393

charismatic authority 410

communism 393

contingent worker 397

deindustrialization 395

democracy 408

division of labor 399

dual labor market 400

economic restructuring 394

economy 392

emotional labor 398

glass ceiling 400

global assembly line 394

global economy 393

government 414

interest group 411

interlocking directorate 413

multinational corporations 393

nationalism 409

occupational segregation 401

outsourcing 394

pluralist model 411

political action committees (PACs) 412

postindustrial society 393

power 409

power elite model 412

propaganda 408

rational–legal authority 410

sexual harassment 404

socialism 393

state 407

totalitarian 408

traditional authority 410

underemployment 403

unemployment rate 402

work 398

xenophobia 394

CHAPTER 16

ENVIRONMENT, POPULATION, AND SOCIAL CHANGE

In this chapter, you will learn to:

Identify the social dimensions of environmental change

Explain how inequality affects environmental quality for different groups

Understand the basic processes of population change

Explain theories of population growth

Describe the different components and sources of social change

Compare and contrast sociological theories of social change

Analyze the social implications of globalization and modernization

Can we preserve the Earth's resources as we know them? Scientists and others are warning us that the polar ice cap is melting, causing ocean levels to rise. Climate patterns are changing. Although the specific causes and effects of climate change are being debated, the increase in severe storms, wildfires, and extreme heat has raised an alarm bell that climate change is already upon us. In some parts of the world, population growth outpaces the ability to feed people. In other places, including in the United States, water can no longer be assumed to be available or safe to drink. Air pollution is so bad in some parts of the world that people routinely wear masks. Is our current lifestyle sustainable?

Sustainability, in fact, has become an organizing cry—a cry for new social policies that will protect the Earth's environment and the people who live within it. New movements have developed—movements to eat local food, to support the creation of urban community gardens, and to recycle used products by transforming them into something else. These and other developments signal the public's concern with the environment and the related phenomena of population, environmental degradation, and social change.

You might think that studying such things as sustainable energy and environmental pollution is solely the work of scientists and engineers. No doubt these are critical scientific problems, yet as scientists and engineers will tell you, social issues are just as important in understanding how we can preserve the Earth's resources. What lifestyles consume the highest amounts of energy? What social and cultural changes are needed to protect our environment and not deplete the Earth's natural resources? How is social inequality related to the degradation of the environment? Are there just too many people for the world to sustain human society as we know it?

These and other questions drive the substance of this chapter—a chapter that looks at the sociological issues that come from studying population and the environment. *Environmental sociology* now has particular urgency as people become more attuned to the potential crises that our planet faces.

A Climate in Crisis: Environmental Sociology

Human beings, animals, and plants all depend on one another and on the physical environment for their survival. **Environmental sociology** is the scientific study of the interdependencies that exist between humans and our physical environment. A *human ecosystem* is any system of interdependent parts that involves human beings in interaction with one another and the physical environment. As examples, a city is a human ecosystem; so is a rural farmland community. In fact, the entire world is a human ecosystem.

The examination of ecosystems has demonstrated two things:

1. The supply of many natural resources is finite, and
2. If one element of an ecosystem is disturbed, the entire system is affected.

For much of the history of humankind, the natural resources of the Earth were abundant. In the past, given the limited resources that people used, they may as well have been infinite. No more. Some resources, such as certain fossil fuels, are simply nonrenewable and may be gone soon. Other resources, such as timber or seafood, are renewable only if we do not plunder the sources of supply so recklessly that they disappear. Some natural resources are so abundant that they still seem infinite, such as the planet's stock of air and water. But, at this stage of our societal development, we are learning that without more vigilance, we can destroy even seemingly infinite resources (Gore 2006). Understanding social behavior as it affects the environment is critical to thinking about and solving our environmental problems.

Society at Risk: Air, Water, and Energy

An engineer might invent a new way to heat our homes and power our cars, but without understanding the social dimensions of issues like energy, pollution, water usage, and other environmental behaviors, we cannot make progress toward environmental sustainability. The challenges we face in protecting the environment are many. Gaseous wastes are gnawing away at the ozone layer. Buried chemical wastes are trickling into the water table, creating underground pools of poison. Pollution has damaged the Earth's surface water so badly that worldwide underground water reserves are being mined faster than nature can replenish them.

The skies of all major cities around the world are stained with pollution hazes. In cities that rest within geological basins, such as Mexico City and Los Angeles, the concentrations of pollutants can rise so high that pollution-sensitive individuals cannot leave their homes or must wear masks when they go outdoors. For many, alerts about unsafe air quality have become a routine phenomenon.

Rather small changes in the average temperature of the Earth can have dramatic consequences (see ▇ map 16-1). A few degrees of difference can cause greater melting in the Arctic regions, which raises

map 16-1 Viewing Society in Global Perspective: Global Warming: Viewing the Earth's Temperature

Scientists at the National Aeronautics and Space Administration (NASA) can measure and map changes in the Earth's temperature. The average temperature on Earth has increased by 1.4 degrees Fahrenheit since 1980 (0.8 degrees Celsius). A change of just one degree has massive consequences. Earth was plunged into a "little ice age" with only a one- to two-degree drop in the seventeenth century. Twenty thousand years ago, much of North America was buried under a towering mass of ice from a five-degree drop. What must be done to protect us from the pending disaster of climate change? How are social policies and social behaviors involved in such changes?

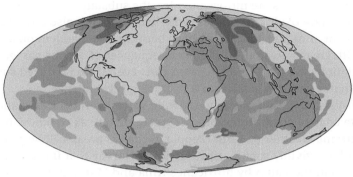

1980–1989

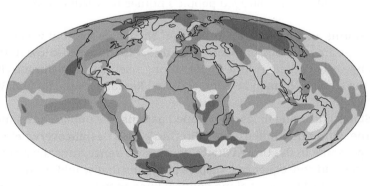

1990–1999

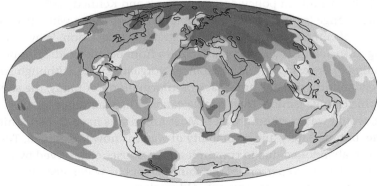

2000–2009

Temperature Anomaly (°C)

−2.5   0   2.5

Source: Carlowicz, Michael. 2014. "Global Temperatures." National Aeronautics and Space Administration. **www.earthobservatory.nasa.gov**

the level of the sea, affecting water, land, and weather systems worldwide. Today, we see images of polar bears drowning because of the breakup of ice floes, that is, the melting of the polar cap. Sea levels are rising because of the melting of the polar ice cap, threatening to drown major urban areas. The rise in ocean temperatures is producing stronger hurricanes.

Human ingenuity has, no doubt, produced inventions that have vastly improved the quality of life. Sometimes, though, these inventions have unintended and possibly dangerous consequences. To illustrate, the invention of the automobile has transformed society. Today, many cannot imagine getting around without a car. We have designed many of our cities and, especially, our suburbs, in ways that require people to drive. But a huge portion of the pollutants released into the air comes from the exhaust pipes of motor vehicles. The major component of this exhaust is carbon monoxide, a highly toxic substance. Also found in exhaust fumes are nitrogen oxides, the substances that give smog its brownish-yellow color. Sunlight causes these oxides to combine with hydrocarbons, also emitted from exhausts, forming a host of health-threatening substances. We have become so dependent on automobiles for transportation that even with greater awareness of the consequences of driving gas-guzzling cars, it is difficult to design transportation systems that rely less on cars. Even if we produced such a design, would people give up their cars? That is as much a social as a technological challenge.

Climate change is the systematic increase in worldwide surface temperatures and the resulting ecological changes. Climate change poses numerous threats to society as we know it. With climate change will come more extreme weather patterns. People in coastal areas will be prone to rising coastal waters and storm surges, all too vividly seen during hurricane Harvey that flooded the city of Houston in the fall of 2017, followed soon thereafter by hurricanes Irma and Maria that devastated the U.S. Virgin Islands and left much of Puerto Rico without potable water or electricity for weeks afterwards. Then in 2018, hurricanes Florence and Michael destroyed people's homes in North Carolina and then the panhandle of Florida.

Were these massive storms an effect of climate change? No one knows, but we do know climate change and its consequences are a serious problem and a threat to life as we know it. Despite the denials of some that climate change is real, there is little doubt among scientists that climate change is happening and that it is largely the result of human activity (National Research Council 2012). What does the public think?

Outside the United States, including in less-developed nations, the public shows more concern about climate change than is true within the United States (Brechin 2003). Despite overwhelming scientific evidence of climate change, some in the United States deny its existence, thus thwarting policy changes that could address its causes and consequences, such as programs that would make us less reliant on fossil fuels. The public is also clearly divided along political lines as to how they think about climate change. Democrats are far more likely than Republicans to think that climate change is occurring and that climate scientists should have a role in policy decisions (Funk and Kennedy 2016).

Denial about climate change is only part of the problem, though. Even when people have information about the potential effects of climate change, they often ignore taking action. Why? The *social organization of denial*, according to sociologist Kari Norgaard, comes from people holding unpleasant emotions— such as the fear of flooding or devastation—at a distance (Norgaard 2006). Although some may deny climate change for purely political reasons or for lack of information, for many, the sheer unpleasantness of facing such catastrophic change is more than people can willingly admit or face.

What, for example, would we do without water? Most Americans have come to think of water as abundant, free, and safe, but the safety and availability of water is now threatened. Thousands of rural water wells have had to be abandoned due to contamination. Households served by municipal water systems are also endangered; fully 20 percent of the country's public water systems do not meet the minimum toxicity standards set by the government. Although many have assumed that the nation's water is plentiful, whether that can still be assumed is questionable (Fishman 2011). In the western and southwestern United States, the groundwater supply is being depleted at a rapid pace, making water one of the causes of political conflict between different states in the region (Espeland 1998).

Threats to our water supply also spill into other issues. As people have become concerned about water quality, drinking water from plastic bottles has increased. Where do the bottles go? Estimates are that about 40 million plastic water bottles go into the nation's trash *every day*—only about 23 percent of which are recycled (Fishman 2011). From production to disposal, water bottles reveal that technical know-how merges with human behavior, creating a complex system of environmental challenges.

The nation's water is also threatened by the chemical pollutants that industries discharge into rivers, lakes, and the oceans, including solid wastes, sewage, nondegradable by-products, synthetic materials, toxic chemicals, and radioactive substances. Add the polluting effects of sewage systems of towns and large cities, detergents, oil spills, pesticide runoff, and runoff from mines, and the enormity of the problem is clear.

The city of Flint, Michigan, with a predominantly African American low-income population, was declared in a state of emergency when the drinking water was contaminated with lead. Several government officials were charged with crimes for negligence in handling this environmental crisis.

Federal and state statutes now prohibit industry from polluting the nation's water, but the pollution continues. Why? The answer is economic, political, and sociological. Industries that contribute to a vigorous economy have traditionally met with little interference from the government. Public awareness and outrage can force the government to crack down on major polluters.

We are racing through our nonrenewable natural resources and destroying much that could be renewable. Addressing this problem also requires looking at some of the inequalities that are revealed when we examine such things as energy usage. On a global scale, the use of natural resources is not evenly shared around the world. The United States, which is a little under 5 percent of the world's population, consumes 20 percent of the world's energy and emits about 20 percent of the carbon dioxide emissions from fossil fuels. China now exceeds the United States in the release of carbon-based emissions, sending 9,377 million metric tons of carbon dioxide emissions into the atmosphere; the United States ranks second in the world, spewing 5,508 million metric tons of carbon dioxide emissions into the atmosphere per year (U.S. Energy Information Administration 2017; see ▲ Figure 16-1). How much is a metric ton? The

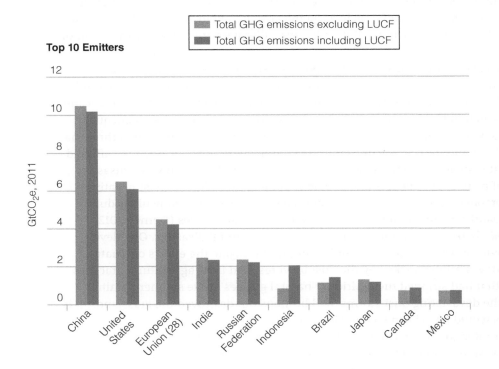

Top 10 Emitters

Legend:
- Total GHG emissions excluding LUCF
- Total GHG emissions including LUCF

Y-axis: GtCO$_2$e, 2011 (scale 0 to 12)

X-axis categories: China, United States, European Union (28), India, Russian Federation, Indonesia, Brazil, Japan, Canada, Mexico

▲ **Figure 16-1   Carbon Dioxide (CO$_2$) Emissions: Top Emitters.** You can vividly see in this graph which nations contribute the most to carbon dioxide emissions. It is a known fact that such emissions contribute to global warming and climate change. What changes would be needed in society to produce a decrease in the future?

Source: World Resources Institute, **http://www.wri.org/blog/2014/05/history-carbon-dioxide-emissions**

The Wasteful Society

For just one day (a full twenty-four-hour period), make a list of everything that you use up or discard. Include everything that you throw away, including garbage, waste from cooking and eating, gasoline in your car, and so forth. At the end of the day, list the things you discarded. Indicate whether there were any alternatives to discarding these things. How might one reduce the amount of waste produced in society generally?

Bottled water (not counting other plastic containers) produces 1.5 million tons of waste every year, only a small percentage of which is recycled.

average car now weighs about two tons, so 5 million metric tons would be the weight of about two and a half million cars sent into the atmosphere, if that were even possible!

Disasters: At the Interface of Social and Physical Life

Even while the normal practices of everyday life threaten the Earth, periodic hazards also come from natural disasters. Floods, fires, earthquakes, tsunamis, hurricanes, tornadoes—these are just a few of the disasters that disrupt communities, families, and public health. Many disasters are forces of nature; some are predictable, some are not. Other disasters, such as chemical spills, explosions, and huge forest fires, are more directly attributable to human actions. But, either way, disasters are not solely the result of physical or natural factors.

According to sociologists who study them, disasters juxtapose physical events, such as floods, fires, hurricanes, earthquakes, and the like with vulnerable populations (Tierney 2007). Although people think of disasters as "nature's wrath," the impact of disasters is often the result of social behavior too. Hurricane Katrina is a good example. Although the hurricane itself emerged from nature, neglect of an inadequate levy system made communities in low-lying areas more vulnerable than others. Likewise, when three major hurricanes struck the Gulf Coast and the Caribbean Islands, including Puerto Rico, in 2017 they affected many, but had a disproportionate impact on poor and lower-income communities. Even when disasters affect hundreds of thousands of people, poor people and people of color are usually the most negatively impacted. They are more vulnerable even before a disaster, but then have the greatest difficulties during recovery from poor access to health care, insurance, housing, and other social services (Tierney 2012).

Time and time again, human behavior is implicated in the impact of natural disasters. Overdevelopment can destroy natural environments, such as barrier islands, that mitigate the effects of a natural disaster. During the 1930s, the devastation wrecked by the dust bowl resulted from agricultural practices (the overproduction of wheat) that had stripped the prairies of natural grasses in the southern plains. Although drought brought on the devastating dust storms, had humans not destroyed the grasslands, it is doubtful that the consequences would have been so dire (Egan 2006).

Social systems are also disrupted in the aftermath of disasters. Hurricane Maria in Puerto Rico in 2017 showed how vulnerable social systems can be to natural disasters as people were left without the ability

---

Debunking Society's Myths

Myth: In the aftermath of a natural disaster, people live in chaos and with a breakdown of the social order.

Sociological Perspective: Even when the ordinary course of life is disrupted following a catastrophic event or disaster, people rely on what disaster researchers call "pro-social behavior"—that is, assisting each other, developing support networks, and organizing informal systems of social control (Tierney et al. 2006).

---

to communicate when cell phone towers went down. For many it was months before they had drinkable water and electric power. In the United States, the massive oil spill in 2010 spewed thousands of gallons of crude oil into the Gulf of Mexico. The spill was so large and disastrous that many shrimpers and fishermen were forced out of business. The spill rapidly polluted major marshes surrounding the Gulf, killing off much flora and fauna, including birds of several species and all varieties of fish, shrimp, and mussels.

Who is most vulnerable during disasters is also shaped by social factors. The poor and the elderly are often the most vulnerable, facing numerous barriers to evacuation. Such vulnerable people also tend to be less visible during disasters and seldom have their opinions heard. They are then less likely to get support—either material or social (Cash, Drotning, and Miller 2017; Tierney et al. 2001).

In a carefully designed research study, sociologist Eric Klinenberg (2002) found that particular social factors influenced the high rate of mortality during the infamous Chicago heat wave of 1993. In this heat wave, temperatures soared above 105 degrees and over 700 people died. Klinenberg found that the isolation of elderly people, retrenchment of social services, and little institutional support in poor neighborhoods meant that those most likely to die from this disaster were the poor, the elderly, African Americans (because of their concentration in poor neighborhoods), and women (because they are more of the older population).

Government responses to disasters also show the consequences of human behavior for understanding the social dimensions of otherwise natural disasters. The slow work of the federal bureaucracy, as well as partisan politics, both play a role in social responses to disasters. Social stereotypes also figure in how victims of disasters are portrayed and how quickly assistance comes to people. Following Hurricane Katrina in New Orleans, African Americans were depicted in media coverage as wild looters and thieves—an image not seen so much when the predominantly White, working- and middle-class communities in New York and New Jersey were so profoundly disrupted following Hurricane Sandy. And, in the more recent hurricanes of 2017, many noted how quickly the federal government responded to people in Houston, but not the U.S. citizens so devastated in Puerto Rico—people who are perceived as people of color (Nixon and Stevens 2017; Rivera and Aranda 2017).

Now, given climate change, people even wonder if disasters will be more frequent. The warming of the Earth's oceans could produce more frequent and more damaging hurricanes. Heat waves could become more frequent, overpowering power supplies as people try to cool their homes. Drought could intensify, fueling political struggles over who controls water supplies in the driest areas of our nation. Some populations will be more vulnerable than others, showing once again how factors such as race, social class, age, and gender shape the impacts and portrayals of so-called natural disasters.

Environmental Inequality and Environmental Justice

Many argue that of all environmental problems facing the United States today, the most urgent is the dumping of hazardous wastes, if only for the sheer noxiousness of the materials being dumped. Since 1970, the production of toxic wastes has increased nine-fold (Weeks 2012). Of course, any degradation in the Earth's well-being affects everyone, but who is most vulnerable to pollutants and toxic waste dumping reveals patterns of social inequality.

© Visual concept by Norman Andersen

▲ **Figure 16-2** Environmental racism refers to the pattern whereby people living in predominantly minority communities are more likely exposed to toxic dumping and other forms of pollution. Nuclear waste and testing in the American Southwest, for example, have been located in areas predominantly inhabited by Native Americans. In other areas, African Americans and Latinos are exposed to the effects of industrial waste.

Environmental racism is the pattern whereby toxic wastes and other pollutants are disproportionately found in minority and poor neighborhoods, a pattern with clear health consequences (Brulle and Pellow 2006). Research has determined that it is virtually impossible that dumps are being placed so often in communities of minority and lower socioeconomic status by chance alone (Bullard and Wright 2009). Is class or race to blame? Race and class both influence toxic waste disposal (Mohai and Saha 2007). Wealthier communities are better able to resist dumping in their neighborhoods, and housing discrimination and other race-related disparities are strongly linked to toxic waste being more present in minority areas, as illustrated in ▲ Figure 16-2.

Studies find that Native American, Hispanic, and particularly African American populations reside disproportionately closer to toxic sources than do Whites. Such patterns are not explainable by social class differences alone. That is, when communities of the same socioeconomic characteristics but different racial–ethnic compositions are compared, Native Americans, Hispanics, and African Americans of a given socioeconomic level live closer to toxic dumps than do Whites of the same socioeconomic level (Bullard and Wright 2010; Mohai and Saha 2007).

Take a look around your own neighborhood. Are there industrial waste sites nearby? Where are toxic products being disposed? For that matter, is there recycling available and to whom? You are likely to find patterns of waste disposal that are significantly linked to the class and racial composition of your neighborhood.

Within minority and poor neighborhoods, many groups have mobilized to protest and stop dumping in their communities. The *environmental justice movement* is the broad term used to refer to the social action that communities have taken to ensure that toxic waste dumping and other forms of pollution do not fall disproportionately on groups because of their race, class, or gender (Pellow 2004).

Environmental justice encompasses a wide array of programs for change. Developing more organic methods of growing food—indeed, encouraging community gardens where people can grow their own food—and other "green" programs are important trends to promote social change for a more sustainable society. But social change is hard to accomplish, in part because it takes more than individual effort. As we will see, social change requires individual action, but it is also collective, that is, a fundamentally *social* process.

Debunking Society's Myths

Myth: Environmental pollution is more common in or near economically poor areas; social class is a more important reason for this than race.

Sociological Perspective: Even when comparing areas of the same low economic status but different racial compositions, the areas with a higher percentage of minorities are on average closer to polluted areas than those with a lower percentage of minorities (Mohai and Saha 2007).

Counting People: Population Studies

Studies of the environment raise fundamental questions about how human societies relate to the physical and natural world. Population growth and density are responsible for some of the challenges we face with our environment: urban overcrowding and sprawl, traffic jams, pollution, and the threat of diminishing or tarnished Earth resources. Can we sustain the current way of living, given the size of the national and world populations? Are there simply too many people for our planet to support?

There are seven billion people living in this world. What do we know about how population is shaped? When a baby is born, what are his or her odds of survival beyond the first year? How many others will be born the following year? Will the population of people born in a given year influence the future of society simply because of the size of this age group?

These questions can be studied through the sociology of population. The scientific study of population is called **demography**. Demography includes studying the size, distribution, and composition of human populations as well as studying population changes over time, both those of the past and those predicted for the future.

Basic population facts drive many of the experiences and attitudes of some people. Young people may feel insecure about their future; decisions about having or not having children may loom; with a large aging population, young people will likely have to care for older people; and, hotly debated topics like immigration are likely to continue to shape national politics. The decisions people make—both personal and national—will ripple forward for years to come. Will there be a need for more senior centers? Will minority children get a good education? If there is a decline in the number of middle-aged people, who will take care of the old? Will environmental resources hold up in such a way as to maintain current lifestyles?

Demography draws on huge bodies of data generated by a variety of sources. One major source is the U.S. Census Bureau. A **census** is a head count of the entire population of a country, usually done at regular intervals. The U.S. census is conducted every ten years, as required by the U.S. Constitution; the 2020 census will provide the most recent data. The census attempts to enumerate every individual and to obtain information such as gender, race, ethnicity, age, education, occupation, and other social factors. The census is updated annually through a much smaller sample of the population that can then be used to track changes more frequently, although in less detail, than the decennial census (conducted every ten years).

The current population of the United States is more than 323.1 million—a milestone when the 300-million mark was passed in 2011. By 2060, the U.S. population is not only predicted to be larger (420 million) but also older and more diverse. White Americans are expected to decline as a percentage of the population; Hispanics and Asians are expected to double in number; African Americans will also increase, though not as much as Hispanics and Asians. As soon as 2030, one in five Americans will be over 65. The older population is expected to almost double in size by mid-century (Ewert 2015).

Even with this detail, however, it is known that the census undercounts particular parts of the country's population. It is simply impossible to have every single person complete the census form that is distributed. Who would be most likely undercounted? You can probably guess. Those most likely to have been undercounted by the census are the homeless, immigrants, minorities living in poor neighborhoods, and others of low social status. In general, the lower your overall social status (such as by income, occupation, race–ethnicity, gender, immigration status, or other measures), the less likely you are to be counted in the U.S. census.

The constitutional requirement for a census was included to ensure fair apportionment of representatives in the federal government. Census data is also the basis for the distribution of various federal benefits and it figures prominently in decisions about where to locate businesses, schools, housing, and so forth. Undercounting specific groups of people thus leaves them underrepresented in government and less served by economic and social development. The Census Bureau itself estimated that Whites were overcounted in the 2010 census by close to 1 percent. African Americans were undercounted by 2 percent; Hispanics, by 1.5 percent. Native Americans living on reservations are estimated to have been undercounted by about 5 percent. Even whether you own or rent your residence affects the likelihood of being counted—renters being more likely to be undercounted; homeowners, overcounted (U.S. Census Bureau 2012). The undercount in the 2020 census remains to be seen.

Although counting people may seem tedious and dry to you, it can actually be a fiercely debated topic. Now (and beginning in the 2000 census), people are allowed to select multiracial (or "mixed race") as a response regarding their racial and ethnic identity on the census questionnaire (see ▲ Figure 16-3). If you are, for example, Hispanic and Black or Black and White, how should the census "count" you in a racial or ethnic category—a category that will later be used to determine such facts as you have seen in this book, such as income distribution by race or the ethnic makeup of neighborhoods? The use of the multiracial response option gives individuals an opportunity to define themselves as mixed race. One argument against this option is that it subtracts from the number of people who would have otherwise indicated only one category, thus further undercounting African Americans, Hispanics, and Native Americans. Currently, only 2.5 percent of people responding to the census indicate a multiracial response, but this is expected to increase substantially in the 2020 census, as well as into the future. Of course, how people identify can change according to the social and political climate of the time, but population experts use these projections to anticipate changes that are likely to occur (Ewert 2015).

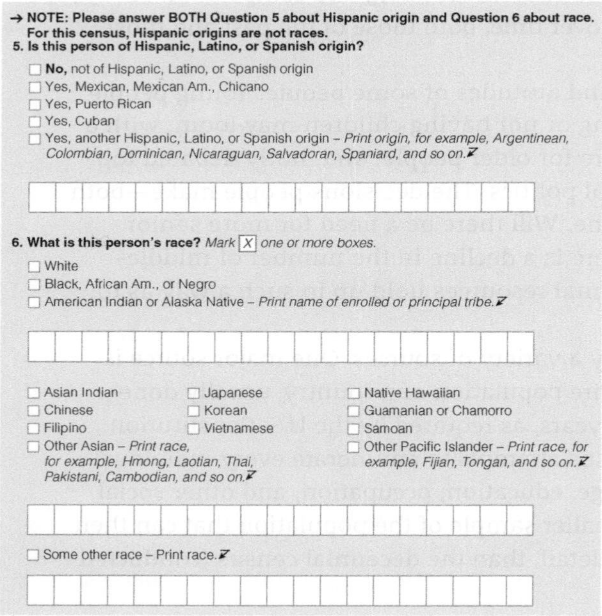
→ NOTE: Please answer BOTH Question 5 about Hispanic origin and Question 6 about race. For this census, Hispanic origins are not races.

5. Is this person of Hispanic, Latino, or Spanish origin?

☐ No, not of Hispanic, Latino, or Spanish origin
☐ Yes, Mexican, Mexican Am., Chicano
☐ Yes, Puerto Rican
☐ Yes, Cuban
☐ Yes, another Hispanic, Latino, or Spanish origin – *Print origin, for example, Argentinean, Colombian, Dominican, Nicaraguan, Salvadoran, Spaniard, and so on.*

6. What is this person's race? *Mark* ☒ *one or more boxes.*

☐ White
☐ Black, African Am., or Negro
☐ American Indian or Alaska Native – *Print name of enrolled or principal tribe.*

☐ Asian Indian ☐ Japanese ☐ Native Hawaiian
☐ Chinese ☐ Korean ☐ Guamanian or Chamorro
☐ Filipino ☐ Vietnamese ☐ Samoan
☐ Other Asian – *Print race, for example, Hmong, Laotian, Thai, Pakistani, Cambodian, and so on.*
☐ Other Pacific Islander – *Print race, for example, Fijian, Tongan, and so on.*

☐ Some other race – Print race.

▲ Figure 16-3 The Census Counts Race. This is the form that the U.S. Census Bureau uses in its decennial census to tally the racial–ethnic composition of the U.S. population. How would you answer? Does this adequately measure your racial identity? If so, why? If not, how might you revise it, and would that be correct for others?

Source: U.S. Census Bureau. 2010. "Overview of Race and Hispanic Origin: 2010." Washington, DC: U.S. Department of Commerce. **www.census.gov**

The world population is currently seven billion people, and it is expected to grow to nine billion by 2050. Most of the growth will be in developing areas of the world, not the already industrialized nations, as you can also see in ▲ Figure 16-4. Many of the nations with the highest rates of population growth also have high rates of poverty. In fact, family size tends to decrease with a rise in income (United Nations Population Fund 2014). Barring some major catastrophe, such as a health epidemic, these nations will have a higher **population density**, defined as the number of people per unit of area, usually per square mile.

The total number of people in a society at any given moment is determined by only three variables: births, deaths, and migrations. These three variables show different patterns for different racial and ethnic groups, different social strata, and both genders. Births add to the total population, and deaths subtract from it. Migration into a society from outside, called **immigration**, adds to the population, whereas **emigration**, the departure of people from a society (also called *out-migration*), subtracts from the population.

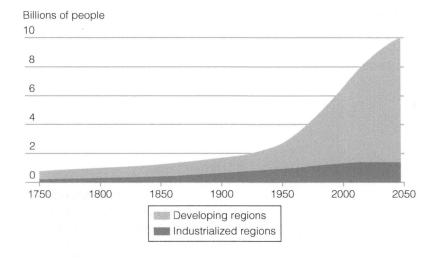

Billions of people

▲ **Figure 16-4   World Population Growth, 1750–2050.** As you can see from this graph, population in the most developed parts of the world is expected to remain somewhat flat or even decline, while population in the less-developed areas will increase dramatically. What implications does this have for feeding the world and protecting people's health?

Data: United Nations Population Division. 2012. "World Population Trends, 2012 Revision." **www.prb.org**

Birthrate

The **crude birthrate** (or **birthrate**) of a population is the number of babies born each year for every 1000 members of the population or, alternatively, the number of births divided by the total population, multiplied by 1000. It is labeled crude because it does not take into account age or sex differences:

$$\text{Crude birth rate (CBR)} = \frac{\text{number of births}}{\text{total population}} \times 1000$$

Nations vary considerably in their birthrates, with the highest being Niger with 48 births per 1000 people and Japan, Italy, and Hong Kong being among the lowest with only births registering only about 8.0 per 1000 people (World Bank 2017). The birthrate for the United States is 12.4 births per 1000 people now, lower than at any other time in U.S. history. By way of comparison, the birthrate in 1880 was 26.7, but has declined rather steadily since then, with the exception of the years just after World War II when the population now called baby boomers was born (Centers for Disease Control 2017; National Center for Health Statistics 2012; Billings 1896).

The effects of birthrates are somewhat cumulative. For example, minorities tend to be overrepresented at the lower end of the socioeconomic scale, compounding the likelihood of a high birthrate. Similarly, religious and cultural differences affect the birthrate. Catholics, for example, have a higher birthrate than non-Catholics of the same socioeconomic status. Hispanic Americans have a high likelihood of being Catholic, another factor that contributes to the higher birthrate among Hispanic Americans. Projections that the United States will have a significantly greater proportion of minorities are based on births, deaths, and migration rates.

Death Rate

The **crude death rate** (or **death rate**) of a population is the number of deaths each year per 1000 people, or the number of deaths divided by the total population, times 1000:

$$\text{Crude death rate (CDR)} = \frac{\text{number of deaths}}{\text{total population}} \times 1000$$

The death rate can be an important measure of the overall standard of living for a population. In general, the higher the standard of living enjoyed by a country, or a group within the country, the lower the death rate. The death rate of a population also reflects the quality of medicine and health care. Poor medical care, which goes along with a low standard of living, will correlate with a high death rate. The death rate can be an important indicator of a population's overall standard of living. In general, the higher the standard of living, the lower the death rate.

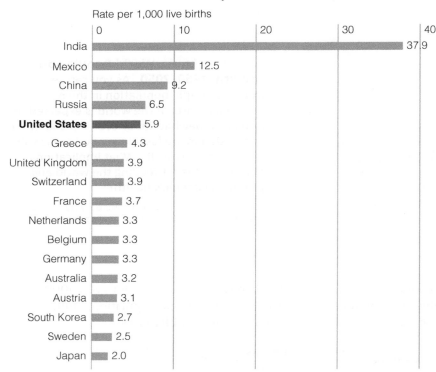

Infant Mortality in Industrial Nations, 2015

Rate per 1,000 live births

Country	Rate
India	37.9
Mexico	12.5
China	9.2
Russia	6.5
United States	5.9
Greece	4.3
United Kingdom	3.9
Switzerland	3.9
France	3.7
Netherlands	3.3
Belgium	3.3
Germany	3.3
Australia	3.2
Austria	3.1
South Korea	2.7
Sweden	2.5
Japan	2.0

▲ **Figure 16-5   Infant Mortality in Industrial Nations.** You may be surprised to see that the United States ranks high in its rate of infant mortality relative to other industrialized nations. What might explain this?

Sources: Organization for Economic Development and Cooperation (OECD). 2017. **https://data.oecd .org/healthstat/infant-mortality-rates.htm;** Centers for Disease Control. 2017. **www.cdc .gov/nchs/pressroom/sosmap/infant_mortality _rates/infant_mortality.htm;** World Data Atlas. 2017. "Infant Mortality." **https://knoema.com /atlas/ranks/Infant-mortality-rate?baseRegion =JP**

In nations with a poor standard of living, infant mortality is typically high. The **infant mortality rate** is measured by the number of deaths per year of infants less than one year old for every 1000 live births. In the United States, the overall infant mortality rate is generally low (5.96 per 1000 live births in 6.1 in 2013), although not compared to other industrialized nations, as you can see in ▲ Figure 16-5.

Infant mortality rates, a measure of the chances of the very survival of members of the population, are important to compare across racial–ethnic groups and across social class strata. The relatively higher rate of infant mortality in the United States stems in large part from the poverty and inequality that exist, especially among racial and ethnic minorities and also the White poor. Infant mortality is a good indicator of the overall quality of life, as well as the survival chances for members of that racial or class group. There are also many other causes of higher infant mortality, such as presence of toxic wastes, malnutrition of the mother, inadequate food, and outright starvation.

Migration

Joining the birthrate and death rate as factors in determining the size of a population is the migration of people into and out of the country. We see the impact of migration in current policy debates about immigration. Who should be allowed into a country? Should those who have immigrated illegally be given amnesty and allowed to stay? Should children who came to the United States at a very young age, but now have never known another country, be allowed to attend college by paying in-state tuition? These questions stem from population changes that are rather dramatically shaping the nation's future.

Migration affects society in many ways. Immigration has ebbed and flowed over the years, but some waves of immigration have, at certain times, had a huge impact on society. Of course, the United States has always been a land of immigrants. Only American Indians and Mexicans, settled in what is now the American Southwest, are indigenous people to this land. Of course, one could hardly call African Americans who came here as slaves "immigrants," as their entry was forced. Still, our nation has become a diverse mix of peoples, given the different origins of our population.

In 1924, the National Origins Quota Act encouraged immigration from northern and western Europe (England, France, Germany, Switzerland, and the Scandinavian countries), but discouraged immigration from eastern and southern Europe (Greece, Italy, Poland, Turkey, and eastern European Jews , among others). Despite this openly discriminatory law, millions of eastern Europeans successfully made the journey to Ellis Island and then the U.S. mainland, only to face prejudice, discrimination, and the accusation that they were taking jobs that would have otherwise gone to the already-present White majority.

Unless there is a change in immigration law as the result of the current national debate about immigration, immigration to the United States is governed by the Hart Celler Act of 1965. This law abolished the national origins quotas that had been mandated since 1920. This meant that the doors were open for immigrants from Asia, Africa, and Latin and Central America—places that had been excluded from the prior policies that favored those from northern and western Europe. Neighborhoods are now invigorated and culturally enriched by mosques or Buddhist temples; by whole neighborhoods of Vietnamese, Koreans, or Asian Indians; or by war refugees from Somalia and Bosnia. Whereas immigrants once settled almost entirely in a small number of cities, now immigration is affecting communities throughout the country (Hirschman and Massey 2008). The simple change in law brought by the Hart Celler Act has had an enormous impact on population diversity, the effects of which we see today.

Diversity and Population Change

The composition of a society's population can reveal a tremendous amount about the society's past, present, and future. To begin with, many nations, including the United States, have a striking imbalance in the number of men and women, with many fewer men than would be expected. The **sex ratio** is the number of males per 100 females, or the number of males divided by the number of females, times 100.

$$\text{Sex ratio} = \frac{\text{number of males}}{\text{number of females}} \times 100$$

A sex ratio above 100 indicates there are more males than females in the population; below 100 indicates there are more females than males. A ratio of exactly 100 indicates the number of males equals the number of females.

In almost all societies, there are more boys born than girls, but because males have a higher infant mortality rate and a higher death rate after infancy, there are usually more females in the overall population. In the United States, approximately 105 males are born for every 100 females, thus giving a sex ratio for live births of 105. After factoring in male mortality, the sex ratio for all ages for the entire country ends up being 94; there are 94 males for every 100 females.

The *age composition* of the U.S. population is presently undergoing major changes. More and more people are entering the sixty-five and older age bracket. This trend is known as the *graying of America*. The elderly are now the largest population category in our society. As our society gets grayer, older people have more influence on national policy and a greater say in matters such as health care and housing.

Population pyramids are graphic depictions of the age and sex distribution of a given population at a point in time (see ▲ Figure 16-6). The bulge in the pyramid for those now in their late sixties and seventies represents the baby boom generation. As the baby boomers age, the bulge will continue rising toward the top of the pyramid, to be replaced underneath by whatever birth trends occur in the coming years. You can also see another bulge for the so-called *millennial generation* (those born after 1980 and reaching adulthood in the twenty-first century). How are these generational structures likely to affect society and its institutions over time as these "bulges" in population move upward?

The bulges you are seeing in Figure 16-6 are *birth cohorts*. A **cohort** consists of all the people born within a given period. A cohort can include all people born within the same year, decade, or other time period. Over time, cohorts either stay the same size or get smaller owing to deaths, but can never grow larger. If we have knowledge of the death rates for this population, we can predict quite accurately the size of the cohort as it passes through the stages of life from infancy to old age. This enables us to predict things such as how many people will enter the first grade in a given period, how many are likely to enroll in college, and how many will arrive at retirement decades down the

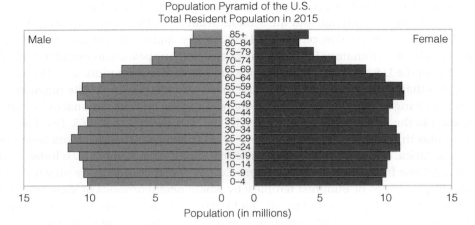

▲ **Figure 16-6   The Age–Sex Pyramids.** A graphic depiction, such as this population pyramid, can capture social realities—such as generational differences—that might drive important issues for public policy. Here, you can see two "bulges" in the U.S. population. How do you think these two populations might have a different stake in national debates about issues such as Medicare, taxes, Social Security, youth programs, and the like?

Source: Rogers, Luke. 2016 (June 23). "America's Age Profile Told through Population Pyramids." Washington, DC: U.S. Census Bureau. **www.census.gov/newsroom/blogs/random -samplings/2016/06/americas-age-profile-told-through-population-pyramids.html**

road. Administrators of social entities such as schools and pension funds can prepare on the basis of cohort predictions.

The United States has long had a diverse population. Even as early as the first U.S. census in 1790, African Americans were 20 percent of the U.S. population, higher than now (at 13 percent). Most were slaves, brought by forced labor, although there was a class of free people (8 percent of the Black population in 1790). Diverse immigrant groups have long characterized the mosaic of the U.S. population, but at no other time has our nation seen as much diversity as now.

What Would a Sociologist Say?

The End of the White Majority?

As the U.S. population is becoming more diverse, many are saying that White people will no longer be the majority. Is this true?

First, even with the population projections indicating that Whites will become a smaller share of the U.S. population, Whites will still constitute the largest racial–ethnic group in the nation.

Second, and perhaps more importantly, from a sociological perspective, the terms *majority* and *minority* do not refer to numbers alone, as you learned in Chapter 10. Sociologically speaking, *majority* refers to a group that holds political, social, and economic power over others—and that group can, indeed, be a small percentage of the population (as we saw in apartheid South Africa when Whites, who had total rule over the whole population, were a mere 10 percent of the population). Even with the increased presence of people of color in social, political, cultural, and economic institutions, Whites are still the dominant group—in terms of power, privilege, and prestige.

Third, White is not a monolithic category. White people are diverse by many social–demographic characteristics—including age, gender, social class, region of residence—and all of these social facts affect the degree to which White people actually hold any power at all! So the next time you hear someone saying that White people are the new minority, you should have the sociological tools to challenge that assertion.

Racial and ethnic diversity is being driven by two major changes, one being immigration. But racial–ethnic groups also vary in basic population matters such as fertility and mortality. Were you to make separate population pyramids for White, Black, Asian, American Indian, and Hispanic people, you would see that these populations are significantly younger and thus more likely to be a larger proportion of the population into the future. Just as examples, the median age of the White, non-Hispanic population is forty-three years of age. For African Americans, it is thirty-three years of age, for Native Americans, thirty-one years, and for Hispanics, twenty-eight years. For Asian Americans, the median age is thirty-six (Gao 2016).

These facts result in what is now being called a population that is "majority minority," realizing that, in a sociological sense, minority refers not just to numbers in a population but also to the unequal treatment of diverse groups. The non-Hispanic White population will peak at about 200 million in 2024 but is predicted to decline by about 10 percent by 2060, comprising 43 percent of the population by then. Meanwhile, the Hispanic population is projected to double such that by 2060, one-third of the U.S. population will be Hispanic. The African American share of the population is projected to be fairly similar in proportional size, reaching a slight increase to about 13 percent by 2060. American Indians and Alaska Natives will remain mostly steady at about 1.5 percent of the population. Native Hawaiians and other Pacific Islanders will likely double, as will Asian Americans who will become 8 percent of the population by 2060. Although non-Hispanic Whites will remain the largest single group, no one of these groups will be a numerical majority (Ewert 2015; see also ▲ Figure 16-7).

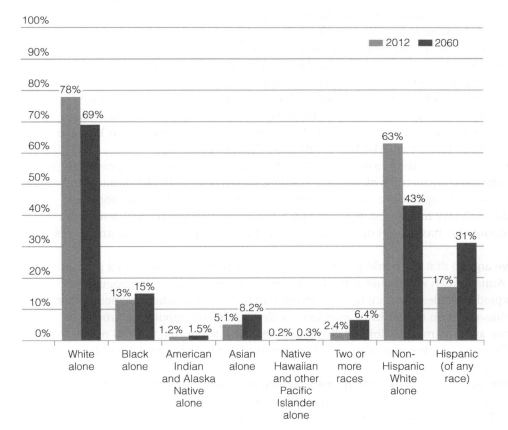

▲ **Figure 16-7   Population Projections: A More Diverse Nation, 2012 and 2060.** This graph shows the projected changes in the population that are predicted by 2060. What social changes do you think these demographic changes might bring?

Source: U.S. Census Bureau. 2012. "U.S. Census Bureau Projections Show a Slower Growing, Older, More Diverse Nation a Half Century from Now." Washington, DC: U.S. Census Bureau. **www.census.gov /newsroom/blogs/random-samplings /2016/06/americas-age-profile-told -through-population-pyramids.html**

These population data will mean significant changes in not just the composition of the U.S. population, but also in the social issues that the nation faces. Will government be more representative? Will intergroup tensions increase or decrease as the nation grapples with a more diverse population? What strains will a younger racial–ethnic population put on school systems? What adjustments must people make in order to work, live, and learn among such population diversity? These and other questions will stem from the diversity in population growth that we are already witnessing.

Population Growth: Are There Too Many People?

Among the major problems facing modern-day civilization is the specter of uncontrolled population growth. Some view overpopulation as a colossal catastrophe about to roll over us like a tidal wave. Others dispute whether the problem exists at all, explaining that there is no scientific consensus on the *carrying capacity* of the planet, the number of people the planet can support on a sustained basis, and that technological advances that have dependably met our needs in the past can be counted on to do so in the future as the number of mouths to feed continues to grow. What can we expect from the future?

A Population Bomb?

Over 250 years ago, Thomas R. Malthus (1798/1926), a Scotch clergyman, predicted disastrous population growth. He thought that populations grow faster than society can sustain. Malthus noted that populations tend to grow not by *arithmetic increase*, adding the same number of new individuals each year, but by *exponential increase*, in which the number of individuals added each year grows, with the larger population generating an even larger number of births with each passing year. Exponential increase, in contrast to arithmetic increase, causes a population to grow ever faster. Malthus predicted widespread catastrophe and famine.

Malthus reasoned that the only checks on population growth were famine, disease, and war. In Malthus's time, disease could reach apocalyptic scales. The outbreak of bubonic plague in Europe from 1334 to 1354 eliminated one-third of the population. A smallpox epidemic in 1707 wiped out three-fourths of the populations of Mexico and the West Indies. Wars took a toll on European men, with deaths in battle producing gaps in the population pyramids of European populations.

Malthusian theory actually predicted rather well the population of many early and agrarian societies. He failed, however, to foresee three revolutionary developments that derailed his predictions of growth and catastrophe. Technological advances have permitted the production of more food, resulting in subsistence levels higher than Malthus would have predicted. Medical science has fought off diseases that Malthus expected to periodically wipe out entire nations. The development of contraceptives has kept the birthrate in many countries at a level lower than Malthus would have thought possible.

Is the specter that Malthus envisioned likely to occur? Viral epidemics warn us that disease can still wipe out huge populations. Heartrending pictures of swollen, starving babies remind us that famine and starvation continue to destroy human populations in some parts of the world just as they have for thousands of years. Overall, Malthus's theory has served as a warning that subsistence and natural resources are limited. The Malthusian doomsday has not yet occurred, but some believe that Malthus's warning was not in error, just premature.

More recent thinkers have argued that the world population cannot continue to expand at its present rates (Ehrlich 1968). Paul and Anne Ehrlich were some of the first modern analysts to argue that population growth is a bomb ready to explode and destroy society if we do not halt it. They and others proposed **zero population growth** (ZPG), achieved when the birthrate matches the death rate and population growth is not influenced by other factors, such as immigration. Ehrlich and others who supported the ZPG movement thought that overpopulation was at the root of many of the country's and world's social problems—poverty, pollution, and violence.

Zero population growth has been reached in some nations. In fact, in some parts of the world, southern and western Europe in particular, population is expected to decline by mid-century. Of course, whether population estimates become true depends in large part on whether there are changes in fertility patterns. The most developed countries have had "below-replacement" levels of fertility for two or three decades already, including, among others, China, the United States, Brazil, the Russian Federation, Japan, Vietnam, Germany, the Islamic Republic of Iran, and Thailand, in order of population size (United Nations 2013). What this will mean for world affairs and the social structures of these societies remains to be seen.

Demographic Transition Theory

Demographic transition theory proposes that countries pass through a consistent sequence of population patterns linked to the degree of development in the society, ending with relatively low birthrates and death rates (Davis 1945). Overall, according to this theory, the population level will eventually stabilize.

Demographic transition theory has three main stages to population change (see ▲ Figure 16-8). Stage 1 is characterized by a high birthrate and high death rate. The United States during its colonial period was in this stage. Women were bearing children at a younger age, and it was not uncommon for a woman to have twelve or thirteen children—a very high birthrate. Infant mortality was also high, as was the overall death rate (both for infants and their mothers), owing to primitive medical techniques and unhealthy sanitary conditions.

Stage 2 in the demographic transition is characterized by a high birthrate but a declining death rate. As a result, the overall level of the population increases. The United States entered stage 2 in the second half of the nineteenth century as industrialization took hold in earnest. The norms of the day continued to encourage large families, thereby causing high birthrates, while advances in medicine and public sanitation whittled away at the infant mortality rate and the overall death rate. Life expectancy increased, and the population grew.

Stage 3 of the demographic transition is characterized by a low birthrate and low death rate. The overall level of the population tends to stabilize in stage 3. Medical advances continue, the general prosperity of the society is reflected in lowered death rates, and cultural changes take place, such as a reduction in family size. The United States entered this stage prior to the Second World War, and with the notable exception of the baby boom, has exhibited stage 3 demographics since then.

Demographic transition theory analyzes the overall population growth in a given society. The different stages may not apply in the same way to diverse population groups within society. High infant mortality rates among the poor, inequality in access to health care, higher death rates among minority populations, and other social factors influencing the experiences of diverse groups mean that population dynamics vary within, as well as between, societies.

It should be apparent by now that population size has an important social dimension. Social forces can

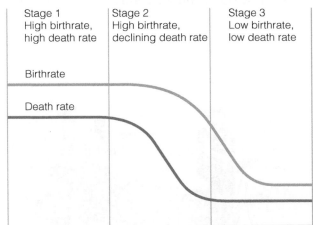

▲ **Figure 16-8   Demographic Transition Theory**

Sources: Davis, Kingsley. 1945. "The World Demographic Transition." *Annals of the American Academy of Political and Social Sciences* 237: 1–11; Coale, Ansley. 1986. "Population Trends and Economic Development." pp. 96–104 in *World Population and the U.S. Population Policy: The Choice Ahead,* edited by J. Menken. New York: W. W. Norton; Weeks, John R. 2012. *Population: An Introduction to Concepts and Issues,* 11th ed. Belmont, CA: Cengage.

cause changes in the size and character of the population, and population changes can likewise trans-form society. In other words, society shapes population, but population also shapes society—a dynamic interaction.

Change: A Multidimensional Process

Population studies focus our attention on the dynamics of social change, but change comes from various sources, not only population change. Currently, we are witnessing extraordinary changes in society—both domestic and global. Consider the following: A gigabyte of information can travel from China to the United States in less time than it takes you to read this paragraph. The reach of electronic communication is amazing, bringing not only innovations in everyday life, but also raising new concerns about privacy and security. Periodic "hacks" into electronic systems have heightened public awareness of the risks—as well as benefits—of technological change. Was all this imaginable fifty years ago? Even thirty years ago? Not really.

Social change is the alteration of social interactions, institutions, stratification systems, and elements of culture over time. Societies are in a constant state of flux. Some changes are rapid, such as the accelerating technological changes like the use of email, Facebook, Twitter, YouTube, and perhaps new systems that will be invented even before this book is published. Other changes are more gradual, such as the increasing urbanization that characterizes the contemporary world. Sometimes people adapt quickly to change, such as the enthusiastic embrace of electronic communication by young people, especially. Other times, people resist change or are slow to adapt to new possibilities. The speed of social change varies from society to society and from time to time within the same society.

Microchanges are subtle alterations in the day-to-day interactions between people. A fad "catching on" is an example of a microchange, such as the flip-flops that are now commonly worn. Not that long ago, flip-flops were just inexpensive sandals, worn largely by poor people. They were introduced into the United States after World War II, modeled on the Japanese sandal, the "zori" (Fortini 2005). Now they have become a common fashion statement.

Macrochanges are gradual transformations that occur on a broad scale and affect many aspects of society. In the process of *modernization*, societies absorb the changes that come with new times and they shed old ways. With modernization, societies develop a more complex division of labor. Inequality may also

Social norms about dress and human activity sometimes make social change more evident when there is historical contrast.

become more multifaceted such that inequality results from multiple social factors, not just a single dimension such as age. Large or small, fast or slow, social change has the following characteristics (Lenski 2005):

1. *Social change is uneven.* The various parts of a society do not all change at the same rate; some parts lag behind others. This is the principle of **culture lag**, a term coined by sociological theorist William F. Ogburn (1922), referring to a delay between social conditions and cultural adjustments to the change. A change in material culture (such as a technological change) may occur more rapidly than happens in nonmaterial culture (meaning the habits and mores of the culture).

2. *The onset and consequences of social change are often unforeseen.* Television pioneers, who envisioned a mode of mass communication more compelling than radio, could not predict television—and now video—would become such a dominant force in determining the interests and habits of people in society.

3. *Social change often creates conflict.* Change often triggers conflicts that may occur along any number of dimensions, such as race and ethnicity, social class, gender, or age. The spread of Western culture into other parts of the world, made possible by the ease of communication, has often resulted in a clash of values between and within nations and can be the basis for international conflict.

4. *The direction of social change is not random.* Change has "direction" relative to a society's history. A populace may want to make a good society better, or it may rebel against a status quo regarded as unsustainable. Whether change is wanted or resisted, when it occurs, it takes place within a specific social and cultural context.

Social change cannot erase the past. As a society moves toward the future, it carries along its past, its traditions, and its institutions. A satisfied populace that strives to make a good society better obviously wishes to preserve its past, but even when a society is in revolt against a status quo that is intolerable, the social change that occurs must be understood in the context of the past as much as the future.

Sources of Social Change

The causes of social change are many and varied but fall into several broad areas, including cultural diffusion; technological innovation; the mobilization of people through social movements and collective behavior; and sometimes, war and revolution. We examine each in the following sections.

Cultural Diffusion

Cultural diffusion (as noted in Chapter 2) is the transmission of cultural elements from one society or cultural group to another. Cultural diffusion can occur by trade, migration, mass communications media, and social interaction. Anthropologist Ralph Linton (1936) long ago alerted us to the fact that much of what many people regard as "American" originally came from other lands—cloth (developed in Asia), clocks (invented in Europe), coins (developed in Turkey), and much more.

Expressions and cultural elements found in the English-speaking United States have been harvested from all over the world. Barbecued ribs, originally eaten by Black slaves in the South after the ribs were discarded by White slave owners who preferred meatier parts of the pig, are now a delicacy enjoyed throughout the United States by many ethnic and racial groups. One theorist, Robert Farris Thompson (1993), points out that an exceptionally large range of elements in material and nonmaterial culture that originated in Africa have diffused throughout many groups and subcultures in the United States, including aspects of language, music, dance, art, dress, decorative styles, and even forms of greeting. For example, the expressions *uh-huh* (yes) and *unh-unh* (no) come from West Africa. Cultural diffusion occurs not only from one place to another (such as from West Africa to the United States) but also across time, such as from a community in the past to many diverse ethnic groups in the present.

The immigration of Latino groups into the United States over time has also dramatically altered U.S. culture by introducing new food, music, language, slang, and many other cultural elements. Popular culture in the United States has diffused into many other countries and cultures: Witness the adoption of American clothing styles, rock, rap, hip-hop, and Big Macs in countries such as Japan, Germany, Russia, and China. The Coca-Cola logo can be found in grocery shops worldwide, from the rain forests of Brazil to the ice floes of Norway.

Technological Innovation and the Cyberspace Revolution

Technological innovations can be strong catalysts of social change. The historical movement from agrarian societies to industrialized societies has been tightly linked to the emergence of technological innovations and inventions (see Chapter 5). Inventions often come about because they answer a need in the society that promises great rewards. The waterwheel promised agrarian societies greater power to raise crops despite dry weather, while also saving large amounts of time and labor. It is possible to trace a timeline from the use of the waterwheel to the use of the large hydroelectric dams that power industrialized societies, and along the way find evidence of how each major advance changed society.

In today's world, the most obvious technological change transforming society is the rise of the computer and the subsequent development of desktop—and now hand-held—computing since the 1980s. The invention and development of the Internet and the resulting communication is now called *cyberspace*, which includes the use of computers for communication between people and communication between people and computers. YouTube videos can go "viral" and reach thousands, perhaps millions, in a very short period of time. A Twitter feed can reach millions in a matter of seconds as people "tweet" and "retweet" messages around the world. Unique in its vastness and lack of a required central location, the Internet has very rapidly become so much a part of human communication and social reality that it pervades and has transformed every social institution—educational, economic, political, familial, and religious.

Social Movements and Collective Behavior

Social change does not develop in the abstract. Change comes from the actions of human beings. A **social movement** is a group that acts with some continuity and organization to promote or resist social change in society. You can see the influence of social movements both now and in the past. What would the United States be like had the civil rights movement not been inspired by Mahatma Gandhi's liberation movement in India? How has the #BlackLivesMatter movement mobilized people now? In what ways has the feminist movement changed your life? We are even witnessing a proliferation of social movements now influenced in part by the ability to mobilize people through the use of social media, such as the #MeToo movement.

Social movements involve groups, sometimes quite large and well organized, that act with some continuity and organization to promote or resist change in society (Turner and Killian 1993). Social movements tend to persist over time, even though they may arise spontaneously. Movements though develop a structure, although they also thrive on spontaneity. Social movements often must swiftly develop new strategies and tactics in their quest for change. During the civil rights movement, students improvised the technique of sit-ins, which quickly spread because they succeeded in gaining attention for the activists' concerns. This is a tactic that has recently been used by demonstrators who have staged "die-ins" to protest the police shootings of young, African American men.

Some social movements aim to change individual behaviors, such as the New Age movement that focuses on personal transformation. Other movements aim to change some aspect of society, such as the gay and lesbian movements that have sought to end discrimination and change public attitudes. Some movements use reform strategies, such as by trying to change laws. Others may be more radical, seeking change in the basic institutions of society. Other social movements are reactionary—that is, organized to resist change or to reinstate an earlier social order that participants perceive to be better.

Doing Sociological Research

Who Cares and Why? Fair Trade and Organic Food

Research Question

There has been a notable increase in the public's use of farmers' markets, a greater presence of organic food sections even in mainstream grocery stores, and other indications of the public's growing concern about where one's food comes from and how it is grown. Social movements to enhance awareness of food production have influenced some of this behavior, as has a corporate response to the public's interest in local, safe, and sustainable food production. Given the widespread growth of organic food options, do people purchase food (and other products) because of their political and ethical values? This is what Philip Howard and Patricia Allen wanted to know.

Research Method

Howard and Allen mailed a survey to 1000 randomly selected respondents, asking them to rate five different reasons why they would select food with different "ecolabels." They identified five different labels: humane (meat, dairy, and eggs coming from animals who have not been treated cruelly); living wage (provides wages to workers above poverty level); locally grown; small-scale (supports small farms or businesses); and "made in the USA." They also collected data on demographic variables, such as age, income, level of education, gender, and place of residence. They analyzed the results using sophisticated statistical techniques of regression analysis.

Research Results

First, the researchers noted that their respondents were more likely to be women, older, White, with a higher income, and better educated than the demographic composition of their random sample. This is an important caveat in interpreting the results because the results are not generalizable to the whole population. One-third of their respondents reported purchasing local foods frequently; many fewer bought organic food regularly.

The three most popular interests in purchasing food were buying local, humane treatment, and providing a living wage for food production workers, but there were differences by demographic group. Buying local was even more important for rural residents. For those who bought organic food, humane reasons topped their preferences. Women were more interested in ecolabeling than men; higher-income people were less likely to care about a living wage than were lower-income respondents. Older respondents were more concerned about the influence of corporations on food production.

Conclusions and Implications

Consumers want the food they buy to reflect their political and ethical judgments. From a sociological perspective, you can also see the influence of demographic variables on the decisions people make about purchasing their food. Although not specifically examined in this study, social movements to "buy local," protect animals, and advocate for food safety have also influenced consumer preferences, meaning that there have been significant changes over time in the food choices that people have.

Questions to Consider

1. Examine your own behavior. What influences what you buy to eat?
2. Do political and ethical values influence your choices?
3. To what degree are your choices influenced by corporations and marketing?
4. Does your social location in particular demographic groups influence your eating habits?

Source: Howard, Philip H., and Patricia Allen. 2010. "Beyond Organic and Fair Trade? An Analysis of Ecolabel Preferences in the United States." *Rural Sociology* 75: 244–269.

Social movements do not typically develop out of thin air. For a movement to begin, there must be a preexisting communication network (Freeman 1983). The importance of a preexisting communication network is well illustrated by the beginning of the civil rights movement, which most historians date to December 1, 1955, the day Rosa Parks was arrested in Montgomery, Alabama, for refusing to give up her

Social movements such as the disability rights movement can raise public awareness and result in new forms of social behavior.

seat to a White man on a municipal bus. Although she is typically understood as simply having been too tired to give up her seat that day, Rosa Parks had been an active member of the movement against segregation in Montgomery. When Rosa Parks refused to give up her seat according to plan, the movement stood ready to mobilize. News of her arrest spread quickly via networks of friends, kin, church, and school organizations (Morris 1999; Robinson 1987).

Celebrities can also advance the cause of social movements by bringing visibility to the movement, such as when NFL players (begun by Colin Kaepernick) kneeled during the national anthem to bring attention to the treatment of African Americans in society. Whether you agree with this protest or not, it has brought attention to important social movements for racial justice as well as sparking a national discussion about free speech and the right to protest. What other examples of celebrities championing social movements can you identify?

As movements develop, they quickly establish an organizational structure. The shape of the movement's organization may range from formal bureaucratic structures to decentralized, interpersonal, and egalitarian arrangements. Many movements combine both. Examples of what are now large bureaucracies are the National Organization for Women (NOW), the National Association for the Advancement of Colored People (NAACP), Amnesty International, Greenpeace, the Jewish Defense League, and the National Rifle Association. As social movements become institutionalized, they are most likely to take on bureaucratic form.

Related to social movements is **collective behavior**, behavior that occurs when the usual conventions that guide social behavior are disrupted for some reason and people establish new, usually sudden, norms in response to an emerging situation (Turner and Killian 1993). Although collective behavior may emerge spontaneously, it can be predicted. Some phenomena defined as collective behavior are whimsical and fun, such as fads, fashions, and certain crowds (flash mobs, for example). Other collective behaviors can be terrifying, as in panics or riots. Whether or not whimsical, collective behavior is innovative, sometimes revolutionary; it is this feature that links collective behavior to social change.

Collective behavior is group, not individual, behavior. A lone gunman, for example, who opens fire in a crowded movie theater is engaged in individual behavior, but the crowd that gathers following this unexpected event is engaged in collective behavior. Collective behavior involves new and emerging relationships that arise in unexpected circumstances. It represents the often novel, but dynamic and changing character of society.

War and Revolution

A **revolution** is the overthrow of state or the total transformation of central state institutions. A revolution thus results in far-reaching social change. Numerous sociologists have studied revolutions and identified the conditions under which revolutions are likely to occur. Revolutions can sometimes break down a nation-state and disenfranchise groups or individuals formerly in power. An array of groups in a society may be dissatisfied with the status quo and organize to replace established institutions. Dissatisfaction alone is not enough to produce a revolution, however. The opportunity must exist for the group to mobilize en masse. Revolutions can result when structured opportunities are created, such as through war or an economic crisis or mobilization through a *social movement*.

Die-ins that came in the aftermath of the police shooting of Michael Brown in Ferguson, Missouri, adopted some of the tactics of the nonviolent campaign of the civil rights movement.

Social structural conditions that often lead to revolution can include a highly repressive state—so repressed that a strong political culture develops out of resistance to state oppression. A major economic crisis can also produce revolution—as can the development of a new economic system, such as capitalism—that transforms the world economy.

War and severe political conflict result in large and far-reaching changes for both the conquering society, or a region within a society (as in civil war), and for the conquered. The conquerors can impose their will on the conquered and restructure many of their institutions, or the conquerors can exercise only minimal changes.

A historic example is the U.S. victory over Japan and Germany in the Second World War. The war transformed the United States into a mass-production economy and affected family structures as well in that women who had not previously been employed joined the labor force during the war with many remaining even after the war's end. Many in the armed forces who returned from the war were educated under a scholarship plan called the GI Bill, resulting in a more highly educated population and encouraging more government investments in higher education.

The war also transformed Germany in countless ways, given the vast physical destruction brought on by U.S. bombs and the worldwide attention brought to anti-Semitism and the Nazi holocaust. The cultural and structural changes in Japan were extensive, as well. The decimation of the Jewish population in Germany and other nations throughout Europe resulted in the massive migration of Jews to the United States. The Vietnam War also resulted in many social changes, including the migration of Vietnamese to the United States. Currently, the various wars in the Middle East have meant more migration of Iraqis and Afghanis to the United States. In countless examples, we can see the impact of war on national and international social change.

Theories of Social Change

How do we explain social change? Some believe that change is cyclical, that is, recurring at regular intervals. Cyclical theories build on the idea that societies have a life cycle, like seasonal plants, or at least a life span, like humans. Arnold J. Toynbee, a social historian and a principal theorist of cyclical social change, argues that societies are born, mature, decay, and sometimes die (Toynbee and Caplan 1972). For at least part of his life, Toynbee believed that Western society was fated to self-destruct because he thought energetic social builders would be replaced by entrenched elites who ruled by force. Then, society would wither under these sterile regimes. Some believe that societies become more decrepit, only to be replaced by more youthful societies.

Sociological theories of change are more nuanced than the idea that there are regular life cycles in society. Each of the major sociological perspectives takes a somewhat different approach to theorizing how social change occurs.

Functionalist Theory

Recall from previous chapters that functionalist theory builds on the postulate that all societies, past and present, possess basic elements and institutions that perform certain functions permitting a society to survive and persist. A *function* is a consequence of a social element that contributes to the continuance of a society. For example, the function of an institution such as the family is to provide the society with sufficient population to assure its continuance.

The early theorists Herbert Spencer (1882) and Emile Durkheim (1964/1895) both argued that as societies move through history, they become more complex. Spencer argued that societies move from "homogeneity to heterogeneity." Durkheim similarly argued that societies move from a state of **mechanical solidarity**, a cohesiveness based on the similarity among its members, to **organic solidarity**, a cohesiveness based on difference; a division of labor that exists among its members joins them together, because each depends on the others to perform specialized tasks (see Chapter 5). Through the creation of specialized roles, structures, and institutions, societies thus move from being less differentiated to greater social differentiation.

According to functional theorists, societies that are structurally simple and homogeneous, such as foraging or pastoral societies, where all members engage in similar tasks, move to societies more structurally complex and heterogeneous, such as agricultural, industrial, and postindustrial societies, where great social differentiation exists in the division of labor among people who perform many specialized tasks. The consequence (or function) of increased differentiation and division of labor is a higher degree of stability and cohesiveness in the society, brought about by mutual dependence (Parsons 1966, 1951a).

Conflict Theory

Karl Marx (1967/1867), the founder of conflict theory, emphasized the role of economic influences in how societies change. He argued that societies could indeed "advance," but that advancement was measured by the movement from a class society to a society with no class structure. Marx believed that, along the way, class conflict was inevitable.

The central notion of conflict theory, as we have seen, is that conflict is inherently built into social relations (Dahrendorf 1959). For Marx, social conflict, particularly between the two major social classes—working class versus upper class, proletariat versus bourgeoisie—was not only inherent in social relations but was indeed the driving force behind all social change. Marx believed that the most important causes of social change were the tensions between social groups, especially those defined along social class lines. Different classes have different access to power, with the relatively lower class carrying less power.

Although Marx was referring specifically to social classes, subsequent interpretations include conflict between any socially distinct groups that receive unequal privileges and opportunities. The central idea of conflict theory is the notion that social groups will have competing interests and that conflict is an inherent part of any society.

Symbolic Interaction Theory

Because symbolic interaction focuses on more micro-level behaviors, it does not explain the wide-scale social changes that other theorists analyze. Symbolic interaction does, though, contribute to understanding social change, especially in how it emphasizes the meaning that people attach to social behavior. Symbolic interaction theorists might, for example, look at the quixotic character of fads and fashion and how people adapt and respond to these behaviors. Fads, for example, emerge, tend to spread rapidly, but then typically vanish (Best 2006). How, for example, did UGG boots become such a symbol of social status? Fads, such as this one, come and go—sometimes passing quickly, even when they are widely popular. Perhaps by the time this book is published, UGGs will be considered passé or even normative and some wholly unpredictable other status symbol will produce changes in how people dress. Can you identify fads from your earlier years that now have vanished from the public scene?

Symbolic interaction theory thus looks at change in terms of how people define social behaviors and how those definitions emerge. Of course, with fads, much of that definition comes through the work of marketing and the sponsorship of the corporate world. Social change, though, also comes from the behaviors of people as they reformulate their ideas and attitudes. Changes in public attitudes and how people think are signs of social change, as symbolic interaction theory would point out.

Globalization and Modernization: Shaping Our Lives

Globalization is the increased interconnectedness and interdependence of numerous societies around the world. No longer can the nations of the world be viewed as separate and independent societies. The irresistible current trend has been for societies to develop deep dependencies on each other, with interlocking economies and social customs. In Europe, this trend proceeded as far as developing a common currency, the *euro*, for all nations participating in the constructed common economy.

As the world becomes increasingly interconnected, does this mean we are moving toward a single, homogeneous culture? In such a culture, electronic communications, computers, and other developments can erase the geographic distances between cultures and, eventually, the cultural differences. As societies become more interconnected, cultural diffusion between them creates common ground, while cultural differences may become more important as the relationships among nations become more intimate. The different perspectives on globalization are represented by three main theories that we will review: modernization theory, world systems theory, and dependency theory, which are included in ◆ Table 16-1.

As societies grow and change, they become more modern in a general sense. **Modernization** is a process of social and cultural change initiated by industrialization and followed by increased social differentiation and division of labor. Societies can, of course, experience social change without industrialization. Modernization has positive consequences, such as improved transportation and a higher gross national product, but also negative consequences, such as pollution, elevated stress, and perhaps social alienation.

Modernization has three general characteristics (Berger et al. 1974):

1. *Modernization is typified by the decline of small, traditional communities.* The individuals in foraging or agrarian societies live in small-scale settlements with their extended families and neighbors. The primary group is prominent in social interaction. Industrialization causes an overall decline in the importance of primary group interactions and an increase in the importance of secondary groups, such as colleagues at work.

2. *With increasing modernization, a society becomes more bureaucratized.* Interactions come to be shaped by formal organizations. Traditional ties of kinship and neighborhood feeling decrease, and members of the society tend to experience feelings of uncertainty and powerlessness.

3. *There is a decline in the importance of religious institutions.* With the mechanization of daily life, people begin to feel that they have lost control of their own lives: People may respond by building new religious groups and communities (Wuthnow 1994).

From Community to Society

The German sociologist Ferdinand Tönnies (1855–1936) formulated a theory of modernization that still applies to today's societies (Tönnies 1963/1887). Tönnies viewed the process of modernization

Table 16-1	Theories of Social Change		
	Functionalist Theory	**Conflict Theory**	**Symbolic Interaction Theory**
How do societies change?	Societies change from simple to complex and from an undifferentiated to a highly differentiated division of labor.	Conflict is inherent in social relations, and society changes from a class-based to a classless society.	Social change occurs when new meaning systems develop around people's behaviors and attitudes.
What is the primary cause of social change?	Technological innovation and globalization make society more differentiated but still stable.	Economic inequality drives social change.	Changes in people's attitudes and beliefs drive social change.
What is the impact of social change on individuals?	Individuals remain integrated into the whole because society seeks equilibrium.	Individuals are faced with conflict, but the powerless may organize to drive social change.	People have to adapt to new understandings that emerge from social change.

as a progressive loss of *gemeinschaft* (German for "community"), a state characterized by a sense of common feeling, strong personal ties, and sturdy primary group memberships, along with a sense of personal loyalty to one another. Tönnies argued that the Industrial Revolution, which emphasized efficiency and task-oriented behavior, destroyed the sense of community and personal ties associated with an earlier rural life. At the crux of this was a society organized on the basis of self-interest, which caused the condition of *gesellschaft* (German for "society"), a kind of social organization characterized by a high division of labor, less prominence of personal ties, the lack of a sense of community among the members of society, and the absence of a feeling of belonging—maladies often associated with modern urban life.

The United States has become a gesellschaft, with social interaction less intimate and less emotional, although certain primary groups such as the family and friendship groups still permit strong emotional ties. Tönnies noted that the role of the family is less prominent in a gesellschaft than in a gemeinschaft. Patriarchy is less prominent, yet more public, and more women are employed outside the home. In the large cities that characterize the gesellschaft, people live among strangers and pass people on the street who are unfamiliar. In a gemeinschaft, people have already seen most of the people they encounter. The level of interpersonal trust is lower in a gesellschaft. Social interaction tends to be even more confined within ethnic, racial, and social class groups. To find personal contact and to satisfy the need for intimate interaction, individuals often join small church groups, training groups, or personal awareness groups or movements.

Urbanization

One consequence of the elaboration of society is the growth of cities. Of course, cities have always been important sites for population density, commerce, and a bustling social life. Scholars locate the development of the first city around 3500 B.C. (Flanagan 1995).

The study of the urban, the rural, and the suburban is the task of *urban sociology*, a subfield of sociology that examines the social structure and cultural aspects of the city compared with rural and suburban centers. These comparisons involve what urban sociologist Gideon Sjøberg (1965) calls the *rural–urban continuum*, those structural and cultural differences that exist as a consequence of differing degrees of urbanization. **Urbanization** is the process by which a community acquires the characteristics of city life and the "urban" end of the rural–urban continuum.

Early German sociological theorist Georg Simmel (1950/1902) argued that urban living had profound social psychological effects on individuals. He was among the early theorists who argued that social structure could affect individuals. Simmel argued that urban life has a quick pace and is stimulating, but as a consequence of this intense style of life, individuals become insensitive to people and events that surround them. Urban dwellers tend to avoid the emotional involvement that, according to Simmel, was more likely found in rural communities. Interaction tends to be characterized as economic rather than social, and close, personal interaction is frowned upon and discouraged. Urban dwelling can, however, increase the likelihood of other ills: Emile Durkheim noted that the suicide rate per 10,000 people was greater in more urbanized areas than in rural areas (Durkheim 1951/1897).

The sociologist Louis Wirth (1928), focusing on Chicago in the 1920s, also argued that the city was a center of distant, cold interpersonal interaction, and as a result, urban dwellers experienced alienation, loneliness, and powerlessness. One positive consequence of all this, according to Wirth and Simmel, was the liberating effect that arises from the relative absence of close, restrictive ties and interactions. City life, in Wirth's thought, offered individuals a certain feeling of freedom.

A contrasting view of urban life is offered by Herbert Gans (1982/1962), who studied people in Boston in the late 1950s and concluded that many city residents develop strong loyalties to others and are characterized by a sense of community. Such subgroupings he referred to as the *urban village*, which is characterized by several "modes of adaptation," among them *cosmopolites*—typically students, artists, writers, and musicians, who together form a tightly knit community and choose urban living to be near the city's cultural facilities. A second category includes the *ethnic villagers*, people who live in ethnically and racially segregated neighborhoods. Such *urban enclaves* tend to develop their own unique identities, such as San Francisco's Chinatown or Miami's Little Havana.

The Cosmopolitan Canopy

Fascinated by the social structure of urban life, sociologist Elijah Anderson defines the *cosmopolitan canopy* as public spaces in urban areas where diverse groups gather and civility is the norm. In such spaces, urban dwellers and visitors interact with grace and respect, different from the usual tensions and distrust that also characterize intergroup relationships.

Anderson studied three neighborhoods in urban Philadelphia—a market, a mall, and a park—with an eye to how diverse groups of people otherwise separated by race, class, and other social truths, interact in public spaces. His careful ethnographic research provides a counterpoint to the usually dreary account of urban blight, poverty, slums, and racial strife.

In cosmopolitan canopies, people interact across racial and class lines with respect and acceptance. Although such interactions are relatively impersonal, they nonetheless reveal the potential for positive group relationships. Anderson asks if people can relate to one another with respect and acceptance in these public spacces, how can such social norms be extended to spaces beyond the cosmopolitan canopy?

Anderson also shows how easily civility can also break down. As Anderson writes, "The promise of the cosmopolitan canopy is challenged by recurring situational racial discrimination or by the occasional racial incident" (p. 157). Still, Anderson is an optimist and believes that some of the civil and respectful behaviors that take place under cosmopolitan canopies can encourage better intergroup relationships.

Perhaps you can think of places in your community that form cosmopolitan canopies. What social interactions occur there? How diverse are the interactions? What would it take for such civil behaviors to be the norm in other social environments?

Source: Anderson, Elijah. 2011. *The Cosmopolitan Canopy: Race and Civility in Everyday Life.* New York: W.W. Norton.

Urban spaces are among some of the nation's most diverse spaces, even while cities remain highly segregated by race and by class. Even with urban racial and class segregation, however, you can observe places in cities where highly diverse groups assemble and interact (see the box, "Understanding Diversity: The Cosmopolitan Canopy").

Social Inequality, Powerlessness, and the Individual

Another product of modernization, along with mass society, is pronounced social stratification, according to theorists such as Karl Marx (1967/1867) and Jurgen Habermas (1970). In their view, the personal feelings of powerlessness that accompany modernization are due to social inequalities related to race, ethnicity, class, and gender stratification. Marx argued that inequalities are the inevitable product of the capitalist system. Habermas argued that inequalities are the cause of social conflict.

The social structural conditions that arise from modernization, such as increased social stratification, affect individual people's lives. Building a stable personal identity is difficult in a highly modernized society that presents individuals with complex and conflicting choices about how to live. Many individuals flounder among lifestyles while searching for personal stability and a sense of self. According to Habermas, individuals in highly modernized environments are more likely than their less modernized peers to experiment with new religions, social movements, and lifestyles in search of a fit with their conception of their own "true self." These individual responses to social structural conditions reveal how the social structure can affect personality.

The influential social theorist Herbert Marcuse (1964) has argued that modernized society fails to meet the basic needs of people, among them the need for a fulfilling identity. In this respect, modern society and its attendant technological advances are not stable and rational, as is often argued, but unstable and irrational. The technological advances of modern society do not increase the feeling of having control over one's life, but instead reduce that control and lead to feelings of powerlessness.

Powerlessness leads to the *alienation* of individuals from society—individuals experience feelings of separation from the group or society. This alienation is most likely to affect people traditionally denied access to power, such as racial minorities, women, and the working class. This alienation from the highly modernized, technological society is, in Marcuse's view, one of the most pressing problems of civilization today. Marcuse argues that, despite the popular view that technology is supposed to yield efficient solutions to the world's problems, it may be more accurate to say that technology is a primary cause of many problems in modern society.

Chapter Summary

What is environmental sociology?

Any society is a human ecosystem with interacting and interdependent forces, consisting of human populations, natural resources, and the state of the environment. *Climate change* is the systematic increase in worldwide surface temperatures and the resulting ecological change. Although some deny that climate change is the result of human behavior, there is little doubt among scientists that climate change is largely the result of human activity.

What is meant by environmental racism?

Environmental racism is the pattern whereby toxic dumps are found more frequently in or very near African American, Hispanic, and Native American communities. Disasters also have a social dimension that can make certain populations more vulnerable than others.

What are the basic dimensions of population development?

Demography is the scientific study of population. The total number of people in a society at any given moment is determined by only three variables: births, deaths, and migrations. Diversity in a population occurs along lines of sex and age, as well as race and ethnicity. Currently, the United States is rapidly becoming more diverse because of immigration, but also fertility and mortality rates that vary among different groups.

How do sociologists explain population growth?

Malthusian theory warns us about the dangers of exponential population growth that only such calamities as famine and war could prevent. Others have also warned that there could be a population bomb, with

the size of the population outpacing the capacity to support people. *Demographic transition theory* postulates that societies pass through sequences of population patterns linked to developmental stages affecting the birthrates and death rates.

What are the different sources of social change?

Social change is that process by which social interaction, the social stratification system, and entire institutions in a society change over time. Sources of social change include *cultural diffusion*, technological innovation, *social movements/collective behavior*, and war and *revolution*.

What sociological theories explain social change?

Functionalist theories explain that societies move or evolve from the structurally simple to the structurally complex. *Conflict theories* predict that social conflict is an inherent part of any social structure and that conflict between social class strata or racial–ethnic groups can bring about social change. *Symbolic interaction theory* studies micro-level processes by which people attribute new meanings to beliefs and behaviors.

What changes are brought about by globalization and modernization?

Globalization is the increased interconnectedness and interdependence of societies around the world. It is being coupled with *modernization*, a process of social and cultural change initiated by industrialization and resulting in increased social differentiation and a strong division of labor. Modernization can produce feelings of powerlessness among individuals as society becomes more stratified and complex.

Key Terms

birthrate (or crude birthrate) 435

census 433

climate change 428

cohort 437

collective behavior 446

cultural diffusion 443

culture lag 443

death rate (or crude death rate) 435

demographic transition

theory 441

demography 433

emigration 434

environmental racism 432

environmental sociology 426

globalization 449

immigration 434

infant mortality rate 436

mechanical solidarity 448

modernization 449

organic solidarity 448

population density 434

population pyramids 437

revolution 446

sex ratio 437

social change 442

social movement 444

urbanization 450

zero population growth 440

GLOSSARY

A

absolute poverty the amount of money needed in a particular country to meet basic needs of food, shelter, and clothing.

achieved status a status attained by effort.

achievement test test intended to measure what is actually learned rather than potential.

adult socialization the process of learning new roles and expectations in adult life.

affirmative action a method for opening opportunities to women and minorities that specifically redresses past discrimination by taking positive measures to recruit and hire previously disadvantaged groups.

Affordable Care Act the nation's health care reform law that has extended some health care insurance to larger segments of the U.S. population.

age cohort an aggregate group of people born during the same time period.

age discrimination different and unequal treatment of people based solely on their age.

age prejudice a negative attitude about an age group that is generalized to all people in that group.

age stereotype preconceived judgments about what different age groups are like.

age stratification the hierarchical ranking of age groups in society.

ageism the institutionalized practice of age prejudice and discrimination.

alienation the feeling of powerlessness and separation from one's group or society.

altruistic suicide the type of suicide that can occur when there is excessive regulation of individuals by social forces.

Americans with Disabilities Act law passed in 1990, stipulating that employers and other public entities must provide "reasonable accommodation" to people with disabilities when they are otherwise qualified for the job or activity.

anomic suicide the type of suicide occurring when there are disintegrating forces in the society that make individuals feel lost or alone.

anomie the condition existing when social regulations (norms) in a society break down.

anticipatory socialization the process of learning the expectations associated with a role one expects to enter in the future.

anti-Semitism the belief or behavior that defines Jewish people as inferior and that targets them for stereotyping, mistreatment, and acts of hatred.

ascribed status a status determined at birth.

assimilation theory the process by which a minority group becomes absorbed into a host society.

attribution error error made in attributing the causes for someone's behavior to their membership in a particular group, such as a racial group.

attribution theory the principle that dispositional attributions are made about others (what the other is "really like") under certain conditions, such as out-group membership.

audit study research technique that poses at least two people, identical in nearly all respects, to test whether discrimination occurs when one of the pair (or teams) present themselves as potential tenants or employees. Such studies test for such things as race, sex, or age discrimination.

authoritarian where power is concentrated in the hands of a very few individuals who rule through centralized power and control.

authority power that is perceived by others as legitimate.

automation the process by which human labor is replaced by machines.

autonomous state model a theoretical model of the state that interprets the state as developing interests of its own, independent of other interests.

aversive racism subtle, nonovert, and nonobvious racism.

B

beliefs shared ideas held collectively by people within a given culture.

bilateral kinship a kinship system where descent is traced through the father and the mother.

biological determinism explanations that attribute complex social phenomena to physical characteristics.

birthrate the number of babies born each year for every 1000 members of the population.

Brown v. Board of Education the 1954 Supreme Court decision that ruled separate but equal public facilities to be unconstitutional.

bureaucracy a type of formal organization characterized by an authority hierarchy, a clear division of labor, explicit rules, and impersonality.

C

capitalism an economic system based on the principles of market competition, private property, and the pursuit of profit.

caste system a system of stratification (characterized by low social mobility) in which one's place in the stratification system is determined by birth.

census a count of the entire population of a country.

charisma a quality attributed to individuals believed by their followers to have special powers.

charismatic authority authority derived from the personal appeal of a leader.

church a formal organization that sees itself and is seen by society as a primary and legitimate religious institution.

Civil Rights Act of 1964 federal law prohibiting discrimination on the basis of race, color, national origin, religion, or sex.

class *see* social class.

class consciousness the awareness that a class structure exists and the feeling of shared identification with others in one's class with whom one perceives common life chances.

class system the organized pattern of social class in society.

climate change the systematic increase in worldwide surface temperatures and the resulting ecological change.

coalition an alliance formed by two or more individuals or groups against another individual or one or more groups to achieve certain ends.

coercive organization organizations for which membership is involuntary; examples are prisons and mental hospitals.

cohort (birth cohort) *see* age cohort.

collective behavior (action)   behavior that occurs when the usual conventions are suspended and people collectively establish new norms of behavior in response to an emerging situation.

collective consciousness   the body of beliefs that are common to a community or society and that give people a sense of belonging.

colonialism   system by which Western nations became wealthy by taking raw materials from other societies (the colonized) and reaping profits from products finished in the homeland.

colorblind racism   ignoring legitimate racial, ethnic, and cultural differences between groups, thus denying the reality of such differences.

coming out   the process of defining oneself as gay or lesbian.

commodity chain   the network of production and labor processes by which a product becomes a finished commodity. By following the commodity chain, it is evident which countries gain profits and which ones are being exploited.

communism   an economic system where the state is the sole owner of the systems of production.

concentrated poverty   refers to geographic areas where large percentages of people are poor.

concept   any abstract characteristic or attribute that has the potential to be measured.

conflict theory   a theoretical perspective that emphasizes the role of power and coercion in producing social order.

conspicuous consumption   the ostentatious display of goods to mark one's social status.

content analysis   the analysis of meanings in cultural artifacts such as books, songs, and other forms of cultural communication.

contingent worker   a person who does not hold a regular job, but whose employment is dependent upon demand.

controlled experiment   a method of collecting data that can determine whether something actually causes something else.

core countries (core nations)   within world systems theory, those nations that are more technologically advanced.

correlation   the degree of positive (direct) or negative (inverse) association between two variables.

countercultures   subcultures created as a reaction against the values of the dominant culture.

covert participant observation   the form of participant observation wherein the observed individuals are not told that they are being studied.

crime   one form of deviance; specifically, behavior that violates criminal laws.

criminology   the study of crime from a scientific perspective.

cross-tabulation   a table that shows how the categories of two variables are related.

crude birthrate   *see* birthrate.

crude death rate   *see* death rate.

cults   religious groups devoted to a specific cause or charismatic leader.

cultural capital   (also known as *social capital*) cultural resources that are socially designated as being worthy (such as knowledge of elite culture) and that give advantages to groups possessing such capital.

cultural diffusion   the transmission of cultural elements from one society or cultural group to another.

cultural hegemony   the pervasive and excessive influence of one culture throughout society.

cultural pluralism   situation wherein different groups inn society maintain distinctive cultures, while coexisting peacefully with the dominant group.

cultural relativism   the idea that something can be understood and judged only in relationship to the cultural context in which it appears.

culture   the complex system of meaning and behavior that defines the way of life for a given group or society.

culture lag   the delay in cultural adjustments to changing social conditions.

culture of poverty   the argument that poverty is a way of life and, like other cultures, is passed on from generation to generation.

culture shock   the feeling of disorientation that can come when one encounters a new or rapidly changed cultural situation.

cybercrime   illegal activities that take place through the use of computers.

D

data   the systematic information that sociologists use to investigate research questions.

data analysis   the process by which sociologists organize collected data to discover what patterns and uniformities are revealed.

death rate   the number of deaths each year per 1000 people.

debriefing   a process whereby a researcher explains the true purpose of a research study to a subject (respondent); usually done after completion of the study.

debunking   looking behind the facades of everyday life.

deductive reasoning   the process of creating a specific research question about a focused point, based on a more general or universal principle.

deindividuation   the feeling that one's self has merged with a group.

deindustrialization   the transition from a predominantly goods-producing economy to one based on the provision of services.

democracy   system of government based on the principle of representing all people through the right to vote.

demographic transition theory   argument that countries pass through a consistent sequence of population patterns linked to the degree of development and ending with a low birth and death rate.

demography   the scientific study of population.

dependency theory   the global theory maintaining that industrialized nations hold less-industrialized nations in a dependent relationship that benefits the industrialized nations at the expense of the less-industrialized ones.

dependent variable   the variable that is a presumed effect (*see also* independent variable).

deviance   behavior that is recognized as violating expected rules and norms.

deviant career   continuing to be labeled as deviant even after the initial (primary) deviance may have ceased.

deviant community   groups that are organized around particular forms of social deviance.

deviant identity   the definition a person has of himself or herself as a deviant.

differential association theory   theory that interprets deviance as behavior one learns through interaction with others.

discrimination   overt negative and unequal treatment of the members of some social group or stratum solely because of their membership in that group or stratum.

diversity   the variety of group experiences that result from the social structure of society.

division of labor   the systematic interrelation of different tasks that develops in complex societies.

doing gender   a theoretical perspective that interprets gender as something accomplished through the ongoing social interactions people have with one another.

dominant culture   the culture of the most powerful group in society.

dominant group   the group that assigns a racial or ethnic group to subordinate status in society.

dual labor market   the division of the labor market into two segments—the primary and secondary labor markets.

dual labor market theory   a theory that contends that the labor market is divided into two segments—the primary and secondary labor markets.

dyad   a group consisting of two people.

E

economic restructuring   contemporary transformations in the basic structure of work that are permanently altering the workplace, including demographic changes, deindustrialization, enhanced technology, and globalization.

economy   the system on which the production, distribution, and consumption of goods and services are based.

educational attainment   the total years of formal education.

egoistic suicide   the type of suicide that occurs when people feel totally detached from society.

emigration (*versus immigration*)   migration of people from one society to another (also called out-migration).

emotional labor work that is explicitly intended to produce a desired state of mind in a client.

empirical refers to something that is based on careful and systematic observation.

Enlightenment the period in eighteenth- and nineteenth-century Europe characterized by faith in the ability of human reason to solve society's problems.

environmental racism the dumping of toxic wastes with disproportionate frequency at or very near areas with high concentrations of minorities.

environmental sociology the scientific study of the interdependencies that exist between humans and our physical environment.

Equal Pay Act of 1963 first legislation requiring equal pay for equal work.

Equal Rights Amendment a constitutional principle, never passed, guaranteeing that equality of rights under the law shall not be denied or abridged on the basis of sex.

estate system a system of stratification in which the ownership of property and the exercise of power is monopolized by an elite or noble class that has total control over societal resources.

ethnic group a social category of people who share a common culture, such as a common language or dialect, a common religion, or common norms, practices, and customs.

ethnocentrism the belief that one's in-group is superior to all out-groups.

ethnomethodology a technique for studying human interaction by deliberately disrupting social norms and observing how individuals attempt to restore normalcy.

eugenics a social movement in the early twentieth century that sought to apply scientific principles of genetic selection to "improve" the offspring of the human race.

evaluation research research assessing the effect of policies and programs.

expressive needs needs for intimacy, companionship, and emotional support.

extended families the whole network of parents, children, and other relatives who form a family unit and often reside together.

extreme poverty the situation in which people live on less than $275 a year, or $1.25 a day.

F

false consciousness the thought resulting from subordinate classes internalizing the view of the dominant class.

family a primary group of people—usually related by ancestry, marriage, or adoption—who form a cooperative economic unit to care for any offspring (and each other) and who are committed to maintaining the group over time.

Family and Medical Leave Act (FMLA) federal law requiring employers of a certain size to grant leave to employees for purposes of family care.

fatalistic suicide suicide that occurs when there is complete over-regulation of people in society.

feminism a way of thinking and acting that advocates a more just society for women.

feminist theory analyses of women and men in society intended to improve women's lives.

feminization of poverty the process whereby a growing proportion of the poor are women and children.

folkways the general standards of behavior adhered to by a group.

formal organization a large secondary group organized to accomplish a complex task or set of tasks.

functionalism a theoretical perspective that interprets each part of society in terms of how it contributes to the stability of the whole society.

G

game stage the stage in childhood when children become capable of taking a multitude of roles at the same time.

gemeinschaft German for *community,* a state characterized by a sense of common feeling among the members of a society, including strong personal ties, sturdy primary group memberships, and a sense of personal loyalty to one another; associated with rural life.

gender socially learned expectations and behaviors associated with members of each sex.

gender apartheid the extreme segregation and exclusion of women from public life.

gender-based violence various forms of violence associated with unequal power relationships between men and women.

gender identity one's definition of self as a woman or man.

gender inequality index measure of three key components of women's lives, including reproductive health, empowerment, and labor market status.

gender segregation the distribution of men and women in different jobs in the labor force.

gender socialization the process by which men and women learn the expectations associated with their sex.

gender stratification the hierarchical distribution of social and economic resources according to gender.

gendered institutions the total pattern of gender relations that structure social institutions, including the stereotypical expectations, interpersonal relationships, and the different placement of men and women that are found in institutions.

generalization applying information obtained on a small sample of units (such as people) to a larger population of the units.

generalized other an abstract composite of social roles and social expectations.

gesellschaft a type of society in which increasing importance is placed on the secondary relationships that are less intimate and more instrumental.

Gini coefficient measure of income distribution within a given population.

glass ceiling popular concept referring to the limits that women and minorities experience in job mobility.

global assembly line an international division of labor where research and development is conducted in an industrial country, and the assembly of goods is done primarily in underdeveloped and poor nations, mostly by women and children.

global culture the diffusion of a single culture throughout the world.

global economy term used to refer to the fact that all dimensions of the economy now cross national borders.

global stratification the differences in wealth, power, and prestige of different societies relative to their position in the international economy.

globalization increased economic, political, and social interconnectedness and interdependence among societies in the world.

government those state institutions that represent the population and make rules that govern the society.

gross national income (GNI) the total output of goods and services produced by residents of a country each year plus the income from nonresident sources, divided by the size of the population.

group a collection of individuals who interact and communicate, share goals and norms, and who have a subjective awareness as "we."

group size effect the effect upon the person of groups of varying sizes.

groupthink the tendency for group members to reach a consensus at all costs.

H

hate crime an assault or other malicious act (including crimes against property) motivated by various forms of bias, including that based on race, religion, sexual orientation, ethnic and national origin, or disability.

Hawthorne effect the effect of the research process itself on the groups or individuals being studied; hence, the act of studying them often itself changes them.

heterosexism institutional structures that define heterosexuality as the only social legitimate sexual orientation.

homophobia the fear and hatred of gays and lesbians.

human capital theory a theory that explains differences in wages as the result of differences in the individual characteristics of the workers.

hypersegregation a pattern of extreme racial, ethnic, and/or social class residential segregation, such that nearly all individuals in an area are of one such group.

hypothesis a statement about what one expects to find in research.

I

ideal type model rarely seen in reality but that defines the principal characteristics of a social form.

identity how one defines oneself.

ideology a belief system that tries to explain and justify the status quo.

imitation stage the stage in childhood when children copy the behavior of those around them.

immigration (*versus emigration*) the migration of people into a society from outside it (also called in-migration).

Immigration Act passed in 1924 (also known as National Origins Act), established ethnic quotas that allowed immigrants to enter the United States only in proportion to their numbers already existing in 1890.

Immigration and Nationality Act (also known as the Hart Celler Act) eliminated the national origins quota of the Immigration Act (or National Origins Act) of 1924.

implicit bias nonconscious form of racism where individuals unconsciously associate negative characteristics with racial–ethnic groups.

impression management a process by which people attempt to control how others perceive them.

income the amount of money brought into a household from various sources during a given year (wages, investment income, dividends, etc.).

independent variable a variable that is the presumed cause of a particular result (*see* dependent variable).

indicator something that points to or reflects an abstract concept.

individualized education program (IEP) programs and services that provide options for students with learning disabilities and physical disabilities.

inductive reasoning the process of arriving at general conclusions from specific observations.

infant mortality rate the number of deaths per year of infants under age 1 for every 1000 live births.

informant in covert participant observation research, a single group member who provides "inside" information about the group being studied.

informed consent a formal acknowledgment by research subjects (respondents) that they understand the purpose of the research and agree to be studied.

institutional racism racism involving notions of racial or ethnic inferiority that have become ingrained into society's institutions.

instrumental needs emotionally neutral, task-oriented (goal-oriented) needs.

interest group a constituency in society organized to promote its own agenda.

interlocking directorate organizational linkages created when the same people sit on the boards of directors of a number of different corporations.

international division of labor system of labor whereby products are produced globally, while profits accrue only to a few.

intersectional theory analytical framework that interprets race, class, and gender as simultaneously overlapping social factors.

intersexed a person born with the physical characteristics of both sexes.

issues problems that affect large numbers of people and have their origins in the institutional arrangements and history of a society.

K

kinship system the pattern of relationships that defines people's family relationships to one another.

L

labeling theory a theory that interprets the responses of others as most significant in understanding deviant behavior.

labor force participation rate the percentage of those in a given category who are employed.

laissez-faire racism maintaining the status quo of racial groups by persistent stereotyping and blaming of minorities for achievement and socioeconomic gaps between groups.

laws the written set of guidelines that define what is right and wrong in society.

liberal feminism a feminist theoretical perspective asserting that the origin of women's inequality is in traditions of the past that pose barriers to women's advancement.

life chances the opportunities that people have in common by virtue of belonging to a particular class.

life course the connection between people's personal attributes, the roles they occupy, the life events they experience, and the social and historical context of these events.

life expectancy the average number of years individuals and particular groups can expect to live.

Lily Ledbetter Fair Pay Act law stating that discrimination claims on the basis of sex, race, national origin, age, religion, and disability accrue with every paycheck.

looking-glass self the idea that people's conception of self arises through reflection about their relationship to others.

M

macroanalysis analysis of the whole of society, how it is organized and how it changes.

mass media channels of communication that are available to very wide segments of the population.

master status some characteristic of a person that overrides all other features of the person's identity.

material culture the objects created in a given society.

matriarchy a society or group in which women have power over men.

matrilineal kinship kinship systems in which family lineage (or ancestry) is traced through the mother.

matrilocal a pattern of family residence in which married couples reside with the family of the wife.

McDonaldization the increasing and ubiquitous presence of the fast-food model in vast numbers of organizations.

mean the sum of a set of values divided by the number of cases from which the values are obtained; an average.

mechanical solidarity unity based on similarity, not difference, of roles.

median the midpoint in a series of values that are arranged in numerical order.

median income the midpoint of all household incomes.

Medicaid a governmental assistance program that provides health care assistance for the poor, including the elderly.

medicalization of deviance explanations of deviant behavior that interpret deviance as the result of individual pathology or sickness.

Medicare a governmental assistance program established in the 1960s to provide health services for older Americans.

meritocracy a system in which one's status is based on merit or accomplishments.

microanalysis analysis of the smallest, most immediately visible parts of social life, such as people interacting.

minority group any distinct group in society that shares common group characteristics and is forced to occupy low status in society because of prejudice and discrimination.

mode the most frequently appearing score among a set of scores.

modernization a process of social and cultural change that is initiated by industrialization and followed by increased social differentiation and division of labor.

modernization theory a view of globalization in which global development is a worldwide process affecting nearly all societies that have been touched by technological change.

monogamy the marriage practice of a sexually exclusive marriage with one spouse at a time.

monotheism   the worship of a single god.

mores   strict norms that control moral and ethical behavior.

multidimensional poverty index   measure of poverty that accounts for health, education, and the standard of living.

multinational corporations   corporations that conduct business across national borders.

multiracial feminism   form of feminist theory noting the exclusion of women of color from other forms of theory and centering its analysis in the experiences of all women.

N

nationalism   the strong identity associated with an extreme sense of allegiance to one's culture or nation.

nativism   extreme beliefs and policies that establish citizens of one nation over all others.

Naturalization Act of 1790   the first law that established rules about who could become a citizen of the United States.

neoliberalism   economic principle promoting the operation of a free market, one with limited (or no) government intervention or subsidies, deregulation, and a shift from public sector spending to the private sector.

net worth   the value of one's financial assets minus debt.

nonmaterial culture   the norms, laws, customs, ideas, and beliefs of a group of people.

nonverbal communication   communication by means other than speech, as by touch, gestures, use of distance, eye movements, and so on.

normative organization   an organization having a voluntary membership and that pursues goals; examples are the PTA or a political party.

norms   the specific cultural expectations for how to act in a given situation.

nuclear family   family in which a married couple resides together with their children.

O

occupational prestige   the subjective evaluation people give to jobs as better or worse than others.

occupational segregation   a pattern in which different groups of workers are separated into different occupations.

organic solidarity   unity based on role differentiation, not similarity.

organizational culture   the collective norms and values that shape the behavior of people within an organization.

organizational ritualism   a situation in which rules become ends in themselves rather than means to an end.

outsourcing   transferring a specialized task or job from one organization to a different organization, usually in another country, as a cost-saving device.

overt participant observation   the form of participant observation wherein the observed individuals are told that they are being studied.

P

panethnicity   process when multiple ethnic groups come together to forge a new, collective identity out of some common purpose.

participant observation   a method whereby the sociologist becomes both a participant in the group being studied and a scientific observer of the group.

patriarchy   a society or group where men have power over women.

patrilineal kinship   a kinship system that traces descent through the father.

patrilocal   a pattern of family residence in which married couples reside with the family of the husband.

peers   those of similar status.

percentage   the number of parts per hundred.

peripheral countries (nations)   poor countries, largely agricultural, having little power or influence in the world system.

personality   the cluster of needs, drives, attitudes, predispositions, feelings, and beliefs that characterize a given person.

play stage   the stage in childhood when children begin to take on the roles of significant people in their environment.

pluralist model   a theoretical model of power in society as coming from the representation of diverse interests of different groups in society.

political action committees (PACs)   groups of people who organize to support candidates they feel will represent their views.

polygamy   a marriage practice in which either men or women can have multiple marriage partners.

polytheism   the worship of more than one deity.

popular culture   the beliefs, practices, and objects that are part of everyday traditions.

population   a relatively large collection of people (or other unit) that a researcher studies and about which generalizations are made.

population density   the number of people per square mile.

population pyramids   graphic depictions of the age and sex distribution of a given population at a point in time.

positivism   a system of thought that regards scientific observation to be the highest form of knowledge.

postindustrial society   a society economically dependent upon the production and distribution of services, information, and knowledge.

poverty line   the figure established by the government to indicate the amount of money needed to support the basic needs of a household.

power   a person or group's ability to exercise influence and control over others.

power elite model   a theoretical model of power positing a strong link between government and business.

preindustrial society   one that directly uses, modifies, and/or tills the land as a major means of survival.

prejudice   the negative evaluation of a social group, and individuals within that group, based upon conceptions about that social group that are held despite facts that contradict it.

prestige   the value with which different groups of people are judged.

primary group   a group characterized by intimate, face-to-face interaction and relatively long-lasting relationships.

profane   that which is of the everyday, secular world and is specifically not religious.

propaganda   information disseminated by a group or organization (such as the state) intended to justify its own power.

Protestant ethic   belief that hard work and self-denial lead to salvation.

psychoanalytic theory   a theory of socialization positing that the unconscious mind shapes human behavior.

Q

qualitative research   research that is somewhat less structured than quantitative research but that allows more depth of interpretation and nuance in what people say and do.

quantitative research   research that uses numerical analysis.

queer theory   a theoretical perspective that recognizes the socially constructed nature of sexual identity.

R

race   a social category, or social construction, that we treat as distinct on the basis of certain characteristics, some biological, that have been assigned social importance in the society.

race-immigration nexus   pattern whereby racial divisions are both shaped and shaped by immigrant arenas and experiences.

racial formation   process by which groups come to be defined as a "race" through social institutions such as the law and the schools.

racial profiling   the use of race alone as a criterion for deciding whether to stop and detain someone on suspicion of having committed a crime.

racial stratification   system of inequality in which race and ethnicity mark differential access to economic, social, political, and cultural resources.

racialization   a process whereby some social category, such as a social class or nationality, is assigned what are perceived to be race characteristics.

racism the perception and treatment of a racial or ethnic group, or member of that group, as intellectually, socially, and culturally inferior to one's own group.

radical feminism feminist theory that locates the source of women's inequality in the power that men hold in society.

random sample a sample that gives everyone in the population an equal chance of being selected.

rate parts per some number (for example, per 10,000; per 100,000).

rational–legal authority authority stemming from rules and regulations, typically written down as laws, procedures, or codes of conduct.

reference group any group (to which one may or may not belong) used by the individual as a standard for evaluating her or his attitudes, values, and behaviors.

reflection hypothesis the idea that the mass media reflect the values of the general population.

relative poverty a definition of poverty that is set in comparison to a set standard.

reliability the likelihood that a particular measure would produce the same results if the measure were repeated.

religion an institutionalized system of symbols, beliefs, values, and practices by which a group of people interprets and responds to what they feel is sacred and that provides answers to questions of ultimate meaning.

religiosity the intensity and consistency of practice of a person's (or group's) faith.

replication study research that is repeated exactly, but on a different group of people at a different point in time.

research design the overall logic and strategy underlying a research project.

residential segregation the spatial separation of racial and ethnic groups in different residential areas.

resocialization the process by which existing social roles are radically altered or replaced.

revolution the overthrow of a state or the total transformation of central state institutions.

risky shift (also polarization shift) the tendency for group members, after discussion and interaction, to engage in riskier behavior than they would while alone.

rite of passage ceremony or ritual that symbolizes the passage of an individual from one role to another.

ritual a symbolic activity that expresses a group's spiritual convictions.

role behavior others expect from a person associated with a particular status.

role conflict two or more roles associated with contradictory expectations.

role modeling imitation of the behavior of an admired other.

role set all roles occupied by a person at a given time.

role strain conflicting expectations within the same role.

S

sacred that which is set apart from ordinary activity, seen as holy, and protected by special rites and rituals.

salience principle categorizing people on the basis of what initially appears prominent about them.

sample any subset of units from a population that a researcher studies.

Sapir–Whorf hypothesis a theory that language determines other aspects of culture because language provides the categories through which social reality is defined and perceived.

schooling socialization that involves formal and institutionalized aspects of education.

scientific method the steps in a research process, including observation, hypothesis testing, analysis of data, and generalization.

secondary group a group that is relatively large in number and not as intimate or long in duration as a primary group.

sect group that has broken off from an established church.

secular the ordinary beliefs of daily life that are specifically not religious.

segregation the spatial and social separation of racial and ethnic groups.

self our concept of who we are, as formed in relationship to others.

self-concept a person's image and evaluation of important aspects of oneself.

self-fulfilling prophecy the process by which merely applying a label changes behavior and thus tends to justify the label.

semiperipheral countries semi-industrialized countries that represent a kind of middle class within the world system.

serendipity unanticipated, yet informative, results of a research study.

sex used to refer to biological identity as male or female.

sex ratio (gender ratio) the number of males per 100 females.

sex tourism practice whereby people travel to engage in commercial sexual activity.

sex trafficking refers to the practice whereby women, usually very young women, are forced by fraud or coercion into commercial sex acts.

sexual harassment unwanted physical or verbal sexual behavior that occurs in the context of a relationship of unequal power and that is experienced as a threat to the victim's job or educational activities.

sexual identity the definition of oneself that is formed around one's sexual relationships.

sexual orientation the attraction that people feel for people of the same or different sex.

sexual politics the link feminists argue exists between sexuality and power, and between sexuality and race, class, and gender oppression.

sexual revolution the widespread changes in men's and women's roles and a greater public acceptance of sexuality as a normal part of social development.

sexual scripts the ideas taught to us about what is appropriate sexual behavior for a person of our gender.

significant others those with whom we have a close affiliation.

social capital see cultural capital.

social change the alteration of social interaction, social institutions, stratification systems, and elements of culture over time.

social class the social structural hierarchical position groups hold relative to the economic, social, political, and cultural resources of society.

social construction perspective a theoretical perspective that explains identity and society as created and learned within a cultural, social, and historical context.

social control the process by which groups and individuals within those groups are brought into conformity with dominant social expectations.

social control agents those who regulate and administer the response to deviance, such as the police or mental health workers.

social control theory theory that explains deviance as the result of the weakening of social bonds.

social fact social pattern that is external to individuals.

social institution an established and organized system of social behavior with a recognized purpose.

social interaction behavior between two or more people that is given meaning.

social learning theory a theory of socialization positing that the formation of identity is a learned response to social stimuli.

social media the term used to refer to the vast networks of social interaction that new media have created.

social mobility a person's movement over time from one class to another.

social movement a group that acts with some continuity and organization to promote or resist social change in society.

social network a set of links between individuals or other social units such as groups or organizations.

social organization the order established in social groups.

social sanction a mechanism of social control that enforces norms.

social stratification a relatively fixed hierarchical arrangement in society by which groups have different access to resources, power, and perceived social worth; a system of structured social inequality.

social structure the pattern of social relationships and social institutions that make up society.

socialism an economic institution characterized by state ownership and management of the basic industries.

socialization the process through which people learn the expectations of society.

socialization agents those who pass on social expectations.

society a system of social interaction, typically within geographical boundaries, that includes both culture and social organization.

socioeconomic status (SES) a measure of class standing, typically indicated by income, occupational prestige, and educational attainment.

sociological imagination the ability to see the societal patterns that influence individual and group life.

sociology the study of human behavior in society.

spurious correlation a false correlation between X and Y, produced by their relationship to some third variable (Z) rather than by a true causal relationship to each other.

state the organized system of power and authority in society.

status an established position in a social structure that carries with it a degree of prestige.

status inconsistency exists when the different statuses occupied by the individual bring with them significantly different amounts of prestige.

status set the complete set of statuses occupied by a person at a given time.

stereotype an oversimplified set of beliefs about the members of a social group or social stratum that is used to categorize individuals of that group.

stereotype interchangeability the principle that negative stereotypes are often interchangeable from one racial group (or gender or social class) to another.

stereotype threat the effect of a negative stereotype about one's self upon one's own test performance.

stigma an attribute that is socially devalued and discredited.

Stockholm syndrome a process whereby a captured person identifies with the captor as a result of becoming inadvertently dependent upon the captor.

structural strain theory a theory that interprets deviance as originating in the tensions that exist in society between cultural goals and the means people have to achieve those goals.

subculture the culture of groups whose values and norms of behavior are somewhat different from those of the dominant culture.

symbolic interaction a theoretical perspective claiming that people act toward things because of the meaning things have for them.

symbol thing or behavior to which people give meaning.

T

taboo behavior that bring the most serious sanctions.

taking the role of the other the process of imagining oneself from the point of view of another.

teacher expectancy effect the effect of a teacher's expectations on a student's actual performance, independent of the student's ability.

Temporary Assistance for Needy Families (TANF) federal program by which grants are given to states to fund welfare.

terrorism premeditated, politically motivated violence perpetrated against targets by groups or individuals who try to achieve their political ends through violence.

Title IX legislation that prohibits schools that receive federal funds from discriminating based on gender.

total institution an organization cut off from the rest of society in which individuals are subject to strict social control.

totalitarian an extreme form of authoritarianism where the state has total control over all aspects of public and private life.

totem an object or living thing that a religious group regards with special awe and reverence.

tracking grouping, or stratifying, students in school on the basis of ability test scores.

traditional authority authority stemming from long-established patterns that give certain people or groups legitimate power in society.

transgender those who deviate from the binary (that is, male or female) system of gender.

transnational family families where one parent (or both) lives and works in one country while the children remain in their country of origin.

triad a group consisting of three people.

troubles privately felt problems that come from events or feelings in one individual's life.

U

underemployment the condition of being employed at a skill level below what would be expected given a person's training, experience, or education.

unemployment rate the percentage of those not working, but officially defined as looking for work.

urban underclass a grouping of people, largely minority and poor, who live at the absolute bottom of the socioeconomic ladder in urban areas.

urbanization the process by which a community acquires the characteristics of city life.

utilitarian organization a profit or nonprofit organization that pays its employees salaries or wages.

V

validity the degree to which an indicator accurately measures or reflects a concept.

values the abstract standards in a society or group that define ideal principles.

variable something that can have more than one value or score.

verstehen the process of understanding social behavior from the point of view of those engaged in it.

W

wealth the monetary value of everything one actually owns.

White privilege the ability for Whites to maintain an elevated status in society that masks racial inequality.

White space the perception Black people have of places where they are typically absent, not expected, or marginalized when present.

work productive human activity that produces something of value, either goods or services.

world cities cities that are closely linked through the system of international commerce.

world systems theory theory that capitalism is a single world economy and that there is a worldwide system of unequal political and economic relationships that benefit the technologically advanced countries at the expense of the less technologically advanced.

X

xenophobia the fear and hatred of foreigners.

Z

zero population growth stable population growth whereby the birthrate is equal to the death rate, without other influences, such as immigration.

References

4th Estate. 2014. "Silenced: Gender Gap in Election Coverage." **www.4thestate.net /female-voices-in-media-infographic/**

Abel, Ernest L. and Michael L. Kruger. 2007. "Gender Related Naming Practices: Similarities and Differences between People and Their Dogs." *Sex Roles: A Journal of Research* 57 (1–2): 15–19.

Abendroth, Anja-Kristin, Matt L. Huffman, and Judith Treas. 2014. "The Parity Penalty in Life Course Perspective: Motherhood and Occupational Status in 13 European Countries." *American Sociological Review* 79 (5): 993.

Abrahamson, Mark. 2006. *Urban Enclaves: Identity and Place in the World,* 2nd ed. New York: Worth.

Acker, Joan. 1992. "Gendered Institutions: From Sex Roles to Gendered Institutions." *Contemporary Sociology* 21 (September): 565–569.

Adorno, T. W., Else Frenkel-Brunswik, D. J. Levinson, and R. N. Sanford. 1950. *The Authoritarian Personality.* New York: Harper and Row.

Ajrouch, K. J., and A. Jamal. 2007. "Assimilate to a White Identity: The Case Of Arab Americans." *International Migration Review* 41 (4): 860–879.

Alba, Richard. 1990. *Ethnic Identity: The Transformation of Ethnicity in the Lives of Americans of European Ancestry.* New Haven, CT: Yale University Press.

Alba, Richard D., and Victor Nee. 2003. *Remaking the American Mainstream: Assimilation and Contemporary America.* Cambridge, MA: Harvard University Press.

Alexander, Michelle. 2010. *The New Jim Crow: Mass Incarceration in the Age of Colorblindness.* New York: The New Press.

Alexander, Stephanie A. C., Katherine L. Frohlich, Blake D. Poland, Rebecca J. Haines, and Catherine Maule. 2010. "I'm a Young Student, I'm a Girl … and for Some Reason They Are Hard on Me for Smoking: The Role of Gender and Social Context for Smoking Behaviour." *Critical Public Health* 20 (3): 323–338.

Allen, Kathleen P. 2012. "Off the Radar and Ubiquitous: Text Messaging and Its Relationship to 'Drama' and Cyberbullying in an Affluent Academically Rigorous U.S. High School." *Journal of Youth Studies* 15 (1): 99–117.

Allison, Anne. 1994. *Nightwork: Sexuality, Pleasure, and Corporate Masculinity in a Tokyo Hostess Club.* Chicago: University of Chicago Press.

Allison, Stuart F. H., Amie M. Schuck, and Kim Michelle Lersch. 2005. "Exploring the Crime of Identity Theft: Prevalence, Clearance Rates, and Victim-Offender Characteristics." *Journal of Criminal Justice* 33 (1): 19–29.

Allport, Gordon W. 1954. *The Nature of Prejudice.* Reading, MA: Addison-Wesley.

Alon, Sigal, and Marta Tienda. 2007. "Diversity, Opportunity, and the Shifting Meritocracy in Higher Education." *American Sociological Review* 72 (August): 487–511.

Altman, Dennis. 2001. *Global Sex.* Chicago: University of Chicago Press.

Amato, Paul R. 2010. "Research on Divorce: Continuing Trends and New Developments." *Journal of Marriage and Family* 72 (3): 650–666.

Amato, Paul R., Alan Booth, Alan R. Johnson, and Stacy J. Rogers. 2007. *Alone Together: How Marriage in America Is Changing.* Cambridge, MA: Harvard University Press.

American Association of University Women. 1998. *Gender Gaps: Where Schools Still Fail Our Children.* Washington, DC: American Association of University Women.

American Psychological Association. 2007. *Report of the APA Task Force on the Sexualization of Girls.* Washington, DC: American Psychological Association.

American Psychological Association. 2008. *Stress in America.* Washington, DC: American Psychological Association. **www.apa.org**

American Sociological Association. 2008. *ASA Code of Ethics.* **www.asanet.org /about/ethics.cfm**

Amott, Teresa L., and Julie A. Matthaei. 1996. *Race, Gender, and Work: A Multicultural History of Women in the United States,* 2nd ed. Boston: South End Press.

Andersen, Margaret L. 2004. "From *Brown* to *Grutter*: The Diverse Beneficiaries of *Brown* vs. *Board of Education*." *Illinois Law Review* 5 (August): 1073–1097.

Andersen, Margaret L. 2015. *Thinking about Women: Sociological Perspectives on Sex and Gender,* 10th ed. Boston: Allyn and Bacon.

Andersen, Margaret L. 2018. *Race in Society: The Enduring American Dilemma.* Lanham, MD: Rowman and Littlefield.

Andersen, Margaret L., and Patricia Hill Collins (eds.). 2016. *Race, Class and Gender: An Anthology,* 9th ed. Belmont, CA: Wadsworth/Cengage.

Anderson, Elijah. 2015. "The White Space." *Sociology of Race and Ethnicity* 1 (1): 10–21.

Anderson, David A., and Mykol Hamilton. 2005. "Gender Role Stereotyping of Parents in Children's Picture Books: The Invisible Father." *Sex Roles: A Journal of Research* 52 (3–4): 145–151.

Anderson, Elijah. 1976. *A Place on the Corner.* Chicago: University of Chicago Press.

Anderson, Elijah. 1990. *Streetwise: Race, Class, and Change in an Urban Community.* Chicago: University of Chicago Press.

Anderson, Elijah. 2011. *The Cosmopolitan Canopy: Race and Civility in Everyday Life.* New York: W. W. Norton.

Anderson, Kathryn Freeman. 2017. "Racial Residential Segregation and the Distribution of Health-Related Organizations in Urban Neighborhoods." *Social Problems* 64 (May): 256–276.

Anderson, Monica, and Andrew Perrin. 2017. "Disabled Americans Are Less Likely to Use Technology." *FactTank: News in the Numbers.* Washington, DC: Pew Research Center. **www .pewresearch.org/fact-tank/2017/04 /07/disabled-americans-are-less -likely-to-use-technology/**

Angwin, Julia. 2010. "The Web's New Gold Mine: Your Secrets." *The Wall Street Journal,* July 30.

Anthony, Dick, Thomas Robbins, and Steven Barrie-Anthony. 2002. "Cult and Anticult Totalism: Reciprocal Escalation and Violence." *Terrorism and Political Violence* 14 (Spring): 211–239.

Apple, Michael W. 1991. "The New Technology: Is It Part of the Solution or Part of the Problem in Education?" *Computers in the Schools* 8 (April– October): 59–81.

Arendt, Hannah. 1963. *Eichmann in Jerusalem: A Report on the Banality of Evil.* New York: Viking Press.

Armstrong, E. A., L. Hamilton, and B. Sweeney. 2006. "Sexual Assault on Campus: Multilevel, Integrative Approach to Party Rape." *Social Problems* 53 (4): 483–499.

Armstrong, Elizabeth A., Laura Hamilton, and Paula England. 2010. "Is Hooking

Up Bad for Young Women?" *Contexts* 9 (Summer): 23–27.

Asch, Solomon. 1951. "Effects of Group Pressure upon the Modification and Distortion of Judgments." In *Groups, Leadership, and Men,* edited by H. Guetzkow. Pittsburgh, PA: Carnegie Press.

Asch, Solomon. 1955. "Opinions and Social Pressure." *Scientific American* 19 (July): 31–35.

Atkins, Celeste. 2011. "Big Black Mammas: The Intersection of Race, Gender and Weight in the U.S." Unpublished manuscript, Pima Community College, AZ.

Aulakh, Ravenna. 2013. "In Uzbekistan, 'Slave Labour' Used to Harvest Cotton." *The Toronto Star,* October 25.

Austin, Algernon. 2013. *High Unemployment Means Native Americans Are Still Waiting for an Economic Recovery.* Economic Policy Institute, December 17. **www.epi.org**

Baca Zinn, Maxine. 1995. "Chicano Men and Masculinity." pp. 33–41 in *Men's Lives,* 3rd ed., edited by Michael S. Kimmel and Michael A. Messner. Boston: Allyn and Bacon.

Baca Zinn, Maxine, and Bonnie Thornton Dill. 1996. "Theorizing Difference from Multiracial Feminism." *Feminist Studies* 22 (Summer): 321–331.

Baca Zinn, Maxine, Pierrette Hondagneu-Sotelo, Michael A. Messner, and Amy Dinissen, eds. 2015. *Gender through the Prism of Difference,* 5th ed. New York: Oxford University Press.

Baca Zinn, Maxine, D. Stanley Eitzen, and Barbara Wells. 2016. *Diversity in Families.* New York: Pearson.

Backman, Maurie. 2016. "Here's How Much the Average American Spends on Child Care." *USA Today* (November 2).

Bailey, Garrick. 2003. *Humanity: An Introduction to Cultural Anthropology,* 6th ed. Belmont, CA: Wadsworth.

Balbo, Nicoletta, and Nicola Barban. 2014. "Does Fertility Behavior Spread among Friends?" *American Sociological Review* 79 (3): 4312–4431.

Bales, Robert F. 1951. *Interaction Process Analysis.* Cambridge: Addison-Wesley.

Bales, Kevin. 2010. *The Slave Next Door: Human Trafficking and Slavery in America Today.* Berkeley, CA: University of California Press.

Ballantine, Jeanne H., and Joan Z. Spade. 2015. *Schools and Society: A Sociological Approach to Education,* 5th ed. Thousand Oaks, CA: Sage Publications.

Bandura, Albert, and R. H. Walters. 1963. *Social Learning and Personality Development.* New York: Holt, Reinhart, and Winston.

Barak, Gregg, Paul Leighton, and Allison Cotton. 2015. *Class, Race, Gender, & Crime: Social Realities of Justice in America,* 4th ed. Lanham, MD: Rowman & Littlefield.

Barnett, Jessica C., and Marina Vornovitsky. 2016. *Health Insurance Coverage in the United States: 2015.* Washington, DC: U.S. Census Bureaa. **www.census.gov**

Barton, Bernadette. 2006. *Stripped: Inside the Lives of Exotic Dancers.* New York: New York University Press.

Barton, Bernadette. 2017. *Stripped: More Stories from Exotic Dancers,* rev. ed. New York: New York University Press.

Baumeister, Roy F., and Brad J. Bushman. 2017. *Social Psychology and Human Nature,* 4th edition. San Francisco, CA: Cengage.

Baxter, Jennifer, Ruth Weston, and Lixia Qu. 2011. "Family Structure, Co-Parental Relationship Quality, Post-Separation Paternal Involvement and Children's Emotional Wellbeing." *Journal of Family Studies* 17 (2): 86–109.

Bean, Frank D., and Marta Tienda. 1987. *The Hispanic Population of the United States.* New York: Russell Sage Foundation.

Bearak, Jonathan, Kristen Burke, and Rachel Jones. 2017. "Disparities and Change Over Time in Distance Women Would Need to Travel to Have an Abortion in the USA: A Spatial Analysis." *The Lancet Public Health.* DOI: **http://dx.doi.org/10.1016/S2468 -2667(17)30158-5**

Becker, Elizabeth. 2000. "Harassment in the Military Is Said to Rise." *The New York Times* (March 10): 14.

Becker, Howard S. 1963. *Outsiders: Studies in the Sociology of Deviance.* New York: Free Press.

Becker, L. B., T. Vlad, W. Kazragis, C. Toledo, and P. Desnoes. 2010. *2010 Annual Survey of Journalism and Mass Communication Graduates.* Athens, GA: James M. Cox, Jr. Center for International Mass Communication Training and Research, Grady College of Journalism and Mass Communication, University of Georgia.

Bell, Myrtle P. 2011. *Diversity in Organizations.* Stamford, CT: Cengage Learning.

Bell, Leslie C. 2013. *Hard to Get: 20-Something Women and the Paradox of Sexual Freedom.* Berkeley, CA: University of California Press.

Bellah, Robert (ed.). 1973. *Emile Durkheim on Morality and Society: Selected Writings.* Chicago: University of Chicago Press.

Beller, Emily, and Michael Hout. 2006. "Intergenerational Social Mobility: The United States in Comparative Perspective." *The Future of Children* 16 (Fall): 19–36.

Ben-Yehuda, Nachman. 1986. "The European Witch Craze of the Fourteenth-Seventeenth Centuries: A Sociologist's Perspective." *American Journal of Sociology* 86: 1–31.

Benedict, Ruth. 1934. *Patterns of Culture.* Boston: Houghton Mifflin.

Bennett, Daniel L., Adam R. Lucchesi, and Richard K Vedder. 2010. *For-Profit Higher Education: Growth, Innovation and Regulation.* Washington, DC: Center for College Affordability and Productivity. **http://files.eric.ed.gov/fulltext /ED536282.pdf**

Berger, Peter L. 1963. *Invitation to Sociology: A Humanistic Perspective.* Garden City, NY: Doubleday Anchor.

Berger, Peter L., Brigitte Berger, and Hansfried Kellner. 1974. *The Homeless Mind: Modernization and Consciousness.* New York: Vintage Books.

Berger, Peter L., and Thomas Luckmann. 1967. *The Social Construction of Reality: A Treatise in the Sociology of Knowledge.* Garden City, NY: Anchor Books.

Best, Joel. 1999. *Random Violence: How We Talk about New Crimes and New Victims.* Berkeley, CA: University of California Press.

Billings, John S. 1896. *Vital and Social Statistics in the United States.* Washington, DC: Government Printing Office. **www.cdc.gov/nchs/data /vsushistorical/vsush_1890_1.pdf**

Bird, Chloe E., and Patricia P. Rieker. 2008. *Gender and Health: The Effects of Constrained Choices and Social Policies.* New York: Cambridge University Press.

Bishaw, Alemayehi. 2011. *Areas with Concentrated Poverty, 2006–2010.* Washington, DC: U.S. Census Bureau.

Black, M. C., K. C. Basile, M. J. Breiding, S. G. Smith, M. I. M. L. Walters, M. T. Merrick, J. Chen, and M. R. Stevens. 2011. *The National Intimate Partner and Sexual Violence Survey: 2010 Summary Report.* Atlanta, GA: National Center for Injury Prevention and Control, Centers for Disease Control and Prevention.

Blair-Loy, Mary, and Amy S. Wharton. 2002. "Employees' Use of Work-Family Policies and the Workplace Social Context." *Social Forces* 80 (3): 813–845.

Blake, C. Fred. 1994. "Footbinding in Neo-Confucian China and the Appropriation of Female Labor." *Signs* 19 (Spring): 676–712.

Blankenship, Kim. 1993. "Bringing Gender and Race in U.S. Employment Discrimination Policy." *Gender & Society* 7 (June): 204–226.

Blassingame, John. 1973. *The Slave Community: Plantation Life in the Antebellum South.* New York: Oxford University Press.

Blau, Peter M., and W. Richard Scott. 1974. *On the Nature of Organizations.* New York: Wiley.

Blee, Kathleen. 2008. *Women of the Klan: Racism and Gender in the 1920s.* Berkeley, CA: University of California Press.

Block, Jason P., Richard A. Scribner, and Karen B. DeSalvo. 2004. "Fast Food, Race/Ethnicity, and Income: A Geographic Analysis." *American Journal of Preventive Medicine* 27 (3): 211–217.

Blumer, Herbert, 1969. *Studies in Symbolic Interaction.* Englewood Cliffs, NJ: Prentice Hall.

Bobo, Lawrence D. 2004. "Inequalities That Endure? Racial Ideology, American Politics, and the Peculiar Role of the Social Sciences." pp. 13–42 in *The Changing Terrain of Race and Ethnicity,* edited by Maria Krysan and Amanda E. Lewis. New York: Russell Sage Foundation.

Bobo, Lawrence D. 2006. "The Color Line, the Dilemma, and the Dream: Race Relations in America at the Close of the Twentieth Century." pp. 87–95 in *Race and Ethnicity in Society: The Changing Landscape,* edited by Elizabeth

Higginbotham and Margaret L. Andersen. Belmont, CA: Wadsworth.

Bobo, Lawrence. 2012 (January). "Post Racial Dreams, American Realities." Paper read before the Department of Sociology, Princeton University, Princeton, NJ.

Bocian, Debbie Gruenstein, Wei Li, and Keith S. Ernst. 2010. *Foreclosures by Race and Ethnicity: The Demographics of a Crisis.* Durham, NC: Center for Responsible Lending.

Bogle, Kathleen. 2008. *Hooking Up: Sex, Dating, and Relationships on Campus.* New York: New York University Press.

Bonilla-Silva, Eduardo. 2004. "From Bi-Racial to Tri-Racial: Towards a New System of Racial Stratification in the USA." *Ethnic and Racial Studies* 27 (6): 931–950.

Bonilla-Silva, Eduardo, and Karen S. Glover. 2004. "'We Are All Americans': The Latin Americanization of Race Relations in the United States." pp. 149–183 in *The Changing Terrain of Race and Ethnicity,* edited by Maria Krysan and Amanda E. Lewis. New York: Russell Sage Foundation.

Bonilla-Silva, Eduardo. 2013. *Racism without Racists: Colorblind Racism and the Persistence of Inequality in America,* 4th ed. Lanham, MD: Rowman and Littlefield.

Bonham, Vence. 2015. "Race" (interview). National Human Genome Research Institute. Accessed on February 9, 2015. **http://www.genome.gov**

Bontemps, Arna (ed.). 1972. *The Harlem Renaissance Remembered.* New York: Dodd, Mead.

Boo, Katherine. 2012. *Behind the Beautiful Forevers: Life, Death, and Hope in a Mumbai Undercity.* New York: Random House.

Boonstra, Heather, Rachel Benson Gold, Cory L. Richards, and Lawrence B. Finer. 2006. *Abortion in Women's Lives.* New York: Guttmacher Institute.

Bourdieu, Pierre. 1984. *Distinction: A Social Critique of the Judgement of Taste,* translated by Richard Nice. Cambridge, MA: Harvard University Press.

Bowles, Samuel, Herbert Gintis, and Melissa Osborn Groves (eds.). 2005. "Introduction." pp. 1–22 in *Unequal Chances: Family Background and Economic Success.* Princeton, NJ: Princeton University Press.

Boyd, Danah. 2014. *It's Complicated: The Social Lives of Networked Teens.* New Haven, CT: Yale University Press.

Boyle, Karen. 2011. "Producing Abuse: Selling the Harms of Pornography." *Women's Studies International Forum* 34 (6): 593–602.

Brame, Robert, Shawn D. Bushway, Ray Paternoster, and Michael G. Turner. 2014. "Demographic Patterns of Cumulative Arrest Prevalence by Ages 18 and 23." *Crime & Delinquency* 60 (3): 471–486.

Branch, Enobong Hannah. 2011. *Opportunity Denied: Limiting Black Women in Devalued Work.* New

Brunswick, NJ: Rutgers University Press.

Branch, Taylor. 2006. *At Canaan's Edge: America in the King Years, 1965–1968.* New York: Simon and Schuster.

Brand, Jennie E., and Yu Xie. 2010. "Who Benefits Most from College? Evidence for Negative Selection in Heterogeneous Economic Returns to Higher Education." *American Sociological Review* 75 (2): 273–302.

Braunstein, Ruth. 2011. "Who Are 'We the People'"? *Contexts* 10 (May): 72–73.

Brechin, Steven R. 2003. "Comparative Public Opinion and Global Climatic Change and the Kyoto Protocol: The U.S. versus the World?" *International Journal of Sociology and Social Policy* 23 (10): 106–134.

Bricker, Jesse, Alice Henriques, Jacob Krimmel, and John Sabelhaus. 2016. "Measuring Income and Wealth at the Top Using Administrative and Survey Data." *Brookings Papers on Economic Activity.* Washington, DC: The Brookings Institute. **https://www.brookings.edu/bpea-articles/measuring-income-and-wealth-at-the-top-using-administrative-and-survey-data/**

Bridges, Ana J., Robert Wosnitzer, Erica Scharrer, Chyng Sun, and Rachael Liberman. 2010. "Aggression and Sexual Behavior in Best-Selling Pornography Videos: A Content Analysis Update." *Violence Against Women* 16 (10): 1065–1085.

Brodkin, Karen. 2006. "How Did Jews Become White Folks?" pp. 59–66 in Elizabeth Higginbotham and Margaret L. Andersen, eds., *Race and Ethnicity in Society: The Changing Landscape.* Belmont, CA: Wadsworth.

Brooke, James. 1998. "Utah Struggles with a Revival of Polygamy." *The New York Times* (August 23): A12.

Brooks-Gunn, Jeanne, Wen-Jui Han, and Jane Waldfogel. 2002. "Maternal Employment and Child Cognitive Outcomes in the First Three Years of Life: The NICHD Study of Early Child Care." *Child Development* 73 (July–August): 1052–1072.

Brown, Cynthia. 1993. "The Vanished Native Americans." *The Nation* 257 (October 11): 384–389.

Brown, Elaine. 1992. *A Taste of Power: A Black Woman's Story.* Pantheon: New York.

Broyard, Bliss. 2007. *One Drop: My Father's Hidden Life.* New York: Little, Brown.

Brulle, Robert J., and David N. Pellow. 2006. "Environmental Justice: Human Health and Environmental Inequalities." *Annual Review of Public Health* 27 (April): 103–123.

Budig, Michelle J. 2002. "Male Advantage and the Gender Composition of Jobs: Who Rides the Glass Escalator?" *Social Problems* 49 (May): 257–277.

Buddel, Neil. 2011. "Queering the Workplace." *Journal of Gay & Lesbian Social Services* 23 (1): 131–146.

Bullard, Robert D., and Beverly Wright. 2009. *Race, Place, and Environmental Justice after Hurricane Katrina.* Boulder, CO: Westview Press.

Burgard, Sarah A., and Lucie Kalousova. 2015. "Effects of the Great Recession: Health and Well-being." *Annual Review of Sociology* 41:181–201.

Burt, Cyril. 1966. "The Genetic Determination of Differences in Intelligence: A Study of Monozygotic Twins Reared Together and Apart." *British Journal of Psychology* 57 (1–2): 137–153.

Cabrera, Nolan Leon. 2014. "Exposing Whiteness in Higher Education: White Male College Students Minimizing Racism, Claiming Victimization, and Recreating White Supremacy." *Race, Ethnicity, and Education* 17 (March): 30–55.

Cannuscio, Carolyn C., Amy Hillier, Allison Karpyn, and Karen Glanz. 2014. "The Social Dynamics of Healthy Food Shopping and Store Choice in an Urban Environment." *Social Science & Medicine* 122 (December): 13–20.

Cantor, Joanne. 2000. "Media Violence." *Journal of Adolescent Health* 27 (August): 30–34.

Carlowicz, Michael. 2014. "*Global Temperatures.*" National Aeronautics and Space Administration. **www.earthobservatory.nasa.gov**

Carmichael, Stokely, and Charles V. Hamilton. 1967. *Black Power: The Politics of Liberation in America.* New York: Vintage.

Carney, Nikita. 2017. "All Lives Matter, But So Does Race: Black Lives Matter and the Evolving Role of Social Media." *Humanity and Society* 40 (2): 180–199.

Carr, Deborah. 2010. "Cheating Hearts." *Contexts* 9 (Summer): 58–60.

Carroll, J. B., 1956. *Language, Thought, and Reality: Selected Writings of Benjamin Lee Whorf.* Cambridge, MA: MIT Press.

Carter, J. S., Mamadi Corra, and Shannon K. Carter. 2009. "The Interaction of Race and Gender: Changing Gender-Role Attitudes, 1974–2006." *Social Science Quarterly* 90(1): 196–211.

Carson, E. Ann, and William J. Sabol. 2012. *Prisoners in 2011.* Washington, DC: Bureau of Justice Statistics, U.S. Department of Justice.

Carson, E. Ann, and Elizabeth Anderson. 2016. *Prisoners in 2015.* Washington, DC: Bureau of Justice Statistics. **www.bjs.gov/content/pub/pdf/p15.pdf**

Case-Levine, Julia. 2016. "All the Ways Psychological Prejudices Help Trump but Hurt Hillary." *Quartz.* **https://qz.com/742723/all-the-ways-that-prejudice-helps-trump-but-hinders-hillary/**

Cash, Hannah, Kelsey Drotning, and Paige Miller. 2017. "Seven Things Social Science Tells Us about Natural Disasters" *Contexts,* October 7. **https://contexts.org/blog/seven-things-social-science-tells-us-about-natural-disasters/**

Cassidy, J., and P. R. Shaver (eds.). 2008. *Handbook of Attachment: Theory,*

Research, and Critical Applications, 2nd ed. New York: Guilford.

Catalano, Shannon. 2012. *Intimate Partner Violence, 1993–2010.* Washington, DC: U.S. Bureau of Justice Statistics. **www.bjs.gov**

Cellini, Stephanie R., and Rajeev Darolia. 2017. "High Costs, Low Resources, and Missing Information: Explaining Student Borrowing in the For-Profit Sector." *Annals of the American Academy of Political and Social Science* 671 (1): 92–112.

Center on Budget and Policy Priorities. 2017. *A Quick Guide to SNAP Eligibility and Benefits.* Washington, DC: Center on Budget and Policy Priorities. **www.cbpp.org**

Center for Responsive Politics. 2014. "Super PACs." **www.opensecrets.org**

Center for Responsive Politics. 2017. "Top PACS." **www.opensecrets.org/pacs /toppacs.php?cycle=2016&Type =E&filter=P**

Center for Responsive Politics. 2018. "Blue Wave of Money Propels 2018 Election to Record-Breaking $5.2 Billion in Spending." **https://www.opensecrets .org/news/2018/10/2018-midterm -record-breaking-5-2-billion/**

Centers for Disease Control and Prevention. 2011. "National Center for Chronic Disease Prevention and Health Promotion, Division of Nutrition, Physical Activity, and Obesity." Atlanta, GA: Centers for Disease Control and Prevention.

Centers for Disease Control and Prevention. 2015. "Hispanic Health." *CDC Vital Signs* (May). **www.cdc.gov/vitalsigns /hispanic-health/**

Centers for Disease Control and Prevention. 2017. "Births and Natality." Atlanta, GA: Centers for Disease Control. **www.cdc .gov/nchs/fastats/births.htm**

Centers, Richard. 1949. *The Psychology of Social Classes.* Princeton, NJ: Princeton University Press.

Chafetz, Janet. 1984. *Sex and Advantage.* Totowa, NJ: Rowman and Allanheld.

Chalfant, H. Paul, Robert E. Beckley, and Chalfant, H. Paul, Robert E. Beckley, and C. Eddie Palmer. 1987. *Religion in Contemporary Society,* 2nd ed. Palo Alto, CA: Mayfield.

Chang, Jung. 1991. *Wild Swans: Three Daughters of China.* New York: Simon & Schuster.

Chen, Elsa, Y. F. 1991. "Conflict between Korean Greengrocers and Black Americans." Unpublished senior thesis, Princeton University.

Cherlin, Andrew J. 2010. "Demographic Trends in the United States: A Review of Research in the 2000s." *Journal of Marriage and Family* 72 (June): 403–419.

Childs, Erica Chito. 2005. *Navigating Interracial Borders: Black-White Couples and Their Social World.* New Brunswick, NJ: Rutgers University Press.

Chou, Rosalind S., and Joe Feagin. 2008. *The Myth of the Model Minority: Asian Americans Facing Racism,* second edition. New York: Paradigm Publishers.

Clawson, Dan, and Naomi Gerstel. 2002. "Caring for Our Young: Child Care in Europe and the United States." *Contexts* 1 (Fall–Winter): 28–35.

Clement, Scott. 2016. "The Most Popular Obama Wasn't the President." *The Washington Post.* **www.washingtonpost .com**

Coakley, Jay. 2014. *Sports in Society: Issues and Controversies,* 11th ed. New York: McGraw Hill.

Coale, Ansley. 1986. "Population Trends and Economic Development." pp. 96–104 in *World Population and the U.S. Population Policy: The Choice Ahead,* edited by J. Menken. New York: W. W. Norton.

Cobb, Jelani. 2016. "The Matter of Black Lives." *The New Yorker* (March 16): 34–40.

Cohen, Andrew. 2015. "How White Users Made Heroin a Public Health Problem." *The Atlantic* (August 12). **www.theatlantic .com/politics/archive/2015/08/crack -heroin-and-race/401015/**

Cole, David. 1999. "The Color of Justice." *The Nation* (October 11): 12–15.

Coles, Roberta, and Charles Green, eds. 2009. *The Myth of the Missing Black Father.* New York: Columbia University Press.

Colley, Ann, and Zazie Todd. 2002. "Gender-Linked Differences in the Style and Content of E-mails to Friends." *Journal of Language and Social Psychology* 21 (December): 380–393.

Collins, Patricia Hill. 1990. *Black Feminist Theory: Knowledge, Consciousness and the Politics of Empowerment.* Cambridge, MA: Unwin Hyman.

Collins, Patricia Hill. 1998. *Fighting Words: Black Women and the Search for Justice.* Minneapolis, MN: University of Minnesota Press.

Collins, Patricia Hill. 2004. *Black Sexual Politics: African Americans, Gender, and the New Racism.* New York: Routledge.

Collins, Patricia Hill, and Silma Bilge. 2016. *Intersectionality.* Malden, MA: Polity Press.

Collins, Patricia Hill, Lionel A. Maldonado, Dana Y. Takagi, Barrie Thorne, Lynn Weber, and Howard Winant. 1995. "On West and Fenstermaker's 'Doing Difference.'" *Gender & Society* 9 (August): 491–505.

Collins, Randall, and Michael Makowsky. 1972. *The Discovery of Society.* New York: Random House.

Collins, S.R., M. Z. Gunja, and H. K. Bhupal. 2017. "New U.S. Census Data Shows the Number of Uninsured Americans Dropped by 1 Million in 2016, with Young Adults Continuing to Show Large Gains." *To the Point.* The Commonwealth Fund (September 12).

Coltrane, Scott, and Melinda Messineo. 2000. "The Perpetuation of Subtle Prejudice: Race and Gender Imagery in 1990s Television Advertising." *Sex Roles* 42 (March): 363–389.

Conley, Dalton, and Jason Fletcher. 2017. *The Genome Factor: What the Social Genomic Revolution Reveals about Ourselves, Our History, and the Future.* Princeton, NJ: Princeton University Press.

Connell, Catherine. 2014. *School's Out: Gay and Lesbian Teachers in the Classroom.* Berkeley, CA: University of California Press.

Connell, Raewyn. 2009. "Accountable Conduct." *Gender & Society* 24: 104–111.

Connell, Catherine. 2010. "Doing, Undoing, or Redoing Gender? Learning from the Workplace Experiences of Transpeople." *Gender & Society* 24 (1): 31–55.

Connell, Raewyn. 2012. "Transsexual Women and Feminist Thought." *Signs* 37 (4): 857–88.

Conner, Thaddieus W. and William A. Taggart. 2013. "Assessing the Impact of Indian Gaming on American Indian Nations: Is the House Winning?" *Social Science Quarterly* 94 (4): 1016–1044.

Conrad, Peter, and Joseph W. Schneider. 1992. *Deviance and Medicalization: From Badness to Sickness,* expanded ed. Philadelphia, PA: Temple University Press.

Cook, S. W. 1988. "The 1954 Social Science Statement and School Segregation: A Reply to Gerard." pp. 237–256 in *Eliminating Racism: Profiles in Controversy,* edited by D. A. Taylor. New York: Plenum.

Cook, Thomas, and John Wall, eds. 2011. *Children and Armed Conflict: Cross-Disciplinary Investigations.* Hampshire, UK: Palgrave Macmillan.

Cooley, Charles Horton. 1902. *Human Nature and Social Order.* New York: Scribner's.

Cooley, Charles Horton. 1967 [1909]. *Social Organization.* New York: Schocken Books.

Coontz, Stephanie. 1992. *The Way We Never Were.* New York: Basic Books.

Cooper, Marianne. 2014. *Cut Adrift: Families in Insecure Times.* Berkeley, CA: University of California Press.

Corak, Miles. 2016. "Economic Mobility." *Pathways.* Stanford, CA: Stanford Center on Poverty and Inequality. **https:// inequality.stanford.edu/sites/default /files/Pathways-SOTU-2016-Economic -Mobility-3.pdf**

Cortes, Patricia. 2008. "The Effect of Low-Skilled Immigration on U.S. Prices: Evidence from CPI Data." *Journal of Political Economy* 116 (3): 381–422.

Coser, Lewis. 1977. *Masters of Sociological Thought.* New York: Harcourt Brace Jovanovich.

Coston, Beverly M., and Michael Kimmel. 2012. "Seeing Privilege Where It Isn't Marginalized Masculinities and the Intersectionality of Privilege." *Journal of Social Issues* 68 (March): 97–111.

Coy, Maddy. 2014. "Pornographic Performances: A Review of Sexualisation and Racism in Music Videos." *End Violence against Women.* **endviolenceagainstwomen.org.uk**

Craig, Lyn, and Killian Mullan. 2013. "Parental Leisure Time: A Gender Comparison in Five Countries." *Social Politics* 20 (3): 329–357.

Crane, Diana (ed.). 1994. *The Sociology of Culture: Emerging Theoretical Perspectives.* Cambridge, England: Blackwell.

Cravey, Tiffany, and Aparna Mmitra. 2011. "Demographics of the Sandwich Generation by Race and Ethnicity in the United States." *Journal of Socio-Economics* 40 (3): 306–311.

Crittenden, Ann. 2002. *The Price of Motherhood: Why the Most Important Job in the World Is the Least Valued.* New York: Holt.

Cronin, Ann, and Andrew King. 2014. "Only Connect? Older Lesbian, Gay and Bisexual (LGB) Adults and Social Capital." *Ageing & Society* 34 (2): 258–279.

Cunningham, Mick. 2001. "The Influence of Parental Attitudes and Behaviors on Children's Attitudes toward Gender and Household Labor in Early Adulthood." *Journal of Marriage and Family* 63 (February): 111–122.

D'Emilio, John. 1998. *Sexual Politics, Sexual Communities: The Making of a Homosexual Minority in the United States, 1940–1970,* rev. ed. Chicago: University of Chicago Press.

Dahrendorf, Rolf. 1959. *Class and Class Conflict in Industrial Society.* Stanford, CA: Stanford University Press.

Dalmage, Heather M. 2000. *Tripping on the Color Line: Black-White Multiracial Families in a Racially Divided World.* New Brunswick, NJ: Rutgers University Press.

Dalton, Susan E., and Denise D. Bielby. 2000. "'That's Our Kind of Constellation': Lesbian Mothers Negotiate Institutionalized Understandings of Gender within the Family." *Gender & Society* 14 (February): 36–61.

Daniels, Jessie. 1997. *White Lies: Race, Class, Gender and Sexuality in White Supremacist Discourse.* New Brunswick. NJ: Rutgers University Press.

Darling-Hammond, Linda. 2010. *The Flat World and Education: How America's Commitment to Equity Will Determine Our Future.* New York: Teachers College Press.

Dart, Richard C., Hilary L. Surratt, Theodore J. Cicero, Mark W. Parrino, S. Geoff Severtson, Becki Bucher-Bartelson, and Jody L. Green. 2015. "Trends in Opioid Analgesic Abuse and Mortality in the United States." *New England Journal of Medicine* 372 (January 15): 372: 241–248.

Davies, James B., Susanna Sandstrom, Anthony Shorrocks, and Edward N. Wolff. 2008. "The World Distribution of Household Wealth." Helsinki, Finland: UNU-WIDER, World Institute for Development Economics Research.

Daugherty, Jill, and Casey Copen. 2016. "Trends in Attitudes about Marriage, Childbearing, and Sexual Behavior: United States 2002, 2006–2010, 2011–2013." *National Health Statistics Reports* 92 (March 17): 1–11.

Davies, Michelle, Jennifer Gilston, and Paul Rogers. 2012. "Examining the Relationship between Male Rape Myth Acceptance, Female Rape Myth Acceptance, Victim Blame, Homophobia, Gender Roles, and Ambivalent Sexism." *Journal of Interpersonal Violence* 27 (14): 2807–2823.

Davis, Georgiann. 2015. *Contesting Intersex: The Dubious Diagnosis.* New York: New York University Press.

Davis, Kingsley. 1945. "The World Demographic Transition." *Annals of the American Academy of Political and Social Sciences* 237 (April): 1–11.

Davis, Kingsley, and Wilbert E. Moore. 1945. "Some Principles of Stratification." *American Sociological Review* 10 (April): 242–247.

DeCamp, Whitney, and Christopher J. Ferguson. 2017. "The Impact of Degree of Exposure to Violent Video Games, Family Background, and Other Factors on Youth Violence." *Journal of Youth and Adolescence* 46 (February): 388–400.

Deegan, Mary Jo. 1988. "W. E. B. DuBois and the Women of Hull-House, 1895–1899." *The American Sociologist* 19 (Winter): 301–311.

Deepwater Horizon Study Group. 2011 (March 1). *Final Report on the Investigation of the Macondo Well Blowout.* **http://ccrm.berkeley.edu/pdfs _papers/bea_pdfs/dhsgfinalreport -march2011-tag.pdf**

Dellinger, Kristen, and Christine L. Williams. 1997. "Makeup at Work: Negotiating Appearance Rules in the Workplace." *Gender & Society* 11 (April): 151–177.

Deming, Michelle E., Eleanor K. Covan, Suzanne C. Swan, and Deborah L. Billings. 2013. "Exploring Rape Myths, Gendered Norms, Group Processing, and the Social Context of Rape among College Women: A Qualitative Analysis." *Violence Against Women* 19 (4): 465–485.

Demos. 2014. *Student Loan Debt by Race and Ethnicity.* **www.demos.org**

DeNavas-Walt, Carmen, and Bernadette D. Proctor. 2014. *Income and Poverty in the United States: 2013.* Washington, DC: U.S. Census Bureau. **www.census.gov**

Denissen, Amy M., and Abigail C. Saguy. 2014. "Gendered Homophobia and the Contradictions of Workplace Discrimination for Women in the Building Trades." *Gender & Society* 28 (3): 381–403.

Desmond, Matthew. 2016. *Evicted: Poverty and Profit in the American City.* New York: Crown.

DeSilver, Drew. 2017. "U.S. Students' Academic Achievement Still Lags That of Their Peers in Many Other Countries." Washington, DC: Pew Research Organization. **www.pewresearch.org /fact-tank/2017/02/15/u-s-students -internationally-math-science/**

DeVault, Marjorie. 1991. *Feeding the Family: The Social Organization of Caring as Gendered Work.* Chicago: University of Chicago Press.

Devine-Eller, Audrey. 2012. "Timing Matters: Test Preparation, Race, and Grade Level." *Sociological Forum* 27 (2): 458–480.

Diallo, Yacouba, Alex Etienne, and Fahrad Mehran. 2013. *Global Child Labour Trends 2008–2012.* Geneva: International Labour Organization.

Dickinson, Timothy. 2014. "Inside the Koch Brothers Toxic Empire." *Rolling Stone,* September 24.

Dill, Bonnie Thornton. 1988. "Our Mothers' Grief: Racial Ethnic Women and the Maintenance of Families." *Journal of Family History* 13 (October): 415–431.

DiMaggio, Anthony. 2011. *The Rise of the Tea Party: Political Discontent and Corporate Media in the Age of Obama.* New York: Monthly Review Press.

DiMaggio, Paul J., and Walter W. Powell. 1991. "Introduction." pp. 1–38 in *The New Institutionalism in Organizational Analysis,* edited by W. W. Powell and P. J. DiMaggio. Chicago: University of Chicago Press.

Dines, Gail, and Jean M. Humez. 2010. *Gender, Race, and Class in Media,* 3rd ed. Thousand Oaks, CA: Sage.

DiPrete, Thomas A., and Claudia Buchmann. 2014. *The Rise of Women: The Growing Gender Gap in Education and What It Means for American Schools.* New York: Russell Sage Foundation.

Dirks, Danielle, and Jennifer C. Mueller. 2010. "Racism and Polar Culture." pp. 115–129 in *Handbook of the Sociology of Racial and Ethnic Relations,* edited by Hernán Vera and Joe R. Feagin. New York: Springer.

Doan, Long, Annalise, Loehr, and Lisa R. Miller. 2014. "Formal Rights and Informal Privileges for Same-Sex Couples: Evidence from a National Survey Experiment." *American Sociological Review* 79 (December): 1172–1195.

Dolan, Jill. 2006. "Blogging on Queer Connections in the Arts and the Five Lesbian Brothers." *GLQ : A Journal of Lesbian and Gay Studies* 12 (3): 491–506.

Domhoff, G. William. 2013. *Who Rules America? The Triumph of the Corporate Rich.* New York: McGraw Hill.

Dowd, James J., and Laura A. Dowd. 2003. "The Center Holds: From Subcultures to Social Worlds." *Teaching Sociology* 31 (January): 20–37.

Drier, Peter. 2012. "The Battle for the Republican Soul: Who Is Drinking the Tea Party?" *Contemporary Sociology* 41 (November): 756–762.

DuBois, W. E. B. 1901. "The Freedmen's Bureau." *Atlantic Monthly* 86 (519): 354–365.

DuBois, W. E. B. 1903. *The Souls of Black Folk: Essays and Sketches.* Chicago: A. C. McClurg and Co.

Dubrofsky, Rachel. 2006. "The Bachelor: Whiteness in the Harem." *Critical Studies in Media Communication* 23 (1): 39–56.

Due, Linnea. 1995. *Joining the Tribe: Growing Up Gay & Lesbian in the '90s.* New York: Doubleday.

Duffy, Mignon. 2011. *Making Care Count: A Century of Gender, Race, and Paid Care Work.* New Brunswick, NJ: Rutgers University Press.

Duneier, Mitchell. 1999. *Sidewalk.* New York: Farrar, Strauss and Giroux.

Dunne, Gillian A. 2000. "Opting into Motherhood: Lesbians Blurring the Boundaries and Transforming the Meaning of Parenthood and Kinship." *Gender & Society* 14 (February): 11–35.

Duggan, Maeva, Nicole B. Ellison, Cliff Lampe, Amanda Lenhart, and Mary Madden. 2015, January 9. *Social Media Update 2014.* Washington, DC: Pew Research Center. **www.pewresearch .org**

Durkheim, Emile. 1951 [1897]. *Suicide.* Glencoe, IL : Free Press.

Durkheim, Emile. 1947 [1912]. *Elementary Forms of Religious Life.* Glencoe, IL: Free Press.

Durkheim, Emile. 1950 [1938]. *The Rules of Sociological Method.* Glencoe, IL: Free Press.

Durkheim, Emile. 1964 [1895]. *The Division of Labor in Society.* New York: Free Press.

Eagan, Kevin, et al. 2013. *The American Freshman: National Norms Fall 2011.* Higher Education Research Institute. Los Angeles, CA: University of California, Los Angeles.

Eberhardt, Jennifer L. 2010. "Enduring Racial Associations: African Americans, Crime, and Animal Imagery." pp. 439–457 in *Doing Race: 21 Essays for the Twenty-First Century,* edited by Paula M. Moya and Hazel Rose Markus. New York: Oxford University Press.

Eberhardt, Jennifer L., Phillip Atiba Goff, Valerie J. Purdie, and Paul G. Davies. 2004. "Seeing Black: Race, Crime, and Visual Processing." *Journal of Personality and Social Behavior* 87 (6): 876–893.

Eckler, Petya, Yusuf Kalyango, and Ellen Paasch. 2017. "Facebook Use and Negative Body Image among U.S. College Women." *Women & Health* 57(2): 249–267.

Economic Mobility Project. 2014. *Women's Work: The Economic Mobility of Women across a Generation.* Washington, DC: Pew Charitable Trusts. www.pewtrusts.org/en/research-and-analysis/reports/2014/04/01/womens-work-the-economic-mobility-of-women-across-a-generation

Economic Policy Institute. 2017a. *State of Working America Data Library.* Washington, DC: Economic Policy Institute. **www.epi.org/data**

Economic Policy Institute. 2017b. "Wages by Percentile." *State of Working America.* Washington, DC: Economic Policy Institute. **www.epi.org**

Edin, Kathryn. 2000. "What Do Low-Income Single Mothers Say about Marriage?" *Social Problems* 47 (February): 112–133.

Edin, Kathryn, and Maria Kefalas. 2005. *Promises I Can Keep: Why Poor Women Put Motherhood before Marriage.* Berkeley, CA: University of California Press.

Edin, Kathryn, and Laura Lein. 1997. *Making Ends Meet: How Single Mothers Survive Welfare and Low-Wage Work.* New York: Russell Sage Foundation.

Edin, Kathryn, and Timothy J. Nelson. 2013. *Doing the Best I Can: Fatherhood in the Inner City.* Berkeley, CA: University of California Press.

Egan. Timothy. 2006. *The Worst Hard Time: The Story of Those Who Survived the Great American Dust Bowl.* Boston: Houghton Mifflin.

Ehrenreich, Barbara, and Deirdre English. 2011. *Complaints and Disorders: The Sexual Politics of Sickness,* 2nd ed. Old Westbury, NY: The Feminist Press.

Ehrlich, Paul. 1968. *The Population Bomb.* New York: Ballantine Books.

Ehrlich, Paul R., and Jianguo Liu. 2002. "Some Roots of Terrorism." *Population and Environment* 24 (2): 183–192.

Eitzen, D. Stanley. 2016. *Fair and Foul: Beyond the Myths and Paradoxes of Sport.* Lanham: MD: Rowman and Littlefield.

Eitzen, D. Stanley, and Maxine Baca Zinn. 2011. *Globalization: The Transformation of Social Worlds,* 3rd ed. San Francisco: Cengage.

Eitzen, D. Stanley, Maxine Baca Zinn, and Kelly Eitzen. 2016. *In Conflict and Order,* 14th ed. Upper Saddle River: Pearson.

Elizabeth, Vivienne. 2000. "Cohabitation, Marriage, and the Unruly Consequences of Difference." *Gender & Society* 14 (February): 87–110.

Emerson, Michael O., and David Hartman. 2006. "The Rise of Religious Fundamentalism." *Annual Review of Sociology* 32: 127–144.

England, Paula. 2010. "The Gender Revolution: Uneven and Stalled." *Gender & Society* 24 (April): 149–166.

England, Paula, and Jonathan Bearak. 2014. "The Sexual Double Standard and Gender Differences in Attitudes toward Casual Sex among U.S. University Students." *Demographic Research* 30 (46): 1327–1338.

Enloe, Cynthia. 2001. *Bananas, Beaches, and Bases: Making Feminist Sense of International Politics,* updated ed. Berkeley, CA: University of California Press.

Erikson, Eric. 1980. *Identity and the Life Cycle.* New York: W. W. Norton.

Erikson, Kai. 1966. *Wayward Puritans: A Study in the Sociology of Deviance.* New York: Wiley.

Ermann, M. David, and Richard J. Lundman. 2001. *Corporate and Governmental Deviance.* New York: Oxford University Press.

Escobar, Edward J. 1993. "The Dialectics of Repression: The Los Angeles Police Department and the Chicano Movement, 1968–1971." *The Journal of American History* 79 (March): 1483–1514.

Espeland, Wendy Nelson. 1998. *The Struggle for Water: Politics, Rationality, and Identity in the American Southwest.* Chicago: University of Chicago Press.

Espiritu, Yen Le. 1992. *Asian American Panethnicity: Bridging Institutions and Identities.* Philadelphia: Temple University Press.

Etzioni, Amatai. 1975. *A Comparative Analysis of Complex Organization: On Power, Involvement, and Their Correlates,* rev. ed. New York: Free Press.

Etzioni, Amatai, John Wilson, Bob Edwards, and Michael W. Foley. 2001. "A Symposium on Robert D. Putnam's Bowling Alone: The Collapse and Revival of American Community." *Contemporary Sociology* 30 (May): 223–230.

Ewert, Stephanie. 2015. *U.S. Population Trends: 2000 to 2060.* Washington, DC: U.S. Department of Commerce. **www.ncsl.org/Portals/1/Documents/nalfo/USDemographics.pdf**

Ewing, Walter A., Daniel E. Martinez, and Rubén G. Rumbaut. 2015. *The Criminalization of Immigration in the United States: July 2015.* Washington, DC: American Immigration Council. **www.immigrationimpact.com**

Farrell, Albert D., Erin L. Thompson, and Krista R. Mehari. 2017. "Dimensions of Peer Influences and Their Relationship to Adolescents' Aggression, Other Problem Behaviors and Prosocial Behavior." *Journal of Youth and Adolescence* 46(6): 1351–1369.

Fausto-Sterling, Anne. 1992. *Myths of Gender: Biological Theories about Women and Men.* New York: Basic Books.

Fausto-Sterling, Anne. 2000. *Sexing the Body: Gender Politics and the Construction of Sexuality.* New York: Basic Books.

Feagin, Joe R. 2000. *Racist America: Roots, Future Realities, and Racial Reparations.* New York: Routledge.

Feagin, Joe R. 2007. *Systemic Racism: A Theory of Oppression.* New York: Routledge.

Federal Bureau of Investigation. 2011. *Uniform Crime Reports.* Washington, DC: U.S. Department of Justice. **www.fbi.gov**

Federal Bureau of Investigation. 2013. *Uniform Crime Reports.* Washington, DC: U.S. Department of Justice.

Federal Bureau of Investigation. 2017. *2017 Hate Crime Statistics.* Washington, DC: U.S. Department of Justice. **www.fbi.gov**

Federal Reserve. 2017. "Consumer Credit: Outstanding Student Debt." **www.federalreserve.gov/releases/g19/current/default.htm**

Federal Trade Commission. 2003. *Identity Theft Survey Report.* **www.ftc.gov** (accessed March 15, 2005).

Fein, Melvyn L. 1988. "Resocialization: A Neglected Paradigm." *Clinical Sociology* 6 (1): 88–100.

Ferber, Abby. 1998. *White Man Falling: Race, Gender, and White Supremacy.* Lanham, MD: Rowman and Littlefield.

Ferdinand, Peter. 2000. *The Internet, Democracy, and Democratization.* London: Frank Cass Publishers.

Ferguson, Christopher J. 2013. *Adolescents, Crime, and the Media: A Critical Analysis.* New York: Springer.

Ferguson, Priscilla Parkhurst. 2014. "Inside the Extreme Sport of Competitive Eating." *Contexts* 13 (Summer): 26–31.

Festinger, Leon, Stanley Schachter, and Kurt Back. 1950. *Social Pressures in Informal Groups: A Study of Human Factors in Housing.* Stanford, CA: Stanford University Press.

Fetner, Tina. 2012. "The Tea Party: Manufactured Dissent or Complex Social Movement?" *Contemporary Sociology* 41 (November): 762–766.

Finkel, David. 2014. *Thank You for Your Service.* New York: Picador.

Finley, Erin P., Mary Bollinger, Polly H. Noël, Megan E. Amuan, Laurel A. Copland, Jacqueline A. Pugh, Albana Dassori, Raymond Palmer, Craig Bryan, and Mary Jo V. Pugh. 2015. "A National Cohort Study of the Association between the Polytrauma Clinical Trial and Suicide-Related Behavior among U.S. Veterans Who Served in Iraq and Afghanistan." *American Journal of Public Health* 105 (2): 380–387.

Fischer, Claude S., Michal Hout, and Martin Sánchez Jankowski. 1996. *Inequality by Design: Cracking the Bell Curve Myth.* Princeton, NJ: Princeton University Press.

Fischer, Nancy L. 2003. "Oedipus Wrecked? The Moral Boundaries of Incest." *Gender & Society* 17 (1): 92–110.

Fishbein, Allan J., and Patrick Woodall. 2006. "Women are Prime Targets for Subprime Lending: Women Are Disproportionately Represented in High-Cost Mortgage Market." Washington, DC: Consumer Federation of America.

Fisher, Bonnie S., Francis T. Cullen, and Michael G. Turner. 2000. *Sexual Victimization of College Women.* The National Criminal Justice Reference Service. **www.ncjrs.gov/pdffiles1 /nij/182369.pdf**

Fishman, Charles. 2011. *The Big Thirst: The Secret Life and Turbulent Future of Water.* New York: Free Press.

Fiske, Susan T., and Federika Durante. 2016. "Stereotype Content across Culture: Variation on a Few Themes." *Handbook of Advances in Culture and Psychology, Volume 6,* edited by Michele J. Gelfand, Chi-yue Chiu, and Ying-yi Hong. Oxford Scholarship Online.

Flanagan, William G. 1995. *Urban Sociology: Images and Structure.* Boston: Allyn and Bacon.

Fleury-Steiner, Ben. 2012. *Disposable Heroes: The Betrayal of African American Veterans.* Lanham, MD: Rowman and Littlefield.

Flippen, Chenoa A. 2014. "Intersectionality at Work: Determinants of Labor Supply among Immigrant Latinas." *Gender & Society* 28 (3): 404–434.

Folbre, Nancy. 2001. *The Invisible Heart: Economics and Family Values.* New York: New Press.

Fontenot, Kayla, Jessica Semega, and Melissa Kollar. 2018. *Income and Poverty in the United States: 2017.* Washington, DC: U.S. Census Bureau. **www.census .gov**

Fortini, Amanda. 2005. "The Great Flip-Flop Flap." *Slate,* July 22. **www .slate.com**

Foucault, Michael. 1995. *Discipline and Punish.* New York: Knopf Doubleday Publishing Group.

Fraga, Luis Ricardo. 2012. *Latinos in the New Millennium: An Almanac of Opinion, Behavior, and Policy Preferences.* New York: Cambridge University Press.

Frankenberg, Erica, and Chungmei Lee. 2002. *Race in American Public Schools: Rapidly Resegregating School Districts.* Los Angeles: Civil Rights Project, UCLA.

Frazier, E. Franklin. 1957. *The Black Bourgeoisie.* New York: Collier Books.

Freedle, Roy O. 2003. "Correcting the SATs Ethnic and Social Class Bias: A Method of Re-Estimating SAT Scores." *Harvard Educational Review* 73 (Spring): 1–43.

Freedman, Estelle B., and John D'Emilio. 1988. *Intimate Matters: A History of Sexuality in America.* New York: Harper & Row.

Freeman, Jo. 1983. "A Model for Analyzing the Strategic Options of Social Movement Organizations." pp. 193–210 in *Social Movements of the Sixties and Seventies,* edited by Jo Freeman. New York: Longman.

Frey, William H. 2015. *Diversity Explosion: How New Racial Demographics Are Remaking America.* Washington, DC: Brookings Institution Press.

Friedman, Thomas L. 1999. *The Lexus and the Olive Tree.* New York: Farrar, Strauss, and Giraux.

Friedman, Thomas L. 2016. *Thank You for Being Late.* New York: Farrar, Straus, and Giroux.

Fry, Richard. 2012. *A Record One-in-Five Households Now Owe Student Loan Debt.* Washington, DC: Pew Research Center. **www.pewsocialtrends.org**

Fryberg, Stephanie, and Alisha Watts. 2010. "We're Honoring You Dude: Myths, Mascots, and American Indians." pp. 458–480 in *Doing Race: 21 Essays for the 21st Century,* edited by Hazel Rose Markus and Paula M. L. Moya. New York: W. W. Norton.

Frye, Marilyn. 1983. *The Politics of Reality.* Trumansburg, NY: The Crossing Press.

Funk, Cary, and Brian Kennedy. 2016. "The Politics of Climate." Washington, DC: Pew Research Center. **www .pewinternet.org/2016/10/04/the -politics-of-climate/**

Gaertner, Samuel L., and John F. Dovidio. 2005. "Understanding and Addressing Contemporary Racism: From Aversive Racism to the Common In-group Identity Model." *Journal of Social Issues* 61 (3): 615–639.

Gallagher, Charles A. 2013. "Color-Blind Privilege." pp. 91–95 in *Race, Class and Gender: An Anthology,* 7th ed., edited by Margaret L. Andersen and Patricia Hill Collins. Belmont CA: Wadsworth/ Cengage.

Gallup Poll. 2008. "Homosexual Relations." *The Gallup Poll.* Princeton, NJ: Gallup Organization. **www.gallup.com**

Gallup Poll. 2012. "Guns." *The Gallup Poll.* Princeton, NJ: Gallup Organization. **www.gallup.com**

Gallup Poll. 2016a. "Confidence in Institutions." *The Gallup Poll.* Princeton, NJ: Gallup Organization. **www.gallup.com**

Gallup Poll. 2016b. "Religion." Princeton, NJ: *Gallup Poll.* **http://news.gallup.com /poll/1690/religion.aspx**

Gallup. 2017. "Gay and Lesbian Rights." Princeton, NJ: The Gallup Organization. **http://news.gallup.com/poll/1651 /gay-lesbian-rights.aspx**

Gamson, Joshua. 1998. *Freaks Talk Back: Tabloid Talk Shows and Sexual Nonconformity.* Chicago: University of Chicago Press.

Gans, Herbert. 1982 [1962]. *The Urban Villagers: Group and Class in the Life of Italian Americans.* New York: Free Press.

Gao, George. 2016. "Biggest Share of Whites in U.S. are Boomers, But for Minority Groups It's Millennials or Younger." Washington, DC: Pew Research Organization. **http:// www.pewresearch.org/fact-tank /2016/07/07/biggest-share-of -whites-in-u-s-are-boomers-but -for-minority-groups-its-millennials -or-younger/**

Garfinkel, Harold. 1967. *Studies in Ethnomethodology.* Englewood Cliffs, NJ: Prentice Hall.

Garner, Steve, and Saher Selod. 2014. "The Racialization of Muslims." *Critical Sociology* 41 (1): 9–19.

Gates, Henry Louis, Jr., and Nellie Y. McKay, eds. 1997. *The Norton Anthology of African American Literature.* New York: W. W. Norton.

Geiger, Abigail. 2018. "About Six-in-Ten Americans Favor Marijuana Legislation." Washington, DC: Pew Research Center. **http://www.pewresearch.org/fact-tank /2018/01/05/americans-support -marijuana-legalization/**

Genovese, Eugene. 1972. *Roll, Jordan, Roll: The World the Slaves Made.* New York: Pantheon.

Gerschick, T. J., and A. S. Miller. 1995. "Coming to Terms: Masculinity and Physical Disability." pp. 183–204 in *Men's Health and Illness: Gender, Power, and the Body,* Research on Men and Masculinities Series, vol. 8, edited by D. Sabo and D. F. Gordon. Thousand Oaks, CA: Sage Publications.

Gerson, Kathleen. 2010. *The Unfinished Revolution: How a New Generation Is Shaping Family, Work, and Gender in America.* New York: Oxford University Press.

Gerstel, Naomi. 2000. "The Third Shift: Gender and Care Work Outside the Home." *Qualitative Sociology* 23 (4): 467–483.

Gerstel, Naomi, and Sally Gallagher. 2001. "Men's Caregiving: Gender and the Contingent Character of Care." *Gender & Society* 15 (April): 197–217.

Gerth, Hans, and C. Wright Mills (eds.). 1946. *From Max Weber: Essays in Sociology.* New York: Oxford University Press.

Giddings, Paula. 1994. *In Search of Sisterhood: Delta Sigma Theta and the Challenge of the Black Sorority Movement.* New York: William Morrow.

Giddings, Paula. 2008. *A Sword among Lions: Ida B. Wells and the Campaign against Lynching.* New York: Amistad.

Gilbert, D. T., and P. S. Malone. 1995. "The Correspondence Bias." *Psychological Bulletin* 117 (1): 21–38.

Gile, K.G., L.G. Johnston, and M.J. L Salganik. 2015. "Diagnostics for Respondent-driven Sampling." *Journal of the Royal Statistical Society* 178 (1): 241–269.

Gilkes, Cheryl Townsend. 2000. "*If It Wasn't for the Women ... Black Women's Experience and Womanist Culture in*

Church and Community. Maryknoll, NY: Orbis Books.

Gimlin, Debra. 1996. "Pamela's Place: Power and Negotiation in the Hair Salon." *Gender & Society* 10 (October): 505–526.

Gimlin, Debra. 2002. *Body Work: Beauty and Self-Image in American Culture*. Berkeley, CA: University of California Press.

Gitlin, Todd. 2007. *Media Unlimited: How the Torrent of Images and Sounds Overwhelms Our Lives*, revised edition. New York: Metropolitan Books.

Glassner, Barry. 1999. *Culture of Fear: Why Americans Are Afraid of the Wrong Things*. New York: Basic Books.

Glaze, Lauren E., and Laura M. Maruschak. 2008. "Parents in Prison and Their Minor Children." Washington, DC: U.S. Bureau of Justice Statistics. **www.ojp.usdoj.gov**

Glenn, Evelyn Nakano. 1986. *Issei, Nisei, War Bride: Three Generations of Japanese American Women in Domestic Service*. Philadelphia, PA: Temple University Press.

Glenn, Evelyn Nakano. 2002. *Unequal Freedom: How Race and Gender Shaped American Citizenship and Labor*. Cambridge, MA: Harvard University Press.

Glock, Charles, and Rodney Stark. 1965. *Religion and Society in Tension*. Chicago: Rand McNally.

Goffman, Alice. 2009. "On the Run: Wanted Men in a Philadelphia Ghetto." *American Sociological Review* 74 (June): 339–357.

Goffman, Erving. 1959. *The Presentation of Self in Everyday Life*. Garden City, NY: Doubleday.

Goffman, Erving. 1961. *Asylums: Essays on the Social Situation of Mental Patients and Other Inmates*. Garden City, NY: Anchor.

Goffman, Erving. 1963. *Stigma: Notes on the Management of Spoiled Identity*. Englewood Cliffs, NJ: Prentice Hall.

Golash-Boza, Tanya, and William Darity Jr. 2008. "Latino Racial Choices: The Effects of Skin Colour and Discrimination on Latinos' and Latinas' Racial Self-Identifications." *Ethnic and Racial Studies* 31 (5): 899–934.

Goldberg, Abbie, Deborah Kashy, and JuilAnna Smith. 2012. "Gender-Typed Play Behaviors in Early Childhood: Adopted Children with Lesbian, Gay, and Heterosexual Parents." *Sex Roles* 67 (9–10): 503–515.

Goncalo, Jack A., Jennifer A. Chatman, Michelle M. Duguid, and Jessica A. Kennedy. 2015. "Creativity from Constraint? How the Political Correctness Norm Influences Creativity in Mixed-Sex Work Groups." *Administrative Science Quarterly* 60 (1): 1–30.

Gonella, Catalina. 2017. "Survey: 20 Percent of Millennials Identify as LGBTQ." *NBC News*, March 31. **www.nbcnews.com /feature/nbc-out/survey-20-percent -millennials-identify-lgbtq-n740791**

Gonzales, Roberto G. 2016. *Lives in Limbo: Undocumented and Coming of Age in America*. Oakland, CA: University of California Press.

Gordon, Linda. 1977. *Woman's Body/ Woman's Right*. New York: Penguin.

Gore, Albert. 2006. *An Inconvenient Truth: The Planetary Emergency of Global Warming and What We Can Do about It*. Emmaus, PA: Rodale.

Gottesdiener, Laura. 2013. "The Great Eviction: Black American and the Toll of the Foreclosure Crisis." *Mother Jones* (August 1). **www.motherjones.com /politics/2013/08/black-america -foreclosure-crisis/**

Gottfredson, Michael R., and Travis Hirschi. 1990. *A General Theory of Crime*. Stanford, CA: Stanford University Press.

Gottfredson, Michael R., and Travis Hirschi. 1995. "National Crime Control Policies." *Society* 32 (January–February): 30–36.

Gould, Stephen Jay. 1999. "The Human Difference." *The New York Times* (July 2).

Grall, Timothy. 2016. *Custodial Mothers and Fathers and Their Child Support: 2013*. Washington, DC: U.S. Census Bureau. **www.census.gov/content/dam/Census /library/publications/2016/demo/P60 -255.pdf**

Gramsci, Antonio. 1971. *Selections from the Prison Notebooks of Antonio Gramsci*, edited by Quintin Hoare and Geoffrey Nowell. London: Lawrence and Wishart.

Granger, Kristen L., Laura D. Hanish, Olga Kornienko, and Robert H. Bradley. 2017. "Preschool Teachers' Facilitation of Gender-Typed and Gender-Neutral Activities during Free Play." *Sex Roles* 76(7–8): 498–510.

Granovetter, Mark. 1973. "The Strength of Weak Ties." *American Journal of Sociology* 78 (May): 1360–1380.

Granovetter, Mark. 1974. *Getting a Job: A Study of Contacts and Careers*. Cambridge, MA: Harvard University Press.

Grant, Don Sherman, II, and Ramiro Martínez Jr. 1997. "Crime and the Restructuring of the U.S. Economy: A Reconsideration of the Class Linkages." *Social Forces* 75 (March): 769–799.

Graves, Joseph L. 2004. *The Race Myth: Why We Pretend Race Exists in America*. New York: Dutton.

Greenbaum, Susan D. 2015. *Blaming the Poor: The Long Shadow of the Moynihan Report on Cruel Images about Poverty*. New Brunswick, NJ: Rutgers University Press.

Greenebaum, Jessica B. 2012. "Managing Impressions: 'Face-Saving' Strategies of Vegetarians and Vegans." *Humanity & Society* 36 (4): 309–325.

Greenstone, Michael, Adam Looney, Jeremy Patashnik, and Muxin Yu. 2013. *Thirteen Economic Facts about Social Mobility and the Role of Education*. Washington, DC: Brookings Institute. **www.brookings .edu/research/thirteen-economic-facts -about-social-mobility-and-the-role-of -education/**

Greenwood, Shannon, Andrew Perrin, and Mauve Duggan. 2016. *Social Media Update 2016*. Washington, DC: Pew Research Organization. **www.pewresearch.org**

Grigoryeva, Angelina. 2017. "Own Gender, Sibling's Gender, Parent's Gender: The Division of Elderly Parent Care among Adult Children." *American Sociological Review* 82: 116–146.

Grimal, Pierre (ed.). 1963. *Larousse World Mythology*. New York: Putnam.

Grimsley, Edwin. 2012. "What Wrongful Convictions Teach Us about Racial Inequality." New York: The Innocence Project. **www.innocenceproject.org**

Grindstaff, Laura. 2002. *The Money Shot: Trash Class, and the Making of TV Talk Shows*. Chicago: University of Chicago Press.

Guinier, Lani. 2016. *The Tyranny of the Meritocracy: Democratizing Higher Education in America*. Boston: Beacon Press.

Gurin, Patricia, E. L. Dey, Sylvia Hurtado, and G. Gurin. 2002. "Diversity and Higher Education: Theory and Impact on Educational Outcomes." *Harvard Educational Review* 72 (3): 330–366.

Guttierez y Muhs, Gabriella, Yolanda Flores Niemann, Carmen G. Gonzalez, and Angela P. Harris (eds.). 2012. *Presumed Incompetent: The Intersections of Race and Class for Women in Academia*. Logan, UT: Utah State University Press.

Habermas, Jürgen. 1970. *Toward a Rational Society: Student Protest, Science, and Politics*. Boston: Beacon Press.

Haley-Lock, Anna, and Stephanie Ewert. 2011. "Serving Men and Mothers: Workplace Practices and Workforce Composition in Two U.S. Restaurant Chains and States." *Community, Work & Family* 14 (4): 387–404.

Hall, Edward T. 1966. *The Hidden Dimension*. New York: Doubleday.

Halpern-Meekin, Kathryn Edin, Laura Tach, and Jennifer Sykes. 2015. *It's Not Like I'm Poor: How Working Families Make Ends Meet in a Post-Welfare World*. Berkeley, CA: University of California Press.

Hamer, Jennifer. 2001. *What It Means to Be Daddy*. New York: Columbia University Press.

Hamilton, Laura, and Elizabeth A. Armstrong. 2009. "Gendered Sexuality in Young Adulthood: Double Binds and Flawed Options." *Gender & Society* 23 (October): 589–616.

Hamilton, Mykol C. 1988. "Using Masculine Generics: Does Generic He Increase Male Bias in the User's Imagery?" *Sex Roles* 19 (December): 785–799.

Handlin, Oscar. 1951. *The Uprooted*. Boston: Little, Brown.

Haney, C., C. Banks, and P. G. Zimbardo. 1973. "Interpersonal Dynamics in a Simulated Prison." *International Journal of Criminology and Penology* 1: 69–97.

Haney, Lynne. 1996. "Homeboys, Babies, Men in Suits: The State and the Reproduction of Male Dominance." *American Sociological Review* 61 (October): 759–778.

Hardie, Jessica H., and Judith A. Seltzer. 2016. "Parent-Child Relationships at the Transition to Adulthood: A Comparison of Black, Hispanic, and White Immigrant and Native-Born Youth." *Social Forces* 95(1): 321–353.

Harnois, Catherine E. 2010. "Race, Gender, and the Black Women's Standpoint." *Sociological Forum* 25 (1): 68–85.

Harp, Dustin, and Mark Tremayne. 2006. "The Gendered Blogosphere: Examining Inequality Using Network and Feminist Theory." *Journalism & Mass Communication Quarterly* 83 (2): 247–264.

Harris, Angel L. 2006. "I Don't Hate School: Revisiting "'Oppositional Culture' Theory of Blacks' Resistance to Schooling." *Social Forces* 85 (2): 797–834.

Harris, Marvin. 1974. *Cows, Pigs, Wars, and Witches: The Riddles of Culture.* New York: Vintage.

Harris, Richard J., Juanita M. Firestone, and Mary Bollinger. 2000. "Gender Role Attitudes among Native Americans in Comparative Perspective." *Free Inquiry in Creative Sociology* 28 (2): 63–76.

Harrison, Roderick. 2000. "Inadequacies of Multiple Response Race Data in the Federal Statistical System." Manuscript. Joint Center for Political and Economic Studies and Howard University, Department of Sociology, Washington, DC.

Hartley, D. 2014. "Rural Health Disparities, Population Health, and Rural Culture." *American Journal of Public Health* 94 (10): 1675–1678.

Hayes, Rebecca M., Rebecca L. Abbott, and Savannah Cook. 2016. "It's Her Fault." *Violence Against Women* 22(13): 1540–1555.

Hayes-Smith, Rebecca. 2011. "Gender Norms in the *Twilight* Series." *Contexts* 10 (Spring): 78–79.

Hays, Sharon. 2003. *Flat Broke with Children: Women in the Age of Welfare Reform.* New York: Oxford University Press.

Hearnshaw, Leslie. 1979. *Cyril Burt: Psychologist.* Ithaca, NY: Cornell University Press.

Henderson, Richard. 2015. "Industry Employment and Output Projections to 2024." *Monthly Labor Review* (December). **www.bls.gov/**

Hendy, Helen M., Cheryl Gustitus, and Jamie Leitzel-Schwalm. 2001. "Social Cognitive Predictors of Body Images in Preschool Children." *Sex Roles* 44 (May): 557–569.

Henry, Kathy. 2007. "Warrior Woman— Ida B. Wells Barnett." *Ezinearticles.* **http:// ezinearticles.com/?Warrior-Woman— Ida-B.-Wells-Barnett&id=446100**

Henry, P. J. and Geoffrey Wetherell. 2017. "Countries with Greater Gender Equality Have More Positive Attitudes and Laws Concerning Lesbians and Gay Men." *Sex Roles* 77 (7–8): 523–532.

Henslin, James M. 1993. "Doing the Unthinkable." pp. 253–262 in *Down to Earth Sociology*, 7th ed., edited by James M. Henslin. New York: Free Press.

Herring, Cedric. 2009. "Does Diversity Pay?: Race, Gender, and the Business Case for Diversity." *American Sociological Review* 74 (2): 208–224.

Herrnstein, Richard J., and Charles Murray. 1994. *The Bell Curve: Intelligence and Class Structure in American Life.* New York: Free Press.

Hertz, Rosanna. 2006. *Single by Chance, Mothers by Choice: How Women Are Choosing Parenthood without Marriage and Creating the New American Family.* New York: Oxford University Press.

Higginbotham, A. Leon, Jr. 1978. *In the Matter of Color: Race and the American Legal Process.* New York: Oxford University Press.

Higginbotham, Elizabeth, and Margaret Andersen, eds. 2012. *Race and Ethnicity in Society: The Changing Landscape*, 3rd edition. Belmont, CA: Wadsworth/ Cengage.

Hill, Catherine. 2010. *Why So Few? Women in Science, Technology, Engineering, and Mathematics.* Washington, DC: American Association of University Women.

Hill, Lori Diane. 2001. "Conceptualizing Educational Attainment Opportunities of Urban Youth: The Effects of School Capacity, Community Context and Social Capital." PhD diss., University of Chicago.

Hirschi, Travis. 1969. *Causes of Delinquency.* Berkeley: University of California Press.

Hirschman, Charles. 1994. "Why Fertility Changes." *Annual Review of Sociology* 20: 203–223.

Hirschman, Charles, and Douglas S. Massey. 2008. "Places and Peoples: The New American Mosaic." pp. 1–21 in *New Faces in New Places: The Changing Geography of American Immigration*, ed. by Douglass Massey. New York: Russell Sage Foundation.

Hochschild, Arlie Russell. 1983. *The Managed Heart: Commercialization of Human Feelings.* Berkeley, CA: University of California Press.

Hochschild, Arlie Russell. 1997. *The Time Bind: When Work Becomes Home and Home Becomes Work.* New York: Metropolitan Books.

Hochschild, Arlie Russell. 2003. *The Commercialization of Intimate Life: Notes from Home and Work.* Berkeley: University of California Press.

Hochschild, Arlie Russell. 2016. *Strangers in Their Own Land: Anger and Mourning on the American Right.* New York: The New Press.

Hochschild, Arlie Russell, with Anne Machung. 1989. *The Second Shift: Working Parents and the Revolution at Home.* New York: Viking.

Hofstra, Bas, Rense Corten, Frankvan Tubergen, and Nicole B. Ellison. 2017. "Sources of Segregation in Social Networks: A Novel Approach Using Facebook." *American Sociological Review* 82 (May): 625–656.

Holyoke, Thomas T. 2009. "Interest Group Competition and Coalition Formation." *American Journal of Political Science* 53 (April): 360–375.

Hondagneu-Sotelo, Pierrette. 2001. *Doméstica: Immigrant Workers Cleaning and Caring in the Shadows of Affluence.* Berkeley, CA: University of California Press.

Hong, Jun S., Jungup Lee, Dorothy L. Espelage, Simon C. Hunter, Desmond U. Patton, and Tyrone Rivers. 2016. "Understanding the Correlates of Face-to-Face and Cyberbullying Victimization among U.S. Adolescents: A Social-Ecological Analysis." *Violence and Victims* 31(4): 638–663.

Hopkins, Andrew. 2012. *Disastrous Decisions: The Human and Organisational Causes of the Gulf of Mexico Blowout.* Sydney: CCH Australia.

Horovitz, Bruce, and Julie Appleby. 2017. "Prescription Drug Costs are Up; So Are TV Ads Promoting Them." *USA Today*, March 16.

Howard, Philip H., and Patricia Allen. 2010. "Beyond Organic and Fair Trade? An Analysis of Ecolabel Preferences in the United States." *Rural Sociology* 75 (2): 244–269.

Huesmann, L. R., J. Moise, C. D. Podolski, and J. D. Eron. 2003. "Longitudinal Relations between Childhood Exposure to Media Violence and Adult Aggression and Violence, 1977–1992." *Developmental Psychology* 39 (2): 201–221.

Hughes, Langston. 1967. *The Big Sea.* New York: Knopf.

Hull, Kathleen E., Ann Meier, and Timothy Ortyl. 2010. "The Changing Landscape of Love and Marriage." *Contexts* 9 (Spring): 32–37.

Hunt, Ruth. 2005. "Subjective Rating by Skin Color Gradation, Gender, and Other Characteristics." Junior Project, Princeton University, manuscript.

Hurtado, Aida, and Mrinal Sinha. 2008. "More Than Men: Latino Feminist Masculinities and Intersectionality." *Sex Roles* 59: 337–349.

Ignatiev, Noel. 1995. *How the Irish Became White.* New York: Routledge.

Imoagene, Onoso. 2012. "Being British vs. Being American: Identification Among Second-Generation Adults of Nigerian Descent in the US And UK." *Ethnic and Racial Studies* 35 (December): 2153–2173.

Insight. 2013 (August). *The Relationship between Alcohol and Sexual Assault.* **www.nvcc.edu**

Institute of Medicine. 2011. *The Health of Lesbian, Gay, Bisexual, and Transgender People: Building a Foundation for Better Understanding.* Washington, DC: The National Academies Press.

Institute of Medicine of the National Academies. 2010. *Returning Home from Iraq and Afghanistan: Preliminary Assessment of Readjustment Needs of Veterans, Service Members, and Their Families.* Committee on the Initial Assessment of Readjustment Needs of Military Personnel, Veterans, and Their Families. Washington, DC: National Academies Press.

Inter-Parliamentary Union. 2017. *Women in National Parliaments 2017.* **www.ipu.org**

International Energy Agency. 2014. *Balances of OECD and Non-OECD Statistics.*

International Labour Organization. 2017. *What is Child Labor.* **www.ilo.org**

Irwin, Katherine. 2001. "Legitimating the First Tattoo: Moral Passage through

Informal Interaction." *Symbolic Interaction* 24 (March): 49–73.

Itakura, Hiroko. 2014. "Femininity in Mixed-Sex Talk and Intercultural Communication." *Pragmatics and Society* 5 (3): 455–483.

Iyengar, Shanto. 2010. "Race in the News: Stereotypes, Political Campaigns, and Market-Based Journalism." pp. 251–273 in *Doing Race: 21 Essays for the Twenty-First Century*, edited by Paula M. Moya and Hazel Rose Marcus. New York: Oxford University Press.

Jakubczyk, A., A. Klimkiewicz, A. Krasowska, Kopera, A. Slawinska-Ceran, K. J. Brower, and M. Wojnar. 2014. "History of Sexual Abuse and Suicide Attempts in Alcohol-Dependent Patients." *Child Abuse & Neglect* 38 (9): 1560–1568.

Jack, Anthony Abraham. 2014. "Culture Shock Revisited: The Social and Cultural Contingencies to Class Marginality." *Sociological Forum* 28 (June): 453–475.

Jackson, Sarah J. 2016. "(Re)Imagining Intersectional Democracy from Black Feminism to Hashtag Activism." *Women's Studies in Communication* 39(4): 375–379.

Jacobs, David, Zhenchao Qian, Jason T. Charmichael, and Stephanie L. Kent. 2007. "Who Survives Death Row: An Individual and Contextual Analysis." *American Sociological Review* 72 (August): 610–632.

Jacobs, Jerry A., and Kathleen Gerson. 2004. *The Time Divide: Work, Family, and Gender Inequality.* New York: Oxford University Press.

Jacobs, Jerry A., and Kathleen Gerson. 2015. "Unpacking Americans' Views of the Employment of Mothers and Fathers Using National Vignette Data." *Gender & Society* 30 (3): 413–441.

Janis, Irving L. 1982. *Groupthink: Psychological Studies of Policy Decisions and Fiascos,* 2nd ed. Boston: Houghton Mifflin.

Jargowsky, Paul A. 2015. *Concentration of Poverty in the New Millennium: Changes in Prevalence, Composition, and Location of High Poverty Neighborhoods.* New Brunswick, NJ: The Century Foundation and Rutgers Center for Urban Research.

Jenness, Valerie, and Sarah Fenstermaker. 2014. "Agnes Goes to Prison: Gender Authenticity, Transgender Inmates in Prisons for Men, and Pursuit of 'The Real Deal.'" *Gender & Society* 28 (February): 5–31.

Jiménez, Tomás. 2009. *Replenished Ethnicity: Mexican Americans, Immigration, and Identity.* Berkeley, CA: University of California Press.

Johnson, Jennifer A. 2011. "Mapping the Feminist Political Economy of the Online Commercial Pornography Industry: A Network Approach." *International Journal of Media and Cultural Politics* 7 (2): 189–208.

Johnstone, Ronald I. 1992. *Religion in Society: A Sociology of Religion,* 4th ed. Englewood Cliffs, NJ: Prentice Hall.

Jones, Diane Carlson. 2001. "Social Comparison and Body Image: Attractiveness Comparison to Models and Peers among Adolescent Girls and Boys." *Sex Roles* 45 (November): 645–664.

Jones, James M., John F. Dovidio, and Deborah L. Vietze. 2013. *The Psychology of Diversity.* Malden, MA: Wiley Blackwell.

Jones-Correa, M., and D. L. Leal. 1996. "Becoming 'Hispanic': Secondary Panethnic Identification among Latin American-Origin Populations in the United States." *Hispanic Journal of Behavioral Sciences* 18 (2): 214–254.

Jordan, Catheleen, and David Cory. 2010. "Boomers, Boomerangs, and Bedpans." *National Social Science Journal* 34 (1): 79–84.

Jordon, Winthrop D. 1968. *White over Black: American Attitudes toward the Negro 1550–1812.* Chapel Hill, NC: University of North Carolina Press.

Jussim, Lee, and Kent D. Harber. 2005. "Teacher Expectations and Self-Fulfilling Prophesies: Knowns and Unknowns, Resolved and Unresolved Controversies." *Personality and Social Psychology Review* 9 (2): 131–155.

Kaiser Family Foundation. 2017. "Poll: Public Divided on Repealing Obamacare, but Few Want It Repealed Without Replacement Details." Menlo Park, CA: Kaiser Family Foundation.

Kalleberg, Arne. 2013. *Good Jobs, Bad Jobs: The Rise of Polarized and Precarious Employment Systems in the United States, 1970s to 2000s.* New York: Russell Sage Foundation.

Kalof, Linda. 1999. "The Effects of Gender and Music Video Imagery on Sexual Attitudes." *Journal of Social Psychology* 139 (June): 378–385.

Kamin, Leon J. 1974. *The Science and Politics of IQ.* Potomac, MD: Lawrence Erlbaum.

Kane, Emily. 2012. *The Gender Trap: Parents and the Pitfalls of Raising Boys and Girls.* New York: New York University Press.

Kang, Miliann. 2010. *The Managed Hand: Race, Gender and the Body in Beauty Service Work.* Berkeley, CA: University of California Press.

Kanter, Rosabeth Moss. 1977. *Men and Women of the Corporation.* New York: Basic Books.

Kaplan, Elaine Bell. 1996. *Not Our Kind of Girl: Unraveling the Myths of Black Teenage Motherhood.* Berkeley, CA: University of California Press.

Kaufman, Gayle. 1999. "The Portrayal of Men's Family Roles in Television Commercials." *Sex Roles* 41 (September): 439–458.

Kennedy, Randall. 2003. *Interracial Intimacies: Sex, Marriage, Identity and Adoption.* New York: Pantheon.

Kephart, W. H. 1993. *Extraordinary Groups: An Examination of Unconventional Life,* rev. ed. New York: Martin's Press.

Kerr, N. L. 1992. "Issue Importance and Group Decision Making." pp. 68–88 in *Group Process and Productivity,* edited by S. Worchel, W. Wood, and J. A. Simpson. Newbury Park, CA: Sage.

Kessler, Suzanne J. 1990. "The Medical Construction of Gender: Case Management of Intersexed Infants." *Signs* 16 (Autumn): 3–26.

Khan, Saher. 2016. "Cotton Farming in Uzbekistan Is Modern Slavery." **https://muftah.org/?s=Cotton+Farming+in+Uzbekistan+Is+Modern+Slavery**

Khashan, Hilal, and Lina Kreidie. 2001. "The Social and Economic Correlates of Islamic Religiosity." *World Affairs* 1654 (Fall): 83–96.

Kibria, Nazli, Cara Bowman, and Megan O'Leary. 2014. *Race and Immigration.* Malden, MA: Polity.

Kiley, Jocelyn. 2017 (June 23). "Public Support for 'Single Payer' Health Coverage Grows, Driven by Democrats." Washington, DC: Pew Research Center. **www.pewresearch.org/fact-tank/2017/06/23/public-support-for-single-payer-health-coverage-grows-driven-by-democrats/**

Kim, Elaine H. 1993. "Home Is Where the Han Is: A Korean American Perspective on the Los Angeles Upheavals." pp. 215–235 in *Reading Rodney King/Reading Urban Uprising,* edited by Robert Gooding-Williams. New York: Routledge.

Kimmel, Michael. 2000. "Saving the Males: The Sociological Implications of Virginia Military Institute and the Citadel." *Gender & Society* 14 (August): 494–516.

Kimmel, Michael. 2008. *Guyland: The Perilous World Where Boys Become Men.* New York: Harper.

Kimmel, Michael S., and Michael A. Messner. 2012. *Men's Lives,* 9th ed. Boston: Allyn and Bacon.

King, Eden B., Jonathan J. Mohr, Chad Peddie, Kristen P. Jones, and Matt Kendra. 2014. "Predictors of Identity Management: An Exploratory Experience-Sampling Study of Lesbian, Gay, and Bisexual Workers." *Journal of Management* 20 (June): 1–27.

Kistler, Michelle E., and Moon J. Lee. 2010. "Does Exposure to Sexual Hip-Hop Music Videos Influence the Sexual Attitudes of College Students?" *Mass Communication and Society* 12 (January): 67–86.

Kitano, Harry. 1976. *Japanese Americans: The Evolution of a Subculture,* 2nd ed. New York: Prentice Hall.

Kleinfeld, Judith. 1999. "Student Performance: Males versus Females." *National Affairs* 134 (Winter): 3–20.

Klinenberg, Eric. 2002. *Heat Wave: A Social Autopsy of Disaster in Chicago.* Chicago: University of Chicago Press.

Klinenberg, Eric. 2012. *Going Solo: The Extraordinary Rise and Surprising Appeal of Living Alone.* New York: The Penguin Press.

Klinger, Lori J., James A. Hamilton, and Peggy J. Cantrell. 2001. "Children's Perceptions of Aggression and Gender-Specific Content in Toy Commercials." *Social Behavior and Personality* 29 (1): 11–20.

Kloer, Amanda. 2010. "Sex Trafficking and HIV/AIDS: A Deadly Junction for Women and Girls." *Human Rights Magazine* 37 (Spring).

Kneebone, Elizabeth. 2014. "The Growth and Spread of Concentrated Poverty, 2000 to 2008–2012." Washington, DC: Brookings Institute. **www.brookings .edu/research/interactives/2014 /concentrated-poverty#/M10420**

Kneebone, Elizabeth, and Natalie Holmes. 2014. *New Census Data Show Few Metro Areas Made Progress Against Poverty in 2013*. Washington, DC: Brookings. **www .brookings.edu**

Ko, Eunjeong, Sunhee Cho, Ramona Perez, Younsook Yeo, and Helen Palomino. 2013. "Good and Bad Death: Exploring the Perspectives of Older Mexican Americans." *Journal of Gerontological Social Work* 56 (1): 6–25.

Kochhar, Rakesh, and Richard Fry. 2014. *Wealth Inequality Has Widened along Racial Ethnic Lines Since End of Great Recession*. Washington, DC: Pew Research Organization. **www .pewresearch.org/fact-tank/2014/12/12/ racial-wealth-gaps-great-recession/**

Kochhar, Rakesh, Richard Fry, and Paul Taylor. 2011. "Wealth Gaps Rise to Record Highs between Whites, Blacks, Hispanics." Washington, DC: Pew Research Center. **www .pewsocialtrends.org**

Kocieniewski, David, and Robert Hanley. 2000. "Racial Profiling Was Routine, New Jersey Says." *The New York Times* (November 28): 1.

Komter, Martha L. 2006. "From Talk to Text: The Interactional Construction of a Police Record." *Research on Language and Social Interaction* 39 (3): 201–228.

Kozol, Jonathan. 2012. *Savage Inequalities: Children in America's Schools*. New York: Broadway Books.

Krcmar, Marina. 2014. "Can Infants and Toddlers Learn Words from Repeat Exposure to an Infant Directed DVD?" *Journal of Broadcasting & Electronic Media* 58 (2): 196–214.

Kreager, David, and Jeremy Staff. 2009. "The Sexual Double Standard and Adolescent Acceptance." *Social Psychology Quarterly* 72 (June): 143–164.

Krebs, Christopher, Christine Lindquist, Tara Warner, Bonnie Fisher, and Sandra Martin. 2007. *The Campus Sexual Assault (CSA) Study: Final Report*. The National Criminal Justice Reference Service. **www .ncjrs.gov/pdffiles1/nij/grants/221153.pdf**

Krogstad, Jens Manuel, and Jeffrey S. Passel. 2015 (July). "5 Facts about Illegal Immigration in the United States." Washington, DC: Pew Research Center. **www.pewresearch.org**

Krogstad, Jens Manuel, Jeffrey S. Passel, and D'Vera Cohn. 2017. "5 Facts about Illegal Immigration in the U.S." Washington, DC: Pew Research Center. **www.pewresearch.org**

Kroll, Luisa, and Kerry A. Dolan. 2012. "The Forbes 400: The Richest People in America." *Forbes* (September 19).

Kurz, Demie. 1995. *For Richer for Poorer: Mothers Confront Divorce*. New York: Routledge.

Laakso, Janice. 2013. "Flawed Policy Assumptions and HOPE VI." *Journal of Poverty* 17 (1): 29–46.

Lachs, M., and K. Pillemer. 2015. "Elder Abuse." *New England Journal of Medicine* 373: 1947–1956.

Lacy, Karen R. 2007. *Blue-Chip Black: Race, Class and Status in the New Black Middle Class*. Berkeley, CA: University of California Press.

LaFlamme, Darquise, Andree Pomerrleau, and Gerard Malcuit. 2002. "A Comparison of Fathers' and Mothers' Involvement in Childcare and Stimulation Behaviors during Free- Play with Their Infants at 9 and 15 Months." *Sex Roles* 11–12 (December): 507–518.

Lamanna, Mary Ann, Agnes Riedman, and Susan D. Stewart. 2018. *Marriage and Families: Making Choices in a Diverse Society*, 13th ed. Belmont, CA: Wadsworth.

Lamont, Michèle. 1992. *Money, Morals, and Manners: The Culture of the French and the American Upper- Middle Class*. Chicago: University of Chicago Press.

Langman, Lauren, and Douglas Morris. 2002. "Internetworked Social Movements: The Promises and Prospects for Global Justice." Paper presented at the International Sociological Association, Brisbane, Australia.

Langton, Lynn, and Sofi Sinozich. 2014. *Rape and Sexual Assault among College-Age Females, 1995–2013*. Washington, DC: U.S. Bureau of Justice Statistics. **www.bjs.gov**

Lara-Millan, Armando. 2014. "Public Emergency Room Overcrowding in the Era of Mass Imprisonment." *American Sociological Review* 79 (5): 866–887.

Lareau, Annette. 2011. *Unequal Childhoods: Class, Race, and Family Life*, 2nd ed. Berkeley, CA: University of California Press.

Laumann, Edward O., John H. Gagnon, Robert T. Michael, and Stuart Michaels. 1994. *The Social Organization of Sexuality: Sexual Practices in the United States*. Chicago: University of Chicago Press.

Laumann, Edward O., Anthony Paik, Dale B. Glasser, Jeong-Han Kang, Tianfu Wang, Bernard Levinson, Edson D. Moreira, Alfredo Nicolosi, and Clive Gingell. 2006. "A Cross-National Study of Subjective Sexual Well-being among Older Women and Men: Findings from the Global Study of Sexual Attitudes and Behaviors." *Archives of Sexual Behavior* 35 (2): 145–161.

Lavigne, Heather J., Katherine G. Hanson, and Daniel R. Anderson. 2015. "The Influence of Television Coviewing on Parent Language Directed at Toddlers." *Journal of Applied Developmental Psychology* 36: 1–10

Lawson, Richard. 2015. "The Reagan Administration's Unearthed Response to the AIDS Crisis is Chilling." *Vanity Fair* (December 1). **www.vanityfair.com**

Ledger, Kate. 2009. "Sociology and the Gene." *Contexts* 8 (3): 16–20.

Lee, Donghoon. 2013. *Household Debt and Credit. Student Debt*. New York: Federal Reserve Bank of New York Consumer Credit Panel. **www.newyorkfed.org**

Lee, Hangwoo. 2006. "Privacy, Publicity, and Accountability of Self Presentation in an On-line Discussion Group." *Sociological Inquiry* 76 (1): 1–22.

Lee, Latoya. 2017. "Black Twitter: A Response to Bias in Mainstream Media." *Social Sciences* 6 (26). **www.mdpi .com/2076-0760/6/1/26**

Lee, M. A., Joachim Singelmann, and Anat Yom-Tov. 2008. "Welfare Myths: The Transmission of Values and Work among TANF Families." *Social Science Research* 37 (2): 516–529.

Lee, Richard M., Harold D. Grotevant, Wendy L. Hellerstedt, and Megan R. Gunnar. 2006. "Cultural Socialization in Families with Internationally Adopted Children." *Journal of Family Psychology* 20 (4): 571–580.

Lee, Stacey J. 2009. *Unraveling the "Model Minority" Stereotype: Listening to Asian American Youth*, 2nd ed. New York: Teacher's College Press.

Legerski, Elizabeth Miklya, and Marie Cornwall. 2010. "Working-Class Job Loss, Gender, and the Negotiation of Household Labor." *Gender & Society* 24 (August): 447–474.

Leidner, Robin. 1993. *Fast Food, Fast Talk: Service Work and the Routinization of Everyday Life*. Berkeley, CA: University of California Press.

Lemert, Edwin M. 1972. *Human Deviance, Social Problems, and Social Control*. Englewood Cliffs, NJ: Prentice Hall.

Lengermann, P. M., and Niebrugge-Brantley, J. 1998. *The Woman Founders: Sociology and Social Theory, 1830–1930*. Boston: McGraw-Hill.

Lenski, Gerhard. 2005. *Ecological-Evolutionary Theory: Principles and Applications*. Boulder, CO: Paradigm Publishers.

Levi, Michael, Alan Doig, Rajeev Gundur, David Wall, and Matthew Williams. 2017. "Cyberfraud and the Implications for Effective Risk-Based Responses: Themes from UK Research." *Crime, Law and Social Change* 67(1): 77–96.

Levitt, Justin. 2017. *The Truth about Voter Fraud*. New York: Brennan Center, New York University School of Law. **http://www.brennancenter.org/**

Levy, Ariel. 2005. *Female Chauvinist Pigs: Women and the Rise of Raunch Culture*. New York: Free Press.

Levy, Becca R., Pil H. Chung, Talya Bedford, and Kristina Navrazhina. 2013. "Facebook as a Site for Negative Age Stereotypes." *The Gerontologist* 54 (2): 172–176.

Lewin, Tamar. 2002. "Study Links Working Mothers to Slower Learning." *The New York Times* (July 17): A14.

Lewin, Tamar. 2010. "Baby Einstein Founder Goes to Court." *The New York Times* (January 13): A15.

Lewis, Amanda E., and John B. Diamond. 2015. *Despite the Best Intentions: How Racial Inequality Thrives in Good Schools*. New York: Oxford University Press.

Lewis, Oscar. 1960. *Five Families: Mexican Case Studies in the Culture of Poverty*. New York: Basic Books.

Lewis, Oscar. 1966. "The Culture of Poverty." *Scientific American* 215 (October): 19–25.

Lewontin, Richard. 1996. *Human Diversity.* New York: W. H. Freeman.

Lichter, Daniel T., Domenico Parisi, and Michael C. Taquino. 2015. "Toward a New Macro-Segregation? Decomposing Segregation Within and Between Metropolitan Cities and Suburbs." *American Sociological Review* 80 (4): 843–873.

Ling, Haping 2009. *Emerging Voices: Experiences of Underrepresented Asian Americans.* New Brunswick, NJ: Rutgers University Press.

Linton, Ralph. 1936. *The Study of Man.* New York: Appleton Century Crofts.

Lipka, Michael. 2017. "Muslims and Islam: Key Findings in the U.S. and Around the World." Washington, DC: Pew Research Organization. **www.pewresearch.org /fact-tank/2017/08/09/muslims-and -islam-key-findings-in-the-u-s-and -around-the-world/**

Litt, Dana M., Melissa A. Lewis, Henriettae Stahlbrandt, Perry Firth, and Clayton Neighbors. 2012. "Social Comparison as a Moderator of the Association between Perceived Norms and Alcohol use and Negative Consequences among College Students." *Journal of Studies on Alcohol and Drugs* 73(6): 961–967.

Lloren, Anouk, and Lorena Parini. 2017. "How LGBT-Supportive Workplace Policies Shape the Experience of Lesbian, Gay Men, and Bisexual Employees." *Sexuality Research & Social Policy* 14(3):289–299.

Locklear, Erin M. 1999. "Where Race and Politics Collide: The Federal Acknowledgement Process and Its Effects on Lumsee and Pequot Indians." Unpublished senior thesis, Princeton University.

Lonsway, K. A., J. Archambault, and D. J. Lisak. 2009. "False Reports: Moving beyond the Issue to Successfully Investigate and Prosecute Non-stranger Sexual Assault." *The Voice,* 3 (1): 1–11. The National District Attorneys Association. **www.ndaa.org/**

Lopez, Mark. 2009. "Dissecting the 2008 Electorate: Most Diverse in History." Pew Research Center. **www.pewresearch.com**

Lorber, Judith. 1994. *Paradoxes of Gender.* New Haven, CT: Yale University Press.

Lorber, Judith and Lisa Jean Moore. 2002. *Gender and the Social Construction of Illness.* Thousand Oaks: Sage.

Lorenz, Conrad. 1966. *On Aggression.* New York: Harcourt Brace Jovanovich.

Lovejoy, Meg. 2001. "Disturbances in the Social Body: Differences in Body Image and Eating Problems among African American and White Women." *Gender & Society* 15 (April): 239–261.

Lu, Melody C. 2005. "Commercially Arranged Marriage Migration: Case Studies of Cross-Border Marriages in Taiwan." *Indian Journal of Gender Studies* 12 (2–3): 275–303.

Lucal, Betsy. 1994. "Class Stratification in Introductory Textbooks: Relational or Distributional Models?" *Teaching Sociology* 22 (April): 139–150.

Luker, Kristin. 1975. *Taking Chances: Abortion and the Decision Not to Contracept.* Berkeley, CA: University of California Press.

Luker, Kristin. 1984. *Abortion and the Politics of Motherhood.* Berkeley, CA: University of California Press.

Luker, Kristin. 1996. *Dubious Conceptions: The Politics of Teenage Pregnancy.* Cambridge, MA: Harvard University Press.

Lyons, Anthony, Marian Pitts, and Jeffrey Grierson. 2013. "Growing Old as a Gay Man: Psychosocial Well-being of a Sexual Minority." *Research on Aging* 35 (3): 275–295.

Lyons, Linda. 2002. "Teen Attitudes Contradict Sex-Crazed Stereotype." *The Gallup Poll.* Princeton, NJ: The Gallup Organization. January 29. **www.gallup .com**

Ma, Jennifer, and Sandy Baum. 2016. "Trends in Community Colleges: Enrollment, Prices, Student Debt, and Completion." *Research Brief: College Board Research.* New York: The College Board. **www.collegeboard.org**

Machel, Graca. 1996. *Impact of Armed Conflict on Children.* New York: UNICEF/ United Nations.

MacKinnon, Catherine. 1983. "Feminism, Marxism, Method, and the State: An Agenda for Theory." *Signs* 7 (Spring): 635–658.

MacKinnon, Catherine A. 2006. "Feminism, Marxism, Method, and the State: An Agenda for Theory." pp. 829–868 in *The Canon of American Legal Thought,* edited by D. Kennedy and W. F. Fisher. Princeton: Princeton University Press.

Mackintosh, N. J. 1995. *Cyril Burt: Fraud or Framed?* Oxford, England: Oxford University Press.

Malamuth, Neil M., Gert M. Hald, and Mary Koss. 2012. "Pornography, Individual Differences in Risk and Men's Acceptance of Violence against Women in a Representative Sample." *Sex Roles* 66 (7–8): 427–439.

Malcomson, Scott L. 2000. *One Drop of Blood: The American Misadventure of Race.* New York: Farrar, Strauss, and Giroux.

Maldonado, Lionel, A. 1997. "Mexicans in the American System: A Common Destiny." In *Ethnicity in the United States: An Institutional Approach,* edited by William Velez. Bayside, NY: General Hall.

Malinauskas, Brenda et al. 2006. "Dieting Practices, Weight Perceptions, and Body Composition: A Comparison of Normal Weight, Overweight, and Obese College Females." *Nutrition Journal* 5 (March 31): 5–11.

Maltby, Lauren E., M. Elizabeth L. Hall, Tamara L. Anderson, and Keith Edwards. 2010. "Religion and Sexism: The Moderating Role of Participant Gender." *Sex Roles* 62: 615–622.

Malthus, Thomas Robert. 1926 [1798]. *First Essay on Population 1798.* London: Macmillan.

Mandel, Daniel. 2001. "Muslims on the Silver Screen." *Middle East Quarterly* 8 (Spring): 19–30.

Mantsios, Gregory. 2010. "Media Magic: Making Class Invisible." pp. 386–394 in *Race, Class, and Gender: An Anthology,* edited by Margaret L. Andersen and Patricia Hill Collins. Belmont, CA: Wadsworth.

Marcos, Christina. 2016. "115th Congress Will Be Most Racially Diverse in History." The Hill, November 17. **www.thehill.com**

Marcuse, Herbert. 1964. *One-Dimensional Man.* Boston: Beacon Press.

Marín, Rebecca, Andrew Christensen, and David C. Atkins. 2014. "Infidelity and Behavioral Couple Therapy: Relationship Outcomes Over 5 Years Following Therapy." *Couple and Family Psychology: Research and Practice* 3 (1): 1–12.

Marks, Carole. 1989. *Farewell, We're Good and Gone: The Great Black Migration.* Bloomington, IN: Indiana University Press.

Marks, Carole, and Deana Edkins. 1999. *The Power of Pride: Stylemakers and Rulebreakers of the Harlem Renaissance.* New York: Crown.

Martell, Luke. 2017. *The Sociology of Globalization,* 2nd ed. New York: Polity Press.

Martin, Joyce A., Brady E. Hamilton, Michelle J. K. Osterman, Anne K. Driscoll, and T. J. Mathews. 2017. "Births: Final Data for 2015." *National Vital Statistics Reports.* 66 (1).

Martin, Karin A. 2005. "William Wants a Doll. Can He Have One? Feminists, Child Care Advisors, and Gender- Neutral Child Rearing." *Gender & Society* 19 (August): 456–479.

Martin, Karin A., and Emily Kazyak. 2009. "Hetero-romantic Love and Heterosexiness in Children's G-Rated Films." *Gender & Society* 23 (June): 315–336.

Martin, Patricia Yancey. 2015. "The Rape Prone Culture of Academic Contexts: Fraternities and Athletics." *Gender & Society* (October): 30–43.

Martin, Patricia Yancey, and Robert Hummer. 1989. "Fraternities and Rape on Campus." *Gender & Society* 3 (December): 457–473.

Martineau, Harriet. 1837. *Society in America.* London: Saunders and Otley.

Martineau, Harriet. 1838. *How to Observe Morals and Manners.* London: Charles Knight and Co.

Martinez, Gladys M., and Joyce C. Amba. 2015. "Sexual Activity, Contraceptive Use, and Childbearing of Teenagers Aged 15–19 in the United States." Atlanta, GA: Centers for Disease Control. **www.cdc .gov/nchs/products/databriefs/db209 .htm**

Martinez, Michael E., and Robin A. Cohen. 2014. *Health Insurance Coverage: Early Release of Estimates from National Health Interview Survey, January–June 2014.* Centers for Disease Control and Prevention, Hyattsville, MD: U.S. Department of Health and Human Services, National Center for Health Statistics.

Martinez, Ramiro, Jr. 2014. *Latino Homicide: Immigration, Violence, and Community*, 2nd ed. New York: Routledge.

Marx, Anthony. 1997. *Making Race and Nation: A Comparison of the United States, South Africa, and Brazil*. New York: Cambridge University Press.

Marx, Karl. 1967 [1867]. *Capital*. F. Engels (ed.). New York: International Publishers.

Marx, Karl. 1972 [1843]. "Contribution to the Critique of Hegel's *Philosophy of Right*." pp. 11–23 in *The Marx-Engels Reader*, edited by Robert C. Tucker. New York: W. W. Norton.

Masci, David, Anna Brown, and Jocelyn Kiley. 2017. "5 Facts about Same-Sex Marriage." Washington, DC: Pew Research Center. **www.pewresearch .org/fact-tank/2017/06/26/same -sex-marriage/**

Massey, Douglas S. 2005. "Five Myths about Immigration: Common Misconceptions about U.S. Border Enforcement Policy." *Immigration Policy in Focus* 4 (August): 1–11.

Massey, Douglas S. 2005. *Strangers in a Strange Land: Humans in an Urbanizing World*. New York: Norton.

Massey, Douglas S., and Nancy A. Denton. 1993. *American Apartheid: Segregation and the Making of the Underclass*. Cambridge, MA: Harvard University Press.

Massey, Douglas S., and Magaly Sanchez R. 2012. *Brokered Borders: Creating Immigrant Identity in Anti-Immigrant Times*. New York: Russell Sage.

Massey, Douglas S., and Jonathan Tannen. 2015. "A Research Note on Trends in Black Hypersegregation." *Demography* 52 (June): 1025–1034.

Matthews, D. 2015. "Sociology in Nursing 5: The Effect of Ageing on Health Inequalities." *Nursing Times* 11 (45): 18–21.

Mayer, Jane. 2016. *Dark Money: The Hidden History of the Billionnaires behind the Rise of the Radical Right*. New York: Doubleday.

McCabe, Janice. 2005. "What's in a Label? The Relationship between Feminist Self-Identification and 'Feminist' Attitudes among U.S. Women and Men." *Gender & Society* 19 (4): 480–505.

McCabe, Janice, Emily Fairchild, Liz Grauerholz, Bernice A. Pescosolido, and Daniel Tope. 2011. "Gender in Twentieth Century Children's Books: Patterns of Disparity in Titles and Central Characters." *Gender & Society* 25 (April): 197–225.

McCartney, Suzanne, Alemayehi Bishaw, and Kayla Fontenot. 2013 (February). *Poverty Rates for Selected Detailed Race and Hispanic Groups by State and Place: 2007–2011*. Washington, DC: U.S. Census Bureau. **www.census.gov**

McClelland, Susan. 2003. "A Grim Toll on the Innocent." *Maclean's* (May 12): 20.

McClintock, Elizabeth Aura. 2010. "When Does Race Matter? Race, Sex, and Dating at an Elite University." *Journal of Marriage and the Family* 72 (February): 45–72.

McClintock, Elizabeth Aura. 2011. "Handsome Wants as Handsome Does: Physical Attractiveness and Gender Differences in Revealed Sexual Preferences." *Biodemography and Social Biology* 57 (2): 221–257.

McDade-Montez, Elizabeth, Jan Wallander, and Linda Cameron. 2017. "Sexualization in U.S. Latina and White Girls' Preferred Children's Television Programs." *Sex Roles* 77 (July): 1–15.

McDonald Paula. 2012. "Workplace Sexual Harassment 30 Years On: A Review of the Literature." *International Journal of Management Reviews* 14: 1–17.

McGrath, Timothy. 2014. "What People around the World Are Saying about Ferguson." *USA Today*, November 25.

McIntyre, Robert S., Matthew Gardner, and Richard Phillips. 2014 (February). *The Sorry State of Corporate Taxes*. Washington, DC: Citizens for Tax Justice and the Institute of Taxation and Economic Policy. **www.itep.org**

McKernan, Signe-Mary, Caroline Ratcliffe, C. Eugene Steuerle, Emma Kalish, and Caleb Quakenbush. 2015. *Nine Charts about Wealth Inequality in America*. Washington, DC: The Urban Institute. **http://datatools.urban.org/Features /wealth-inequality-charts/**

McLaughlin, Heather, Christopher Uggen, and Amy Blackstone. 2012. "Sexual Harassment, Workplace Authority, and the Paradox of Power." *American Sociological Review* 77: 625–647.

McLaughlin, Heather, Christopher Uggen, and Amy Blackstone. 2017. "The Economic and Career Effects of Sexual Harassment on Working Women." *Gender & Society* 31 (3): 333–358.

McNamee, Stephen, and Robert K. Miller. 2009. *The Meritocracy Myth*. Lanham, MD: Rowman and Littlefield.

McVeigh, Ricky. 2012. "Making Sense of the Tea Party." *Contemporary Sociology* 41 (November): 766–769.

Mead, George Herbert. 1934. *Mind, Self, and Society*. Chicago, IL: University of Chicago Press.

Meier, Barry. 2013. "Maker Hid Data about Design Flaw in Hip Implant." *The New York Times* (January 13): B1 and B6.

Mendes, Elizabeth, Lydia Saad, and Kyley McGeeney. 2012. "Stay at Home Moms Report More Depression, Sadness, Anger." *The Gallup Poll*. Princeton, NJ: The Gallup Organization, May 18. **www.gallup.com**

Meredith, Martin. 2003. *Elephant Destiny: Biography of an Endangered Species in Africa*. New York: HarperCollins.

Merkow, Carla H. and Eugenia Costa-Giomi. 2014. "Infants' Attention to Synthesised Baby Music and Original Acoustic Music." *Early Child Development and Care* 184 (1): 73–83

Mernissi, Fatema. 2011. *Beyond the Veil: Male-Female Dynamics in Modern Muslim Society*. London: Seqi Books.

Merton, Robert K. 1957. *Social Theory and Social Structure*. New York: Free Press.

Merton, Robert K. 1968. "Social Structure and Anomie." *American Sociological Review* 3 (5): 672–682.

Merton, Robert, and Alice K. Rossi. 1950. "Contributions to the Theory of Reference Group Behavior." pp. 279–334 in *Continuities in Social Research Studies, Scope and Method of "The American Soldier*," edited by Robert K. Merton and Paul F. Lazarsfeld. New York: Free Press.

Messina-Dysert, Gina, and Rosemary Ruether, eds. 2014. *Feminism and Religion in the Twenty-First Century: Technology, Dialogue and Expanding Borders*. New York: Routledge.

Messner, Michael A. 2002. *Taking the Field: Women, Men, and Sports*. Minneapolis, MN: University of Minnesota Press.

Messner, Michael A., and Michela Musto. 2016. *Child's Play: Sport in Kids' Worlds*. New Brunswick, NJ: Rutgers University Press.

Messner, Steven F. 2011. *Crime and the American Dream*. Belmont, CA: Wadsworth/Cengage.

Meyer, B. D., and J. X. Sullivan. 2012. "Identifying the Disadvantaged: Official Poverty, Consumption Poverty, and the New Supplemental Poverty Measure." *Journal of Economic Perspectives* 26 (3): 111–136.

Mezey, Nancy J. 2008. *New Choices, New Families: How Lesbians Decide about Motherhood*. Baltimore, MD: Johns Hopkins University Press.

Mickelson, Roslyn Arlin (ed.). 2000. *Children on the Streets of the Americas: Globalization, Homelessness, and Education in the United States, Brazil, and Cuba*. New York: Routledge.

Milanovic, Brandon. 2010. *The Haves and Have Nots: A Brief and Idiosyncratic History of Global Inequality*. New York: Basic Books.

Milgram, Stanley. 1974. *Obedience to Authority: An Experimental View*. New York: Harper & Row.

Milkie, Melissa A., Suzanne M. Bianchi, Marybeth J. Mattingly, and John P. Robinson. 2002. "Gendered Division of Childrearing: Ideals, Realities, and the Relationship to Parental Well-being." *Sex Roles* 47 (July): 21–38.

Miller, Susan L. 1997. "The Unintended Consequences of Current Criminal Justice Policy." Talk presented at Research on Women Series, University of Delaware, Newark, DE.

Mills, C. Wright. 1956. *The Power Elite*. New York: Oxford University Press.

Mills, C. Wright. 1959. *The Sociological Imagination*. New York: Oxford University Press.

Miner, Horace. 1956. "Body Ritual among the Nacirema." *American Anthropologist* 58 (3): 503–507.

Mirandé, Alfredo. 1979. "Machismo: A Reinterpretation of Male Dominance in the Chicano Family." *The Family Coordinator* 28 (4): 447–449.

Mirandé, Alfredo. 1985. *The Chicano Experience*. Notre Dame, IN: Notre Dame University Press.

Misra, Joy, Stephanie Moller, and Maria Karides. 2003. "Envisioning Dependency: Changing Media Depictions of Welfare in the 20th Century." *Social Problems* 50 (November): 482–504.

Moen, Phyllis. 2003. *It's about Time: Couples and Careers.* Ithaca, NY: Cornell University Press.

Moen, Phyllis, Jungmeen E. Kim, and Heather Hofmeister. 2001. "Couples' Work/Retirement Transitions, Gender, and Marital Quality." *Social Psychology Quarterly* 64 (March): 55–71.

Mohai, Paul, and Robin Saha. 2007. "Racial Inequality in the Distribution of Hazardous Waste: A National-Level Reassessment." *Social Problems* 54 (3): 343–370.

Mohanty, Jayashree. 2013. "Ethnic and Racial Socialization and Self-Esteem of Asian Adoptees: The Mediating Role of Multiple Identities." *Journal of Adolescence* 36(1): 161–170.

Moore, Joan. 1976. *Hispanics in the United States.* Englewood Cliffs, NJ: Prentice Hall.

Moore, Valerie A. 2001. "'Doing' Racialized and Gendered Age to Organize Peer Relations: Observing Kids in Summer Camp." *Gender & Society* 15 (December): 835–858.

Mora, Christina. 2009. *DeMuchos, Uno: The Institutionalization of Latino Panethnicity in the United States, 1960–1990.* PhD diss., Princeton University, Princeton, NJ.

Morales, Lymari. 2010. "A Scientific Measure of Religious Prejudice." *The Gallup Poll.* Princeton, NJ: The Gallup Organization.

Morgan, Marcyliena. 2009. *The Real Hiphop: Battling for Knowledge, Power, and Respect in the LA Underground.* Durham, NC: Duke University Press.

Morgan, Marcyliena. 2010. (With Dawn-Elissa Fischer). "Hiphop and Race: Blackness, Language and Creativity" pp. 509–527 in *Doing Race: 21 Essays for the 21st Century,* edited by Hazel Rose Markus and Paula M. L. Moya. New York: W. W. Norton.

Morin, Rich, and Renee Stepler. 2016. "The Racial Confidence Gap in Police Performance." Washington, DC: Pew Research Organization. **www.pewsocialtrends.org**

Morin, Rich, Kim Parker, Renee Stepler, and Andrew Mercer. 2017. *Behind the Badge.* Washington, DC: Pew Research Organization. **http://assets.pewresearch.org/wp-content/uploads/sites/3/2017/01/06171402/Police-Report_FINAL_web.pdf**

Morning, Ann. 2011. *The Nature of Race: How Scientists Think and Teach about Human Differences.* New York: New York University Press.

Morris, Aldon. 1984. *The Origins of the Civil Rights Movement: Black Communities Organizing for Change.* New York: Free Press.

Morris, Aldon D. 1999. "A Retrospective on the Civil Rights Movement: Political and Intellectual Landmarks." *Annual Review of Sociology* 25: 517–539.

Morris, Aldon. 2015. *The Scholar Denied: W. E. B. DuBois and the Birth of Modern Sociology.* Berkeley, CA: University of California Press.

Mulder, Mark T., Aida I. Ramos, and Gerardo Martí. 2017. *Latino Protestants in America: Growing and Diverse.* Lanham, MD: Rowman and Littlefield.

Mullen, Ann L. 2010. *Degrees of Inequality: Culture, Class, and Gender in American Higher Education.* Baltimore, MD: Johns Hopkins University Press.

Mullings, Leith, and Amy J. Schulz, eds. 2005. *Gender, Race, Class and Health: Intersectional Approaches.* New York: Jossey-Bass.

Murphy, Joseph. 2014. "The Social and Educational Outcomes of Homeschooling." *Sociological Spectrum* 34 (May): 244–272.

Myers, Steven Lee. 2000. "Survey of Troops Finds Antigay Bias Common in Service." *The New York Times* (March 24): 1.

Myers, Walter D. 1998. *Amistad Affair.* New York: NAL/Dutton.

Nagel, Joane. 1996. *American Indian Ethnic Renewal: Red Power and the Resurgence of Identity and Culture.* New York: Oxford University Press.

Nagel, Joane. 2003. *Race, Ethnicity, and Sexuality: Intimate Intersections, Forbidden Frontiers.* New York: Oxford University Press.

Nanda, Serena. 1998. *Neither Man Nor Woman: The Hijras of India.* Belmont, CA: Wadsworth.

National Cancer Institute. 2017. "Cancer Health Disparities." Washington, DC: National Institute of Health. **www.cancer.gov/about-nci/organization/crchd/cancer-health-disparities-fact-sheet#q5**

National Center for Education Statistics. 2013. *The Condition of Education, 2012.* Washington, DC: U.S. Department of Education.

National Center for Education Statistics. 2013. *Private School Universe Survey 1995–1996 through 2011–2012.* Washington, DC: National Center for Education Statistics, U.S. Department of Education. Table 205.20.

National Center for Education Statistics. 2016. *Digest of Education Statistics.* Washington, DC: U.S. Department of Education. **https://nces.ed.gov/programs/digest/current_tables.asp** Hyattsville, MD: U.S. Department of Health and Human Services.

National Center for Health Statistics. 2016. *Health United States 2016.* Atlanta, GA: Centers for Disease Control. **www.cdc.gov**

National Center for Health Statistics. 2015. *Health United States 2015.* Atlanta, GA: Centers for Disaese Control. **www.cdc.gov**

National Center for Health Statistics. 2017. *Health United States 2016.* Atlanta, GA: Centers for Disease Control. **www.cdc.gov/nchs**

National Center on Elder Abuse. 2015. **www.ncea.aoa.gov**

National Coalition for the Homeless. 2017. "Fact Sheet." **www.nationalhomeless.org**

National Human Genome Research Institute. *Fact Sheet: The Human Genome Project.* National Institutes of Health. **www.genome.gov**

National Institutes of Health. 2007. *National Center for Complementary and Alternative Medicine.* Hyattsville, MD: U.S. Department of Health and Human Services.

National Institute of Mental Health. 2016. "Eating Disorders." **www.nimh.nih.gov/health/topics/eating-disorders/index.shtml**

National Research Council. 2012. *Advancing the Science of Climate Change.* Washington, DC: The National Academies Press. **www.nas-sites.org**

Nawaz, Maajid. 2013. *Radical: My Journey Out of Islamist Extremism.* Guilford, CT: Lyons Press.

Nee, Victor. 1973. *Longtime Californ': A Documentary Study of an American Chinatown.* New York: Pantheon Books.

Negrón-Muntaner, Frances with Chelsea Abbos, Luis Figueroa, and Samuel Robson. 2014. *The Latino Media Gap: A Report on the Status of U.S. Latinos in U.S. Media.* New York: Columbia University Center for the Study of Race and Ethnicity. **www.columbia.edu**

Newheiser, Anna-Kaisa, Manuela Barreto, and Jasper Tiemersma. 2017. "People Like Me Don't Belong Here: Identity Concealment is Associated with Negative Workplace Experiences." *The Journal of Social Issues* 73 (2): 341–358.

Newman, Katherine S. 2012. *The Accordion Family: Boomerang Kids, Anxious Parents, and the Private Toll of Global Competition.* Boston: Beacon Press.

Newman, Katherine S., Cybelle Fox, David Harding, Jal Mehta, and Wendy Roth. 2006. *Rampage: Social Roots of School Shootings.* New York: Basic Books.

Newport, Frank. 2015. "Fewer Americans Identify as Middle Class in Recent Years." *The Gallup Poll.* Princeton, NJ: Gallup Organization.

Newport, Frank. 2012a. "Bias against Mormon Presidential Candidate Same as in 1967." *The Gallup Poll.* Princeton, NJ: The Gallup Organization. **www.gallup.com**

Newport, Frank, 2012b. "Seven in 10 Americans Are Moderately or Very Religious." *The Gallup Poll.* Princeton, NJ: The Gallup Organization. **www.gallup.com**

Newport, Frank, 2013. "In U.S. 4 in 10 Report Attending Church in Last Week." Princeton, NJ: *The Gallup Poll.* Princeton, NJ: The Gallup Organization. **www.gallup.com**

Newport, Frank. 2017a. "Five Key Findings on Religion in the US." Princeton, NJ: Gallop Poll.

Newport, Frank. 2017b. "Middle-Class Identification at Pre-Recession Levels." *The Gallup Poll.* Princeton, NJ: Gallup Organization.

Newport, Frank, and Joseph Carroll. 2005. "Another Look at Evangelicals in America Today." *The Gallup Poll.* Princeton, NJ:

The Gallup Organization. **www.gallup .com**

Ngai, Mae M. 2012. "Impossible Subjects: Illegal Aliens and the Making of Modern America." pp. 192–196 in *Race and Ethnicity in Society: The Changing Landscape*, 3rd ed., edited by Elizabeth Higginbotham and Margaret L. Andersen. Belmont, CA: Wadsworth/ Cengage.

Nielsen, Søren Beck. 2014. "Medical Record Keeping as Interactional Accomplishment." *Pragmatics and Society* 5 (2): 221–242.

Nielsen. 2016. *The Nielsen Total Audience Report; Q1*. **www.nielsen.com**

Nixon, Ron, and Matt Stevens. 2017. "Harvey, Irma, Maria: Trump Administration's Response Compared." *The New York Times*, September 27.

Noah, Timothy. 2012. *The Great Divergence: America's Growing Inequality Crisis and What We Can Do about It*. New York: Bloomsbury.

Norgaard, Kari M. 2006. "'People Want to Protect Themselves a Little Bit': Emotions, Denial, and Social Movement Nonparticipation." *Sociological Inquiry* 76 (3): 372–396.

Norris, Pippa, and Ronald Inglehart. 2002. "Islamic Culture and Democracy: Testing the 'Clash of Civilizations' Thesis." *Comparative Sociology* 1: 44 (1) 235–263.

Norris, Tina, Paula L. Vines, and Elizabeth M. Hoeffel. 2012. "The American Indian and Alaska Native Population: 2010." *U.S. Census Briefs*. Washington, DC: U.S. Census Bureau.

Nunley, John M., Adam Pugh, Nicholas Romero, and R. Alan Seals. 2015. "Racial Discrimination in the Labor Market for Recent College Graduates: Evidence from a Field Experiment." *The B. E. Journal of Economic Analysis & Policy* 15 (3): 1093–1125.

Oakes, Jeannie. 2005. *Keeping Track: How Schools Structure Inequality*, 2nd ed. New Haven, CT: Yale University Press.

Offer, Shira. 2014. "Time with Children and Employed Parents' Emotional Well-being." *Social Science Research* 47 (September): 192–203.

Ogburn, William F. 1922. *Social Change with Respect to Cultural and Original Nature*. New York: B. W. Huebsch.

Ogden, Cynthia L., Margaret D. Carrol, Brian K. Kit, and Katherine Flegal. 2015. "Prevalence of Childhood and Adult Obesity in the United States, 2011–2012." *JAMA* 311 (8): 806–815.

Okamoto, Dina. 2014. *Redefining Race: Asian American Panethnicity and Shifting Racial Boundaries*. New York: Russell Sage Foundation.

Okamoto, Dina, and G. C. Mora. 2014. "Panethnicity." *Annual Review of Sociology* 40: 219–239.

Oliver, Melvin L., and Thomas M. Shapiro. 2006. *Black Wealth/White Wealth: A New Perspective on Racial Inequality*, 10th ed. New York: Routledge.

Oliver, Melvin L., and Thomas M. Shapiro. 2008. "Sub-Prime as Black Catastrophe." *The American Prospect*, September 22.

Oliviera, Voctor. 2017. *The Food Assistance Landscape: FY 2016 Annual Report*. Washington, DC: U.S. Department of Agriculture. **https://www.ers.usda.gov /publications/pub-details/?pubid=82993**

Olson, Laura R., Wendy Cadge, and James T. Harrison. 2006. "Religion and Public Opinion about Same-Sex Marriage." *Social Science Quarterly* 87 (2): 340–360.

Olson, Peter, and Louise Sheiner. 2017. "The Hutchins Center Explains: Prescription Drug Spending." Washington, DC: The Brookings Institution. **www.brookings. edu/blog/up-front/2017/04/26/the -hutchins-center-explains-prescription -drug-spending/**

Omi, Michael, and Howard Winant. 2014. *Racial Formation in the United States*, 3rd ed. New York: Routledge.

O'Neil, John. 2002. "Parent Smoking and Teenage Sex." *The New York Times*, Sept. 3, p. F7.

Orfield, Gary, Erica Frankenberg, and Laurie Russman. 2014. "School Resegregation and Civil Rights Challenges for the Obama Administration." Los Angeles, CA: The Civil Rights Project/ Proyecto Derechos Civiles, University of California Los Angeles. **www.civilrightsproject .ucla.edu**

Organization for Economic Co-operation and Development (OECD). 2014. *Society at a Glance—OECD Social Indicators*. **www.oecd.org/els/social/indicators/SAG**

Ortman, Jennifer, M., and Christine E. Guarneri. Nd. *United States Population Projections: 2000 to 2050*. Washington, DC: U.S. Census Bureau. **https://www .census.gov/content/dam/Census /library/working-papers/2009/demo /us-pop-proj-2000-2050/analytical -document09.pdf**

Ortiz, Isabel, and Matthew Cummins. 2011. "Global Inequality: Beyond the Bottom Billion." *Social and Economic Policy Working Paper*. UNICEF, April. **www.unicef.org**

Ousey, Graham C., and Charis E. Kubrin. 2014. "Immigration and the Changing Nature of Homicide in US Cities, 1980–2010." *Journal of Quantitative Criminology* 30 (3): 453–483.

Padavic, Irene, and Barbara Reskin. 2002. *Women and Men at Work*, 2nd ed. Thousand Oaks, CA: Sage.

Padgett, Tim, Anthony Esposito, and Aaron Nelson. 2011. "The Chilean Miners." *Time* (January 3): 105–111.

Page, Scott E. 2007. *The Difference: How the Power of Diversity Creates Better Groups, Firms, Schools, and Societies*. Princeton, NJ: Princeton University Press.

Pager, Devah. 2007. *Marked: Race, Crime, and Finding Work*. Chicago: University of Chicago Press.

Paik, Anthony, Kenneth J. Sanchagrin, and Karen Heimer. 2016. "Broken Promises: Abstinence Pledging and Sexual and Reproductive Health." *Journal of Marriage and Family*. 78 (2): 546–561.

Pain, Emil. 2002. "The Social Nature of Extremism and Terrorism." *Social Sciences* 33: 55–68.

Painter, Nell Irvin. 2011. *The History of White People*. New York: W. W. Norton.

Parcel, Toby L., Joshua A. Hendrix, and Andrew J. Taylor. 2016. "The Challenge of Diverse Public Schools." *Contexts* 15 (1): 42–47.

Park, Robert E., and Ernest W. Burgess. 1921. *Introduction to the Science of Society*. Chicago: University of Chicago Press.

Parker, Kim. 2012. *The Boomerang Generation: Feeling OK about Living with Mom and Dad*. Washington, DC: Pew Research Center. **ww.pewsocialtrends .org**

Parreñas, Rhacel Salazar. 2001. *Servants of Globalization: Women, Migration, and Domestic Work*. Stanford, CA: Stanford University Press.

Parreñas, Rhacel. 2005. *Children of Global Migration: Transnational Families and Gendered Woes*. Stanford, CA: Stanford University Press.

Parsons, Talcott (ed.). 1947. *Max Weber: The Theory of Social and Economic Organization*. New York: Free Press.

Parsons, Talcott. 1951a. *The Social System*. Glencoe, IL: Free Press.

Parsons, Talcott. 1951b. *Toward a General Theory of Action*. Cambridge, MA: Harvard University Press.

Parsons, Talcott. 1966. *Societies: Evolutionary and Comparative Perspectives*. Englewood Cliffs, NJ: Prentice Hall.

Pascoe, C. J. 2011. *Dude, You're a Fag: Masculinity and Sexuality in High School*. Berkeley, CA: University of California Press.

Paternoster, Ray, Jean Marie McGloin, Holly Nguyen, and Kyle J. Thomas. 2013. "The Causal Impact of Exposure to Deviant Peers: An Experimental Investigation." *Journal of Research in Crime and Delinquency* 50 (4): 476–503.

Pattillo, Mary. 2013. *Black Picket Fences: Privilege and Peril among the Black Middle Class*, 2nd ed. Chicago: University of Chicago Press.

Pattillo, Mary. 2015. "Everyday Politics of School Choice in The Black Community." *Du Bois Review* 12(1): 41–71.

Paulus, P. B., T. S. Larey, and M. T. Dzindolet. 2001. "Creativity in Groups and Teams." pp. 319–338 in *Groups at Work: Theory and Research*, edited by E. Turner. Mahwah, NJ: Erlbaum.

Payne, Brian K. 2016. "Editor's Introduction: Special Issue on Cybersecurity and Criminal Justice." *Criminal Justice Studies* 29 (2): 89–91.

Pedraza, Silvia. 1996. "Cuba's Refugees: Manifold Migrations." pp. 263–279 in *Origins and Destinies: Immigration, Race, and Ethnicity in America*, edited by Silvia Pedraza and Rubén Rumbaut. Belmont, CA: Wadsworth.

Pelham, Brett, and Steve Crabtree. 2009. "Religiosity and Perceived Intolerance of Gays and Lesbians." Princeton, NJ: Gallup Poll, March 10. **www.gallup.com**

Pellow, David N. 2004. "The Politics of Illegal Dumping: An Environmental Justice Framework." *Qualitative Sociology* 27 (4): 511–525.

Pempek, Tiffany A., Lindsay B. Demers, Katherine G. Hanson, Heather L. Kirkorian, and Daniel R. Anderson. 2011. "The Impact of Infant-Directed Videos on Parent-Child Interaction." *Journal of Applied Developmental Psychology* 32 (1): 10–19

Peralta, Katherine. 2014. "Native Americans Left Behind in the Economic Recovery." *U.S. News and World Report*, November 17. **www.usnews.com**

Peri, Giovanni. 2014. "Does Immigration Hurt the Poor?" *Pathways* (Summer): 15–18.

Peri, Giovanni, and Chad Sparber. 2009. "Task Specialization, Immigration, and Wages." *American Economic Journal: Applied Economics* 1 (July): 135–169.

Perlmutter, David. 2008. *Blogwars: The New Political Battleground.* New York: Oxford University Press.

Perrow, Charles. 2007. *Organizing America: Wealth, Power, and Origins of American Capitalism.* Princeton: Princeton University Press.

Pershing, Jana L. 2003. "Why Women Don't Report Sexual Harassment: A Case Study of an Elite Military Institution." *Gender Issues* 21 (4): 3–30.

Peterson-Withorn, Chase. 2016. "Forbes 400: The Full List of the Richest People in America 2016." *Forbes* (October 4). **www .forbes.com/sites/chasewithorn /2016/10/04/forbes-400-the-full-list -of-the-richest-people-in-america -2016/#a82839322f4b**

Pettigrew, Thomas F. 1992. "The Ultimate Attribution Error: Extending Allport's Cognitive Analysis of Prejudice." pp. 401–419 in *Readings about the Social Animal,* edited by Elliott Aronson. New York: Freeman.

Pew Charitable Trusts. 2016. *Portrait of Financial Security.* **www.pewtrusts .org/en/multimedia/data -visualizations/2016/portrait-of -financial-security**

Pew Forum on Religion & Public Life. 2012. *Asian Americans: A Mosaic of Faiths.* Washington, DC: Pew Research Center. **www.pewforum.org**

Pew Research Center. 2007. "Muslim Americans: Middle Class and Mostly Mainstream." Washington, DC: Pew Research Center. **www.pewresearch.org**

Pew Research Center. 2009. *Dissecting the 2008 Election: Most Diverse in U.S. History.* Washington, DC: Pew Research Center. **www.pewresearch.org**

Pew Research Center. 2010. *Marrying Out: One-in-Seven New U.S. Marriages Is Interracial or Interethnic.* Washington, DC: Pew Research Center. **www .pewsocialtrends.org**

Pew Research Center. 2011. *Muslim Americans: No Signs of Growth in Alienation or Support for Extremism.* Washington, DC: Pew Research Center. **www.people-press.org**

Pew Research Center. 2014 (December 8). "Sharp Divisions in Reaction to Brown, Garner Decisions." Washington, DC: Pew Research Center. **www.pewresearch.org**

Pew Research Center. 2015 (June 8). "Changing Attitudes on Gay Marriage." Washington, DC: Pew Research Center. **www.pewresearch.org**

Pew Research Center. 2015. *America's Changing Religious Landscape.* Washington, DC: Pew Research Organization. **www.pewforum .org/2015/05/12/americas-changing -religious-landscape/**

Pew Research Center for the People and the Press. 2014. "For 2016 Hopefuls, Washington Experience Could Do More Harm Than Good." Washington, DC: Pew Research Center, May 14. **www .pewresearch.org**

Pew Research Center. 2016a. *A Portrait of Jewish Americans.* Washington, DC: Pew Research Center. **www.peopleforum.org**

Pew Research Center. 2016b. *On Immigration Policy, Partisan Differences But Also Some Common Ground.* Washington, DC: Pew Research Center. **www.peopleforum.org**

Pew Research Global Attitudes Project. 2017. *Emerging and Developing Economies Much More Optimistic than Rich Countries about the Future.* Washington, DC: Pew Research Center. **www.pewglobal.org**

Pew Research Internet Project. 2014. *Mobile Technology Fact Sheet.* **www .pewinternet.org**

Pew Research Religion & Public Life Project. 2014. "The Shifting Religious Identity of Latinos in the United States." Washington, DC: Pew Research Center. **www.pewforum.org**

Pfeffer, Fabian T., Sheldon Danziger, and Robert F. Shoeni. 2014. *Wealth Levels, Wealth Inequality, and the Great Recession.* New York: Russell Sage Foundation. **www.russellsage.org**

Piketty, Thomas. 2014. *Capital in the Twenty-First Century.* Cambridge, MA: Harvard University Press.

Polce-Lynch, Mary, Barbara J. Myers, Wendy Kliewer, and Christopher Kilmartin. 2001. "Adolescent Self- Esteem and Gender: Exploring Relations to Sexual Harassment, Body Image, Media Influence, and Emotional Expression." *Journal of Youth and Adolescence* 30 (April): 225–244.

Popenoe, David. 2001. "Today's Dads: A New Breed?" *The New York Times* (June 19): A22.

Portes, Alejandro. 2002. "English-Only Triumphs, but the Costs Are High." *Contexts* 1 (February): 10–15.

Portes, Alejandro, and Rubén G. Rumbaut. 1996. *Immigrant America: A Portrait,* 2nd ed. Berkeley, CA: University of California Press.

Portes, Alejandro, and Rubén G. Rumbaut. 2001. *Legacies: The Story of the Immigrant Second Generation.* Berkeley, CA: University of California Press.

Potok, Mark. 2017 (February). "The Year in Hate and Extremism." *Intelligence Report.* Montgomery, AL: Southern Poverty Law Center. **www.splcenter.org**

Potts, Monica. 2012. "The Collapse of Black Wealth." *The American Prospect*

(November 21). **http://prospect.org /article/collapse-black-wealth**

Poushter, Jacob. 2014. "What's Morally Acceptable? It Depends on Where in the World You Live." Fact Tank. Washington, DC: Pew Research Center. **www .pewresearch.org**

Powell, Brian, Catherine Bolzendahl, Claudia Geist, and Lala Carr Steelman. 2010. *Counted Out: Same-Sex Relations and Americans' Definitions of Family.* New York: Russell Sage Foundation.

Press, Andrea. 2002. "The Paradox of Talk." *Contexts* 1 (Fall–Winter): 69–70.

Preves, Sharon E. 2003. *Intersex and Identity: The Contested Self.* New Brunswick, NJ: Rutgers University Press.

Prokos, Anastasia. 2011. "An Unfinished Revolution." *Gender & Society* 25 (1): 75–80.

Prostitutes Education Network. 2009. **www.bayswan.org**

Puffer, Phyllis. 2009. "Durkheim Did Not Say 'Normlessness': The Concept of Anomic Suicide for Introductory Sociology Courses." *Southern Rural Sociology* 24 (2): 200–222.

Putnam, Robert D. 2015. *Our Kids: The American Dream in Crisis.* New York: Simon and Schuster.

Quadagno, Jill. 2005. *One Nation, Uninsured: Why the U.S. Has No National Health Insurance.* New York: Oxford University Press.

Quiroz, Pamela A., and Vernon Lindsay. 2015. "Selective Enrollment, Race, and Shifting the Geography of Educational Opportunity." *Humanity & Society* 39 (4): 376–393.

Rainie, Lee. 2013 (June 6). "Cell Phone Ownership Hits 91% of Adults." Washington, DC: Pew Research Center, Internet and Technology. **www .pewresearch.org**

Raley, Sara B., Marybeth J. Mattingly, and Suzanne M. Bianchi. 2006. "How Dual Are Dual-Income Couples? Documenting Change from 1970 to 2001." *Journal of Marriage and Family* 68 (1): 11–28.

Rampersad, Arnold. 1986. *The Life of Langston Hughes: Vol. I: 1902–1941. I, Too, Sing America.* New York: Oxford University Press.

Rampersad, Arnold. 1988. *The Life of Langston Hughes: Vol. II: 1941–1967. I Dream A World.* New York: Oxford University Press.

Rashid, Ahmed. 2000. *Taliban: Militant Islam, Oil, and Fundamentalism in Central Asia.* New Haven, CT: Yale University Press.

Rasmussen Report. 2017. "Most Support Temporary Ban on Newcomers from Terrorist Havens." **www .rasmussenreports.com**

Read, Jen'nan Ghazal. 2003. "The Sources of Gender Role Attitudes among Christian and Muslim Arab-American Women." *Sociology of Religion* 64 (Summer): 207–222.

Read, Piers Paul. 1974. *Alive: The Story of the Andes Survivors.* Philadelphia, PA: Lippincott.

Reich, Robert. 2010. "The Root of Economic Fragility and Political Anger." **ww.robertreich.org**

Reid, T. R. 2010. *The Healing of America: A Global Quest for Better, Cheaper, and Fairer Health Care*. New York: Penguin Books.

Reiman, Jeffrey H. 2012. *The Rich Get Richer and the Poor Get Prison*, 10th ed. Boston: Allyn and Bacon.

Reiman, Jeffrey, and Paul Leighton. 2012. *The Rich Get Richer and the Poor Get Prison: A Reader*, 10th ed. Upper Saddle River, NJ: Pearson.

Renzetti, Claire, Jeffrey L. Edieson, and Raquel K. Bergen. 2010. *Sourcebook on Violence against Women*. Thousand Oaks, CA: Sage Publications.

Reskin, Barbara. 1988. "Bringing the Men Back In: Sex Differentiation and the Devaluation of Women's Work." *Gender & Society* 2 (March): 58–81.

Reyns, Bradford W. 2013. "Online Routines and Identity Theft Victimization: Further Expanding Routine Activity Theory beyond Direct-Contact Offenses." *Journal of Research in Crime and Delinquency* 50 (2), 216–238.

Rheault, Magail, and Dalia Mogahed. 2008. "Moral Issues Divide Westerners from Muslims in the West." *The Gallup Poll*. Princeton, NJ: Gallup Organization. **www.gallup.com**

Rich, Adrienne. 1980. "Compulsory Heterosexuality and Lesbian Existence." *Signs* 5 (Summer): 631–660.

Ridgeway. Cecilia L. 2011. *Framed by Gender: How Gender Inequality Persists in the Modern World*. New York: Oxford University Press.

Riegle-Crumb, Catherine, and Melissa Humphries. 2012. "Exploring Bias in Math Teachers' Perceptions of Students' Ability by Gender and Race/Ethnicity." *Gender & Society* 26 (April): 290–322.

Riffkin, Rebecca. 2014. "Americans Favor Ban on Smoking in Public, but not Total Ban." *The Gallup Poll*, July 30. Princeton, NJ: The Gallup Organization. **www .gallup.com**

Risman, Barbara, and Pepper Schwartz. 2002. "After the Sexual Revolution: Gender Politics in Teen Dating." *Contexts* 1 (Spring): 16–24.

Ritzer, George. 2010. *The McDonaldization of Society*, 6th ed. Thousand Oaks, CA: Sage Publications.

Rivas-Drake, Deborah. 2011. "Ethnic- Racial Socialization and Adjustment among Latino College Students: The Mediating Roles of Ethnic Centrality, Public Regard, and Perceived Barriers to Opportunity." *Journal of Youth and Adolescence* 40 (May): 606–619.

Rivera, Fernando I., and Elizabeth Aranda. 2017. "When the U.S. Sneezes, Puerto Rico Already Has a Cold." *Contexts* October 5. **https://contexts.org/articles/puerto -rico-already-has-a-cold/**

Roberts, Dorothy. 1997. *Killing the Black Body: Race, Reproduction and the Meaning of Liberty*. New York: Vintage Books.

Roberts, Dorothy. 2012. *Fatal Invention: How Science, Politics, and Big Business Re-Create Race in the Twenty-First Century*. New York: New Press.

Robinson, Jo Ann Gibson. 1987. *The Montgomery Bus Boycott and the Women Who Started It*. Knoxville, TN: The University of Tennessee Press.

Roda, Allison, and Amy Stuart Wells. 2013. "School Choice Policies and Racial Segregation: Where Parents' Good Intensions, Anxiety, and Privilege Collide." *American Journal of Education* 119 (February): 261–293.

Rodriguez, Clara E. 1989. *Puerto Ricans: Born in the U.S.A.* Boston: Unwin Hyman.

Rodriguez, Clara E. 2009. "Changing Race." pp. 22–25 in *Race and Ethnicity in Society: The Changing Landscape*, 3rd ed., edited by Elizabeth Higginbotham and Margaret L. Andersen. Belmont, CA: Wadsworth.

Ropelato, Jerry. 2007. "Internet Pornography Statistics." *Top Ten Reviews*. **http://www.ministryoftruth.me.uk /wp-content/uploads/2014/03 /IFR2007.pdf**

Roper Organization. 2017. "How Groups Voted 2016." Ithaca, NY: Roper Center, Cornell University. **https://ropercenter. cornell.edu/polls/us-elections/how -groups-voted/groups-voted-2016/**

Rose, Stephen J. 2014. *Social Stratification in the United States: The American Profile Poster*. New York: The New Press.

Rosenbaum, Janet Elise. 2009. "Patient Teenagers? A Comparison of the Sexual Behavior of Virginity Pledgers and Matched Nonpledgers." *Pediatrics* 123 (January): 110–120.

Rosenfeld, Michael J., and Reuben J. Thomas. 2012. "Searching for a Mate: The Rise of the Internet as a Social Intermediary." *American Sociological Review* 77 (August): 523–547.

Rosengarten, Danielle. 2000. "Modern Times." *Dollars & Sense* (September): 4.

Rosenhan, David L. 1973. "On Being Sane in Insane Places." *Science* 179 (January 19): 250–258.

Rosenthal, Robert, and Lenore Jacobson. 1968. *Pygmalion in the Classroom: Teacher Expectations and Pupils' Intellectual Development*. New York: Holt, Rinehart and Winston.

Rospenda, Kathleen M., Judith A. Richman, and Candice A. Shannon. 2009. "Prevalence and Mental Health Correlates of Harassment and Discrimination in the Workplace. Results from a National Study." *Journal of Interpersonal Violence* 24 (5): 819–843.

Rossi, Alice S., and Peter H. Rossi. 1990. *Of Human Bonding: Parent-Child Relations across the Life Course*. New York: Aldine de Gruyter.

Royster, Deirdre. 2003. *The Invisible Hand: How White Networks Exclude Black Men from Blue-Collar Jobs*. Berkeley, CA: University of California Press.

Rudman, Laurie A., Peter Glick, Tahnee Marquardt, and Janell C. Fetterolf. 2017. "When Women Are Urged to Have Casual Sex More Than Men Are: Perceived Risk Moderates the Sexual Advice Double Standard." *Sex Roles* 77: 409–418.

Rudski, Jeffrey M. 2014. "Treatment Acceptability, Stigma, and Legal Concerns of Medical Marijuana Are Affected by Method of Administration." *Journal of Drug Issues* 44 (3): 308–320.

Rueschmeyer, Dietrich, and Theda Skocpol. 1996. *States, Social Knowledge, and the Origins of Modern Social Policies*. Princeton, NJ: Princeton University Press.

Rugh, Jacob S., and Douglas S. Massey. 2010. "Racial Segregation and the American Foreclosure Crisis." *American Sociological Review* 75 (5): 629–651.

Rugh, Jacob S., and Douglas S. Massey. 2010. "Racial Segregation and the American Foreclosure Crisis." *American Sociological Review* 75 (October): 629–651.

Rumbaut, Rubén. 1996. "Origins and Destinies: Immigration, Race, and Ethnicity in American History." pp. 1–20 in *Origins and Destinies: Immigration, Race, and Ethnicity in America*, edited by Silvia Pedraza and Rubén G. Rumbaut. Belmont, CA: Wadsworth.

Rumbaut, Rubén G., and Walter A. Ewing. 2007. *The Myth of Immigrant Criminality and the Paradox of Assimilation*. Washington, DC: American Immigration Law Foundation. **www .immigrationpolicy.org**

Rupp, Leila J., and Verta Taylor. 2003. *Drag Queens at the 801 Cabaret*. Chicago: University of Chicago Press.

Rupp, Leila J., and Verta Taylor. 2010. "Straight Girls Kissing." *Contexts* 9 (Summer): 28–33.

Rury, John L. 2015. *Education and Social Change: Contours in the History of Schooling*, 5th ed. New York: Routledge.

Rust, Paula. 1995. *Bisexuality and the Challenge to Lesbian Politics*. New York: New York University Press.

Rust, Paula C. 1993. "'Coming Out' in the Age of Social Constructionism: Sexual Identity Formation among Lesbian and Bisexual Women." *Gender & Society* 7 (March): 50–77.

Rutter, Virginia, and Pepper Schwartz. 2011. *The Gender of Sexuality: Exploring Sexual Possibilities*. Lanham, MD: Rowman and Littlefield.

Ruvolo, Julie. 2011. "How Much of the Internet is Actually for Porn?" *Forbes*, September 7. **www.forbes.com**

Ryan, Joelle Ruby. 2016. "From Transgender to Trans." *Introducing the New Sexuality Studies*, edited by Nancy L. Fischer, and Steven Seidman. New York: Routledge.

Ryan, Kathryn M. 2011. "The Relationship between Rape Myths and Sexual Scripts: The Social Construction of Rape." *Sex Roles: A Journal of Research* 65 (11–12): 774–782.

Ryan, Paul. 2014. "Interview with Bill Bennett." Bill Bennett: *Morning in America*, (radio) March 12.

Ryan, William. 1971. *Blaming the Victim*. New York: Pantheon.

Saad, Lydia. 2011 (August 8). "Plenty of Common Ground Found in Abortion Debate." *The Gallup Poll*. Princeton,

NJ: The Gallup Organization. **www .gallup.com**

Saad, Lydia. 2012. "In U.S., Half of Women Prefer a Job Outside the Home." *The Gallup Poll*. Princeton, NJ: The Gallup Organization, September 7. **www.gallup .com**

Saad, Lydia. 2014. "U.S. Still Split on Abortion: 47% Pro-Choice, 46% Pro-Life." *The Gallup Poll*. Princeton, NJ: The Gallup Organization. **www .gallup.com**

Sadker, Myra, and David Sadker. 1994. *Failing at Fairness: How America's Schools Cheat Girls*. New York: Scribner's.

Saez, Emmanuel, and Gabriel Zucman. 2014. "Exploding Wealth Inequality in the United States." Washington, DC: Washington Center for Equitable Growth. **www.equitablegrowth.org**

Saguy, Abigail. 2003. *What Is Sexual Harassment? From Capitol Hill to the Sorbonne*. Berkeley, CA: University of California Press.

Sanchez-Jankowski, Martin. 1991. *Islands in the Street: Gangs and American Urban Society*. Berkeley: University of California Press.

Sanday, Peggy. 2002. *Women at the Center: Life in a Modern Matriarchy*. Ithaca, NY: Cornell University Press.

Sandberg, Sheryl. 2013. *Lean In: Women, Work, and the Will to Lead*. New York: Knopf.

Sanders, Clinton R., and D. Angus Vail. 2008. *The Art and Culture of Tattooing*. Philadelphia, PA: Temple University Press.

Santelli, John, et al. 2007. "Explaining Recent Declines in Adolescent Pregnancy in the United States: The Contribution of Abstinence and Increased Contraceptive Use." *American Journal of Public Health* 97 (1): 150–156.

Sapir, Edward. 1921. *Language: An Introduction to the Study of Speech*. New York: Harcourt Brace.

Sara, Siddharth. 2010. *Sex Trafficking: Inside the Business of Modern Slavery*. New York: Columbia University Press.

Sarkasian, Natalia, and Naomi Gerstel. 2012. *Nuclear Family Values, Extended Family Lives: The Power of Race, Class, and Gender*. New York: Russell Sage Foundation.

Saunders, Jessica F. and Leslie D. Frazier. 2017. "Body Dissatisfaction in Early Adolescence: The Coactive Roles of Cognitive and Sociocultural Factors." *Journal of Youth and Adolescence* 46 (6): 1246–1261.

Sayer, Liana C., and Leigh Fine. 2011. "Racial-Ethnic Differences in U.S. Married Women's and Men's Housework." *Social Indicators Research* 101 (2): 259–265.

Shaefer, H. Luke, and Kathryn Edin. 2014. "The Rise of Extreme Poverty in the United States." *Pathways* (Summer): 28–32.

Schaffer, Kay, and Song Xianlin. 2007. "Unruly Spaces: Gender, Women's Writing and Indigenous Feminism in China." *Journal of Gender Studies* 16 (1): 17–30.

Schalet, Amy. 2010. "Sex, Love, and Autonomy in the Teenage Sleepover." *Contexts* 9 (Summer): 16–21.

Schilt, Kristen. 2011. *Just One of the Guys? Transgender Men and the Persistence of Gender Inequality*. Chicago: University of Chicago Press.

Schlosser, Eric. 2001. *Fast Food Nation: The Dark Side of the All-American Meal*. New York: Houghton Mifflin.

Schmitt, Eric. 2001. "Segregation Growing among U.S. Children." *The New York Times* (May 6): 28.

Schmitt, Frederika E., and Patricia Yancey Martin. 1999. "Unobtrusive Mobilization by an Institutionalized Rape Crisis Center: 'It Comes From the Victims.'" *Gender & Society* 13 (3): 364–384.

Schriever, Norm. 2017. "Ranking the Biggest Industries in the US Economic—With a Surprise #1!" Blue Water Credit. **http:// bluewatercredit.com/ranking-biggest -industries-us-economy-surprise-1/**

Schwartz, John, and Matthew L. Wald. 2003. "NASA's Failings Go Far beyond Foam Hitting Shuttle, Panel Says." *The New York Times* (June 7): 1 and 12.

Scott, Ellen K., Andrew S. London, and Nancy A. Myers. 2002. "Dangerous Dependencies: The Intersection of Welfare Reform and Domestic Violence." *Gender & Society* 16 (December): 878–897.

Segal, David R., and Mady Wechsler Segal. 2004. "America's Military Population." *Population Bulletin* 59 (December): 1–44.

Seidman, Steven. 2014. *The Social Construction of Sexuality*, 3rd ed. New York: W. W. Norton.

Semega, Jessica L., Kayla R. Fontenot, and Melissa A. Kollar. 2017. *Income and Poverty in the United States: 2016*. Washington, DC: U.S. Census Bureau. **www.census.gov**

Sen, Amartya. 2000. "Population and Gender Equity." *The Nation* (July 24–31): 16–18.

Sentse, Miranda, Noona Kiuru, René Veenstra, and Christina Salmivalli. 2014. "A Social Network Approach to the Interplay between Adolescents' Bullying and Likeability Over Time." *Journal of Youth and Adolescence* 43(9): 1409–1420.

Shapiro, Thomas. 2017. *Toxic Inequality: How America's Wealth Gap Destroys Mobility, Depends the Racial Divide, & Threatens Our Future*. New York: Basic Books.

Shelley, Louise. 2010. *Human Trafficking: A Global Perspective*. Cambridge, MA: Cambridge University Press.

Sherman, Jennifer. 2009. "Bend to Avoid Breaking: Job Loss, Gender Norms, and Family Stability in Rural America." *Social Problems* 56 (November): 599–620.

Shibutani, Tomatsu. 1961. *Society and Personality: An Interactionist Approach to Social Psychology*. Englewood Cliffs, NJ: Prentice Hall.

Shilts, Randy. 2007. *And the Band Played On: Politics, People, and the AIDS Epidemic*, 20th anniversary ed. New York: St. Martin's.

Short, Natalie N. 2015. "Christopher J. Ferguson: Adolescents, Crime, and the Media: A Critical Analysis." *Journal of Youth and Adolescence* 44 (3): 773–776.

Silva, Jennifer M. 2013. *Coming Up Short: Working-Class Adulthood in an Age of Uncertainty*. New York: Oxford University Press.

Silva, Jennifer M. 2014. "Working-Class Growing Pains." *Contexts* 13 (2): 26–31.

Silva, Jennifer M. 2015. *Coming Up Short: Working-Class Adulthood in an Age of Uncertainty*. New York: Oxford University Press.

Silver, Alexandra. 2010. "Brief History of the U.S. Census." *Time* (February 8): 16.

Silverthorne, Zebulon A., and Vernon L Quinsey. 2000. "Sexual Partner Age Preferences of Homosexual and Heterosexual Men and Women." *Archives of Sexual Behavior* 29 (1): 67–76.

Simmel, Georg. 1950 [1902]. "The Number of Members as Determining the Sociological Form of the Group." *The American Journal of Sociology* 8 (July): 1–46.

Simon, David R. 2011. *Elite Deviance*, 10th ed. Boston: Allyn and Bacon.

Singer, Peter W. 2007. *Corporate Warriors: The Rise of the Privatized Military Industry*. Ithaca, NY: Cornell University Press.

Sinozich, Sofi, and Lynn Langton. 2014. *Rape and Sexual Assault Victimization Among College-Age Females, 1995–2013*. Washington, DC: Bureau of Justice Statistics. **www.bjs.gov/content/pub /pdf/rsavcaf9513.pdf**

Sjøberg, Gideon. 1965. *The Preindustrial City: Past and Present*. New York: Free Press.

Skocpol, Theda. 1992. *Protecting Soldiers and Mothers: The Origins of Social Policy in the United States*. Cambridge, MA: Belknap Press.

Skocpol, Theda, and Vanessa Williamson. 2012. *The Tea Party and the Remaking of Republican Conservatism*. New York: Oxford University Press.

Smelser, Neil J. 1992. "Culture: Coherent or Incoherent." pp. 3–28 in *Theory of Culture*, edited by R. Münch and N. J. Smelser. Berkeley, CA: University of California Press.

Smith, Aaron. 2012 (November 30). "The Best (and Worst) of Mobile Connectivity." Washington, DC: Pew Research Center, Internet and Technology. **www .pewresearch.org**

Smith, Aaron, and Dana Page. 2015. *Smartphone Usage in 2015*. Washington, DC: Pew Research Organization. **www .pewresearch.org**

Smith, Andrew. 2009. "Nigerian Scam E-Mails and the Charms of Capital." *Cultural Studies* 23 (1): 27–47.

Smith, Brad W., and Malcolm D. Holmes. 2014. "Police use of Excessive Force in Minority Communities: A Test of the Minority Threat, Place, and Community Accountability Hypotheses." *Social Problems* 61 (1): 83–104.

Smith, M. Dwayne, Joel A. Devine, and Joseph F. Sheley. 1992. "Crime and Unemployment: Effects Across Age and Race Categories." *Sociological Perspectives* 35 (Winter): 551–572.

Smith, Stacy L., Marc Choueiti, Ashley Prescott, and Katherine Pieper. 2013. *Gender Roles & Occupation: A Look at Character Attributes and Job-Related Aspirations in Film and Television.* **www.seejane.org**

Smith, Tom W., Peter V. Marsden, Michael Hout, and Jibum Kim. *General Social Surveys, 1972–2012.* Chicago: National Opinion Research Center.

Smith, Tom, and Jaesok Son 2013. *Trends in Public Attitudes about Sexual Morality.* Chicago: National Opinion Research Center.

Smith, R. Tyson, and Gala True. 2014. "Warring Identities: Identity Conflict and the Mental Distress of American Veterans of the Wars in Iraq and Afghanistan." *Society and Mental Health* 4 (2): 147–161.

Smolak, L, and M. Levine, eds., 1996. *The Developmental Psychopathology of Eating Disorders: Implications for Research, Prevention, and Treatment.* Hillsdale, NJ: Lawrence Erlbaum Associates Inc.

Snipp, C. Matthew. 1989. *American Indians: The First of This Land.* New York: Russell Sage Foundation.

Snipp, C. Matthew. 1996. "The First Americans: American Indians." pp. 390–404 in *Origins and Destinies: Immigration, Race, and Ethnicity in America,* edited by Sylvia Pedraza and Rubén G. Rumbaut. Belmont, CA: Wadsworth.

Snipp, Matthew. 2007. "An Overview of American Indian Populations." pp. 38–48 in *American Indian Nations: Yesterday, Today, and Tomorrow,* edited by George Horse Capture, Duane Champaign, and Chandler Jackson. Walnut Creek, CA: Altamira Press.

Solomon, Brittany C., and Simine Vazire. 2014. "You Are So Beautiful . . . To Me: Seeing beyond Biases and Achieving Accuracy in Romantic Relationships." *Journal of Personality and Social Psychology* 107 (3): 516–528.

Spencer, Herbert. 1882. *The Study of Sociology.* London: Routledge.

Spencer, Ranier. 2011. *Reproducing Race: The Paradox of Generation Mix.* Las Vegas, Nevada: Lynne Rienner.

Spitzer, Brenda L., Katherine A. Henderson, and Marilyn T. Zivian. 1999. "A Comparison of Population and Media Body Sizes for American and Canadian Women." *Sex Roles* 700 (7/8): 545–565.

Stacey, Judith, and Timothy J. Bibliarz. 2001. "(How) Does the Sexual Orientation of Parents Matter?" *American Sociological Review* 66 (April): 159–183.

Stack, Carol. 1974. *All Our Kin: Strategies for Survival in a Black Community.* New York: Harper Colophon Books.

The Statistics Portal. 2017. New York: Statista. **https://www.statista.com/statistics/264810/number-of-monthly-active-facebook-users-worldwide/**

Steele, Claude M. 1997. "A Threat in the Air: How Stereotypes Shape Intellectual Identity and Performance." *American Psychologist* 52 (6): 613–629.

Steele, Claude. M. 2010. *Whistling Vivaldi and Other Clues to How Stereotypes Affect Us.* New York: W. W. Norton.

Steele, Claude M., and Joshua Aronson. 1995. "Stereotype Threat and the Intellectual Test Performance of African Americans." *Journal of Personality and Social Psychology* 69 (5): 797–811.

Steger, Manfred B. 2017. *Globalization: A Very Short Introduction,* 3rd ed. New York: Oxford University Press.

Stein, Arlene, and Marcy Westerling. 2012. "The Politics of Broken Dreams." *Contexts* 11 (Summer): 8–10.

Stengers, Jean, and Van Neck, Anne. 2001. *Masturbation: The History of a Great Terror.* New York: St. Martin's.

Stern, Jessica. 2003. *Terror in the Name of God: Why Religious Militants Kill.* New York: Ecco.

Sternheimer, Karen. 2007. "Do Video Games Kill?" *Contexts* 6 (Winter): 13–17.

Sternheimer , Karen. 2011. *Celebrity Culture and the American Dream: Stardom and Social Mobility.* New York: Routledge.

Sternthal, Michelle J., Natalie Slopen, and David R. Williams. 2011. "Racial Disparities in Health." *Du Bois Review* 8 (1): 95–113.

Stewart, Susan D. 2001. "Contemporary American Stepparenthood: Integrating Cohabiting and Nonresident Stepparents." *Population Research and Policy* 20 (August): 345–364.

Stombler, Mindy, and Irene Padavic. 1997. "Sister Acts: Resisting Men's Domination in Black and White Fraternity Little Sister Programs." *Social Problems* 44 (May): 257–275.

Stone, Michael E. 1993. *Shelter Poverty: New Ideas on Housing Affordability.* Philadelphia: Temple University Press.

Stoner, J. A. F. 1961. "A Comparison of Individual and Group Decisions Involving Risk." Unpublished M. A. thesis, MIT.

Stryker, Susan, and Stephen Whittle (eds.). 2006. *The Transgender Studies Reader.* New York: Routledge.

Su, Dejun, Chad Richardson, and Guangzhen Wang. 2010. "Assessing Cultural Assimilation of Mexican Americans: How Rapidly Do Their Gender-Role Attitudes Converge to the U.S. Mainstream?" *Social Science Quarterly* 91 (3): 762–776.

Sullivan, Maureen. 1996. "Rozzie and Harriet? Gender and Family Patterns of Lesbian Coparents." *Gender & Society* 12 (December): 747–767.

Sullivan, Nikki. 2003. *A Critical Introduction to Queer Theory.* New York: New York University Press.

Sullivan, Patrick F. 1995. "Mortality in Anorexia Nervosa." *American Journal of Psychiatry* 152 (July): 1073–1074.

Suls, Rob. 2017. *Less Than Half the Public Views Border Wall as an Important Goal for U.S. Immigration Policy.* Washington, DC: Pew Research Center. **www.pewresearch.org**

Sumner, William Graham. 1906. *Folkways: A Study of the Sociological Importance of Usages, Manners, Customs, Mores, and Morals.* Boston: Ginn.

Sun, Lena H. 2017. "Rural Americans Are More Likely to Die from the Top 5 Causes of Death." *The Washington Post,* January 12. **www.washingtonpost.com**

Sussman, N. M., and D. H. Tyson. 2000. "Sex and Power: Gender Differences in Computer-Mediated Interactions." *Computers in Human Behavior* 16 (July): 381–394.

Sutherland, Edwin H. 1940. "White Collar Criminality." *American Sociological Review* 5 (February): 1–12.

Sutherland, Edwin H., and Donald R. Cressey. 1978. *Criminology,* 10th ed. New York: Lippincott.

Swidler, Ann. 1986. "Culture in Action: Symbols and Strategies." *American Sociological Review* 51 (April): 273–286.

Swift, Art. 2017. "Majority in US Still Say Religion Can Answer Most Problems." Princeton, NJ: Gallup Poll. **http://news.gallup.com/poll/211679/majority-say-religion-answer-problems.aspx**

Switzer, J. Y. 1990. "The Impact of Generic Word Choices: An Empirical Investigation of Age- and Sex-Related Differences." *Sex Roles* 22 (1): 69–82.

Tach, Laura, and Sarah Halpern-Meekin. 2009. "How Does Premarital Cohabitation Affect Trajectories of Marital Quality?" *Journal of Marriage and Family* 71 (May): 298–317.

Takaki, Ronald. 1989. *Strangers from a Different Shore: A History of Asian Americans.* New York: Penguin.

Tanner, Lindsay. 2014. "New Study Suggests Genetic Link for Male Homosexuality." *The Huffington Post,* November 17.

Tatum, Beverly. 1997. *Why Are All the Black Kids Sitting Together at the Cafeteria?* New York: Basic Books.

Taylor, Howard F. 1980. *The IQ Game: A Methodological Inquiry into the Heredity-Environment Controversy.* New Brunswick, NJ: Rutgers University Press.

Taylor, Howard F. 1992. "The Structure of a National Black Leadership Network: Preliminary Findings." Unpublished manuscript, Princeton University.

Taylor, Howard F. 2012. "Defining Race." pp. 7–13 in *Race and Ethnicity in Society: The Changing Landscape,* 3rd ed., edited by in Elizabeth Higginbotham and Margaret L. Andersen. Belmont, CA: Wadsworth/Cengage.

Taylor, Paul. 2012. "The Growing Electoral Clout of Blacks Is Driven by Turnout, Not Demographics." *Pew Social & Demographic Trends.* Washington, DC: Pew Research Center. **www.pewresearch.org**

Taylor, Shelley E., Letitia Anne Peplau, and David O. Sears. 2016. *Social Psychology,* 14th ed. Upper Saddle River, NJ: Prentice-Hall.

Telles, Edward E. 2004. *Race in Another America: The Significance of Skin Color in Brazil.* Princeton, NJ: Princeton University Press.

Telles, Edward E., and Vilma Ortiz. 2008. *Generations of Exclusion: Mexican Americans, Assimilation, and Race.* New York: Russell Sage Foundation.

Telles, Edward, Mark Q. Sawyer, and Gaspar Rivera-Salgado, eds. 2011. *Just Neighbors? Research on African American and Latino Relations in the United States.* New York: Russell Sage Foundation.

Tenenbaum, Harriet R. 2009. "'You'd Be Good at That': Gender Patterns in Parent–Child Talk about Courses." *Social Development* 18 (2): 447–463.

Thoits, Peggy A. 2009. "Sociological Approaches to Mental Illness." In *A Handbook for the Study of Mental Health,* edited by Teresa L. Scheid and Tony L. Brown. Cambridge: Cambridge University Press.

Thomas, A. J., J. D. Hacker, and D. Hoxha. 2011. "Gendered Racial Identity of Black Young Women." *Sex Roles* 64: 530–542.

Thomas, William I. 1931. *The Unadjusted Girl.* Boston: Little, Brown.

Thomas, William I., with Dorothy Swaine Thomas. 1928. *The Child in America.* New York: Knopf.

Thompson, Robert Farris. 1993. *Face of the Gods: Art and Alters of Africa and the African Americas.* New York: Random House.

Thornton, Russell. 1987. *American Indian Holocaust and Survival: A Population History.* Norman, OK: University of Oklahoma Press.

Thornton, Russell. 2001. "Trends among American Indians in the United States." pp. 135–169 in *America Becoming: Racial Trends and Their Consequences,* vol. I, edited by Neil J. Smelser, William Julius Wilson, and Faith Mitchell. Washington, DC: National Academies Press.

Thui, Prak Chan. 2016. "Cambodia Raises Minimum Wage for Textile Industry Workers." **www.reuters .com/article/cambodia-garment -idUSL3N1C51OD**

Tichenor, Veronica. 2005. "Maintaining Men's Dominance: Negotiating Identity and Power When She Earns More." *Sex Roles* 53 (3–4): 191–205.

Tierney, Kathleen J. 2007. "From the Margins to the Mainstream? Disaster Research at the Crossroads." *Annual Review of Sociology* 33: 503–525.

Tierney, Kathleen. 2012 (November 11). "After Hurricane Sandy: Understanding a Hurricane's Worst Impacts: Q&A." *Star-Ledger.* **blog.nj.com**

Tierney, Kathleen J., Michael K. Lindell, and Ronald W. Perry. 2001. *Facing the Unexpected: Disaster Preparedness and Response in the United States.* Washington, DC: National Academies Press.

Tierney, Kathleen, Christine Bevc, and Erica Kuglikowski. 2006. "Metaphors Matter: Disaster Frames, Media Myths, and Their Consequences in Hurricane Katrina." *Annals of the American Association of Political and Social Science* 604 (March): 57–81.

Timmerman, Kelsey. 2012. *Where Am I Wearing: A Global Tour to the Countries, Factories, and People that Make our Clothes.* Hoboken, NJ: Wiley.

Tjaden, Patricia, and Nancy Thoennes. 2000. *Extent, Nature, and Consequences of Intimate Partner Violence.* Washington, DC: National Institute of Justice and the Centers for Disease Control and Prevention.

Tönnies, Ferdinand. 1963 [1887]. *Community and Society (Gemeinschaft and Gesellschaft).* New York: Harper & Row.

Toossi, Mitra. 2015. "Labor Force Projections to 2024: The Labor Force is Growing, But Slowly." *Monthly Labor Review* (December): 1–33.

Towne, Samuel D., Janice C. Probst, James W. Hardin, Bethany A. Bell, and Saundra Glover. 2017. "Health and Access to Care among Working-Age Lower Income Adults in the Great Recession: Disparities Across Race and Ethnicity and Geospatial Factors." *Social Science & Medicine* 182: 30–44.

Toynbee, Arnold J., and Jane Caplan. 1972. *A Study of History.* New York: Oxford University Press.

TransAtlantic Slave Trade Database. 2014. **www.slavevoyages.org**

Travers, Jeffrey, and Stanley Milgram. 1969. "An Experimental Study of the Small World Problem." *Sociometry* 32 (4): 425–443.

Treiman, Donald J. 2001. "Occupations, Stratification and Mobility." pp. 297–313 in *The Blackwell Companion to Sociology,* edited by Judith R. Blau. Malden, MA: Blackwell.

Tsoukalas, I. 2007. "Exploring the Microfoundations of Consciousness." *Culture and Psychology* 13 (1): 39–81.

Tuan, Yi-Fu. 1984. *Dominance and Affection: The Making of Pets.* New Haven, CT: Yale University Press.

Tuchman, Gaye. 1979. "Women's Depiction by the Mass Media." *Signs* 4 (Spring): 528–542.

Tumin, Melvin M. 1953. "Some Principles of Stratification." *American Sociological Review* 18 (August): 387–393.

Turner, Ralph, and Lewis Killian. 1993. *Collective Behavior,* 4th ed. Englewood Cliffs, NJ: Prentice Hall.

Turner, Terence. 1969. "Tchikrin: A Central Brazilian Tribe and Its Symbolic Language of Body Adornment." *Natural History* 78 (October): 50–59.

Tyson, Alec, and Shiva Maniam. 2016. "Behind Trump's Victory: Divisions by Race, Gender, Education" Washington, DC: Pew Research Organization. **www.pewreserch.org**

UNICEF. 2000. *Child Poverty in Rich Nations.* Florence, Italy: United Nations Children's Fund. New York: United Nations. **www.unicef-icdc.org**

United Nations. 2006. *In Depth Study on All Forms of Violence against Women.* New York: United Nations.

United Nations. 2010. *The Gender Inequality Index.* New York: United Nations. **www .undp.org**

United Nations. 2012. *WomenWatch.* New York: United Nations. **www.un .org/womenwatch**

United Nations. 2013. *World Population Prospects: The 2012 Revision.* New York: United Nations. **www.unpopulation .org**

United Nations. 2014. *Statistics on Literacy.* **www.unesco.org**

United Nations. 2015. *International Migration Report 2015.* **www.un.org/en /development/desa/population /migration/publications /migrationreport/docs /MigrationReport2015_Highlights.pdf**

United Nations Population Fund. 2014. *Population and Poverty.* **www.unfpa.org /resources/population-and-poverty**

U.S. Administration for Children and Families. 2015. *Child Maltreatment 2015.* Washington, DC: Children's Bureau. **www.acf.hhs.gov/cb/resource/child -maltreatment-2015**

U.S. Bureau of Justice Statistics. 2012. *National Crime Victimization Survey.* Department of Justice. **www.bjs.gov**

U.S. Bureau of Justice Statistics. 2013. *Arrest Data.* Department of Justice. **www.bjs.gov**

U.S. Bureau of Justice Statistics. 2016. *NCVS Victimization Analysis Tool.* **www .bjs.gov**

U.S. Bureau of Labor Statistics. 2010. "Labor Force Participation Rates among Mothers." Washington, DC: U.S. Department of Labor. **www.bls.gov**

U.S. Bureau of Labor Statistics. 2012a. Employment Projections: Civilian Labor Force Participation Rates by Age, Sex, Race, and Ethnicity. Washington, DC: U.S. Department of Labor. **www.bls.gov**

U.S. Bureau of Labor Statistics. 2012b. *Employment and Earnings.* Washington, DC: U.S. Department of Labor. **www.bls.gov**

U.S. Bureau of Labor Statistics. 2012c. "A Profile of the Working Poor, 2008." Washington, DC: U.S. Department of Labor. **www.bls.gov**

U.S. Bureau of Labor Statistics. 2014. *Highlights of Women's Earnings in 2013,* Report 1051. BLS Reports. **www .bls.gov**

U.S. Bureau of Labor Statistics. 2016a. *Employment and Earnings.* Washington, DC: Department of Labor. **www.bls.gov**

U.S. Bureau of Labor Statistics. 2016b. *Employer-Provided Quality-of-Life Benefits, March 2016.* Washington, DC: U.S. Department of Labor. **www.bls.gov /opub/ted/2016/employer-provided -quality-of-life-benefits-march-2016 .htm**

U.S. Bureau of Labor Statistics. 2017a. *A Profile of the Working Poor, 2015.* Washington, DC: U.S. Department of Labor.

U.S. Bureau of Labor Statistics. 2017b. *Consumer Expenditure Survey.* Washington, DC: U.S. Department of Labor. **www.bls.gov/cex/**

U.S. Bureau of Labor Statistics. 2017c. *Employment and Earnings.* Washington, DC: U.S. Department of Labor. **www.bls .gov**

U.S. Bureau of Labor Statistics. 2017d. *Employment Characteristics of Families Summary.* Washington, DC: US. Department of Labor. **www.bls.gov**

U.S. Census Bureau. 2003. "Racial and Ethnic Classification Used in Census 2000 and Beyond." Washington, DC: U.S. Census Bureau. **www.census.gov**

U.S. Census Bureau. 2012a. "U.S. Census Bureau Projections Show a Slower Growing, Older, More Diverse Nation A Half Century Years from Now." Washington, DC: U.S. Census Bureau. **www.census.gov**

U.S. Census Bureau. 2012b. *Census Bureau Releases Estimate of Undercount and Overcount in the 2010 Census.* Washington, DC: U.S. Census Bureau. **https://www.census.gov/newsroom /releases/archives/2010_census /cb12-95.html**

U.S. Census Bureau. 2012c. *Current Population Reports: Annual Social and Economic Supplement.* Washington, DC: U.S. Census Bureau. **www.census.gov**

U.S. Census Bureau. 2014. *Foreign-Born: 2014 Current Population Survey Detailed Tables.* Washington, DC: U.S. Census Bureau. **www.census.gov/data/tables /2014/demo/foreign-born/cps-2014 .html**

U.S. Census Bureau. 2016a. Estimated Median Age at First Marriage: 1890 to Present, Table MS-2. *America's Families and Living Arrangements: 2016.* Washington, DC: U.S. Census Bureau.

U.S. Census Bureau. 2016b. "Living Arrangements of Children under 18 Years and Marital Status of Parents, By Age, Sex, Race, and Hispanic Origin and Selected Characteristics of the Child for all Children: 2016. Table C3." *America's Families and Living Arrangements: 2016.* Washington, DC: U.S. Census Bureau. **www.census.gov/data/tables/2016 /demo/families/cps-2016.html**

U.S. Census Bureau. 2016c. "Marital Status of People 15 Years and Over, by Age, Sex, and Personal Earnings: 2016, Table A1." *America's Families and Living Arrangements: 2016.* Washington, DC: U.S. Census Bureau. **www.census.gov /data/tables/2016/demo/families /cps-2016.html**

U.S. Census Bureau. 2016d. "PINC-03 Educational Attainment—People 25 Years Old and Over, by Total Money Earnings in 2016, Work Experience in 2016, Age, Race, Hispanic Origin, and Sex." Washington, DC: U.S. Census Bureau. **www.census.gov/data/tables/time -series/demo/income-poverty /cps-pinc/pinc-03.html#par_ textimage_54**

U.S. Census Bureau. 2017a. *American Fact Finder.* Washington, DC: U.S. Census Bureau. **www.census.gov**

U.S. Census Bureau. 2017b. *Detailed Income Tables: Selected Characteristics of People 15 Years and Over, by Total Money Income, Work Experience, Race, Hispanic Origin, and Sex, PINC-01.* Washington, DC: U.S. Census Bureau. **www.census.gov**

U.S. Department of Agriculture. 2014. *Supplemental Nutrition Assistance Program (SNAP) Characteristics of Supplemental Nutrition Assistance Program Households: Fiscal Year 2014.* Washington, DC: USDA Food and Nutrition Program.

U.S. Department of Defense. 2012. *2012 Demographics: A Profile of the Military Community.* Washington, DC: U.S. Department of Defense.

U.S. Department of Justice. 2015. *Investigation of the Ferguson Police Department.* Washington, DC: U.S. Department of Justice. **www.justice.gov**

U.S. Department of Labor. 2018. *Employment and Earnings.* Washington, DC: U.S. Department of Labor. **www.bls.gov**

U.S. Department of State. 2011. *Annual Report on Intercountry Adoption.* Washington, DC: U.S. Department of State.

U.S. Department of State. 2014. *Trafficking in Persons Report: June 2014.* Washington, DC: U.S. Department of State. **www.state .gov**

U.S. Department of State. 2017. *Trafficking in Persons Report 2017.* Washington, DC: U.S. Department of State. **www.state .gov/j/tip/rls/tiprpt/2017/**

U.S. Energy Information Administration. 2017. *U.S. Energy Related Carbon Dioxide Emissions 2011.* Washington, DC: U.S. Department of Energy. **www.eia.gov**

U.S. Energy Information Administration. 2012. Washington, DC: U.S. Department of Energy. **https://www.eia.gov**

U.S. Energy Information Administration (EIA). 2015. *International Energy Statistics Database.* **www.eia.gov/ies**

U.S. State Department. 2012. *Trafficking in Persons Report, June 2012.* Washington, DC: U.S. State Department. **www.state .gov**

VA Suicide Prevention Program. 2016. "Facts about Veteran Suicide 2014." Washington, DC: Veterans Administration. **www.va.gov/opa /publications/factsheets/Suicide _Prevention_FactSheet_New_VA _Stats_070616_1400.pdf**

Valentine, Kathyrn, Mary Prentice, Monica F. Torres, and Eduardo Arellano. 2012. "The Importance of Cross-Racial Interactions as Part of College Education: Perceptions of Faculty." *Journal of Diversity in Higher Education* 5 (December): 191–206.

Valian, Virginia. 1999. *Why So Slow? The Advancement of Women.* Cambridge, MA: MIT Press.

Vallejo, Jody. 2013. *From Barrios to Burbs: The Making of the Mexican American Middle Class.* Stanford, CA: Stanford University Press.

Valles, Mike. 2014. "Alternative Medicine and Your Health Insurance." *The Simple Dollar,* September 16. **thesimpledollar.com**

Van Ausdale, Debra, and Joe R. Feagin. 1996. "The Use of Racial and Ethnic Concepts by Very Young Children." *American Sociological Review* 61 (October): 779–793.

van Dijck, José. 2013. "'You Are One Identity': Performing the Self on Facebook and LinkedIn." *Media, Culture, & Society* 35 (2): 199–215.

Vanneman, Reeve, and Lynn Weber Cannon. 1987. *The American Perception of Class.* Philadelphia: Temple University Press.

Vaughan, Diane. 1996. *The Challenger Launch Decision: Risky Technology, Culture, and Deviance at NASA.* Chicago: The University of Chicago Press.

Veblen, Thorstein. 1994 [1899]. *Theory of the Leisure Class.* New York: Penguin.

Veblen, Thorstein. 1953 [1899]. *The Theory of the Leisure Class: An Economic Study of Institutions.* New York: The New American Library.

Veliz, Philip, and Sohaila Shakib. 2012. "Interscholastic Sports Participation and School Based Delinquency: Does Participation in Sport Foster a Positive High School Environment?" *Sociological Spectrum* 32 (6): 558–580.

Venkatesh, Sudhir. 2008. *Gang Leader for a Day: A Rogue Sociologist Takes to the Streets.* New York: Penguin.

Vespa, Jonathan. 2009. "Gender Ideology Construction." *Gender & Society* 23 (June): 363–387.

Vidmar, Neil, and Valerie P. Hans. 2007. *American Juries: The Verdict.* Amherst, NY: Prometheus Books.

Vijayasiri, Ganga. 2008. "Reporting Sexual Harassment: The Importance of Organizational Culture and Trust." *Gender Issues* 25 (1): 43–61.

Villareal, Andres. 2010. "Stratification by Skin Color in Contemporary Mexico." *American Sociological Review* 75 (6): 652–678.

Volscho, Thomas W., and Nathan J. Kelly. 2012. "The Rise of the Super-Rich: Power Resources, Taxes, Financial Markets, and the Dynamics of the Top 1 Percent, 1949–2008." *American Sociological Review* 77 (October): 679–699.

Voss, Georgina. 2012. "'Treating It as a Normal Business': Researching the Pornography Industry." *Sexualities* 15 (3–4): 391–410.

Waldfogel, Jane, Wen-Jui Han, and Jeanne Brooks-Gunn. 2002. "The Effects of Early Maternal Employment in Child Cognitive Development." *Demography* 39 (May): 369–392.

Walker, Renee E., Christopher R. Keane, and Jessica G. Burke. 2010. "Disparities and Access to Healthy Food in the United States: A Review of Food Deserts Literature." *Health & Place* 16 (5): 876–884.

Walker, Samuel, Cassia Spohn, and Miriam Delone. 2012. *The Color of Justice: Race, Ethnicity, and Crime in America.* Belmont, CA: Wadsworth.

Wallerstein, Immanuel M. 1974. *The Modern World System: Capitalist Agriculture and the Origins of the European World Economy in the Sixteenth Century.* New York: Academic Press.

Wallerstein, Immanuel M. 1980. *The Modern World-System II.* New York: Academic Press.

Walters, Suzanna Danuta. 1999. "Sex, Text, and Context: (In) Between Feminism and Cultural Studies." pp. 222–260 in *Revisioning Gender,* edited by Myra Marx Ferree, Judith Lorber, and Beth B. Hess. Thousand Oaks, CA: Sage.

Wang, Wendy. 2015. *Interracial Marriage: Who is "Marrying Out?"* Washington, DC: Pew Research Center. **www.pewresearch.org/fact -tank/2015/06/12/interracial -marriage-who-is-marrying-out/**

Warren, Cortney S., Andrea Schoen, and Kerri J. Schafer. 2010. "Media Internalization and Social Comparison as Predictors of Eating Pathology among Latino Adolescents: The Moderating Effect of Gender and Generational Status." *Sex Roles* 63 (9–10): 712–724.

Washington Post. 2017. "Data: Police Shootings." **https://github.com /washingtonpost/data-police -shootings/blob/master/fatal-police -shootings-data.csv**

Wasserman, Stanley, and Katherine Faust (eds.). 1994. *Social Network Analysis: Methods and Applications.* Cambridge, MA: Cambridge University Press.

Waters, Mary C. 1990. *Ethnic Options: Choosing Identities in America.* Berkeley, CA: University of California Press.

Waters, Mary C., and Peggy Levitt (eds.). 2002. *The Changing Face of Home: The Transnational Lives of the Second Generation.* New York: Russell Sage Foundation.

Weber, Max. 1958 [1904]. *The Protestant Ethic and the Spirit of Capitalism.* New York: Scribner's.

Weber, Max. 1962 [1913]. *Basic Concepts in Sociology.* New York: Greenwood.

Weber, Max. 1978 [1921]. *Economy and Society: An Outline of Interpretive Sociology,* edited by Guenther Roth and Claus Wittich. Berkeley, CA: University of California Press.

Weeks, John R. 2012. *Population: An Introduction to Concepts and Issues,* 12th ed. Belmont, CA: Wadsworth Publishing Co.

Weiss, Gregory L., and Lynne E. Lonnquist. 2017. *The Sociology of Health, Healing, and Illness,* 9th ed. New York: Routledge.

Weitz, Rose. 2016. *The Sociology of Health, Illness, and Health Care: A Critical Approach,* 7th ed. San Francisco: Cengage.

Weitzer, Ronald. 2009. "Sociology of Sex Work." *Annual Review of Sociology* 35 (February): 213–234.

West, Candace, and Sarah Fenstermaker. 1995. "Doing Difference." *Gender & Society* 9 (February): 8–37.

West, Candace, and Don Zimmerman. 1987. "Doing Gender." *Gender & Society* 1 (June): 125–151.

West, Cornel. 1994. *Race Matters.* New York: Vintage.

West, Cornel R. 2004. *Democracy Matters: Winning the Fight against Imperialism.* New York: The Penguin Press.

Westbrook, Laurel, and Kristen Schilt. 2014. "Doing Gender, Determining Gender: Transgender People, Gender Panics, and the Maintenance of the Sex/ Gender/ Sexuality System." *Gender & Society* 28 (February): 32–57.

Western, Bruce. 2007. *Punishment and Inequality in America.* New York: Russell Sage Foundation.

Western, B. (2014). "Incarceration, Inequality, and Imagining Alternatives." *The Annals of the American Academy of Political and Social Science* 651 (1): 302–306.

White, Deborah Gray. 1999. *Ar'n't I a Woman? Female Slaves in the Plantation South.* New York: W. W. Norton.

White, Jonathan R. 2013. *Terrorism and Homeland Security.* Belmont, CA: Wadsworth.White, Kevin. 2017. *An Introduction to the Sociology of Health.* Thousand Oaks, CA: Sage.

Whitehead, Andrew L., and Joseph O. Baker. 2012. "Homosexuality, Religion, and Science: Moral Authority and the Persistence of Negative Attitudes." *Sociological Inquiry* 82 (4): 487–509.

White House. 2012. *Critical Issues Facing Asian Americans and Pacific Islanders.* Washington DC: Initiative on Asian Americans and Pacific Islanders. **www.whitehouse.gov**

Whiten, A., J. Goodall, W. C. McGrew, T. Nishida, V. Reynolds, Y. Sugiyama, C. E. G. Tutin, R. W. Wrangham, and C. Boesch. 1999. "Cultures in Chimpanzees." *Nature* 399 (June 17): 682–684.

Whorf, Benjamin. 1956. *Language, Thought, and Reality: Selected Writings.* Cambridge, MA: Technology Press, MIT.

Whyte, William F. 1943. *Street Corner Society.* Chicago: University of Chicago Press.

Wilchins, Riki. 2014. *Queer Theory/Gender Theory.* New York: Riverdale.

Wilder, Esther I., and Toni Terling Watt. 2002. "Risky Parental Behavior and Adolescent Sexual Activity at First Coitus." *The Milbank Quarterly* 80 (September): 481–524.

Wilkins, Amy. 2012. "Stigma and Status: Interracial Intimacy and Intersectional Identities among Black College Men." *Gender & Society* 26 (April): 165–189.

Williams, Christine L. 1992. "The Glass Escalator: Hidden Advantages for Men in the 'Female' Professions." *Social Problems* 39 (August): 253–267.

Williams, Christine L. 1995. *Still a Man's World: Men Who Do Women's Work.* Berkeley, CA: University of California Press.

Williams, Christine L., Patti A. Giuffre, and Kirsten Dellinger. 2009. "The Gay-Friendly Closet." *Sexuality Research and Social Policy: Journal of NSRC* 6 (1): 29–45.

Williams, David R., and Selina A. Mohammed. 2013. "Racism and Health II: A Needed Research Agenda for Effective Interventions." *American Behavioral Scientist* 57 (8): 1200–1226.

Willie, Charles Vert. 1979. *The Caste and Class Controversy.* Bayside, NY: General Hall.

Wilson, John F. 1978. *Religion in American Society: The Effective Presence.* Englewood Cliffs, NJ: Prentice Hall.

Wilson, William Julius. 1978. *The Declining Significance of Race: Blacks and Changing American Institutions.* Chicago: University of Chicago Press.

Wilson, William Julius. 1987. *The Truly Disadvantaged: The Inner City, the Underclass and Public Policy.* Chicago: University of Chicago Press.

Wilson, William Julius. 1996. *When Work Disappears: The World of the New Urban Poor.* New York: Knopf.

Wilson, William Julius. 2009. *More Than Just Race: Being Black and Poor in the Inner City.* New York: W. W. Norton.

Winnick, Louis. 1990. "America's Model Minority." *Commentary* 90 (August): 222–229.

Wirth, Louis. 1928. *The Ghetto.* Chicago: University of Chicago Press.

Wolcott, Harry F. 1996. "Peripheral Participation and the Kwakiutl Potlatch." *Anthropology and Education Quarterly* 27 (December): 467–492.

Wolf, Stephen. 2014. "Your Guide to 2014 Election's Results and the 114th Congress Members and Their Districts." *DailyKos,* December 4.

Wolff, Edward N. 2014. "Wealth Inequality." *Pathways:* Special Issue: 34–41.

Wolfinger, Nicholas. 2017. *America's Generation Gap in Extramarital Affairs.* Charlottesville, VA: Institute for Family Studies. **www.ifstudies.org**

Wong, Janelle, S. Karthick Ramakrishnan, Taeku Lee, and Jane Junn. 2011. *Asian American Political Participation: Emerging Constituents and Their Political Identities.* New York: Russell Sage.

Woo, Deborah. 1998. "The Gap between Striving and Achieving." pp. 247–256 in *Race, Class, and Gender: An Anthology,* 3rd ed., edited by Margaret L. Andersen and Patricia Hill Collins. Belmont, CA: Wadsworth.

Wood, Julia. 2013. *Gendered Lives: Communication, Culture, and Gender.* Belmont, CA: Wadsworth/Cengage.

Woodhams, Jessica, Claire Cooke, Leigh Harkins, and Teresa da Silva. 2012. "Leadership in Multiple Perpetrator Stranger Rape." *Journal of Interpersonal Violence* 27 (4): 728–752.

World Bank. 2017a. *Health Expenditure, Total (% of GDP).* New York: The World Bank. **www.worldbank.org**

World Bank. 2017b. GNI Per Capita. **http://data.worldbank.org/indicator /NY.GNP.PCAP.PP.KD**

World Bank. 2017c. *Water.* **www .worldbank.org/en/topic/water /overview**

World Bank. 2017d. *Poverty.* **www .worldbank.org/en/topic/poverty /overview**

World Bank. 2017e. Birth Rate, Crude (per 1,000 People). Washington, DC: The World Bank. **https://data.worldbank .org/indicator/SP.DYN.CBRT.IN**

Wright, Beth. 2016. Gap Reveals List of Factory Names and Locations."

www.just-style.com/news/gap-reveals-list-of-factory-names-and-locations_id128784.aspx

Wright, Erik Olin. 1979. *Class Structure and Income Determination*. New York: Academic Press.

Wright, Erik Olin. 1985. *Classes*. London: Verso.

Wuthnow, Robert (ed.). 1994. *I Come Away Stronger: How Small Groups Are Shaping American Religion*. Grand Rapids, MI: Eerdmans.

Wuthnow, Robert. 1998. *After Heaven: Spirituality in America since the 1950s*. Berkeley, CA: University of California Press.

Wuthnow, Robert. 2010. *Boundless Faith: The Outreach of American Churches*. Berkeley: University of California Press.

Yardi, Sarita, and Dana Boyd. 2010. "Dynamic Debates: An Analysis of Group Polarization over Time on Twitter." *Bulletin of Science, Technology and Society* 30 (5): 316–327.

Young, Alford A. 2003. *The Minds of Marginalized Black Men: Making Sense of Mobility, Opportunity, and Future Life Chances*. Princeton, NJ: Princeton University Press.

Zewde, Naomi, and Terceira Berdahl. 2017. "Children's Usual Source of Care: Insurance, Income, and Racial/Ethnic Disparities, 2004–2014." *Medical Expenditure Panel Survey*. Washington, DC: Agency for Healthcare Research and Quality. **https://meps.ahrq.gov/data_files/publications/st501/stat501.shtml**

Zhou, Min, and Carl L. Bankston. 2000. "Immigrant and Native Minority Groups, School Performance, and the Problem of Self-Esteem." Paper presented at the Southern Sociological Society, April, New Orleans, LA.

Zimbardo, Phillip G., Ebbe B. Ebbesen, and Christina Maslach. 1977. *Influencing Attitudes and Changing Behavior*. Reading, MA: Addison-Wesley.

Zippel, Kathrin S. 2006. *The Politics of Sexual Harassment: A Comparative Study of the United States, the European Union, and Germany*. Cambridge: Cambridge University Press.

Zola, Irving Kenneth. 1989. "Toward the Necessary Universalizing of a Disability Policy." *Milbank Quarterly* 67 (Suppl. 2): 401–428.

Zola, Irving Kenneth. 1993. "Self, Identity and the Naming Question: Reflections on the Language of Disability." *Social Science and Medicine* 36 (January): 167–173.

Zoroya, Gregg. 2012. "Homeless, At-Risk Vets Double." *USA Today* (December 27).

Zuckerman, Phil. 2002. "The Sociology of Religion of W. E. B. DuBois." *Sociology of Religion* 63 (July): 239–253.

Zweigenhaft, Richard L., and G. William Domhoff. 2006. *Diversity in the Power Elite: How It Happened, Why It Matters*. Lanham, MD: Rowman and Littlefield.

Zweigenhaft, Richard L., and G. William Domhoff. 2011. *The New CEOs: Women, African Americans, Latino, and Asian American Leaders of Fortune 500 Companies*. Lanham, MD: Rowman and Littlefield.

Zwick, Rebecca (ed.). 2004. *Rethinking the SAT: The Future of Standardized Testing in University Admissions*. New York: Routledge.

Name Index

Subject Index

Note: Page numbers in **bold** indicate definitions/explanations of key terms; page numbers in *italics* indicate boxed material or illustrations.